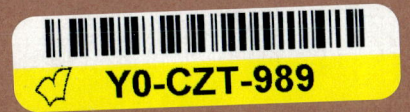

ELECTRIC MACHINERY
AND TRANSFORMERS

PRENTICE-HALL INTERNATIONAL, INC., *London*
PRENTICE-HALL OF AUSTRALIA PTY. LTD., *Sydney*
PRENTICE-HALL OF CANADA, LTD., *Toronto*
PRENTICE-HALL OF INDIA PRIVATE LIMITED, *New Delhi*
PRENTICE-HALL OF JAPAN, INC., *Tokyo*

ELECTRIC MACHINERY AND TRANSFORMERS

IRVING L. KOSOW, PH.D.

Staten Island Community College
City University of New York

PRENTICE-HALL, INC., ENGLEWOOD CLIFFS, NEW JERSEY

© 1972 by
PRENTICE-HALL, INC.
Englewood Cliffs, N.J. 07632

All rights reserved. No part of this book may be reproduced in any form or by any means without permission in writing from the publisher.

10

ISBN: 13-247205-8

Library of Congress catalog card number: 79-164665

Printed in the United States of America

To my wife
RUTH
and my children
SONIA, MARTIN, and JULIA

preface

This work is an outgrowth of the author's earlier *Electric Machinery and Control*, originally published in 1964. In revising, supplementing and updating that work, it became clear that two volumes were necessary to present the material properly and keep pace with the state-of-the-art. A variety of reasons dictated this choice. The original work was already fairly large (over 700 pages) and the contemplated new material would inevitably result in a most unwieldy and expensive volume.

A logical division between electric machinery theory and control applications of electric machinery already exists in the literature. Numerous works already exist in separate volumes in these areas, so there is a precedent for such a dichotomy. The student who requires a background in the theory of electric machines and their characteristics should be introduced to the subject in a way that is different from that required by the practicing engineer and technician in the field. The latter, primarily, are interested in the control and commercial applications of electric machinery covered in the second volume, although reference to this first volume may occasionally be required.

This first volume, therefore, is a text which reflects the feedback from teachers and students who used the earlier *Electric Machinery and Control*. In response to numerous requests, a new chapter on Transformers has been added. In addition, questions have been added to each chapter to sharpen the reader's qualitative comprehension of the material. The language of the text has been rewritten, in part, to clarify important theoretical distinctions, facilitate comprehension, and more importantly, to enable self-study. New problems and illustrative examples have been added. Unit abbreviations have been revised to reflect IEEE standards.

The earlier rationale for the study of electric machinery cited in the preface of *Electric Machinery and Control* has been accentuated by two major worldwide problems: pollution (of our lands, waters and atmosphere) and overpopulation. The latter has resulted in tremendously increased demands for power and personalized transportation, along with consumer goods of a wide variety, concomitant with a rising standard of living, and this inevitably has produced the former. As a consequence, engineers and scientists are taking a new look at electric power generation, energy conversion, and the use of electrical (pollution free or relatively low pollution) traction techniques for rail and automotive transportation. The electric car, cited by the author as a possibility in the earlier volume, is rapidly becoming a reality, as a result. The brownouts and blackouts of the late sixties are a direct consequence of man's insatiable need for electric power, generally; and extended reliance on electric machinery, specifically. And the seventies inevitably will see an intensified interest in electrical energy conversion and machinery, on the part of governments, educational institutions, and industry, in response to these pressing global problems.

A strong attempt has been made to unify the subject matter and its method of presentation, as begun in the earlier work. Chapter 1 conveys the unifying principle that generator and motor action simultaneously occur in all rotating machines. Chapter 2 treats windings on the basis of similarities rather than differences between dc and ac dynamos. Chapters 5 and 7 treat armature reaction and parallel operation, respectively, in a similar unified way, leading to generalizations regarding the effects of excitation and armature reaction on all dynamos. Chapters 8 and 9 stress the distinctions between synchronous and asynchronous dynamos, always directed to the increased understanding of the characteristics of alternators, synchronous motors, induction motors and generators, and various single-phase motors. Chapter 11 on specialized dynamos includes selsyns, servomotors, and multifield exciters as well as other cross-field machines, essential for a study of servomechanisms. Dynamo efficiency is treated in Chapter 12 as a unified topic in electromechanical conversion, in which dc and ac dynamo efficiency and the underlying theory of basic tests are closely related. This chapter also gives particular attention to the rating, selection, speed control, and maintenance of electric machinery. The final chapter on transformers is closely related and referred to previous chapters on alternators and efficiency to stress similarities and unify the presentation. This chapter also includes higher order polyphase conversions for high power dc requirements.

As noted earlier, the emphasis of the writing, based on the author's quarter century of teaching experience, is directed toward self study. This has resulted in somewhat more detail in text material, illustrative examples indicating solution of problems, and many specific questions designed to motivate reading. It also has the advantages of decreasing the teacher's work load and placing more responsibility on the student in the learning process. Consequently, this frees the teacher to place more stress on those aspects of the subject he feels requires emphasis or in-depth study, and on those particular topics which students require help. Further, because of its self-study aspect, the work lends itself to either a two-semester or one-semester course on the subject. In the latter case, the teacher may assign specific chapters and/or sections of chapters as representing the course outline, with the preliminary injunction to the student to read whatever peripheral explanatory material he may require in other chapter sections, to broaden and enhance his understanding.

Thanks and appreciation is expressed to the Prentice-Hall staff, generally, and particularly to Steven Bobker for his careful supervision of the production of the manuscript and the many helpful suggestions which resulted in the present format of the book. The author also acknowledges the support and help of Mr. Matthew Fox, Executive Editor and Mr. Edward Francis, Editor, Electronic Technology.

As in the case of my other books and editorial work, my wife, Ruth, has made significant contributions directly in the proofreading and indexing of this entire ms and indirectly by her encouragement, patience and understanding through the many days of loneliness and isolation required to produce this work.

IRVING L. KOSOW

New York City, 1971

contents

1 ELECTROMECHANICAL FUNDAMENTALS   1

     1-1   Electromagnetic Energy Conversion, 2
     1-2   Relation between Electromagnetic Induction and Electromagnetic Force, 3
     1-3   Faraday's Law of Electromagnetic Induction, 4
     1-4   Factors Affecting Magnitude of Induced emf, 5
     1-5   Direction of Induced Voltage—Fleming's Rule, 9
     1-6   Lenz's Law, 10
     1-7   Elementary Generators, 12
     1-8   Proof of Fleming's Right-hand Rule by Means of Lenz's Law, 13
     1-9   Polarity of an Elementary Generator, 13
     1-10   Sinusoidal emf Generated by a Coil Rotating in a Uniform Magnetic Field at Constant Speed, 14
     1-11   Rectification by Means of a Split-ring Commutator, 15
     1-12   The Gramme-ring Winding, 18
     1-13   Dynamo Voltage, Current, and Power Ratings, 23
     1-14   Average emf Generated in a Quarter-revolution, 24

1-15 Fundamental dc Generator Voltage Equation for emf between Brushes, 25
1-16 Electromagnetic Force, 26
1-17 Factors Affecting Magnitude of EM Force, 27
1-18 Direction of EM Force and Left-hand Rule, 28
1-19 Counter emf, 29
1-20 Comparison of Motor Action vs Generator Action, 29

2 DYNAMO CONSTRUCTION AND WINDINGS 40

2-1 Dynamo Possibilities, 40
2-2 Direct-current (dc) Dynamo Construction, 41
2-3 Synchronous Dynamo (Stationary Field) Construction, 43
2-4 Rotating Field Synchronous Dynamo Construction, 44
2-5 Asynchronous Induction Dynamo Construction, 45
2-6 dc Dynamo Magnetic Fields and Circuits, 46
2-7 Armature Reactance, 47
2-8 ac Dynamo Magnetic Fields and Circuits, 47
2-9 Magnetic Flux Calculations, 48
2-10 Armature Windings, 51
2-11 Lap and Wave Windings—Similarities and Differences, 52
2-12 Summary—Windings, 57
2-13 ac Synchronous Dynamo Armature Windings, 58
2-14 Half-coil and Whole-coil Windings, 59
2-15 Chorded or Fractional-pitch Windings, 60
2-16 Distribution or Belt Factor—Distributed Windings, 62
2-17 Effect of Fractional Pitch and Distribution of Coils on Waveform, 64
2-18 Generated emf in an ac Synchronous Dynamo, 66
2-19 Frequency of ac Synchronous Dynamo, 68

3 DC DYNAMO VOLTAGE—DC GENERATORS 76

3-1 General, 76
3-2 dc Generator Types, 77
3-3 Schematic Diagram and Equivalent Circuit of a Shunt Generator, 77
3-4 Schematic Diagram and Equivalent Circuit of a Series Generator, 79
3-5 Schematic Diagram and Equivalent Circuit of a Compound Generator, 80
3-6 The Separately Excited Generator, 82
3-7 No-load Voltage Characteristic of dc Generator, 82
3-8 Self-excited Generator-field Resistance Lines, 86
3-9 Build-up of Self-excited Shunt Generator, 87
3-10 Critical Field Resistance, 88
3-11 Reasons for Failure of Self-excited Shunt Generator to Build up Voltage, 88

3-12 Effect of Load in Causing a Shunt Generator to Unbuild, 89
3-13 Load-voltage Characteristics of a Shunt Generator, 90
3-14 Effect of Speed on No-load and Load Characteristics of a Shunt Generator, 93
3-15 Voltage Regulation of a Generator, 94
3-16 Series Generator, 96
3-17 Compound Generator, 97
3-18 Cumulative Compound Generator Characteristics, 98
3-19 Adjusting the Degree of Compounding of Cumulative Compound Generators, 100
3-20 Differential Compound Generator Characteristic, 101
3-21 Comparison of Generator Load-voltage Characteristics, 102
3-22 Effect of Speed on Load-voltage Characteristics of Compound Generators, 103

4 DC DYNAMO TORQUE RELATIONS—DC MOTORS 110

4-1 General, 110
4-2 Torque, 111
4-3 Fundamental Torque Equation for a dc Dynamo, 115
4-4 Counter emf or Generated Voltage in a Motor, 117
4-5 Motor Speed as a Function of Counter emf and Flux, 117
4-6 Counter emf and Mechanical Power Developed by a Motor Armature, 119
4-7 Relation between Torque and Speed of a Motor, 121
4-8 Starters for dc Motors, 122
4-9 Electromagnetic Torque Characteristics of dc Motors, 125
4-10 Speed Characteristics of dc Motors, 127
4-11 Speed Regulation, 133
4-12 External Torque, Rated Horsepower, and Speed, 134
4-13 Reversal of Direction of Rotation, 135
4-14 Effect of Armature Reaction on Speed Regulation of all dc Motors, 136

5 ARMATURE REACTION AND COMMUTATION IN DYNAMOS 144

5-1 General, 144
5-2 Magnetic Field Produced by Armature Current, 145
5-3 Effect of Armature Flux on Field Flux, 146
5-4 Shift of Neutral Plane in Generator vs Motor, 149
5-5 Compensating for Armature Reaction in dc Dynamos, 150
5-6 The Commutation Process, 154
5-7 Reactance Voltage, 157
5-8 Armature Reaction in the ac Dynamo, 158
5-9 Summary of Armature Reaction in Dynamos, 161

6 AC DYNAMO VOLTAGE RELATIONS—ALTERNATORS 169

- 6-1 General, 169
- 6-2 Construction, 170
- 6-3 Advantages of Stationary Armature and Revolving Field Construction, 170
- 6-4 Prime Movers, 173
- 6-5 Equivalent Circuit for a Single-phase and a Polyphase Synchronous Dynamo, 174
- 6-6 Comparison between Separately Excited dc Generator and Separately Excited Synchronous Alternator, 176
- 6-7 Relation between Generated and Terminal Voltage of an Alternator at Various Load Power Factors, 176
- 6-8 Voltage Regulation of an ac Synchronous Alternator at Various Power Factors, 180
- 6-9 Synchronous Impedance, 182
- 6-10 The Synchronous Impedance (or emf Method) for Predicting Voltage Regulation, 183
- 6-11 Assumptions Inherent in the Synchronous Impedance Method, 188
- 6-12 Short-circuit Current and Use of Current-limiting Reactors, 189

7 PARALLEL OPERATION 197

- 7-1 Advantages of Parallel Operation, 197
- 7-2 Voltage and Current Relations for Sources of emf in Parallel, 198
- 7-3 Parallel Operation of Shunt Generators, 201
- 7-4 Conditions Necessary for Parallel Operation of Shunt Generators, 202
- 7-5 Parallel Operation of Compound Generators, 203
- 7-6 Conditions Necessary for Parallel Operation of Compound Generators, 204
- 7-7 Procedure for Paralleling Generators, 206
- 7-8 Conditions Necessary for Paralleling Alternators, 206
- 7-9 Synchronizing Single-phase Alternators, 207
- 7-10 Effects of Synchronizing (Circulating) Current between Single-phase Alternators, 211
- 7-11 Load Division between Alternators, 218
- 7-12 Hunting or Oscillation of Alternators, 220
- 7-13 Synchronizing Polyphase Alternators, 222
- 7-14 Synchroscopes, 224
- 7-15 Phase-sequence Indicator, 226
- 7-16 Summary of Procedure for Paralleling Polyphase Alternators, 226

8 AC DYNAMO TORQUE RELATIONS—SYNCHRONOUS MOTORS 234

8-1 General, 234
8-2 Construction, 236
8-3 Operation of the Synchronous Motor, 236
8-4 Starting Synchronous Motors, 239
8-5 Starting a Synchronous Motor as an Induction Motor by Means of its Damper Windings, 239
8-6 Starting a Synchronous Motor under Load, 241
8-7 Synchronous Motor Operation, 242
8-8 Effect of Increased Load at Normal Excitation of Synchronous Motor, 247
8-9 Effect of Increased Load at Conditions of Underexcitation, 249
8-10 Effect of Increased Load at Conditions of Overexcitation, 250
8-11 Summary of the Effect of Increased Load (Neglecting Effects of Armature Reaction) under Constant Excitation, 250
8-12 Effect of Armature Reaction, 251
8-13 Power Factor Adjustment of Synchronous Motor at Constant Load, 253
8-14 V-curve of a Synchronous Motor, 256
8-15 Computation of Torque Angle and Generated Voltage Per Phase for a Polyphase Synchronous Motor, 260
8-16 Use of the Synchronous Motor as a Corrector of the Power Factor, 268
8-17 Developed Electromagnetic Torque Per Phase of a Synchronous Motor, 270
8-18 Synchronous Motor Ratings, 274
8-19 Synchronous Capacitors, 274
8-20 Economic Limit to Improvement of Power Factor, 276
8-21 Computation of Synchronous Motor Power Factor Improvement Using the kW–kvar Method, 277
8-22 Use of the Synchronous Capacitor as a Synchronous Reactor, 280
8-23 Use of the Synchronous Motor as a Frequency Changer, 281
8-24 The Supersynchronous Motor, 282
8-25 Special Types of Synchronous Motors Which Do Not Employ dc Field Excitation, 283
8-26 The Synchronous-induction Motor, 284
8-27 Reluctance Motor, 285
8-28 Hysteresis Motor, 286
8-29 Subsynchronous Motor, 287
8-30 Solid-state dc Field Supplies–Static Supplies, 288
8-31 Brushless Synchronous Motors, 289

9 POLYPHASE INDUCTION (ASYNCHRONOUS) DYNAMOS 300

9-1 General, 300
9-2 Construction, 301
9-3 Production of a Rotating Magnetic Field by Application of Polyphase Alternating Current to Stator Armature, 302
9-4 Induction Motor Principle, 305
9-5 Rotor Conductors, Induced emf, and Torque; Rotor Stalled, 307
9-6 Maximum Torque, 314
9-7 Operating Characteristics of an Induction Motor, 315
9-8 Running Characteristics of an Induction Motor, 317
9-9 Effect of Change in Rotor Resistance, 319
9-10 Starting Characteristics with Added Rotor Resistance, 320
9-11 Running Characteristics with Added Rotor Resistance, 326
9-12 Induction Motor Torque and Developed Rotor Power, 327
9-13 Measurement of Slip by Various Methods, 332
9-14 Induction Motor Starting, 335
9-15 Reduced Voltage—Autotransformer Starting, 336
9-16 Reduced Voltage, Primary Resistor or Reactor Starting, 338
9-17 Wye-delta Starting, 338
9-18 Part-winding Starting, 340
9-19 Wound-rotor Starting, 340
9-20 Double-cage Rotor Line-starting Induction Motor, 341
9-21 Commercial Induction Motor Classification, 342
9-22 Induction Generator, 347
9-23 Induction Frequency Converters, 348

10 SINGLE-PHASE MOTORS 359

10-1 General, 359
10-2 Construction of the Single-phase Induction Motor at Standstill, 361
10-3 Balanced Torque of a Single-phase Induction Motor at Standstill, 361
10-4 Resultant Torque of a Single-phase Induction Motor as a Product of Rotor Rotation, 363
10-5 Split-phase (Resistance-start) Induction Motor, 366
10-6 Split-phase (Capacitor-start) Motor, 369
10-7 Permanent-split (Single-value) Capacitor Motor, 371
10-8 Two-value Capacitor Motor, 373
10-9 Shaded-pole Induction Motor, 375
10-10 Reluctance-start Induction Motor, 379
10-11 Single-phase Commutator Motors, 380
10-12 The Repulsion Principle, 380
10-13 Commercial Repulsion Motor, 384
10-14 Repulsion-start Induction Motor, 385

- 10-15 Repulsion-induction Motor, 386
- 10-16 Universal Motor, 388
- 10-17 ac Series Motor, 389
- 10-18 Summary of Types of Single-phase Motors, 391

11 SPECIALIZED DYNAMOS 401

- 11-1 General, 401
- 11-2 Diverter-pole Generator, 402
- 11-3 Third-brush Generator, 403
- 11-4 Homopolar or Acyclic Dynamo, 405
- 11-5 Dynamotor, 406
- 11-6 Single-phase Rotary Converter, 408
- 11-7 Polyphase Rotary Converter, 411
- 11-8 Three-wire-system Generators, 417
- 11-9 Effect of Line Resistance and Unbalanced Loads in Three-wire Systems, 419
- 11-10 Induction Phase Converters, 423
- 11-11 Synchronizing (Selsyn) Divices, 424
- 11-12 Power Selsyns and Synchro Tie Systems, 432
- 11-13 dc Servomotors, 434
- 11-14 ac Servomotors, 437
- 11-15 Rosenberg Generator, 439
- 11-16 The Amplidyne, 441
- 11-17 Multiple-field Exciters—Rototrol and Regulex, 444

12 POWER AND ENERGY RELATIONS; EFFICIENCY, RATINGS SELECTION AND MAINTENANCE OF ROTATING ELECTRIC MACHINERY 462

- 12-1 General, 462
- 12-2 Dynamo Power Losses, 464
- 12-3 Power-flow Diagrams, 467
- 12-4 Determination of losses, 468
- 12-5 dc Dynamo Efficiency, 469
- 12-6 Maximum Efficiency, 471
- 12-7 Duplication of Flux and Speed, 475
- 12-8 ac Synchronous Dynamo Efficiency, 476
- 12-9 Ventilation of Alternators, 478
- 12-10 ac Synchronous Dynamo Efficiency by the Calibrated dc Motor Method, 479
- 12-11 Asynchronous Induction Dynamo Efficiency, 480
- 12-12 Equivalent Resistance of Induction Motor, 481
- 12-13 Induction Motor Efficiency from Open-circuit and Short-circuit (Locked-rotor) Tests, 483
- 12-14 Induction Motor Efficiency from AIEE Load-slip Equivalent-circuit Method, 486

12-15 Efficiency of Single-phase Motors, 489
12-16 Factors Affecting Ratings of Machines, 489
12-17 Temperature Rise, 490
12-18 Voltage Rating, 492
12-19 Effect of Duty Cycle and Ambient Temperature on Rating, 493
12-20 Types of Enclosures, 493
12-21 Speed Rating; Classifications of Speed and Reversibility, 494
12-22 Factors Affecting Generator and Motor Selection, 497
12-23 Maintenance, 498

13 TRANSFORMERS 514

13-1 Fundamental Definitions, 514
13-2 Ideal Transformer Relations, 517
13-3 Reflected Impedance, Impedance Transformation, and Practical Transformer, 524
13-4 Equivalent Circuits for a Practical Power Transformer, 529
13-5 Voltage Regulation of a Power Transformer, 533
13-6 Voltage Regulation from the Short Circuit Test, 536
13-7 Assumptions Inherent in the Short Circuit Test, 539
13-8 Transformer Efficiency from the Open Circuit and Short Circuit Test, 540
13-9 All-day Efficiency, 546
13-10 Phasing, Identification and Polarity of Transformer Windings, 547
13-11 Connecting Transformer Windings in Series and Parallel, 551
13-12 The Autotransformer, 554
13-13 Autotransformer Efficiency, 562
13-14 Three Phase Transformation, 565
13-15 Transformer Harmonics, 572
13-16 Importance of Neutral and Means for Providing It, 574
13-17 V-V Transformer Relations—Open Delta System, 576
13-18 T-T Transformer Relations, 578
13-19 Three Phase to Two Phase Transformation—Scott Connection, 582
13-20 Three Phase to Six Phase Transformations, 585
13-21 Use of Polyphase Transformations for Power Conversion, 592

APPENDIX 611

INDEX 629

ONE

electromechanical fundamentals

For a number of years the fields of electric power generation and conversion have occupied a subordinate place in the public mind in comparison to the more glamorous fields of electron-tube and solid-state electronics. Electrical engineers, scientists, professors, and their students have considered electric power a rather sterile field of study, generally lacking opportunity, challenge, or excitement. Yet a number of studies, national and international, which have estimated our fossil fuel reserves (coal, gas, and petroleum accounting for 96 per cent of our energy supply), our population growth, and our rising standard of living, predict an optimistic estimated fuel reserve of about 230 years and a pessimistic estimated reserve of 23 years*. New sources of energy as well as improved methods of energy conversion are indicated. Man's insatiable explorations into the ocean depths and outer space have begun to stimulate investigation of other means of energy conversion (solar, biochemical, chemical, and

* J. A. Hutcheson, "Engineering for the Future," *Journal of Engineering Education* (April 1960), pp. 602–607.

nuclear). But whatever the method of generation of energy, it appears that, since electricity is the only form of energy that is relatively easy to use, control, and convert to other energy forms, it will probably continue to be the major energy form used by man. It would appear, then, that here is a field which should continue to prove both challenging and rewarding.

This text is concerned mainly with the use, control, and conversion of electromechanical energy, as well as its transmission and distribution. Whatever exotic or sophisticated methods of electric generation may be discovered in the future, the principles of the conversion and utilization of electricity must still be advanced, analyzed, and developed.

1-1. ELECTROMAGNETIC ENERGY CONVERSION

The first inkling of the possibility of interchange between mechanical and electric energy was advanced by Michael Faraday in 1831. This discovery has been hailed by some as the greatest single advance in the progress of science toward the ultimate betterment of mankind. It has given birth to the electric generator and motor, the microphone, the loudspeaker, the transformer, the galvanometer, and in fact, practically all of the devices whose principles and characteristics will be considered in this volume. (See Sec. 1-3.)

Electromagnetic energy conversion, as we have come to know it today, relates the electric and magnetic forces of the atom to mechanical force applied to matter and to motion. As a result of this relationship, mechanical energy may be *converted* to electric energy, and vice versa, by means of *dynamos*. Although this conversion may also produce other energy forms such as heat and light, for most practical purposes the state-of-the-art has been advanced to a degree where such energy losses are held to a minimum and relatively direct conversion is accomplished in either direction. Thus the mechanical energy of a waterfall is readily converted to electric energy by an *alternator*; the electric energy produced is *transformed* by electromagnetic energy conversion to a higher voltage for transmission over long distances, and at some terminal point is *transformed* once again for distribution at a load-center substation where the electric energy is again distributed to specific farms, factories, residences, and commercial establishments. In these individual occupancies, the electric energy may be *converted* once more to mechanical energy by means of motors, to heat energy through the use of electric ovens, to light energy through the use of electric lamps, and to chemical energy through the use of electrochemical techniques and processes; or it may be converted to other forms of electric energy through the use of rotary converters, rectifiers, and frequency changers. The electric energy produced by means of such electromechanical energy conversion may be reconverted several times by the devices considered in this volume before that energy is ultimately converted into *useful work* in other energy forms.

This chapter will concern itself only with the fundamental and pertinent relations between electric and mechanical energy.

1-2. RELATION BETWEEN ELECTROMAGNETIC INDUCTION AND ELECTROMAGNETIC FORCE

A number of natural electromagnetic phenomena have been discovered which relate electric and mechanical energy. The relative ease with which such energy conversion is accomplished is due, indeed, to a knowledge of these relations. For most practical purposes, the conversion from electric to mechanical energy and vice versa may be considered a reversible reaction.

The extent to which the process is not completely reversible and other undesirable energy forms are produced in the process (such as heat, light, and chemical energy), results in a loss of energy from the electromechanical energy system. The subject of energy loss and efficiency is covered in Chapter 12. The following discussion of electromagnetic phenomena assumes complete electromechanical energy conversion.

Perhaps the most important electromagnetic effects are those relating mechanical force applied to a body (i.e., a mass consisting of charged particles, chiefly protons and electrons, in motion, resulting in motion of that body) in the presence of electric and magnetic fields.*

There are four such effects considered here. The first two are mentioned, briefly. The last two are the subject of this chapter. These phenomena involved in electromechanical energy conversion are:

1. A force of attraction that exists between the (oppositely) charged plates of a capacitor. This force is mechanical in nature, for should a dielectric sample be placed between these plates, it would tend to be moved into that part of the electric field which has the greatest density. The electric field thus acts in such a way on a dielectric sample to maintain an electrostatic (electric) field of maximum density. For this reason, if the sample is irregular in shape, it will align itself with its longest axis (or maximum mass) parallel to the field. Thus, particles of mica sprinkled on a surface are aligned in the presence of an electric field.
2. The reluctance principle: mechanical force is exerted on a sample of magnetic material located in a magnetic field. The force tends to act on the material in such a way as to bring the material into that portion of the magnetic field which has the greatest density. If the sample is irre-

---

* A number of thermoelectric, galvanomagnetic, and thermomagnetic effects (e.g., the Hall, Ettingshausen, Nernst, or Righi-Leduc effects) do not involve a force applied to a conductive body or motion of that body, yet serve to relate electric and magnetic fields with properties of matter. In the Hall effect, for example, a longitudinal electric current (produced by a longitudinal electric field) will produce, in the presence of an orthogonal magnetic field, a transverse electric field. Since no motion is involved, this effect is not considered an electromechanical energy phenomenon. These effects, excluding the principle of the transformer, are outside the scope of this text and will not be considered.

gularly shaped, it will tend to be aligned in such a way as to produce minimum magnetic reluctance and, consequently, maximum flux density. Thus, particles of iron filings are aligned, in the presence of a magnetic field, parallel to the field direction. (Sec. 8-17 and 8-27.)
3. Electromagnetic induction (see Section 1-3).
4. Electromagnetic force (see Section 1-16).

1-3. FARADAY'S LAW OF ELECTROMAGNETIC INDUCTION

Prior to Faraday's discovery, a voltage was generated in a circuit by means of chemical action such as that which occurs in a dry cell or storage battery. The unique contribution of Faraday's discovery in 1831 was the generation of a voltage because of relative motion between a magnetic field and a conductor of electricity. Faraday called this voltage an "induced" voltage because it occurred only when there was relative motion between the conductor and a magnetic field without actual "physical" contact between them. Faraday's actual device resembles that shown in the footnote to Section 1-11, but the principle of electromagnetic induction is perhaps better understood from the diagram shown in Fig. 1-1.

The general statement of Faraday's law may be given as follows.

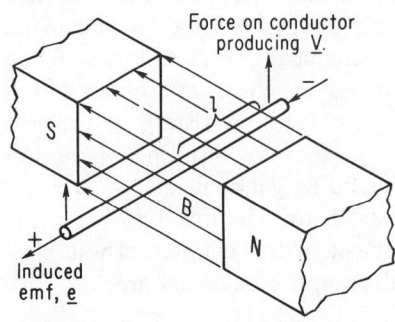

Figure 1-1

Conductor of length *l* moving through a magnetic field *B*, to generate an emf.

> The magnitude of voltage induced in a single turn of wire is proportional to the rate of change of lines of force passing through (or linked with) that turn.

Neumann, in 1845, quantified this statement in an equation in which the magnitude of a generated induced electromotive force (emf) was directly proportional to the rate of change of flux linkages:

$$E_{av} = \frac{\phi}{t} \text{ abvolts} = \frac{\phi}{t} \times 10^{-8} \quad \text{V} \tag{1-1}$$

where E_{av} is the average voltage generated in a single turn (volts/turn)
  ϕ is the number of maxwells or lines of magnetic force linked by the turn during
  t the time in seconds in which ϕ lines are "linked"
  10^8 is the number of lines a single turn must link each second in order to induce a voltage of 1 volt

From the above statement and equation, it is fairly evident that an induced generated voltage may be increased by increasing the magnetic field strength (i.e., the number of flux lines in motion relative to the conductor) or by decreasing the time during which the change in flux linkages occurs (i.e., an increase in speed or relative motion between conductor and magnetic field).

1-4. FACTORS AFFECTING MAGNITUDE OF INDUCED EMF

Neumann's quantification of Faraday's law, as stated in Eq. (1-1), holds true only when the magnetic circuit is physically the same at the end as at the beginning and during the period of change of flux linkages. In rotating electric machinery, however, the change of flux linking each individual turn because of rotation (of either the armature or the field) is not clearly defined or easily measured. It is more convenient, therefore, to express this rate of change in terms of an (assumed constant) average flux density and the relative velocity between this field and a single conductor moving through it. In Fig. 1-1, for the conductor of *active* length l, the instantaneous induced emf may be expressed as*

$$e = Blv\, 10^{-8} \quad \text{V} \qquad (1\text{-}2)$$

where B is the flux density in gauss (lines/cm²) or in lines/in²
l is the length of the active portion of the conductor linking the flux in cm or in inches
v is the relative velocity between the conductor and the field in cm/s or in/s

In terms of English or practical units, Eq. (1-2) can be expressed as

$$e = \tfrac{1}{5} Blv\, 10^{-8} \quad \text{V} \qquad (1\text{-}3)$$

where B is the flux density in lines/in²
l is the length in inches of the active portion of the conductor linking the flux
v is the velocity in ft/min
10^8 is the number of lines a single conductor must link each second in order to induce a voltage of 1 V

If both the flux density, B, and the relative velocity of either conductor or field are uniform and constant, then the instantaneous and average values of induced emf are the *same*. Either Eq. (1-1) or Eq. (1-3) may be used with the same results, as shown by Ex. 1-1.

* Equation (1-2) may be derived from Eq. (1-1) in the following way: If the conductor of Fig. 1-1 moves a distance ds in the time dt, the change in flux linkage may be expressed as $d\phi = -Bl\,ds$. But since $e = -(d\phi/dt)10^{-8}$ V, by substitution $e = Bl(ds/dt)10^{-8}$ V. But since ds/dt is the same as the velocity v of the conductor with respect to the magnetic field, $e = (Blv)10^{-8}$ V.

EXAMPLE 1-1: A single conductor 18 inches long is moved by a mechanical force perpendicularly to a uniform magnetic field of 50,000 lines/in², covering a distance of 720 inches in 1 second. Calculate:
a. The instantaneous induced emf using Eq. (1-3)
b. The average induced emf using Eq. (1-1).

Solution:

a. $e_{\text{inst}} = \dfrac{1}{5} Blv \times 10^{-8}$ V (1-3)

$= \dfrac{1}{5}\left(50{,}000 \,\dfrac{\text{lines}}{\text{in}}\right)(18 \text{ in})\left(720 \,\dfrac{\text{in}}{\text{s}} \times 60 \,\dfrac{\text{s}}{\text{min}} \times \dfrac{1}{12}\,\dfrac{\text{ft}}{\text{in}}\right) 10^{-8}$ V

$= 6.48$ V

b. $\phi = BA = (50{,}000 \text{ lines/in}^2)(720 \text{ in} \times 18 \text{ in}) = 6.48 \times 10^8$ lines

$e_{\text{av}} = \dfrac{\phi}{t} \times 10^{-8} \text{ V} = \dfrac{6.48 \times 10^8 \text{ lines}}{1 \text{ s}} \times 10^{-8} \text{ V} = 6.48$ V (1-1)

The preceding equations are subject to a number of qualifications, however, that cannot be overlooked. In Fig. 1-1, it is assumed that

1. The field B is of uniform flux density.
2. The force applied to move either the field or the conductor, or both, will produce uniform relative motion between them.
3. The conductor, the field, and the direction in which the conductor moves with respect to the field, are mutually perpendicular (orthogonal).

For the most part, commercial dynamos are designed in such a way that the first two assumptions may be adhered to for all practical purposes. Even where there is a change in load condition, once the change has occurred it may be assumed that the new flux density and speed will remain constant as long as the given load condition is constant. The third assumption is discussed briefly below. Before considering it, however, it would be well to give some attention to certain implications of Eq. (1-3). If, for example, any term in this equation is reduced to zero, say either the flux density B or the velocity v, then the induced voltage in a given conductor l is also zero. In order to induce an emf in a given conductor it is, therefore necessary that there be a *continuous change* of flux linkages, i.e., some motion is required so that "new" lines of force link the conductor, or vice versa.

For a given active length of conductor, the product Bv in Eq. (1-3) represents the rate of change of flux linkage upon which the magnitude of induced emf in a given conductor of length l depends. Increasing either the flux density or the relative velocity (or both) will increase the rate of change of flux linkages and the induced emf in a given conductor, in turn. It is also fairly evident that increasing the over-all length of the conductor will not increase the emf since the *active* length of the conductor is unchanged. The active length *l* of the conductor may be increased (1) by using larger field poles or more of them, or (2) by turning the conductor back on itself so that several active lengths are connected in series and presented to the magnetic field in such a way that they are all moving in the same direction. It will later be seen that both these methods are employed in commercial dynamos.

It has been stated that Eq. (1-3), which represents the induced emf for a single conductor shown in Fig. 1-1 and illustrated in the above example, is a special or ideal case of induced emf (recall Assumption 3). Equation (1-3) does not take into account the fact that the conductor may not be perpendicular to the magnetic field and that its motion may not be perpendicular to the magnetic field. Both of these possibilities arise in the operation of the commercial dynamo, and their consequent effect on the magnitude of the induced emf should be considered.

The special case of a conductor moving at right angles to a magnetic field is shown in Fig. 1-1, and this mutually orthogonal condition is also represented in Fig. 1-2a. For any given magnetic field strength, active

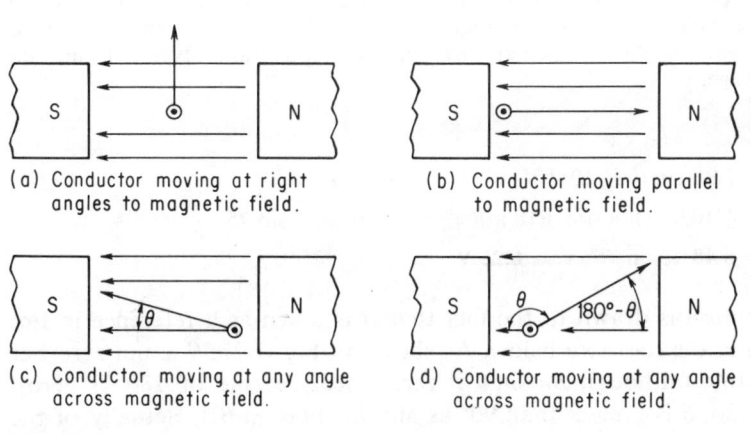

(a) Conductor moving at right angles to magnetic field.

(b) Conductor moving parallel to magnetic field.

(c) Conductor moving at any angle across magnetic field.

(d) Conductor moving at any angle across magnetic field.

Figure 1-2

Effect of change of flux linkages on induced emf in a conductor.

length of conductor, and speed of conductor, the voltage induced in the conductor shown in Fig. 1-2a is expressed by Eqs. (1-1) and (1-3).

Consider, however, the conductor shown in Fig. 1-2b moving at the same speed in a field of equal strength in a direction *parallel* to the magnetic field. The induced voltage in this conductor is zero because the rate of change of flux linkage is zero, i.e., the conductor does not link any new lines of force when moving parallel to a magnetic field. Since the product Bv in Eq. (1-3) represents the rate of change of flux linkage, it is evident that this expression must equal zero. But the flux density and the velocity, respectively, are the same in both Figs. 1-2a and b, yet in the former case the voltage is a maximum and in the latter the voltage is zero. Since the active length of the conductor is assumed unchanged, it is obvious that the product Bv must be multiplied by some factor which takes into account the difference in rate of change of flux linkage produced by a change in direction of the conductor. One might almost intuitively infer that this factor is a sine function, since it is zero at zero degrees and maximum at 90°. For the emf of *any* conductor moving in *any* direction with respect to the (reference) magnetic field,* as shown in Fig. 1-2c,

$$e = \tfrac{1}{5}[Blv \sin (B, v)]10^{-8} \quad \text{or} \quad \tfrac{1}{5}(Blv \sin \theta)10^{-8} \quad \text{V} \qquad (1\text{-}4)$$

where all quantities are the same as in Eq. (1-3) and where θ is the angle formed between B and v, taking B as a reference.

EXAMPLE 1-2: The conductor of Example 1-1 is moved by the prime mover at the same velocity but at an angle of 75° with respect to the same field (instead of 90°). Calculate the instantaneous (and average) induced voltage.

Solution:

$$e = \tfrac{1}{5} Blv \sin \theta \times 10^{-8} \text{ V} \qquad (1\text{-}3)$$
$$= \tfrac{1}{5}(50{,}000 \text{ lines/in}^2)(18 \text{ in})(\tfrac{720}{12} \times 60 \text{ ft/min}) \sin 75° \times 10^{-8} \text{ V}$$
$$= 6.48 \times \sin 75° \text{ V} = \mathbf{6.25 \text{ V}}$$

Attention is drawn to the fact that B is taken as a reference in Eq. (1-4) and in the footnote below. As shown in Fig. 1-2b, θ is not zero but actually 180° (although $Bv \sin \theta$ is zero since the sine of 180° is zero). In Fig. 1-2d, θ is greater than 90° as shown, but $\sin \theta$ is actually of the same value as $\sin (180° - \theta)$.

The case of a conductor moving perpendicular to the magnetic field but lying at some angle ϕ (other than 90°) with respect to the magnetic field, is treated the same as Ex. 1-2, above. See Eq. (1-4) and footnote to same.

---

* A problem may arise in which none of the factors B, l, and v, are mutually perpendicular. Equation (1-3) may be multiplied by the sine of the angle between pairs of quantities, B, v and B, l, using B as a reference:

$$e = Blv \sin (B, v) \sin (B, l) = Blv \sin \theta \sin \phi$$

where θ is the angle between B and v and
ϕ is the angle between B and l

1-5. DIRECTION OF INDUCED VOLTAGE—FLEMING'S RULE

It should be noted that when a conductor is moved in an *upward* direction, as shown in Fig. 1-2c, from the lower right to the upper left, so that θ is less than 90°, the induced voltage e will have the same direction (and polarity) as that shown in Fig. 1-2d where θ is greater than 90°. Since $\sin \theta$ is positive for all angles between 0 and 180°, e of Eq. (1-4) is positive for all directions with respect to B from 0° to 180°, i.e., for a generally upward motion of the conductor. Similarly, if the force applied to the conductor tends to move the conductor *downward*, as shown in Fig. 1-3b, the direction of induced voltage will be opposite to that shown in Fig. 1-2. Since $\sin \theta$ is negative for all angles between 180° and 360°, θ of Eq. (1-4) is negative for all directions generally downward. If the magnetic field were reversed, however, the polarities would reverse. Thus, the basic reference for polarity and for angle θ in Eq. (1-4) is the direction of the magnetic field.

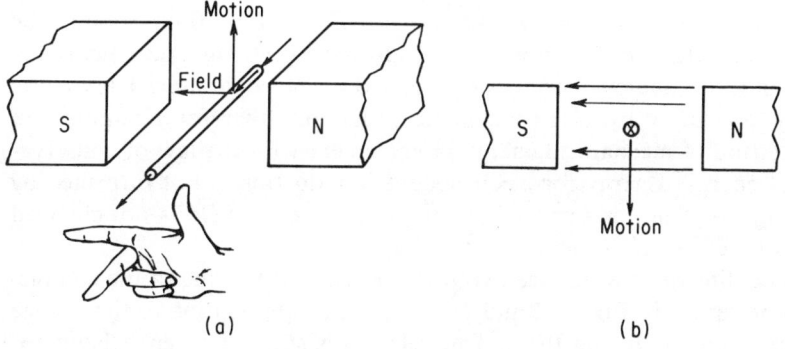

Figure 1-3

Fleming's right-hand rule for direction of induced emf (conventional current).

The relation between the directions of induced emf, magnetic field, and motion of a conductor, is conveniently represented and remembered by Fleming's rule, shown in Fig. 1-3a. When conventional current is employed* for determining the direction of generated emf, it may be called Fleming's "right-hand" rule, as shown in Fig. 1-3a.

Fleming's right-hand rule assumes that the field is stationary and that the conductor is moving with respect to the stationary (reference) field. Since induced voltage depends on the relative motion between

---

* Throughout this text, conventional current is used. All left-hand or right-hand rules must therefore be reversed by the reader desiring to use the direction of electron flow.

conductor and field, it may be applied in the case of a stationary conductor and moving field, but with the assumption that the conductor is moving in the opposite direction. Since the thumb in Fig. 1-3a shows the direction of relative upward motion of the conductor only, the direction of induced emf in the figure would represent downward motion of a field with respect to the stationary conductor. Using the thumb to represent conductor motion, the forefinger (index finger) to represent the direction of the magnetic field, and the middle (third) finger to represent induced emf, the reader may verify the direction of induced emf of Fig. 1-3b, which is opposite to that of Fig. 1-3a because its direction has been reversed.

1-6. LENZ'S LAW

By way of summary, it should be noted that Faraday's law of electromagnetic induction is but one of the electromechanical effects which relate the mechanical force applied to a body with an electromagnetic field as discussed in Section 1-2. Whereas the preceding paragraphs stressed motion and the direction of motion, it must be emphasized that the motion of a conductor in a magnetic field is the result of a mechanical force (work) applied to the conductor. The electric energy produced by electromagnetic induction, therefore, requires an expenditure of mechanical energy in agreement with the law of conservation of energy. Energy for electromagnetic induction is not furnished by the magnetic field, as some might suppose, since the field is not changed or destroyed in the process.

The directions of induced voltage and current flowing in the conductor, represented in Figs. 1-2 and 1-3, bear a definite relation to the change in flux linkage inducing them. This relation is stated by Lenz's law:*

> In all cases of electromagnetic induction, an induced voltage will cause a current to flow in a closed circuit in such a direction that its magnetic effect will oppose the change that produces it.

The above statement of Lenz's law implies both (1) a cause and (2) an effect opposing a cause. The implied cause is not necessarily conductor motion resulting from a mechanical force, but a *change in flux linkages*. The implied effect is a current (due to an induced voltage) whose field

---

* In 1833, Heinrich Lenz reported that "the electrodynamic action of an induced current opposes equally the mechanical action inducing it." It should be noted that Lenz's law is, in reality, an extension of Le Chatelier's principle. The latter states that natural forces exist in an equilibrium in such a manner so as to oppose any change in the equilibrium. Newton's third law of motion is equally derived from this principle: to every action, there is an equal and opposite reaction. Further, the law of conservation of energy is implied in Lenz's law, since it requires mechanical energy to produce electric energy by electromagnetic action. Thus, it is only when a force overcomes resistance that energy is expended.

opposes the cause. Thus, in all cases of electromagnetic induction, whenever a change in flux linkage occurs, a voltage is induced that tends to set up a current in such a direction as to produce a field which opposes the change in flux linking the turns of the circuit. If viewed in this manner, a concept of Lenz's law will emerge which satisfies all cases of induced emf, including transformer and induction motor action as well as induced emf in dc motors and generators.

It can also be shown that the property of inductance is an effect and a result of Lenz's law (which implies that the voltage generated in a conductor by a change of flux linkages will establish a current whose surrounding magnetic field tends to oppose the change of flux linking that conductor). In fact, when a circuit or component possesses the property of opposing any change of current in itself, that property is called *inductance*, and the induced emf is termed an *emf of self-induction*. This is discussed in greater detail in Chapter 2, Eq. (2-2), and in Chapter 13.

Consider the conductor shown in Fig. 1-4a as an elementary

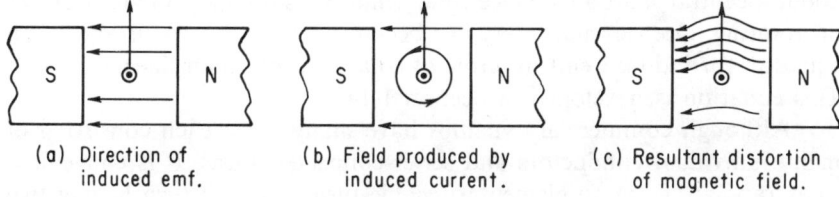

(a) Direction of induced emf.   (b) Field produced by induced current.   (c) Resultant distortion of magnetic field.

Figure 1-4

Illustration of Lenz's law.

generator which is being moved by some prime mover in the upward direction as shown. If some electric load is connected to this elementary generator, current will tend to flow in the conductor, in the same direction as the emf, producing a magnetic field around the conductor as shown in Fig. 1-4b. The counter-clockwise magnetic field surrounding the conductor repels the magnetic field above it and attracts the magnetic field below it (i.e., the induced current produces a field which opposes the motion that caused it). The tendency of the magnetic field is, therefore, of such a nature, by Lenz's law, as to oppose the upward motion of the conductor.

In the case of an elementary generator, electric energy is consumed only when a load completes the path so that a current flows due to the induced emf. But the field produced by this load current acts in such a way so as to react with the magnetic field of the generator and thus oppose the prime mover driving the generator. As more electric energy is drawn by the load, the conductor current produces a stronger field in opposition to the motion of the prime mover driving the generator. The greater the electric energy drawn from the generator, therefore, the greater the opposition produced by field interaction, and more mechanical energy is required to turn the generator. Conversely, if no load current is drawn from the elementary generator, no field is produced around the conductor since there is no induced current; and, theoretically, no energy is required from the prime mover. Again, in accordance with the law of conservation of energy, work is done only in overcoming a resistance.

1-7.
ELEMENTARY
GENERATORS

It goes without saying that the generators shown in Figs. 1-1 through 1-4 and discussed in the preceding paragraphs are impractical for a number of reasons. One of these reasons is that such generators would require a prime mover which imparts linear or reciprocating motion to the conductor. Commercial prime movers provide rotary motion to commercial electric generators (including those, such as a steam engine, that produce reciprocating motion). The conductors of most commercial generators, therefore, rotate about a central shaft axis. Since rotary motion is involved in all instances of rotating electric machinery, it becomes necessary to establish an equation for induced emf in terms of rotary (rather than linear) motion. This equation is developed in Section 1-14.

Although commercial dynamos have many coils, each consisting of many individual conductors and series-connected turns, it is convenient to extrapolate from an elementary single-turn coil (one turn having two conductors), rotating clockwise in a bipolar field, shown in Fig. 1-5a. The direction of emf induced in each conductor or coil-side may be determined by Fleming's right-hand rule or by means of Lenz's law as described in Section 1-8. The polarity of the emf of an elementary generator will be defined in Section 1-9, and the nature of the emf waveform determined in Section 1-10.

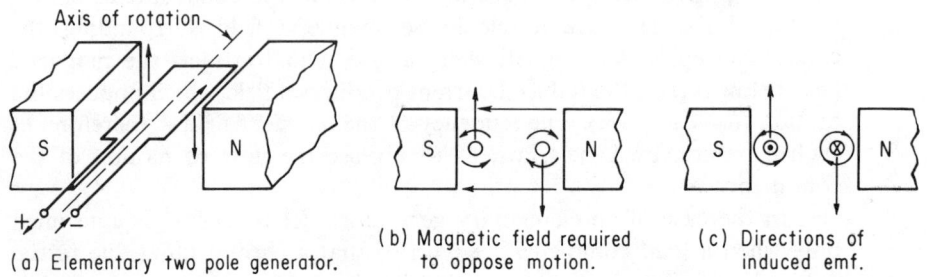

(a) Elementary two pole generator.

(b) Magnetic field required to oppose motion.

(c) Directions of induced emf.

Figure 1-5

Proof of Fleming's right-hand rule by means of Lenz's law.

**1-8.
PROOF OF
FLEMING'S
RIGHT-HAND RULE
BY MEANS OF
LENZ'S LAW**

It is interesting to note that the direction of induced emf for a specific conductor moving in a given magnetic field may also be verified by means of Lenz's law. This technique requires use of the "right-hand corkscrew rule" for direction of flux around a current-carrying conductor,* as well as Lenz's law.

The upward motion of the conductor on the *left-hand* side, shown in Fig. 1-5b, should produce an emf and current whose magnetic field would oppose the upward motion of the conductor. This method of verification asks: "What type of magnetic field will oppose the motion of the conductor?" Reason indicates that a *counterclockwise* magnetic field will oppose the conductor motion, because such a field produces repulsion above the conductor and attraction below the conductor. Lines of force in the same direction produce repulsion, and in the opposite direction produce attraction.

In the case of the conductor which is on the *right-hand* side shown in Fig. 1-5b, since the conductor is moving downward, the field around the conductor would require attraction above and repulsion below the conductor to oppose conductor motion by Lenz's law. This is accomplished by means of a *clockwise* magnetic field around the conductor on the right. Note that Fig. 1-5c agrees with Fleming's right-hand rule in determining the direction of induced emf. Note also that since both conductors are in the *same* magnetic field but traveling in *opposite* directions, the emfs and resultant magnetic fields produced by the conductor current are reversed with respect to each other.

**1-9.
POLARITY OF
AN ELEMENTARY
GENERATOR**

It should be noted that the polarity of the elementary two-pole generator in Fig. 1-5a shows the left-hand conductor as positive and the right-hand conductor as negative. This polarity designation may cause some confusion, since in conventional current flow it is assumed that current flows from the positive to the negative terminal. There is no inconsistency in this designation, however, if the reader realizes that the conductor is to be treated as a *source* of emf, i.e., a battery. If an external load were connected to the terminals, shown in Fig. 1-5a, current would flow from the positive terminal through the load and back to the negative terminal of the source. Since a generator coil, and indeed a complete generator, is a *source* of emf, its polarity will always be determined by the direction of current flow it will produce in an external load.

---

* H. W. Jackson, *Introduction to Electric Circuits*, 3rd. ed., (Englewood Cliffs, N.J.: Prentice-Hall, Inc., 1970), Sec. 8-4.

1-10.
SINUSOIDAL EMF GENERATED BY A COIL ROTATING IN A UNIFORM MAGNETIC FIELD AT CONSTANT SPEED

If the single-turn coil of Fig. 1-5 is rotated in a uniform magnetic field at a constant speed, as shown in Fig. 1-6a, the emf induced in a given coil side will vary as the coil moves through the various positions 0 through 7, as shown in the figure.

Using coil side *ab* as a reference, note that when this coil side is in position 0 shown in Fig. 1-6a, the emf induced in the coil is zero, since conductor *ab* (and conductor *cd*, as well) is moving *parallel* to the magnetic field and experiencing no change in flux linkages. When conductor *ab* moves to position 1,

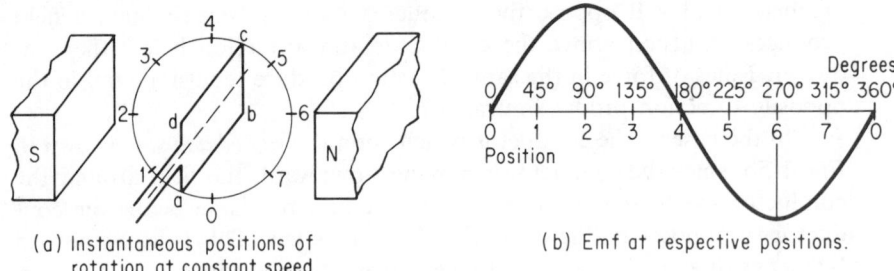

(a) Instantaneous positions of rotation at constant speed.

(b) Emf at respective positions.

Figure 1-6

EMF generated by a coil moving in a uniform field.

rotating in a *clockwise* direction, it cuts the uniform magnetic field at an oblique angle of 45°. The emf induced in this upward-moving conductor with respect to an external load will be positive (by the method described in Section 1-9) and will amount to approximately 70.7 per cent of the maximum induced voltage [by Eq. (1-4) where θ is 45°]. The change in voltage is shown graphically in Fig. 1-6b, where the emf is positive at position 1 and has the approximate value given. When the coil reaches 90°, position 2, conductor *ab* has the maximum number of flux linkages, since it is moving perpendicularly to the magnetic field, and has the maximum positive value shown in the previous figure and in Fig. 1-6b. Position 3, corresponding to 135° of rotation, yields an emf in coil side *ab* identical to that produced at position 1 [sin 135° = sin 45° in Eq. (1-4)] with a positive polarity since the conductor is still moving upward but the change in flux linkage occurs at a lower rate than at position 2. When conductor *ab* reaches 180°, position 4, the induced emf is again zero, since no change in flux linkage occurs when a conductor is moving parallel to a magnetic field. In position 5, corresponding to 225°, the emf induced in conductor *ab* is reversed since conductor *ab* is now moving downward in the same uniform magnetic field. The emf induced increases to a negative maximum at 270°, position 6, and finally decreases through position 7 back to zero at position 0.

14

CHAP. ONE / *Electromechanical Fundamentals*

It should be noted that the nature of the induced emf in a conductor rotating in a magnetic field is *both* sinusoidal and alternating. It will later be seen that an alternating emf is produced in the conductors of all dynamos, dc as well as ac. Observe that during this process no emf is induced in coil sides *bc* or *ad* since these conductors experience no change in flux linkages. Even if an emf were produced in these coil sides, they would not contribute to the emf of the coil because they are moving in the *same* direction through the *same* field and hence would produce equal emfs in opposition. Coil sides *ab* and *cd*, however, *aid* each other, and the total emf produced by the coil is *twice* the magnitude represented in Fig. 1-6b. It should be noted that no emf is produced in positions 0 and 4, known as the *interpolar* or *neutral zones* of the dynamo.

It should be emphasized that a *sinusoidal waveshape* is produced by a conductor rotating in a *theoretically uniform field*, represented in Fig. 1-6, in which the air gap is not constant because of the straight pole sides. If the pole tips are curved so that the pole face produces a more uniform gap and flux density (except in the interpolar regions), the waveshape of the induced emf will tend to be flat-topped, more nearly approaching a square wave than a sine wave. (See Fig. 1-8).

1-11. RECTIFICATION BY MEANS OF A SPLIT-RING COMMUTATOR

All rotating electric machines, regardless of their type or purpose, generate alternating current (ac), with the single exception of the *homopolar dynamo*.* The truth of this statement emerges from consideration of Fig. 1-6 and the fact that commercial dynamos employ many conductors which move relative to field poles of alternate magnetic polarity, N–S–N–S–N, etc. Each time a conductor moves in the same direction but under a pole of opposite polarity, the direction of induced emf reverses. Thus, if the ends

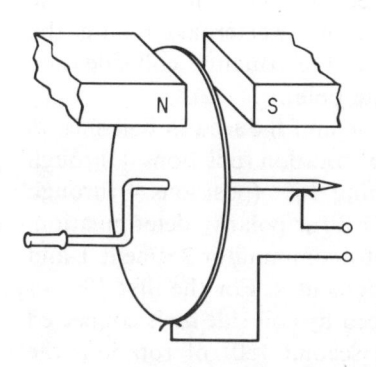

---

* It is rather ironic to note that the first electric generator discovered by Michael Faraday and reported in his diary of October 28, 1831, was in fact a *true* dc generator, known as *Faraday's disc*. This dynamo is the single exception to all of the various rotating machines developed since Faraday's first discovery of the homopolar generator and motor. The dynamo consisted of a disc of copper rotating in a permanent magnetic field (bipolar). When the disc was manually rotated, an emf of pure direct current was generated between the center of the disc and its extreme outer edge. Conversely, if a dc voltage is applied to the same terminals, the disc rotates as a homopolar motor. In the author's personal experience as a teacher, several of his students have independently "discovered" the homopolar principle. It is of final interest to note that the principle of magnetohydrodynamics (MHD) is, in reality, homopolar generation in which the moving conductor is an ionized plasma. This type of generation appears to lend itself to generation by nuclear fusion at extremely high temperatures. Thus, it would seem that from the first discovery by Faraday, the cycle has come full around and we are back to the homopolar principle once again! See Section 11-4 for a further discussion of the homopolar machine.

of the coil sides in Fig. 1-6a were connected to some external circuit by means of slip rings, an alternating voltage would flow through that circuit, as shown in Fig. 1-6b.

In order to convert the alternating voltage (ac) to unidirectional current (dc), it is necessary to employ a mechanical switching device which is actuated by the mechanical rotation of the dynamo shaft. The simple split-ring *commutator*, shown in Fig. 1-7a, is such a switching

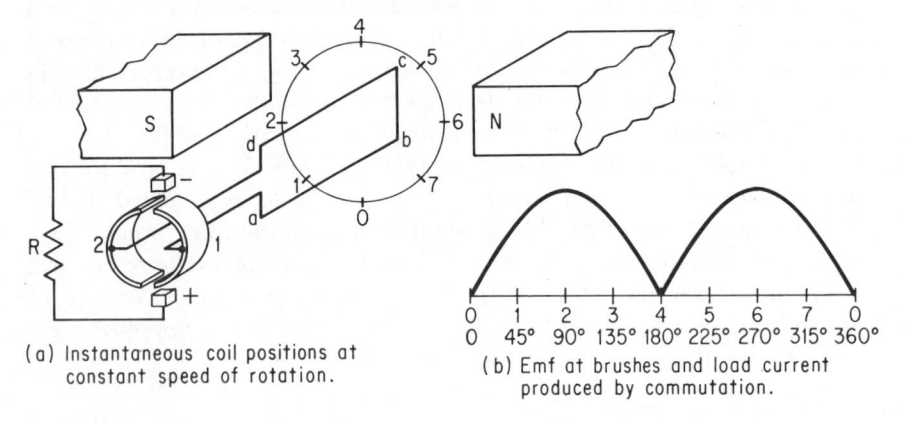

(a) Instantaneous coil positions at constant speed of rotation.

(b) Emf at brushes and load current produced by commutation.

Figure 1-7

Bipolar generator with two segment commutator.

device. The commutator shown consists of two segments, secured to and insulated from the armature shaft and from each other as well. Each conducting commutator segment is connected, respectively, to a coil side. Since both coil side and commutator segment are fastened mechanically to the same shaft, the action of the mechanical rotation is to reverse the armature coil and connections to the stationary external circuit at the same instant that the induced emf reverses in the armature coil side (i.e., when the coil side moves under an opposite pole).

As shown in both Figs. 1-6 and 1-7, the emf induced in coil side *ab* causes a positive polarity for the first 180° of rotation (positions 0 through 4), and a negative polarity for the remaining 180° (positions 4 through 0), using the method developed in Section 1-8 for polarity determination. But, in Fig. 1-7, coil side *ab* is connected to commutator segment 1 and coil side *cd* is connected to commutator segment 2. For the first 180° of rotation, therefore, the positive emf produced by coil side *ab* is connected to the stationary positive brush. For the second 180° of rotation, the negative emf produced by coil side *ab* is connected to the stationary negative brush. The same effect occurs in reverse order for coil side *cd*. In effect, therefore, the action of the commutator is to reverse connections to the external circuit simultaneously and at the same instant that the direction of emf reverses in each of the coil sides. Each brush, positive or negative, respectively, is always maintained, therefore, at the *same* polarity. Figure 1-7b shows the emf (and current) waveform produced as a result of the above process for one full cycle (or 360°) of rotation.

The split-ring commutator and conductors shown in Fig. 1-7(a) are represented in *cross section* in Fig. 1-8a. The position illustrated in Fig. 1-8a is displaced 90° in the direction of rotation from that shown in

(a) Elementary generator.

(b) Emf at brushes produced by coil rotating in a uniform magnetic field.

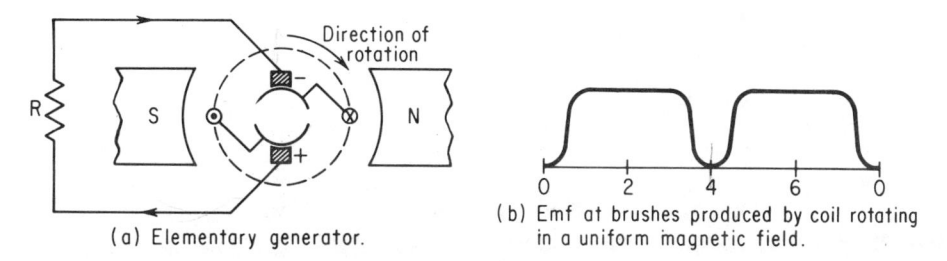

Figure 1-8

Elementary dc generator.

Fig. 1-7a. Note that the brushes in both figures are *perpendicular* to the pole axis, so that *commutation* of the conductor occurs when the coil is *perpendicular* to the magnetic field (in the so-called *neutral zone* or *interpolar* space) or in the position shown in Fig. 1-7a, where no emf is induced, as at points 0 and 4 of Fig. 1-7b. Thus, in Fig. 1-7a, no current flows in load resistor R at the instant illustrated, whereas in Fig. 1-8a, the maximum current is flowing in the external load R. Note that the pole faces of Fig. 1-8a are *curved* in order to produce a more uniform magnetic field and thus improve the wave form, producing a dc voltage (and current) of higher average value; in effect, maintaining the maximum for a longer period of the cycle.

The pulsating unidirectional current, which has a zero value twice each cycle as shown in Figs. 1-7b and 1-8b, is hardly suitable for commercial dc use. The output emf may be made less pulsating by using a large number of coils and commutator segments. The effect of increasing the number of coil sides and segments is shown in Fig. 1-9a, and the resultant waveform is shown in Fig. 1-9b. With only two brushes and four segments, there are now four commutations, shown as *a, b, c, d* of Fig. 1-9b, in one full cycle of rotation (time t to t'). Although the resultant emf is less pulsating, this particular type of winding (half-coil, open winding) is not employed for several reasons. The main disadvantage is that the emfs generated in each coil are not added because the coils are not connected in series. Note that Coil 2 passes out of contact with the brushes at *a*, and Coil 1 passes out of contact with the brushes at *b*.

(a) Cross-sectional view.

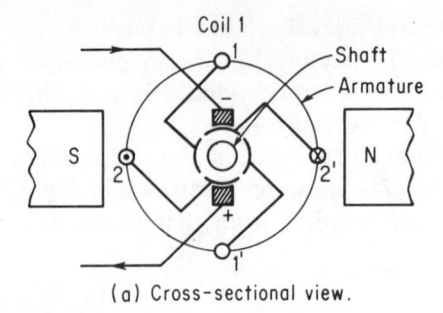

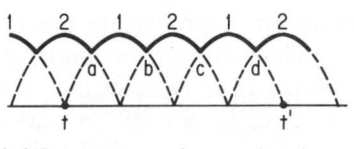

(b) Resultant waveform at brushes.

Figure 1-9

Effect of four conductors and segments on output waveform.

| 1-12. THE GRAMME-RING WINDING | One of the earliest types of armature windings designed to produce series-connected conductors was the *gramme-ring* winding. Although obsolete for many years, it is shown here |

because it illustrates the modern commercial armature winding so well. The *armature* is the structure which supports and protects the current-carrying conductors; and, since it is part of the magnetic circuit, as shown in Fig. 1-9a, it is constructed of (laminated) iron to reduce magnetic reluctance. The simplest type of armature construction and method of "winding" the conductors is employed, as shown in Fig. 1-10a. The armature is a laminated iron cylinder which provides two low-reluctance paths for the flux linking the conductors. The armature winding is wound axially and spirally around the cylinder, with taps from the commutator segments to equally spaced portions of the winding, as shown in Fig. 1-10b. Unlike the open winding shown in Fig. 1-9a, the gramme-ring winding is known as a *closed* winding because all the coils connected between the brushes are in series and the winding is *re-entrant*, i.e., it closes on itself, as shown in Fig. 1-10a. For the clockwise direction of rotation shown, the induced emfs will produce the brush polarities given in Fig. 1-10b, using the method described in Section 1-8.

As with our elementary generator described in Section 1-10, those conductors located under a given pole will all have the same direction of emf, which is opposite to that for conductors under a pole of opposite polarity, also as shown in Fig. 1-10b. Each of the coil sides under a given pole will have some emf induced in it; and, since they are all connected in series and the directions of induced emf are the same, the emfs are additive. The equivalent armature circuit representing the directions of induced emf for the conductors under the two poles, respectively, is shown in Fig. 1-10c. Assuming that the armature is perfectly centered between the poles so that the air gaps are identical, the sum of the induced

CHAP. ONE / *Electromechanical Fundamentals*

(a) Bipolar spiral winding and flux paths.

(b) Cross section showing conductors, induced voltages (and currents) and commutator connections.

(c) Equivalent armature circuit.

(d) Resultant waveform

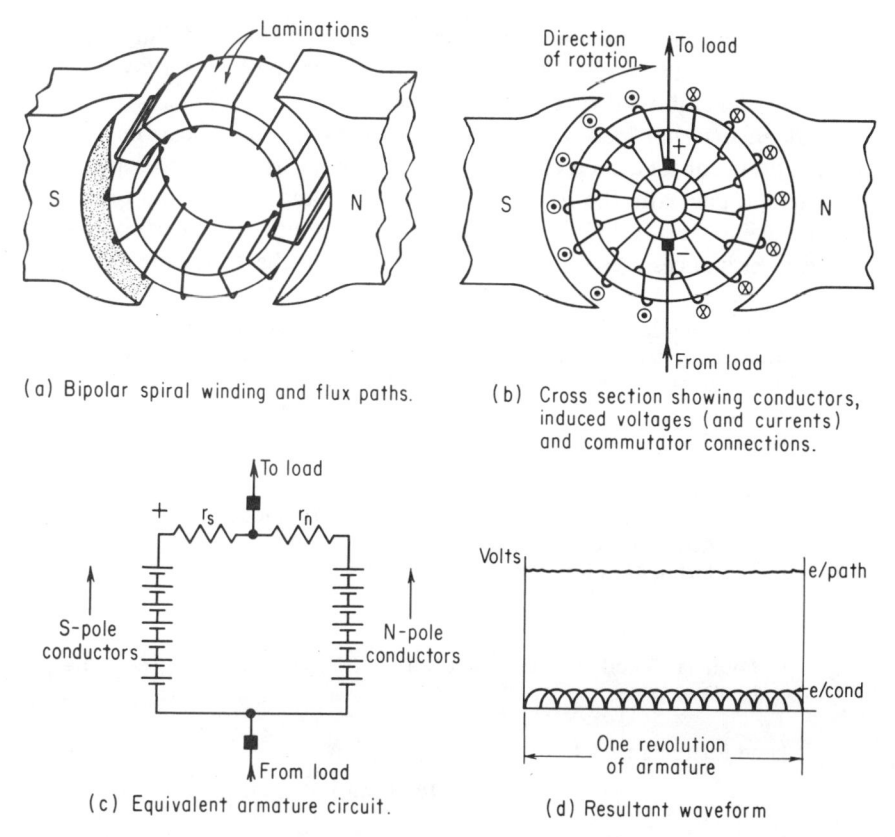

Figure 1-10

Gramme-ring winding.

emfs under each pole will be equal and opposite, producing no circulating current in the two paths of the equivalent battery shown in Fig. 1-10c. This is true even if the brushes (located in the interpolar axis) are lifted off the commutator. Observe that the two brushes are located at the points of maximum negative and positive polarity, and, when they are connected to an external load, serve to produce two parallel conductor paths. Those conductors that are immediately adjacent to the brushes may not have as high an induced voltage as those directly under the center of each pole; but, since the total voltage between the brushes is the sum of all the induced voltages, and since both paths have equal total emfs, this is of little concern. The resultant waveform, produced by summing the individual flat-topped waves, contains very little ripple (the ac component is practically zero), as shown in Fig. 1-10d.

Since one conductor immediately takes the place of another during each revolution of the armature, the cross section shown in Fig. 1-10b is a dynamic representation of a constant or static condition. We can

SEC. 1-12. / The Gramme-Ring Winding

consider that the total number of conductors producing voltage at any instant is constant, and that the resultant emf per path, for a given speed and flux density, is also relatively constant.

If an external load were connected across the brushes, the current which flows through each of the two generator paths would be determined by the emf per path, the resistance of the load, and the internal resistance of the armature winding paths, r_s and r_n, the south and north pole conductor resistances, respectively, as indicated by the following example:

EXAMPLE 1-3: A two-pole dc generator has an armature containing a total of 40 conductors connected in two parallel paths. The flux per pole is 6.48×10^8 lines, and the speed of the prime mover is 30 rpm. The resistance of each conductor is 0.01 ohm, and the current-carrying capacity of each conductor is 10 A. Calculate:
 a. The average generated voltage per path and the generated armature voltage
 b. The armature current delivered to an external load
 c. The armature resistance
 d. The terminal voltage of the generator.

Solution:

a. Total ϕ linked in one revolution $= P \times \phi/\text{pole} = 2 \text{ poles} \times 6.48 \times 10^8 \text{ lines/pole}$.

Time for one revolution, $t/\text{rev} = \frac{1}{30}$ min/rev
$$= (60 \text{ s/min}) \times \tfrac{1}{30} \text{ min/rev} = 2 \text{ s/rev}$$

From Eq. (1-1),
$$e_{av}/\text{cond} = \frac{\phi}{t} \times 10^{-8} \text{ V}$$
$$= \frac{2 \times 6.48 \times 10^8 \text{ lines}}{2 \text{ s/rev}} \times 10^{-8} \text{ V}$$
$$= \mathbf{6.48 \text{ V/conductor}}$$

Generated voltage per path,
$$E_g = (\text{voltage/cond}) \times \text{no. of cond/path}$$
$$= (6.48 \text{ V/cond}) \times 40 \text{ cond/2 paths}$$
$$= \mathbf{129.6 \text{ V/path}}$$

Generated armature voltage,
$$E_g = \text{generated voltage/path}$$
$$= \mathbf{129.6 \text{ V}}$$

b. $I_a = (I/\text{path}) \times 2 \text{ paths} = (10 \text{ A/path}) \times 2 \text{ paths} = \mathbf{20 \text{ A}}$

c. $R_a = \dfrac{r \text{ per path}}{\text{no. of paths}} = \dfrac{0.01 \text{ ohm/cond}}{2 \text{ paths}} \times 20 \text{ cond} = \mathbf{0.1 \ \Omega}$

d. $V_t = E_g - I_a R_a = 129.6 \text{ V} - [20 \text{ A} \times 0.1 \ \Omega] = \mathbf{127.6 \text{ V}}$ \hfill (1-10)

The gramme-ring winding is no longer used because the conductors on the inside of the ring are *inactive*, i.e., no change in flux linkages are produced in them because they are magnetically shielded by the armature. In order to save copper and to reduce the weight of the armature winding, drum-wound armatures are employed in which the conductors all lie on the outer surface of the armature, placed in slots in such a way that a good portion of the conductor is active. A theoretical advantage of the gramme-ring winding, however, is that *any* given armature is adapted to any number of poles. The effect of increasing the number of poles on the equivalent circuit and emf is shown in Fig. 1-11.

The four-pole armature winding of Fig. 1-11 is still a closed winding; and, since there are now four interpolar zones in which commutation can take place, four brushes are required. For the sake of clarity, these

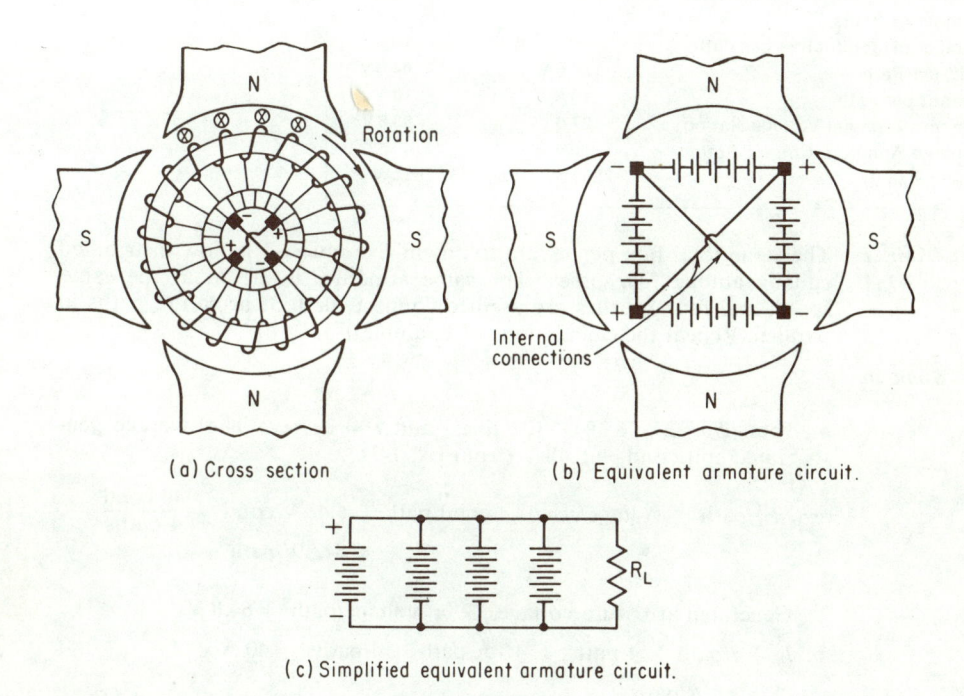

(a) Cross section

(b) Equivalent armature circuit.

(c) Simplified equivalent armature circuit.

Figure 1-11

Four-pole gramme-ring winding.

brushes are drawn on the inside of the commutator in Fig. 1-11a. By means of Fleming's right-hand rule, the direction of induced emf and the polarity may be verified. The two brushes of positive polarity and the two of negative polarity, respectively, are internally connected as shown in Figs. 1-11a and b. A simplified, redrawn form of the equivalent circuit is shown in Fig. 1-11c. Note that the two-pole winding of Fig. 1-10 required two brushes and produced two armature paths. The four-pole winding of Fig. 1-11 requires four brushes and produces four armature paths. The effect of increasing the number of poles on the voltage, current, and power developed by the dynamo is illustrated in Example 1-4 and Table 1-1.

TABLE 1-1.
EFFECT OF THE NUMBER OF POLES ON VOLTAGE, CURRENT, AND POWER RELATIONS OF A DYNAMO

PARAMETER	NUMBER OF POLES 2	NUMBER OF POLES 4
Number of Armature Conductors	40	40
Number of Paths	2	4
Number of Conductors per Path	0	10
EMF per Path	129.6 V	64.8 V
Current per Path	10 A	10 A
Dynamo Terminal Voltage Rating	127.6 V	63.8 V
Dynamo Armature Current Rating	20 A	40 A
Power Rating	2552 W	2552 W

EXAMPLE 1-4: The same total flux per revolution as in Example 1-3 is now distributed equally among *four* poles. The same armature is driven at the same speed, and four brushes are used to connect the four armature paths in parallel. Repeat the calculations of Example 1-3.

Solution:

a. Total $\phi = 2 \times 6.48 \times 10^8$ lines, and $t = 2$ s/rev. The average generated emf/cond = 6.48 V (from Ex. 1-3).

$$E_g/\text{path} = \text{voltage/cond} \times \text{cond/path} = 6.48 \text{ V/cond} \times \frac{40 \text{ cond}}{4 \text{ paths}}$$
$$= \textbf{64.8 V/path}$$

Generated armature voltage, E_g = voltage/path = **64.8 V**

b. $I_a = I/\text{path} \times 4 \text{ paths} = 10 \text{ A/path} \times 4 \text{ paths} = \textbf{40 A}$

c. $R_a = \dfrac{R/\text{path}}{\text{no. of paths}} = 0.01$ ohm/cond $\times$ 10 cond/4 paths $= \textbf{0.025 } \Omega$

d. $V_t = E_g - I_a R_a = 64.8 \text{ V} - [40 \text{ A} \times 0.025 \text{ }\Omega] = \textbf{63.8 V}$ (1-10)

Note that the terminal voltage and the generated voltage are each reduced to one-half of their original values, but that the armature current has doubled. Examples 1-3 and 1-4 are summarized in Table 1-1.

**1-13.
DYNAMO VOLTAGE,
CURRENT, AND
POWER RATINGS**

Table 1-1 serves to illustrate a fundamental relationship that applies to all modern dynamo armature windings. In the commercial generator, a large number of conductors are employed to link the flux with one or more pairs of poles (the number of poles is always an even number). As illustrated by the preceding problems and Table 1-1, commercial armatures may have two or more parallel paths (the number of paths is also always an even number). Each path consists of a series-connected group of coils, each coil possessing an allowable voltage rating (in the case of a motor) or a generated voltage rating (for rated flux and speed, in the case of a generator). The voltage rating of the dynamo, therefore, is determined only by the approximate equal number of series-connected coils per path and not by the number of paths in parallel.*

The major factor in the current rating of the dynamo is the *current-carrying capacity* of the individual coil or conductor in each path, or the series-connected group of coils. As the number of paths is increased, the current rating of the dynamo is increased. It is most important to realize, however, that the number of paths and the current rating in a given dynamo may be increased *only* at the expense of the voltage rating, since the total number of conductors or coils is fixed for a given armature.

The implication of the last statement refers back to a fundamental relationship that applies to cells and batteries. A battery may consist of a series-parallel group of cells. The power rating of each cell determines, in effect, the power rating of the battery, *regardless of the method of connection*. For a given number of cells, therefore, the power rating of any battery is fixed, although its voltage and current ratings may vary with the series-parallel connections employed.

The same situation that exists with respect to voltage, current, and power ratings of cells and batteries applies to the conductors and the armature windings of a dynamo. The power rating of a given armature is actually fixed by the current and voltage rating of its individual coil in a given path. The *only* way to *increase* the power rating of a dynamo, based on the above consideration, is to employ a *larger* armature having *more* conductors and coils. Thus, physical size is an approximate indication of the power rating of electric machinery.

* For any given number of armature conductors, however, increasing the number of parallel paths must of course reduce both the number of series-connected coils per path and the voltage.

Table 1-2 illustrates that power rating is independent of the manner in which armature conductors are connected. For the sake of simplicity, the voltage and current rating of each conductor of a given armature are 10 V and 10 A respectively. The armature contains 120 conductors, connected in various numbers of parallel paths. Observe that the power rating (120 conductors × 100 W/conductor) is fixed, but that the voltage rating decreases in the same proportion that the current rating increases.

TABLE 1-2.
EFFECT OF INCREASING THE NUMBER OF
PARALLEL PATHS IN AN ARMATURE

PATHS IN PARALLEL	VOLTAGE RATING VOLTS	CURRENT RATING AMPERES	POWER RATING WATTS
2	600	20	12,000
4	300	40	12,000
6	200	60	12,000
8	150	80	12,000
10	120	100	12,000
20	60	200	12,000

**1-14.
AVERAGE EMF
GENERATED IN
A QUARTER-
REVOLUTION**

The preceding paragraphs have stated (repeatedly) that the emf between the brushes of multicoil armatures is produced by many coils connected in series, in which each coil consists of many conductors of wire. In order to calculate the resultant emf between brushes, it is first necessary to determine the average emf induced in a single conductor (see Example 1-3) in one quarter of a revolution (i.e., 90 electrical degrees), in which the conductor moves from a position at the center of the interpolar zone to a position directly under the center of a given pole. As shown in Figs. 1-6b and 1-7b, a single coil side rotates from position 0 to position 2 in a quarter of a revolution (i.e., from a position where there are zero flux linkages to a position of maximum flux linkage). The average emf induced in each conductor may be derived in the following manner. Assume that the total flux produced between the poles of Figs. 1-6a and 1-7a consists of ϕ lines, and that t is the time required for one quarter of a revolution (i.e., 90 electrical degrees). Since the flux linkages have gone from zero to maximum in one quarter of a revolution, the average emf induced in a single turn of two coil sides during this period (see Sec. 1-3) is

$$E_{av} = \frac{\phi}{t} \times 10^{-8} \quad \text{V} \tag{1-1}$$

But, since the time t for one quarter of a revolution is $1/4s$, where s is the number of revolutions of the coil per second, the average induced emf per turn is, by substitution

$$E_{av} = 4s\phi \times 10^{-8} \quad \text{V}$$

For an armature coil consisting of N turns, the average induced emf per coil is

$$E_{av/coil} = 4\phi Ns \times 10^{-8} \quad \text{V} \tag{1-5}$$

where ϕ is the number of lines or maxwells per pole
N is the number of turns per coil
s is the relative speed in revolutions per second (rps) between the coil of N turns and the magnetic field

Note that the derivation of Eq. (1-5) follows exactly the procedure used in the solution of Example 1-3, with the exception that *turns* are used instead of conductors. There are two coil sides (or two conductors) per single-turn coil.

EXAMPLE 1-5: Compute the average emf per coil and per conductor (per coil side) for the one-turn coil of Example 1-3, using Eq. 1-5.

Solution:

$$E_{av/coil} = 4\phi Ns \times 10^{-8} \text{ V} \tag{1-5}$$
$$= 4(6.48 \times 10^8 \text{ lines/pole})(1 \text{ turn})(30 \text{ rev/min} \times \tfrac{1}{60} \text{ min/s})$$
$$\times 10^{-8} \text{ V}$$
$$= \mathbf{12.96 \text{ V}}$$

$$E_{av/coil\ side} = 12.96 \text{ V/coil} \times \tfrac{1}{2} \text{ coil/coil side} = \mathbf{6.48 \text{ V/coil side}}$$

1-15. FUNDAMENTAL DC GENERATOR VOLTAGE EQUATION FOR EMF BETWEEN BRUSHES

Equation (1-5) makes it possible to calculate the average voltage rating of a single coil (having one or more turns) rotating at a given speed (rps) under a given pole of known field strength. But the discussion of Section 1-13 considered voltage between brushes in terms of the total number of conductors and paths on a given armature in combination with a given number of poles. The average induced emf between brushes may be derived, as follows.

If Z is the total number of armature conductors or coil sides, and if a is the number of parallel coil paths between brushes of opposite polarity, then the total number of turns N per armature circuit is $Z/2a$. Further, if, speed S is stated in rpm, then $s = S/60$. Finally, since Eq. (1-5) is derived for a bipolar machine, if a machine has P poles, the result must be multiplied by $P/2$. The total average induced emf between brushes, therefore, is

$$E_g = 4\phi Ns \times 10^{-8} = 4\phi\left(\frac{P}{2}\right)\left(\frac{Z}{2a}\right)\left(\frac{S}{60}\right)10^{-8} = \left(\frac{\phi ZSP}{60a}\right)10^{-8} \quad \text{V} \quad (1\text{-}6)$$

where ϕ is the flux per pole
P is the number of poles
Z is the number of armature conductors (twice the total armature turns)
a is the number of parallel armature paths
S is the speed in rpm

EXAMPLE 1-6: Compute (a) the average induced emf between brushes for the data of Example 1-4 using Eq. 1-6 and (b) the applied voltage required to overcome counter emf and armature resistance.

Solution:

a. $E_g = \dfrac{\phi ZSP}{60a} \times 10^{-8}$ V (1-6)

$= \left(\dfrac{2 \times 6.48 \times 10^8 \text{ lines}}{4 \text{ poles}}\right)\left(\dfrac{40 \text{ cond}}{4 \text{ paths}}\right)\left(\dfrac{30 \text{ rpm}}{60 \text{ sec/min}}\right) \times 4 \text{ poles} \times 10^{-8}$ V

$= 64.8$ V

b. $V_t = E_g + I_a R_a = 64.8 \text{ V} + [40 \text{ A} \times 0.025 \text{ }\Omega] = 65.8$ V (1-9)

Example 1-6 illustrates the fundamental unity of Eqs. (1-1) through (1-6), all of which are derived from Eq. (1-1), Neumann's quantification of Faraday's law.

Equation (1-6) applies to a dc motor as well as to a dc generator. In a motor, conductors of the armature are rotating past a magnetic pole, and an emf will be induced in these conductors, in accordance with Eq. (1-6), called a *counter emf*. The following example will serve to illustrate Eq. (1-6) and the relation between counter emf and applied voltage in a motor (see Example 1-4).

1-16. ELECTROMAGNETIC FORCE

It was shown in Section 1-2 that electromechanical energy conversion of practically all rotating electric machinery depends on two basic electromagnetic principles which are closely interrelated, namely (1) electromagnetic induction and (2) electromagnetic force. The basic principles of electromagnetic induction were discussed in the preceding sections, and we may now consider electromagnetic force and its relation to electromagnetic induction.

Figure 1-12a shows a current-carrying conductor located in a uniform magnetic field. *An electromagnetic force will exist between the conductor and the field whenever a current-carrying conductor is located in a magnetic field in such a position that there is a component of the active length of the conductor perpendicular to the field.* Thus, if a conductor is inserted or lies in a magnetic field, and a voltage is applied to the conductor so that current flows through the conductor, a force will be devel-

(a) Conductor carrying current in a magnetic field.

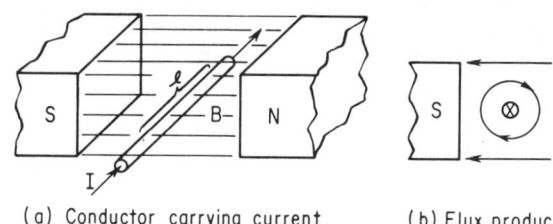

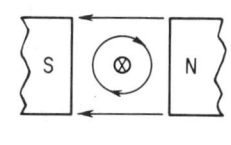

(b) Flux produced by conductor with respect to field.

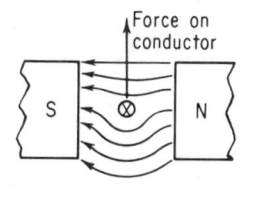

(c) Resultant distortion of magnetic field.

Figure 1-12

Conductor of length *l*, carrying a current *I*, in a magnetic field *B*, developing a result force *F*.

oped, and the conductor will tend to move with respect to the field or vice versa. This principle is sometimes called "motor action."

1-17. FACTORS AFFECTING MAGNITUDE OF EM FORCE

The definition above implicitly contains three requirements which affect the magnitude of the electromagnetic (EM) force, namely: a magnetic field (*B*), the length of active conductor (*l*), and the quantity of current (*I*) flowing in the conductor. Thus, if any or all of these three factors are varied, the EM force, *F*, will vary directly and in the same proportion. If, as in Section 1-4, the factors *B* and *l* are perpendicular, an orthogonal force *F* is developed.

$$F = \frac{BIl'}{10} \text{ dynes} \qquad (1\text{-}7)$$

where *B* is the flux density in lines per square centimeter
  I is the current in amperes (absolute)
  l' is the active conductor length in centimeters

In terms of English or practical units, the force *F*, in pounds, becomes*

$$F = \frac{(B/6.45 \text{ cm}^2/\text{in}^2) \times (I)(l \times 2.54 \text{ cm/in})}{10 \times 980 \text{ dynes/gram} \times 453.6 \text{ grams/lb}} = \frac{BIl}{1.13} \times 10^{-7} \text{ lb} \qquad (1\text{-}8)$$

where *B* is the flux density in lines per square inch
  I is the current in amperes
  l is the active length of the conductor in inches

---

* Since it is usually assumed that the current flow is in the same direction as the axial length of the conductor, Eq. (1-8) is not subject to the same conditions of orthogonality as Eq. (1-4) (see Section 1-4). If a problem arises, therefore, in which *B* and *l* are *not* perpendicular, Eq. (1-8) may be multiplied by the sine of the angle between the quantities *B* and *l*, as in Example 1-8.

SEC. 1-17. *Factors Affecting Magnitude of EM Force*

EXAMPLE 1-7: A single conductor 18 in. long carries a current of 10 A and is perpendicular to a uniform magnetic field of 50,000 lines/in². Calculate the EM force developed by the current-carrying conductor in pounds.

Solution:

$$F = \frac{BIl}{1.13} \times 10^{-7} \text{ lb} = \left(\frac{50{,}000 \text{ lines/in}^2 \times 10 \text{ A} \times 18 \text{ in}}{1.13}\right) 10^{-7} \text{ lb} \quad (1\text{-}8)$$

$$= \mathbf{0.797 \text{ lb}}$$

EXAMPLE 1-8: Repeat Example 1-7 with the conductor lying at an angle of 75° with respect to the same field (instead of 90°).

Solution:

$$F = \frac{BIl}{1.13} \times \sin(B, l) \times 10^{-7} \text{ lb} = 0.797 \text{ lb} \times \sin 75° = \mathbf{0.77 \text{ lb}}$$

1-18. DIRECTION OF EM FORCE AND LEFT-HAND RULE

Sections 1-16 and 1-17 above describe the magnitude and nature of the force developed orthogonal to the mutually perpendicular current-carrying conductor and magnetic field, as shown in Fig. 1-12a. It is possible to predetermine the direction of the EM force by the method shown in Figs. 1-12b and c. Figure 1-12b shows the clockwise magnetic field produced by the current-carrying conductor. Observe that the magnetic field produced by the conductor causes attraction of the main field above and repulsion below the conductor shown in Fig. 1-12b. The resultant distortion of the main magnetic field created by the field of the current-carrying conductor is shown in Fig. 1-12c. The tendency of the interaction of the two fields, therefore, is to force the conductor in an upward direction as shown in the figure. The relations between the direction of current in the conductor, the direction of the magnetic field, and the direction of the developed

(a) Left-hand motor rule.

(b) Right-hand generator rule.

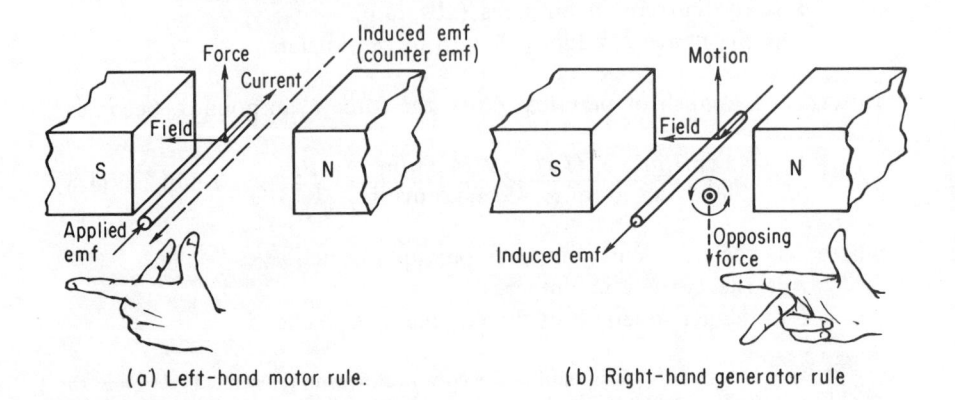

Figure 1-13

Comparison of motor and generator action.

CHAP. ONE / *Electromechanical Fundamentals*

force on the conductor may also be conveniently remembered and determined by means of the left-hand* or motor rule, as shown in Fig. 1-13a, for the same direction conditions as the previous figure. As in the case of Fleming's right-hand rule (see Fig. 1-3) for generator action, the index finger also indicates the direction of the field (N to S), the third or middle finger indicates the direction of current flow (or applied emf), and the thumb points to the direction of force developed on the conductor or the resultant motion.

1-19. COUNTER EMF

In Fig. 1-13 we have an opportunity (finally) to unify some of the relationships occurring in electromechanical energy conversion. Figure 1-13a shows motor action as described in the preceding section and its associated figure. For the direction of the field and armature current shown, the force developed on the conductor is in an upward direction. But the force developed on the conductor causes the conductor to move through the magnetic field, resulting in a change of flux linkage about this conductor. An emf is induced in the "motor" conductor of Fig. 1-13a. The direction of this induced emf is shown in Fig. 1-13b, for the same motion and field direction. Applying this induced emf to the conductor of Fig. 1-13a, observe that it opposes or is developed counter to the direction of current flow (and emf) which created the force or motion; hence, it is termed *counter emf*. Note that the development of a counter emf, shown as a dotted line in Fig. 1-13a, is an application of and in accordance with Lenz's law in that the direction of induced voltage opposes the applied emf which created it. Thus, *whenever motor action occurs, generator action is simultaneously developed*, as shown in Fig. 1-13a.

1-20. COMPARISON OF MOTOR ACTION VS. GENERATOR ACTION

If, whenever motor action occurs, generator action is also developed, the question may be raised as to whether the converse is also the case. Generator action is shown in Fig. 1-13b, where a mechanical force moves a conductor in an upward direction inducing an emf in the direction shown. When a current flows as a result of this emf, there is a current-carrying conductor existing in a magnetic field; hence, motor action occurs. Shown as a dotted line in Fig. 1-13b, the force developed as a result of motor action opposes the motion which produced it. It then may be stated categorically that *generator action and motor action occur simultaneously*

---

* A convenient mnemonic device is to picture an M-G (motor-generator) set with the motor on the left and the generator on the right. Thus, the left-hand rule is used for motor action, and the right-hand rule for generator action. The term "M-G set" is commonly used and not easily forgotten. These rules presuppose conventional current direction.

in rotating electric machinery. Hence, the same dynamo may be operated either as a motor or a generator, or *both.**

A more graphic representation in terms of rotational elements is presented in Fig. 1-14, which compares the elementary motor and genera-

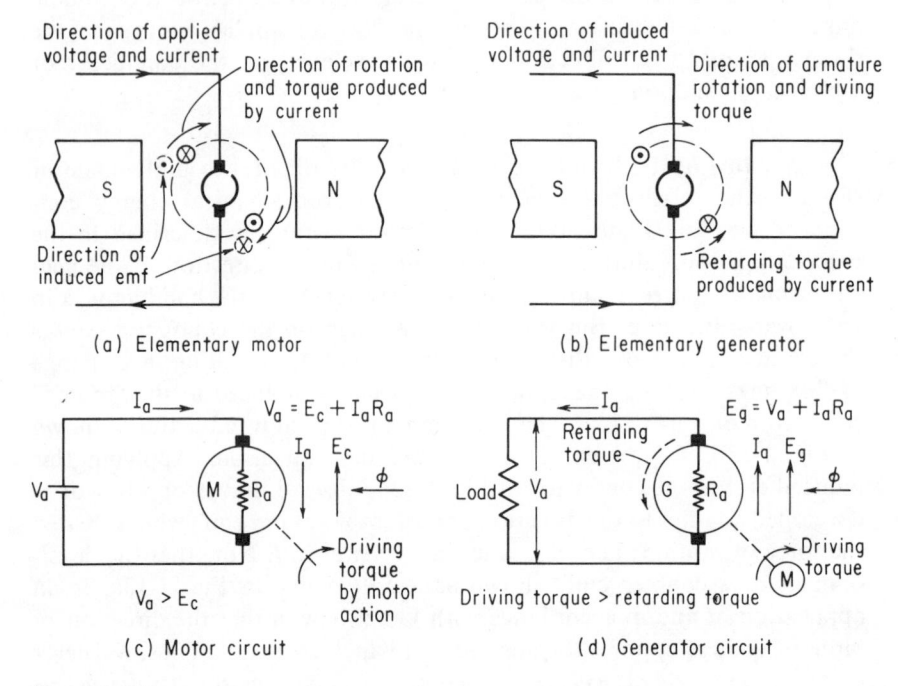

Figure 1-14

Elementary motor action vs generator action.

tor for the same direction of rotation and shows the electric circuits of each. The reader should study this figure very carefully because it is the *key* to an understanding of electromechanical energy conversion. Given the direction of applied voltage and current shown in Fig. 1-14a, the motor action that results produces a clockwise rotational force on both conductors. The direction of induced counter emf is also shown as opposed to the applied voltage, both in Fig. 1-14a and in the motor circuit of Fig. 1-14c. Observe that in order for the current to produce clockwise rotation and to have the direction shown in Fig. 1-14c, it is necessary that the applied armature terminal voltage, V_a, be greater than the developed counter emf, E_c. Thus, *when a dynamo is operating as a motor,* the generated counter emf is always *less* than the terminal voltage (that produces motor action) and it *opposes* the armature current.

Assuming that the elementary generator conductors of Fig. 1-14b

* As in the dynamotor or synchronous converter.

30

CHAP. ONE / *Electromechanical Fundamentals*

are rotated in a clockwise direction, an emf is induced in the direction shown in the figure. When connected to a load as shown in Fig. 1-14d, the resultant armature current that flows will produce the *retarding torque* shown by dots in both generator figures. Observe that, in the generator circuit of Fig. 1-14d, for the same direction of conductor rotation and magnetic field, the current flow is reversed. Note also that the retarding torque developed by the current flow opposes the driving torque of the prime mover. Thus, *when a dynamo is operating as a generator,* the armature current is in the *same* direction as the generated emf, and the generated emf E_g *exceeds* the armature terminal voltage V_a applied across the load.

This distinction between generator and motor, in which the armature-generated voltage aids or opposes the armature current, respectively, gives rise to the basic armature circuit equations shown in Fig. 1-14 and summarized, as follows:

For a motor, $$V_a = E_c + I_a R_a \tag{1-9}$$
For a generator, $$E_g = V_a + I_a R_a \tag{1-10}$$

where V_a is the applied voltage (measurable terminal voltage) across the armature
 E_c is the generated counter emf developed in the motor armature
 E_g is the generated emf developed in the generator armature
 $I_a R_a$ is the armature voltage drop due to a flow of armature current through an armature of given resistance, R_a

It should be noted that for a given dynamo, E_c and E_g may be evaluated precisely in terms of the factors expressed in Eq. 1-6. Also note that when the armature current, I_a, is flowing, V_a is a quantity which may be measured by means of a voltmeter, whereas E_g and E_c are inferred quantities determined only by computation from Eq. (1-6), (1-9), or (1-10).

EXAMPLE 1-9: The armature of a motor has a resistance of 0.25 ohm, and, when connected to a 125 V dc bus, draws an armature current of 60 A. Calculate the counter emf generated in the armature conductors of the motor.

Solution:
$$E_c = V_a - I_a R_a = 125 - (60 \text{ A} \times 0.25 \, \Omega) = 110 \text{ V} \tag{1-9}$$

The preceding example indicates that it is possible to calculate the generated emf of a motor from *external* measurements. The following example shows how it is possible also to determine the flux per pole, in the same manner, as an inferred quantity rather than by direct measurement.

EXAMPLE 1-10: The armature of a 110 V dc generator delivers a current of 60 A to a load. The armature circuit resistance is 0.25 ohm. The generator has 6 poles and 12 paths, and a total of 720 armature conductors rotating at a speed of 1800 rpm. Calculate:
a. The generated emf in the armature
b. The flux/pole.

Solution:

a. $E_g = V_a + I_a R_a = 110 \text{ V} + (60 \text{ A} \times 0.25 \text{ }\Omega) = \mathbf{125 \text{ V}}$ (1-10)

b. $\phi = \dfrac{E_g(60a)}{(ZSP)10^{-8}} = \left(\dfrac{125 \times 60 \times 12 \times 10^8}{720 \times 1800 \times 6}\right)$

$= \mathbf{1.16 \times 10^6 \text{ lines/pole}}$ (1-6)

The fundamental electromechanical dynamo relations which distinguish dynamo motor operation from generator operation, discussed above, may be summarized as follows

Motor Action	Generator Action
1. Electromagnetic torque produces (aids) rotation.	1. Electromagnetic torque (developed in the current-carrying-conductor) opposes rotation (Lenz's law).
2. Generated voltage opposes armature current (Lenz's law).	2. Generated voltage produces (aids) armature current.
3. $E_c = V_a - I_a R_a$ (1-9)	3. $E_g = V_a + I_a R_a$ (1-10)

BIBLIOGRAPHY

Alger, P. L. and E. Erdelyi. "Electromechanical Energy Conversion," *Electro-Technology*, September 1961.

Carr, C. C. *Electrical Machinery*, New York: John Wiley & Sons, Inc., 1958.

Crosno, C. D. *Fundamentals of Electromechanical Conversion*, New York: Harcourt, Brace, Jovanovich, Inc, 1968.

Daniels. *The Performance of Electrical Machines*, New York: McGraw-Hill, 1968.

Fitzgerald, A. E. and Kingsley, C. *The Dynamics and Statics of Electromechanical Energy Conversion*, 2nd Ed., New York: McGraw-Hill, 1961.

Fitzgerald, A. E., C. Kingsley, Jr. and Kusko, A. *Electric Machinery*, 3rd Ed. New York: McGraw-Hill, Inc., 1971.

Gemlich, D. K. and Hammond, S. B. *Electromechanical Systems*, New York: McGraw-Hill, 1967.

Hindmarsh, J. *Electrical Machines*, Elmsford, N. Y.: Pergamon Press, 1965.

Jones, C. V. *The Unified Theory of Electrical Machines*, New York: Plenum Publishing Corp., 1968.

Koenig, H. E. and W. A. Blackwell. *Electromechanical System Theory*, New York: McGraw-Hill, Inc., 1961.

Ku, Y. H. *Electrical Energy Conversion*, New York: The Ronald Press Company, 1959.

Levi, E. and Panzer, M. *Electromechanical Power Conversion*, New York: McGraw-Hill, 1966.

Majmudar, H. *Introduction to Electrical Machines*, Boston: Allyn and Bacon, 1969.

Meisel, J. *Principles of Electromechanical Energy Conversion*, New York: McGraw-Hill, 1966.

Nasar, S. A. *Electromagnetic Energy Conversion Devices and Systems*, Englewood Cliffs, N. J.: Prentice-Hall, Inc., 1970.

O'Kelly and Simmons. *An Introduction to Generalized Electrical Machine Theory*, New York: McGraw-Hill, 1968.

Robertson, B. L. and L. J. Black. *Electric Circuits and Machines*, 2nd Ed., Princeton, N. J.: D. Van Nostrand Co., Inc., 1957.

Rotating Machinery. (Group 10.) (ASA C42.10.) New York: American Standards Association.

Schmitz, N. L., and Novotny, D. W. *Introductory Electromechanics*, New York: Ronald Press, 1965.

Seely, S. *Electromechanical Energy Conversion*, New York: McGraw-Hill, 1962.

Selmon. *Magnetoelectric Devices: Transducers, Transformers and Machines*, New York: Wiley/Interscience, 1966.

Siskind, C. S. *Direct-Current Machinery*, New York: McGraw-Hill, 1952.

Skilling, H. H. *Electromechanics: A First Course in Electromechanical Energy Conversion*, New York: Wiley/Interscience, 1962.

Thaler, G. J. and Wilcox, M. L. *Electric Machines: Dynamics and Steady State*, New York: Wiley/Interscience, 1966.

Walsh, E. M. *Energy Conversion—Electromechanical, Direct, Nuclear*, New York: Ronald Press, 1967.

White, D. C. and H. H. Woodson. *Electromechanical Energy Conversion*, New York: John Wiley & Sons, Inc., 1959.

QUESTIONS

1-1. Describe four (4) electromechanical energy conversion effects.

1-2. State Faraday's law of electromagnetic induction
   a. in your own words
   b. in terms of an equation, giving all factors in the equation.

1-3. a. What is the name of the scientist and his law given in Quest. 1-2b?
   b. Does this law hold for instantaneous or average flux conditions? Explain.

1-4. Give an equation which may be used for computing the instantaneous magnitude of induced emf when a constant flux density is known. Give all the factors in the equation, including units, expressed in the cgs system

1-5. a. Repeat Question 1-4 for the equation in the English system
   b. Repeat for the RMKS system.

1-6. Give three reservations which apply to the equations given in Questions 1-4 and 1-5.

1-7. a. In the equation $e = Blv \sin(B, v) \sin(B, l)$ what factor is taken as the reference?
   b. Draw a diagram illustrating the situation in which l, B and v are all mutually perpendicular (orthogonal)
   c. Draw another diagram illustrating the equation given in 7a above.

1-8. a. Draw a diagram illustrating Fleming's rule.
   b. What does Fleming's rule show?
   c. What is meant by "conventional" current as opposed to "electron" flow?
   d. Draw a diagram showing Fleming's rule if used for determining direction of electron flow produced by electromagnetic induction.

1-9. a. State Lenz's law
   b. Show that both Lenz's law and Newton's third law of motion are related to Le Chatelier's principle.
   c. Draw a diagram showing that the direction of induced emf of a conductor moving in a magnetic field produces a current which in turn produces a flux opposing the motion.
   d. Develop Fleming's rule from the diagram you have drawn in (c).

1-10. Draw a diagram for a single-turn coil rotating in a uniform magnetic magnetic field. Show:
   a. Direction of induced emf in each coil side
   b. Direction of current flow if a load is connected across the terminals
   c. Polarity of the terminals with respect to the load.

1-11. For the diagram drawn in Quest. 1-10 explain; starting with the positive terminal:
   a. direction of current flow within the coil
   b. direction of current flow in the load

CHAP. ONE / *Electromechanical Fundamentals*

- c. compare this to current flow within and external to a battery supplying a load and explain.

1-12. Explain why:
- a. ac must be induced in a conductor rotating in a bipolar magnetic field.
- b. the waveform produced is sinusoidal.

1-13. For the single-turn coil shown in Fig. 1-6a explain:
- a. why no emf is induced in those portions labelled *ad* and *bc*
- b. assuming that the flux is not parallel to the coil sides in (a) above, why may the voltage produced in these coil sides be disregarded?
- c. under what conditions the waveform produced is not sinusoidal?

1-14. "All rotating electric machines tend to generate ac, regardless of type or purpose."
- a. Describe one exception to this statement
- b. Explain why ac is not generated in this particular rotating machine.

1-15. If "all" rotating dynamos are ac generators, explain:
- a. how dc may be produced when a single-turn coil rotates in a bipolar field
- b. how a positive polarity is always maintained at one terminal of the external circuit.

1-16. Explain:
- a. why a sinusoidal half-waveform is produced in Fig. 1-7 but a flat-topped wave is produced in Fig. 1-8
- b. what accounts for the difference between the two waveforms?
- c. why both waveforms are called "pulsating unidirectional current"
- d. under what conditions the output may be made less "pulsating".

1-17. Compare the waveform shown in Fig. 1-9b with that shown in Fig. 1-10d and explain:
- a. why the voltage at the brushes of the latter is the sum of the voltages per conductor in each parallel path
- b. the advantages of a closed winding over an open winding (list three).

1-18. For the diagram shown in Fig. 1-10c, explain:
- a. why the voltage induced in each coil side can be treated as an individual electric cell
- b. the meaning of the symbols r_s and r_n
- c. why no circulating current is produced between the two paths, even if the brushes are lifted off the commutator.

1-19. If an external load is connected across the brushes of a gramme-ring generator,
- a. list four factors which determine the current which flows in the load
- b. what is the relationship between the current/path and the load current?

1-20. Give one theoretical advantage and three disadvantages of the gramme-ring winding as compared to the modern drum-wound armature.
- a. Why is the gramme-ring winding known as a re-entrant or closed winding?

b. Why is it possible to use a ring (or pipe) rather than a solid cylinder in Fig. 1-10a?

1-21. Why does the resultant output waveform shown in Fig. 1-10d contain a small ac component, even though the individual coils produce a square wave (rich in harmonics and ac) for each revolution of rotation?

1-22. a. Using Tables 1-1 and 1-2, explain why the power rating of each coil determines the power rating of the dynamo, regardless of the method of connection.
b. Explain why physical size is an approximate indication of power rating of electrical machinery.

1-23. a. Rewrite Eq. 1-6 in an algebraic form for determining the number of paths.
b. Repeat (a) above for the number of poles, P.
c. If, for any *given* dynamo already constructed, the number of conductors, Z, the number of poles, P, and the number of paths, a, are fixed, rewrite Eq. 1-6 in terms of the variables involved.

1-24. State the equation expressing the relation between electromagnetic force on a current carrying conductor in a magnetic field in
a. cgs system of units
b. English units.

1-25. Using Fig. 1-13 as an illustration, show that Lenz's law applies to:
a. motor action
b. generator action
In each of the above cases, indicate both the cause and the opposition to it.

1-26. Using Fig. 1-14 as an illustration, explain the universality of the statements that motor action is always accompanied by generator action and generator action is always accompanied by motor action.

1-27. a. Using Eq. 1-9, explain why it is impossible for the counter emf to equal the applied voltage in a motor.
b. Using Eq. 1-10, explain under what conditions the generated voltage, E_g and the voltage across the armature, V_a, are the same for a generator.

PROBLEMS

1-1. A flux of 6.5×10^6 lines links a loop of one turn. The flux collapses to zero in 0.125 s. The closed loop has a resistance of 0.05 Ω. Calculate:
a. The average value of voltage generated in the loop
b. The average circulating current in the loop

1-2. A single conductor 1 m long moves perpendicularly to a uniform magnetic field of 25,000 gauss (maxwells/cm^2) at a uniform velocity of 25 m/s. Calculate:
a. The instantaneous emf induced in the conductor
b. The average voltage induced in the conductor

1-3. A conductor 24 in long moves at a velocity of 12 in/min through the gap in a U-shaped permanent magnet having a flux of 50,000 lines. The poles of the magnet measure 4 in square (not 4 in^2!). Assuming no fringing of flux, calculate:
   a. The emf induced in the conductor when it moves perpendicularly to the magnetic field (at an angle of 90°)
   b. The emf induced in the conductor when it moves at an angle of 75° with respect to the magnetic field

1-4. Michael Faraday's homopolar generator (See footnote figure, Section 1-11) is a disc 12 in in diameter in a field of 80,000 lines/in^2. The disc is rotated manually at 60 rpm. The shaft is 1 in in diameter. Calculate the voltage induced between the outer edge of the shaft and the rim of the disc. (Hint: compute the *average* linear velocity.)

1-5. The *vertical* component of the earth's magnetic field is 0.645 gauss in the vicinity of a locomotive traveling due south at a speed of 60 mi/hr. The locomotive rails and the axle of the locomotive span 6 ft. Calculate:
   a. The emf induced in the axles of each set of wheels
   b. The average emf measured at the rails produced in Problem 1-5a
   c. The polarity of the rails, east and west, respectively
   d. Draw the scale of an electrical speedometer from zero to a maximum speed of 80 mi/hr using a millivoltmeter
   e. Consider the practicality of a device to measure aircraft velocity and altitude using this principle. Discuss pros and cons.

1-6. A coil measuring 12 in × 18 in lies with its coil axis *parallel* to a uniform magnetic field of 5000 lines/in^2. The coil has 20 turns and its axis is through the center of its smaller dimension. The coil is rotated about its axis so that it is *perpendicular* to the uniform magnetic field (90° rotation) in 0.1 s. Calculate:
   a. Average induced emf for one-quarter turn (0 to 90°)
   b. Instantaneous induced emf at the zero degree (original) position, at instant coil is set in motion
   c. Instantaneous induced emf at the 90° position (axis perpendicular to the magnetic field)
   d. Average induced emf if the coil is continuously rotated at a speed of 20 rps

1-7. The voltage induced in a conductor moving in a uniform magnetic field is 25 V when the velocity is 60 cm/s. Calculate the induced emf when
   a. The field flux is increased by 15 per cent
   b. The velocity is reduced by 30 per cent
   c. The velocity is increased by 20 per cent and the flux reduced by 10 per cent

1-8. The flux per pole of a two-pole generator is 10×10^6 lines. It is driven at a speed of 1500 rpm. In order to induce a voltage of 20 V/coil, calculate:
   a. The time to complete one revolution and one-quarter revolution. (Time to go from zero to maximum flux per pole)
   b. The number of turns in series per coil using Eq. (1-1)
   c. Verify Problem 1-8b by using Eq. (1-5)

1-9. The flux per pole of a four-pole generator is 10×10^6 lines. It is driven at a speed of 1500 rpm. In order to induce a voltage of 20 V/coil, calculate:
  a. The time to complete one-eighth of a revolution (time to go from zero to maximum flux per pole)
  b. The number of series turns using Eqs. (1-1) and (1-5), respectively
  c. Explain the difference in series turns required for Problems 1-8 and 1-9, respectively
  d. The number of *conductors* required between brushes to generate 120 V

1-10. Given a generator having 1 turn/coil, four poles, four paths, a flux per pole of 10×10^6 lines and a speed of 1500 rpm, calculate:
  a. The number of series-connected conductors on the entire armature required to produce a voltage of 120 V between brushes
  b. The number of series-connected conductors per path (compare to Problem 1-9d)
  c. Distinguish between Eqs. (1-5) and (1-6) on the basis of the comparison in Problem 1-10b

1-11. Given the following information regarding a generator: active conductor length of 14 in, armature 12 in in diameter, flux density 66,000 lines/in^2. The pole faces cover 80 per cent of the armature surface and the speed is 1600 rpm. Assuming uniform flux density under the pole, calculate:
  a. The instantaneous induced emf per conductor when moving directly under the center of a pole
  b. The average induced emf per conductor taking into account the lack of flux in the interpolar region
  c. The average emf between brushes assuming a total of 40 conductors/path

1-12. An eight-pole generator has a total of 480 conductors connected in 16 parallel paths. The flux per pole is 1.6×10^7 lines and the speed is 1200 rpm. If the pole faces cover 75 per cent of the armature surface, calculate the voltage generated between brushes.

1-13. The armature of the generator in Problem 1-12 is replaced with one having four parallel paths. Calculate:
  a. The voltage developed between brushes
  b. The per cent of change in either original flux or speed in order to develop the same voltage as in Problem 1-12

1-14. Each conductor of the generator in Problem 1-11 is carrying a current of 20 A when connected to a load. Calculate:
  a. The counterforce developed (opposing the motion) by the conductor, when directly under the center of a pole
  b. The average counterforce developed by a conductor taking into account lack of flux and no useful torque in the interpolar region

1-15. If the flux density of the generator in Problem 1-14 is increased 10 per cent and the load is increased by 15 per cent, calculate the average counterforce developed by each conductor of the generator.

1-16. The axial length of the armature of a dc motor is 9 in, the poles have a flux density of 72,000 lines/in^2 and cover 72 per cent of the armature surface.

Calculate the force developed by each conductor when carrying a current of 25 A

1-17. A dynamo runs at a speed of 1200 rpm. Its armature has a total resistance of 0.04 Ω, a length of 16 in, a total of 630 conductors and 6 paths. The diameter of the armature is 18 in and the air gap is 0.100 in. The six poles cover 80 per cent of the total armature circumference. Rated dynamo current (per path) per conductor is 25 A. When operated at rated speed and flux, the generated voltage per path is 120 V. Calculate:
a. The flux per pole and the flux density
b. The terminal voltage across the armature when operating as a generator
c. The force per conductor developed by motor action
d. The applied voltage across the armature required to develop a generated voltage of 120 V when operated as a motor

ANSWERS

1-1(a) 0.52 V (b) 10.4 A   1-2(a) 62.5 V (b) 62.5 V   1-3(a) 25 μV (b) 24.2 μV
1-4 0.083 V   1-5(a) 3.16 mV (b) 3.16 mV (c) East(+), West(−) (d) 4.21 mV
1-6(a) 2.16 V (b) 3.392 V (c) 0 (d) 0.458 V   1-7 (a) 28.75 V (b) 17.5 V (c) 27 V
1-8(a) 0.01 s (b) 2 turns/coil (c) 2 turns/coil   1-9(a) 5 ms (b) 1 turn (d) 12 conductors   1-10 (a) 48 (b) 12   1-11(a) 9.3 V (b) 7.44 V (c) 297.6 V   1-12 577 V
1-13(a) 2308 V (b) 75 per cent reduction 1-14 (a) 1.635 lb (b) 1.308 lb   1-15 1.655 lb   1-16 1.03 lb   1-17 (a) 7.82 × 10^3 lines/in² (b) 114 V (c) 0.2765 lb (d) 126 V

TWO

dynamo construction and windings

**2-1.
DYNAMO
POSSIBILITIES**

The preceding chapter established that a dynamo is a rotating electric machine capable of converting mechanical energy into electric energy (a *generator*) or electric energy into mechanical energy (a *motor*). For the generator, rotary motion is supplied by a prime mover (a source of mechanical energy) to produce relative motion between the conductors and the magnetic field of the dynamo in order to generate electric energy. For the motor, electric energy is supplied to the conductors and the magnetic field of the dynamo in order to produce relative motion between them and thus produce mechanical energy. In *both* cases, we have *relative motion* between a magnetic field and the conductors in a dynamo. This gives rise to several interesting possibilities and choices in determining which shall be the *rotor* (that part of the dynamo which rotates) and the *stator* (that part of the dynamo which is stationary). There are specific, sound engineering reasons that dictate the choice as to whether the armature conductors or the field coils which provide the magnetic flux shall serve as the rotor or the stator.

The various types of dynamo possibilities which will be discussed are

1. The *direct-current* (dc) dynamo, which has a rotating armature and a stationary field.
2. The *synchronous* (ac) dynamo with a rotating armature and a stationary field.
3. The *synchronous* (ac) dynamo with a rotating field and a stationary armature.
4. The *asynchronous* (ac) dynamo, which has both stationary and rotating armature windings.

2-2. DIRECT-CURRENT (DC) DYNAMO CONSTRUCTION*

Figure 2-1a shows a cross section of a typical commercial "dc" dynamo, simplified for emphasis of the major portions. The rotor of the dynamo consists of*

1. The *armature shaft*, which imparts rotation to the armature core, winding, and commutator. Mechanically joined to the shaft is
2. The *armature core*, constructed of laminated layers of dynamo steel, providing a low-reluctance magnetic path between the poles. The laminations serve to reduce eddy currents in the core, and the dynamo steel used is of such a grade as to produce a low hysteresis loss. The core contains axial slots in its periphery for
3. The *armature winding*, consisting of insulated coils, insulated from each other and from the armature core, embedded in the slots and electrically connected to
4. The *commutator*, which, by virtue of the shaft rotation, provides the necessary switching for the commutation process. The commutator consists of copper segments, individually insulated from each other and from the shaft, electrically connected to the armature winding coils.

The rotor armature of the dc dynamo performs four major functions: (1) permits rotation for mechanical generator action or motor action; (2) by virtue of rotation, it produces the switching action necessary for commutation; (3) contains the conductors which induce a voltage or provide an electromagnetic torque; and (4) provides a low-reluctance magnetic flux path.

The stator of the dynamo consists of

1. A *yoke* or cylindrical frame of cast or rolled steel. Not only does the yoke serve as a support for the portions described below, but also it provides a flux return path for the magnetic circuit created by

---

* It is beyond the scope or intent of this work (as the author has stated in the preface) to give a detailed description of machine construction. An interested reader who desires more detail than given here may consult either G. C. Blalock, *Direct Current Machinery* (New York: McGraw-Hill, Inc., 1947), Chapter 2; or Kloeffler, Kerchner, and Brenneman, *Direct Current Machinery* (New York: The Macmillan Company, 1948), Chapter 1. Also see references at end of chapter.

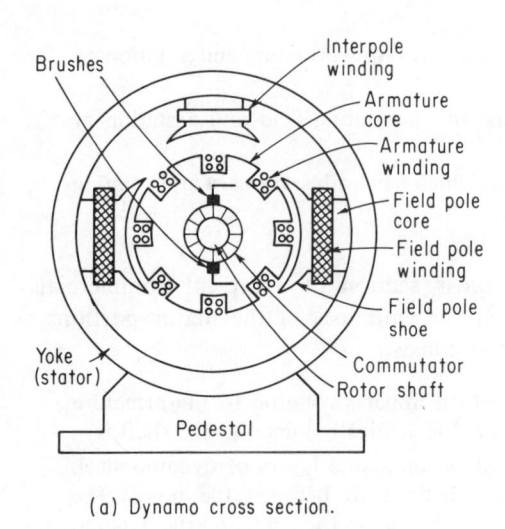

Figure 2-1

"DC" dynamo construction and electrical circuits (shunt and series).

2. The *field windings*, consisting of a few turns of wire for a series field or many turns of fine wire for a shunt field. Essentially, the field coils are electromagnets whose ampere-turns provide a magnetomotive force adequate to produce, in the air gap, the flux needed to generate an emf or electromotive force. The field windings are supported on

3. The *field poles*, constructed of laminated steel and bolted or welded to the yoke after the assembly of field windings have been inserted on them. The pole shoe is curved, and it is wider than the pole core to spread the flux more uniformly.

4. The *interpole* and its winding are also mounted on the yoke of the dynamo. These are located in the interpolar region between the main poles, and are generally smaller in size. The interpole winding is composed of a few turns of heavy wire, since it is connected in series with the armature circuit so that its magnetomotive force (mmf) is proportional to the armature current.

5. *Compensating windings* (not shown) are optional; they are *connected* in the *same* manner as the interpole windings but are *located* in axial slots of the field pole shoe. (See Fig. 5-7.)

6. *Brushes and brush rigging*, like interpoles and compensating windings, are part of the armature circuit. The brushes are composed of carbon

42

CHAP. TWO / *Dynamo Construction and Windings*

and graphite, supported from the stator structure by a rigging, and held in brush holders by means of springs so that the brushes will maintain firm contact with the commutator segments. The brushes are always instantaneously connected to a segment in contact with a coil located in the interpolar zone.

7. *Mechanical details.* Mechanically connected to the yoke are *end-bells* containing bearings in which the armature shaft is supported, as well as the brush rigging on some machines. These details are not shown in Fig. 2-1 or 2-2.

The electrical connections of the dc dynamo are shown in Figs. 2-1b and c. The former shows the shunt-field connection in which the field winding shunts the armature circuit. The latter shows the series-field connection in which the heavy winding of a few turns is located on the main field poles and connected in series with the armature circuit. Note that the compensating and interpole windings, if employed, are always part of the armature circuit, as are the brushes. Note also that the shunt-field connection of Fig. 2-1b employs its field rheostat in the field circuit.

It is most important to observe that the dc dynamo of Fig. 2-1c may be used universally and will operate either as a dc or ac dynamo, or both, as in the case of a universal motor (Section 10-16). Actually, the term "dc dynamo" is a misnomer. Most dc generators and motors employ the construction shown in Fig. 2-1a and the electric connections of Figs. 2-1b and c. It will later be shown that some dynamos combine the field connections shown in these figures, creating *compound-wound dynamos.*

2-3. SYNCHRONOUS DYNAMO (STATIONARY FIELD) CONSTRUCTION

Precisely the same stator construction is employed in the synchronous (stationary field) dynamo as in the dc dynamo, and the field winding is excited by a dc source. The rotor armature winding is brought out to slip rings as well as to a commutator, as shown in the construction cross section of Fig. 2-2b. Such a dynamo will function either as a revolving-armature synchronous generator or as a motor, depending on whether the input is dc to the brushes, or polyphase (or single-phase) alternating current to the rotor slip rings. This type of dynamo finds its greatest application in the synchronous or rotary converter, shown in Fig. 2-2b, which is used to convert direct current to alternating current or vice versa (see Sections 11-6 and 11-7). If direct current is applied to the brushes, the dynamo operates as a dc motor and ac alternator, simultaneously. If alternating current is applied to the slip rings, the dynamo functions as an ac motor and dc generator, simultaneously. It has already been demonstrated (Section 1-20) that generator action and motor action always occur simultaneously, and rotary converters are perhaps the best example of the simultaneous employment of both. This type of construction is also used in the three-wire Dobrowolsky generator (shown in Fig. 11-9d).

(a) Synchronous dynamo cross section

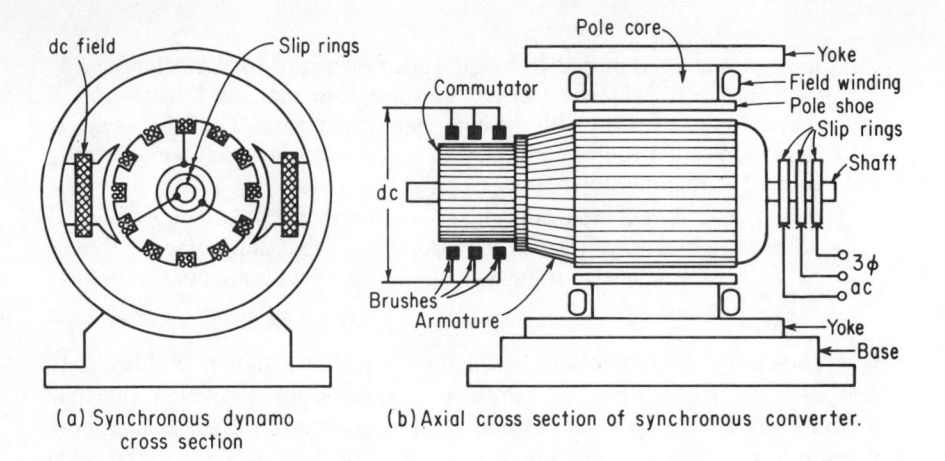

(b) Axial cross section of synchronous converter.

Figure 2-2

Synchronous dynamo, revolving armature, salient field pole type.

2-4. ROTATING FIELD SYNCHRONOUS DYNAMO CONSTRUCTION

The synchronous dynamo construction of Section 2-3, in which the field is stationary and the armature rotates, is relatively limited (for reasons outlined in Section 6-3) in comparison to the *rotating-field synchronous dynamo* shown in Fig. 2-3. In this dynamo, the field winding is supplied from a dc source through two slip rings and the armature is connected directly to a polyphase ac source or load.

(a) Salient pole synchronous dynamo.

(b) Nonsalient pole four-pole synchronous dynamo showing stator armature connections.

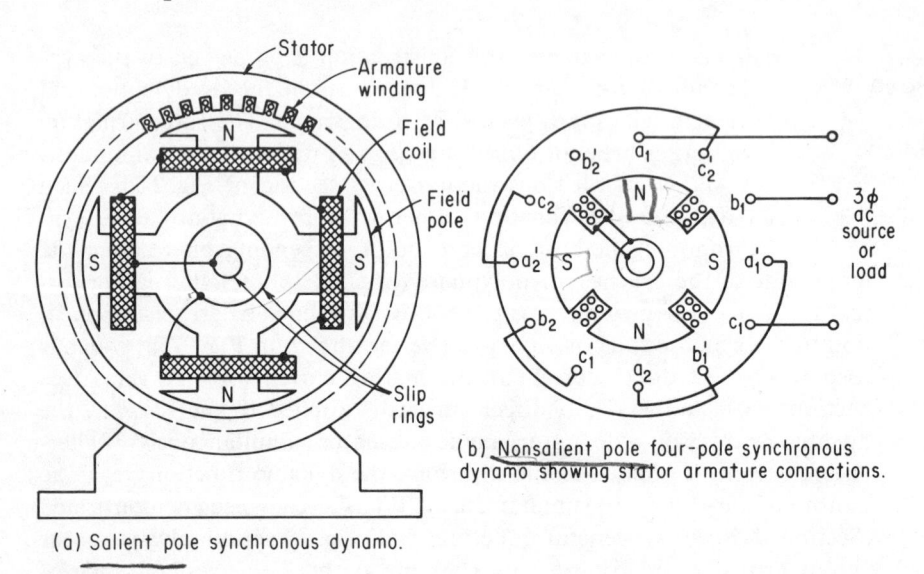

Figure 2-3

Synchronous revolving-field dynamo.

44

CHAP. TWO / *Dynamo Construction and Windings*

If the stator armature is connected to a single-phase or polyphase ac supply, the dynamo will function as a *synchronous motor*, and the rotor will rotate at a synchronous speed in synchronism with the rotating field developed by the stator winding as determined by the number of poles and the frequency of the supply. If the rotor, either *salient* as shown in Fig. 2-3a or *nonsalient*, as shown in Fig. 2-3b, is rotated at a synchronous speed by a prime mover, the dynamo will function as an *alternator*, either single-phase or polyphase, depending on the armature connections. The armature stator connections shown in Fig. 2-3b are those required for obtaining a three-phase output using a four-pole rotor.

**2-5.
ASYNCHRONOUS
INDUCTION
DYNAMO
CONSTRUCTION**

The *asynchronous* induction dynamo shown in Fig. 2-4a has the identical *stator* construction as that described for the synchronous dynamo in Section 2-4. The stator armature winding, therefore, may be connected to a single-phase or polyphase ac supply. The rotor is not (dc) separately excited, as described below. The induction dynamo becomes an asynchronous *induction generator* (Section 9-22) when the rotor is turned by a prime mover at a speed exceeding synchronous speed.

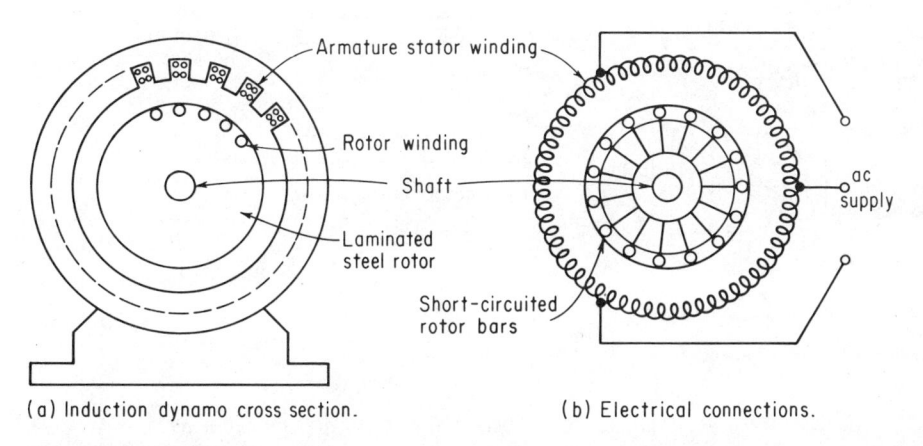

(a) Induction dynamo cross section.   (b) Electrical connections.

Figure 2-4

Induction (asynchronous) dynamo.

If the stator armature is connected to a single-phase or polyphase ac supply, the dynamo will function normally as an *induction motor*. Single-phase induction motors require auxiliary starting devices, but polyphase induction motors are inherently self-starting. It should be noted that, whether it is operated as a generator or as a motor, the asynchronous induction dynamo requires that the stator armature is connected to an ac supply. Like the dc and ac synchronous dynamo, it is *doubly excited* (see Section 9-1), but alternating current flows in *both* the stator and the rotor windings.

The rotor winding, which carries alternating current produced by induction from the directly connected stator winding, consists of copper or aluminum conductors embedded or cast in a laminated steel rotor. Short-circuiting end rings are placed at both ends in the squirrel-cage type, or effectively at one end in the wound-rotor type.

The asynchronous dynamo receives its name from the fact that it operates as a generator or a motor at a speed other than the synchronous speed of its rotating magnetic field.

2-6.
DC DYNAMO
MAGNETIC FIELDS
AND CIRCUITS

As demonstrated by the preceding sections, all dynamos, regardless of type or purpose, require: (1) a winding whose function is to produce a magnetic field; (2) a winding consisting of current-carrying conductors; and (3) a means of providing relative motion between (1) and (2). Figure 2-5a shows the flux distribution produced by a four-pole stator in the rotating armature of a

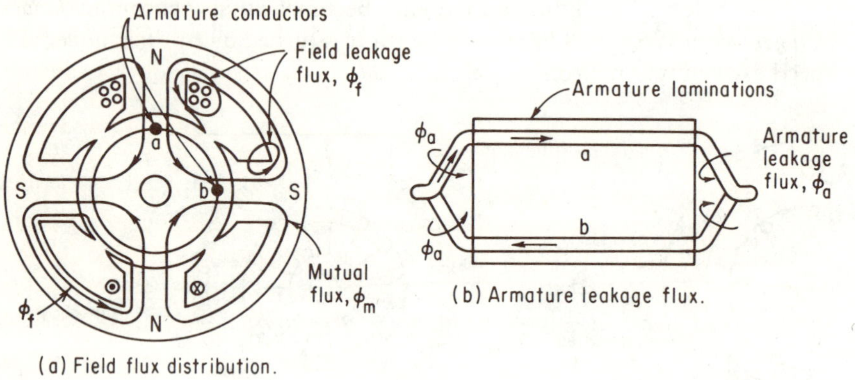

(a) Field flux distribution.

(b) Armature leakage flux.

Figure 2-5

DC dynamo mutual and leakage fluxes.

dc dynamo. Regardless of whether the dynamo operates as a motor or as a generator, the only useful flux for either generator action or motor action is that which links *both* the field and the armature conductors simultaneously. This flux, called the mutual flux ϕ_m, is shown in Fig. 2-5a as produced by the field winding around the field poles. Complete magnetic circuit loops are formed, passing from a north pole through the air gap to link the armature conductors, back through the air gap to a south pole, and through the yoke back to the original north pole. Since there is a reluctance of twice the air gap in the mutual flux circuit (and the air gap varies in commercial dynamos from $\frac{1}{16}$ to $\frac{1}{4}$ in), a possibility exists for a shorter magnetic circuit (or nonmutual leakage flux) which does *not* link both the field and armature simultaneously. A leakage flux path may be set up directly

CHAP. TWO / *Dynamo Construction and Windings*

from the north to the south field pole, or from a given pole to the yoke, as shown by the leakage flux designated ϕ_f in Fig. 2-5a.

The rotating current-carrying conductors, because of the nature of the armature winding, also tend to produce an armature leakage flux, ϕ_a, shown in Fig. 2-5b, particularly in that portion of the coil which is not embedded in the armature iron. Thus, both the field windings and the armature windings tend to produce leakage fluxes which are independent of the mutual or air-gap flux.

2-7. ARMATURE REACTANCE

Of the two leakage fluxes, the field leakage flux is of lesser significance. The loss of field flux created by field leakage can be compensated by *increasing* the field strength, as described in Section 2-9. In addition to reducing the mutual flux slightly, field leakage flux only affects dynamo operation during transient periods such as starting or change of load conditions.

The armature leakage flux is of greater importance, since it is responsible for the effect termed *armature leakage reactance* or, briefly, *armature reactance*. Armature reactance varies, in a given machine, only with the armature current, since it is produced by the armature current and links the armature conductor only. Armature reactance is simply an effect due to *self-inductance* of the armature conductor, and it is observable, by definition, *only* when the current through the conductor is *changing*. In the dc dynamo, the current in the armature conductor (ignoring transients) changes only during the commutation period when the direction of current in the conductor reverses. The emf of self-induction created by commutation of the conductor gives rise to the need for *commutating poles* or interpoles in the dc dynamo. Leakage reactance in ac dynamos is discussed in the following section.

2-8. AC DYNAMO MAGNETIC FIELDS AND CIRCUITS

The mutual flux distribution for a four-pole ac synchronous dynamo is shown in Figs. 2-6a and b. As in the case of the dc dynamo, the mutual or useful air-gap flux, ϕ_m, is that which links simultaneously both the field and armature conductors. Similarly, some field leakage flux, ϕ_f, is produced which links only the field conductors; but the effect of this field leakage flux is identical to the dc dynamo and can be overcome by increasing the field current.

In the case of armature leakage flux, ϕ_a, shown in Fig. 2-6, produced by current-carrying armature conductors in which the current is continuously alternating, an *inductive armature reactance*, X_a, is produced. This armature reactance in quadrature combination with the armature resistance is a factor in the over-all *armature impedance*; and, as will be shown later, it plays an important part in determining the voltage regulation of an alternator or the power factor adjustment of a synchronous motor.

The reader should not confuse armature reactance with *armature*

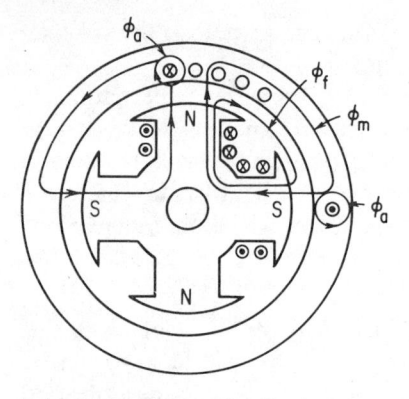

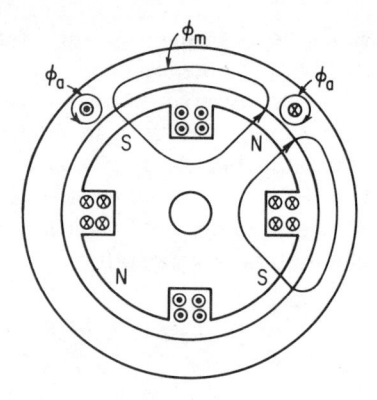

(a) Salient pole flux distribution.   (b) Nonsalient pole flux distribution.

Figure 2-6

AC dynamo mutual and leakage fluxes.

reaction whenever either term is employed. The effects of the former have been described for dc and ac dynamos above. Armature reaction (Chapter 5) is the effect of the magnetomotive force (mmf) produced by the armature ampere-turns ($I_a N_a$) in varying and distorting the mutual flux, ϕ_m, of the field. In the dc and ac dynamo, armature reaction and its effects are continuously present, whereas armature reactance affects the conductors of the dc dynamo only when the current is changing (as during commutation resulting in sparking between the commutator and the brushes).

2-9. MAGNETIC FLUX CALCULATIONS

The detailed procedure for the design of field coils required to set up a given flux density in a dynamo is beyond the scope of this volume.* Summarized briefly, it consists of dividing the magnetic circuit, shown in Fig. 2-6a, into its component parts: pole core, pole shoe, air gaps, yoke, armature teeth, and armature core. A table is then set up listing the dimensions (average length and cross-sectional area), material, total flux, flux density, the magnetomotive force per portion, and the total ampere turns per pole. Each of the above component parts is treated as a series element in a magnetic circuit based on a relation known as the "Circuital Law of the Magnetic Field":†

The magnetic field strength of a closed path is the sum of the ampere turns with which this path is linked.

---

* The procedure is outlined in Kloeffler, Kerchner, and Brenneman, *Direct Current Machinery*, revised ed. (New York: The Macmillan Company, 1948), pp. 25–29.

† Sometimes called Kirchhoff's "Law of the Magnetic Circuit," in which the analog of total applied voltage is total ampere-turns required to set up a desired magnetic flux in a given series-magnetic circuit containing various reluctances.

Stated in terms of an equation, the above law becomes:

$$\phi = \frac{0.4\pi IN}{\mathscr{R}} = 0.4\pi\left(\frac{IN_1}{\mathscr{R}_1} + \frac{IN_2}{\mathscr{R}_2} + \frac{IN_3}{\mathscr{R}_3} + \cdots + \frac{IN_n}{\mathscr{R}_n}\right) \quad (2\text{-}1)$$

where ϕ is the desired magnetic field strength required to produce torque or to generate an emf in a dynamo in lines or maxwells

$\mathscr{R}$ is the total reluctance of the closed path

$\mathscr{R}_1, \mathscr{R}_2$, etc.,   are the individual reluctances of the component parts cited above

IN is the total ampere-turns or *magnetomotive force* to be produced by the field winding

IN_1, IN_2, etc.,   are the individual mmf's required to overcome reluctances $\mathscr{R}_1, \mathscr{R}_2$, etc., in order to produce ϕ total flux lines

As stated in Section 2-6, a certain amount of leakage flux is produced which tends to reduce the useful mutual or air-gap flux. In the above computation, it is customary to increase the value of the desired ϕ by a *leakage* or *dispersion coefficient*. In very small dynamos and in older dynamos, this coefficient may be as high as 1.25; in the modern dynamos having shorter pole cores and improved pole-shoe designs, the coefficient may be as low as 1.05, particularly for dynamos of larger capacity.

Field coils for given dynamos, whether stationary as on dc dynamos, or rotating as on ac types, have the same number of turns per pole and the coils are always connected in series to ensure the same air-gap mmf ($I_f N_f$) and flux under each pole. Field coils are designed with an attempt to provide a proper balance between the field-copper loss and the amount of copper used for the turns. A large number of field turns with a relatively high resistance will produce the required mmf and result in low field-copper loss. But such a design may be at the expense of higher construction costs involving more copper, larger poles, and increased physical size. Thus a given field coil on a dynamo represents an *optimum balance* between field-copper loss and the number of turns and copper current-carrying capacity which will result in the most economical size and lowest material cost.

Field windings, whether self-excited or separately excited, require direct current for their operation. Since they are generally constructed of many turns and are placed on iron cores of relatively low reluctance, they produce a *highly inductive* circuit. When a field circuit is energized, the current i in the circuit rises in accordance with the general voltage equation

$$V_{dc} = iR + L\left(\frac{di}{dt}\right) \quad (2\text{-}2)$$

where V_{dc} is the applied voltage across the field circuit
i is the instantaneous current in the field circuit
L is the inductance of the field winding circuit
R is the field circuit resistance
di/dt is the rate of rise of current in the field circuit

Solving Eq. 2-2 for i (by calculus) yields the same equation in the form

$$i = \frac{V_{dc}}{R}(1 - \epsilon^{-tR/L}) \qquad (2\text{-}3)$$

where ϵ is 2.718 ... (the base of Naperian or natural logarithms) and all other terms are previously defined.

Once the current rises to a steady value (at the end of approximately five times the value of L/R seconds), current through the field winding is constant and is limited only by the field-circuit resistance. The inductive effect of the field circuit occurs only during the transient conditions of: (1) fluctuations in the power applied to the field; (2) a change in the field current by means of a field rheostat or potentiometer rheostat; or (3) opening (or removing power from) a field circuit.

The last-named transient is by far the most serious, for if the field circuit is suddenly opened, the rate of change of current is high and the term $L(di/dt)$ in Eq. (2-2) operates independently of the supply voltage since it is no longer connected to it. The result is an emf of self-induction which may be several times the supply voltage. This voltage may damage instruments connected across the field, puncture the insulation of the field windings, or cause severe arcing at the switch contacts supplying power to the field circuit. Unless properly protected, the dc field winding should *not* be opened while the field is energized.

When it becomes necessary to open the field circuit, for laboratory study or commercial operation, a special *field-discharge switch* and resistor are employed, shown in Fig. 2-7a. Such a switch permits a *field-discharge resistor, R,* to be placed in parallel with the field circuit at the instant of

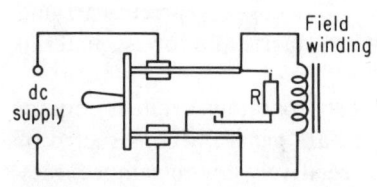

(a) Field discharge resistor and switch.

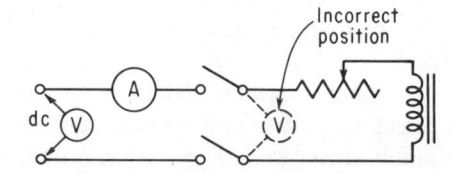

(b) Correct position of voltmeter in measurement of field circuit resistance or field circuit power loss.

Figure 2-7

Field discharge circuit and field measurements.

opening the field circuit. The energy of the magnetic field ($\frac{1}{2}LI^2$) is thus dissipated in the low-resistance field-discharge resistor rather than across the switch contacts or the voltmeter across the field. In the laboratory, if no field-discharge switch and resistor are available, the connection of Fig. 2-7b may be used, preferably with circuit breakers rather than open knife switches. The connection shown not only protects the voltmeter but also yields a better measurement in terms of instrument sensitivity.

2-10. ARMATURE WINDINGS*

As shown in Figs. 2-1 through 2-4, representing the four basic dynamo types, the armature windings, whether on the rotor or the stator, are always of the *nonsalient* type and are distributed equally in slots adjacent to the air gap around the periphery of the armature. Essentially, there are two kinds, depending on the type of closure or re-entrance of the winding: closed-circuit windings, employed in dc dynamos; and open-circuit windings, employed usually in ac dynamos.

Regardless of the type or application, most armature windings consist of diamond-shaped pre-formed coils, as shown in Fig. 2-8a, which are inserted into the armature slots and connected in a manner to produce a complete winding. Each coil consists of many turns of fine silk-covered, cotton-covered, or enamel-covered wire, individually taped, lacquer-dipped, and insulated from the armature slot. The number of *conductors* [Z in Eq. (1-6)] in any given coil will be *twice* the number of *turns* making up the coil, i.e., two conductors per turn.

In general, armature coils span 180 electrical degrees, i.e., from the center of a given pole to the center of a pole of *opposite* polarity, which may nevertheless be physically adjacent, as shown in Figs. 2-8b and c. If a coil covers a span of 180 electrical degrees, it is called a *full-pitch coil*, whereas one that spans less than 180° is called a *fractional-pitch coil*. An armature wound with a fractional pitch is called a *chorded winding*. Chorded windings require the use of less copper than full-pitch coils, but they have approximately the same characteristics because the shorter front-end and back-end spans are inactive. A coil which spans 150 electrical degrees would have a *pitch factor*, p, of $150°/180° = 0.833$, or 83.3 per cent. In general, pitch factors of less than 80 per cent are avoided.

Most armature windings are *two-layer* windings, i.e., two coil sides are inserted in each slot. In winding a two-layer armature, one coil side p is placed in the bottom of a slot, as shown in Fig. 2-8d for coil 1, in which the right-hand side of the coil is inserted and the other side is not. The

---

* Although the subject of windings and winding calculations is beyond the intention of this text, this section is included here to discuss the principles that will be developed later. For a fairly comprehensive coverage of the subject, the reader is referred to Liwschitz-Garik and Whipple, *Electrical Machinery* (Princeton, N.J.: D. Van Nostrand Company, Inc., 1946), Vol. I, Chapter 6; Vol. II, Chapter 4.

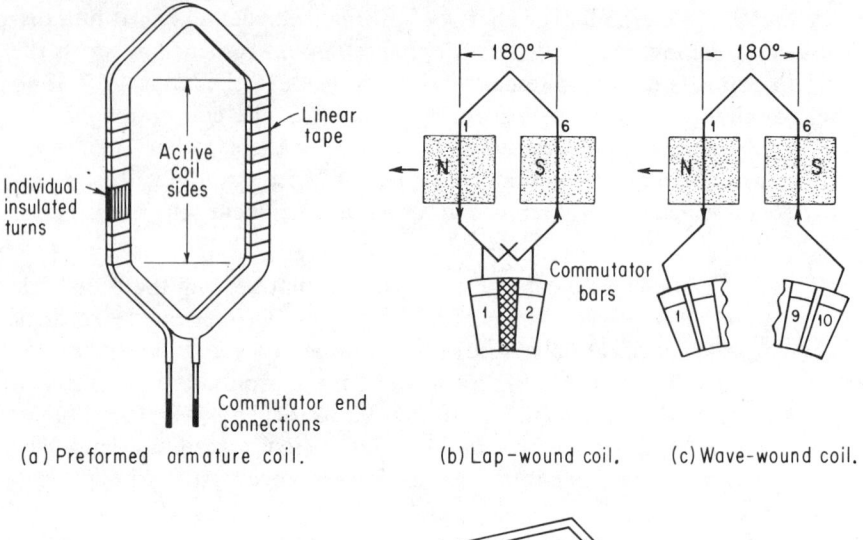

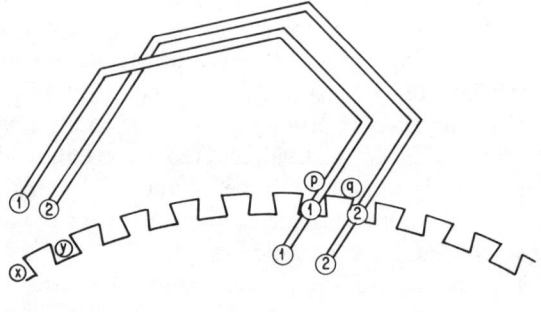

Figure 2-8

Types of dc dynamo armature coils and end connections.

(a) Preformed armature coil.
(b) Lap-wound coil.
(c) Wave-wound coil.
(d) Two-layer windings.

second coil side is not inserted until all the other armature coils have been inserted in bottom slots. When coil-side x has been inserted, only then coil-side 1 is inserted; when coil-side y has been inserted in a bottom slot, only then is coil-side 2 inserted; and so on. The purpose of this procedure is to assure both strength against centrifugal forces and nearly perfect equality in the size, shape, and weight of all the coils.

2-11. LAP AND WAVE WINDINGS—SIMILARITIES AND DIFFERENCES

In Section 1-13, it was stated that armatures consisted of series-connected coils having two or more current paths. Two types of end connections are employed to ensure that the emf's induced in the series-connected coil sides will aid each other, namely, the *lap-wound* and *wave-wound* connections shown in Figs. 2-8b and c. Observe that in both types of connections, one active coil side is under a north pole and the other under a south pole; assuming that the field is moving to the left (conductor motion to the right), the emf induced in the coil side inserted in slot 1 is in the same direction as that inserted in slot 6, regardless of the end connection. Note that the lap-wound coil end adds emf's in such a manner that the coil ends must be series-connected to adjacent commutator bars. Since the dc dynamo winding must close on itself, the last coil end in the winding is connected to the first coil side of an adjacent coil at the first commutator bar where the winding began. The winding described is also shown in Fig. 2-9a where the *upper* coil side is represented by a *solid* line and the *lower* coil side by a *dotted* line.

Figure 2-9a also serves to show the brush connections with respect to the coils. Observe that the positive brush is connected to commutator 3, and at this instant is connected to coil 3. Since the brushes are stationary, for the particular position of each of the four brushes, they are *always* in

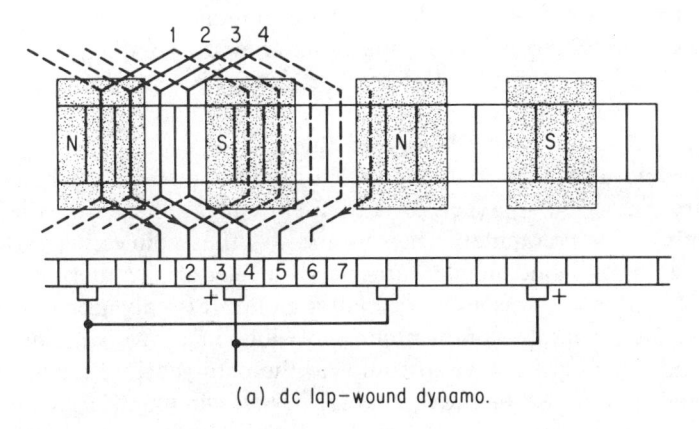

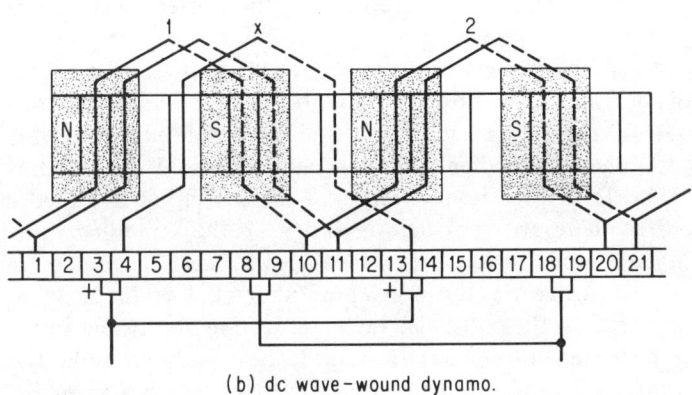

(a) dc lap-wound dynamo.

(b) dc wave-wound dynamo.

Figure 2-9

Developed views of dc dynamo armatures.

contact with a coil undergoing commutation, i.e., a coil lying in the interpolar space. At the same time, the sum of the emf's from coil sides to the left and to the right of the positive brush are such that they tend to send current into the brush from both directions or paths [a, in Eq. (1-6)]. Thus, each positive brush in the lap winding shown in Fig. 2-9a receives current from two paths, making a total of four paths for the four-pole dynamo shown.

A wave-wound coil is shown in Fig. 2-8c, and the wave winding in Fig. 2-9b. As stated previously, the windings differ only in the manner in which the coil ends are connected to the commutator. In the wave winding, the series connections of coil sides between brushes are created by going several times around the armature before one path between brushes is completed. Thus, starting at commutator 1, coil 1 enters upper slot 1 under a north pole, then goes to lower slot 6 under the adjacent south pole to commutator 10, then to coil 2 at slot 11 under the subsequent north pole and slot 16 under the next south pole to commutator 20, and so on. Thus, coil 1 is in series with coil 2 and, unlike the lap winding, these series-connected coils are under a different pair of poles. In a four-pole machine, coil 3 will now be under the first set of poles and coil 4 under the second set, and so on, until all slots are filled.

Wave windings also differ from lap windings in the manner in which the brushes are connected together. In Fig. 2-9b, coil X, located in the interpolar zone, spans commutator bars 4 and 14, thus connecting the positive brushes. In the same manner, other coils located in the interpolar space connect the negative brushes to a common point. It is only necessary to have two brushes, regardless of the number of poles in a wave winding, because the coils themselves serve to complete the paths back to a common brush point. It can also be shown that for a wave winding which uses all the coils no the armature to form a closed path, there are only two paths regardless of the number of poles.

There are many winding variations as well as other types of windings which are employed on dc dynamos to yield high current or voltage characteristics. In one of these types, called a multiplex winding, there are several sets of completely closed and independent windings. If there is only one set of coils to form the closed winding, such a winding is called a *simplex winding*. If there are two such windings on the same armature, it is called a *duplex*, and so on. The multiplicity affects the number of possible paths in the armature. For a given number of slots and coils, as the multiplicity increases, the paths also increase, raising the current rating but decreasing the voltage rating, as shown in Section 1-13 and Table 1-2.

In computing the average generated emf between brushes from Eq. (1-6), the number of paths, a, for lap and wave windings may be determined by the following simple relations.

For a lap winding, $\quad a = mP \quad$ (2-4)

For a wave winding, $\quad a = 2m \quad$ (2-5)

where a is the number of parallel paths in the armature
m is the multiplicity of the armature
P is the number of poles

EXAMPLE 2-1: a. A triplex lap-wound armature is used in a 14-pole machine with fourteen brush sets, each spanning three commutator bars. Calculate the number of paths in the armature.
b. Repeat (a) for a triplex wave-wound armature having two such brush sets and 14 poles.

Solution:

a. $a = mP = 3 \times 14 = $ **42 paths** $\quad$ (2-4)

b. $a = 2m = 2 \times 3 = $ **6 paths** $\quad$ (2-5)

EXAMPLE 2-2: Calculate the generated emf in each of the above problems if the flux per pole is 4.2×10^6 lines, the generator speed is 60 rpm, and there are 420 coils on the armature, each coil having 20 turns.

Solution:

a. $Z = 420$ coils $\times$ 20 turns/coil $\times$ 2 conductors/turn $= 16{,}800$ conductors.

From Eq. (1-6):

$$E_g = \frac{\phi ZSP}{60a} \times 10^{-8} \text{ V} = \frac{4.2 \times 10^6 \times 16{,}800 \times 60 \times 14}{60 \times 42} \times 10^{-8}$$

$$= 235.2 \text{ V}$$

b. $E_g = \dfrac{4.2 \times 10^6 \times 16{,}800 \times 60 \times 14}{60 \times 6 \times 10} = 1646.4 \text{ V}$

The paths for the four-pole lap-wound simplex dynamo, shown in Fig. 2-9a, are represented in Fig. 2-10. Note that there are four poles and four paths, each path carrying one-fourth of the total current and generating a voltage per path of e_p. The total power generated by the dynamo operating as a generator is $e_p I$. Figure 2-10c shows the equivalent circuit of this armature operating as a motor at the same speed and flux density. The counter emf generated per path is e_p, and the resistance of the winding in each path is r_w. In the motor mode, the applied line voltage V_L exceeds e_p by the voltage drop across the resistance of the windings, $Ir_w/4$ (assuming no brush voltage drop).

The paths for the four-pole wave-wound simplex dynamo shown in Fig. 2-9b are represented in Fig. 2-11. Note that there are four poles and

(a)

(b)

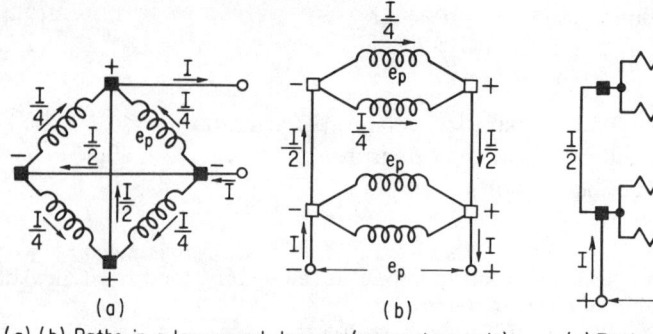

(a),(b) Paths in a lap-wound dynamo (generator mode).

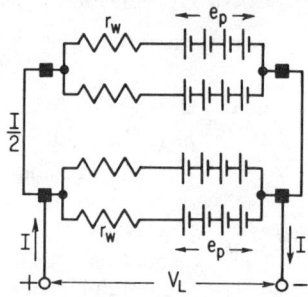

(c) Equivalent circuit of a lap-wound dynamo (motor mode).

Figure 2-10

Armature paths and equivalent circuit for a four-pole lap-wound simplex dynamo.

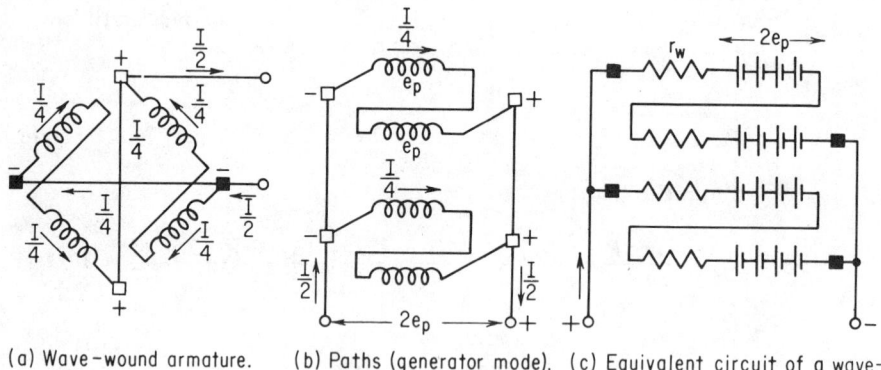

(a) Wave-wound armature.   (b) Paths (generator mode).   (c) Equivalent circuit of a wave-wound dynamo (motor mode).

Figure 2-11

Equivalent capacity wave-wound dynamo showing paths and equivalent circuit for a four-pole simplex wave winding.

two paths, since the paths are independent of the number of poles in a wave winding, as stated in Eq. 2-5. For conductors of the same current-carrying capacity as the lap winding above, i.e., $I/4$, the current per path in the wave winding is $I/4$. The total current, since there are but two paths, is $I/2$. But now that there are only two paths, the total number of conductors per path is doubled, and the emf per path is $2e_p$. The total power generated by the dynamo operating in the generator mode is $2e_p I/2$, or still $e_p I$. This is also still in agreement with the statements of Section 1-13 and Table 1-2.

56

CHAP. TWO / *Dynamo Construction and Windings*

The wave-wound dynamo in the motor mode is shown in Fig. 2-11c. Note that the resistance per path and the voltage per path are each twice that of the equivalent lap-wound armature, because there are twice as many conductors in series producing a counter emf and resistance in opposition to the applied voltage.

2-12.
SUMMARY—
WINDINGS

Based on the preceding material, the following summary may be made regarding each of the windings studied.

2-12.1
Lap Winding

There are always as many paths in parallel as the product of the multiplicity and the number of poles [Eq. (2-4)]. Each path at any instant contains a series group of coils, N_c/a, where N_c is the total number of armature coils. The current carried by each armature coil is the total terminal armature current (current entering or leaving the armature) divided by the number of paths, I/a. The winding requires as many brushes as there are poles.

2-12.2
Wave Winding

The number of paths in the armature is twice the multiplicity, $2m$, and independent of the number of poles [Eq. (2-5)]. Each path at any instant contains a series group of coils, N_c/a, similar to the lap winding above. The current carried by each armature coil is the total terminal armature current divided by the number of paths, I/a, also similar to the lap winding above. The wave winding requires only two brushes, regardless of the number of poles, although, in large dynamos, it may use as many sets of brushes as poles to reduce the current carried per brush.

In comparing the relative merits of lap and wave windings, it is fairly obvious that lap winding lends itself to high-current, low-voltage, dc dynamos. For a given current-carrying capacity per coil, say 100 A, a 12-pole triplex lap-wound armature could deliver a rated current of 3600 A or 36 times the current per coil. On the other hand, the wave-wound dynamo lends itself to high-voltage, low-current applications in which the dynamo is operating at fairly high speeds. The voltage rating of such dynamos is limited by the possibility of flash-over* between adjacent commutator segments and the insulation quality of the armature coils. The wave winding, because of the possibility of using only two sets of brushes, lends itself to those applications, such as railway traction service, where the complete periphery of the motor may be inaccessible for brush maintenance and replacement.

* A general rule-of-thumb is a maximum allowable value of 15 V between adjacent commutator segments. Flash-over may be avoided by designing and operating the dynamo at speeds and voltages well below this value.

**2-13.
AC SYNCHRONOUS
DYNAMO
ARMATURE
WINDINGS**

Unlike the dc dynamo in which closed (re-entrant) windings are employed, the ac dynamo may use either closed or open windings; but the vast majority of applications employ open windings. Diamond-shaped preformed coils of either the wave or lap variety are used on open-type ac dynamo windings. As stated in Section 2-3 and its accompanying figure, no commutator is needed to convert the alternating current generated in the individual coil into direct current. Since *no commutation is required*, it is not necessary to employ closed windings or to rotate the armature. As in the dc armature, groups of coils are connected in series for purposes of increasing voltage or torque. Such a series-connected group, whose ends are brought out of the armature, is called a *phase*. If all the coils wound in all the slots on a stator armature are connected in series, the ac synchronous dynamo winding is a *single-phase winding*. If two separate and insulated series-connected windings are placed in slots on the stator armature and are mechanically displaced with respect to each other, the ac synchronous dynamo winding is a *two-phase winding*; and so on.

A *three-phase winding*, in extremely simplified form, is shown in Fig. 2-12a. The start and finish of all the coils in phase A are designated, re-

(a) Concentrated three-phase, half-coil wave winding with one slot per pole per phase (one coil side per slot).

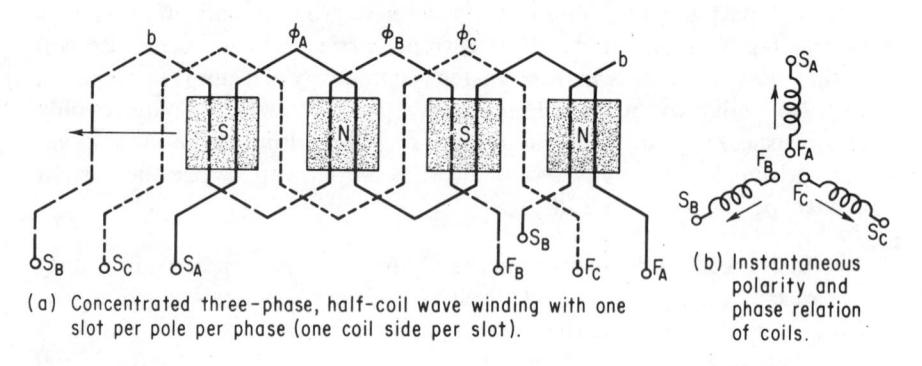

(b) Instantaneous polarity and phase relation of coils.

Figure 2-12

Open windings used on ac synchronous dynamo.

spectively, as S_A and F_A. Phase A is shown as a solid line in the figure, phase B as a dashed line, and phase C as a dotted line. Note that each winding does not start and finish under the same pole. If we assume that the poles on the rotor are moving to the left as shown, then the relative motion of the armature conductors (see Section 1-5) is to the right, producing a *phase sequence* of $ACBACBA$, etc. If the distance between two adjacent corresponding points on the poles is 180 electrical degrees, we can see that the distance between the coil side at the start of A and that at the start of C is approximately 120 electrical degrees. Thus, the leading

CHAP. TWO / *Dynamo Construction and Windings*

pole tip of a unit north pole moving to the left in Fig. 2-12a will induce identical voltages in corresponding coil sides *A*, *C*, and *B*, respectively, 120 electrical degrees apart. By Fleming's right-hand rule, given the above conductor motion with respect to the field, a north pole will induce a corresponding voltage from the finish to the start of the coils, displaced as shown in Fig. 2-12b. Note that phase *B* lags phase *A* by 240 electrical degrees, or leads phase *A* by 120.

The winding employed in Fig. 2-12 is an *open* winding, since both ends of the windings have been brought out for suitable connections. It is a *wave* winding since it progresses from pole to pole. It is a *concentrated* winding because all the coils of one phase are concentrated in the same slot under one pole. It is a *half-coil* winding because there is only one-half of a coil (one coil side) in each slot.

2-14. HALF-COIL AND WHOLE-COIL WINDINGS

Half-coil or single-layer windings are sometimes used in small induction motor stators and in the rotors of small wound-rotor induction motors. A cross section of a half-coil, single-layer winding is shown on the left side of Fig. 2-13b. Like the dc dynamo armature windings, the preponderance of commercial armatures for ac synchronous dynamos are of the full or whole-coil two-

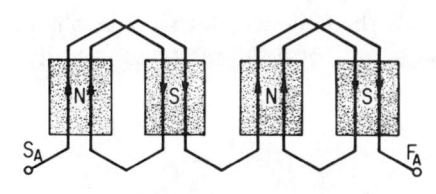

(a) Single-phase distributed half-coil lap winding, two slots per pole-phase (one coil side per slot).

(b) Single vs. double layer.

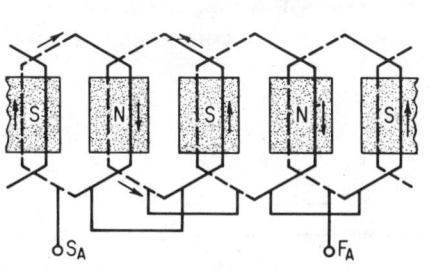

(c) Single-phase, two-layer concentrated whole coil, lap winding, one slot per pole-phase (two coil sides per slot).

Figure 2-13

Distributed and concentrated half-coil and whole-coil windings.

layer type, shown in cross section at the right of Fig. 2-13b. The whole-coil two-layer winding derives its name from the fact that there are two coil sides (one coil) per slot. Fig. 2-13a shows a single-layer half-coil lap winding, and Figs. 2-13b and c show a double-layer full-coil lap winding.

2-15. CHORDED OR FRACTIONAL-PITCH WINDINGS

Whereas most single-layer windings are full-pitch windings, the two-layer whole-coil winding is generally designed on an armature as a chorded or fractional-pitch winding (see Section 2-10). This common practice stems from the fact that the primary advantage of the whole-coil winding is that it permits the use of fractional-pitch coils. Section 2-10 pointed out that dc dynamos employed fractional-pitch in order to save copper. As will be shown later, fractional-pitch windings, when used in ac synchronous and asynchronous dynamo armatures, in addition to saving copper, (1) reduce the mmf harmonics produced by the armature winding, and (2) reduce the emf harmonics induced in the winding, without reducing the magnitude of the fundamental emf wave to any extent. For the three reasons cited, two-layer windings are almost universally employed in ac synchronous dynamo armatures.

Although fractional pitch is employed in dc dynamos, it is *not* necessary to apply a *pitch factor*, k_p, to the computation of induced emf between brushes of Eq. 1-6. The average emf induced in each coil of a dc dynamo may be assumed to be in phase with all the other coils in any given path between brushes. In the case of an ac dynamo using a full-pitch coil, such as that shown in Fig. 2-12a, the two coil sides span a distance exactly equal to the pole pitch of 180 electrical degrees. As a result, the emf's induced in a full-pitch coil are such that the coil-side emf's are in phase, as shown in Fig. 2-14a. The pitch factor, k_p, of a full-pitch coil is unity; and the total coil voltage, E_c, is $2E_1 \times k_p$ or $2E_1$.

(a) Full-pitch coil (b) Fractional-pitch coil

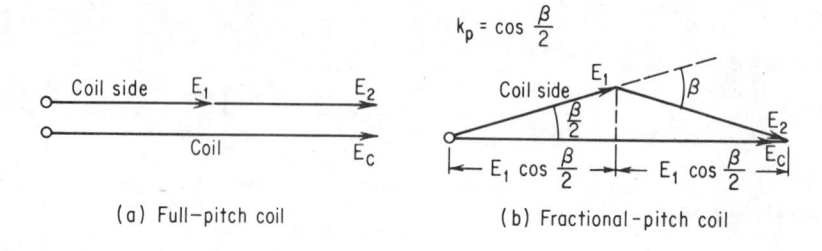

Figure 2-14

Coil emf in terms of coil side emfs for full- and fractional-pitch coils.

In the case of the two-layer winding shown in Fig. 2-13c, it will be noted that the coil span of a single coil is *less* than the pole span of 180 electrical degrees. The emf induced in each coil side is *not* in phase, and the resultant coil voltage, E_c, would be less than the arithmetic sum of each coil side, or less than $2E_1$. It is obvious that $2E_1$ must be multiplied by a factor which is less than unity or $2E_1 k_p$ to produce the proper coil voltage E_c. From the equality of the previous sentence, the pitch factor, k_p is

$$k_p = \frac{E_c}{2E_1} = \frac{\text{phasor sum of the two coil sides}}{\text{arithmetic sum of the two coil sides}} \quad (2\text{-}6)$$

The above relationship, in terms of voltages, is of interest in helping to understand the concept; but it is hardly useful, since we have no way of predicting what the voltage change would be if the coil were stretched or compressed. If we assume that the induced emf's of two coils, E_1 and E_2, are out of phase with respect to each other by some angle β, as shown in Fig. 2-14b, then the angle between E_1 and the resultant coil voltage E_c is $\beta/2$. The resultant coil voltage E_c is from Eq. (2-6) and Fig. 2-14b:

$$E_c = 2E_1 \cos \frac{\beta}{2} = 2E_1 k_p$$

and, therefore,

$$k_p = \cos \frac{\beta}{2} \quad (2\text{-}7)$$

where β is 180° minus the number of electrical degrees spanned by the coil.

Since β is the supplementary angle of the coil span, the pitch factor, k_p may also be expressed as

$$k_p = \sin \frac{p^\circ}{2} \quad (2\text{-}8)$$

where p° is the span of the coil in electrical degrees.

EXAMPLE 2-3: A 72-slot stator armature having four poles is wound with coils spanning 14 slots (slot 1 to slot 15). Calculate:
  a. The full-pitch coil span (pole span).
  b. The span of the coil in electrical degrees.
  c. The pitch factor, using Eq. 2-7.
  d. The pitch factor, using Eq. 2-8.

Solution:

a. Full-pitch coil span $= \dfrac{72 \text{ slots}}{4 \text{ poles}} = \dfrac{18 \text{ slots}}{\text{pole}}$

or 18 slots/180 electrical degrees

b. $p^\circ = \frac{14}{18} \times 180° = \mathbf{140°}$

c. $k_p = \cos \dfrac{\beta}{2} = \cos \dfrac{180° - 140°}{2} = \cos 20° = \mathbf{0.94}$ \quad (2-7)

d. $k_p = \sin \dfrac{p^\circ}{2} = \sin \dfrac{140°}{2} = \sin 70° = \mathbf{0.94}$ \quad (2-8)

It is sometimes convenient to speak of an armature coil span as having a fractional pitch expressed as a fraction, e.g., a $\frac{5}{6}$ pitch, or an $\frac{11}{12}$ pitch, etc. In such a case, the electrical degrees spanned, p, is $(\frac{5}{6}) \times 180°$, or 150°;

or $(\tfrac{11}{12}) \times 180°$, or $165°$; etc.; and the pitch factor, k_p, is still computed, as in Eq. (2-8), and Ex. 2-4 below.

EXAMPLE 2-4: A 96-slot, six-pole armature is wound with coils having $\tfrac{13}{16}$ fractional pitch. Calculate the pitch factor.

Solution:

$$k_p = \sin\left(\frac{p°}{2}\right) = \sin\left[\frac{(\tfrac{13}{16}) \times 180°}{2}\right] = \sin 73.2° = \mathbf{0.957} \qquad (2\text{-}8)$$

2-16.
DISTRIBUTION OR BELT FACTOR— DISTRIBUTED WINDINGS

The windings shown in Figs. 2-12a and 2-13c are called *concentrated windings* because all the coil sides of a given phase are concentrated in a single slot under a given pole. For Fig. 2-12a, in determining the induced ac voltage per phase, it would only be necessary to multiply the voltage induced in any given coil by the number of series-connected coils in each phase. This is true for the winding shown in Fig. 2-12a because the conductors of each coil, respectively, lie in the same position with respect to the N and S poles as other series coils in the same phase. Since these individual coil voltages are induced in phase with each other, they may be added arithmetically, or a given coil voltage may be multiplied by the number of series-connected coils per phase to obtain the induced voltage per phase.

Concentrated windings, in which all the conductors of a given phase per pole are concentrated in a single slot, are not commercially used and have numerous disadvantages. They fail to use the entire inner periphery of the stator iron efficiently, and they make it necessary to use extremely deep slots where the windings are concentrated, thus increasing armature leakage and reactance (see Section 2-7). Finally, they result in low copper-to-iron ratios (for a given weight of dynamo iron, the more copper concentrated in the slots, the greater the capacity and output of the dynamo) since they do not efficiently use the mutual air-gap flux in the stator armature core. It will also be shown later that *distributed windings* (like fractional-pitch coils) reduce harmonics in the output waveform. It is much more effective to distribute the armature slots around the inner periphery of the stator, using a uniform spacing between slots, than to concentrate the windings in a few deep slots.

When the slots are distributed around the armature uniformly, the winding which is inserted is a *distributed winding*. A distributed lap winding is shown in Fig. 2-13a. Note that two coils in phase belt *A* are displaced with respect to each other. The induced voltages of each of these coils will be displaced by the same degree to which the slots have been distributed, and the total voltage induced in any phase will be the phasor sum of the individual coil voltages. The four individual coil voltages, shown in Fig. 2-13a, are represented vectorially, shown in Fig. 2-15, as displaced by some angle α, the number of electrical degrees between adjacent slots. Voltages E_{c1}, E_{c2}, etc., are the individual coil voltages, and n is the number of coils in a given phase belt. The *belt or distribution factor* by which the arithmetic sum of the individual coil voltages must be multiplied in order to yield the phasor sum is

$$E_\phi = nE_c \times k_d$$

or

$$k_d = \frac{E_\phi}{nE_c} = \frac{\text{phasor sum of coil emf's per phase}}{\text{arithmetic sum of coil emf's per phase}} \quad (2\text{-}9)$$

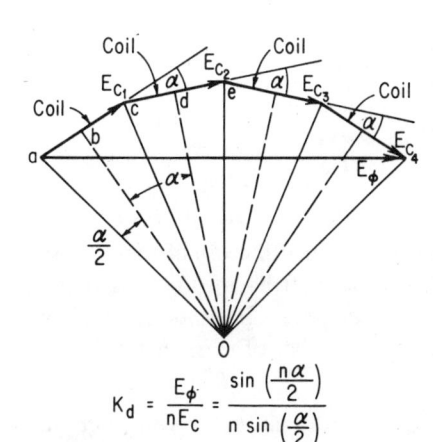

As in the case of Eq. (2-6), the computation of k_d in terms of voltages (theoretical or actual) is impractical. The construction of Fig. 2-15, in which perpendiculars have been drawn to the center of each of the individual coil voltages to a common center of radius, serves to indicate that $\alpha/2$ is the angle bOa. Coil-side voltage ab equals $(Oa) \sin \alpha/2$, and chord E_{c1} is $2(Oa) \sin \alpha/2$. For n coils in series per phase, chord E_ϕ is also $2(Oa) \sin n\alpha/2$, and the distribution or belt factor, k_d, is

$$K_d = \frac{E_\phi}{nE_c} = \frac{\sin\left(\frac{n\alpha}{2}\right)}{n \sin\left(\frac{\alpha}{2}\right)}$$

$$k_d = \frac{E_\phi}{nE_c} = \frac{2Oa \sin n\alpha/2}{n \times 2Oa \sin \alpha/2} = \frac{\sin(n\alpha/2)}{n \sin(\alpha/2)} \quad (2\text{-}10)$$

Figure 2-15
Determination of distribution factor.

where n is the number of slots per pole per phase (slots/pole-phase)
α is the number of electrical degrees between adjacent slots

EXAMPLE 2-5:
a. Calculate the distribution factor, k_d, for a four-pole three-phase armature having:
 1. 12 slots
 2. 24 slots
 3. 48 slots
 4. 84 slots
b. Tabulate n, α, and k_d for ready reference and comparison.

Solution:

180°/pole × 4 poles = 720 electrical degrees.
a. 1. α = 720 electrical degrees/12 slots = 60°/slot
n = 12 slots/(4 poles × 3 phases) = 1 slot/pole-phase

$$k_d = \frac{\sin(1 \times \frac{60}{2})}{1 \times \sin(\frac{60}{2})} = 1.0 \quad (2\text{-}9)$$

63
SEC. 2-16. / *Distribution or Belt Factor—Distributed Windings*

2. $\alpha = 720$ electrical degrees/24 slots $= 30°$/slot
$n = 24$ slots/12 pole-phase $= 2$ slots/pole-phase

$$k_d = \frac{\sin[(2 \times 30)/2]}{2 \times \sin(\frac{30}{2})} = 0.966 \qquad (2\text{-}9)$$

3. $\alpha = 720$ electrical degrees/48 slots $= 15°$/slot
$n = 48$ slots/12 pole-phase $= 4$ slots/pole-phase

$$k_d = \frac{\sin[(4 \times 15)/2]}{4 \times \sin(\frac{15}{2})} = 0.958 \qquad (2\text{-}9)$$

4. $\alpha = 720$ electrical degrees/48 slots $= 8\frac{4}{7}°$/slot
$n = 84$ lots/12 pole-phase $= 7$ slots/pole-phase

$$k_d = \frac{\sin[7(\frac{60}{7})(\frac{1}{2})]}{7 \times \sin(\frac{60}{7 \times 2})} = 0.955 \qquad (2\text{-}9)$$

b.

n	α	k_d
1	60°	1.0
2	20°	0.966
3	15°	0.958
4	8°	0.955

It should be noted from Example 2-5 that the distribution factor, k_d, for any fixed or given number of phases, is a sole function of the number of distributed slots under a given pole. As the distribution of coils (slots/pole) increases, the distribution factor, k_d, decreases. It is *not* affected by the type of winding, lap or wave, nor by the number of turns per coil, etc.

**2-17.
EFFECT OF
FRACTIONAL PITCH
AND DISTRIBUTION
OF COILS ON
WAVEFORM**

Section 2-15 showed that with a fractional-pitch winding, the emf in each coil side must be added vectorially to obtain the emf per coil. Section 2-16 showed that the emf's per coil must be added vectorially to obtain the belt emf or emf per phase. Although both of these factors result in a slightly smaller resultant emf per phase, the use of fractional-pitch coil sand distributed windings in armatures of ac synchronous and asynchronous dynamos is almost universal.

In Section 2-15, it was stated that one of the advantages of fractional-pitch coils was the reduction of harmonics. The fractional-pitch coil of Fig. 2-14b is shown in Fig. 2-16, where the coil sides E_1 and E_2 are out of phase by some angle β, and the resultant coil voltage is E_c. As shown in Fig. 1-8, when the field poles are curved and the flux density made more uniform, the resultant waveshape will be more nearly a square wave than a sine wave. Such a wave is very rich in odd harmonics, in phase with the fundamental, and has an instantaneous equation whose approximate value is represented by the Fourier series of

$$e = E_m \sin \omega t + \frac{E_m}{3} \sin 3\omega t + \frac{E_m}{5} \sin 5\omega t + \cdots + \frac{E_m}{n} \sin n\omega t \qquad (2\text{-}11)$$

(a) Fundamental (b) Third harmonic (c) Fifth harmonic (d) Seventh harmonic

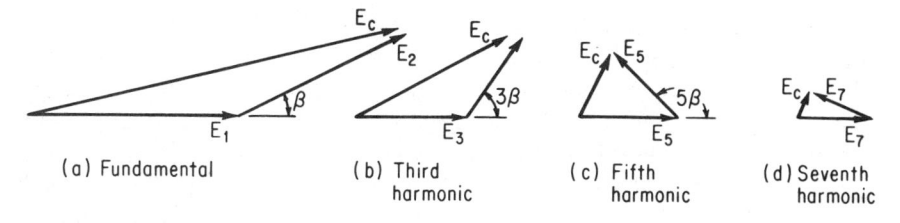

Figure 2-16

Effect of fractional pitch on harmonic voltages generated.

As shown in Fig. 2-16, if the instantaneous voltage in Eq. 2-11 is induced in each coil side, and if the coil-sides are displaced by an angle β, the third harmonic must be displaced by 3β, the fifth harmonic by 5β, and so on. Note that in the case of the fifth harmonic, a component of the harmonic coil-side voltage subtracts from the fundamental to reduce the resultant harmonic voltage. The resultant seventh, ninth, etc., harmonics will produce coil voltages which will decrease the harmonic still more. *Any one harmonic may be completely eliminated* by choosing a fractional pitch that will yield a pitch factor of zero for that harmonic. For example, a $\frac{4}{5}$ pitch (a coil span of 144 electrical degrees) will eliminate the fifth harmonic, or a $\frac{5}{6}$ pitch (a coil span of 150 electrical degrees) will greatly reduce both the fifth and the seventh harmonics, as shown in Figs. 2-16c and d.

The effect of using distributed windings on the waveform (see Section 2-16) is shown in Fig. 2-17. The ac dynamo flux distribution, and the resultant waveshape of emf induced per coil side, are shown in Figs. 2-17a and

(a) Flux distribution at stator.

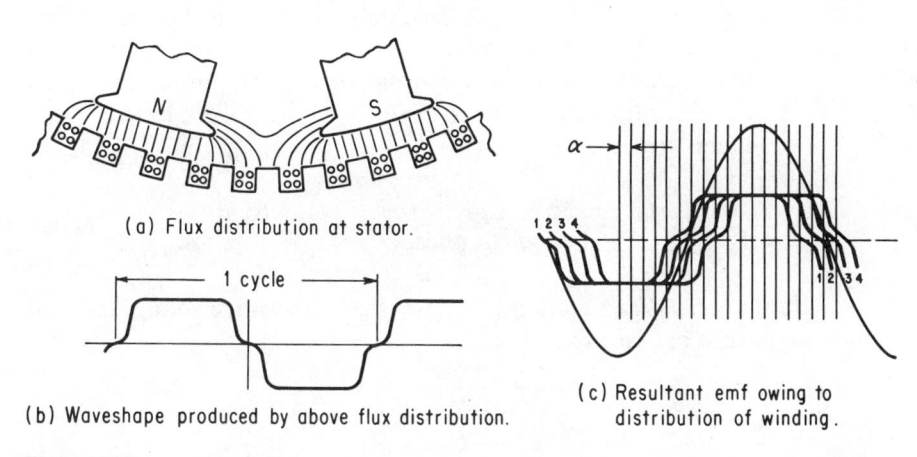

(b) Waveshape produced by above flux distribution.

(c) Resultant emf owing to distribution of winding.

Figure 2-17

Effect of using distributed windings on the waveform.

SEC. 2-17. / *Effect of Fractional Pitch and Distribution of Coils on Waveform*

(b), respectively. For a given constant length and relative velocity of conductors with respect to the field, the emf wave has the same shape as the flux density curve [see Eq. (1-2)], i.e., e is proportional to B. For series-connected coils whose emf's are displaced by some angle α in any given phase belt, as shown in Figs. 2-15 and 2-17c, the resultant emf is the phasor or graphical sum of the individual emf's (both coil side and coil emf). The graphical sum of the individual coil emf's is shown in Fig. 2-17c. Note that although the individual coil emf's are almost square waves, the resultant phase emf is a sine wave. Since a sine wave contains no harmonics, it is fairly obvious that the harmonics of Eq. (2-11) have been canceled out by using a distributed winding, as well as a fractional-pitch coil.

It is fairly obvious then, why commercial ac synchronous dynamos employ distributed windings having fractional-pitch coils. One might well ask, why it is necessary that the output ac voltage per phase of a polyphase alternator should have a waveform which most nearly approaches a sinusoidal wave? Why not a square wave, a triangular wave, or a sawtooth wave?

Perhaps the two most important reasons are: (1) since a sine wave contains no harmonics of higher frequency, the losses resulting from eddy currents and hysteresis are reduced, resulting in increased efficiency; and (2) all electric machines, transformers, and appliances (clocks, etc.) are designed with the assumption that the waveform furnished by the utility to operate them is sinusoidal. This assumption greatly simplifies all design calculations for present and future electric apparatus.

2-18. GENERATED EMF IN AN AC SYNCHRONOUS DYNAMO

It is now possible to derive the computed or expected emf per phase generated in an ac synchronous dynamo. Let us assume that this dynamo has an armature winding consisting of a total number of coils, C, each coil having a given number of turns, N_C. Then the total number of turns in any given phase of a dynamo armature is

$$N_p = \frac{CN_C}{P} = \frac{\text{total armature coils} \times \text{turns/coil}}{\text{number of phases}} = \frac{\text{total turns}}{\text{phase}} \quad (2\text{-}12)$$

But Faraday's law, Section 1-3, states that the average voltage induced in a single turn of two coil sides is

$$E_{\text{av}} = \frac{\phi}{t} \times 10^{-8} \text{ V} \quad (1\text{-}1)$$

Furthermore, when a coil consisting of N turns rotates in a uniform magnetic field, at a uniform speed, it was shown (Section 1-14) that the average voltage induced in an armature coil is,

$$E_{av/coil} = 4\phi N_c s \times 10^{-8} \text{ V} \tag{1-5}$$

where ϕ is the number of lines of maxwells per pole
N_c is the number of turns per coil
s is the relative speed in revolutions/second (rps) between the coil of N_c turns and the magnetic field ϕ

Equation (1-5) was derived for a two-pole machine, as shown in Fig. 1-6a, generating a sine wave in one complete revolution of 360 electrical and mechanical degrees. Thus, in Eq. (1-5), a speed s of 1 rps will produce a frequency f of 1 Hz. Since f is directly proportional and equivalent to s, replacing the latter in Eq. (1-5), for all the series turns in any phase

$$E_{av/phase} = 4\phi N_p f \times 10^{-8} \text{ V} \tag{2-13}$$

But in the preceding section we discovered that the voltage per phase is made more completely sinusoidal by intentional distribution of the armature winding. The effective rms value of a sinusoidal ac voltage is 1.11 times the average value. The effective ac voltage per phase is

$$E_{eff} = 4.44\phi N_p f \times 10^{-8} \text{ V} \tag{2-14}$$

But Eq. (2-14) is still not representative of the effective value of the phase voltage generated in an armature in which fractional-pitch coils and a distributed winding are employed. Taking the pitch factor k_p and the distribution factor k_d into account, we may now write the equation for the effective value of the voltage generated in each phase of an ac synchronous dynamo, as

$$E_{gp} = 4.44\phi N_p f k_p k_d \times 10^{-8} \text{ V} \tag{2-15}$$

where ϕ is the flux per pole in lines or maxwells
N_p is the total number of turns per phase [Eq. (2-12)]
f is the frequency in hertz [Eq. (2-16)]
k_p is the pitch factor [Eq. (2-8)]
k_d is the distribution factor [Eq. (2-10)]

EXAMPLE 2-6: A 72-slot three-phase stator armature is wound for six poles, using double-layer lap coils having 20 turns per coil with a $\frac{5}{8}$ pitch. The flux per pole is 4.8×10^6 lines, and the rotor speed is 1200 rpm. Calculate:
a. The generated effective voltage per coil of a full-pitch coil.
b. The total number of turns per phase.
c. The distribution factor.
d. The pitch factor.
e. The total generated voltage per phase from (a), (c), and (d) above, and by Eq. (2-15).

Solution:

a. $E_{g/coil} = 4.44\phi N_c f \times 10^{-8}$ V (2-14)
$= 4.44(4.8 \times 10^6)(20)\left(\dfrac{6 \times 1200}{120}\right) \times 10^{-8}$
$= $ **256 V/coil**

b. $N_p = 72$ coils/3 phase $\times$ 20 turns/coil
$= $ **480 turns/phase** [from (Eq. 2-12)]

c. $k_d = \dfrac{\sin(n\alpha/2)}{n \sin(\alpha/2)}$, where $n = 72$ slots/(3 phase $\times$ 6 poles) (2-9)
$= 4$ slots/pole-phase
and $\alpha = $ (6 poles $\times$ 180°/pole)/72 slots $= 15°$/slot

$k_d = \dfrac{\sin[(4 \times 15)/2]}{4 \sin(\frac{15}{2})} = \dfrac{\sin 30°}{4 \sin 7.5°} = $ **0.958**

d. $k_p = \sin \dfrac{p°}{2} = \sin\left(\dfrac{5}{6} \times \dfrac{180}{2}\right) = \sin 75° = $ **0.966** (2-8)

e. $E_{gp} = 4.44 \times 480$ turns/phase $\times 4.8 \times 10^6$
$\times 60 \times 0.966 \times 0.958 \times 10^{-8}$
$= $ **5680 V** [from Eq. (2-15)]

$E_{gp} = 256$ V/coil $\times 24$ coils/phase $\times 0.966 \times 0.958$
$= $ **5680 V/phase** [from (a), (c), and (d)]

2-19.
FREQUENCY OF AC SYNCHRONOUS DYNAMO

Commercial ac synchronous dynamos have many poles and may rotate at various speeds, either as alternators or as synchronous or induction motors. Equation 2-15 above was derived for a two-pole device in which the generated emf in the stationary armature winding changes direction every half-revolution of the two-pole rotor. One complete revolution will produce one complete positive and negative pulse each cycle. The frequency in cycles per second (Hz) will, as stated previously, depend directly on the speed or number of revolutions per second (rpm/60) of the rotating field.

If the ac synchronous dynamo is multipolar (having, say, two, four, six, or eight poles), then for a speed of one revolution per second (1 rpm/60) the frequency per revolution will be one, two, three, or four cycles per revolution, respectively. The frequency per revolution is, therefore, equal to the number of pairs of poles. Since the frequency depends directly on the speed (rpm/60) and also on the number of pairs of poles ($P/2$) we may combine these into a single equation in which

$$f = \dfrac{P}{2} \times \dfrac{\text{rpm}}{60} = \dfrac{PS}{120} \quad (2\text{-}16)$$

where P is the number of poles
S is the speed in rpm
f is the frequency in hertz

Table 6-1 illustrates this relation for three of the most commonly used frequencies.

EXAMPLE 2-7: An ac generator has eight poles and operates at a speed of 900 rpm. Calculate:
a. The frequency of the generated voltage.
b. The prime mover speed required to generate frequencies of 50 Hz and 25 Hz.

Solution:

a. $f = \dfrac{PS}{120} = \dfrac{8 \times 900}{120} =$ **60 Hz** (2-16)

b. $S = \dfrac{120f}{P} = \dfrac{120 \times 50}{8} =$ **750 rpm** (to generate 50 Hz)

$= \dfrac{120 \times 25}{8} =$ **375 rpm** (to generate 25 Hz)

BIBLIOGRAPHY

Alger, P. L. and Erdelyi, E. "Electromechanical Energy Conversion," *Electro-Technology*, September 1961.

Bewley, L. V. *Alternating Current Machinery*, New York: The Macmillan Company, 1949.

Carr, C. C. *Electrical Machinery*, New York: John Wiley & Sons, Inc., 1958.

Crosno, C. D. *Fundamentals of Electromechanical Conversion*, New York: Harcourt, Brace, Jovanovich, Inc, 1968.

Daniels. *The Performance of Electrical Machines*, New York: McGraw-Hill, Inc., 1968.

Fitzgerald, A. E. and Kingsley, C. *The Dynamics and Statics of Electromechanical Energy Conversion*, 2nd Ed., New York: McGraw-Hill, 1961.

Fitzgerald, A. E., Kingsley, Jr. C., and Kusko, A. *Electric Machinery*, 3rd Ed. New York: McGraw-Hill, Inc., 1971.

Gemlich, D. K. and Hammond, S. B. *Electromechanical Systems*, New York: McGraw-Hill, 1967.

Hindmarsh, J. *Electrical Machines*, Elmsford, N. Y.: Pergamon Press, 1965.

Jones, C. V. *The Unified Theory of Electrical Machines*, New York: Plenum Publishing Corp., 1968.

Kloeffler, S. M., Kerchner, R. M. and Brenneman, J. L. *Direct Current Machinery*, Rev. ed., New York: The Macmillan Company, 1948.

Koenig, H. E. and Blackwell, W. A. *Electromechanical System Theory*, New York: McGraw-Hill, 1961.

Liwschitz, M. M., Garik, M. and Whipple, C. C. *Alternating Current Machines*, Princeton, N. J.: D. Van Nostrand Co., Inc., 1946.

Selmon. *Magnetoelectric Devices: Transducers, Transformers and Machines*, New York: Wiley/Interscience, 1966.

Siskind, C. S. *Direct-Current Machinery*, New York: McGraw-Hill, 1952.

Skilling, H. H. *Electromechanics: A First Course in Electromechanical Energy Conversion*, New York: Wiley/Interscience, 1962.

Thaler, G. J., and Wilcox, M. L. *Electric Machines: Dynamics and Steady State*, New York: Wiley/Interscience, 1966.

White, D. C., and Woodson, H. H. *Electromechanical Energy Conversion*, New York: Wiley/Interscience, 1959.

QUESTIONS

2-1. Give four possible types of dynamo construction, listing for each
   a. the name of the particular type
   b. choice of rotor (rotating element)
   c. choice of stator (stationary element).

2-2. For the commercial dc dynamo, list and define
   a. four distinct parts of the rotor
   b. four distinct *functions* of the armature
   c. seven distinct parts of the stator.

2-3. Show, by means of a diagram, the relation between the electric circuits of
   a. a shunt-wound dc dynamo
   b. a series-wound dc dynamo.

2-4. Show, by means of a construction diagram, the major parts of a stationary field revolving armature synchronous dynamo.

2-5. Repeat Quest. 2-4 for a synchronous dynamo having a rotating field and stationary armature.

2-6. Repeat Quest. 2-4 for an *asynchronous* dynamo.

2-7. Show the magnetic circuit of a 6-pole dc commercial dynamo tracing the following flux paths:
   a. mutual flux paths (a total of six)
   b. leakage flux between adjacent poles
   c. leakage flux between a given pole and the yoke.

2-8. With reference to your diagram drawn for Quest. 2-7 explain, in terms of magnetic circuit theory,

a. why the mutual field flux linking the armature is mainly confined to the surface of the armature

b. why the armature flux, produced by current in the armature conductors embedded in the armature slots, links the armature laminations primarily.

2-9. Repeat Quest. 2-8 for a salient pole ac synchronous dynamo.

2-10. During the period of commutation in a dc dynamo, the current in an armature conductor undergoing commutation varies considerably. This change in conductor flux should induce a voltage in accordance with Neumann's law. For the coil shown in Fig. 2-5(b) explain:

a. where the voltage is induced by that portion of the coil embedded in iron

b. where the voltage is induced by that portion of the coil NOT embedded in iron.

2-11. Explain:

a. which of the induced voltages in Quest. 2-10 gives rise to armature reactance

b. which of the induced voltages in Quest. 2-10 contributes to iron losses.

2-12. Explain:

a. why all field coils placed on field poles have the same number of turns/pole and are always connected in series

b. why all shunt field coils are designed to produce the required mmf using a large number of turns and low current

c. why all field coils require dc for their operation, even in ac dynamos.

2-13. a. Under what three conditions is the inductive effect of a field coil produced?

b. Give the equation for the voltage produced by this inductive effect.

2-14. a. Under what conditions may the induced voltage, $L\,di/dt$ in Eq. 2-2 exceed E_{dc}? Explain.

b. What method is usually employed to absorb the energy of the magnetic field?

2-15. a. Assuming that the dynamo shown in Fig. 2-1b is a motor energized from a dc supply, explain what happens to the energy of the magnetic field produced by the shunt field winding when the motor is disconnected from the line.

b. Why is it unnecessary to take into account the field energy ($\frac{1}{2}LI^2$) of series field windings or interpole and compensating windings? (*Hint:* compare the inductance of these windings with that of a shunt field.)

2-16. a. Distinguish between open and closed armature windings and describe their application to dc and ac dynamos.

b. Name two types of dc armature winding.

c. Explain why only 2 brushes are needed on wave windings, regardless of the number of poles.

2-17. Define the following terms:

a. Coil span (in terms of pole pitch)

      b. Full-pitch coil
      c. Fractional-pitch coil (in terms of pitch factor)
      d. Chorded winding
      e. Two-layer winding
      f. Coil side
      g. Number of conductors/coil side
      h. Multiplicity
      i. Parallel paths
      j. Degree of reentrancy
      k. Full-coil winding
      l. Half-coil winding
      m. Concentrated winding
      n. Distributed winding
      o. Phase.

2-18. a. Give a major difference between dc and ac dynamo armature windings as to reentrancy.
      b. Give one good reason for this difference in the case of ac dynamos.

2-19. a. Give one advantage of the use of fractional-pitch windings in dc dynamos.
      b. Repeat (a) for two additional advantages in ac dynamos.

2-20. a. Why is it unnecessary to take the phasor sum of induced voltages in series-connected coils of a dc dynamo, using fractional-pitch coils, to obtain the voltage per path between brushes?
      b. Why is it necessary to use a pitch factor to determine the induced emf of ac dynamos using fractional-pitch coils?
      c. Define pitch factor in terms of three separate equations.

2-21. a. What is the distribution factor of a concentrated winding?
      b. Give three advantages of distributed over concentrated windings.
      c. Define distribution factor in terms of two separate equations.

2-22. Give two reasons affecting output waveform to explain why commercial ac synchronous dynamos employ distributed windings and fractional-pitch coils.

2-23. Give the equation expressing the generated voltage per phase of an ac synchronous polyphase dynamo which takes pitch and distribution factor into account.

2-24. Give the equation for the frequency of an ac synchronous polyphase dynamo.

PROBLEMS

2-1. A six-pole generator has field coils each having an inductance of 25 H/pole. If the field current drops from 3 A to zero in 15 ms when the generator's field circuit is opened, calculate:
      a. The average induced voltage per pole
      b. The total induced voltage across the field circuit switch

2-2. A dynamo has a total of 8000 field turns. When carrying 2.5 A, a total flux of 5.2×10^6 lines is produced. Calculate:
a. Self-inductance of the field coils ($L = \phi N/I \times 10^8$)
b. Average voltage generated if the current drops to zero in 10 ms.

2-3. The field coils of a dynamo have an inductance of 8 H, a resistance of 60 Ω and are connected to a 120-V dc source. Calculate:
a. The value of the discharge resistor to be connected across the field coils if the voltage across the field circuit must never exceed 150 V
b. The time required to discharge the energy stored in the magnetic field through the field discharge resistor
c. The total energy discharged when the field circuit is disconnected from the supply

2-4. A 220-V dc motor has a field circuit resistance of 220 Ω and an inductance of 55 H. Calculate:
a. The initial field current at the instant the motor is connected to its rated dc supply
b. The instantaneous rate of change of current at the instant the field circuit is connected to the 220-V supply
c. The final steady-state current
d. The time required to reach 63.2 per cent of its steady-state value
e. The time required to reach its steady-state value
f. The instantaneous current one second after the switch is closed
g. The time required to reach a field current of 0.75 A
h. The energy stored in the magnetic field once the current reaches a steady-state value
i. The time required to dissipate the field energy in a discharge resistor at a rate of 100 W

2-5. A dynamo has a rated armature current of 250 A and 12 poles. Calculate the current per path if the armature is
a. Wave wound
b. Lap wound
(Assume simplex unless otherwise indicated.)

2-6. Calculate the number of parallel paths in the following armatures when inserted in an 18-pole field structure
a. Triplex wave wound
b. Triplex lap wound
c. Duplex lap wound
d. Quadruplex wave wound

2-7. A 12-pole 120-V dc generator has a triplex lap-wound armature having 80 coils of 9 turns/coil and runs at a speed of 3600 rpm. Calculate:
a. The flux per pole required to produce rated generated voltage
b. The current per path if the rated output is 60 kW (neglect field current)
c. The minimum number of brushes required and commutator segments spanned

2-8. A duplex double-layer lap winding is wound on an armature having 48 slots with 1 coil/slot; each coil has a total of 60 turns. The armature is to be used in a 250-V, 1200-rpm, four-pole, 50-kW generator. Calculate

a. Current per conductor when the generator is delivering rated load
b. Flux per pole required to produce generated voltage.

2-9. Repeat Problem 2-8 using a simplex wave-wound armature.

2-10. A 16-pole alternator is driven at a speed of 3000 rpm. Calculate the frequency generated in the stator armature.

2-11. The rotor of a six-pole 60 Hz alternator produces a field flux of 5×10^6 lines per pole. Calculate
a. The speed at which the alternator must be driven to produce rated frequency
b. The average generated voltage induced in each stator coil having 200 turns
c. The effective voltage per phase for a single-phase stator armature having 60 coils equally distributed on the stator

2-12. A stator to be used for a three-phase half-coil six-pole armature winding has 144 slots. Each coil spans 20 slots. Calculate
a. Pitch factor
b. Number of coils per phase
c. Distribution factor
d. Effective voltage per phase if the generated voltage per coil is 30 V rms

2-13. A 72-slot three-phase whole-coil stator armature has a coil span of 10 slots, 10 turns/coil. The armature is wye connected. The six-pole rotor has a flux of 5.2×10^6 maxwells/pole and is driven at a speed of 120 rpm. Calculate:
a. Effective phase voltage
b. Alternator line voltage

2-14. Repeat Problem 2-13 for a two-phase (three-wire) alternator.

2-15. A 24-pole 60 Hz, three-phase wye-connected alternator has 6 slots/pole and a full-pitch two-layer lap winding in which there are 8 conductors/slot. The air-gap flux is 6×10^6 lines per pole. Calculate:
a. Number of conductors per phase
b. Distribution factor
c. Pitch factor
d. Field pole rpm
e. Induced emf per phase and per line.

2-16. Repeat Problem 2-15e for a coil span of slots 1 to 6 ($\frac{5}{6}$ pitch).

ANSWERS

2-1(a) 5 kV (b) 30 kV   2-2(a) 166.5 H (b) 41.6 kV   2-3(a) 15 Ω (b) 533 ms (c) 16 W-s   2-4(a) 0 (b) 4 A/s (c) 1 A (d) 0.25 s (e) 1.25 s (f) 0.982 A (g) 0.346 s (h) 27.5 W (i) 0.275 s   2-5(a) 125 A (b) 20.8 A   2-6(a) 6 (b) 54 (c) 36 (d) 8.   2-7(a) 4.167×10^5 (b) 13.88 A (c) 12.   2-8(a) 25 A/conductor (b) 4.33×10^5 lines   2-9(a) 100 A/conductor   (b) 1.08×10^5 lines   2-10. 400 Hz

2-11(a) 1200 rpm (b) 2400 V (c) 160 kV   2-12(a) 0.966 (b) 24 coils/phase (c) 0.954 (d) 664 V   2-13(a) 3080 V (b) 5330 V 2-14(a) 4360 V (b) 6170 V   2-15(a) 384 conductors/phase (b) 0.966 (c) 1 (d) 300 rpm (e) 2970 V, 5140 V   2-16 2870 V, 4970 V

THREE

dc dynamo voltage relations—dc generators

3-1.
GENERAL

In comparing dc dynamo generator action with motor action, Section 1-20 concluded with a summary of the fundamental differences between them. This chapter is devoted to the dc dynamo used as a dc generator, and is concerned primarily, therefore, with dc dynamo voltage relations since a generator is a source of voltage. The summary of Section 1-20 stated for generator action:

1. The electromagnetic torque (developed in the current-carrying conductor) *opposes* prime mover rotation (Lenz's law).
2. The generated voltage (induced in the armature) aids and produces armature current.
3. The generated voltage, $E_g = V_a + I_a R_a$ (1-10)

The general construction of the dc dynamo was discussed in Section 2-2 and shown in Fig. 2-1. For purposes of commutation, it was seen that the armature, containing the current-carrying conductors in which voltages are induced, of necessity *must* rotate to perform the functions outlined in Section 2-2. A discussion of the stator comprising the dc magnetic field and some of its design considerations was outlined in Section 2-9. A discussion of some of the design considerations affecting the dc dynamo armature was developed in Sections 2-10, 2-11, and 2-12.

The generated armature voltage, E_g, of Eq. (1-10), for the total average induced emf between brushes, was given as

$$E_g = \frac{\phi ZSP}{60a} \times 10^{-8} \text{ V} \qquad (1\text{-}6)$$

where, depending on the nature of the winding, the number of paths, a, in the armature, is determined by these equations:

For a lap winding, $\qquad a = mP \qquad (2\text{-}4)$
For a wave winding, $\qquad a = 2m \qquad (2\text{-}5)$

The reader should review the equations and sections cited above since they are fundamental and apply to all the commercial dc generator types and characteristics discussed below.

**3-2.
DC GENERATOR
TYPES**

The three basic types of dc generators employing dc dynamo construction are the *shunt*, *series*, and *compound* generators. The differences among these types emerge from the manner in which the dc stator field winding's excitation is produced. The purpose of the generator is to produce a dc voltage by conversion of mechanical to electric energy, and a portion of this dc voltage is employed to *excite* the stationary magnetic field winding.

**3-3.
SCHEMATIC
DIAGRAM AND
EQUIVALENT
CIRCUIT OF A
SHUNT GENERATOR**

When the excitation is produced by a field winding connected across the complete (or almost complete) line voltage produced between the brushes of the armature, the dc dynamo is called a *shunt generator*. The complete schematic circuit diagram of the shunt generator is shown in Fig. 3-1a.

The rotor armature is represented (enclosed in a rectangle of dashed lines) as consisting of: a source of emf, E_g, generated in accordance with Eq. (1-6); a resistance, R_w, of the armature winding; and a resistance, R_b, of the carbon brushes and the brush contact resistance made with the rotating armature. The entire *armature circuit* consists of the armature (enclosed in the rectangle of dashed lines) and two optional windings, the compensating winding, R_c and the interpole winding, R_i, located on the stator. Thus, the portion of the armature circuit which rotates is shown enclosed in the rectangle, and that portion of the armature circuit which is fixed on the stator is outside the rectangle.

(a) Complete schematic circuit diagram of shunt generator.

(b) Equivalent circuit of shunt generator.

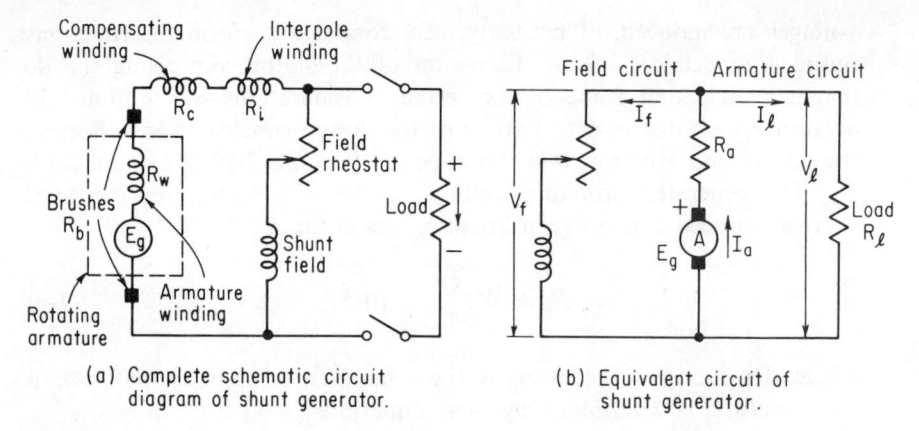

Figure 3-1

Shunt generator; schematic and equivalent circuit.

For purposes of simplicity, all series resistances in the armature circuit may be added and lumped together under a single resistance, R_a, called the armature circuit resistance. In the equivalent circuit of a shunt generator shown in Fig. 3-1b, the armature circuit consists of a source of emf, E_g, and an armature circuit resistance, R_a, hereafter designated merely as the armature resistance.

The field circuit of a shunt generator is in parallel with the armature circuit and, as shown in Figs. 3-1a and b, consists of the shunt field winding wound on the stationary field poles and a field rheostat.

Note that the shunt generator, when loaded, is composed of three parallel circuits: (1), the armature circuit; (2), the field circuit; and (3), the load circuit. Since the basic source of emf and current is the armature, the equivalent circuit of Fig. 3-1b yields the following current relation:

$$I_a = I_f + I_l \tag{3-1}$$

where I_a is the armature current produced in the same direction as the generated voltage, E_g (Eq. 1-6)
I_f is the field current (V_f/R_f) in the field circuit
I_l is the load current, V_l/R_l

For the three circuits in parallel, by definition, the same voltage exists across the armature, field, and load circuits, respectively, or

$$V_a = V_f = V_l \tag{3-2}$$

where V_a is the voltage across the armature, i.e.,

$$V_a = E_g - I_a R_a \tag{1-10}$$

CHAP. THREE / DC Dynamo Voltage Relations—DC Generators

and where V_f is the voltage across the field circuit
V_l is the voltage across the load

EXAMPLE 3-1: A 150 kW, 250 V shunt generator has a field circuit resistance of 50 ohms and an armature circuit resistance of 0.05 ohm. Calculate:
a. The full-load line current flowing to the load.
b. The field current.
c. The armature current.
d. The full-load generated voltage.

Solution:

a. $I_l = \dfrac{\text{kW} \times 1000}{V_l} = \dfrac{150 \times 1000 \text{ W}}{250 \text{ V}} = 600 \text{ A}$

b. $I_f = \dfrac{V_l}{R_f} = \dfrac{250 \text{ V}}{50 \text{ }\Omega} = 5\text{A}$

c. $I_a = I_f + I_l = 5 + 600 = 605 \text{ A}$ \hfill (3-1)

d. $E_g = V_a + I_a R_a = 250 \text{ V} + 605 \times 0.05 = 280.25 \text{ V}$ \hfill (1-10)

3-4. SCHEMATIC DIAGRAM AND EQUIVALENT CIRCUIT OF A SERIES GENERATOR

When the excitation is produced by a field winding connected in series with the armature in such a way that the flux produced by the series-connected field winding is a function of the current in the armature and the load, the dc dynamo is called a *series generator*. The complete schematic diagram of a series generator is shown in Fig. 3-2a. The series field is excited only when a load is connected to complete the circuit. Since this field must carry the full or rated current of the armature, it is constructed of *few turns of heavy wire*. As in the previous case, the compensating winding R_c located on the field poles, and the interpole winding, R_i, are included in series with the armature winding, R_w of the rotating armature which

(a) Complete schematic.   (b) Equivalent circuit.

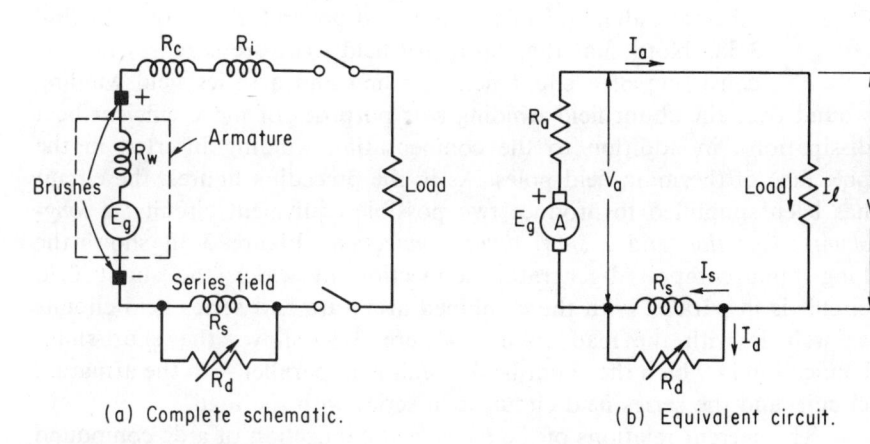

Figure 3-2

Series generator; schematic and equivalent circuit.

produces a generated emf, E_g, in accordance with Eq. (1-6). The equivalent circuit (under load) of a series generator is shown in Fig. 3-2b. Note that the current in the series field winding I_s is controlled by a diverter, R_d, which serves to provide some adjustment of excitation of the series field in much the same way as the rheostat in the shunt generator. It should be noted that, unlike the shunt generator, whose field excitation is virtually (for purposes of comparison) independent of load, the series field excitation depends primarily on the *magnitude* of *resistance* of the *load*. Thus, the diverter serves only to provide minor adjustment of series field excitation in a series generator.

The current relations of a series generator are

$$I_a = I_l = I_s + I_d \tag{3-3}$$

The voltage relations of a series generator, as shown in the equivalent circuit of Fig. 3-2b, may be summarized as

$$V_a = V_l + I_s R_s \tag{3-4}$$

where V_a is the voltage across the armature, or $E_g - I_a R_a$ [from Eq. (1-10)]
V_l is the voltage across the load
$I_s R_s$ is the voltage drop across the series field

3-5. SCHEMATIC DIAGRAM AND EQUIVALENT CIRCUIT OF A COMPOUND GENERATOR

When the field excitation is produced by a combination of the two types of windings discussed above, namely (1) a series field winding excited by the armature or line current, and (2) a shunt field winding excited by the voltage across the armature, the dc dynamo is called a *compound generator*. The complete schematic diagram of a compound generator is shown in Fig. 3-3a. Note that the stationary field structure is represented as consisting of a shunt field winding and a series field winding wound over the shunt field winding (for purposes of more efficient heat dissipation), in addition to the compensating winding inserted in the pole face of the main field poles. As in the preceding figures, the circuit has been simplified to produce two possible equivalent circuits: a *long-shunt connection* and a *short-shunt connection*. Figure 3-3b shows the long-shunt compound-generator connection in which the shunt field circuit is in parallel with the combined armature and series field circuits as well as with the load circuit. Figure 3-3c shows the short-shunt connection in which the shunt field circuit is in parallel with the armature circuit, and the series field circuit is in series with the load.

The current relations of the *long-shunt* connection of a dc compound generator are

$$I_a = I_f + I_l = I_s + I_d \tag{3-5}$$

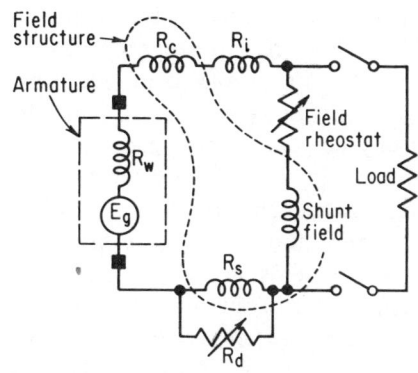

(a) Complete schematic, long shunt connection.

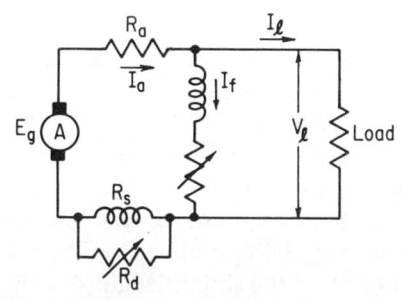

(b) Equivalent circuit, long shunt compound generator.

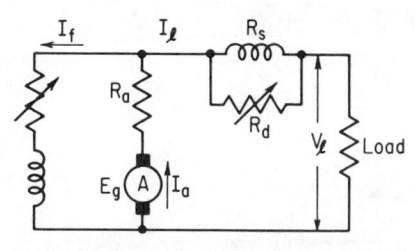

(c) Equivalent circuit, short shunt compound generator.

Figure 3-3

Compound generator; schematic and equivalent long and short shunt connections.

The current relations of the *short-shunt* connection of a dc compound generator are

$$I_a = I_f + I_l$$

and

$$I_l = I_s + I_a \qquad (3\text{-}6)$$

From Eqs. (3-5) and (3-6), it may be noted that the essential difference between long-shunt and short-shunt connections is that, in the long-shunt connection, the *armature* current excites the series field; whereas, in the short-shunt connection, the *load* current excites the series field.

EXAMPLE 3-2: A long-shunt compound generator, rated at 100 kW and 500 V dc, has an armature resistance of 0.03 ohm, a shunt field resistance of 125 ohms, and a series field resistance of 0.01 ohm. The diverter carries 54 amperes. Calculate:
a. The diverter resistance at full load.
b. The generated voltage at full load.

Solution:

a. $I_l = \dfrac{\text{kW} \times 1000}{V_l} = \dfrac{100 \times 1000}{500 \text{ V}}$

$= 200 \text{ A}$

$I_f = \dfrac{V_f}{R_f} = \dfrac{500 \text{ V}}{125 \, \Omega} = 4 \text{ A}$

$I_a = I_f + I_l = 4 + 200 = 204 \text{ A}$ \hfill (3-1)

$I_s = I_a - I_d = 204 \text{ A} - 54 \text{ A}$
$= 150 \text{ A}$ \hfill (3-5)

Since diverter and series field are in parallel, $I_d R_d = I_s R_s$ and

$$R_d = \dfrac{I_s R_s}{I_d} = \dfrac{150 \times 0.01}{54} = 0.0278 \, \Omega \qquad (3\text{-}11)$$

b. $E_g = V_l + I_a R_a + I_s R_s = 500 + (204 \times 0.03) + (150 \times 0.01)$
$= 507.62 \text{ V}$

3-6. THE SEPARATELY EXCITED GENERATOR

One method of classification of dc generators is based on the manner in which the field winding is excited in order to produce the necessary ampere-turns and mmf per pole [ϕ in Eq. (1-6)] required to generate a voltage [E_g in Eq. (1-6)]. Thus, it would appear that it is possible for any dc generator to produce a dc voltage and current of sufficient magnitude to excite its own field, and such excitation is called *self-excitation*. The generators shown in Figs. 3-1, 3-2, and 3-3 are *self-excited shunt, series,* and *compound generators,* respectively. When, however, one or more fields are connected to a separate dc voltage supply which is independent of the armature voltage of the generator, the generator is termed a *separately excited generator*.

Two separately excited generators are shown in Figs. 3-4a and b. The circuit arrangement of Fig. 3-4a shows the (shunt) field connected to a potentiometer and a dc source which is independent of the armature voltage, V_a. Since the field is no longer excited by armature voltage, the armature current, I_a, is the same as the load current I_l. In the same way, the armature voltage V_a is the same as the load voltage, $I_l R_l$, assuming transmission lines of zero resistance.

Note that the potentiometer connection of Fig. 3-4a permits zero adjustment of the (shunt) field current as a minimum, whereas the rheostat connection of Fig. 3-4b permits minimum current adjustment but not zero. The separately excited generator of Fig. 3-4b combines self-excitation of the series field and separate excitation of the shunt field, providing the advantages of compound operation with the advantages of separate field excitation. The armature current relations of this generator are the same as that for the series generator, given in Eq. 3-3.

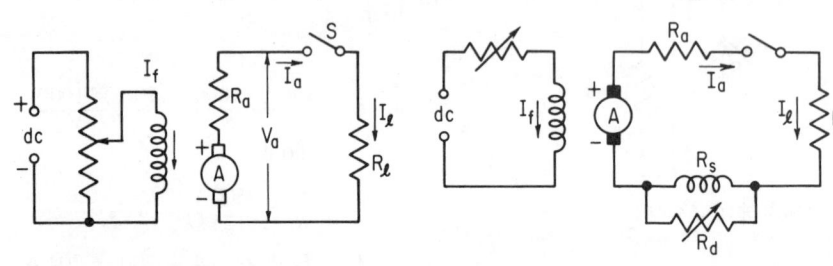

(a) Separate excitation, shunt field using potentiometer rheostat.

(b) Separate excitation, shunt field, compound operation.

Figure 3-4

Separately excited generators.

3-7. NO-LOAD VOLTAGE CHARACTERISTIC OF DC GENERATOR

The circuit of Fig. 3-4a is commonly used in machinery laboratories to investigate both no-load and load characteristics of shunt generators. With switch S open, the generator is driven by a prime mover at an approximately constant rate of speed. An ammeter is connected in the potentiometer circuit to

(a) Connections

(b) Magnetization curve.

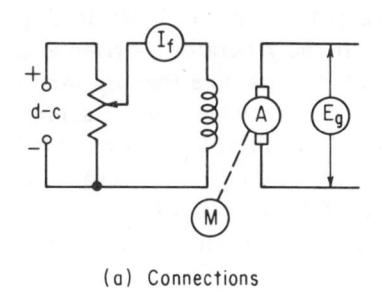

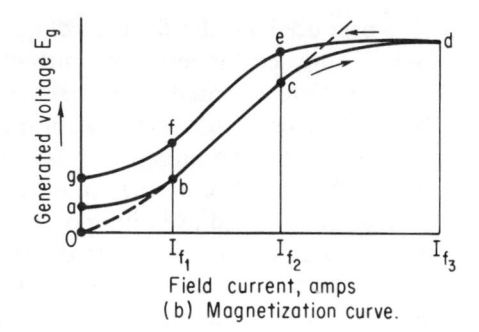

Figure 3-5

Separately excited no-load connections for obtaining saturation curve.

record the current drawn by the field, I_f, and a high-sensitivity voltmeter is connected across the armature to record the generated voltage, E_g, as shown in Fig. 3-5a* and expressed by Eq. (1-6):

$$E_g = \frac{\phi ZSP}{60a} \times 10^{-8} \text{ V} \tag{1-6}$$

For any given generator, the number of poles, P, the total number of armature conductors, Z, and the number of paths, a, may be determined from the armature winding data and the use of Eqs. (2-4) or (2-5). Thus, for a given armature in a given generator, P, Z, and a in the above equation are *fixed*. Equation (1-6) may be written as

$$E_g = K\phi S \tag{3-7}$$

where $K = \dfrac{PZ \times 10^{-8}}{60a}$

ϕ is the flux per pole (mutual air-gap flux)
S is the speed in rpm

Since the prime mover of Fig. 3-5a is being driven at an approximately constant speed, the generated emf, E_g, of Eq. (3-7) is

$$E_g = K'\phi \tag{3-8}$$

It would appear, on the basis of Eq. (3-8), that the voltmeter reading in Fig. 3-5a is solely and purely a function of the mutual air-gap

---

* A high-sensitivity voltmeter is commonly used because it draws negligible current from the armature. Under these circumstances, V_a may be assumed equal to E_g since the $I_a R_a$ armature voltage drop is negligible. If an electronic voltmeter or an electrometer is employed, the value obtained may be considered, for all practical purposes, the value of E_g in Eq. (1-6).

SEC. 3-7. / *No-Load Voltage Characteristic of DC Generator*

flux produced by the field winding. If the potentiometer shown in Fig. 3-5a is adjusted for zero field current, and if the generator is driven at a constant rate of speed, one might be tempted to assume that E_g would be zero. Such is not the case, however, and even when the field mmf $(I_f N_f)$ is zero, the airgap flux is *not* zero. A small voltage is recorded across the armature by the voltmeter when the field current is zero. This voltage is indicated as point *a* on the curve of Fig. 3-5b, where the field current is zero and the generated voltage, E_g, is some small value, a few volts. The voltage at *a* is due to the *retentivity* of the field poles and is proportional to the amount of *residual magnetism* that was left in the dynamo iron when the generator was last turned off.

If the field current is increased by means of the potentiometer so that a field current is recorded of I_{f_1}, the voltage will rise to point *b* in Fig. 3-5b. If the current is increased in the same direction so that the ammeter records a field current of I_{f_2}, the generated voltage will rise to point *c* on Fig. 3-5b. Thus, the generated induced voltage rises in proportion to the air-gap mmf produced by the field current $(I_f N_f)$. It should be noted that the portion of the curve *a* to *b* is *nonlinear*, since it is composed of a *fixed* residual mmf and a *variable* field-current mmf. The portion *b* to *c* is linear, however, since the residual mmf is now negligible in comparison to the mmf produced by the field current, and the generated voltage varies directly with the variation in field current. Beyond point *c* (the knee of the curve), an increase in field current does *not* produce a proportional increase in generated voltage. Here the iron of the field poles and the surrounding core of the magnetic circuit approaches saturation. Beyond point *c*, therefore, any increase in mmf above the knee of the saturation curve will fail to produce a proportionate increase in flux, and the magnetization curve from *c* to *d* is again nonlinear, this time because of the effect of *magnetic saturation*.

If the field current is now reduced by means of the potentiometer in Fig. 3-5a from the value I_{f_3} to the value I_{f_2}, the generated voltage decreases from *d* to *e*. Note that the voltage at *e* is greater than at *c*, and that further field current decreases produce higher generated voltages than those produced when the field current was increasing. This action is identical to that produced in any magnetic circuit which contains a ferromagnetic material; it is a property of the material called *hysteresis*.*

It can be seen, then, that the shape of the magnetization curve (E_g vs. I_f) is no different from the shape of the saturation curve (B vs. H) obtained for any ferromagnetic material. As a matter of fact, if the machine were not rotating, and if measurements were made of the air-gap flux vs. the magnetizing force, the *B-H* curve would be identical to that

---

* See Eq. (12-3), Section 12-2, of this book; also, for additional information on the subject, see H. W. Jackson, *Introduction to Electric Circuits*, 3rd ed., Englewood Cliffs, N.J.: Prentice-Hall, Inc., 1970 Sec. 8-16.

shown in Fig. 3-5b. Since $E_g = K\phi S$, rotation of the armature conductors at a constant speed produces a voltage directly proportional to the air-gap flux (at all times) and *not* necessarily proportional to the field current!

In obtaining a generator magnetization curve in the laboratory, therefore, care should be taken to increase the field current to a maximum and decrease the field current to a minimum, moving in one direction *only*, as readings are taken. If this is not done, minor hysteresis loops are produced, yielding erroneous results. In addition, care should be taken to maintain absolutely *constant* speed, since Eq. (3-8) is predicated on the assumption that the speed is, in fact, constant. The magnetization curve of Fig. 3-5b is a graphic representation of this equation. If the speed is recorded at the same instant that the field current and voltage readings are taken, then it is a simple matter to *correct for any variations* of speed that may occur, as shown by Examples 3-3 and 3-4, using the *ratio* method.

EXAMPLE 3-3: Assuming constant field excitation, calculate the no-load voltage of a separately excited generator whose armature voltage is 150 V at a speed of 1800 rpm, when
   a. The speed is increased to 2000 rpm.
   b. The speed is reduced to 1600 rpm.

Solution:

From Eq. 3-7, $E_g = K''S$ at constant field excitation, and therefore

$$\frac{E_{\text{final}}}{E_{\text{orig}}} = \frac{S_{\text{final}}}{S_{\text{orig}}}$$

a. $E_{\text{final}} = (E_{\text{orig}}) \dfrac{S_{\text{final}}}{S_{\text{orig}}} = (150 \text{ V}) \dfrac{2000}{1800} = 166.7 \text{ V}$

b. $E_{\text{final}} = (150 \text{ V}) \dfrac{1600}{1800} = 133.3 \text{ V}$

EXAMPLE 3-4: In obtaining a magnetization curve at the constant speed of 1200 rpm, the following values of voltage were recorded when the speed was simultaneously noted as varying from 1200 rpm
   a. 64.3 V at 1205 rpm.
   b. 82.9 V at 1194 rpm.
   c. 162.3 V at 1202 rpm.
   What corrections must be made in the data before the curve is plotted?

Solution:

a. $E_1 = (64.3 \text{ V}) \dfrac{1200}{1205} = 64.0 \text{ V at 1200 rpm}$

b. $E_2 = (82.9 \text{ V}) \dfrac{1200}{1194} = 83.3 \text{ V at 1200 rpm}$

c. $E_3 = (162.3 \text{ V}) \dfrac{1200}{1202} = 162.0 \text{ V at 1200 rpm}$

3-8.
SELF-EXCITED GENERATOR-FIELD RESISTANCE LINES

In the separately excited generator shown in Fig. 3-5a, the field circuit was excited independently of the voltage across the armature. Thus, the current in the field circuit, I_f, was independent of the generated voltage, E_g, although the reverse was *not* the case. When, however, the field circuit (consisting of the field winding and the field rheostat) is connected across the armature as shown in Fig. 3-6, the field current, I_f, is no longer independent of the generated voltage. For the connection of Fig. 3-6, the field current as

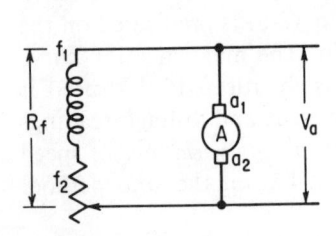

Figure 3-6
Self-excited shunt generator.

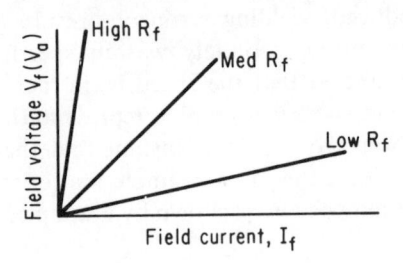

Figure 3-7
Field resistance lines.

given in Eq. (3-1) depends on the ratio V_f/R_f, where V_f is the same as the voltage across the armature, V_a. The field current at any instant, therefore, is a function of *two* variables: (1) the armature voltage, which varies with the air-gap mmf; and (2) the field resistance, which varies with the setting

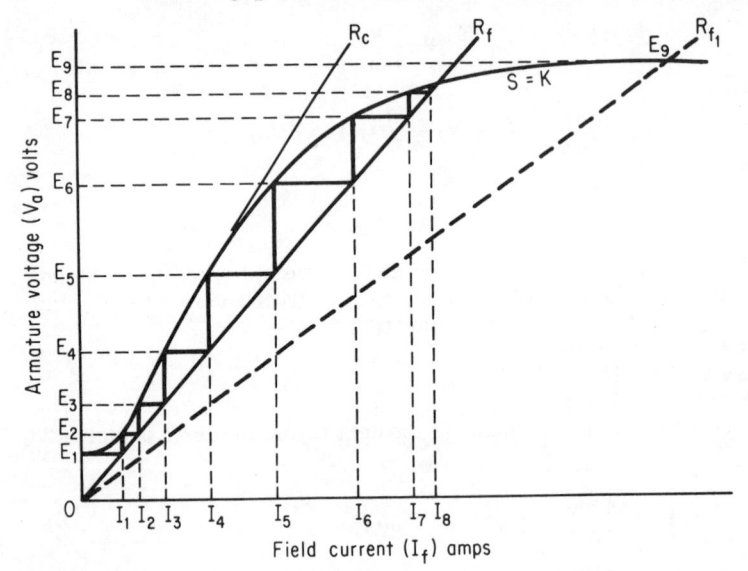

Figure 3-8
Build-up a of self-excited shunt generator.

86
CHAP. THREE / DC Dynamo Voltage Relations—DC Generators

of the field rheostat shown in Fig. 3-6. In order to express the field and armature current which may flow at any instant in the circuit of Fig. 3-6, it is necessary to represent, graphically, a *family of field resistance lines*. If the field resistance is assumed to be linear and constant, a given family of field resistances may be plotted as shown in Fig. 3-7. Thus, according to Ohm's law, a high field resistance (i.e., one with a high slope) is one which will produce a small field current for a fairly high value of field voltage, as shown in Fig. 3-7. Conversely, as shown in the figure, a low field resistance (i.e., one with a low slope) will produce a fairly high field current for a fairly low value of field voltage. The slope of the field resistance line is, therefore, an indication of the field resistance, V_f/I_f.

Since the generator of Fig. 3-6 is supplying a relatively small current (in proportion to its rated full-load current) to excite its own field circuit, we may assume (for the moment) that the internal $I_a R_a$ drop is negligible and that the ordinates of Fig. 3-5b and Fig. 3-7 are the same, i.e., V_a is equal to E_g. It is now possible to represent both the field resistance lines and the dynamo magnetization curve on a common axis. Such a representation is shown in Fig. 3-8.

3-9.
BUILD-UP OF
SELF-EXCITED
SHUNT GENERATOR

The dynamo magnetization curve for the separately excited generator of Fig. 3-5 and a particular shunt field resistance line, R_f, are shown in Fig. 3-8, for the identical generator connected as a self-excited shunt generator, originally shown in Fig. 3-6. As shown in Figs. 3-6 and 3-8, since the field circuit is connected directly across the armature, the ordinate of the field resistance line, R_f, is the terminal voltage of the generator, V_a. The manner in which a self-excited generator manages to excite its own field and build a dc voltage across its armature is described with reference to Fig. 3-8 in the following steps:

1. Assume that the generator starts from rest, i.e., prime-mover speed is zero. Despite a residual magnetism, the generated emf, E_g, is zero.
2. As the prime mover rotates the generator armature and the speed approaches rated speed, the voltage due to residual magnetism and speed ($E = K\phi S$) increases.
3. At rated speed, the voltage across the armature due to residual magnetism is small, E_1, as shown in the figure. But this voltage is also across the field circuit whose resistance is R_f. Thus, the current which flows in the field circuit I_1, is also small.
4. When I_1 flows in the field circuit of the generator of Fig. 3-6, an increase in mmf results (due to $I_f N_f$) which aids the residual magnetism in increasing the induced voltage to E_2, as shown in Fig. 3-8.
5. Voltage E_2 is now impressed across the field, causing a larger current I_2 to flow in the field circuit. $I_2 N_f$ is an increased mmf, which produces generated voltage E_3.

6. E_3 yields I_3 in the field circuit, producing E_4. But E_4 causes I_4 to flow in the field, producing E_5; and so on, up to E_8, the maximum value.

7. The process continues until that point where the field resistance line crosses the magnetization curve in Fig. 3-8. Here the process stops. The induced voltage produced, when impressed across the field circuit, produces a current flow that in turn produces an induced voltage of the same magnitude, E_8, as shown in the figure.

3-10. CRITICAL FIELD RESISTANCE

The above description for the build-up of a self-excited shunt generator, shown in Fig. 3-8, used a particular value of field resistance, R_f. If the field resistance were reduced by means of adjusting the field rheostat of Fig. 3-6 to a lower value, say R_{f_1}, shown in Fig. 3-8, the build-up process would take place along field resistance line R_{f_1}, and build up a somewhat higher value than E_8, i.e., the point where R_{f_1} intersects the magnetization curve, E_9. Since the curve is extremely saturated in the vicinity of E_9, reducing the field resistance (to its limiting field winding resistance) will not increase the voltage appreciably. Conversely, increasing the field rheostat resistance and the field circuit resistance (to a value having a higher slope than R_f in the figure) will cause a reduction of the maximum value to which build-up can possibly occur.

The resistance of the field rheostat may be increased until the field circuit reaches a critical field resistance. Field circuit resistance above the critical field resistance will fail to produce build-up. The critical field circuit resistance, R_c, is shown as a tangent to the saturation curve passing through the origin, O, of the axes of the curve of Fig. 3-8. Thus a field circuit resistance higher than R_c will produce an armature voltage of E_1 approximately and no more.

3-11. REASONS FOR FAILURE OF SELF-EXCITED SHUNT GENERATOR TO BUILD UP VOLTAGE

There are four specific (electrical) reasons for the failure of a self-excited unloaded shunt generator to build up voltage. They are

1. Lack of (or low) residual magnetism. Since the build-up process as described in Section 3-9 requires some residual magnetism for its initiation, it is evident that an extremely low value or complete loss of residual magnetism will inhibit build-up. Residual magnetism may be lost as a result of conditions which would tend to demagnetize the field poles: mechanical jarring in shipment, excessive vibration, extreme heat, alternating current inadvertently connected across the field winding, inactivity for long periods, etc. Such a failure may be remedied by *flashing the field*, i.e., by applying direct current to the highly inductive field [see Eq. (2-2) and the discussion following it] and then removing it, which produces an accompanying inductive spark. The residual magnetism should then be regained, and the self-excited generator should build up voltage.

2. Field circuit connections reversed with respect to the armature circuit. In Section 3-9, Step 4, it was stated that the current which flows in the field circuit should produce an mmf which aids the residual magnetism, i.e., the flux produced by the field coil must be of the same magnetic polarity as the residual mmf. If the field connections are reversed with respect to the armature, the resulting field flux will tend to buck or decrease the residual flux, thus decreasing the net flux and the generated emf, E_g, when the field circuit is closed. A simple test for this condition is to open the field circuit of a running generator and observe the voltmeter across the armature. If the voltage across the armature *rises* when the field circuit is opened, then the field circuit connections are *reversed* with respect to the armature. This failure may be remedied merely by reversing $f_1 - f_2$ with respect to $a_1 - a_2$, as shown in Fig. 3-6. It should be noted, parenthetically, that reversing the prime mover direction of rotation will accomplish the same purpose, where possible, since it reverses the polarity of the armature. A possible and really identical cause of failure is the use of the wrong direction of rotation, which produces reversed armature connections with respect to the field.*

3. Field circuit resistance higher than critical field resistance. An open connection in the field circuit windings, rheostat, or connections will result in a resistance higher than critical field resistance. This will prevent build-up, as described in Section 3-10. This trouble may be checked by means of an ohmmeter or a series-connected voltmeter to a dc source.

4. Open connection or high resistance in the armature circuit. An open connection in the armature circuit, a dirty commutator, a loose brush, or the lack of brushes will tend to act in the same manner as a high resistance in the field circuit, because they reduce field current and tend to prevent a voltage higher than the residual magnetism. The resistance of connections a_1–a_2 to the armature should also be checked by means of an ohmmeter (for a comparatively low resistance). A high armature circuit resistance indicates an open connection in the armature circuit.

3-12. EFFECT OF LOAD IN CAUSING A SHUNT GENERATOR TO UNBUILD

Since the load is a resistance which is generally low in comparison to the shunt field of a shunt generator, it should be remembered that if too great a load is connected across a shunt generator and the prime mover brought up to speed, the generator may fail to build up voltage.† The reason here is that most of the armature current is diverted to the load rather than the

---

* The reversal of direction of the prime mover is hardly recommended as a means of eliminating failure to build up. Most mechanical prime movers, such as steam engines and turbines, are designed to run in one direction only. A dc motor is similarly designed, since its brushes are pitched in a particular direction. (Only the induction and synchronous motors, when used as prime movers, may be reversed with impunity).

Similarly, excessively low speed as a cause for failure to build up is also discounted. It is assumed that the prime mover is running at the proper speed in the proper direction.

Only electrical reasons are listed above. Otherwise one might list as a cause for failure to build up, the fact that the generator is not rotating at all!

† This situation is usually not included among causes for failure to build up because it is assumed that the generator is not under load during the build-up process. In effect, a high (low resistance) load is a short-circuit across the armature.

89
SEC. 3-12. / *Effect of Load in Causing a Shunt Generator to Unbuild*

field, and little additional field current is available to produce the additional mmf required to start the build-up process. Thus, in order to achieve build-up, it is necessary that the shunt generator is not connected to a load until its voltage is brought up to its rated value by the build-up process described.

This raises an interesting question as to whether a generator will "unbuild" with the application of additional load (i.e., less equivalent load resistance because of more load resistors in parallel). It will be shown in this section that the effect of additional load does in fact reduce the armature voltage and, hence, the field current excitation. The student might be tempted to reason, therefore, that, if application of load reduces armature voltage, the reduction in field current should in turn reduce armature voltage, which in turn should reduce the field current still more, and so on, until the machine is back to its residual voltage. An examination of the expanded saturated portion of the magnetization curve is shown in Fig. 3-9. Note that in the saturated portion of the curve, the field current may be reduced from I_{f_4} to I_{f_3} (a sizeable reduction) with only a small change in armature voltage from E_4 to E_3.

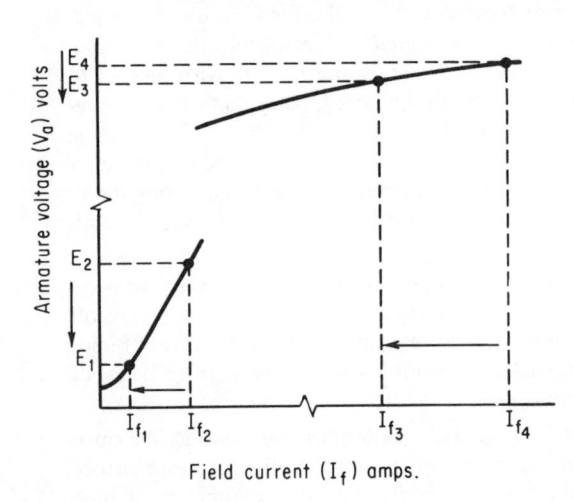

Figure 3-9
Effect of decreasing field current on armature voltage.

In the unsaturated portion of the magnetization curve, however, a smaller reduction of field current from I_{f_2} to I_{f_1} will produce a large drop in armature voltage from E_2 to E_1. It would appear, then, insofar as loading is concerned, that the shunt generator should be operated in the saturated portion of its magnetization curve. If it is operating below the knee of the magnetization curve (in the linear or unsaturated portion), it *may unbuild* with application of load, as shown in Fig. 3-10.

3-13. LOAD-VOLTAGE CHARACTERISTICS OF A SHUNT GENERATOR

It was stated in an earlier section that the effect of application of load across the armature terminals is to reduce the generated and armature voltage. There are three reasons for this drop in voltage: (1) an internal armature voltage drop produced by the armature circuit resistance, R_a; (2) the effect of armature reaction on the air-gap flux; and (3) the reduction in field current caused by the two preceding factors. Let us consider each of these factors in turn.

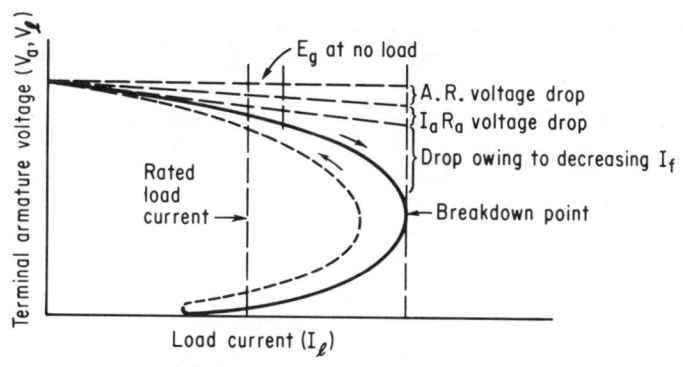

(a) Shunt generator load characteristic.

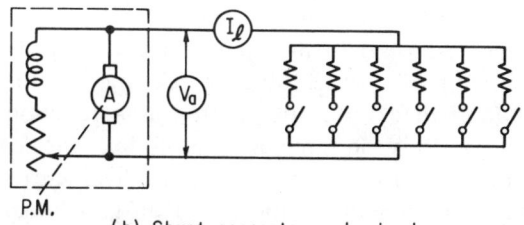

(b) Shunt generator under load.

Figure 3-10

External load voltage characteristic of a shunt generator.

1. Armature circuit voltage drop. The various resistive components of the armature circuit were described in Section 3-3 and shown in Fig. 3-1a. The equivalent armature circuit shown in Fig. 3-1b verified the equation representing the terminal armature voltage

$$V_a = E_g - I_a R_a \tag{1-10}$$

For a separately excited generator, shown in Fig. 3-5, which is not supplying load current, I_a is zero, and V_a equals E_g. For a loaded self-excited shunt generator, shown in Fig. 3-1b, as the load current, I_l increases, the armature current I_a also increases [Eq. (3-1)], as does the armature circuit voltage drop, $I_a R_a$, in Eq. (1-10). Thus, the terminal voltage across the shunt generator armature, V_a, *decreases with application of load*, as shown in Fig. 3-10a.

2. Armature reaction. The individual current-carrying conductors of the armature furnish a load current in the same direction as the induced voltage (Section 3-1). Since these conductors are embedded in an iron armature, an armature mmf is produced in proportion to the load current. It will be shown (Chapter 5) that the effect of this armature flux is to distort and reduce the air-gap flux produced by the field. The reduction in mutual field flux ϕ_m reduces the generated and terminal armature voltages,

E_g and V_a, respectively. Thus, as the armature current, I_a, increases, the effect of armature reaction is a progressive reduction of ϕ_m, E_g and V_a as shown in Fig. 3-10a.

3. Reduction in field current. The terminal voltage V_a drops as a function of the load current as a result of (1) armature reaction and (2) the internal armature circuit voltage drop. This drop in V_a results in a decreased field current and excitation produced by the field poles. Decreased excitation $(I_f N_f)$ results in decreased air-gap flux and reduced E_g and V_a. If the generator field current and speed are such that the field poles are unsaturated, the machine will rapidly unbuild (see Fig. 3-10a). Note that this cause of voltage drop does not occur in a separately excited dc generator; and for this reason, the same generator operated with separate excitation always has improved regulation.

The effect of each of the preceding three factors is shown in Fig. 3-10, which shows the (external) load-voltage characteristic of a shunt generator. For the circuit of Fig. 3-10b, the readings of the voltage across the armature (and load), V_a, are plotted as a function of load current, I_l. The voltage, V_a, is the same as E_g at no load (neglecting the $I_a R_a$ and armature reaction drop produced by the field current). The effects of armature reaction (in reducing the mutual air-gap flux), armature circuit voltage drop, and decrease in field current are all shown with progressive increases in load. Note that both the armature reaction and the $I_a R_a$ drop are shown as dashed straight lines, representing theoretically linear voltage decreases directly proportional to the increase in load current. The drop owing to decreased field current is a curved line, since it depends on the degree of saturation existing in the field at that value of load.

In general, the external load-voltage characteristic decreases with application of load only to a small extent up to its rated load (current) value. Thus, the shunt generator is considered as having a fairly constant output voltage with application of load, and, in practice, is rarely operated beyond the rated load current value continuously for any appreciable time. As shown in Fig. 3-10a, further application of load causes the generator to reach a breakdown point, beyond which further load causes it to "unbuild" as it operates on the unsaturated portion of its magnetization curve (Section 3-12, Fig. 3-9). This unbuilding process continues until the terminal voltage is zero, at which point the load current is of such magnitude that the internal armature circuit voltage drop equals the emf generated on the unsaturated or linear portion of its magnetization curve. This is illustrated by the following example:

EXAMPLE 3-5: A 125 V dc generator having an armature resistance of 0.15 ohm is loaded progressively until the voltage across the load is zero. If the load current is 96 A and the field current 4 A what is the voltage generated in the armature?

Solution:

$$I_a = I_f + I_l = 4 + 96 = 100 \text{ A} \tag{3-1}$$
$$E_g = V_a + I_a R_a = 0 + (100 \times 0.15) = 15 \text{ V} \tag{3-10}$$

The preceding example serves to indicate the possible extent of unbuilding of the generator. It also serves to indicate that the generated emf at no load is *not* the same as the generated emf at any given load condition, since the generator is no longer operating on the same portion of the magnetization curve, *due primarily* to the *unbuilding of the generator*.

Finally, it should be noted that if the external load is decreased (an increase of external load resistance), the generator will tend to build up gradually along the dashed line shown in Fig. 3-10a. Note that for any value of load current, the terminal or armature voltage is less (as the voltage increases) compared to the solid lines which yield a higher voltage (as the voltage decreases). This difference is due to hysteresis (Section 3-7) and is in agreement with Fig. 3-5b, where *increasing* values of voltage are *less* than decreasing values of voltage *for any given fixed excitation*.

3-14. EFFECT OF SPEED ON NO-LOAD AND LOAD CHARACTERISTICS OF A SHUNT GENERATOR

The above discussion is predicated on the assumption that the speed of the prime mover is constant with application of load. But since (Section 3-1) the electromagnetic torque (developed in the current-carrying conductor of the generator) opposes rotation (Lenz's law), this opposition has a tendency to decrease the speed of the prime mover with application of load.* What effect does a change in speed produce on the load characteristic of a shunt generator?

Equation (3-7) states that the generated voltage, E_g, of any given generator is proportional to both flux and speed ($E_g = K\phi S$). For a given constant mutual air-gap flux, an increase in speed will produce an increase in voltage, and an infinite speed will produce an infinite voltage. Unfortunately, we have no way of holding the air-gap flux constant (except to use a permanent magnetic field, in which case $E_g = K'S$, which is the principle of the tachometer generator), but we *can* maintain a constant field current. The effect of constant field current on saturation is shown in Fig. 3-11a for two different values of speed, S_1 and S_2. For the same field current, I_{f_1}, the higher speed will produce less saturation, since the slope at point 2 is more vertical than the slope at point 1. But the less saturated a given shunt generator is, the more rapidly it will

---

* The discussion of prime movers such as gasoline, diesel, and steam engines and hydro or steam turbines is beyond the scope of this volume. In general, while it is assumed that these are equipped with speed-regulating devices that will maintain a constant speed with an increased application of load, the possibility of a drop in prime-mover speed due to load is *inherent* in the nature of any drooping-load prime mover. It will be shown (Chapter 7) that such a drooping-load characteristic is, in fact, essential to parallel operation of compound generators for purposes of stability.

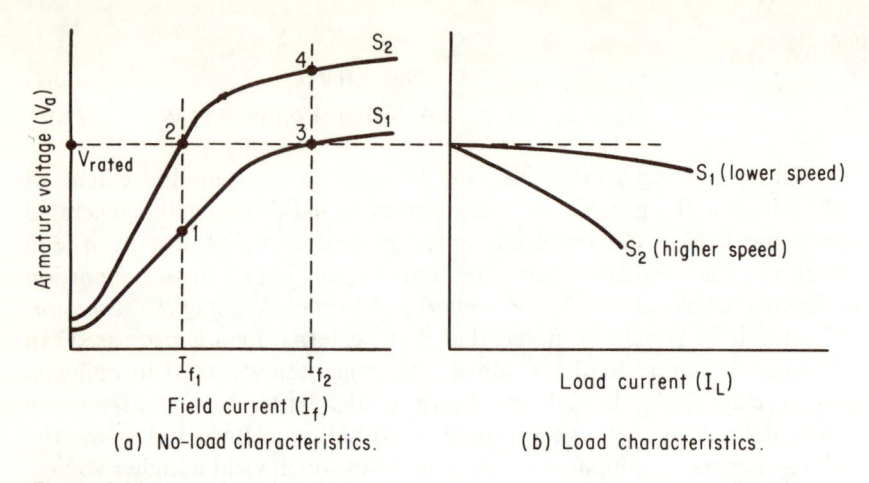

Figure 3-11

Effect of speed on saturation and voltage of a shunt generator.

unbuild (Section 3-12). Thus, we would expect a machine of higher speed to unbuild more rapidly and have a more drooping load characteristic than a slower machine. The effect is verified and even more pronounced in Fig. 3-11b where, instead of maintaining the field current at a constant value, we compare the two speeds of the dynamo at the same rated voltage. At the lower speed, S_1, we require a higher field current (I_{f_2}) to produce rated voltage than at a higher speed, S_2, which requires a field current of I_{f_1}. Thus, at the lower speed, we are operating on the more saturated portion (point 3) of the magnetization curve, whereas at the higher speed, we are operating on the less saturated portion (at point 2) of the magnetization curve. At rated voltage, therefore, as shown in Fig. 3-11b, the lower speed will produce the more satisfactory load-voltage characteristic.

If the prime-mover speed *drops*, therefore, it will tend to *improve* the voltage regulation of the shunt generator. If, in addition, due to the drop in speed and the reduction of terminal voltage, we restore the voltage to its original value by increasing the field current, the voltage regulation is improved still more as a result of *increased saturation* of the field.

3-15. VOLTAGE REGULATION OF A GENERATOR

In the last paragraph above, the term "voltage regulation" was used to indicate the degree of change in armature voltage produced by application of load. If there is little change from no load to full load, the generator or voltage-supplying device is said to possess good voltage regulation. If the voltage changes appreciably with load, it is considered to have poor voltage regulation.

Voltage regulation is defined as the change in voltage from no load to full load, expressed as a percentage of the rated terminal voltage (armature voltage at full load) or

$$\text{VR (per cent voltage regulation)} = \frac{V_{nl} - V_{fl}}{V_{fl}} \times 100 \qquad (3\text{-}9)$$

where V_{fl} is the full-load (rated) terminal voltage
V_{nl} is the terminal voltage at no load

EXAMPLE 3-6: The no-load voltage of a shunt generator is 135 V, and its full-load voltage is 125 V. Calculate VR, the per cent voltage regulation.

Solution:

$$\text{VR (per cent voltage regulation)} = \frac{V_{nl} - V_{fl}}{V_{fl}} \times 100 \qquad (3\text{-}9)$$

$$= \frac{135 - 125}{125} \times 100 = \textbf{8 per cent}$$

EXAMPLE 3-7: The per cent voltage regulation for a 250-volt shunt generator is given as 10.5 per cent. Calculate the no-load voltage of the generator.

Solution:

$$V_{nl} = V_{fl} + (V_{fl} \times \text{VR}) = V_{fl}(1 + \text{VR}) \qquad (3\text{-}9)$$
$$= 250(1 + 0.105) = \textbf{276.3 V}$$

Equation (3-9) serves to indicate that an "ideal" generator would maintain the same voltage from no load to full load and, since the voltage change is zero, would have zero per cent regulation. The generator with the smallest or least change in terminal voltage has the lowest per cent regulation and is thus closest to the ideal generator as a source of constant emf, regardless of load. If additional loads are added or removed, an ideal generator will continue to supply the same voltage across the load terminals. An examination of the curve of Fig. 3-10a indicates that the shunt generator does have a tendency to drop in voltage as load is increased. If the load is located in the immediate vicinity of the generator, the drop in voltage may be compensated by means of automatic voltage regulators which decrease the field resistance (increasing the field current) and restore the voltage to its no-load value.

If the load is remotely located from the generator, however, the problem of maintaining constant voltage at the load is further complicated by the voltage drop in the transmission lines. The line-voltage drop varies in direct proportion to the load current drawn from the generator, as does each of the three causes of voltage drop described in Section 3-13 and shown in Fig. 3-10a. Because of the possibly great distance between the generator and the load, it becomes impractical to reflect (feed back) changes in load voltage by means of wiring to a voltage regulator located at the shunt generator. Furthermore, since all of the causes of voltage drop, and transmission-line voltage drop as well, vary with the load, it is best to employ some type of generator whose *magnetization* will be *controlled* or affected, in part, *by a change in load*. Both the series and the compound generator possess this characteristic.

**3-16.
SERIES GENERATOR**

The complete schematic diagram of the series generator is shown in Fig. 3-2a, and the equivalent circuit is represented in Fig. 3-2b. The series generator on open circuit (i.e., without a load connection) is *incapable* of building up. Thus, when the load current is zero, the generated and terminal armature voltages, E_g and V_a, are identical, and both are due to the residual magnetic flux, shown as E_1 in Fig. 3-12. If a load is connected across the series generator armature, as shown in Fig. 3-2b, a common armature and load current, I_L, will flow through the series field, creating additional mmf (which aids the residual flux) to produce a higher generated voltage. Automatic buildup will commence, since additional voltage produces additional current in the load; this, in turn, produces additional series field $(I_s N_s)$ mmf. But the action of the series generator is somewhat more complex than that of the shunt generator. There are now *two* voltage drops which limit the voltage, V_l, across the load [see Eq. (3-4)], and Eq. (3-4) may be rewritten as

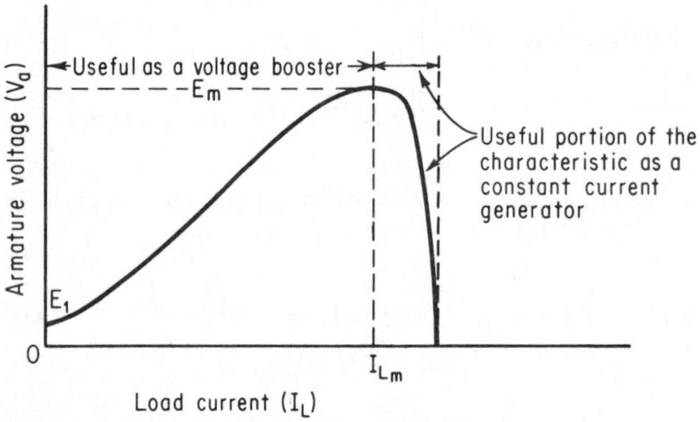

$$V_l = E_g - (I_s R_s + I_a R_a) \tag{3-4a}$$

Figure 3-12
Series generator load characteristic.

In addition to these voltage drops, the generated voltage E_g is also reduced by the effects of armature reaction, so that the voltage across the load, V_l (producing the magnetized current, I_s) represents a *resultant* of two forces: (1) the factors tending to *decrease* the voltage, V_l; and (2) the magnetized current (I_s) tending to *increase* the generated voltage E_g. As a result, for a given prime-mover speed, a maximum voltage E_m, is produced, as shown in Fig. 3-12, which represents that critical point at which build-up ceases and no additional current is automatically produced. At this value of load current, I_{L_m}, the voltage drops of the series field and the armature, as well as the armature reaction drop, exactly counterbalance the increased mmf produced in the series field; and the terminal voltage, V_l, remains constant.

The useful portion of the series-generator characteristics as a *constant-current* generator is shown in Fig. 3-12, where further application of load beyond the critical maximum voltage point produces a *sharp drop* in load voltage. The sharp decline of load voltage is due to the combined factors of increased armature reaction and the increased voltage drops of Eq. (3-4), which now decrease the load voltage at a faster rate than the generated voltage, E_g, is increased by the load current. This sharply *drooping* characteristic lends itself to *welding generators* (where the current must be relatively constant to produce the same heating effect, I^2R, over the wide range of voltage drops produced by the electric arc). The *rising* portion of the magnetization curve, on the other hand, lends itself to *voltage boosters* in ground dc trolley or railway systems to reduce the electrolytic action between the ground-return rails and the conduit or other underground structures. It has also been employed extensively in Europe for series boosting of high-voltage dc transmission lines to compensate for transmission line drop, as in the Thury system. The rising or linear portion of the series generator characteristic is also used in a multifield exciter known as the Rototrol (Section 11-17).

3-17. COMPOUND GENERATOR

The voltage regulation of the series generator is obviously very poor, as seen from an examination of the combined load and magnetization characteristic shown in Fig. 3-12. But the ability of the series field to produce additional useful magnetization in response to increased load cannot be denied. This useful characteristic of the series field, combined with the relatively constant voltage characteristic of the shunt generator, led to the compound generator, whose construction and circuitry is discussed in Section 3-5 and shown in Fig. 3-3. The current relations for the long-shunt connection are given in Eq. (3-5), and for the short-shunt connection in Eq. (3-6). An examination of Figs. 3-3b and c will reveal that regardless of the method of connection, the terminal voltage, V_l, of the short- or long-shunt compound generator is the same as Eq. (3-4a) (Section 3-16) for a *series* generator (!); that is,

$$V_l = E_g - (I_s R_s + I_a R_a) \tag{3-4a}$$

The generated voltage, E_g, of a compound generator is the result of the combination of mmf's produced by the series $(I_s N_s)$ and shunt $(I_f N_f)$ ampere-turns, due to current which flows in their field windings. In a compound generator, the *shunt field predominates* and is much the stronger of the two. When the series field mmf *aids* the shunt field mmf, the generator is said to be *cumulatively* compounded. When the series field mmf *opposes* the shunt field mmf, the generator is said to be *differentially* compounded.

**3-18.
CUMULATIVE
COMPOUND
GENERATOR
CHARACTERISTICS**
There are three types of load characteristics possible for the cumulative compound generator (whether long-shunt or short-shunt), depending on the relative additional aiding mmf produced by the series field. These types are called (1) overcompound; (2) flat-compound; and (3) undercompound.

A compound generator whose terminal voltage *rises* with the application of load so that its full-load voltage exceeds its no-load voltage (negative regulation) is called an *overcompound generator*; see Fig. 3-13.

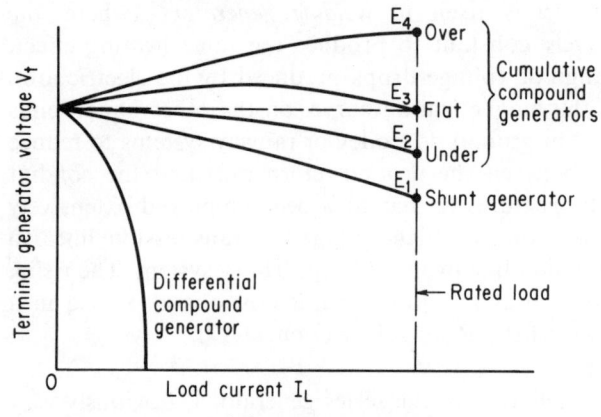

(a) Load characteristics.

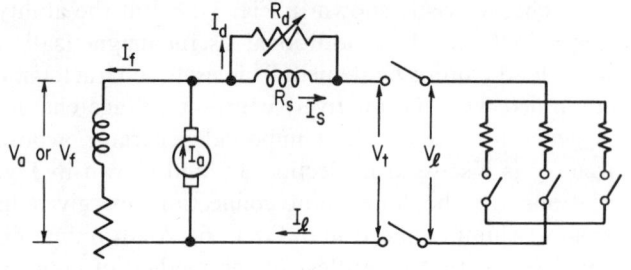

(b) Currents and voltages in compound generators.

Figure 3-13

External load voltage characteristic of cumulative and differential compound generators.

A *flat-compound generator* has a load-voltage characteristic in which the no-load and full-load voltages are equal (zero per cent regulation); refer again to Fig. 3-13.

An *undercompound generator* has a load-voltage characteristic in which the full-load voltage is somewhat less than the no-load voltage, but whose aiding series-field ampere-turns cause its characteristic to have better regulation than an equivalent shunt generator; see Fig. 3-13.

98

CHAP. THREE / DC Dynamo Voltage Relations—DC Generators

Most commercial compound dc dynamos (whether used as generators or motors) are *normally supplied by the manufacturer as overcompound machines*. As shown in Figs. 3-3b and c, the degree of compounding (over, flat, or under) may be adjusted by means of a diverter which shunts the series field (see Examples 3-2 and 3-8).

EXAMPLE 3-8: A long-shunt compound generator has a shunt field winding of 1000 turns per pole, and a series field winding of 4 turns per pole. In order to obtain the same (rated) voltage at full load as at no load, when operated as a shunt generator, it is necessary to increase the field current 0.2 A. The full-load armature current of the compound generator is 80 A and the series field resistance is 0.05 ohm. Calculate:
a. The number of series field ampere-turns (At) required for flat-compound operation.
b. The diverter resistance required for flat-compound operation.

Solution:

a. $\delta I_f N_f = 0.2\,\text{A} \times 1000\,\text{turns} = 200\,\text{At} = I_s N_s$ for flat-compound operation (3-10)

b. $I_s = \dfrac{I_s N_s}{N_s} = \dfrac{200\,\text{At}}{4\,\text{t}} = 50\,\text{A}$, required in series field winding for flat-compound operation

$$I_d = I_a - I_s = 80\,\text{A} - 50\,\text{A} = 30\,\text{A}$$

$$R_d = \dfrac{I_s R_s}{I_d} = \dfrac{50\,\text{A} \times 0.05\,\Omega}{30\,\text{A}} = 0.0833\,\Omega \quad (3\text{-}11)$$

As mentioned in Section 3-15, the overcompound generator is the type best suited for the transmission of direct-current electric energy where the load is remotely located from the generator. The rising-voltage characteristic of this generator is more than sufficient to compensate for the line drop in the transmission line. A diverter is used to control and produce a sufficient voltage rise at the generator to compensate for the voltage drop in the lines at full load. Since the line drop and the adjusted voltage rise produced by the series field are both proportional to the load current, the voltage at the remotely located load will be substantially constant from no load to full load, thus obviating the need for voltage regulators. An overcompound generator *always* has a *negative* voltage regulation. (see Eq. 3-9.)

The flat-compound generator finds similar application for a constant-voltage generator where line drop is negligible and the load is located in the immediate vicinity of the generator. It should be noted from Fig. 3-13, however, that the voltage of a flat-compound generator is not necessarily "flat" but has a negative "regulation" at the half-load point* and zero per cent regulation at full load.

* Since regulation is defined by the ASA in terms of no-load and full-load conditions only, it is technically improper to speak of the regulation of a voltage generating device under any other load conditions than at full load or no load.

The undercompound generator, shown in Fig. 3-13, has a somewhat drooping characteristic, similar to the shunt generator, but with improved regulation. If a short circuit is connected across the series field ($R_d = 0$), the cumulative overcompound generator will act as a shunt generator. If the diverter resistance is increased so that some small current flows through the series field, any cumulative compound generator will act as an undercompound generator. It is for this reason that manufacturers supply only overcompound generators and expect the consumer to adjust the degree of compounding, using a *diverter*.

3-19. ADJUSTING THE DEGREE OF COMPOUNDING OF CUMULATIVE COMPOUND GENERATORS

The diverter shown in Fig. 3-13 is represented as a variable resistor. In actual practice, a diverter is usually a fixed, high-resistivity, constant-temperature resistor, wound with manganin or constantan. In extremely large compound generators of high capacity, the diverter may actually be a cable or copper conductor of a given length. It is necessary, therefore, to determine in advance the desired diverter resistance to accomplish the degree of compounding desired, rather than to employ a variable resistance which may be adjusted and left at a particular setting. The degree of compounding depends, as shown in Fig. 3-13, on the voltage rise required at the rated load, from E_1 to any particular value up to E_4. The method employed is as follows:

The compound generator is operated as a shunt generator (series field not connected) at rated speed and rated load. The field current, load current, and terminal voltage, E_1, are noted. At rated load, if a flat-compound generator is desired, it will be necessary to raise the voltage from E_1 to E_3. In order to raise the full-load voltage, it is necessary to increase the field current. Thus, the increased field ampere-turns is also an indication of the series ampere-turns (as well as the diverter resistance) needed to produce a desired degree of compounding; or

$$\delta I_f N_f = I_s N_s \tag{3-10}$$

and since the diverter and series-field are connected in parallel,

$$R_d = \frac{I_s R_s}{I_d} \tag{3-11}$$

where δI_f is the increased shunt field current at full load required to produce a desired compound generator terminal voltage

N_f is either the total number of shunt field turns or the number of turns per pole

N_s is either the total number of series field turns or the turns per pole

R_s is the resistance of the series field

I_s is the desired current in the series field required to produce the voltage rise

I_d is the necessary current in the diverter required to produce I_s

R_d is the diverter resistance required for the desired regulation

EXAMPLE 3-9: A 60 kW, 240 V short-shunt compound generator, operated as a shunt generator, required an increase in field current of 3 A, to provide an overcompounded voltage of 275 V, at rated load current of 250 A. The shunt field has 200 turns per pole and the series field 5 turns per pole, with resistances of 240 ohms and 0.005 ohm, respectively. Calculate:
a. The required diverter resistance.
b. If the no-load voltage of the compound generator is also 240 V, calculate the total air-gap mmf per pole at no load and at full load.

Solution:

a. $\delta I_f N_f = 3 \text{ A} \times 200 \text{ turns} = 600 \text{ At} = I_s N_s$ (3-10)

$$I_s = \frac{I_s N_s}{N_s} = \frac{600 \text{ At}}{5 \text{ turns}} = 120 \text{ A}$$

$$I_d = 250 - 120 = 130 \text{ A}$$

$$R_d = \frac{I_s R_s}{I_d} = \frac{120 \times 0.005}{130} = 0.00462 \, \Omega \quad (3\text{-}11)$$

b. No-load mmf $= I_f N_f = \dfrac{240 \text{ V}}{240 \, \Omega} \times 200 \text{ turns} = \mathbf{200 \text{ At/pole}}$

Full-load mmf $= I_f N_f + I_s N_s = 200 \text{ At} + 600 \text{ At} = \mathbf{800 \text{ At/pole}}$

3-20. DIFFERENTIAL COMPOUND GENERATOR CHARACTERISTIC

The *differential compound* generator is defined as that compounding produced when the series field mmf *opposes* the shunt field mmf. The difference in current direction of the two windings is shown in Fig. 3-14a, where, for the sake of clarity, the series field winding is shown above (rather than directly around) the shunt field winding. The load characteristic of the differential compound generator is shown in Fig. 3-14b. Without load, the differential compound generator builds up and self-excites its shunt field in much the same manner as the shunt generator. When a load is applied, however, the generated voltage, E_g, is now reduced by the reduction in the main field flux created by the opposing mmf of the series field. This reduction in E_g occurs in addition to the armature and series circuit volt drop, the armature reaction, and the reduction in field current produced by reduction of the armature voltage, V_a, as described in Section 3-13. The result is a sharp drop in the terminal voltage with load, as shown in Figs. 3-13a and 3-14b, as the field is brought below saturation and rapidly unbuilds. The differential compound generator is used as a constant-current generator for the same constant-current applications as the series generator. (Sec. 3-16.)

(a) Series and shunt field mmf's.

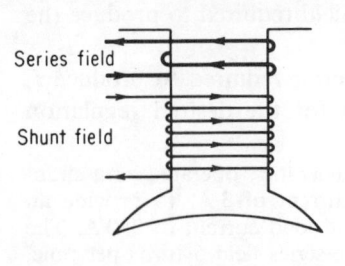

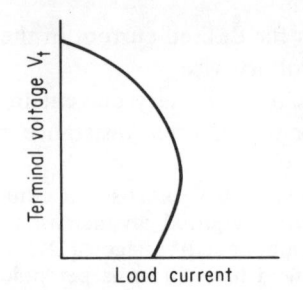

(b) Differential load characteristic.

Figure 3-14

Differential compound generator.

**3-21.
COMPARISON
OF GENERATOR
LOAD-VOLTAGE
CHARACTERISTICS**

If a compound generator is tested in the laboratory to determine and compare the various characteristics discussed, it is customary to adjust the generator for rated voltage at rated load and rated speed. If the load is decreased in increments, and if readings are taken of the load voltage and current at each step, it is possible to compare the characteristics for various connections at the same rated load, voltage, and speed. Such a comparison, using the same dc dynamo, is shown in Fig. 3-15.

Note that all the characteristics droop with increased application of load with the exception of the overcompound generator. Only the latter has negative voltage regulation. With full (or maximum) shunt field mmf, it is impossible to adjust the differential compound generator to yield rated voltage at rated speed and rated load. Would raising the speed make it possible to get a higher voltage and to obtain rated voltage and current for the differential compound generator? What *is* the effect of speed on the load-voltage characteristics of compound generators?

Figure 3-15

Comparison of dc generator load-voltage characteristics at constant speed.

CHAP. THREE / DC Dynamo Voltage Relations—DC Generators

3-22.
EFFECT OF SPEED ON LOAD-VOLTAGE CHARACTERISTICS OF COMPOUND GENERATORS

The effect of a change in speed on both shunt generator and cumulative generator characteristics is shown in Fig. 3-16, using the same dc dynamo for all the tests at rated voltage and load. We have previously seen (Fig. 3-11) that the effect of a reduction in speed is to increase the saturation of the magnetic circuit. The increase in saturation at lower speeds tends to improve the voltage regulation of both the shunt and compound generators. In the case of the cumulative compound generator, shown in Fig. 3-16, since the field circuit is more saturated at a lower speed, the resultant effect of additional mmf produced by the series field is less pronounced. *Increasing the speed*, therefore, would result in a *less saturated* condition, as shown by the magnetization curve of Fig. 3-11a. We can now answer the question related to the differential compound generator. If the magnetic circuit is less saturated, the opposing effect of the series field would be even more pronounced, and the net field flux approaches zero even more rapidly with the application of load!

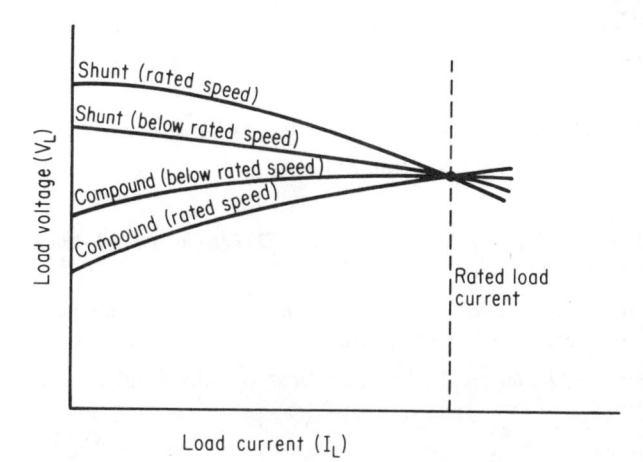

Figure 3-16
Effect of speed on load-voltage characteristics.

One might infer from the above that it is always better procedure to run generators at reduced speeds with higher excitations to produce more saturation and better regulation.

While lower speeds and increased excitation are beneficial for voltage regulation they are not for efficiency. Increased excitation results in higher field-copper and core losses, while the reduced speed results in overheating as a result of less efficient ventilation of the dynamo. It is best, therefore, to operate generators at the rated speed recommended by the manufacturer and stated on the generator nameplate.

BIBLIOGRAPHY

Alger, P. L. and Erdelyi. E. "Electromechanical Energy Conversion," *Electro-Technology*, September 1961.

Carr, C. C. *Electrical Machinery*, New York: John Wiley & Sons, Inc., 1958.

Crosno, C. D. *Fundamentals of Electromechanical Conversion*, New York Harcourt, Brace, Jovanovich, Inc, 1968.

Daniels. *The Performance of Electrical Machines*, New York: McGraw-Hill, Inc., 1968.

D-C Generators and Motors (ASA 050.4.) New York: American Standards Association.

Emunson, B. M. and Ward. A. J. "An Evaluation of the New 'Industrial' d-c Motors and Generators," *Electrical Manufacturing*, June 1958.

Fitzgerald, A. E. and Kingsley, C. *The Dynamics and Statics of Electromechanical Energy Conversion*, 2nd Ed., New York: McGraw-Hill, 1961.

Fitzgerald, A. E., Kingsley, Jr. C., and Kusko, A. *Electric Machinery*, 3rd Ed. New York: McGraw-Hill, Inc., 1971.

Gemlich, D. K. and Hammond, S. B. *Electromechanical Systems*, New York: McGraw-Hill, 1967.

Hindmarsh, J. *Electrical Machines*, Elmsford, N.Y.: Pergamon Press, 1965.

Jones, C. V. *The Unified Theory of Electrical Machines*, New York: Plenum Publishing Corp., 1968.

Kloeffler, S. M., Kerchner R. M., and Brenneman. J. L. *Direct Current Machinery*, Rev. ed., New York: The Macmillan Company, 1948.

Liwschitz, M. M., Garik M., and Whipple. C. C. *Direct Current Machines*, Princeton, N. J.: D. Van Nostrand Co., Inc., 1947.

Majmudar, H. *Introduction to Electrical Machines*, Boston: Allyn and Bacon, 1969.

Nasar, S. A. *Electromagnetic Energy Conversion Devices and Systems*, Englewood Cliffs, N.J.: Prentice-Hall, Inc., 1970.

Robertson, B. L. and Black, L. J. *Electric Circuits and Machines*, 2nd Ed., Princeton, N. J.: D. Van Nostrand Co., Inc., 1957.

Seely, S. *Electromechanical Energy Conversion*, New York: McGraw-Hill, 1962.

Selmon. *Magnetoelectric Devices: Transducers, Transformers and Machines*, New York: Wiley/Interscience, 1966.

Siskind, C. S. *Direct-Current Machinery*, New York: McGraw-Hill, 1952.

Skilling, H. H. *Electromechanics: A First Course in Electromechanical Energy Conversion*, New York: Wiley/Interscience, 1962.

Thaler, G. J., and Wilcox, M. L. *Electric Machines: Dynamics and Steady State*, New York: Wiley/Interscience, 1966.

White, D. C., and Woodson, H. H. *Electromechanical Energy Conversion*, New York: Wiley/Interscience, 1959.

QUESTIONS

3-1. a. Name three basic types of dc generators using dc dynamo construction.
      b. What accounts for the difference in construction between these types?

3-2. Draw a schematic and equivalent circuit for each of the dc generator types cited in Quest. 3-1a, labelling all currents and voltages. (Note: a total of four sets of circuits should be shown because of possible connections.)

3-3. Under each equivalent circuit drawn in Quest. 3-2, write the equations for
a. the current relation
b. the voltage relation
c. the relationship between generated and terminal load voltage.

3-4. Define:
a. self excitation
b. separate excitation.

3-5. a. Draw the schematic for obtaining the magnetization or saturation curve of a dc generator.
b. Draw a typical magnetization curve showing the variation of E_g vs I_f for both increasing and decreasing values of I_f.
c. If a series field winding was connected in series with the armature (in the schematic drawn in answer to Quest. 3-5a), would this affect the magnetization curve obtained? Explain.

3-6. With reference to the saturation curve drawn in Quest. 3-5b, explain:
a. why the curve does not usually begin at the origin
b. under what circumstances it might begin at the origin
c. why the curve is non-linear at extremely low values of voltage
d. why the curve is non-linear at extremely high values of voltage
e. why the curve is linear at moderate values of voltage
f. why decreasing values of field current produce higher voltage values than increasing values of field current.

3-7. Explain:
a. why the shape of the saturation or magnetization curve is identical to the B-H curve for the mutual flux density in the air gap of the dynamo
b. what precautions must be taken in obtaining a smooth magnetization curve and why
c. why the equation $E_g = k\phi$ represents the magnetization curve rather than $E_g = kI_f$.

3-8. a. For a separately excited generator, explain why the field current is independent of the generated voltage and yet the generated voltage is dependent upon the field current. Illustrate by schematic diagram.
b. For a self-excited shunt generator, explain why the field current is dependent upon the generated voltage. Illustrate by schematic diagram.
c. In addition to the generated voltage, what other variable determines field current?
d. What is meant by a "family of field resistance lines?"

3-9. With reference to the curve for build-up of a self-excited generator, explain:
a. critical field resistance, R_c
b. field resistance which is higher than R_c and its effect on build-up
c. field resistance much lower than R_c and its effect on build-up.

3-10.   a. State four specific reasons why a self-excited shunt generator will not build up voltage.
b. If a low resistance load is connected across the armature terminals, will a self-excited shunt generator build-up? Explain.
c. Repeat (b) for a series generator.

3-11.   Explain what is meant by "increasing generator load" in terms of:
a. load resistance
b. load current.

3-12.   If application of load causes the terminal voltage of a shunt generator to decrease, explain under what conditions a shunt generator may tend to unbuild, using a magnetization curve to illustrate your explanation.

3-13.   a. Give three reasons why the terminal voltage of a self-excited shunt generator will decrease with application of increased load.
b. Repeat (a) for a separately excited generator (two reasons).
c. What conclusions would you draw regarding the voltage regulation of a self-excited shunt generator compared to the same generator operated under conditions of separate excitation?

3-14.   Define voltage regulation
a. in terms of an equation
b. in your own words.

3-15.   What is meant by
a. good voltage regulation
b. poor voltage regulation
c. negative voltage regulation
d. positive voltage regulation
e. zero voltage regulation.

3-16.   Explain, for a series generator
a. why it is incapable of build-up without load
b. under what conditions the terminal voltage rises with load
c. under what conditions the terminal voltage drops sharply with load
d. what prevents the generated voltage from rising indefinitely with load.

3-17.   Draw a series generator load characteristic and explain:
a. which portion is useful as a voltage booster and why
b. which portion is useful as a constant current generator and why
c. four applications of dc series generators indicating which portion of the voltage load characteristic is used and why.

3-18.   Explain for a compound generator, using illustrations where necessary,
a. long-shunt connection
b. short-shunt connection
c. cumulative compound
d. differential compound
e. overcompound
f. flat compound
g. undercompound
h. which of the above compoundings produces negative regulation
i. repeat (h) for zero percent regulation.

3-19. Give one application for each of the following compound generators:
a. overcompound
b. flat compound
c. undercompound
d. differential compound.

3-20. "Most commercial compound dc dynamos are normally supplied as overcompound machines." Explain:
a. why this is the case
b. how the degree of compounding may be adjusted
c. how the diverter resistance is determined to produce a desired degree of compounding

3-21. Using Fig. 3-15, explain:
a. why it is normally impossible for a differential compound generator to deliver *rated* load current
b. how it is possible for a differential compound generator to deliver a value of load current when its terminal voltage is zero.

3-22. For a given value of load current, what is the effect of increasing the prime mover speed on the terminal voltage of
a. a shunt generator
b. a cumulative compound generator
c. an undercompound generator
d. a differential compound generator
e. a series generator operating on its constant current characteristic
f. a series generator operating on its voltage-booster characteristic.

PROBLEMS

3-1. A 50 kW, 250 V dc shunt generator has a field circuit resistance of 62.5 Ω, a brush voltage drop of 3 V, and an armature resistance of 0.025 Ω. When delivering rated current at rated speed and voltage, calculate:
a. Load, field and armature currents
b. Generated armature voltage.

3-2. A 10 kW, 125 V dc series generator has a 2 V brush drop, an armature circuit resistance of 0.1 Ω and a series field resistance of 0.05 Ω. When delivering rated current at rated speed, calculate:
a. Armature current
b. Generated armature voltage.

3-3. A 600 V, 100 kW, long-shunt compound generator has a brush volt drop of 5 V, a series field resistance of 0.02 Ω, a shunt field resistance of 200 Ω and an armature resistance of 0.04 Ω. When rated current is delivered at rated speed of 1200 rpm, calculate
a. Armature current
b. Generated armature voltage.

3-4. A separately excited generator has a no-load voltage of 125 V at a field current of 2.1 A when driven at a speed of 1600 rpm. Assuming that it is

operating on the straight line portion of its saturation curve, calculate
a. Generated voltage when the field current is increased to 2.6 A
b. Generated voltage when the speed is reduced to 1450 rpm and the field current is increased to 2.8 A.

3-5. Assuming that a 100 per cent increase in field current produces a 70 per cent increase in flux, repeat Problems 3-4a and 3-4b.

3-6. A compound generator has a no-load voltage of 125 V and a full-load voltage of 150 V. Calculate the per cent voltage regulation of the generator.

3-7. A 125-V shunt generator has a voltage regulation of 5 per cent. Calculate the voltage at no load.

3-8. The shunt field current of a 125-V, 60-kW dc generator has to be increased from 3.5 to 4.0 A to produce flat compounding from no load to full load, respectively. Each field pole has 1500 turns. Calculate:
a. The number of series field turns per pole, assuming a short-shunt connection.
b. Repeat (a) assuming long-shunt connection.

3-9. A 50 kW, dc generator has 2000 turns/pole in its shunt field winding. A shunt field current of 1.20 A is required to generate 125 V at no load and 1.75 amp is required to generate 140 V at full load. Calculate
a. The minimum number of series field-ampere turns per pole required to furnish the required no-load and full load voltages as a compound generator (assume a short-shunt connection)
b. If the machine is equipped with a series field having 5 turns/pole and a resistance of 0.02 Ω, calculate the resistance of the diverter required to produce the desired compounding
c. Voltage regulation of the compound generator.

3-10. A 250 V compound generator (long shunt) is rated at 25 kW. The shunt and series field resistances are 125 Ω and 0.05 Ω, respectively, and are wound with 1000 and 10 turns/pole, respectively. It is found that with the series field short-circuited, the field current must be increased from 2.0 to 2.3 A to produce flat compounding. Calculate:
a. The diverter resistance required to produce flat compounding
b. The number of ampere turns per pole produced by both shunt and series fields at full load
c. Ampere turns per pole produced by both shunt and series fields at no load.

3-11. A 30 kW, 250 V shunt generator produces a generated armature voltage of 265 V in order to develop rated output when its field excitation is 1.5 A. Calculate:
a. Field circuit resistance to produce rated terminal voltage
b. Armature circuit resistance (neglect brush voltage drop).

3-12. A 30 kW, 250 V shunt generator is separately excited to determine its armature reaction voltage drop. Its armature resistance is 0.1235 Ω. Assume a 3 V brush drop. Calculate:
a. The armature circuit voltage drop at full load, at rated voltage and speed

b. The voltage drop due to armature reaction if the no-load voltage is 275 V at rated speed.

3-13. A 125-V dc compound generator operates as a flat compound generator at its rated speed of 1200 rpm. Assuming no change in prime mover speed, discuss the effect on compounding if
   a. The no-load voltage is increased to 150 V
   b. The no-load voltage is reduced to 100 V.

3-14. A 125 V, dc compound generator operates as a flat compound generator at its rated speed of 1200 rpm. Assuming no change in excitation, discuss the effect on compounding if
   a. The speed is increased to 1500 rpm
   b. The speed is reduced to 1000 rpm.

ANSWERS

3-1(a) 204 A (b) 258.1 V   3-2(a) 80 A (b) 139 V   3-3(a) 169.7 A (b) 615.2 V
3-4(a) 154.8 V (b) 151 V   3-5(a) 145.8 V (b) 140 V   3-6. −16.7 per cent   3-7. 131.3 V   3-8(a) 1.56 turns/pole (b) 1.545 turns/pole.   3-9(a) 3.08 turns/pole (b) 0.0321 Ω (c) −10.7 per cent   3-10(a) 0.0208 Ω (b) 300 At/pole (c) 5.87 At/pole
3-11(a) 166.7 Ω (b) 0.1235 Ω   3-12(a) 14.8 V (b) 7.2 V

FOUR

dc dynamo torque relations—dc motors

4-1. GENERAL

In comparing dc dynamo motor action vs. generator action, Section 1-20 concluded with a summary of the fundamental differences between them. This chapter is devoted to the dc dynamo used as a dc motor. It is concerned, therefore, with dc dynamo *torque* relations and the characteristics of the dc motor as a means of producing electromagnetic torque. The summary of Section 1-20 stated for motor action:

1. The developed electromagnetic torque produces (aids) rotation.
2. The voltage generated in the current-carrying conductors (counter emf) opposes the armature current (Lenz's law).
3. The counter emf may be expressed by the equation

$$E_c = V_a - I_a R_a \tag{1-9}$$

and is less than the applied voltage causing a given armature current flow, I_a.

Equation (1-9) may be rewritten in terms of the armature current, I_a, produced for a given applied voltage and load:

$$I_a = \frac{V_a - E_c}{R_a} \qquad (1\text{-}9)$$

It was also shown in Section 1-20 that the three factors which determine the magnitude and which are required to produce electromagnetic force on a given current-carrying armature conductor (a force orthogonal to B and I) may be expressed by

$$F = \left(\frac{BIl}{1.13}\right) 10^{-7} \text{ lbs} \qquad (1\text{-}8)$$

and finally, the direction of electromagnetic force developed by such a current-carrying conductor in a given magnetic field may be determined by the left-hand rule (Section 1-18).

The reader should review the above relations and Sections 1-16 through 1-20, since they are fundamental and apply to all commercial motor types and the characteristics discussed below.

4-2. TORQUE

The terms electromagnetic *force* and electromagnetic *torque* have been used in the above summary of the motor relations covered in Chapter 1. These terms are not synonymous, but they are related. The relation between the force on a conductor [developed in accordance with Eq. (1-8) above] and the torque produced is shown in Fig. 4-1.

A single-turn coil (supported on a structure capable of rotation) is carrying current in a magnetic field, as shown in Fig. 4-1a. In accordance with Eq. (1-8) and the left-hand rule, an orthogonal force f_1 is developed in coil-side 1, and a similar force f_2 is developed in coil-side 2, as shown in Fig. 4-1b. Forces f_1 and f_2 are developed in such a direction that they

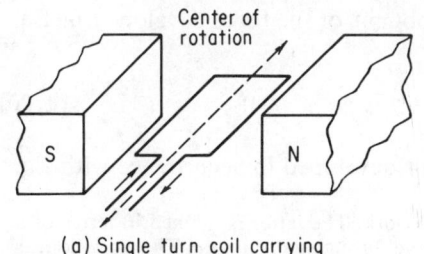

(a) Single turn coil carrying current in a magnetic field.

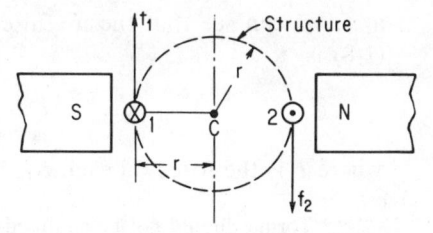

(b) Definition of torque developed.

Figure 4-1

Production of torque in a single-turn coil.

tend to produce a clockwise rotation of the structure supporting the conductors about the center of rotation C.

Torque is defined as the tendency of a mechanical coupling (of a force and its radial distance to the axis of rotation) to produce rotation. It is expressed in units of force and distance, such as lb-ft, gram-cm, ounce-inches, etc.,* to distinguish it from *work*, which is expressed in ft-lbs, cm-gram, etc. The torque acting on the structure of Fig. 4-1b is the sum of the products $f_1 r$ and $f_2 r$, i.e., the total sum of the torques acting on or produced by the individual conductors which tend to produce rotation. It should be noted that forces f_1 and f_2 are equal in magnitude since they lie in a field of the same magnetic strength and carry the same current. This is true for the forces developed by all conductors carrying the same current in a uniform magnetic field; but the torques developed, by definition, are not the same for each of these conductors.

The distinction between the force developed on the various armature conductors and the useful torque developed by these conductors to produce rotation is shown in Fig. 4-2. It has been shown in Chapters 1 and 2 that there is essentially no difference in construction between a generator armature and a motor armature. A dc dynamo, as described in Section 1-20, may be considered a motor when it fulfills the three conditions summarized in Section 4-1.

An armature and a field of a two-pole motor are shown in Fig. 4-2. Note that all of the conductors carrying current in the same direction develop the same force. This is true because they carry the same current and lie perpendicularly in the same field. But, since torque is defined as the product of a force and its perpendicular distance from the axis, we can see that the useful component of the force developed in Eq. (1-8) is

$$f = F \sin \theta \qquad \text{lbs} \qquad (4\text{-}1)$$

where F is the force on each conductor developed in accordance with Eq.

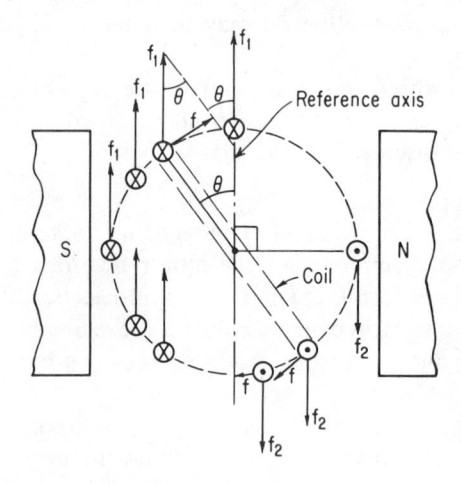

Figure 4-2

Useful torque for rotation.

---

* Torque should not be confused with work. The latter is defined in terms of a force f, acting on a body and causing it to move through a distance d. The work done is the product of that component of force f acting in the same direction in which the body moves (to overcome resistance) for some distance d. If there is a force applied but no motion results, no work is done. Conversely, a force may exist on a body tending to produce rotation (a torque) and, even if the body does not rotate, the torque exists as the product of that force and the radial distance to the center of the axis of rotation.

(1-8), and θ is the complement of the angle created by the force developed on the conductor and the useful force f tangential to the armature periphery; and thus the torque developed by *any* conductor, T_c, on the armature surface is

$$T_c = fr = (F \sin \theta) r \qquad \text{lb-ft} \qquad (4\text{-}2a)$$

where f is the force in pounds [Eq. (4-1)] perpendicular to r, and r is the radial distance to the axis of rotation in feet.

EXAMPLE 4-1: The single-turn coil in Fig. 4-2 lies on an armature 18 in. in diameter having an axial length of 24 inches, in a field with a density of 24,000 lines/in². When the coil carries a current of 26 A calculate
a. The force developed on each conductor.
b. The useful force at the instant the coil lies at an angle of 60° with respect to the interpolar reference axis.
c. The torque developed in lb-ft.

Solution:

a. $F = \dfrac{BIl}{1.13} \times 10^{-7}$ lbs $= \dfrac{24{,}000 \times 26 \times 24}{1.13 \times 10^7} = $ **1.325 lbs** $\qquad (1\text{-}8)$

b. $f = F \sin \theta = 1.325 \sin 60° = $ **1.145 lbs** $\qquad (4\text{-}1)$

c. $T_c = fr = 1.145 \text{ lb} \times \left(9 \text{ in} \times \dfrac{1 \text{ ft}}{12 \text{ in}}\right) = $ **0.858 lb-ft** $\qquad (4\text{-}2)$

Note that the conductors lying in the interpolar region of Fig. 4-2 develop (theoretically) just as much force as those lying directly under the pole; but that the component of *useful* force, f, tangential to the armature is zero. Furthermore, if the coil of Fig. 4-2 is free to rotate in the direction of the developed torque *without* undergoing commutation, the current directions in the conductors would remain *unchanged* but the force developed on them would *reverse*, as shown in Fig. 4-3.

The necessity for commutation to *reverse* the current in a conductor as it moves under a pole of *reversed* polarity is just as fundamental for a dc motor as it is for a dc generator. Finally, since no *useful* torque is produced by conductors lying in the interpolar region, very little torque

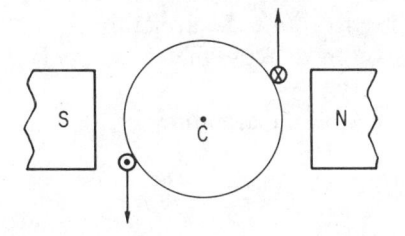

Figure 4-3
Necessity for commutation in a dc motor.

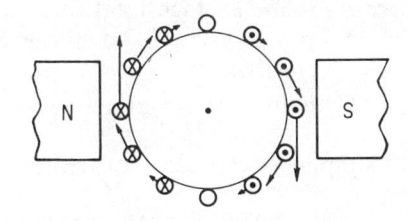

Figure 4-4
Reversal of conductor current required to produce continuous rotation.

SEC. 4-2. / *Torque*

is lost by those conductors undergoing commutation. This is shown in Fig. 4-4, where the components of *useful* force and their magnitudes are indicated, as well as the current reversal in the conductor required to produce uniform and continuous rotation.

The preceding relations were developed for an armature having straight field poles and a fairly appreciable neutral interpolar zone. As shown in Fig. 4-5, in a commercial armature having many poles, slots, and armature conductors, the difference between the useful force developed directly under the pole and that developed almost at the pole tip is relatively small. It is customary, instead, to consider only that percentage of conductors *directly under the pole* which contribute useful torque, and to assume that an average or common torque is produced by each conductor.

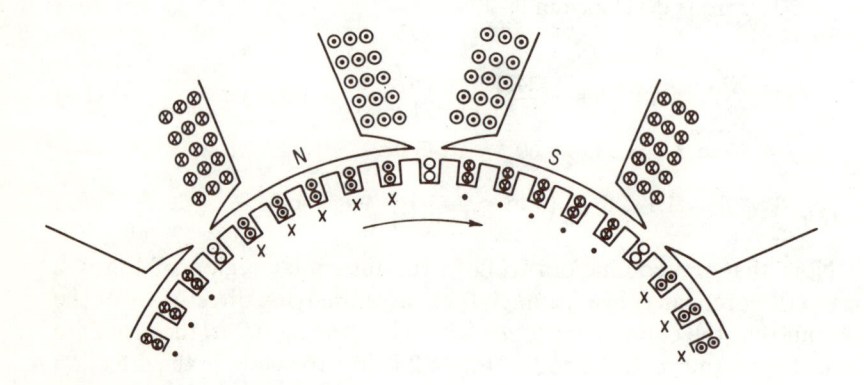

Figure 4-5

Direction of force, current flow, and counter emf in a commercial dc motor.

These assumptions lead to the simple relation

$$F_{av} = F_c \times Z_a \tag{4-2b}$$

where F_{av} is the average total force tending to rotate the armature
F_c is the average force per conductor directly under the pole (Eq. 1-8)
Z_a is the number of active conductors on the armature

This simplifies calculation of the total torque developed by the armature, since

$$T_{av} = F_{av} \times r = F_c \times Z_a \times r \tag{4-2c}$$

where all terms have been defined above.

EXAMPLE 4-2: The armature of a dc motor contains 700 conductors and has a diameter of 24 in and an axial length of 34 in. If 70 per cent of the conductors lie directly under the poles with a flux density of 50,000 lines/in² and carry a current of 25 A, calculate
a. The average total force tending to rotate the armature
b. The armature torque in lb-ft

Solution:

From Eq. (4-2b):
a. $F_{av} = F_c \times$ No. active conductors

$$= \left(\frac{50,000 \times 25 \times 34}{1.13 \times 10^{+7}}\right)(700 \times 0.7) = \mathbf{1860\ lbs} \quad (1\text{-}8)$$

b. $T_{av} = F_{av} \times r = 1860\ \text{lbs} \times 1\ \text{ft} = \mathbf{1860\ lb\text{-}ft} \quad (4\text{-}2c)$

EXAMPLE 4-3: Find the total external armature current of a motor which has the following specifications: 120 armature slots, 6 conductors per slot; 60,000 lines per sq in flux density; 28 in armature diameter, 14 in armature axial length; 4 armature paths in parallel; the pole arcs span 72 per cent of the armature surface; and the torque developed by the armature is 1500 lb-ft.

Solution:

$$F_{av} = \frac{T_{av}}{r} = \frac{1500\ \text{lb-ft}}{14\ \text{in}} \times 12 \frac{\text{in}}{\text{ft}} = 1285\ \text{lbs} \quad (4\text{-}2c)$$

$$I_a/\text{cond} = I_a/\text{path} = \frac{F_{av} \times 1.13 \times 10^7}{B \times l \times Z_{active}}$$

$$= \frac{1285 \times 1.13 \times 10^7}{60{,}000 \times 14(120 \times 6 \times 0.72)} = 33.4\ \text{A/path} \quad (1\text{-}8)$$

$$I_a = I_a/\text{path} \times a = 33.4\ \text{A/path} \times 4\ \text{paths} = \mathbf{133.6\ A}$$

4-3. FUNDAMENTAL TORQUE EQUATION FOR A DC DYNAMO

The preceding discussion and the above problems indicate that the torque developed by the armature of any given dynamo may be computed in terms of the number of poles, paths, conductors, and flux per pole linking the armature conductor, etc. As an exercise for the reader, Example 4-3 may be recomputed in terms of the following equation:*

$$T = 0.1173 \left(\frac{P}{a}\right) Z I_a \phi \times 10^{-8}\ \text{lb-ft} \quad (4\text{-}3)$$

where P is the number of poles
a is the number of paths
Z is the number of active conductors on the surface of the armature, each producing useful average torque
I_a is the total armature current entering the armature
ϕ is the flux per pole linking the conductors

* See Problem 4-3 at the end of the chapter.

For any given dc dynamo, however, the number of paths, poles, and armature conductors is fixed or constant, and, therefore, the equation for electromagnetic torque developed by any given armature may be written only in terms of its possible variables as

$$T = k\phi I_a \text{ lb-ft} \tag{4-4}$$

where ϕ and I_a are the same as in Eq. (4-3) and $k = 0.1173(P/a)Z \times 10^{-8}$ for any given dynamo.

Note that Eq. (4-4) is but another form of Eq. (1-8) ($F = k'BIl$), where the variables B and I for any given dynamo determine the value of the electromagnetic force which produces the motor torque.

Also note that this electromagnetic torque opposes rotation in a generator and aids (is in the same direction as) rotation in a motor. Since the torque is a function of the flux and the armature current, it is *independent of speed* in the case of either a generator or a motor. It will be seen later that the *speed* of a motor *does*, in fact, *depend on torque* (but not vice versa). The terms torque and speed should *not*, however, be used synonymously, for a motor that is stalled may tend to develop appreciable torque but no speed.

A change in flux may produce a change in armature current, and also produce a change in torque, as shown by the following example.

EXAMPLE 4-4: A motor develops a torque of 150 lb-ft and is subjected to a 10 per cent reduction in field flux, which produces a 50 per cent increase in armature current. Find the new torque produced as a result of this change.

Solution:

	ϕ	I_a	T
Original Condition	1.0	1.0	150 lb-ft
New Condition	0.9	1.5	?

$$T = K\phi I_a \tag{4-4}$$

Using the ratio method, the new torque is the product of 2 new ratio changes:

$$T = 150\left(\frac{0.9\phi}{1.0\phi}\right) \times \left(\frac{1.5I_a}{1.0I_a}\right) = 202.5 \text{ lb-ft}$$

Finally, it should be noted that the electromagnetic torque developed by the armature in accordance with Eqs. (4-3) and (4-4) is customarily called the *developed torque*. The developed torque, developed by the armature conductors, is somewhat analogous to generated emf, E_g, in that it is developed *internally*, within the armature. The *torque available at the pulley or shaft* of a motor is somewhat *less* than the developed torque because of specific rotational losses that require and consume some portion of the developed torque during motor action (Fig. 12-1, Sec. 12-3).

4-4. COUNTER EMF OR GENERATED VOLTAGE IN A MOTOR

We are already aware that when a dc dynamo operates as a motor, generator action simultaneously occurs, since the conductors are moving in a magnetic field. The current-carrying conductors that produce a clockwise torque are shown in the armature slots of Fig. 4-5. The opposing direction of induced emf is shown below the conductors (left-hand vs. right-hand rules, respectively) in the figure. The counter emf generated in the armature conductors is expressed in Eq. (1-6) for a given armature. The current which flows through the armature is limited by (1) the armature resistance and (2) the counter emf, in accordance with Eq. (1-9) as rewritten in Section 4-1, i.e., $I_a = (V_a - E_c)/R_a$. It is fairly evident that the counter emf can never equal the voltage applied across the armature terminals because, as shown in Fig. 4-5, the direction in which the current flow *first* occurs determines the direction of rotation and, in turn, creates the counter emf. Clearly, then, the counter emf, like the armature resistance, is a current *limiting* factor. The nature of the counter-emf in limiting the current may best be understood by Ex. 4-5 which also includes brush voltage drop, BD, as a limiting factor, as well.

EXAMPLE 4-5: A dc shunt motor having an armature resistance of 0.25 ohm and a brush contact volt drop of 3 V, receives an applied voltage across its armature terminals of 120 V. Calculate the armature current when
  a. The speed produces a counter emf of 110 V at a given load.
  b. The speed drops (due to application of additional load) and the counter emf is 105 V.
  c. Compute the per cent change in counter emf and in armature current.

Solution:

a. $I_a = \dfrac{V - (E_c + BD)}{R_a} = \dfrac{120 - (110 + 3)}{0.25} = 28$ A   (1-9)

b. At increased load, $I_a = \dfrac{120 - (105 + 3)}{0.25} = 48$ A

c. $\delta E_c = \dfrac{110 - 105}{110} \times 100 = 4.53\%$; $\delta I_a = \dfrac{28 - 48}{28} \times 100 = 71.5\%$

In the above problem it should be noted that a *small* change in counter emf and speed (4.53 per cent) has resulted in a *substantial* change (71.5 per cent) in armature current. Consequently, changes in motor speed, however slight, are reflected by correspondingly larger changes in motor current. For this reason, in some types of servomechanism transducer devices, motor current is employed as an indication of motor load and speed.

4-5. MOTOR SPEED AS A FUNCTION OF COUNTER EMF AND FLUX

The value of counter emf given in Example 4-5 may be computed readily from Eq. (1-6) (see Example 1-6, Section 1-15). For any given dc dynamo, either Eq. (1-5) or Eq. (1-6) may be rewritten in terms of its variables, and the counter emf of a motor may be expressed by

$$E_c = k\phi S \tag{4-5}$$

where ϕ is the flux per pole
k is $(ZP/60a)10^{-8}$ for any given dynamo
S is the speed of rotation of the motor in rpm

But the counter emf of a motor, including volt drop across brushes, BD, is

$$E_c = V_a - (I_a R_a + BD) \tag{1-9}$$

and substituting $k\phi S$ for E_c from Eq. (4-5), and solving for S, the motor speed, results in

$$S = \frac{V_a - (I_a R_a + BD)}{k\phi} \tag{4-6}$$

where all the terms have been previously defined.

Equation 4-6 has been called the *fundamental dc motor speed equation*, since it permits prediction of dc motor performance so readily. For example, if the field flux of a dc motor is weakened considerably, the motor will *run away*. If the denominator of Eq. (4-6) approaches zero, the speed approaches infinity. Similarly, if the load current and flux are held constant, while the voltage impressed across the motor armature is increased, the speed will increase in the same proportion. Finally, if the field flux and the voltage across the armature are fixed, and the armature current is increased because of increased load, the motor speed will drop in the same proportion as the decrease in counter emf [Eq. (4-5)].

EXAMPLE 4-6: A 120 V dc shunt motor having an armature circuit resistance of 0.2 ohm and a field circuit resistance of 60 ohms, draws a line current of 40 A at full load. The brush volt drop is 3 V and rated, full load speed is 1800 rpm. Calculate:
  a. The speed at half load.
  b. The speed at an overload of 125 per cent.

Solution:

a. at full load

$$I_a = I_l - I_f = 40 \text{ A} - \frac{120 \text{ V}}{60 \text{ }\Omega} = 38 \text{ A};$$

$$E_c = V_a - (I_a R_a + BD) = 120 - (38 \times 0.2 + 3) = 109.4 \text{ V} \tag{1-9}$$

at the rated speed of 1800 rpm

$$E_c = 109.4 \text{ V} \quad \text{and} \quad I_a = 38 \text{ A (full load)}$$

half-load speed

$$I_a = \frac{38 \text{ A}}{2} = 19 \text{ A};$$

$$E_c = V_a - (I_a R_a + BD) = 120 - (19 \times 0.2 + 3) = 113.2 \text{ V}$$

Using the ratio method, half-load speed

$$S = S_{orig} \frac{E_{final}}{E_{orig}} = 1800 \frac{113.2}{109.4} = 1860 \text{ rpm} \qquad (4\text{-}5)$$

b. at $1\frac{1}{4}$ load

$$I_a = \tfrac{5}{4} 38 \text{ A} = 47.5 \text{ A};$$

$$E_c = V_a - (I_a R_a + BD) = 120 - (47.5 \times 0.2 + 3) = 107.5 \text{ V}$$

$$S_{5/4} = 1800 \frac{107.5}{109.4} = 1765 \text{ rpm} \qquad (4\text{-}5)$$

The above results are tabulated in Ex. 4-8, below.

EXAMPLE 4-7: The dc motor of Example 4-6 is loaded to a line current of 66 A (temporarily), but in order to produce the necessary torque, the field flux is increased by 12 per cent by decreasing the field circuit resistance to 50 ohms. Calculate the speed of the motor

Solution:

$$I_a = I_l - I_f = 66 - \frac{120}{50} = 63.6 \text{ A}$$

$$E_c = V_a - (I_a R_a + BD) = 120 - (63.6 \times 0.2 + 3) = 104.3 \text{ V}$$

from Eq. (4-6)

$$S = \frac{K E_c}{\phi} = 1800 \frac{104.3}{109.4} \times \frac{1.0}{1.12} = 1535 \text{ rpm} \qquad (4\text{-}5)$$

Note that in the preceding solution, the method employed is the *ratio method*. The original full-load speed of 1800 rpm is affected by *two* factors, counter emf and flux. The counter emf has decreased and since speed varies *directly* as counter emf, the speed is multiplied by a *decreasing* ratio. Similarly, the flux has increased, but an increase in ϕ produces a *decrease* in speed. Therefore, the speed is again multiplied by a *decreasing* ratio. This calculation technique is more economical and useful than proportions. The reader should study it carefully by solving Exs. 4-6 and 4-7 independently.

**4-6.
COUNTER EMF
AND MECHANICAL
POWER DEVELOPED
BY A MOTOR
ARMATURE**

In general, it may be noted from the preceding problems that the full-load counter emf is less than the counter emf at a lighter value of load. As a function of the armature voltage across the armature circuit, the *full-load* counter emf will vary from approximately 80 per cent in the smaller dynamos to about 95 per cent of the armature applied voltage in the larger dc motors. The counter emf, E_c, as a percentage of the armature voltage, V_a, is an important ratio in determining the relative efficiency and mechanical power developed by a given armature. The mechanical power developed by the armature may be derived in the following way.

The voltage drop across the armature, ignoring brush drop BD, is

$$I_a R_a = V_a - E_c \qquad (1\text{-}9)$$

and the power lost in the armature when V_a is impressed across it and I_a flows is [multiplying both sides of Eq. (1-9) by I_a]

$$(I_a R_a) I_a = I_a (V_a - E_c)$$

or

$$I_a^2 R_a = V_a I_a - E_c I_a$$

Solving for $E_c I_a$ we get

$$E_c I_a = V_a I_a - I_a^2 R_a \qquad (4\text{-}7)$$

The significance of Eq. (4-7) is that when electrical power, $V_a I_a$, is supplied to the motor armature circuit to produce rotation, a certain amount of power is dissipated in the various components making up the armature circuit resistance; this dissipation is termed an armature copper loss, $I_a^2 R_a$. The remaining power, $E_c I_a$, is required by the armature for the production of developed or internal torque (cf. Fig. 12-1). The ratio of *power developed by* to *power supplied to* the armature, $E_c I_a / V_a I_a$, is the same as the ratio E_c / V_a. Thus, the higher the percentage of counter emf to voltage across the armature, the higher the motor efficiency. Further, for a given load current, it is fairly evident that, when the counter emf is a maximum, the motor will develop the greatest possible power for that value of armature current, I_a.

The last sentence bears some reflection for it would appear from Eq. (4-5) ($E_c = K\phi S$) that, in order to develop the maximum counter emf possible, it is only necessary to increase the field current and flux to a maximum (without overheating the field winding) and at the same time "operate" the motor at very high speeds. But Eq. (4-6) shows that when the field flux is increased, the speed decreases (Example 4-7). Furthermore, both the speed and the counter emf are, in part, determined by the mechanical load on the motor. But it is fairly certain that, for a given mechanical load and resulting line and armature current, there is a particular speed and field rheostat setting that should produce maximum power.

EXAMPLE 4-8: Calculate the armature power developed for each of the loads of Examples 4-6 and 4-7 and tabulate them for reference and comparison.

Solution:

Example	I_a	E_c	Speed	$P_d (E_c I_a)$
4-6a	38	109.4	1800	4160 W at full load
	19	113.2	1860	2510 W at $\frac{1}{2}$ load
	47.5	107.5	1765	5110 W at $1\frac{1}{4}$ load
4-7	63.6	104.3	1535	6640 W at overload

Note from Ex. 4-8 that the small reduction in counter emf results in a proportionately large increase in armature current, with the result that the power developed increases as the counter emf drops because of load.

4-7. RELATION BETWEEN TORQUE AND SPEED OF A MOTOR

Let us assume that in the basic speed equation, Eq. (4-6) that the brush voltage drop, BD, is zero. In the derivation and discussion of the basic speed equation $S = (V_a - I_a R_a)/k\phi$, the reader may have noticed what may appear to be an obvious inconsistency between this equation and Eq. (4-4), $T = k\phi I_a$.

Since torque is defined as a force tending to produce rotation, according to Eq. (4-4), increasing the field flux would tend to increase the torque and (possibly) the speed. On the other hand, increasing the field flux in Eq. (4-6) would reduce the speed. Is there an inconsistency and is it possible to reconcile the two equations?

Actually, there is no inconsistency; and with the help of Eq. (1-9), $I_a = (V_a - E_c)/R_a$, it is possible to give both a qualitative and a quantitative explanation of what happens when the field flux is reduced. Qualitatively, the steps are:

1. The field flux of a shunt motor is reduced by decreasing the field current.
2. The counter emf, $E_c = k\phi S$, drops instantly (the speed remains constant as a result of the inertia of the large and heavy armature).
3. The decrease in E_c causes an increase in the armature current, I_a; refer to Eq. (1-9), cited above.
4. But Example 4-4 showed that a small reduction in field flux produces a large increase in armature current.
5. In Eq. (4-4), therefore, where $T = k\phi I_a$, the *small decrease* in flux is more than counterbalanced by a *large increase* in armature current. Note that the *torque has increased more than the flux was reduced!*
6. This increase in torque produces an *increase in speed*.

Since the speed of a running machine is determined by the torque developed, the question arises, then, is it possible to increase the field flux and, at the same time, increase the speed? The answer is that it *is* possible, but *only* if the armature current is held *constant* ($T = K\phi I_a$). This is actually done in the dc servomotor, shown in Fig. 4-6, in which I_a is constant because the armature is connected to a constant-current source (a series or differential compound generator; see Section 3-20). With no dc voltage impressed on the separately

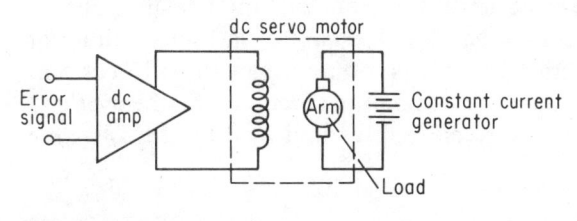

Figure 4-6

Separately excited dc motor.

excited field, there is no torque [Eq. (4-4)]. When a small dc voltage is applied to the field, a small torque is developed and the armature rotates slowly in accordance with Eq. (4-4). Since the armature current is constant at all times, the torque and speed are therefore proportional only to the field flux. A field flux of zero produces zero, *not infinite*, speed. The separately excited dc servo-motor (Section 11-13) does not violate the basic motor equations. On the contrary, it proves them!

A final question frequently asked (and a tempting one for students in the laboratory) is: What would occur if the field circuit of a loaded shunt motor is suddenly opened? Would the motor, if unprotected by fuses, gain speed to a point where it would destroy itself? We already know that any small decrease in flux produces a large increase in armature current and torque. A loaded motor with an open field draws an abnormally high armature current as it races to higher speeds and in turn produces higher mechanical loads and centrifugal forces on its armature conductors.

The answer to the question lies in the nature of the source and the lines supplying the armature. Given a source capable of supplying an infinite current, and given feeder lines of zero resistance, an open field will cause higher speed, more load, more armature current, more torque, and, in turn, higher speed. The motor speed will be almost infinite (ultimately) and the motor will, indeed, be destroyed by the centrifugal forces acting on its armature conductors. But, happily for most students in the laboratory, the supply lines have resistance, the voltage supply is limited as to the current it can deliver, and, fortunately, a circuit breaker or fuse opens the circuit before too much damage is done to the motor by excessive armature current and speed.

Summarizing then, in attempting to predict the effect of changes in armature current and flux, on either torque or speed, there is no inconsistency between Eqs. (4-4) and (4-6). The reader must bear in mind that when armature current (I_a) is *not* held constant, a decrease in flux produces correspondingly larger increases in armature current, torque and speed.

**4-8.
STARTERS FOR
DC MOTORS**

At the instant of applying a voltage, V_a, across the armature terminals in order to cause a motor to rotate, the motor armature is not producing any counter emf since the speed is zero [Eq. (4-5)]. The only current-limiting factors are the armature brush volt drop and the resistance of the armature circuit, R_a. Since neither of these, under normal conditions, amounts to more than 10 or 15 per cent of the applied voltage, V_a, across the armature (Section 4-6), the overload is many times the rated armature current, as indicated by the following example.

EXAMPLE
4-9:
A 120 V dc shunt motor has an armature resistance of 0.2 ohm and a brush volt drop of 2 V. The rated full-load armature current is 75 A. Calculate the current at the instant of starting, and the per cent of full load.

Solution:

$$I_{st} = \frac{V_a - BD}{R_a} = \frac{120 - 2}{0.2} = 590 \text{ A (counter emf is zero)}$$

$$\text{Per cent full load} = \frac{590 \text{ A}}{75 \text{ A}} \times 100 = 786 \text{ per cent}$$

Example 4-9 serves to illustrate that damage may be done to a motor unless the starting current is limited by means of a *starter*.*

The current in the above problem is excessive because of a lack of counter emf at the instant of starting. Once rotation has begun, counter emf is built up in proportion to speed. What is required, then, is a device, usually a tapped or variable resistor, whose purpose is to limit the current during the starting period and whose resistance may be progressively reduced as the motor gains speed. Given an external resistor, R_s, in series with the armature, Eq. (1-9) must be modified for computing the armature current.

$$I_a = \frac{V_a - (E_c + BD)}{R_a + R_s} \tag{4-8}$$

where all terms have been defined above.

The value of the starting resistor at zero speed or any step along the way may be computed from Eq. (4-8), as illustrated by the following example:

EXAMPLE 4-10: Calculate the various values (taps) of starting resistance to limit the current in the motor of Example 4-9 to
a. 150 per cent rated load at the instant of starting.
b. A counter emf which is 25 per cent of the armature voltage, V_a, at 150 per cent rated load.
c. A counter emf which is 50 per cent of the armature voltage at 150 per cent rated load.
d. Find the counter emf at full load, without starting resistance.

Solution:

$$R_s = \frac{V_a - (E_c + BD)}{I_a} - R_a \quad \text{[from Eq. (4-8)]}$$

a. At starting, E_c is zero; $R_s = \frac{V_a - BD}{I_a} - R_a = \frac{120 - 2}{1.5 \times 75} - 0.2 = 1.05 - 0.2 = \mathbf{0.85 \ \Omega}$

b. $R_s = \frac{V_a - (E_c + BD)}{I_a} - R_a = \frac{120 - 30 - 2}{1.5 \times 75} - 0.2 = 0.782 - 0.2$

$= \mathbf{0.582 \ \Omega}$

c. $R_s = \frac{120 - (60 + 2)}{1.5 \times 75} - 0.2 = 0.516 - 0.2 = \mathbf{0.316 \ \Omega}$

d. $E_c = V_a - (I_a R_a + BD) = 120 - [(75 \times 0.2) + 2] = \mathbf{103 \text{ V}}$

* The subject of commercial motor starters, both manual and automatic, is covered in detail in Kosow, *Control of Electric Machines*, Prentice-Hall, 1972, and Kosow, *Electric Machinery and Control*, Prentice-Hall, 1964, Ch. 14.

Note that, in Example 4-10, a progressively decreasing value of starting resistance is required as the motor develops an increased counter emf owing to acceleration. This is the principle of the armature resistance motor starter.

The manner in which a starter is used in conjunction with the three basic types of dc dynamos, used as motors, is shown in Fig. 4-7. The techniques shown here for motor starting are *schematic* diagrams only; as stated previously, commercial forms of manual and automatic starters and controllers differ somewhat from these.

The shunt and compound motors are started with *full field excitation* (i.e., the full line voltage is impressed across the field circuit) in order to develop maximum starting torque ($T = k\phi I_a$). In all three types of dynamos, the armature starting current is limited by a high-power series connected variable starting resistor. In commercial practice, the initial inrush of armature current is generally limited to a higher value than the full-load current, as given in Example 4-10, again to develop greater starting torque, particularly in the case of large motors which have great inertia and which come up to speed slowly.

With the starting arm at position 1 in Fig. 4-7a, the maximum series resistance will limit the armature current on starting to about 150 per cent of its rated value. As the motor slowly increases its speed, the armature develops counter emf and the armature current drops to approximately full load. If the starting arm were left at position 1, the armature current would drop somewhat and the speed would stabilize at some value well below the rated speed. In order to accelerate the motor armature once more, it is necessary to move the arm to position 2. Again, there is an inrush of armature current and the motor rises in speed. This process is continued until the motor armature attains its rated speed, where the counter emf at that speed and flux is sufficient to limit the armature current without the need for a series armature resistance.

It should be noted that all three types (series, shunt, and compound motor), if started with a mechanical load coupled to the armatures, as shown in Fig. 4-7, will accelerate more

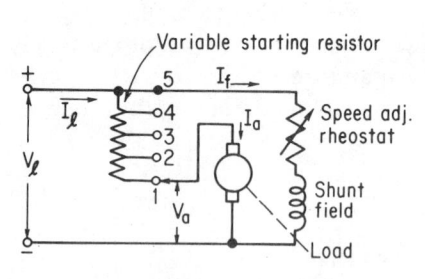

(a) Shunt motor starter (schematic form).

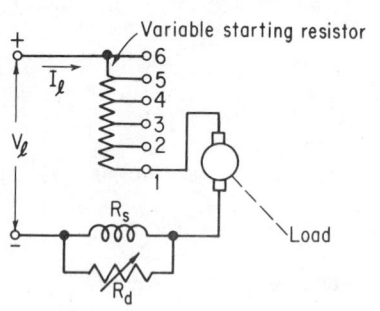

(b) Series motor starter (schematic form).

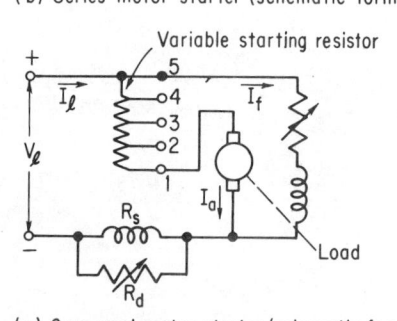

(c) Compound motor starter (schematic form).

Figure 4-7

Schematic starter connections for shunt, series, and compound motors.

slowly than if started without load. The series motor, moreover, should *never* be started and accelerated without a load coupled to its armature (see Section 4-10), although the shunt and compound motors may be started with or without mechanical load.

4-9. ELECTROMAGNETIC TORQUE CHARACTERISTICS OF DC MOTORS

The fundamental torque equation, Eq. (4-4), in which $T = k\phi I_a$, provides a means of predicting how the torque of each of the three types of motors shown in Fig. 4-7 will vary with application of load (i.e., with armature current). The torque-load characteristic of each motor type will be taken up in turn.

It is now assumed that each motor has been properly started and accelerated so that its armature is connected directly across the line terminals, V_l, in Fig. 4-7. What is the effect of increased load on the torque of dc motors?

4-9.1 Shunt Motor

During the starting and the running periods, the current in the shunt field circuit, as shown in Fig. 4-7a, is essentially constant for a given setting of the field rheostat, and consequently the flux (for the present) is also essentially constant. As the mechanical load is increased, the motor slows down somewhat, causing decreased counter emf and increased armature current.* In the basic torque equation, therefore, if the flux is essentially constant and if the armature current increases directly with the application of mechanical load, the torque equation for the shunt motor may be expressed as a perfectly linear relation $T = k'I_a$, shown in Fig. 4-8 (for the shunt motor).

4-9.2 Series Motor

If the shunt field coils were removed from the above dc dynamo and replaced with a full series field winding, the identical armature and construction would produce the torque curve shown in Fig. 4-8 for the series motor. In a series motor, the armature and series field currents are the same (ignoring the effects of a diverter), and the flux produced by the series field, ϕ, is at all times proportional to the armature current, I_a. The basic torque equation for series motor operation therefore becomes $T = k''I_a^2$. To the extent that the field core is unsaturated (i.e., on the linear portion of its magnetization curve) the relation between series motor torque and load current is exponential, as shown in Fig. 4-8. It should be noted that the series motor torque at extremely light loads (low values of I_a) is less than the shunt motor because it develops less flux. For the same armature current at full load, however, its torque is greater, as evidenced by a comparison of the two equations, respectively, shown in Fig. 4-8.

---

* The effect of increased armature current produces an armature mmf called "armature reaction" which, depending on the degree of saturation of the field, will tend to demagnetize and reduce the field flux somewhat. Armature reaction will be covered in detail in Chapter 5.

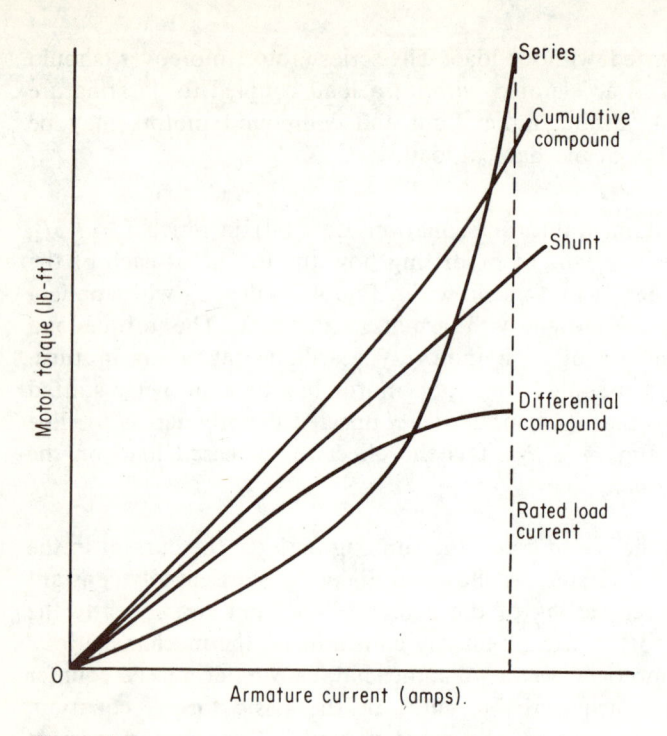

Figure 4-8

Comparison of torque-load characteristics for a given dc dynamo.

4-9.3 Compound Motors

When a combined shunt and series field winding is installed on the poles of the same dc dynamo used above, the series field may be cumulative or differentially compounded. Regardless of compounding, however, the current in the shunt field circuit and the field flux, ϕ_f, during starting or running, is essentially constant. The current in the series field is a function of the load current drawn by the armature.

The basic torque equation for cumulative compound motor operation is $T = k(\phi_f + \phi_s)I_a$, where the series field flux ϕ is a function of the armature current I_a. Starting with a flux equal to the shunt field flux at no load and one which increases with armature current, the cumulative compound motor produces a torque curve which is *always* higher than the shunt motor for the *same* armature current as shown in Fig. 4-8.

For the *differential compound* motor, however, the above torque equation may be written as $T = k(\phi_f - \phi_s)I_a$ where ϕ_s is still a function of I_a and ϕ_f is (presumably) constant. Starting with a flux equal to the shunt field flux at no load, any value of armature current will produce a series field mmf which reduces the total air-gap flux and hence the torque. Thus, the differential compound motor produces a torque curve which is always less than the shunt motor.

EXAMPLE 4-11: A cumulative compound motor is operated as a shunt motor (series field disconnected) and develops a torque of 160 lb-ft when the armature current is 140 A and the field flux is 1.6×10^6 lines. When reconnected as a cumulative compound motor at the same current, it develops a torque of 190 lb-ft. Find
a. The flux increase due to the series field, in per cent.
b. The torque when the compound motor load increases by 10 per cent (assume operation on the *linear* portion of the saturation curve).

Solution:

The data given in the Ex. 4-11 is arranged in tabular form below for convenient reference.

	Torque in lb-ft	Armature Current I_a, in amperes	Field Flux ϕ_f, in lines
Original	160	140	1.6×10^6
Added flux	190	140	$\phi_{f'}$
Final torque	$T_{f'}$	154	$1.1 \times 1.9 \times 10^6$

From the data given, the calculations are worked out as follows:

a. $\phi_{f'} = \phi_{orig}\left(\dfrac{T_{final}}{T_{orig}}\right) = 1.6 \times 10^6 \times \left(\dfrac{190}{160}\right) = 1.9 \times 10^6$ lines

Per cent flux increase $= \dfrac{1.9 \times 10^6}{1.6 \times 10^6} \times 100 - 100 = 118.8 - 100$

$= $ **18.8 per cent**

b. The final field flux is $1.1 \times 1.9 \times 10^6$ lines (due to the 10% increase in load)

The final torque, $T_{f'} = 190 \text{ lb-ft} \left(\dfrac{154 \text{ A}}{140 \text{ A}}\right) \times \left(\dfrac{1.1 \times 1.9 \times 10^6}{1.0 \times 1.9 \times 10^6}\right)$ (4-4)

$= $ **230 lb-ft**

EXAMPLE 4-12: A series motor draws a current of 25 A and develops a torque of 90 lb-ft. Calculate:
a. The torque when the current rises to 30 A if the field is *unsaturated*.
b. The torque when the current rises to 50 A and the increase in current produces a 60 per cent increase in flux.

Solution:

a. $T = kI_a^2 = 90 \text{ lb-ft} \left(\dfrac{30}{25}\right)^2 = $ **129.5 lb-ft**

b. $T = k\phi I_a = 90 \text{ lb-ft} \left(\dfrac{50}{25}\right)\left(\dfrac{1.6}{1.0}\right) = $ **288 lb-ft** (4-4)

4-10. SPEED CHARACTERISTICS OF DC MOTORS

The fundamental speed equation, Eq. (4-6), in which $S = (V_a - I_a R_a)/k\phi$, provides a means of predicting how the speed of each of the motors shown in Fig. 4-7 will vary with application of load. The speed-load characteristic of each motor will be taken up in turn. To simplify the discussion, it is assumed that the brush voltage drop, BD, is zero.

4-10.1
Shunt Motor

Assume that the shunt motor of Fig. 4-7a has been brought up to rated speed and is operating at no load. Since the field flux of the motor (ignoring armature reaction) may be considered *constant*, the speed of the motor may be expressed in terms of the basic speed equation

$$S = \frac{E}{k'\phi_f} = k\frac{V_a - I_a R_a}{\phi_f} \tag{4-6}$$

As mechanical load is applied to the armature shaft, the counter emf decreases and the speed decreased proportionately. But since the counter emf from no load to full load is a change of approximately 20 per cent (i.e., from 0.75 V_a at full load to approximately 0.95 V_a to no load), the motor speed is essentially constant, as shown in Fig. 4-9.

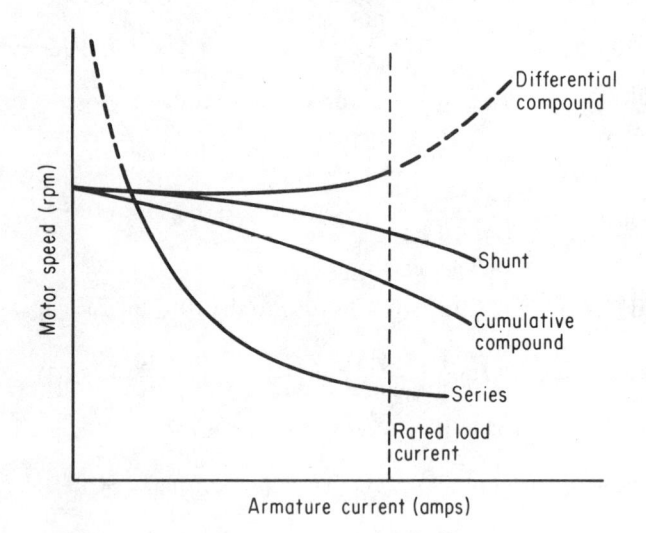

Figure 4-9

Comparison of speed-load characteristics for a given dc dynamo.

4-10.2
Series Motor

The basic speed equation, Eq. (4-6), as modified for the circuit of the series motor, is clearly

$$S = \frac{V_a - I_a(R_a + R_s)}{k\phi} \tag{4-9}$$

where V_a is the voltage applied across the motor terminals; and, since the air-gap flux produced by the series field is proportional to the armature current, only, the speed may be written as

$$S = K'\frac{V_a - I_a(R_a + R_s)}{I_a} \tag{4-10}$$

128
CHAP. FOUR / DC Dynamo Torque Relations—DC Motors

Equation (4-10) gives us an indication of the speed-load characteristic of a series motor. If a relatively small mechanical load is applied to the shaft of the armature of a series motor, the armature current I_a is small, making the numerator of the fraction in Eq. (4-10) large and its denominator small, resulting in an unusually high speed. At no load, therefore, with little armature current and field flux, the speed is extremely excessive. For this reason, series motors are always operated coupled or geared to a load, as in hoists, cranes, or dc traction (railway) service. As the load increases, however, the numerator of the fraction in Eq. (4-10) decreases faster than the denominator increases (the numerator decreases by a product of I_a, compared to the denominator which increases directly with I_a), and the speed drops rapidly, as shown in Fig. 4-9. The dotted line represents that lightly loaded portion of the characteristic in which series motors are not operated.

As shown in Fig. 4-9, excessive speed for a series motor does *not* result in a high armature current (as with shunt and compound motors) which will open a fuse or a circuit breaker and disconnect the armature from the line. Some other method of protection against runaway must be used. Series motors are usually equipped with centrifugal switches, normally closed in the operating range, which open at speeds of approximately 150 per cent of the rated speed.

4-10.3 Cumulative Compound Motor

The basic speed equation for the cumulative compound motor may be written as

$$S = K\left[\frac{V_a - I_a(R_a + R_s)}{\phi_f + \phi_s}\right] \quad (4\text{-}11)$$

and further simplified to

$$S = K\left(\frac{E}{\phi_f + \phi_s}\right) \quad (4\text{-}12)$$

On comparing Eq. (4-12) for the cumulative compound motor with $S = KE/\phi_f$ for the shunt motor, it is evident that, as the load and the armature current increase, the flux produced by the series field also increases, while the counter emf decreases. The denominator therefore increases while the numerator decreases proportionately more than for a shunt motor. The result is that the speed of the *cumulative* compound motor will drop at a faster rate than the speed of the shunt motor with application of load, as shown in Fig. 4-9.

4-10.4 Differential Compound Motor

Equation (4-12) for the cumulative compound motor may be modified slightly to show the effect of the *opposing* field mmf's and the speed is

$$S = \frac{KE}{\phi_f - \phi_s} = k\frac{V_a - I_a(R_a + R_s)}{\phi_f - \phi_s} \quad (4\text{-}13)$$

As the load and I_a increase, the numerator of the fraction in Eq. (4-13) decreases somewhat but the denominator decreases more rapidly. The speed may drop slightly at light loads; but as the load increases, the speed increases. This condition sets up a dynamic instability. As the speed increases, most mechanical loads increase automatically (since *more work is done at a higher speed*) causing an increase in current, a decrease in total flux, and a higher speed, producing still more load. Because of this inherent instability, differential motors are rarely used. In a machinery laboratory where these motors are tested, the student may occasionally observe a condition where a differential motor begins to race away and suddenly drops in speed and reverses direction! This may be explained using Eq. (4-13) and Fig. 4-9. As the counter emf decreases due to decreased mutual flux, the armature current and torque increase is so excessive that the series field flux *exceeds* the shunt field flux, and the motor reverses direction (in accordance with the left-hand rule). It is for this reason that, when starting a differential motor, for testing purposes in the laboratory, care should be taken to short out the series field so that the high starting and armature current will not start the motor in the reverse direction.

The curves of Figs. 4-8 and 4-9 were developed for the same dc dynamo operating from the same no-load point. But, since all electrical machinery is specified in terms of rated (full-load) values, the comparison of torque-load and speed-load characteristics should be made at rated load. If one were to compare dc motors of the same voltage horsepower output and speed rating, the curves of Fig. 4-10 would be obtained. The reader should compare the curves of Fig. 4-10 with those of Figs. 4-8 and 4-9, to verify the characteristics.

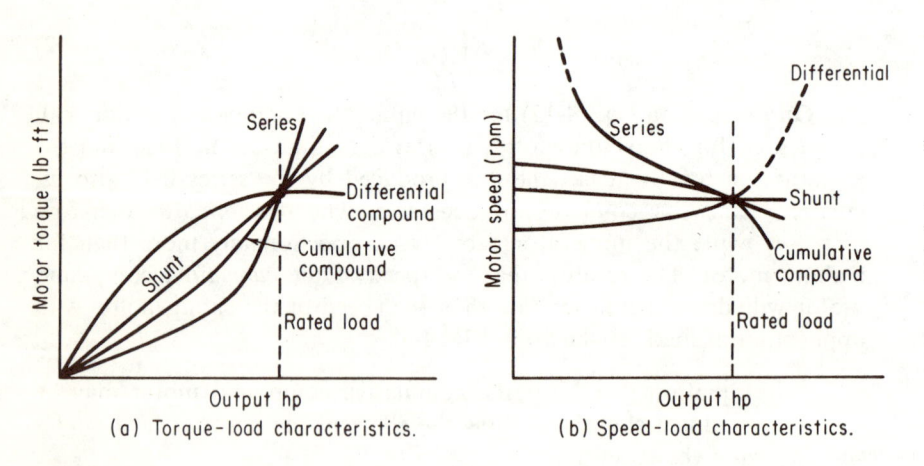

Figure 4-10

Comparison of torque and speed-load characteristics at rated load.

EXAMPLE 4-13: A 230 V, 10 hp, 1250 rpm compound motor has an armature resistance of 0.25 ohm, a combined compensating winding and interpole resistance of 0.25 ohm and a brush volt drop of 5 V. The resistance of the series field is 0.15 ohm and the shunt field resistance is 230 ohms. When connected as a shunt motor, the line current rated load is 55 A and the no-load line current is 4 A. The no-load speed is 1810 rpm. Neglecting armature reaction at rated voltage, calculate
a. Speed at rated load.
b. Internal power in watts and horsepower.

Solution:

a. $I_a = I_l - I_f = 4 \text{ A} - 1 \text{ A} = 3 \text{ A}$

no-load $E_c = V_a - (I_a R_a + BD) = 230 - (3 \times 0.5 + 5) = 223.5 \text{ V}$ at a speed of 1810 rpm

full-load $E_c = V_a - (I_a R_a + BD) = 230 - (54 \times 0.5 + 5) = 198 \text{ V}$;

$$S_{fl} = 1810 \frac{198}{223.5} = \mathbf{1600 \text{ rpm}} \tag{4-5}$$

b. $P_d = E_c I_a = 198 \text{ V} \times 54 \text{ A} = \mathbf{10{,}700 \text{ W}}$

$$\text{hp} = \frac{10{,}700 \text{ W}}{746 \text{ W/hp}} = \mathbf{14.35 \text{ hp}} \tag{4-15}$$

EXAMPLE 4-14: The motor of Example 4-13 is reconnected as a long shunt cumulative compound motor. At rated load (55 A), the compound winding increases the flux per pole 25 per cent. Calculate:
a. The speed at no load (4 A line current)
b. The speed at full load (55 A line current)
c. Internal torques at full load with and without the series field. Use Eq. (4-15.)
d. Internal horsepower of the compound motor based on above flux increase
e. Explain the difference between internal and rated horsepower

Solution:

a. No-load $E_c = V_a - (I_a R_a + I_a R_s + BD)$ (4-8)
$= 230 - [(3 \times 0.5) + (3 \times 0.15) + 5]$
$= \mathbf{223.05 \text{ V}}$

$$S_{nl} = 1810 \left(\frac{223.05}{223.5} \right) = \mathbf{1805 \text{ rpm}} \tag{4-5}$$

b. Full-load $E_c = 230 - [(54 \times 0.5) + (54 \times 0.15) + 5]$ (4-8)
$= \mathbf{190 \text{ V}}$

$$S_{fl} = K \left(\frac{\delta E}{\delta \phi} \right) = 1805 \text{ rpm} \left(\frac{190}{223.05} \right) \times \left(\frac{1.0}{1.25} \right) \tag{4-5}$$
$= \mathbf{1231 \text{ rpm}}$

c. Internal torque of shunt motor [Eq. (4-15)] at full load:

$$T_{\text{shunt}} = \frac{\text{hp} \times 5252}{S} = \frac{14.35 \times 5252}{1600} = \mathbf{47.2 \text{ lb-ft}} \tag{4-15}$$

SEC. 4-10. / *Speed Characteristics of DC Motors*

$$T_{comp} = T_{shunt}\left(\frac{\phi_2}{\phi_1}\right) \times \left(\frac{I_{a2}}{I_{a1}}\right) = 47.2\left(\frac{1.25}{1.0}\right) \times \left(\frac{54}{54}\right)$$
$$= 59.1 \text{ lb-ft}$$

d. Horsepower $= \dfrac{E_c I_a}{746} = \dfrac{190 \times 54}{746} = 13.8$ hp

e. The internal horsepower exceeds the rated horsepower because the power developed in the motor must also overcome the internal mechanical rotational losses (see Fig. 12-1).

Note that the shunt motor develops slightly more horsepower than the compound motor because it is running at a speed higher than rated [Eq. (4-15)]. Note also that the torque of the compound motor is greater than the shunt motor torque because of the added series field flux. It is precisely because of the additional flux that the speed of the compound motor drops [Eq. (4-12)].

EXAMPLE 4-15: The armature circuit resistance of a 25 hp, 250 V series motor is 0.1 ohms, the brush volt drop is 3 V and the resistance of the series field is 0.05 ohms. When the series motor takes 85 A, the speed is 600 rpm. Calculate:
a. The speed when the current is 100 A
b. The speed when the current is 40 A
  Neglect armature reaction and assume that the machine is operating on the linear portion of its saturation curve at all times
c. Recompute speeds in (a) and (b), using a 0.05 ohm diverter at these speeds.

Solution:

a. $E_{c2} = V_a - I_a(R_a + R_s) - BD = 250 - 100(0.15) - 3$ (4-8)
 $= 232$ V when $I_a = 100$ A

$E_{c1} = 250 - 85(0.15) - 3 = 234.3$ V at a speed of 600 rpm when $I_a = 85$ A

$S = K\dfrac{E}{\phi}$, assuming ϕ is proportional to I_a (on the linear portion of saturation curve)

$S_2 = S_1 \dfrac{E_2}{E_1} \times \dfrac{\phi_1}{\phi_2} = 600 \dfrac{232}{234.3} \times \dfrac{85}{100} = \mathbf{506 \text{ rpm}}$

b. $E_{c3} = V_a - I_a(R_a + R_s) - BD = 250 - 40(0.15) - 3 = 241$ V at 40 A

$S_3 = S_1\left(\dfrac{E_{c3}}{E_{c1}}\right) \times \dfrac{\phi_1}{\phi_3} = 600\left(\dfrac{241}{234.3}\right) \times \left(\dfrac{85}{40}\right) = \mathbf{1260 \text{ rpm}}$

c. The effect of the diverter is to reduce the series field current (and flux) to half their previous values.

$E_{c2} = V_a - I_a(R_a + R_{sd}) - BD = 250 - 100(0.125) - 3$
$= 234.5$ V at 100 A

$$E_{c3} = V_a - I_a(R_a + R_{sd}) - BD = 250 - 40(0.125) - 3$$
$$= 242 \text{ V at } 40 \text{ A}$$

$$S_2 = S_1\left(\frac{E_{c2}}{E_{c1}}\right) \times \frac{\phi_1}{\phi_2} = \frac{234.5}{234.3} \times \frac{85 \text{ A}}{\left(\frac{100}{2}\right) \text{ A}} = 1022 \text{ rpm}$$

$$S_3 = S_1\left(\frac{E_{c3}}{E_{c1}}\right) \times \frac{\phi_1}{\phi_3} = 600\left(\frac{242}{234.3}\right) \times \frac{85 \text{ A}}{\left(\frac{40}{2}\right) \text{ A}} = 2630 \text{ rpm}$$

Observe that the 50 per cent reduction of the series field current, the effect of diverting current away from the field, has resulted in a sharp rise in speed of approximately 200 per cent of the original values (computed without the diverter).

4-11. SPEED REGULATION

The speed regulation of a motor is defined as:* *the change in speed from rated to zero load, expressed in per cent of rated load speed.* In equation form, the speed regulation becomes

$$\text{Per cent speed regulation} = \frac{S_{nl} - S_{fl}}{S_{fl}} \times 100 \quad (4\text{-}14)$$

From an examination of the curves of Fig. 4-10b, it is evident that shunt motors may be classified as motors of fairly constant speed, whose speed regulation is good (a small percentage). The speed regulation of the cumulative compound motor is poorer than the shunt motor, and its speed regulation is a higher percentage. The series motor speed regulation is extremely poor (since it has an infinite no-load speed). Both the cumulative and series motors are considered variable speed motors (see Section 12-20). The differential compound motor has a negative speed regulation, which may always be associated with load instability.

EXAMPLE 4-16: Compute the per cent speed regulation for the motors of:
a. Example 4-13.
b. Example 4-14.
c. Example 4-15 (assume the 40 A current as no load, and the 100 A current as full load).

Solution:

a. Per cent SR (shunt) $= \dfrac{S_{nl} - S_{fl}}{S_{fl}} \times 100 = \dfrac{1810 - 1600}{1600} \times 100$

$= \textbf{13.12 per cent}$ \quad (4-14)

b. Per cent SR (compound) $= \dfrac{1805 - 1231}{1231} \times 100 = \textbf{46.6 per cent}$

c. Per cent SR (series) $= \dfrac{1260 - 506}{506} \times 100 = \textbf{149 per cent}$

* ASA Standard C50, *Rotating Electrical Machinery*. Note the similarity between this definition and the definition of voltage regulation, Eq. (3-9).

**4-12.
EXTERNAL TORQUE,
RATED
HORSEPOWER,
AND SPEED**

It should be noted that, in comparing the motors of Fig. 4-10, they were compared in terms of output horsepower values (i.e., 1 hp = 33,000 ft-lb/min) as a measure of the motor's rate of doing mechanical work. Rated output values of horsepower for a motor are more significant than the internal horsepower ($E_c I_a/746$ watts per hp) developed by the armature in Eq. (4-7), for reasons outlined in Section 4-6. In specifying and selecting motors, therefore, a question may arise as to the amount of external torque available at the pulley or the motor shaft to perform useful work at any rated speed. The equation expressing the relationship between external torque, horsepower, and speed is derived as follows:

Let F equal the useful force developed by all the conductors on the armature producing electromagnetic torque
r equal the radius of the armature, in feet
n equal the number of revolutions of the armature
t equal the time (1 min) for the armature to rotate n times

The work done per revolution of the armature is then

$$W = F \times 2\pi r \text{ ft-lb/revolution}$$

and the power or rate of doing work of a revolving armature is, in ft-lb/min,

$$P = \frac{W}{t} = (2F\pi r \text{ ft-lb/rev}) \times \frac{n}{t} \text{ rpm}$$

But since torque $T = Fr$ [Eq. (4-2)], and speed $S = n/t$ by definition, then

$$P = \frac{W}{t} = 2\pi T S \text{ ft-lbs/min}$$

and

$$\text{hp} = \frac{2\pi T S \text{ ft-lbs/min}}{33,000 \text{ ft-lbs/min/hp}}$$

or

$$\text{hp} = \frac{TS}{5252} \qquad (4\text{-}15)$$

The above relation permits computation of the internal electromagnetic torque [determined by Eq. (4-3)] at a given speed from Eq. (4-7) in combination with Eq. (4-15). It also permits computation of the torque available at the pulley, given the rated speed and horsepower of a motor. Both of these computations are illustrated by the following example:

EXAMPLE 4-17: From the calculated values of rated speed and internal power computed in Example 4-13, calculate
a. The internal torque
b. The external torque from rated hp and speed given in Example 4-13
c. Account for the differences

Solution:

a. $T_{\text{int.}} = \dfrac{\text{hp} \times 5252}{S} = \dfrac{14.35 \times 5252}{1595} = \mathbf{47.25 \text{ lb-ft}}$ (4-15)

b. $T_{\text{ext.}} = \dfrac{\text{hp} \times 5252}{S} = \dfrac{10 \times 5252}{1250} = \mathbf{42.0 \text{ lb-ft}}$ (4-15)

c. Internal horsepower is developed as a result of electromagnetic torque produced by energy conversion. Some of the mechanical energy is used internally to overcome mechanical losses of the motor, reducing the torque available at its shaft to perform work

4-13. REVERSAL OF DIRECTION OF ROTATION

In order to reverse the direction of rotation of any d-c motor, it is necessary to reverse the direction of current through the armature with respect to the direction of the magnetic field. For either the shunt or series motor, this is done simply by reversing either the armature circuit with respect to the field circuit or vice versa. Reversal of *both* circuits will produce the *same* direction of rotation.

It might appear that since the field circuit carries less current than the armature circuit, the former would be the one selected for reversal. However, in designing automatic starters and control equipment, the *armature* circuit is usually selected for reversal because: (1) the field is a highly inductive circuit (see Fig. 2-7), and frequent reversals produce high induced emf's and pitting of the switch contacts which serve to accomplish field circuit reversal; (2) if the shunt field is reversed, the series field must also be reversed, otherwise a cumulative compound motor will be differentially connected; (3) the armature circuit connections are normally opened for purposes of dynamic, regenerative, or plugging braking, and since these connections are available, they may as well be used for reversal; and (4) if the reversing switch is defective and field circuit fails to close, the motor may "run away."

In the case of the compound motor, therefore, reversing only the armature connections achieves a reversal of direction of rotation for either the long-shunt or short-shunt connections, as shown in Fig. 4-11, without changing the direction of the current in the fields.

For the above reasons therefore, reversal of rotation dictates reversal of armature connections *only*, as shown in Figs. 4-11a and b.

(a) Short shunt connection.

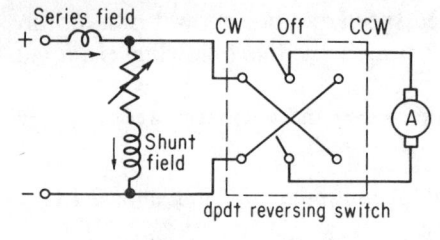

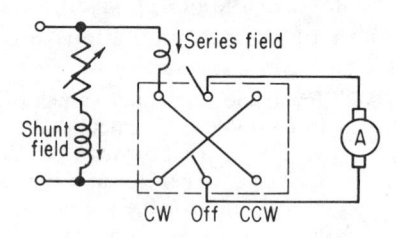

(b) Long shunt connection.

Figure 4-11

Reversal of direction of long or short shunt compound motor.

4-14.
EFFECT OF ARMATURE REACTION ON SPEED REGULATION OF ALL DC MOTORS

Armature reaction (as defined in Section 2-8 and in Chapter 5) is the effect of the mmf produced by the armature conductors ($I_a N_a$) in reducing and distorting the mutual air-gap flux ϕ_m produced by the field winding (series and shunt fields). The fundamental speed equation, Eq. (4-6), indicates that a reduction of the field flux in the denominator of this equation will cause an increase in speed. It will be shown in the next chapter that the extent and effect of armature reaction varies directly with the load or with the armature current, I_a. As *any* dc motor (regardless of type) is loaded, the effect of armature reaction is to reduce the air-gap flux and (depending on the degree of saturation) tend to *increase the motor speed*. An examination of the speed-load curves shown in Fig. 4-9 indicates that the speed regulation of *each* of the commercial motor types (shunt, series, and cumulative compound) would be *improved*, somewhat, by this effect (if not too pronounced so as to cause negative speed regulation). In the case of the shunt motor, for example, since armature reaction increases with load, the decrease in flux and increase in speed with load may increase the load to such an extent that its characteristic may tend to be the same as that of the differential compound motor, shown in Fig. 4-9. A shunt motor operating with a weak field and without some means of compensating for armature reaction (as discussed in Chapter 5) is particularly susceptible to load instability and runaway.

BIBLIOGRAPHY

Ahlquist, R. W. "Equations Depicting the Operation of the dc Motor," *Electrical Engineering*, April 1955.

Alger, P. L. and Erdelyi. E. "Electromechanical Energy Conversion," *Electro-Technology*, September 1961.

Carr, C. C. *Electrical Machinery*, New York: John Wiley & Sons, Inc., 1958.

Crosno, C. D. *Fundamentals of Electromechanical Conversion*, New York, Harcourt, Brace, Jovanovich, Inc, 1968.

Daniels. *The Performance of Electrical Machines*, New York: McGraw-Hill Inc., 1968.

D-C Generators and Motors (ASA 050.4.) New York: American Standards Association.

Emunson, B. M. and Ward. A. J. "An Evaluation of the New 'Industrial' dc Motors and Generators," *Electrical Manufacturing*, June 1958.

Fitzgerald, A. E. and Kingsley, C. *The Dynamics and Statics of Electromechanical Energy Conversion*, 2nd Ed., New York: McGraw-Hill, 1961.

Fitzgerald, A. E., Kingsley, Jr. C., and Kusko, A. *Electric Machinery*, 3rd. Ed. New York: McGraw-Hill, Inc., 1971.

Gemlich, D. K. and Hammond, S. B. *Electromechanical Systems*, New York: McGraw-Hill, 1967.

Hindmarsh, J. *Electrical Machines*, Elmsford, N.Y.: Pergamon Press, 1965.

Jones, C. V. *The Unified Theory of Electrical Machines*, New York: Plenum Publishing Corp., 1968.

Kloeffler, S. M., Kerchner R. M., and Brenneman. J. L. *Direct Current Machinery*, Rev. ed., New York: The Macmillan Company, 1948.

Liwschitz, M. M., Garik M., and Whipple. C. C. *Direct Current Machines*, 2d. ed., Princeton, N. J.: D. Van Nostrand Co., Inc., 1947.

Majmudar, H. *Introduction to Electrical Machines*, Boston: Allyn and Bacon, 1969.

Millermaster, R. A. *Harwood's Control of Electric Motors*, 4th Ed., New York: Wiley/Interscience, 1970.

Puchstein, A. F. *The Design of Small Direct Current Motors*, New York: Wiley/Interscience, 1961.

Robertson, B. L. and Black, L. J. *Electric Circuits and Machines*, 2nd Ed., Princeton, N. J.: C. Van Nostrand Co., Inc., 1957.

Seely, S. *Electromechanical Energy Conversion*, New York: McGraw-Hill, 1962.

Selmon. *Magnetoelectric Devices: Transducers, Transformers and Machines*, New York: Wiley/Interscience, 1966.

Siskind, C. S. *Direct-Current Machinery*, New York: McGraw-Hill, 1952.

Skilling, H. H. *Electromechanics: A First Course in Electromechanical Energy Conversion*, New York: Wiley/Interscience, 1962.

Smeaton, Motor *Applications and Maintenance Handbook*, New York: McGraw-Hill, 1969.

Thaler, G. J. and Wilcox, M. L. *Electric Machines: Dynamics and Steady State*, New York: Wiley/Interscience, 1966.

Veniott. *Fractional and Subfractional Horsepower Electric Motors*, 3rd Ed., New York: McGraw-Hill, 1970.

White, D. C. and Woodson, H. H. *Electromechanical Energy Conversion*, New York: Wiley/Interscience, 1959.

QUESTIONS

4-1. Using the equation, $I_a = (V_a - E_c)/R_a$, explain:
 a. why it is impossible for E_c to equal V_a
 b. what proportion of V_a is normally represented by E_c and $I_a R_a$, respectively, at full load.

4-2. a. What is the relation between electromagnetic force and electromagnetic torque?
 b. What is the relation between torque and work?

4-3. Using Fig. 4-3, a conductor located exactly midway in the interpolar region, give two reasons why it will develop no useful torque.

4-4. a. Distinguish between torque and speed of a motor.
 b. What two factors determine motor torque?
 c. Distinguish between developed torque and torque available at the pulley. Which is greater and why?

4-5. a. Explain why a small change in motor speed and counter emf will produce correspondingly larger changes in armature current.
 b. If the speed is increased, what effect does this have on
  1. the counter emf? (Why?)
  2. the armature current?
 c. Why is armature current often employed as an indication of motor load and speed?

4-6. Using Eq. (4-6), explain the effect on speed of a shunt motor when
 a. armature current is increased
 b. counter emf is decreased
 c. field flux (field current) is increased.

4-7. For a given applied voltage across the armature, V_a, explain:
 a. the significance of a high ratio of E_c to V_a on efficiency
 b. why the power developed ($E_c I_a$) is not merely a function of load (I_a).

4-8. Using the same steps outlined in Sec. 4-7, explain the effect, qualitatively, of an *increase* in field current on the speed of a motor.

4-9. Since the armature current, I_a, of a shunt motor is affected by the change in field current and flux, is it possible to control motor torque in a way which is independent of armature current? Explain.

4-10. Assuming a shunt motor is connected (unfused) to zero resistance lines capable of supplying an infinite current, if the field circuit is opened, what would happen?

4-11. Using equation $I_{st} = (V_a - E_c)/R_a$, for the starting current of a shunt motor, explain:

a. why a starting resistance is necessary
b. the development of Eq. (4-8)
c. why starting current is limited using Eq. (4-8).
d. why progressively decreasing values of starting resistance are required.

4-12. Using Fig. 4-7, explain why all dc motors are started:
a. with maximum resistance in series with armature
b. with maximum field excitation.

4-13. Explain why the series motor must be started with a mechanical load coupled to its armature.

4-14. For the series, cumulative compound, shunt and differential compound motor:
a. state the torque equation for the particular motor
b. show how this torque equation produces the characteristic shown in Fig. 4-8.

4-15. For the differential compound, shunt, cumulative compound and series motor:
a. state the speed equation for each particular motor
b. show how this equation produces the speed characteristic obtained in Fig. 4-9.

4-16. Compare the family of curves shown in Fig. 4-10a and b with those of Figs. 4-8 and 4-9, and
a. explain the advantages of the former over the latter
b. show where starting torque and starting speed should appear on these curves.

4-17. Define:
a. starting torque
b. full load torque
c. no load torque
d. speed regulation
e. internal torque
f. external torque.

4-18. Why are motors specified on the basis of rated speed and rated output hp rather than no load speed and internal hp?

4-19. Give four reasons why the armature connections are selected for reversal of motor direction rather than the field connections of dc motors.

4-20. What is the effect of armature reaction on the speed regulation of all dc motors?

PROBLEMS

4-1. The conductors of a dynamo armature have an axial length of 12 in. When carrying a current of 80 A, the field flux density is adjusted to 61,000 lines/in^2. Calculate:
a. The force developed by each current-carrying conductor
b. The total force developed, given a total of 60 active conductors on the armature
c. The total torque developed if the armature diameter is 18 in.

4-2. A dc motor armature has 48 slots, a two-layer full-coil simplex lap winding (1 coil/slot) in which each coil has 42 turns. The four field poles span 78 per cent of the armature circumference and produce a uniform flux density of 56,000 lines/in². The armature core has a diameter of 14 in and an axial length of 16 in, but the slots are skewed at an angle of 20° with respect to the shaft. The current per conductor is 20 A. Calculate:
a. The number of active conductors
b. The active length of each conductor
c. The total electromagnetic force developed by the armature conductors
d. The torque tending to produce rotation.

4-3. Calculate the torque in Problem 4-2 assuming the flux density is increased by 10 per cent and the current is reduced by 20 per cent.

4-4. Derive Eq. (4-3) from an examination of Example 4-3.

4-5. Solve Problem 4-2 using Eq. (4-3) as verification of the validity of Example 4.3.

4-6. Solve illustrative Example 4-3 in the text by using Eq. (4-3) to determine total torque developed.

4-7. A six-pole shunt motor has an armature containing a simplex *wave* winding which draws a current of 80 A. Each pole produces a flux density of 52,000 lines/in.² and an area of 19.25 in.². The armature has a total of 300 active conductors. Calculate:
a. The total average electromagnetic torque
b. The torque if the armature is wound as a simplex lap winding.

4-8. A 220 V dc shunt motor has a 5 V brush drop, an armature resistance of 0.2 Ω and a rated armature current of 40 A. Calculate:
a. The voltage generated in the armature under these conditions of load applied to the armature shaft
b. Power developed by the armature in watts
c. Mechanical power developed by the armature in hp.

4-9. A 125 V, dc shunt motor has a 2 V brush drop, an armature resistance of 0.1 Ω and a counter emf of 118 V when rated load is applied to the shaft of the motor armature. Calculate:
a. The rated load current drawn by the armature
b. The total armature circuit voltage drop.

4-10. A 10 hp motor has an armature resistance of 0.05 Ω, a 4.1-V brush drop and develops a mechanical power of 12 hp at rated armature current of 80 A at 120 V. Calculate:
a. Counter emf from mechanical power developed in the armature
b. Counter emf from armature circuit voltage drops
c. Account for discrepancies between (a) and (b).

4-11. A 220 V shunt motor has a speed of 1200 rpm, an armature resistance of 0.2 Ω and a 4 V brush drop. The motor draws an armature current of 20 A when connected to a rated supply voltage for a given load. As the mechanical load is increased, the field flux is increased by 15 per cent, causing the metered armature current to rise to 45 A. Calculate:
a. Counter emf at the 20 A load

b. Counter emf at the 45 A load
c. Speed at the 45 A load
d. Internal hp developed by the armature at the 20 A and 45 A loads, respectively.

4-12. A 50 hp, 230 V shunt motor has a brush volt drop of 5 V and an armature resistance of 0.05 Ω. The field circuit resistance is 115 Ω. At no load, the motor draws 12 A and has a speed of 1300 rpm. Calculate:
   a. Motor speed at rated line current (see Appendix Table A-3 and correct line current accordingly)
   b. Motor speed at an armature current which is half rated
   c. Speed regulation
   d. Mechanical power developed by the armature at rated load and output hp
   e. Compare computed full-load hp with rated hp (50 hp) and account for the differences.

4-13. Using the developed hp computed at full load in Problem 4-12, calculate
   a. The developed torque in lb-ft
   b. The half-load hp and torque using any data from Problem 4-12
   c. The no-load hp and torque, using data from Problem 4-12
   d. Tabulate the following parameter for full-, half- and no-load conditions: speed, hp, and torque developed by the armature. Account for differences in variation between these parameters.

4-14. A shunt motor has a shunt field resistance of 600 Ω, an armature circuit resistance of 0.1 Ω and a brush voltage drop of 5 V. The motor nameplate rating values are 600 V, 1200 rpm, 100 hp; full-load efficiency is 90 per cent. At these rated values, calculate:
   a. Motor line current
   b. Motor line current from Appendix Table A-3 data. Account for differences
   c. Field and armature currents using line current computed in (a)
   d. Counter emf at rated speed
   e. Internal power and the internal torque developed
   f. Output torque
   g. Ratio of output to internal torque (compare with efficiency and account for differences).

4-15. Calculate the motor speed at one-half rated hp output if the efficiency in Problem 4-14 is 85 per cent (at one-half rated hp output).

4-16. A 10 hp, 240 V *Series* motor has a line current of 38 A and a rated speed of 600 rpm. The armature circuit and series field resistances, respectively, are 0.4 and 0.2 Ω. The brush voltage drop is 5 V. Assume that the motor is operating on the linear portion of its saturation curve below rated armature current. Calculate:
   a. Speed when the load current drops to 20 A at half rated load
   b. The no-load speed when the line current is 1 A
   c. The speed at 150 per cent rated load when the line current is 60 A and the series field flux is 125 per cent of the full-load flux due to saturation.

4-17. Repeat Problem 4-16 using a 0.2 Ω diverter across the series field. Tabulate results of Problems 4-16 and 4-17 for ready reference and comparison with and without diverter at no load, 50 per cent, 100 per cent, and 150 per cent rated load. Draw inferences as to effects of diverter. (Hint: first compute full-load speed using diverter.)

4-18. A 15 hp, 240 V, 500 rpm *Series* motor develops an internal torque of 170 lb-ft at its rated line current of 55 A at rated speed. Assume a straight line saturation curve at currents below rated load and calculate the internal torques when the armature current drops to
  a. 40 A
  b. 25 A
  c. 10 A
  d. Calculate the internal torque at 125 per cent rated load if the increase in armature current causes a 60 per cent increase in series field flux

4-19. Given the following measurements for the motor of Problem 4-18: armature resistance 0.25 Ω, series field resistance 0.1 Ω, brush voltage drop 3 V. Calculate for each of the values of load given in Problem 4-18
  a. Motor speed (at each load)
  b. Internal hp developed at each load
  c. For load currents of 10 A, 25 A, 40 A, 55 A and 68.75 A, *tabulate* in summary form the internal torque, speed, and internal hp computed in Problems 4-18 and 4-19
  d. Since hp is the product of torque and speed, explain why the hp output and developed internal hp *increases* with increased load current, despite the high no-load and light-load speeds
  e. Explain the discrepancy between internal hp developed at rated load and the rated output hp (15 hp)

4-20. A 10 hp, 1800 rpm, 120 V dc shunt motor has an armature circuit resistance of 0.05 Ω and a shunt field circuit resistance of 60 Ω. The brush voltage drop is 2 V. Calculate:
  a. The line current if the motor were connected directly across the 120 V mains without a protective armature starting resistance
  b. The resistance of the starting resistor which will limit the current to a 50 per cent armature current overload (Appendix Table A-3) at the instant of starting (zero counter emf)
  c. Full load counter emf
  d. Three resistance taps on the starting resistor if each step is to be limited to $\frac{1}{4}$, $\frac{1}{2}$ and $\frac{3}{4}$ of the full-load counter emf and speed, respectively (assume that at each step the armature current increases to a maximum of 1.5 times rated current and must be limited to that value by the resistor)
  e. Speed at each of the above values
  f. Tabulate resistance steps vs. acceleration speed from zero to full load computed above

4-21. A 25 hp, 600 rpm, 240 V dc series motor has a rated line current of 89 A, an armature resistance of 0.08 Ω, a series field resistance of 0.02 Ω and a brush voltage drop of 4 V. It is desired that the starting torque of the motor shall be 225 per cent of the rated load torque. Calculate:

a. Resistance of the first step of the accelerating resistor (assuming a linear saturation curve in which series field flux is proportional to series field current)
b. The rated-load motor speed if the starting resistance (computed above) is inserted in series with the armature when the motor is running.

ANSWERS

4-1(a) 5.18 lb/conductor (b) 311 lb (c) 233 lb-ft 4-2(a) 3140 conductors (b) 15.05 in. (c) 4680 lb (d) 2730 lb-ft 4-3. 2400 lb-ft 4-5. 2740 lb-ft 4-6. 1500 lb-ft 4-7(a) 84.5 lb-ft (b) 28.17 lb-ft 4-8(a) 207 V (b) 8280 W (c) 11.1 hp 4-9(a) 50 A (b) 5 V 4-10(a) 111.8 V (b) 111.9 V (c) same 4-11(a) 212 V (b) 207 V (c) 1020 rpm (d) 5.675 hp 12.48 hp 4-12(a) 1250 rpm (b) 1275 rpm (c) 4 per cent (d) 51.6 hp 4-13(a) 216.2 lb-ft (b) 26.4 hp; 109 lb-ft (c) 3.01 hp, 12.2 lb-ft 4-14(a) 138 A (b) 136.5 A (c) 137 A (d) 581.3 V (e) 466 lb-ft (f) 437.5 lb-ft (g) 0.9 4-15 1191 rpm 4-16(a) 1200 rpm (b) 25,185 rpm (c) 450 rpm 4-17(a) 2420 rpm (b) 50,400 rpm (c) 735 rpm 4-18(a) 90 lb-ft (b) 35.2 lb-ft (c) 5.62 lb-ft (d) 244 lb-ft 4-19(a) 705, 1152, 2720, and 432 rpm (b) 16.2, 12.1, 7.73, 2.91, 20.1 lb-ft 4-20(a) 2402 A (b) 1.012 Ω (c) 114.3 V (d) 0.756 Ω, 0.498 Ω, 0.141 Ω (e) 4.50, 900, and 1350 rpm 4-21(a) 1.665 Ω (b) 209 rpm

FIVE

armature reaction and commutation in dynamos

5-1.
GENERAL

The emphasis of presentation used in this volume is that all types of generators and motors have a great deal in common, whether motor or generator, whether dc or ac. Perhaps the most dramatic evidence of this is the fact that a single electric machine may be constructed to operate as a dc or ac generator or motor.* This unifying principle was stressed in Chapter 1, Electromechanical Fundamentals, and Chapter 2, Dynamo Construction and Windings, where it was shown that the only modifications necessary to change an electric dynamo from dc to ac operation are in the external connections of the

* First described by Brown, Kusko, and White as "A New Educational Program in Energy Conversion," *Electrical Engineering*, Feb. 1956, pp. 180–185. Other articles describing this machine and its laboratory use in teaching are: (a), D. C. White and A. Kusko, "A Unified Approach to the Teaching of Electromechanical Energy Conversion," *Electrical Engineering*, November, 1956, pp. 1028–1033; and (b), D. C. White and H. H. Woodson, "A New Electromechanical Energy Conversion Laboratory," *Transactions AIEE*, 1957, Paper 57-603.

windings and in the use of a commutator or of slip-rings.* In the spirit of this principle, armature reaction will be discussed from a point of view of application to all machines, followed by commutation as it applies to dc and ac commutator machines.

5-2. MAGNETIC FIELD PRODUCED BY ARMATURE CURRENT

All armatures, whether rotating or stationary, carry ac current*. In machines of large capacity, the current in the armature conductors is appreciable. In all dynamos, the armature conductors are embedded in slots in an iron armature core where they produce a flux or mmf in proportion to the amount of current they carry. In both dc and ac generators, relative motion is produced between the conductors and the magnetic field such that the direction of induced emf and current in conductors lying under a given pole is *opposite* to that of conductors lying under an *opposite* pole. The same holds true for dc and ac motors in order to produce continuous rotation in the same direction. The current which flows in the armature conductors (as a result of the voltage applied to the motor) must be *reversed* as the conductor moves under a pole of *opposite* polarity. In the case of the dc motor, this is accomplished by the commutator, which converts the dc applied to the brushes to ac in the armature conductors. In the case of the ac motor, this is accomplished by the impressed sinusoidal voltage.

A two-pole *universal* dynamo armature is shown in Fig. 5-1a in which the **mmf** produced by the instantaneous direction of current flow in the armature conductors is shown. Each series-connected current-

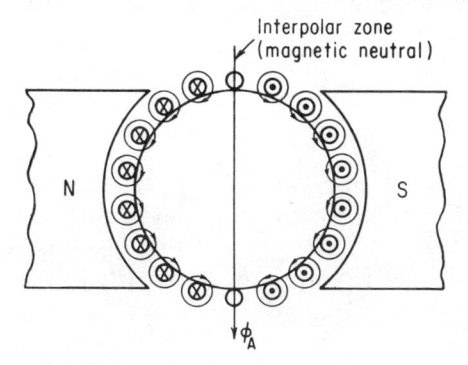

(a) Mmf produced by current carrying armature conductors.

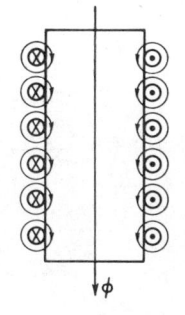

(b) Equivalent iron core electromagnet.

Figure 5-1

Armature flux in universal dynamo.

* The only exception to this statement, oddly enough, is the first generator discovered by Faraday, the homopolar machine (see footnote, Section 1-11), and MHD generators based on this principle (Section 11-4).

carrying conductor produces the same **mmf** under a given pole and an opposite **mmf** under an opposite pole. The net effect of the individual **mmf**s is the production of a resultant armature flux in the direction shown in the figure. The armature flux produced is analogous to that produced in the equivalent iron-core electromagnet shown in Fig. 5-1b, and the direction of the magnetic field is in accordance with the right-hand corkscrew rule.

The resultant armature flux produced in Fig. 5-1a lies in the so-called *interpolar zone* or *magnetic neutral* between the poles, *perpendicular* to the main field flux, If the field poles of Fig. 5-1a were rotated clockwise, the magnetic neutral would shift clockwise to the same degree since, by definition, it is *always perpendicular* to the magnetic field.

5-3. EFFECT OF ARMATURE FLUX ON FIELD FLUX

There are, of course, two primary **mmf**s and fluxes operating in the dynamo shown in Fig. 5-1a.* One is the armature flux, described above, and the other is the field flux produced by the field windings around the N and S poles of the bipolar machine.

The *interaction* of the two fluxes is shown in Fig. 5-2. The armature flux is shown in Fig. 5-2a with its resultant magnetic field phasor ϕ_a, produced by the armature mmf $(I_a N_a)$. The main field flux is shown in Fig. 5-2b with its phasor ϕ_f, produced by the field **mmf** $(I_f N_f)$. The phasor sum of the two **mmf**s is shown in Fig. 5-2c as a resultant flux, ϕ_r. Note that in this figure the field flux entering the armature is not only *shifted* but also *distorted*. The shift has caused the magnetic neutral to be shifted *clockwise* (but still perpendicular to the resultant field flux). The distortion shown in Fig. 5-2c has produced crowding of the flux

(a) Armature flux. (b) Field flux. (c) Resultant distortion of field flux produced by armature flux.

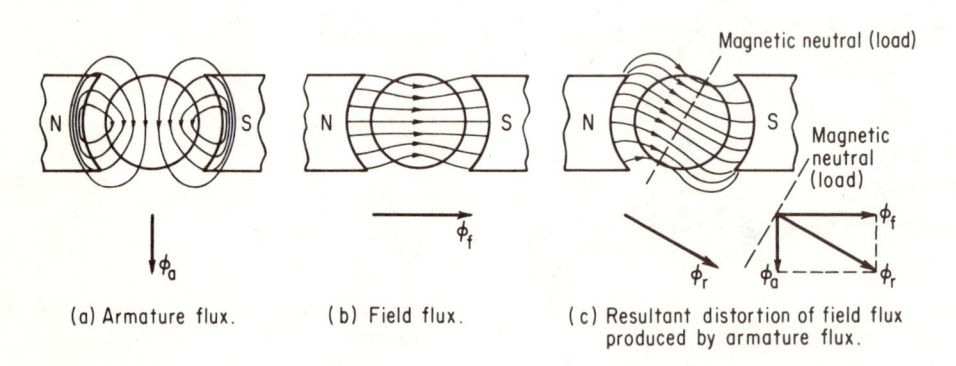

Figure 5-2

Interaction of armature and field flux to shift magnetic neutral.

* This follows from the concept that practically all dynamos are doubly excited electric machines which differ only in the nature of their excitation.

(increase of flux density) in one pole tip and a reduction of the flux (decreased flux density) in the other tip of the same pole.

The phasor diagram of Fig. 5-2(c) might give the impression that the resultant flux, ϕ_r, is now greater than the original field flux, ϕ_f, having been augmented by the perpendicular armature flux, ϕ_a. This is not true, however, because of the effect of saturation of one of the pole tips of each pole. Assuming that the poles were normally saturated, the effect of a shift of the magnetic neutral is to create a path of *greater* reluctance to the resultant flux, ϕ_r, and increased saturation of part of each of the poles. The net effect of armature reaction is therefore twofold: (1) a *distortion* of the main field flux in which the mutual air-gap flux is no longer uniformly distributed under the poles and the neutral plane is shifted; and (2) a *reduction* of the main field flux.*

A universal multipolar dynamo is shown in Fig. 5-3, where the armature conductors are shown moving with respect to the magnetic field, or vice versa. The direction of current in the armature conductors shown is the same as that for Fig. 5-1. Figure 5-3a shows the flux distribution produced by the field flux ϕ_f under the N and S poles, respectively. The armature flux, ϕ_a, produced by the armature current-carrying conductors is shown in Fig. 5-3b. Note that the armature flux is a maximum in the magnetic neutral planes and that its maximum is displaced from the field

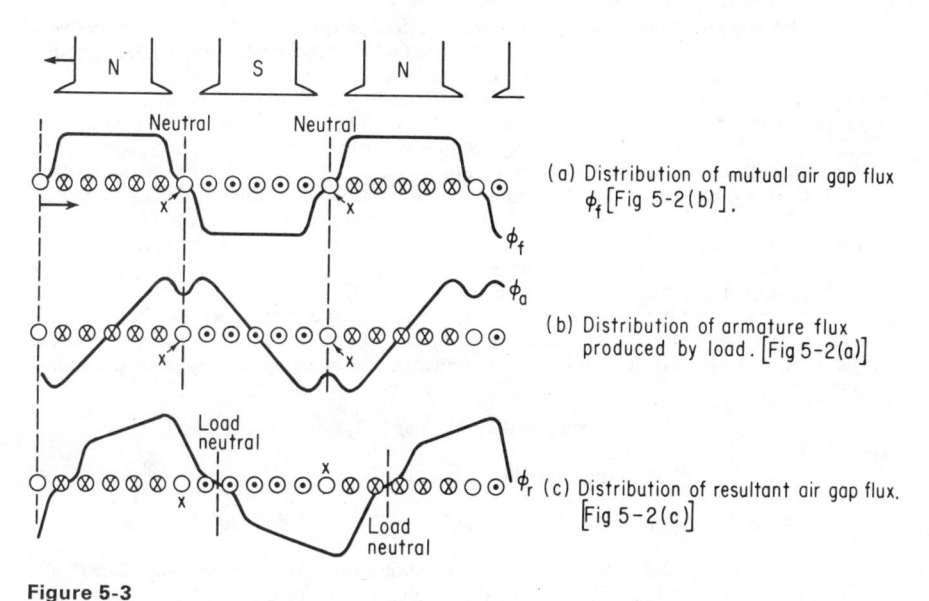

(a) Distribution of mutual air gap flux ϕ_f [Fig 5-2(b)].

(b) Distribution of armature flux produced by load. [Fig 5-2(a)]

(c) Distribution of resultant air gap flux. [Fig 5-2(c)]

Figure 5-3

Distortion of air-gap flux in multipolar machines.

* The reduction in field flux is responsible, in part, for the drop in voltage of a generator (Section 3-13) with increased load, and the increase in speed of a motor (Section 4-14) with increased load.

flux by 90°. For this reason, the armature flux is sometimes called the *quadrature* or *cross-magnetizing* flux.* This terminology may be verified in Fig. 5-2 also, where the armature flux is at right angles to and cross-magnetizes the field flux. The graphical sum of the two waveforms is shown in Fig. 5-3c where the distortion of the resultant field waveshape is evident. The (almost) square-wave appearance of the main field flux distribution shown in Fig. 5-3a has been distorted by the (almost) triangular wave appearance of the armature flux to produce a flux distribution which is no longer uniform under the poles. Instead, it tends to crowd toward the right-hand side of each pole. Since this increase of saturation or flux density on the right-hand side of each pole is produced by the direction of the armature current only, it is independent of (1) the direction of rotation, (2) whether the dynamo is a motor or a generator, and (3) whether the dynamo is ac or dc.

A harmonic analysis† of the waveforms would also indicate a reduction or *subtractive effect* even produced by the armature flux over the main field flux, ignoring the effects of saturation. This same resultant flux ϕ_r waveshape may be demonstrated in the laboratory, using a special genera-

---

* Students occasionally ask the following question, particularly in regard to ac dynamos (although it applies equally to dc machines): "If ac flows in the armature, why isn't there any transformer action developed, especially in view of the closed iron circuits of both the armature and the field?"

The answer to this question is a verification of the quadrature relation between the fields of the stator and the rotor. The mere fact that these fields *are* in space quadrature implies little or no coupling between the field and armature circuits when ac flows in either or both. In effect, there is zero mutual flux and this is verified continuously in actual machine operation. If coupling did in fact exist between the two circuits (see Transformers, Ch. 13), each time a change occurred in the armature current of a dc generator or motor due to load, it would produce a change of current in the field circuit. This does not occur either in dc dynamos or synchronous ac dynamos. It *does* occur, however, in single- and polyphase *asynchronous* dynamos (induction types) for the simple reason that the rotor and stator fluxes are *not* in quadrature. It is precisely for this reason that transformer theory is sometimes employed in explaining polyphase and single-phase motor operation (see Chs. 9 and 10).

† A complete harmonic analysis is beyond the scope of this text. For our purposes here, it may be assumed that the curve of ϕ_f is a square wave and that of ϕ_a is a triangular wave, displaced 90° from the square wave. The equation for ϕ_f is of the square wave form

$$\phi_f = \phi_m \sin \omega t + \frac{\phi_m}{3} \sin 3\omega t + \frac{\phi_m}{5} \sin 5\omega t, \text{ etc.}$$

The equation for ϕ_a is of the triangular wave form

$$\phi_a = \phi_m \sin(\omega t + 90°) - \frac{\phi_m}{9} \sin(3\omega t + 90°) - \frac{\phi_m}{25} \sin(5\omega t + 90°), \text{ etc.}$$

Since both waveforms contain odd harmonics only, the resultant waveform must contain odd harmonics. The fact that the resultant waveform exhibits "mirror-image" symmetry indicates the absence of even harmonics. The fact that the resultant waveform does not exhibit Z-axis symmetry is an indication of the presence of odd harmonics no longer at 0 or 180° with respect to the fundamental (obviously due to the 90° displacement of the armature flux with respect to the field flux). Because the negative harmonic terms in ϕ_a are subtracted from the positive harmonic terms of ϕ_f, the resultant flux, ϕ_r, is decreased.

tor containing an exploring armature coil whose ends have been brought out to slip rings. The nature of the voltage induced in this coil under load may be viewed on an oscilloscope. Laboratory measurements indicate the reduction in total mutual air-gap flux is approximately one to five per cent, from no load to full load, as a result of armature reaction.

5-4. SHIFT OF NEUTRAL PLANE IN GENERATOR VS. MOTOR

The shift of the load neutral, as shown in Fig. 5-3c, from the original neutral plane shown in Figs. 5-1, 5-2, and 5-3a and b may have serious effects on both dc generator and dc motor operation. In the case of a dc generator, for example, the coil whose conductors are marked x-x, is in the neutral plane (Fig. 5-3) originally, and hence *not* undergoing a change in flux linkages. As a result, this coil is normally being short circuited by the brushes. In Fig. 5-3c, however, if the brushes remain on the original neutral plane, the coil which is being commutated (short-circuited) is undergoing the *greatest* change in flux linkage compared to any other coil under the pole. If the conductors cut across the flux while being short circuited by the brushes, the voltage induced in the conductors may be sufficient to produce a heavy circulating current and sparking at the brushes each time a new coil comes up to take the place of coil x-x. Furthermore, since the brushes of a dc generator were placed [Section 2-11, Fig. 2-9a] at a point of minimum coil flux but maximum path voltage, it is obvious that they must be shifted (to the new magnetic neutral to obtain maximum voltage); but in which direction? Since direction of rotation was not indicated in the above discussion, the question arises as to how the brushes must be shifted for a motor or a generator in terms of the direction of rotation.

A dc generator is represented in Fig. 5-4a in which the armature conductors are rotated clockwise by a prime mover. Using the right-hand rule, the direction of induced emf in the armature conductors is as shown. Under

(a) dc generator load neutral.

(b) dc motor load neutral.

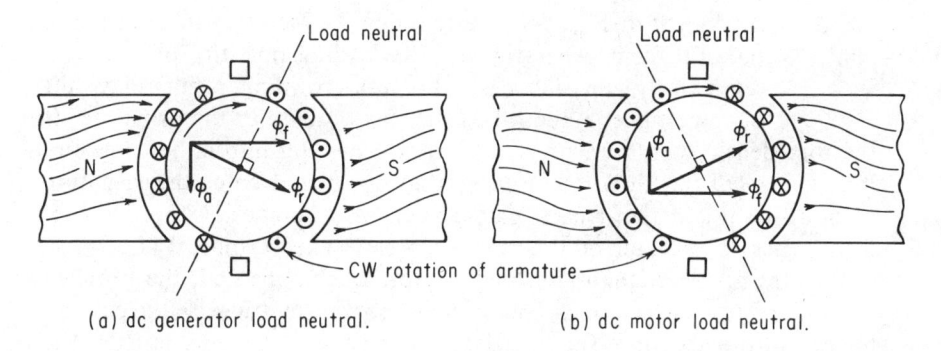

Figure 5-4

Comparison of shift of load neutral for generator and motor, same direction of rotation.

load, the armature mmf would produce a resultant flux, as shown, and the load neutral would also be shifted clockwise, *in the direction* of rotation indicated in the figure.

A dc motor is represented in Fig. 5-4b, and the current direction produced by the armature voltage in the armature conductors is designed to produce clockwise direction of rotation (left-hand rule). The armature flux produced by these armature conductors (right-hand corkscrew rule), will produce an effect on the field flux such that the resultant flux and its perpendicular load neutral are shifted counterclockwise or *opposite to the direction* of rotation, as shown in Fig. 5-4b. Also note that in the case of the *generator*, the flux is always crowding in the *trailing pole tip* (i.e., the pole tip last encountered by a conductor on a moving armature), whereas for a *motor*, the flux density is higher in the *leading pole tip*.

We are now on the horns of a dilemma. If a dc dynamo is operated as a *generator*, it is necessary to *advance* the brushes in the direction of rotation as the load is increased. If it is to be operated as a *motor*, it is necessary to move the brushes *against* the direction of rotation, as the load is increased. The manufacturer is well aware that a consumer might use a given dc dynamo for either motor or generator applications. The consumer, furthermore, is not inclined to be bothered with details for shifting brush neutrals.

5-5. COMPENSATING FOR ARMATURE REACTION IN DC DYNAMOS

It is obvious that one cannot stand continuously at a dynamo and shift the brushes in accordance with load variations and applications (as a generator or motor). Some *automatic* method is necessary in which the effects of armature reaction are compensated for, or the factors that cause it are neutralized. Some of the various methods of compensating for armature reaction effects are discussed below.

5-5.1 High-Reluctance Pole Tips

As shown in Fig. 5-4, the flux density increases in the trailing pole tip for a generator and the leading pole tip for a motor, respectively. In each case, this flux crowding results in a shift of the magnetic neutral as the flux entered in the armature. If the flux density could be prevented from crowding into either pole tip, it might solve the situation because the flux would enter the armature without a shift in the neutral plane. Such a technique is shown in Fig. 5-5a, where the center of the pole is closer to the peripheral circumference of the armature than are the pole tips. In this method, the rounded circumferential surface of the pole shoe is not concentric with the circumference of the armature surface (the former greater than the latter). The greater reluctance at the tips, because of the greater air gap, forces the field flux to be confined to the center of each of the field pole cores.

A similar result is also accomplished as shown in Fig. 5-5b, where,

(a) High reluctance pole tips.

(b) Chamfered pole laminations.

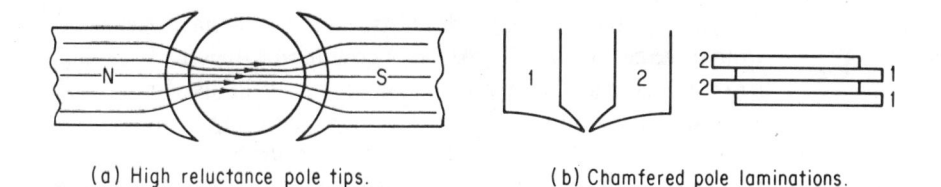

Figure 5-5

Use of chamfered laminations to counteract effects of armature reaction.

in assembling the field pole laminations, the same punched laminations are alternately reversed. The effect is to produce a cross section, as shown, in which the center of the pole core has more iron than the leading or trailing pole tips and, hence, less center reluctance. This produces a similar effect as that shown in Fig. 5-5a, in preventing the main field flux and neutral plane from shifting.

5-5.2
Reduction in Armature Flux

Another constructional technique is to attempt a reduction of armature flux without effectively reducing the main field flux. The method employed here is to create a high reluctance in the quadrature, cross-magnetizing, armature flux path without materially affecting the main field flux path. The use of slotted, punched, field-pole laminations, as shown in Fig. 5-6, introduces several air gaps into the armature flux magnetic path without affecting the field flux path materially. Thus the armature flux is reduced considerably but the field flux remains substantially the same. By chamfering the field lamination, as shown in Fig. 5-6a, the combination of the two methods described above may be used most effectively and inexpensively, since it requires only a slightly more complex die for punching the field laminations.

(a) Field lamination

(b) Armature magnetic flux path.

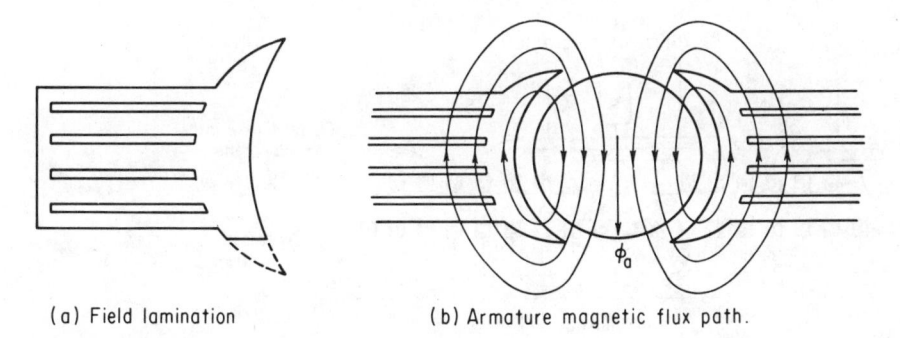

Figure 5-6

Use of slotted laminations to reduce armature flux and armature reaction.

**5-5.3
Compensating
Windings**

The two mechanical techniques discussed above have one major disadvantage in that they do not counteract the effects of high armature currents and **mmf**s due to heavy loads. In large dc dynamos, even those with chamfered pole tips and slotted field laminations, the high armature currents produce sufficient magnetic flux to produce air-gap flux distortion and consequent shifting of the magnetic neutral. On large dynamos, therefore, an *electrical* method is employed in which the armature flux is neutralized or counterbalanced by a winding in the armature circuit whose effect varies with the armature current. The winding is called respectively a *compensating winding*, a *pole-face winding*, or a *Thomson-Ryan winding* after its designers. As shown in Fig. 5-7, the winding is inserted in slots in the face of the stationary pole shoe. It is *not* necessary to have the same number of slots or conductors in the compensating winding as there are conductors on the surface of the armature. The important fact is that the number of conductors in each pole face times the armature line current must equal the number of armature conductors under each pole times the armature conductor current per path. This equivalency in **mmf** is expressed by

$$Z_p I_l = Z_a I_a = \frac{Z_a I_l}{a} \qquad (5\text{-}1)$$

where Z_p is the number of pole-face conductors in each pole
Z_a is the number of active armature conductors under each pole
a is the number of parallel paths in the armature
I_l is the load or total current entering the armature
I_a is the current carried by each armature conductor

Solving the above equation for Z_p, the number of pole face conductors per pole, we get

$$Z_p = \frac{Z_a}{a} \qquad (5\text{-}2)$$

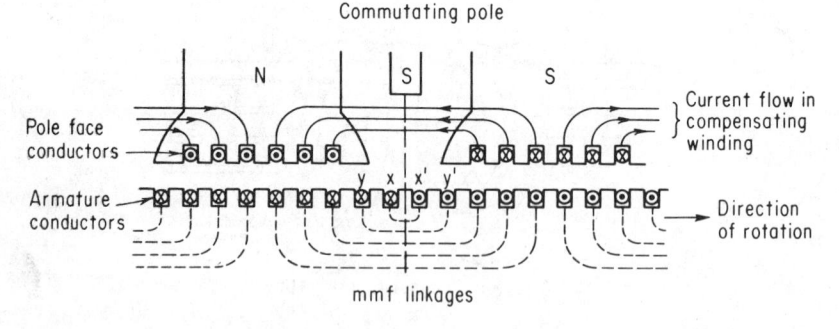

Figure 5-7

Use of compensating winding to neutralize armature magnetomotive force.

Since two conductors equal one turn, in effect Eq. (5-1) states that the *pole face* mmf ($I_l N_p$) is counterbalancing the armature conductor mmf ($I_a N_a$). The number of armature compensating pole face conductors may be determined either from Eqs. (5-1) or (5-2).

EXAMPLE 5-1: A simplex lap wound dc dynamo has 800 conductors on its armature, a rated armature current of 1000 A and 10 poles. Calculate the number of pole face conductors per pole to give full armature reaction compensation, if the pole face covers 70 per cent of the pitch.

Solution:

Using Eq. (5-1):
$$Z = \frac{800}{10} = 80 \text{ cond/path under each pole}$$

Active conductors/pole,
$$Z_a = 80 \text{ cond/path} \times 0.7 = 56 \text{ conductors/pole}$$

Solving for Z_p in Eq. (5-1)
$$Z_p \times 1000 \text{ A} = 56 \text{ cond} \times 1000 \text{ A}/10 \text{ paths}$$
$$= \mathbf{5.6 \text{ or } 6 \text{ conductors/pole}}$$

Using Eq. 5-2, instead
$$Z_p = \frac{Z_a}{a} = \frac{56}{10} \cong \mathbf{6 \text{ conductors/pole}}$$

as shown in Fig. 5-7.

EXAMPLE 5-2: For the armature given in Ex. 5-1, calculate
a. The cross-magnetizing ampere-conductors/pole and ampere-turns/pole, respectively, with the brushes set on the geometric neutral
b. The demagnetizing ampere-turns/pole, if the brushes are shifted 5 electrical degrees from the geometric neutral
c. The cross-magnetizing ampere-turns/pole with the brushes shifted as in (b) above

Solution:

a. With the brushes on the geometric neutral, the entire armature reaction effect is completely *cross-magnetizing*.
The cross-magnetizing ampere-conductors per pole are
$$\frac{ZI_l}{Pa} = \frac{8000}{10} \times \frac{1000}{10} = 80 \frac{\text{conductors}}{\text{pole}} \times \frac{100 \text{ A}}{\text{path}}$$
$$= 800 \text{ ampere-conductors/pole}$$

and since there are 2 conductors/turn, the cross-magnetizing ampere-turns/pole are
$$\frac{1 \text{ turn}}{2 \text{ conductors}} \times \frac{8000 \text{ ampere-conductors}}{\text{pole}} = 4000 \text{ At/pole}$$

b. Let α = the number of electrical degrees that the brushes are shifted. Then the total number of *demagnetizing* electrical degrees are 2α
While the (remaining) *cross-magnetizing* electrical degrees, $\beta = 180° - 2\alpha$

153
SEC. 5-5. / Compensating for Armature Reaction in DC Dynamos

The *ratio* of demagnetizing to cross-magnetizing ampere-turns is always $2\alpha/\beta$

The *fraction* of demagnetizing ampere-turns/pole is

$$\frac{2\alpha}{180°} \times \frac{IN \text{ total}}{\text{pole}} = \frac{2 \times 5°}{180°} \times 4000 \text{ At/pole} = \textbf{22.2 At/pole}$$

c. Since $\beta = 180° - 2\alpha = 180° - 10° = 170°$, the cross-magnetizing ampere-turns/pole are

$$\frac{\beta}{180°} \times \frac{IN \text{ total}}{\text{pole}} = \frac{170°}{180°} \times 4000 \text{ At/pole} = \textbf{3778 At/pole}$$

The action of the pole face conductors, as stated above, and as shown in Fig. 5-7, is to produce an mmf which is equal and opposite to the armature mmf. In effect, the compensating winding demagnetizes or *neutralizes* the armature flux produced by armature conductors lying *under* the poles. If the load increases or decreases, the current in the armature circuit and in the compensating winding will vary in exact proportion to the armature mmf so that the latter is (theoretically) effectively neutralized for all load conditions.

Note in Fig. 5-7 that the compensating winding produces an equal and opposite mmf to that produced by the armature conductors *except in the interpolar zone*. Thus, the mmf produced by conductors y-y' and x-x' is not compensated by pole face conductors. It will later be shown (Sec. 5-7) that some of the flux produced by the interpoles is also useful in reducing armature reaction in the interpolar zone.

Finally, for reasons given in Sec. 5-8, compensation of armature reaction is *only* confined to dc dynamos. Armature reaction is *never* compensated in ac machines, because it may worsen performance under certain conditions of load and power factor. The reader should appreciate that both compensating windings and interpoles (Sec. 5-7) are devices used to improve the the performance of (larger) dc dynamos, only.

5-6.
THE
COMMUTATION
PROCESS

Unfortunately, while the compensating winding neutralizes the armature mmf produced by these conductors lying directly under the poles, it does not neutralize the mmf produced by these conductors lying in the interpolar region (x-x' and y-y' shown in Fig. 5-7). These conductors still produce a cross-magnetizing armature flux which is uncompensated. The effects of sparking at and overheating of the brushes would still be noticeable in large dynamos using pole-face compensating windings only. These effects are due to: (1) the uncompensated armature conductors in the interpolar region and (2) the commutation process, itself, which we will not consider.

As shown in Fig. 5-7 for all dynamos, the armature conductors under a given pole will have a particular current direction; and, as the conductors move under an opposite pole, the direction of the current is reversed. All dc dynamos, and some ac dynamos as well, are equipped with *com-*

mutators. The purpose of the commutator and its associated brushes are: (1) in the case of a generator, to change the generated alternating current to external direct current; or, in the case of a motor, to change the external applied direct current to alternating current as the conductors alternately move under opposite poles (to produce rotation in the same direction); and (2) to accomplish a transfer of current between a *moving* armature and *stationary* brushes.

The transfer of current between the coils, the commutator bars, and a brush is shown in Fig. 5-8 for a dc generator. This particular dc generator

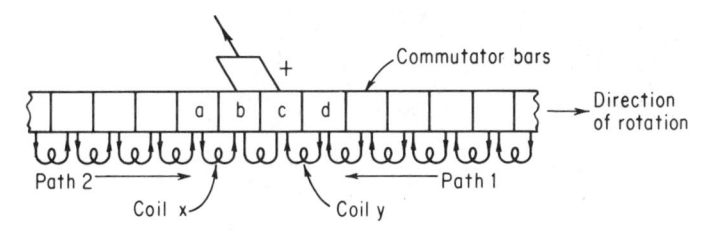

Figure 5-8

Current paths in a dc generator.

has an armature winding which produces two paths [i.e., a two-pole simplex lap winding, or a simplex wave with any even number of poles; see Eqs. (2-4) and (2-5)]. The series-connected coils of path 1 and path 2 carry the induced emf and the current drawn by the load to the positive brush. Thus, the coils of path 2 carry current which enters the positive brush at commutator bar b, and the coils of path 1 carry current in the opposite direction which enters the positive brush at commutator bar c. Since the drawing of Fig. 5-8 represents a dynamic process, it is fairly evident that during a short time interval, a given coil in path 2 carrying current in one direction (after undergoing commutation and passing the brush) becomes a specific series-connected coil in path 1, carrying current in the opposite direction.

Let us consider the commutation process as it affects a particular coil, coil x, located in path 2, as shown in Fig. 5-8, about to undergo commutation. Coil x in Fig. 5-8 is connected to commutator bars a and b and is, therefore, carrying the full path emf and current from a to b, I_{a-b}.

Figure 5-9 is an incremental graphic description of coil x undergoing commutation, and the changes in emf at each instant of time are described below:

Instant t_1: Coil x is now the last coil in the series string carrying the full path emf
(Fig. 5-9a): and current of path 2 to the positive brush, and its direction of current is still I_{a-b}, as indicated above and also shown in Fig. 5-10 at time t_1.

Instant t_2: Coil x, whose coil sides are still undergoing some change in flux linkages
(Fig. 5-9b): at the (trailing) tips of an N and an S pole, respectively, is being partially

SEC. 5-6. / *The Commutation Process*

(a) Instant t_1. Path 2 / Path 1

(b) Instant t_2. Path 2 / Path 1

(c) Instant t_3. Path 2 / Path 1

(d) Instant t_4. (zero current) Path 2 / Path 1

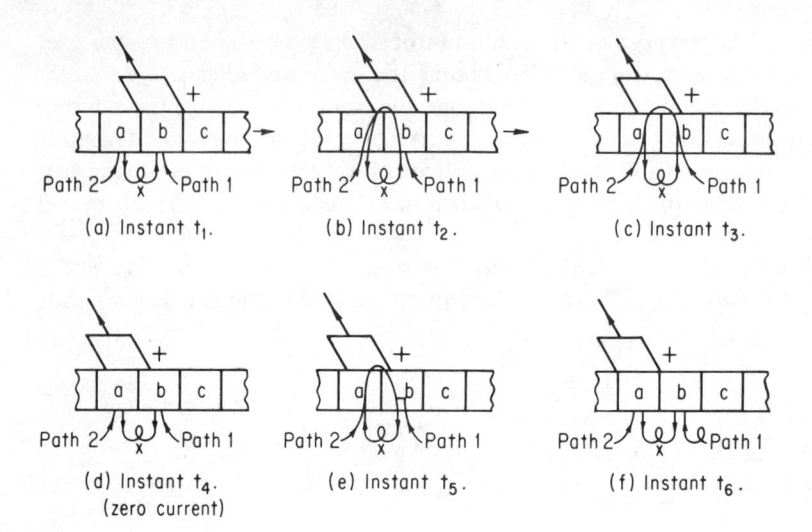

(e) Instant t_5. Path 2 / Path 1

(f) Instant t_6. Path 2 / Path 1

Figure 5-9

Changes in emf and current in a coil undergoing commutation.

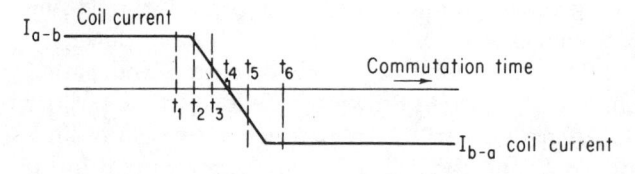

Figure 5-10

Reversal of current direction in any coil undergoing commutation.

short circuited by the positive brush, which causes a circulating current through the brush, bars a and b, and coil x in the direction shown. Current from path 2 begins to enter the brush through bar a, reducing the path current in coil x as shown in Fig. 5-10.

Instant t_3:   Coil x is now being fully short circuited by the positive brush. Due
(Fig. 5-9c):   to the short circuit at the previous instant, it is still carrying some current (since, by Lenz's law, an emf of self-induction is produced which opposes the decay of current in the coil) in accordance with the exponential decay of current in a circuit containing resistance and inductance. The current which flows, therefore, is still in the same direction (sustained by the emf of self-induction).

Instant t_4:   Current from path 2 is flowing to the positive brush through bar a, and
(Fig. 5-9d);   current from path 1 is flowing to the positive brush through bar b. The path emfs are equal and opposite in coil x; and, consequently, no current flows in coil x. Note however, that there is less resistance between the commutator and the brush for bar a than for bar b, as shown in Fig. 5-9d.

Instant t_5:   The fairly high resistance between bar b and the brush will cause current
(Fig. 5-9e):   from path 1 to flow through coil x and bar a to the positive brush. Since the brush still short circuits bars a and b, a small circulating current will be set up in coil x in the opposite direction, as shown in Fig. 5-9e and

156

CHAP. FIVE / *Armature Reaction and Commutation in Dynamos*

indicated in Fig. 5-10 at t_5. Note that previously the current in coil x was counterclockwise, and now the short-circuit current is clockwise, i.e., I_{b-a}.

Instant t_6: (Fig. 5-9f): Coil x is now carrying the full path current of path 1, I_{b-a}, in the opposite direction, as shown in Fig. 5-9f and Fig. 5-10, and will continue to do so until it reaches a negative brush. At the negative brush, the same process is repeated, except that the current enters the commutator instead of leaving it.

5-7. REACTANCE VOLTAGE

Two conditions have occurred in the process described above, both of which prevent smooth commutation:

1. An emf self-induction was created which opposes the sudden reversal of current in the coil, shown in Fig. 5-10. Since each and every coil undergoes commutation consecutively and continually, this constant voltage is called a reactance voltage because it "reacts" against the current reversal in each coil undergoing commutation.
2. A potentially high short-circuit current is developed due to voltage existing in the coil sides of the coil undergoing commutation during periods t_2 to t_4, when the resistances of the short-circuit paths are lowest.

> For a dc generator, both of these conditions may be lessened somewhat by shifting the brush axis *in the direction* of the armature rotation, or, for a dc motor, by rotating the brush axis *against the direction* of the armature rotation.

Using the dc generator as an example, shifting the brushes in the direction of rotation will cause commutation to occur, as shown in Fig. 5-8, at coil y, i.e., at a coil in which a voltage is generated in the direction in which the reverse current is about to flow. Thus, the counterclockwise circulating currents sustained by the reactance voltage and shown in Fig. 5-9 will be reversed and become clockwise more rapidly.

But shifting the brushes, in effect, is equivalent to *moving the neutral plane*. As shown in Fig. 5-7, shifting the load neutral in the direction of rotation would bring conductors x-x' under an S pole. We have already observed, however, that, since the shift is *opposite* for dc generator compared to dc motor operation, and since the degree or extent of the shift depends on the *load* current, an electrical technique similar to that used for armature reaction compensation is required. In this case the solution is to interpose a pole between the main poles whose mmf also *varies with load*. Such a pole, called an *interpole* or *commutating pole*, is shown in Fig. 5-7.

In practice as well as in theory, the *interpole winding* is designed to have *more* ampere turns than are required to generate sufficient voltage in those conductors lying in the interpolar region to accomplish a smooth reversal of current and eliminate the effects of reactance voltage. There are two reasons for this: (1) to produce neutralization of those armature conductors lying in the interpolar region which are not neutralized by the compensating winding (Secs. 5-5, 6); and (2) the complex nature of the transients involved in the commutation process requires assumptions in the design equations and calculations which do not yield a precise determination of the number of interpole turns required. By making the interpole stronger than necessary, nonmagnetic (and magnetic) shims may be used experimentally between the yoke frame (to which the interpole is bolted) to provide the exact reluctance for the interpole magnetic path. Adjustments are made until smooth and effective commutation and interpolar armature reaction compensation is achieved over a range of load.

It should be pointed out that when both compensating and commutating pole windings are employed in series with the armature and connected in the armature circuit, as shown in Figs. 3-1, 3-2, and 3-3, it is: (1) *not* necessary to shift the brushes from the no-load neutral plane; and (2) that the dc dynamo will operate as a motor or a generator equally well without the adverse effects of armature reaction or commutation difficulties. Finally, it should be noted that the use of compensating and commutating windings are peculiar to dc dynamos primarily or to some ac dynamos employing commutators.

The ac synchronous and asynchronous dynamos do not require commutators; hence, they do not require a commutating winding. Furthermore, while armature reaction does occur in all dynamos since the above discussion of armature reaction does occur in all dynamos (since the above discussion of armature reaction was based on the universal dynamo), we shall see that because of the possibility of current lead or lag with respect to voltage in an ac dynamo, it may produce beneficial, as well as deleterious effects, as will be shown below. The net result is that the ac synchronous and asynchronous induction dynamos are *never compensated*, either for armature reaction or commutation.

5-8.
ARMATURE REACTION IN THE AC DYNAMO

The distribution of armature flux due to current in the armature conductors, and the consequent distortion of air-gap flux shown in Fig. 5-3, applies to the universal dynamo but with some qualifications. In the case of the ac dynamo, the current in the individual armature conductors (1) is not uniform but varies sinusoidally, and (2) may lead or lag with respect to the field flux at leading or lagging power factors. As a result, the waveform shown in Fig. 5-3b may be shifted to the right or left with respect to the mutual air-gap flux shown in Fig. 5-3a.

The manner in which various power-factor load conditions affect the ac dynamo may be determined from the phase relation between the induced emf and the current which flows in the armature conductors, as shown in Fig. 5-11. The four sets of conductors lying under the field poles represent, in effect, the range of conditions which may occur in the single set of armature conductors (shown in Fig. 5-3) because of armature reaction. The first set of armature conductors shown in Fig. 5-11 represent the emf induced in the armature conductors as a result of relative motion between the poles and the armature of an ac dynamo. This emf may be either the induced emf E_g of an ac alternator which produces the terminal voltage, or the counter emf (E_c) of an ac synchronous motor which, in part, serves to limit the armature current drawn from the lines to produce motor action. This induced emf is no different from that shown in Fig. 5-3 for the universal dynamo and in Fig. 5-7 for the dc dynamo. Power factor (for purposes of this discussion) will be defined as the phase relation between the emf induced in the armature conductors, E_g per phase, and the armature current which flows in them, I_a per phase, as a result of generator or motor action. The effect of power factor on the field flux and the generated emf will be discussed below for the extreme conditions of power-factor variation, namely, unity power factor, zero power factor lagging, and zero power factor leading.

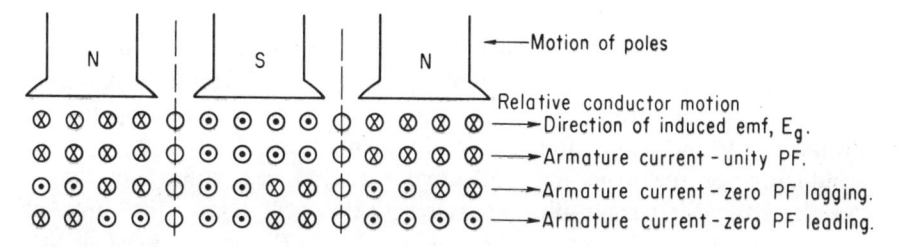

Figure 5-11

Displacement of armature current with respect to field flux owing to power factor in ac dynamos.

5-8.1 Unity Power Factor

At unity power factor, the ac armature phase current is in phase with the ac armature induced phase voltage (as it is in the dc dynamo). This is shown in the second set of armature conductors in Fig. 5-11, where the instantaneous conductor currents match the instantaneous induced voltages in the armature. The armature mmf produced by these armature conductors (by the right-hand corkscrew rule) will produce a cross-magnetizing flux. This flux is a maximum in the interpolar region, and it lags the primary air-gap flux by 90°.

This relation is no different from that shown in Fig. 5-2 and is the same as the cross-magnetizing armature reaction occurring in a dc dynamo (since, in a dc circuit, current is always in phase with voltage). The phasor diagram showing the relation between the field flux, ϕ_f, taken as a reference, and the quadrature cross-magnetizing armature current and flux, I_a and ϕ_a, respectively, is shown in Fig. 5-12a.

Note that it requires 90 electrical degrees for the primary air-gap flux, ϕ_f, to produce the ac armature generated voltage E_g and its consequent (in phase) armature current, I_a. The field surrounding a current-

(a) Unity PF   (b) Zero PF lagging.   (c) Zero PF leading.

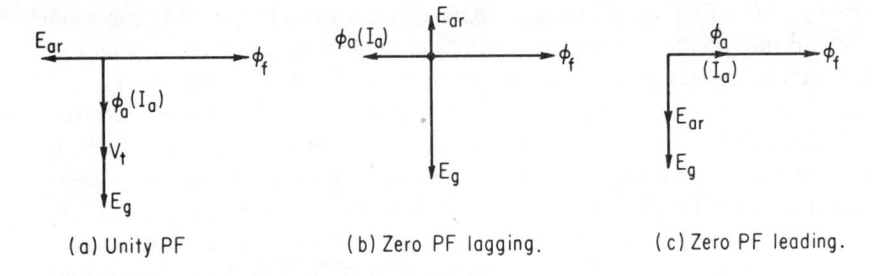

Figure 5-12

Phasor diagrams showing effect of armature reaction at various power factor conditions of ac dynamos.

carrying armature conductor, moreover, depends directly on the current in that conductor, and the cross-magnetizing armature flux, ϕ_a, is always in phase with the armature current, I_a. Thus, an ac armature reaction (cross magnetizing) flux, ϕ_a, lagging the field flux, ϕ_f, is set up by the ac armature current in the interpolar region. This ac armature flux links the conductors of the armature and creates an induced voltage in them. As stated in this paragraph, it requires 90 electrical degrees for a flux to produce a voltage. An armature reaction voltage, E_{ar}, is produced in the armature, therefore, which lags E_g, the generated voltage, by 90° at unity power factor, as shown in Fig. 5-12a. It will later be shown that this voltage drop plays a part in the voltage regulation of the ac synchronous alternator (Chapter 6).

5-8.2 Zero Power Factor Lagging

At zero power factor lagging, the ac armature phase current, I_a, lags the ac armature induced phase voltage, E_g, by 90°, as defined above. As shown in Fig. 5-11, the instantaneous conductor currents, produced by the ac armature current, create an ac armature mmf and flux ϕ_a, which opposes the primary air-gap flux ϕ_f (right-hand corkscrew rule). This effect is shown in the phasor diagram of Fig. 5-12b, where the air-gap flux induces voltage E_g, in 90 electrical degrees, the armature current lags E_g by 90 electrical degrees, and ϕ_a opposes ϕ_f. The armature flux at zero power factor lagging is now demagnetizing rather than cross magnetizing as at unity PF. The ac armature flux ϕ_a links the armature conductors to produce an ac armature reaction voltage, E_{ar}, in 90 electrical degrees. Note that since the armature flux is displaced from the field flux by 180°, the armature reaction voltage, E_{ar}, is displaced from the induced voltage, E_g, by 180°. Voltage E_{ar} *reduces* the generated voltage, E_g, at zero power factor lagging, tending to produce poor voltage regulation with the application of load.

5-8.3
Zero Power Factor Leading

At zero power factor leading, the ac armature current per phase, I_a, leads the ac armature induced phase voltage E_g by 90 electrical degrees, in accordance with the above definition. As shown in Fig. 5-11, the instantaneous conductor currents, produced by the ac armature current, create an ac armature mmf and flux, ϕ_a, which aids the primary air-gap flux, ϕ_f (right-hand corkscrew rule). This effect is shown in the phasor diagram of Fig. 5-12c, where the air gap induces voltage E_g in 90 electrical degrees. The armature current I_a leads E_g by 90 electrical degrees, and ϕ_a aids ϕ_f. The armature flux at zero power factor leading is *magnetizing* rather than cross magnetizing as at unity PF. The ac armature flux ϕ_a links the armature conductors to produce an ac armature reaction voltage E_{ar} in 90 electrical degrees. Note that the armature reaction voltage E_{ar} produced is in phase with the generated voltage E_g and thus tends to *increase* E_g, thereby improving the voltage regulation of the ac synchronous alternator.

5-9.
SUMMARY OF ARMATURE REACTION IN DYNAMOS

The effects of armature reaction discussed in this chapter may be summarized by the following statements:

1. In dc dynamos and in ac single-phase and polyphase dynamos at unity power factor, where the phase current in the armature is in phase with the induced armature voltage per phase, the armature reaction is *cross magnetizing*, and the induced armature reaction voltage lags the generated voltage by 90 electrical degrees.

2. In ac single-phase and polyphase dynamos at zero power factor lagging, where the phase current in the armature *lags* the induced armature voltage per phase by 90 electrical degrees, the armature reaction is *demagnetizing*, and the induced armature reaction voltage lags the generated voltage by 180 electrical degrees.

3. In ac single-phase and polyphase dynamos, where the phase current in the armature lags the induced armature voltage per phase by some angle *between* zero and ninety electrical degrees, the armature reaction is *partly demagnetizing* and *partly cross magnetizing*. If θ represents the angle by which the phase current lags the induced phase voltage in the armature, the demagnetizing component is sine θ and the cross-magnetizing component is cosine θ of the armature reaction flux. Note that θ equals 0° at unity power factor and equals 90° at zero power factor.

4. In ac single-phase and polyphase dynamos at zero power factor leading, where the phase current in the armature *leads* the induced armature voltage per phase by 90 electrical degrees, the armature reaction is *magnetizing*, and the induced armature reaction voltage is in phase with the voltage generated in the armature by the air-gap flux.

5. In ac single-phase and polyphase dynamos where the phase current in the armature leads the induced armature voltage per phase by some angle between zero and ninety electrical degrees, the armature reaction is partly magnetizing and partly cross magnetizing. If θ represents

the angle by which the phase current leads the induced phase voltage in the armature, the magnetizing component is sine θ and the cross-magnetizing component is cosine θ of the armature reaction flux. Note that θ equals 0° at unity power factor, and equals 90° at zero power factor.

6. The summarized statements are subjected to the following qualifications:*
    a. Armature reaction in single-phase dynamos is pulsating, whereas in polyphase and dc dynamos the armature reaction flux is constant.
    b. The resultant constant armature reaction flux is produced as a result of balanced three-phase loads whose armature currents are equal and displaced by 120 electrical degrees. If the loads are unbalanced, angle θ above, is meaningless, as is the distribution of cross-magnetizing, demagnetizing, and magnetizing components of the armature flux.

7. Armature reaction is compensated, and its effects are neutralized, only in some dc dynamos and rarely in ac dynamos. Indeed, armature reaction is used to good advantage in the third-brush generator, Rosenberg generator, amplidyne, and multifield exciters (see Chapter 11).

BIBLIOGRAPHY

Alger, P. L. *The Nature of Polyphase Induction Machines*, New York: John Wiley & Sons, Inc., 1951.

Bewley, L. V. *Tensor Analysis of Electrical Circuits and Machines*, New York: Ronald Press, 1961.

Carr, C. C. *Electrical Machinery*, New York: John Wiley & Sons, Inc., 1958.

Crosno, C. D. *Fundamentals of Electromechanical Conversion*, New York Harcourt, Brace, Jovanovich, Inc, 1968.

Daniels. *The Performance of Electrical Machines*, New York: McGraw-Hill, Inc., 1968.

Fitzgerald, A. E. and Kingsley, C. *The Dynamics and Statics of Electromechanical Energy Conversion*, 2nd Ed., New York: McGraw-Hill, 1961.

Fitzgerald, A. E., Kingsley, Jr. C., and Kusko, A. *Electric Machinery, 3rd Ed.* New York: McGraw-Hill, Inc., 1971.

Gemlich, D. K. and Hammond, S. B. *Electromechanical Systems*, New York: McGraw-Hill, 1967.

Hindmarsh, J. *Electrical Machines*, Elmsford, N.Y.: Pergamon Press, 1965.

Jones, C. V. *The Unified Theory of Electrical Machines*, New York: Plenum Publishing Corp., 1968.

* For a more complete study of the differences between single-phase and polyphase armature reaction, cf. Dawes, *Electrical Engineering*, New York: McGraw-Hill, Inc., Vol. II, Chapter 7, "Alternating Currents." 1948.

Kloeffler, S. M., Kerchner R. M., and Brenneman. J. L. *Direct Current Machinery*, Rev. ed., New York: The Macmillan Company, 1948.

Koenig, H. E. and Blackwell, W. A. *Electromechanical System Theory*, New York: McGraw-Hill, 1961.

Liwschitz, M. M., Garik M., and Whipple. C. C. *Direct Current Machines*, 2nd Ed., Princeton, N. J.: D. Van Nostrand Co., Inc., 1947.

Majmudar, H. *Introduction to Electrical Machines*, Boston: Allyn and Bacon, 1969.

Meisel, J. *Principles of Electromechanical Energy Conversion*, New York: McGraw-Hill, 1966.

Nasar, S. A. *Electromagnetic Energy Conversion Devices and Systems*, Englewood Cliffs, N.J.: Prentice-Hall, Inc., 1970.

O'Kelly and Simmons. *An Introduction to Generalized Electrical Machine Theory*, New York: McGraw-Hill, 1968.

Puchstein, A. F. *The Design of Small Direct Current Motors*, New York: Wiley/Interscience, 1961.

Robertson, B. L. and Black. L. J. *Electric Circuits and Machines*, 2nd Ed., Princeton, N. J.: D. Van Nostrand Co., Inc., 1957.

Schmitz, N. L., and Novotny, D. W. *Introductory Electromechanics*, New York: Ronald Press, 1965.

Seely, S. *Electromechanical Energy Conversion*, New York: McGraw-Hill, 1962.

Selmon. *Magnetoelectric Devices: Transducers, Transformers and Machines*, New York: Wiley/Interscience, 1966.

Siskind, C. S. *Direct-Current Machinery*, New York: McGraw-Hill, 1952.

Skilling, H. H. *Electromechanics: A First Course in Electromechanical Energy Conversion*, New York: Wiley/Interscience, 1962.

Thaler, G. J., and Wilcox, M. L. *Electric Machines: Dynamics and Steady State*, New York: Wiley/Interscience, 1966.

White, D. C., and Woodson, H. H. *Electromechanical Energy Conversion*, New York: Wiley/Interscience, 1959.

QUESTIONS

5-1. Draw a 4-pole dynamo in cross-section and show direction of:
   a. induced voltage in armature conductors under each pole
   b. current in armature conductors under load
   c. flux produced by armature conductors.

5-2. a. Does the above diagram (Q. 5-1) also apply to a dc motor? Explain.
   b. What relation exists between the direction of current in the armature conductors and the relative polarity of the field?

5-3. Draw a diagram which shows that for each pair of poles a net armature flux is produced by the armature conductors. Show the mmf linkages of the armature with respect to the field flux.
   a. What space relation exists between the field flux and the armature flux?
   b. By means of a diagram, show the interaction between the field and armature flux.
   c. Give two effects of this interaction on the flux entering the armature.

5-4. Give two reasons why the resultant flux entering an armature, whose conductors are carrying full-load current, is less than the no-load flux.

5-5. Compare, for the same direction of rotation, the shift in the load neutral plane (perpendicular to the resultant flux) from the no-load neutral in (a) a generator, (b) a motor.

5-6. a. Using the diagram developed in Quest. 5-5 above, explain why it is impossible for the manufacturer to set the brushes for a dc dynamo.
   b. Is it possible to set the brushes at a particular load netural setting for a dc motor operating at a fixed load? Explain.
   c. Repeat (b) for a generator.

5-7. a. Give three non-electrical construction techniques used to compensate for armature reaction automatically without requiring brush shifting.
   b. What are the advantages of these techniques?
   c. What disadvantages may be noted, particularly for large dc dynamos?

5-8. a. Describe an electrical method of compensating for armature reaction.
   b. What is the advantage of this method over those discussed in Quest. 5-7?
   c. Give an equation expressing the relation between the number of pole-face conductors required to compensate a given number of armature conductors.
   d. Does a compensating winding compensate for armature reaction produced by current in those current-carrying conductors lying in the interpolar space? Explain.

5-9. Give two reasons why sparking is produced at brushes in large dynamos using only pole-face compensating windings.

5-10. As a result of the commutation process (described in Sec. 5-6), describe two conditions which prevent smooth commutation and result in sparking at brushes in contact with the commutator.

5-11. a. Give two functions of an interpole (commutating pole) winding.
   b. Where is it physically located in a dynamo and how is it connected electrically?
   c. Give two reasons why an interpole is designed to produce more mmf and has more ampere-turns than are normally required to eliminate the effects of reactance voltage.

5-12. When both compensating and commutating pole windings are connected in series with the armature is it necessary to shift the brushes from the no-load neutral plane for a dynamo
   a. operating under load as a motor? Explain.
   b. operating under load as a generator? Explain.

5-13. Are commutating pole windings required on synchronous ac dynamos which are not equipped with commutators? Explain.

5-14. With respect to the phasor diagrams shown in Fig. 5-12, explain:
   a. why I_a and ϕ_a are always in phase
   b. why E_g always lags ϕ_f by 90°
   c. why E_{ar} always lags ϕ_a by 90°
   d. why I_a is in phase with E_g at unity PF but lags or leads E_g at other power factors
   e. why E_{ar} is 100% demagnetizing at zero PF lagging
   f. why E_{ar} is 100% magnetizing at zero PF leading
   g. what proportion of E_{ar} is magnetizing and cross magnetizing at a PF of 0.707 leading
   h. what proportion of E_{ar} is demagnetizing and cross magnetizing at a PF of 0.5 lagging.

5-15. a. Why is it unnecessary and even undesirable to compensate for armature reaction in ac dynamos, generally, and even in some dc dynamos, particularly?
   b. Give two types of dynamos in which the armature reaction flux is constant.
   c. Give one type of dynamo in which the armature reaction flux is pulsating.

PROBLEMS

5-1. A 50 kW, 250 V separately excited generator has an armature resistance of 0.05 Ω, a brush drop of 6 V, and an armature reaction voltage drop of 20 V at rated load. Assuming a linear armature reaction effect with load, calculate
   a. Armature circuit voltage drop at full load, $\frac{3}{4}$, $\frac{1}{2}$, $\frac{1}{4}$ and no load
   b. Armature reaction voltage drop at above load conditions
   c. Generated armature voltage at each of the load conditions in (a)
   d. Explain why the generated voltage at no load is different from the generated voltage at full load.

5-2. A 125 V, dc, 5 kW two-pole generator has a total of 1800 armature conductors on the periphery of its armature. The coil sides span exactly 180 electrical degrees on each coil. Neglecting the field current, calculate
   a. Ampere-turns per pole when rated armature current is delivered by the generator
   b. Demagnetizing and cross-magnetizing ampere-turns per pole when the brushes are shifted five electrical degrees from the neutral (no-load) plane
   c. Demagnetizing and cross-magnetizing (distortion component) ampere-turns per pole when the brushes are shifted 10 electrical degrees
   d. If the field flux is 10,000 amp-turns/pole and the no-load generated voltage is 140 V, calculate the armature reaction voltage drop at rated load, where it is necessary to shift the brushes 10 electrical degrees.

5-3. Using the ratio of demagnetizing to cross-magnetizing ampere turns in Problem 5-2, derive a universal equation to express the demagnetizing and cross-magnetizing ampere-turns per pole for an armature having Z active conductors, a armature paths, P poles. (Assume that the brushes may be shifted α electrical degrees/pole and that the remaining electrical degrees, β, are $180 - 2\alpha$.)

5-4. A 12 pole, triplex-lap armature winding having a total of 720 armature, conductors is used in a 120 kW, 600 V generator whose field resistance is 100 Ω. The field is wound with 20 turns/pole. At rated load, it is necessary to shift the brushes six electrical degrees. Calculate
    a. Armature current per path, current per armature conductor, total armature ampere conductors and total ampere-turns per pole pair
    b. Demagnetizing and cross-magnetizing ampere-turns per pole (ampere turns per 180°)
    c. Verify (b) by using equations derived in Problem 5-3
    d. Net mmf per pole at full load as a result of the effects of armature reaction.

5-5. It is desired to compensate for the distortion component of armature reaction in the generator of Problem 5-2 using a compensating winding in the pole face conductors. Calculate
    a. The number of pole-face conductors required to neutralize the cross-magnetizing mmf
    b. Will the pole-face conductors neutralize the degenerative effects of armature reaction completely? Why not?
    c. Will it be necessary to shift the brushes as a result of such compensation? Why?

5-6. In the generator of Problem 5-2, it is desirable to compensate for the armature reaction produced by those conductors in the interpolar zone by adding turns on each commutating pole. Calculate
    a. The number of commutating pole turns to be added to the commutating winding to neutralize the mmf produced by the demagnetizing conductors
    b. If both compensating and commutating pole windings are employed, will it be necessary to shift the brushes for the generator of Problem 5-2? Why? Explain
    c. If the generator is rotating in a clockwise direction, in what direction should the brushes be shifted?

5-7. For the generator of Problem 5-4, calculate
    a. Compensating ampere conductors required to neutralize conductors directly under the poles and number of pole-face conductors per pole
    b. Compensating ampere-turns required to neutralize conductors in the interpolar region, and number of turns on the commutating pole.

5-8. A two-pole dynamo has a speed of 1200 rpm, a commutator diameter of 6.375 in and a total of 50 bars comprising its commutator. The brushes span a linear distance of 0.8 in and the armature circuit full-load current is 50 A. Neglecting the thickness of insulation between commutator bars, calculate

    a. Width of each commutator bar
    b. Current carried by each commutator bar at rated load
    c. Time required for each bar to pass a brush (commutation time)
    d. Average rate of change of current during commutation at rated load
    e. Reactance voltage per coil if the inductance of each coil on the armature is 0.01 H.

5-9. A six-pole, 250 rpm dynamo has a simplex lap-wound armature with a diameter of 24 in commutator consisting of 240 bars. The brushes are $\frac{3}{4}$ in wide. When operating as a generator, the armature delivers 300 A at rated load. Neglecting insulation between adjacent commutator bars, calculate at rated load
    a. Average rate of change of current during commutation
    b. Reactance voltage, if the self inductance of each armature coil is 0.2 mH.

5-10. The self inductance of each coil of an armature is 1.0 mH and the rate of change of current in each coil (coils consist of five turns) is 20,000 A/s whenever a coil is undergoing commutation. If the peripheral speed of the armature is 4000 ft/min and the dimension of the commutating pole *parallel* to the armature shaft is 8 in, calculate
    a. emf of self induction in each coil undergoing commutation
    b. Voltage per conductor induced in each coil under each commutating pole (assume there are as many commutating poles as main poles)
    c. The flux density under the commutating pole

5-11. A duplex lap-wound, 1200 rpm, 250 V, 50 kW generator has six poles, an armature resistance of 0.1 Ω, a field resistance of 250 Ω, a brush voltage drop of 5.9 V, and an armature reaction voltage drop of 12 V at rated load. The commutator diameter is 12 in and there are as many commutator bars as there are coils on the armature. Calculate:
    a. Generated emf between brushes at full load
    b. Number of armature conductors if the flux per pole is 6×10^5 lines
    c. Number of commutator bars if there are five turns per coil
    d. Voltage between each commutator bar and current in each commutator bar
    e. Width of each commutator bar, allowing a mica thickness of $\frac{1}{32}$ inch between bars
    f. Commutation time, assuming a brush width of $\frac{1}{4}$ in.
    g. Reactance voltage per coil if the inductance of each coil is 0.1 mH
    h. The flux density under the commutating pole assuming the dimension of the commutating pole parallel to the armature shaft is 4 in.

5-12. Given the phase currents of a three-phase alternator as

$$i_A = I_m \sin(\omega t + \theta)$$
$$i_B = I_m \sin(\omega t + \theta - 120°)$$
$$I_C = I_m \sin(\omega t + \theta - 240°)$$

where I_m is the instantaneous maximum current and θ is the power factor angle. *Prove* that the armature reaction fluxes produce a *constant* resultant of magnitude $1.5\phi_m$ having a *constant* geometrical relation with respect to the field flux for the following conditions of *time* and *power factor*

a. Case I: time, $\omega t = 0$, power factor of zero, leading, i.e., $\theta = 90°$
b. Case II: time, $\omega t = 90°$, power factor of zero, lagging, i.e., $\theta = -90°$
c. Case III: time, $\omega t = 120°$, power factor of unity, i.e., $\theta = 0°$.

ANSWERS

5-1(a) 2.5 V, 5.0 V, 7.5 V, and 10 V at full load (b) 5.0 V, 10.0 V, 15.0 V, and 20 V (c) 263.5 V, 271 V, 278.5 V, 286 V 5-2(a) 18,000 amp-turns/pole (b) 1000 and 17,000 amp-turns/pole (c) 2000 amp-turns/pole and 16,000 amp-turns/pole (d) 28 V 5-3 Demag amp-turns/pole $= (\alpha/90) \times (ZIa/Pa)$; X-Mag. amp-turns/pole $= (\beta/180) \times (ZIa/Pa)$ 5-4(a) 343 amp-turns/pole (b) 22.9 and 320.1 amp-turns/pole (c) same as (b) (d) 97.1 amp-turns/pole 5-5(a). 800 5-6(a) 500 turns/pole 5-7(a) 3.1 conductor/pole (b) 0.1 turns/pole 5-8(a) 0.4 in (b) 25 A/bar (c) 2 ms (d) 25,000 A/s (e) 250 V 5-9(a) 42×10^3 A/s (b) 8.4 V/coil 5-10(a) 20 V/coil (b) 2 V/conductor (c) 3.13×10^4 lines/in² 5-11(a) 288 V (b) 4800 (c) 480 (d) 7.2 V (e) 0.0464 in (f) 0.104 ms (g) 0.5975 V (h) 1.98×10^4 lines/in² 5-12. $\phi_R = 1.5\phi_{max}$ at all times and power factors

SIX

ac dynamo voltage relations—alternators

6-1.
GENERAL

In comparing generator versus motor operation, Section 1-20 concluded with a summary of the fundamental differences between them. This chapter is devoted to the use of the ac dynamo as an ac generator or (so-called) *alternator*, since it produces an alternating voltage. Since the alternator is a source of voltage, we are concerned primarily with the ac dynamo voltage relations and how they are affected by resistive and reactive loads. The summary of Section 1-20 (modified somewhat to apply specifically to alternator performance) may be stated as:

1. The electromagnetic torque (developed in the current-carrying armature conductor) *opposes* rotation (of the rotor's magnetic field with respect to the armature), in accordance with Lenz's law.
2. The generated (induced) voltage in the armature produces armature current. The *phase* of the armature current with respect to the generated alternator voltage depends on the nature of the electrical load placed across the alternator terminals.

3. The generated voltage per phase E_{gp} of a polyphase or single-phase alternator may be stated by the phasor sum

$$\dot{E}_{gp} = \dot{V}_p + I_p \dot{Z}_p \qquad (6\text{-}1)$$

where V_p is the terminal voltage per phase of the alternator
$I_p Z_p$ is the internal synchronous impedance voltage drop of the alternator

6-2. CONSTRUCTION

The general construction of the ac synchronous dynamo was discussed in Section 2-3 and 2-4 and shown in Figs, 2-2b and 2-3. Section 2-3 and Fig. 2-2b dealt with a synchronous dynamo having a rotating armature and a stationary field. Section 2-4 and Fig. 2-3 discussed a synchronous dynamo having a rotating field and a stationary armature. Although both types of construction may be used in an alternator, the latter type, with a _stationary armature_ and a _revolving field_, is almost universally used for alternating electric power generation, for reasons discussed below in Section 6-3. The former type, with the rotating armature, finds its greatest application as a synchronous or rotary converter (Sections 11-6 and 11-7).

The armature windings used on the stationary armature are generally _lap_ windings (both lap windings and wave windings will produce the same voltage for the same number of coils) because of the _shorter end connections_ (jumpers) required between coils. Sections 2-13 to 2-17 inclusive discussed various types of alternator armature windings. Based on this presentation, the generated voltage per phase was developed in Section 2-18 as

$$E_{gp} = 4.44 \phi N_p f k_p k_d \times 10^{-8} \text{ V} \qquad (2\text{-}15)$$

The frequency of an ac synchronous alternator was also related to its construction, since the frequency varies with the number of salient or nonsalient poles. Frequency was expressed by the equation

$$f = \frac{PS}{120} \qquad (2\text{-}16)$$

The reader should review the equations and sections cited above, since they are fundamental to an understanding of the ac synchronous alternator, discussed below.

6-3. ADVANTAGES OF STATIONARY ARMATURE AND REVOLVING FIELD CONSTRUCTION

The reader, on studying Chapters 2 and 3, might at first react adversely to the idea of making the armature stationary. After all, it should be a simple matter to bring out the alternating current generated by moving armature to slip-rings on one side of a shaft, and the direct current generated by the same moving armature to commutator bars on the other side of the shaft. In this way, we would have universal dynamos which could supply either direct or alternating current, or both simultaneously. Indeed, this *is* done in a synchronous converter, as will be seen in Chapter 11; but there are several compelling reasons for abandoning the idea of a universal dynamo having a rotating ac armature. Once the armature is stationary, we no longer achieve automatic switching from ac to dc by commutation, and only ac is generated. The more significant advantages of the stationary armature and revolving field construction are:

6-3.1 Increased Armature Tooth Strength

Machines of higher capacity require more armature copper and deeper slots in the armature iron than are needed in a machine built for lighter duties. On a stationary armature, as the slots are made *deeper* from $a\text{-}a'$ to $b\text{-}b'$ in Fig. 6-1a, the armature tooth becomes *wider* and stronger. On a rotating armature, however, as the slots are made deeper from $a\text{-}a'$ to $b\text{-}b'$ in Fig. 6-1b, the armature tooth becomes *narrower* and thus *weaker*. It should be noted that, in both instances, the top of the slot must be narrower than the bottom. Cutting the slots in this way tends to prevent the winding from "creeping" out of the slot during vibration. Such creeping would result in damage to the machine.

(a) Stationary armature slots.   (b) Rotating armature slots.

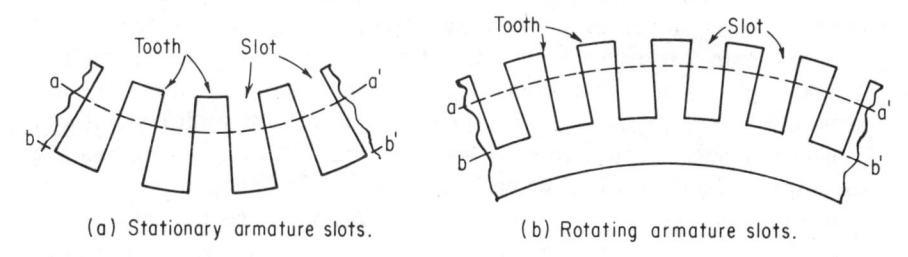

Figure 6-1

Weakened teeth produced by deeper slots on rotating armature.

In a rotating armature, the teeth might be subjected to high centrifugal forces; and in any armature, whether rotating or stationary, the teeth might encounter shocks in either construction or operation. With its sturdier teeth, the stationary armature is less likely to be damaged.

6-3.2 Reduced Armature Reactance

The *mutual* air-gap flux created by the primary field mmf must pass through the armature iron and slots. For the same air-gap width at the bottom of the slot for a given armature coil, the stationary armature provides a *reduced* reluctance to the flux. This is due to an *increased* cross section of iron, as shown in Fig. 6-1. The reduced reluctance also reduces the amount of armature *leakage* flux produced (Section 2-7), because the armature flux path sees increased reluctance, particularly in the case of those armature conductors lying in the bottoms of the slots.

6-3.3 Improved Insulation

High-speed, high-voltage, and high-capacity commercial alternators carry appreciable currents at appreciable voltages, requiring efficient insulation. The shafts, through metallic bearings, are electrically grounded to the stationary frame of the dynamo. It is easier to insulate a stationary member than a rotating member, since size, weight, and quantity of insulation are not as critical for the former. Furthermore, since the rotor is grounded, it is less of a problem to insulate the low voltage dc field on a rotor than a rotating high voltage ac armature.

6-3.4 Construction Advantages

In large polyphase stators, the armature winding is more complex than the field winding. The various coil and phase interconnections may be built more easily on a rigid stationary structure than on a rotor, and the armature winding is braced more securely when built on a rigid frame.

6-3.5 Number of Insulated Slip-Rings Required

If the armature of a polyphase alternator were permitted to rotate, a three-phase alternator would require a minimum of three slip-rings, a six-phase alternator six, etc. The problem of transferring the high induced voltage (in some cases as high as 13,000 V/phase) at high currents from armature slip rings to stationary armature brushes in contact with these rings is not accomplished without difficulty. Insulating the slip-rings from the shaft is one problem. Spacing the slip-rings sufficiently far apart to avoid flashover is another. As the number of phases increases, the insulation problem becomes more complex. A stationary armature presents none of these problems, however, and the voltage per phase is more easily insulated and brought out of a stationary dynamo. <u>Only two slip-rings are required to excite the field winding at a comparatively low voltage, perhaps 300 V of direct current at most.</u>

6-3.6 Reduced Rotor Weight and Inertia

From the preceding discussion, it is fairly obvious that a low-voltage field winding, which uses many turns of fine wire to produce the field **mmf**, hardly requires the weight of copper and the equivalent insulation that are needed for the high-voltage armature winding. It is easier to build rotors for efficient high-speed operation using the low-voltage field winding as the rotating member. Inertia of the rotor plays an important factor in bringing the alternator up to speed; and on extremely high-capacity alternators, even with the dc field as the rotor, it may take several hours to bring the machine up to its rated speed and voltage, particularly with steam turbines as prime movers.

6-3.7
Ventilation
Advantages

Most of the heat is produced in and related to the armature winding and its surrounding iron. With a stationary armature the winding may be cooled more efficiently, because the stator core and its peripheral size have fewer limitations. Thus, the stator core may be enlarged somewhat to permit radial air ducts and ventilation holes for forced air, hydrogen, or other forms of cooling.

In addition to the foregoing, there are advantages in size and total weight for the same capacity of rotating field over rotating armature dynamos. The reasons cited above have compelled designers of ac dynamos to abandon the dream of "the universal dynamo" (having one basic construction and design), except for relatively small rotary converters. Without exception, ac synchronous alternators have *rotating* fields of either *salient* or *nonsalient pole* construction, as shown in Fig. 2-3.

6-4.
PRIME MOVERS

The salient-pole rotor construction lends itself to medium and low-speed alternators having many poles. The *nonsalient, cylindrical* type of rotor is used almost universally on *high-speed*, two-pole (and sometimes four-pole) alternators. Since the number of poles and the speed are also related to frequency, [Eq. (2-16), where $f = PS/120$], it would be of interest to compare the most commonly used commercial frequencies as to the poles and speeds required; such a comparison is shown in Table 6-1.

TABLE 6-1.
SPEED-FREQUENCY RELATIONS FOR VARIOUS NUMBERS OF POLES IN AC SYNCHRONOUS DYNAMOS

NUMBER OF POLES	FREQUENCY DESIRED		
	25 Hz*	50 Hz*	60 Hz*
2	1500 rpm†	3000 rpm†	3600 rpm†
4	750	1500	1800
6	500	1000	1200
8	375	750	900
10	300	600	720
12	250	500	600
14	$214\frac{2}{7}$	$428\frac{4}{7}$	$514\frac{2}{7}$

* Frequency in hertz (cycles per second).
† Speed in revolutions per minute.

Table 6-1 indicates that where the prime mover is essentially a *low-speed* driver, as in the case of a hydroturbine (used at hydroelectric powerhouses), a *large* number of poles is required. Since windage is not a problem at low speeds, a *salient* pole rotor may be employed. Similarly, if the prime mover is a gasoline, diesel, gas, or steam engine, a prime mover of essentially *moderate* speed, anywhere from four to twelve salient poles might be employed. In the case of an essentially *high-speed* prime mover, such as a gas-driven or steamdriven *turbine* (the steam may be obtained from a conventional coal-fired or oil-fired boiler, or from an atomic reactor), usually only two *nonsalient* poles are used. To a large extent, the determination of the type of alternator field construction employed is determined by the kind of fuel or the energy source available in the geographic location where the electricity is to be generated. Fuels such as coal or oil may be shipped by rail or barge; or oil or gas may be piped to the generating site, if that site is neither too remote nor inaccessible. Transportation expense is a factor in the cost of generation per kilowatt hour.

Slow-speed salient-pole alternators require stator armatures with a *large* circumference into which many conductors may be inserted. Such stators require field pole or armature conductors of *short axial* length. On the other hand, *high-speed cylindrical nonsalient-pole* rotors have a *small* circumference requiring field pole and armature conductors of *long axial* length. Thus, by the very marked difference in external appearance, one may easily distinguish between salient and nonsalient-pole synchronous dynamos, without even viewing the rotor, as shown in Fig. 6-2.

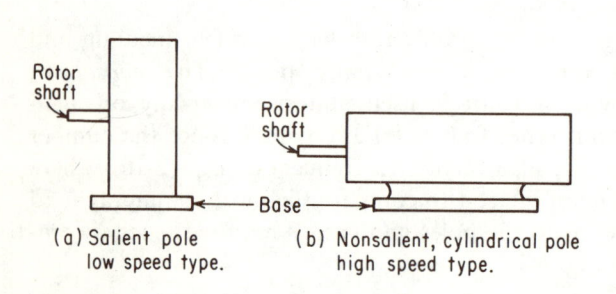

Figure 6-2

General appearance of synchronous dynamo.

(a) Salient pole low speed type.

(b) Nonsalient, cylindrical pole high speed type.

6-5. EQUIVALENT CIRCUIT FOR A SINGLE-PHASE AND A POLYPHASE SYNCHRONOUS DYNAMO

The relation between the terminal and generated voltage of a synchronous dynamo was given in Section 6-1 by Eq. (6-1), and the circuit is represented in Fig. 6-3. For a single-phase or polyphase synchronous alternator, Eq. (6-1) may be expanded and rewritten as follows:

$$\dot{V}_p = \dot{E}_{gp} - \dot{I}_a R_a - \dot{I}_a(jX_a) \pm \dot{E}_{ar} \qquad (6\text{-}2)$$

where V_p is the terminal voltage/phase

E_{gp} is the generated voltage/phase, from Eq. (2-15) and Section 6-2

$I_a R_a$ is the voltage drop across the armature winding, having an effective (ac) resistance of R_a, per phase

$I_a(jX_a)$ is the voltage drop across the reactance of the armature winding due to leakage reactance (Section 2-7 and 6-3) per phase

and E_{ar} is the effect of armature reaction (magnetizing, cross-magnetizing, or demagnetizing as summarized in Section 5-9) per phase

(a) Single phase

(b) 3-phase, wye connected equivalent circuit of a synchronous alternator.

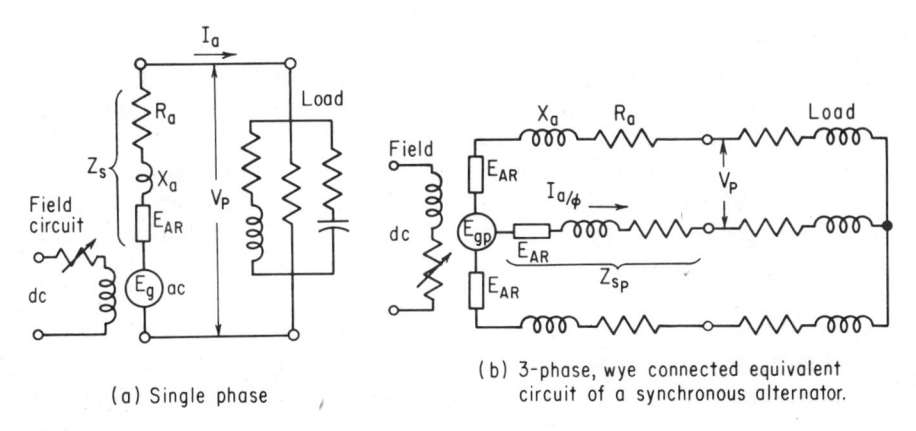

Figure 6-3
Equivalent circuit of a synchronous alternator.

It should be noted that there is essentially little difference between the equivalent circuit of a single-phase ac synchronous alternator and the three-phase ac synchronous alternator shown in Figs. 6-3a and b, respectively. Each phase winding of a three-phase alternator is assumed to have an effective armature resistance per phase of R_a, an armature reactance per phase of X_a, and a generated voltage per phase of E_{gp}. Furthermore, if the load is balanced (as we have seen in Section 5-9) it may be assumed that the voltage drop due to the effect of armature reaction is the same in each phase: hence, E_{ar}, or armature reaction volt drop per phase. The components of Eq. (6-2) apply equally well both to polyphase and to single-phase synchronous alternators. Since the dots over the various components of Eq. (6-2) imply phasor addition (or subtraction), the diagrams for the various power factor conditions will be considered, in turn, in order to predict the voltage relations and to write equations for the voltage regulation of an ac synchronous alternator.

Before doing so, however, it might be well to consider those factors which may account for differences between the no-load generated voltage per phase, E_{gp}, and the terminal voltage per phase, V_p. The equivalent circuit shown in Fig. 6-3 employs separate dc excitation for the rotating field windings of both and single-phase and polyphase synchronous alternators. Consequently, any change in terminal voltage as a result of loading

SEC. 6-5. / *Equivalent Single-Phase and a Polyphase Synchronous Dynamo* does *not* affect the excitation of the field emf. In this respect, the alternator is similar to the separately excited dc generator, and a comparison between the two will reveal similarities as well as differences.

6-6.
COMPARISON BETWEEN SEPARATELY EXCITED DC GENERATOR AND SEPARATELY EXCITED SYNCHRONOUS ALTERNATOR

There are two causes of voltage drop (see Section 3-13) from no-load to full-load in the separately excited dc generators: (1) armature circuit voltage drop; and (2) armature reaction. Equation (6-2) indicates that there are now *three* causes for voltage "drop" in the separately excited synchronous alternator: (1) armature circuit voltage drop; (2) armature reactance; and (3) armature reaction. In addition, for an alternator it would appear that, while the first two factors always tend to reduce the generated voltage, the third factor (armature reaction) may tend to decrease or increase the generated voltage [Eq. (6-2)]. Thus, the voltage regulation of the synchronous alternator differs from the separately excited dc generator in two important respects: (1) there is a voltage drop due to armature reactance; and (2) the armature reaction effect (depending on the power factor of the load) may produce a voltage which aids the generated voltage and tends to increase the terminal voltage [Eq. (6-2)]. Since the nature of the load affects the voltage regulation of the ac synchronous alternator, in addition to the load current, I_a, let us consider the relation between the generated and the terminal voltage in the ac synchronous alternator.

6-7.
RELATION BETWEEN GENERATED AND TERMINAL VOLTAGE OF AN ALTERNATOR AT VARIOUS LOAD POWER FACTORS

As in the dc generator, if there is *no load* on the (separately excited) ac synchronous alternator, the terminal voltage and the generated voltage are the *same*, as shown in Eq. (6-2). The magnitude of the three causes of voltage drop in the synchronous alternator cited in Section 6-6 are solely a function of the load current, I_a. The relation between these magnitudes in affecting alternator voltage characteristics is shown below.

6-7.1
Unity Power Factor Loads

The relationships among the various voltage drops producing a difference between the generated and the terminal voltage are shown in Fig. 6-4a. At unity power factor, the phase current in the armature I_a is in phase with the terminal phase voltage V_p, by definition. The voltage drop per phase across the effective resistance of the armature, $I_a R_a$, is also always in phase with the armature current, I_a. The inductive voltage drop due to armature reactance, $I_a X_a$, is always 90° leading with respect to the current through it (since the current lags the voltage by 90° in a circuit possessing inductive reactance only). At unity power factor, the armature reaction voltage drop, E_{ar} from Section

176

CHAP. SIX / *AC Dynamo Voltage Relations—Alternators*

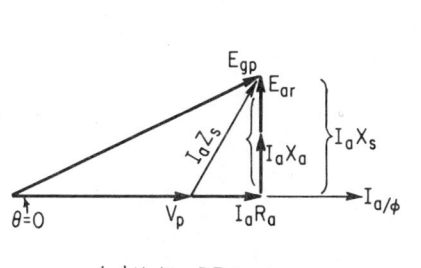

(a) Unity PF loads

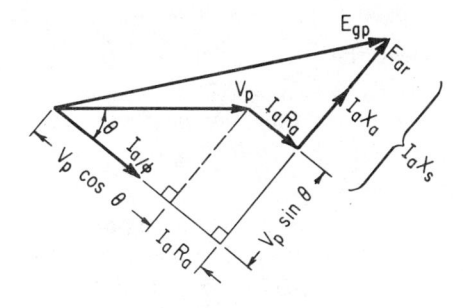

(b) Lagging PF loads.

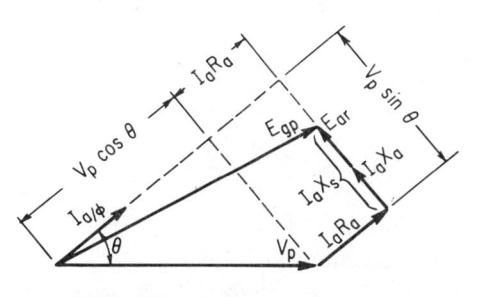

(c) Leading PF loads.

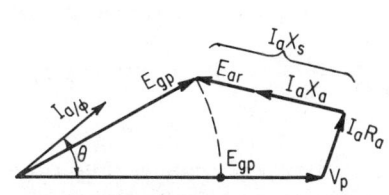

(d) Negative regulation at leading PF loads.

Figure 6-4

Relation between generated (no-load) and terminal (full-load) voltage of a synchronous alternator for three types of load conditions.

5-9 and Fig. 5-12a, leads* the armature current, I_a, which produced it, and is therefore always in phase with the armature reactance voltage drop, $I_a X_a$. The basic generator equation, Eq. (6-1), may now be written for unity power factor loads in complex form as

$$E_{gp} = (V_p + I_a R_a) + j(I_a X_a + E_{ar}) \qquad (6\text{-}3)$$

where all the terms have been defined in Eq. (6-2) above.

From both the diagram of Fig. 6-4a and Eq. (6-3), it may be seen

---

*This is a point which usually causes confusion in the mind of the student. The generated voltage due to armature reaction, as shown in Fig 5-12a, lags the armature current by 90° at unity power factor. The component of the total generated voltage necessary to overcome the effect of generated voltage due to armature reaction must be in the opposite direction. This distinction is between a voltage *generated*, and a voltage *drop* necessary to overcome it. Further, the armature reaction generated voltage always lags the armature current and the flux producing it by 90°. The component of voltage drop necessary to overcome this generated voltage must always lead the armature current by 90°, as does the component of applied voltage necessary to overcome the emf of self induction.

177

SEC. 6-7. / *Relation between Generated and Terminal Voltage of an Alternator*

that, at unity power factor, the terminal voltage per phase, V_p, is always less than the generated voltage per phase by a total impedance drop $I_a(R_a + jX_s)$, where jI_aX_s is the quadrature *synchronous reactance* voltage drop, or the *combined* voltage drop due to armature *reactance* and armature *reaction*.

6-7.2 Lagging Power Factor Loads

If the armature phase current, I_a (by definition), *lags* the terminal phase voltage, V_p, by some angle θ, as a result of an external load (primarily inductive) across the ac synchronous alternator the voltages may be represented by the diagram shown in Fig. 6-4b. The I_aR_a drop is still in phase with the armature phase current, and the quadrature reactance and armature reaction voltage drops lead the armature current by 90°. The relations of Eq. (6-1) still apply to this condition, but it is simpler to indicate the value of E_{gp} in terms of its *horizontal* and *vertical components*.

$$E_{gp} = (V_p \cos\theta + I_aR_a) + j(V_p \sin\theta + I_aX_s) \qquad (6\text{-}4)$$

From both the diagram of Fig. 6-4b and Eqs. (6-3) and (6-4), it would appear that to obtain the same rated terminal voltage per phase, V_p, a *higher* induced voltage per phase, E_{gp}, is required at *lagging* power factors than at unity power factor. This is indicated by the following example.

EXAMPLE 6-1: A 1000 kVA, 4600 V, three-phase, wye-connected alternator has an armature resistance of 2 ohms per phase and a synchronous armature reactance, X_s, of 20 ohms per phase. Find the full-load generated voltage per phase at:
a. Unity power factor
b. A power factor of 0.75 lagging

Solution:*

$$V_p = \frac{V_L}{\sqrt{3}} = \frac{4600 \text{ V}}{1.73} = 2660 \text{ V};$$

$$I_p = \frac{\text{kVA} \times 1000}{3V_p} = \frac{1000 \times 1000}{3 \times 2660} = 125 \text{ A}$$

$$I_aR_a \text{ drop/phase} = 125 \text{ A} \times 2\,\Omega = 250 \text{ V}$$

$$I_aX_s \text{ drop/phase} = 125 \text{ A} \times 20\,\Omega = 2500 \text{ V}$$

a. At unity power factor,

$$E_g = (V_p' + I_aR_a) + jI_aX_s = (2660 + 250) + j2500 \qquad (6\text{-}3)$$
$$= 2910 + j2500 = \mathbf{3845 \text{ V/phase}}$$

*Note that the solution is performed on a per-phase basis because the basic definition of power factor is in these terms.

b. At 0.75 PF lagging,

$$E_g = (V_p \cos \theta + I_a R_a) + j(V_p \sin \theta + I_a X_s) \quad (6\text{-}3)$$
$$= (2660 \times 0.75 + 250) + j(2660 \times 0.676 + 2500)$$
$$= 2250 + j4270 = \mathbf{4820 \text{ V/phase}}$$

6-7.3 Leading Power Factor Loads

If the armature phase current, I_a (by definition), *leads* the terminal phase voltage V_p by some angle θ, as a result of an external load (containing a capacitive component) across the ac synchronous alternator, the voltages may be represented as shown in Fig. 6-4c. The $I_a R_a$ drop is always in phase with the phase current in the armature and quadrature synchronous reactance drop $I_a X_s$ leads the armature current by 90°. Indicating E_{gp} in terms of the horizontal and vertical components we find

$$E_{gp} = (V_p \cos \theta + I_a R_a) + j(V_p \sin \theta - I_a X_s) \quad (6\text{-}5)$$

From both the diagram of Fig. 6-4c and Eq. (6-5), it would appear that for the same rated terminal phase voltage, less generated voltage is required for a leading power factor than for a lagging power factor. This is indicated by the following example:

EXAMPLE 6-2: Repeat Example 6-1 to determine the generated voltage per phase at full load with:
a. A leading load of 0.75 PF
b. A leading load of 0.4 PF

Solution:

From Example 6-1,

$$I_a R_a/\text{phase} = 250 \text{ V}$$
$$I_a X_s/\text{phase} = 2500 \text{ V}$$

a. At 0.75 PF leading

$$E_g = (V_p \cos \theta + I_a R_a) + j(V_p \sin \theta - I_s X_a) \quad (6\text{-}5)$$
$$= [(2660 \times 0.75) + 250] + j[(2660 \times 0.676) - 2500]$$
$$= 2250 - j730 = \mathbf{2360 \text{ V/phase}}$$

b. At 0.40 PF leading

$$E_g = [(2660 \times 0.4) + 250] + j[(2660 \times 0.916) - 2500] \quad (6\text{-}5)$$
$$= 1314 - j40 = \mathbf{1315 \text{ V/phase}}$$

Note that the *generated* voltage is *less* than the terminal voltage at *both* power factors, and *decreases as the power factor becomes more leading*.

6-8. VOLTAGE REGULATION OF AN AC SYNCHRONOUS ALTERNATOR AT VARIOUS POWER FACTORS

The preceding examples serve to illustrate two facets of the effect of leading or lagging loads on alternator-generated voltage and, in turn, voltage regulation, namely: (1) the lower the leading power factor, the greater the voltage *rise* from no load (E_{gp}) to full load (V_p); and (2) the lower the lagging power factor, the greater the voltage *decrease* from no load (E_{gp}) to full load (V_p). This may also be seen in the graphical representation taken from the data of these examples as shown in Fig. 6-5.

The figure also indicates that raising the power factor of a lagging load to unity power factor is still insufficient to produce zero per cent voltage regulation, and that the terminal voltage will still drop as purely resistive load is applied to the alternator. The figure also shows the effects of armature reaction, discussed in Section 5-9. At *leading* loads, the armature reaction is *magnetizing*, and tends to produce *additional* generated voltage as load is applied, producing a *negative* regulation, as in Fig. 6-4d. This high generated voltage is more than sufficient to compensate for the internal resistive voltage drop in the armature. At a specific leading power factor, as shown in Fig. 6-5 the additional magneti-

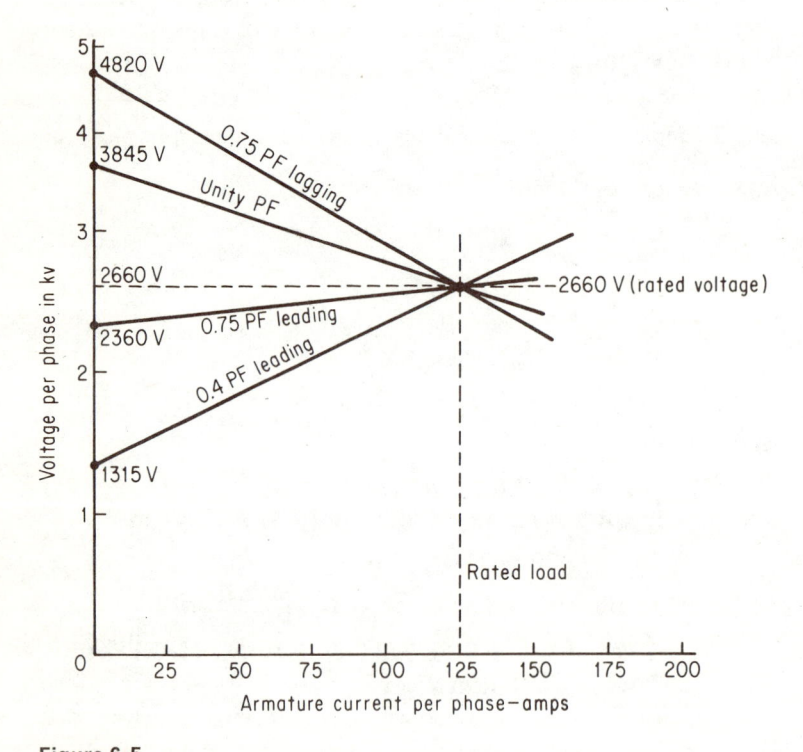

Figure 6-5

Voltage regulation of an alternator at various power factors.

zation produced by armature reaction is just balanced by the internal voltage drops, and the voltage regulation is zero.

At lagging loads, the armature reaction is *demagnetizing* (Section 5-9), and its effect in reducing the generated voltage, coupled with the internal armature resistive and reactive voltage drops, results in a rapid decrease in terminal voltage as load is applied, as shown in Fig. 6-5.

The voltage regulation of an alternator is the same as that for a dc generator, namely:

$$\text{VR (per cent voltage regulation)} = \frac{V_{nl} - V_{fl}}{V_{fl}} \times 100 \qquad (3\text{-}9)$$

and is usually performed on a per-phase basis, although line voltage values may be used with the same results.

EXAMPLE 6-3: Calculate the voltage regulation for the four power factors shown in Fig. 6-5 and computed in Examples 6-1 and 6-2.

Solution:

a. At 0.75 PF lagging, per cent
$$\text{VR} = \frac{4820 - 2660}{2660} \times 100 = \mathbf{81.3 \text{ per cent}}$$

b. At unity PF per cent
$$\text{VR} = \frac{3845 - 2660}{2660} \times 100 = \mathbf{44.4 \text{ per cent}}$$

c. At 0.75 PF leading, per cent
$$\text{VR} = \frac{2360 - 2660}{2660} \times 100 = \mathbf{-11.25 \text{ per cent}}$$

d. At 0.4 PF leading, per cent
$$\text{VR} = \frac{1315 - 2660}{2660} \times 100 = \mathbf{-50.6 \text{ per cent}}$$

It should be noted that the regulation of a separately excited dc generator (whose voltage drops with the application of load because of armature resistance and armature reaction) is inherently better than a separately excited ac synchronous alternator. Since *commercial* electrical loads are generally loads of a *lagging* nature, the voltage of a separately excited ac alternator will drop because of armature resistance, armature reactance, and armature reaction. The effect of armature reaction in a dc generator is, primarily, cross magnetizing and slightly demagnetizing whereas in an alternator its demagnetizing component is the armature flux, $\phi_a \sin \theta$ (Section 5-9).

Furthermore, the effects of armature reaction are *compensated* in a dc dynamo; but, although numerous attempts have been made to produce a compounding action with alternators to offset armature reaction at various power factors, none have been successful. In practice, therefore, the alternator's inherently poor regulation is ignored and its output maintained at a constant terminal voltage by means of external *voltage regulators* which automatically increase or decrease the field excitation from a dc generator (exciter) with changes in electric load and power factor. The *exciter* is usually on the same shaft as the prime mover and the alternator. Its characteristics are usually closely related to the alternator regulation, i.e., if the exciter is to maintain a constant voltage over a wide range of load, the limits of field current, power and exciter rating depend on the amount of field current required by the alternator to maintain good regulation.

6-9. SYNCHRONOUS IMPEDANCE

The difference between the generated voltage, E_{gp}, and the terminal voltage, V_p, per phase of an alternator, as stated by Eq. (6-1) and shown in Fig. 6-4a, is the synchronous impedance voltage drop, $I_a Z_s$. This same difference, in point of fact, exists between V_p and E_{gp} for any power factor and any load, as shown by the various diagrams of Fig. 6-4. The synchronous impedance voltage drop is, at all times, the phasor sum of the effective armature resistance voltage drop per phase and the quadrature equivalent voltage drops due to armature reactance and armature reaction per phase for the same load. All of the voltage drops comprising synchronous impedance, by definition, are taken at full load, I_a. The phasor diagram is shown in Fig. 6-6a, using the armature current, I_a, as a reference. If each of the voltage drops in the phasor diagram are divided by the armature current, an *impedance triangle* is obtained in which the armature effective resistance, quadrature synchronous reactance, and synchronous impedance per phase are represented as shown in Fig. 6-6b.

(a) Vector diagram
(b) Impedance triangle

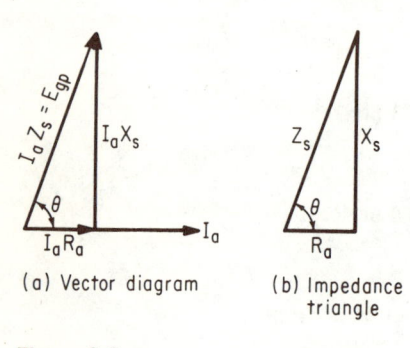

Figure 6-6

Phasor diagram and impedance triangle for synchronous impedance of an alternator.

The concept of an internal equivalent synchronous impedance possessed by an ac alternator is similar to that of an internal equivalent armature circuit resistance possessed by a dc dynamo (Fig. 3-1). Knowing the armature circuit resistance of a dc dynamo, it is possible to compute the terminal voltage of a dc generator or the counter emf of a dc motor for any value of load. In a similar manner, if the effective armature resistance per phase and synchronous reactance per phase are known, it is also possible to compute the generated emf of a synchronous alternator or motor. The advantage of the synchronous impedance concept is that it is possible to treat the quadrature voltage drop necessary to overcome the voltage due to armature reaction, as a reactive impedance component. This is permissible since this voltage is always in quadrature with the armature current, as previously shown.

The synchronous impedance and effective resistance per phase are determined by means of specific tests in a technique called the *synchronous impedance method*. The results give a value of synchronous reactance which, when used in the various voltage equations, yield a voltage regulation for the alternator which is somewhat *greater* than is actually obtained by *direct loading*. For this reason, the synchronous impedance method has been termed a "pessimistic method." But its simplicity, coupled with the assurance that the machine will, in actual performance, produce better regulation, has led to its almost universal use.*

6-10. THE SYNCHRONOUS IMPEDANCE (OR EMF) METHOD FOR PREDICTING VOLTAGE REGULATION

A single commercial alternator may have a capacity as high as 500,000 kVA or 500 million watts at unity power factor. Loading such an alternator electrically to determine its voltage characteristics (and efficiency) is a difficult matter; and to obtain such a load, one would have to "borrow" a sizeable city. Furthermore, if the alternator is built in an electrical plant, and is intended for use with specific steam or hydro turbines, there is no guarantee that sufficiently large prime movers are available in the vicinity of the plant to drive the alternator at its rated load. It is customary, therefore, to test large-capacity dynamos by a "conventional" no-load technique which will duplicate or *simulate* the load conditions. Such a technique uses only a *fraction* of the power for its performance that direct loading would require. The method employed for determining the *effective armature resistance* per phase in shown in Fig. 6-7a. The *synchronous impedance* method, consisting of the *open-circuit* and *short-circuit* tests, is shown in Figs. 6-7b and c. All *three* measurements are necessary for the determination of regulation.

The effective armature resistance per phase may be computed from the dc test shown in Fig. 6-7a. The armature is *assumed* to be connected in wye (even if connected in delta, the assumption that it is wye-connected produces the same results). It is customary to use a low-voltage dc source and the voltmeter-ammeter method rather than an ac source and a wattmeter. Direct current is used because the ac method would include coupled losses in the field pole structure, and its surrounding iron, yielding spurious values. The dc resistance per phase is

$$R_{dc} = \left(\frac{1}{2}\right)\frac{\text{Voltmeter Reading}}{\text{Ammeter Reading}} = \frac{V}{A \times 2}$$

The ac resistance per phase is obtained by multiplying the dc resistance by a factor which varies from 1.2 to about 1.8, depending on the frequency,

* Other methods, such as the mmf method (an optimistic method), the Potier method, the adjusted synchronous reactance method, and the ASA method (employed principally for calculations of field current required for specific operating conditions of load and power factor) are beyond the scope of this volume.

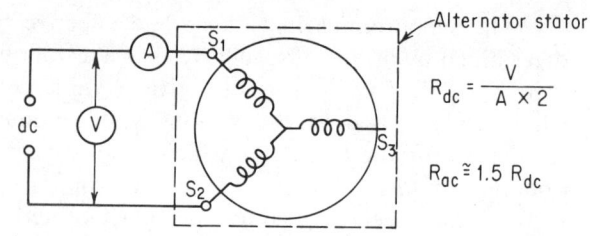

(a) Resistance measurement of dc and ac armature (effective) resistance per phase.

(b) Open-circuit test.

(c) Short-circuit test.

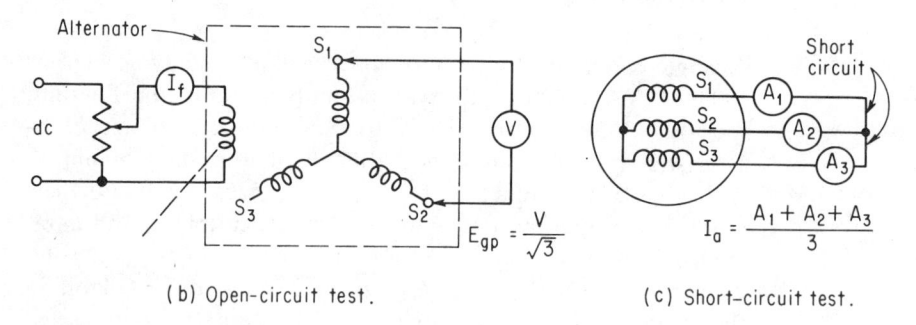

Figure 6-7

Synchronous impedance test-circuit connections.

quality of insulation, size and capacity, etc. For purposes here, we shall use a factor of 1.5 in computing the effective (ac) armature resistance per phase.

As stated above, the synchronous impedance test consists of two parts:

1. The open-circuit test. A separately excited (no-load) magnetization curve is obtained for the alternator, operated at *synchronous speed*. A dc ammeter is connected in the field circuit to record the field current, and an ac voltmeter is connected across any two stator leads to record the line voltage, V_l. A sufficient number of readings are taken, starting with zero field current, both below and above the knee of the curve. In each case, the field current, I_f, and the generated phase voltage, E_{gp}, (i.e., $V_l/\sqrt{3}$), are recorded, and a *saturation curve* is plotted as shown in Fig. 6-8. As with the dc magnetization curve, the results should be taken in one direction to avoid minor hysteresis loops.

2. The short-circuit test. The short-circuit characteristic is taken by connecting ammeters in the line to record the line current (even if the alternator is delta connected). The field current is adjusted to zero, and the aletrnator is brought up to speed. Readings are taken of dc field current versus ac short-circuited armature current. The results are plotted as shown in Fig. 6-8. It should be noted that this curve is completely linear; this is

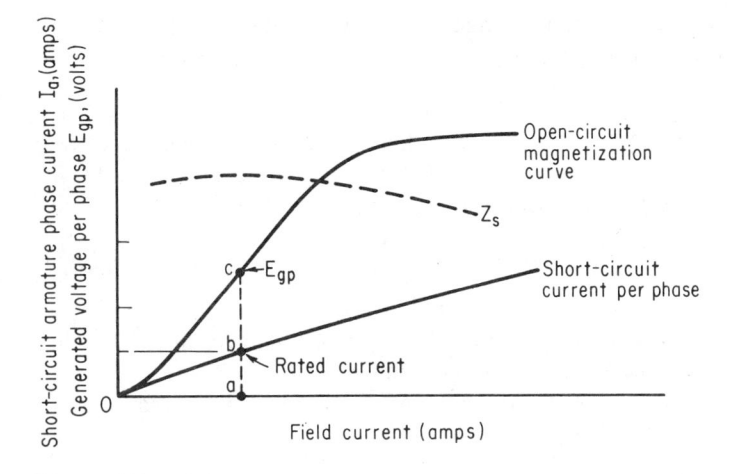

Figure 6-8

Open- and short-circuit characteristics of a synchronous alternator.

evident from both Eq. (6-1) and Fig. 6-6a. On short-circuit, the alternator terminal voltage is zero. The entire generated phase voltage, E_{gp}, is necessary to overcome the internal synchronous impedance drop, $I_a Z_s$, per phase. Since Z_s is almost a constant for a given dynamo, the short-circuit current varies directly with the generated voltage and the field current required to produce it (below saturation). Since the internal impedance is an extremely lagging, low power factor load, the demagnetizing effect is such that it reduces the field flux (and the generated voltage) considerably. Thus, fairly high field currents may be used without producing unusually excessive short-circuit currents.

When the short-circuit armature current per phase is equal to the full or rated load current, and when the alternator speed and frequency correspond to their respective rated values, if one were to remove the short circuit while the field current was held constant, the generated voltage per phase could be measured across the open-circuit armature terminals. The use of the curves permits this technique with less danger to personnel and the machine, which may be rated as high as 13,200 V/phase, or even higher.

As shown in Fig. 6-8, point ob represents the rated armature current per phase and field current; and oa, that excitation required to produce short-circuit current. But this *same* excitation will produce an open-circuit generated voltage of E_{gp}, corresponding to point oc. Since the terminal voltage is zero in Eq. (6-1), we may write

$$E_{gp} = I_a Z_p \quad \text{or} \quad Z_p = E_{gp}/I_a \tag{6-6}$$

where I_a is the full-load or rated current per phase

SEC. 6-10. / *The Synchronous Impedance (or EMF) Method for Predicting Voltage Regulation*

E_{gp} is the open-circuit voltage produced by the same field current that caused the rated short-circuit current per phase

Z_p is the synchronous impedance per phase

The various equations for voltage regulation are stated in terms of the voltage drops produced by effective armature resistance per phase and synchronous reactance per phase, and, therefore,

$$\dot{X}_s = \dot{Z}_p - \dot{R}_a \tag{6-7}$$

where X_s is the synchronous reactance per phase

Z_p is the synchronous impedance per phase, as determined by the short-circuit test and Eq. (6-6) above

R_a is the effective armature resistance per phase as determined by the dc resistance test above

It is also possible to combine Eqs. (6-1) to (6-5), inclusive, into a single general equation that will work equally well for all power factors and load conditions, i.e.,

$$E_{gp} = (V_p \cos\theta + I_a R_a) + j(V_p \sin\theta \pm I_a X_s) \tag{6-8}$$

where all terms are the same as in Eq. (6-5), and where $+$ is used for lagging loads and $-$ is used for leading loads.

EXAMPLE 6-4: A 100 kVA, 1100 V, three-phase alternator is tested in accordance with the above procedure to determine its regulation under various conditions of load and power factor. The data obtained is as follows:

dc resistance test	Open-circuit test	Short-circuit test
E bet. lines = 6 V dc	Field current = 12.5 A dc	Field current = 12.5 A dc
I in lines = 10 A dc	E bet. lines = 420 V ac	Line current = rated value

From the preceding data, assuming that the alternator is wye connected, calculate

a. Effective resistance, synchronous impedance and reactance per phase.
b. Voltage regulation of the alternator at 0.8 lagging and 0.8 leading power factors.

Solution:

Assuming that the alternator is wye connected:

a. I_a rated $= \dfrac{\text{kVA} \times 1000}{V_l \sqrt{3}} = \dfrac{100{,}000}{1100 \times 1.73} = 52.5$ A

$R_{dc} = \dfrac{V_l}{2I_a} = \dfrac{6\text{ V}}{2 \times 10} = 0.3\ \Omega/\text{winding}$ and

$R_{ac} = 0.3 \times 1.5 = 0.45\ \Omega/\text{phase}$

$Z_p = \dfrac{E_{gp}}{I_a} = \dfrac{420}{\sqrt{3} \times 52.5} = 4.62\ \Omega/\text{phase} \tag{6-6}$

$\dot{X}_s = \dot{Z}_p - \dot{R}_a = \sqrt{(4.62)^2 - (0.45)^2} = 4.61\ \Omega/\text{phase}$ [from Eq. (6-7)]

b. $V_p = \dfrac{V_l}{\sqrt{3}} = \dfrac{1100}{\sqrt{3}} = 635$ V/phase

$I_a R_a = 52.5$ A $\times$ 0.45 Ω = 23.6 V/phase
$I_a X_s = 52.5$ A $\times$ 4.61 Ω = 242 V/phase

At 0.8 lagging PF
$$\begin{aligned} E_{gp} &= (V_p \cos\theta + I_a R_a) + j(V_p \sin\theta + I_a X_s) \quad \text{[from Eq. (6-8)]} \\ &= (635 \times 0.8 + 23.6) + j(635 \times 0.6 + 242) \\ &= 530 + j623 = 820 \text{ V/phase} \end{aligned}$$

Per cent voltage regulation $= \dfrac{V_{nl} - V_{fl}}{V_{fl}} \times 100$ (3-9)

$= \dfrac{820 - 635}{635} \times 100 = $ **29.1 per cent**

At 0.8 PF leading
$$\begin{aligned} E_{gp} &= (V_p \cos\theta + I_a R_a) + j(V_p \sin\theta - I_a X_s) \quad \text{(Eq. 6-8)} \\ &= (635 \times 0.8 + 23.6) + j(635 \times 0.6 - 242) \\ &= 530 + j139 = 548 \text{ V/phase} \end{aligned}$$

per cent voltage regulation $= \dfrac{548 - 635}{635} \times 100$ (3-9)

$= $ **−13.65 per cent**

EXAMPLE 6-5: Repeat Example 6-4 assuming that the alternator is delta connected and the same measurements were taken.

Solution:

Assuming that the alternator is delta connected:

$V_l = V_p = 420$ V (from short-circuit test)

$I_p = \dfrac{I_L}{\sqrt{3}} = \dfrac{52.5 \text{ A}}{1.73} = 30.31$ A

$Z_s = \dfrac{420 \text{ V}}{30.31 \text{ A}} = 13.86$ Ω/phase

R_{eff} in Δ = 3 $\times$ R_{eff} in wye = 3 $\times$ 0.45 Ω/phase = **1.35 Ω/phase**

$X_s = \sqrt{Z_s^2 - R_{\text{eff}}^2} = \sqrt{(13.86)^2 - (1.35)^2} = 13.8$ Ω/phase (6-7)

Note that in each instance the equivalent resistance, reactance, and impedance per phase connected in delta is *three times* the value when connected in wye (see Example 6-4).

Rated voltage $V_l = V_p = 1100$ V and $I_p = 30.31$ A/phase

$I_a R_a = 30.31 \times 1.35 = 40.8$ V/phase
$I_a X_s = 30.31 \times 13.8 = 4.19$ V/phase

At 0.8 PF lagging
$$\begin{aligned} E_{gp} &= (V_p \cos\theta + I_a R_a) + j(V_p \sin\theta + I_a X_s) \quad (6\text{-}8) \\ &= (1100 \times 0.8 + 40.8) + j(1100 \times 0.6 + 419) \\ &= 920.8 + j1079 \\ &= 1421 \text{ V/phase} \end{aligned}$$

$$\text{Per cent voltage regulation} = \frac{V_{nl} - V_{fl}}{V_{fl}} \times 100 \tag{3-9}$$

$$= \frac{1421 - 1100}{1100} \times 100$$

$$= \textbf{29.1 per cent} \text{ (as in Example 6-4)}$$

At 0.8 PF leading

$$E_{gp} = (V_p \cos\theta + I_a R_a) + j(V_p \sin\theta - I_a X_s) \tag{6-8}$$

$$= (1100 \times 0.8 + 40.8) + j(1100 \times 0.6 - 419)$$

$$= 920.8 + j241$$

$$= 950 \text{ V/phase}$$

$$\text{Per cent voltage regulation} = \frac{950 - 1100}{1100} \times 100 \tag{3-9}$$

$$= \textbf{-13.65 per cent} \text{ (as in Example 6-3)}$$

Examples 6-3 and 6-4 prove conclusively that, *regardless* of which assumption (delta or wye) is taken for an alternator, if the calculations are carried through *consistently* and *completely*, the *same* results are obtained. The wye assumption is *recommended* because of the relative simplicity of the dc and effective resistance between lines. Moreover, most alternators are, in fact, wye-connected because a neutral connection may be brought out to provide a ground protection circuit. Furthermore, the wye connection automatically produces a higher line voltage for a given phase voltage and is therefore preferred where ac power is transmitted over long distances.

6-11. ASSUMPTIONS INHERENT IN THE SYNCHRONOUS IMPEDANCE METHOD

Examination of Fig. 6-8 and Eq. (6-6) reveals that synchronous impedance is, at all times, the ratio of the open-circuit curve to the short-circuit curve. When the two curves are *linear*, the synchronous impedance is *constant*, i.e., the ratio of two points on straight lines. Above the knee of the saturation curve, however, the synchronous impedance drops as the curves approach each other. As shown in Fig. 6-8, the synchronous impedance is obtained well *below* saturation and is therefore *larger* than under normal operating conditions. What is worse, however (as may be seen from Examples 6-3 and 6-4), is that the armature resistance is negligible compared to the synchronous reactance per phase. Under short-circuit conditions, therefore, the current in the armature lags the generated voltage by almost 90°, and the armature reaction is almost totally *demagnetizing*, much greater than under normal conditions. This demagnetizing effect reduces the degree of saturation still further. The effect due to armature reaction, therefore, is *too* pronounced, and the corresponding values of synchronous impedance and reactance, computed by this method, are *too large*. An attempt is sometimes made to compensate for this by reducing the synchronous reactance by a factor of 0.75 of that calculated.

Another assumption reverts back to Figs. 6-3 and 6-4, where we assumed that the effects of the armature reaction flux induce a voltage which could be added to the armature reactance volt drop, and that this voltage is a function of the load current. Neither of these assumptions is quite correct, since the shift of armature flux varies with the power factor as well as with the load current, and a distortion of the main field flux is produced as shown in Fig. 5-3. Thus, the *armature flux* and its resultant emf (the reason why the method is sometimes called the emf method) *cannot* always be assumed to be *in phase* with the armature reactance.

Still another assumption is that the armature flux path, produced by the armature mmf, is constant through the armature iron, the air gap, and the field poles. As the armature flux shifts with the power factor, however the reluctance of the magnetic circuit varies, particularly for machines with salient poles having large interpolar spaces. This variation in reluctance and armature flux with the power factor rather than with the load current causes some differences between actual (direct loading) and calculated values of regulation by the synchronous impedance method.

Despite the theoretical weaknesses of the synchronous impedance or emf method of determining alternator regulation, it *is* the simplest to perform, compute and comprehend. The "pessimistic" value of synchronous reactance* is not a disadvantage when one is aware of and can make allowances for it.

6-12. SHORT-CIRCUIT CURRENT AND USE OF CURRENT-LIMITING REACTORS

Although alternators are protected by circuit breakers and other overload protection devices, these may require a few cycles before responding to the overload. At the instant of short circuit, only the almost negligible resistance of the alternator limits the current, since it takes the armature reaction several cycles to weaken the field appreciably by demagnetizing the field. This inrush of momentary maximum short-circuit current may be so excessive that even the switch gear, the bus bars, and the machine windings can be severely damaged. It is customary, therefore, to place *current-limiting reactors*, consisting of heavy *turns* of cable or bus bar, in series with each phase of the alternator stator, external to the machine. Without series reactors, the short-circuit current may be as high

---

* A more accurate method, sometimes used, for determining synchronous reactance is the so-called "slip" method which uses the components of the synchronous reactance: X_d, the direct synchronous reactance, and X_q, the quadrature synchronous reactance. Vector diagrams are plotted using the variations in these reactances for the pole position with respect to the stator, and the E_{gp} is determined from the diagrams. The details of this method are beyond the scope of this volume; further information can be obtained by reference to B. L. Robertson and L. J. Black, *Electric Circuits and Machines*, Second Edition, Princeton, N.J.: D. Van Nostrand Company, Inc., 1957, pp. 221–227.

as ten times the full-load current. With series reactors, it is usually limited to about twice the full-load current, the reactors creating an impedance drop (almost pure reactance) as high as twenty per cent of the rated alternator voltage. Within a few cycles after the inrush of current, armature reaction reduces the short-circuit current from maximum to its sustained or steady short-circuit value, where it may be simultaneously interrupted by overload equipment.

It is primarily for reasons of short-circuit protection that *no* attempt is made to compensate for armature reaction in large alternators (see the last sentence in Section 5-7). Instead, it is customary to design alternators with a *high* ratio of armatures synchronous reactance to resistance to reduce the sustained short-circuit current to approximately rated current, as shown by the following example.

EXAMPLE 6-6: A three-phase, wye-connected, 11,000 V, 165,000 kVA alternator has a synchronous reactance of 1.0 ohm and an armature resistance of 0.1 ohm/phase. Calculate
 a. Maximum short-circuit current at instant of short circuit and overload.
 b. Sustained short-circuit current and overload.
 c. Maximum short-circuit current with reactors of 0.8 Ω reactance/phase and negligible resistance.

Solution:

$$\text{Rated } E_p = \frac{E_L}{\sqrt{3}} = \frac{11{,}000}{1.73} = 7040 \text{ V}$$

$$\text{Rated } I_p = \frac{\text{kVA} \times 1000}{3 E_p} = \frac{165{,}000 \times 1000}{3 \times 7040} = 7810 \text{ A}$$

a. $$I_{max} = \frac{E_p}{R_p} = \frac{7040 \text{ V}}{0.1 \text{ Ω}} = 70{,}400 \text{ A}$$

$$\text{overload} = \frac{70{,}400 \text{ A}}{7810 \text{ A}} = 9 \times \text{rated current}$$

b. $$I_{steady} = \frac{E_p}{Z_p} = \frac{7040 \text{ V}}{1 \text{ Ω}} = 7040 \text{ A}$$

$$\text{overload} = \frac{7040 \text{ A}}{7810 \text{ A}} = 0.9 \times \text{rated current}$$

c. $$I_{max} = \frac{E_p}{Z_t} = \frac{7040 \text{ V}}{0.1 + j0.8} = \frac{7040 \text{ V}}{0.814 \text{ Ω}} = 8640 \text{ A}$$

BIBLIOGRAPHY

Alger, P.L. *The Nature of Polyphase Induction Machines*, New York: Wiley Interscience 1951.

Bewley, L. V. *Alternating Current Machinery*, New York: The Macmillan Company, 1949.

Bewley, L. V. *Tensor Analysis of Electrical Circuits and Machines*, New York: Ronald Press, 1961.

Carr, C. C. *Electrical Machinery*, New York: John Wiley & Sons, Inc., 1958.

Crosno, C. D. *Fundamentals of Electromechanical Conversion*, New York: Harcourt, Brace, Jovanovich, Inc, 1968.

Daniels. *The Performance of Electrical Machines*, New York: McGraw-Hill, Inc., 1968.

Fitzgerald, A. E. and Kingsley, C. *The Dynamics and Statics of Electromechanical Energy Conversion*, 2nd Ed., New York: McGraw-Hill, 1961.

Fitzgerald, A. E., Kingsley, Jr. C., and Kusko, A. *Electric Machinery*, 3rd Ed. New York: McGraw-Hill Inc. 1971.

Gemlich, D. K. and Hammond, S. B. *Electromechanical Systems*, New York: McGraw-Hill, 1967.

Hindmarsh, J. *Electrical Machines*, Elmsford, N.Y.: Pergamon Press, 1965.

Jones, C. V. *The Unified Theory of Electrical Machines*, New York: Plenum Publishing Corp., 1968.

Koenig, H. E. and Blackwell, W. A. *Electromechanical System Theory*, New York: McGraw-Hill, 1961.

Liwschitz, M. M., Garik M., and Whipple. C. C. *Alternating Current Machines*, Princeton, N. J.: D. Van Nostrand Co., Inc., 1946.

Majmudar, H. *Introduction to Electrical Machines*, Boston: Allyn and Bacon, 1969.

McFarland, T. E. *Alternating Current Machines*, Princeton, N. J.: D. Van Nostrand Co., Inc., 1948.

Meisel, J. *Principles of Electromechanical Energy Conversion*, New York: McGraw-Hill, 1966.

Nasar, S. A. *Electromagnetic Energy Conversion Devices and Systems*, Englewood Cliffs, N.J.: Prentice-Hall, Inc., 1970.

O'Kelly and Simmons. *An Introduction to Generalized Electrical Machine Theory*, New York: McGraw-Hill, 1968.

Puchstein, A. F., Lloyd, R., and Conrad, A. G. *Alternating Current Machines*, 3rd Ed., New York: Wiley/Interscience, 1954.

Robertson, B. L. and Black. L. J. *Electric Circuits and Machines*, 2nd Ed., Princeton, N. J.: D. Van Nostrand Co., Inc., 1957.

Schmitz, N. L., and Novotny, D. W. *Introductory Electromechanics*, New York: Ronald Press, 1965.

Seely, S. *Electromechanical Energy Conversion*, New York: McGraw-Hill, 1962.

Selmon. *Magnetoelectric Devices: Transducers, Transformers and Machines*, New York: Wiley/Interscience, 1966.

Skilling, H. H. *Electromechanics: A First Course in Electromechanical Energy Conversion*, New York: Wiley/Interscience, 1962.

Synchronous Generators and Motors. (ASA C50.1.) New York: American Standards Association.

Thaler, G. J., and Wilcox, M. L. *Electric Machines: Dynamics and Steady State*, New York: Wiley/Interscience, 1966.

White, D. C., and Woodson, H. H. *Electromechanical Energy Conversion*, New York: Wiley/Interscience, 1959.

QUESTIONS

6-1. For an alternator, state:
   a. The relation between electromagnetic torque developed in the armature conductors and the torque applied by the prime mover driving the alternator
   b. The relation between generated armature voltage per phase, armature current and terminal voltage per phase
   c. The relations expressed in (b) in the form of an equation
   d. The equation for generated voltage per phase in terms of the number of turns per phase.

6-2. A study of dc dynamos shows that ac is generated in the conductors of a rotating armature. This leads to the possible advantage of universal dynamo construction capable of supplying either dc or ac (or both), having rotating armatures and stationary fields. Explain:
   a. Why the universal dynamo is seldom used
   b. Give 7 compelling reasons for use of stationary armature and rotating field on ac dynamos.

6-3. a. Describe two types of rotating field construction used in alternators.
   b. What factors determine the choice of construction type?
   c. How is it possible to distinguish between the two types of construction on the basis of general appearance?

6-4. a. Draw the equivalent circuit of a three-phase, wye-connected synchronous alternator connected to a balanced three-phase reactive load.
   b. Write the equation expressing the relationship between the generated voltage per phase and the terminal voltage per phase, including those factors which are responsible for the difference between the two.

6-5. Draw phasor diagrams showing the relationship between V_p and E_{gp} for loads at:
   a. unity power factor
   b. lagging power factors
   c. leading power factors.

6-6. a. On a phase basis, represent the voltage-load characteristic of an alternator for the three conditions of Quest. 6-5 above.
   b. Under what conditions of load is it possible for an alternator to have a voltage regulation of zero percent?

6-7. a. Compare the inherent voltage regulation of a separately excited dc generator with a separately excited alternator and account for the advantages of the former.

b. Explain why armature reaction is always compensated in the dc generator but never in the ac alternator.

6-8. a. Define synchronous impedance.
b. What are the advantages of this concept and where is it used?

6-9. In performing the synchronous impedance test for determining alternator voltage regulation:
a. draw the test circuit connections
b. describe the precautions required in measuring effective armature resistance and explain why a dc rather than an ac measurement is made
c. explain why a conventional method is used to predict voltage regulation and efficiency rather than direct loading
d. explain what precautions are necessary in performing the open-circuit test
e. describe the short-circuit test.

6-10. For the synchronous impedance method, explain:
a. why it is called a "pessimistic method"
b. why Z_s as calculated from the curves of Fig. 6-8 is not constant but decreases at higher values of field current
c. at least four assumptions of the method which are not equivalent to direct loading and cause differences in calculated regulation
d. why it is used despite its disadvantages.

6-11. a. Explain why alternators are designed intentionally to have a high ratio of armature reactance to resistance
b. Under what conditions are current-limiting reactors used?

6-12. Assuming a delta-connected alternator is tested by the methods shown in Fig. 6-7 and *all* measurements are made of voltage *between* lines and current *in* the lines, explain why:
a. the same results are obtained using the assumption that the alternator is wye-connected
b. it is preferable to assume the alternator is wye-connected
c. it is necessary to carry the assumption completely throughout *all* the calculations.

PROBLEMS

6-1. Calculate:
a. The number of poles required for an alternator driven by a prime mover having a speed of 720 rpm if it is desired to generate ac at a frequency of 60 Hz
b. From two poles to 10 poles, calculate the various prime mover speeds required to generate 25 Hz

c. The frequency produced by a prime mover turning a 10-pole alternator at 800 rpm.

6-2. Given a 60 Hz, four-pole synchronous alternator driven at a speed of 1000 rpm, calculate
a. The effect on its generated voltage
b. The effect on its armature leakage reactance
c. The effect on its armature reaction.

6-3. A 600 kVA, 125 V delta-connected alternator is reconnected in wye. Calculate its new rating in:
a. volts
b. amperes
c. kilovolt amperes.

6-4. A 1000 kVA, 440 V wye-connected alternator is reconnected in delta. Calculate its new rating in
a. volts
b. amperes
c. kilovolt amperes.

6-5. A delta-connected alternator supplies a delta-connected resistive load requiring 150 kW at 550 V. Calculate:
a. The line current
b. The line current to the *same* resistive load and total power dissipated, if the alternator is reconnected in wye and driven at the same speed and excitation as in (a) of this problem.

6-6. A three-phase load of 10 Ω/phase may be connected by means of switches in wye or delta. If connected to a 220-V, three-phase alternator, calculate
a. Power dissipated in wye
b. Power dissipated in delta
c. Ratio of (b) to (a).

6-7. A three-phase alternator delivers 500 kW to a group of induction motors at a power factor of 0.8 lagging. If the alternator is rated at 750 kVA, calculate
a. The number of 100-W lamps which may be supplied in addition to the motors before the alternator is operating at rated load.
b. Repeat (a) if the power factor of the motors drops to 0.7.

6-8. A three-phase, wye-connected, 1500-kVA, 13-kV alternator has an armature resistance of 0.9 Ω and a synchronous reactance of 8.0 Ω. When carrying rated load at rated voltage, calculate generated voltage per phase at loads of
a. unity power factor
b. 0.8 lagging
c. 0.8 leading.
d. Calculate voltage regulation for each of these loads and determine best regulation.

6-9. A three-phase, 13,000 V, 2500 kVA alternator has an armature resistance of 0.3 Ω/phase and a synchronous reactance of 4.0 ohms/phase. The alternator excitation is adjusted in each of the cases to provide rated

voltage at rated load. If, after the adjustment is made, the load is suddenly removed from the alternator, calculate the no-load voltage per phase and line for loads of
a. unity power factor
b. 0.8 lagging
c. 0.8 leading.
d. Calculate voltage regulation for each of the above and determine best regulation.

6-10. A 2300 V, 2500 kVA, three-phase, delta-connected alternator has a resistance of 0.1 Ω/phase and a synchronous reactance of 1.5 Ω/phase The alternator is adjusted to rated voltage *at no load*. Calculate its terminal voltage when carrying rated current at a power factor of 0.6 lagging.

6-11. A 220 V, 100 kVA, wye-connected alternator has an armature resistance per phase of 0.1 Ω and a reactance per phase of 0.5 Ω. Assuming that when it is connected to a 0.4 PF lagging load and delivers rated current, the armature reaction has twice the effect of armature reactance, and neglecting the effect of saturation, calculate
a. No-load voltage when load is disconnected and speed and field current are the same
b. The no-load voltage required to produce rated current assuming the alternator were short circuited.

6-12. The voltage regulation of a 550 V, single-phase alternator rated at 100 kVA is to be determined from the following test data, using the synchronous impedance method. When the alternator armature is short circuited through an ammeter, the alternator delivered 350 A at a dc field excitation current of 12 A. At the same excitation current, when the short circuit was removed, an open circuit voltage of 350 V was recorded across the armature terminals. A Wheatstone Bridge measurement of the armature resistance indicated it to be 0.1 Ω. Assuming this alternator has a ratio of effective to ohmic resistance of 1.25, calculate:
a. Synchronous impedance and reactance
b. Voltage regulation at 0.8 PF leading and lagging.

6-13. A 2300 V, three-phase, 60 Hz alternator rated at 1200 kVA is short circuited, brought up to rated speed, and its field excitation is increased until 1.5 times rated armature current flows. The short circuit is then removed, and with identical field current and speed, the voltage across each of the line terminals is 1000 V. The average of the dc resistance of the armature windings, taken between lines, is 0.225 Ω. Assuming the alternator to be delta-connected and the ratio of effective to ohmic resistance as 1.4, calculate
a. Rated line and phase current and ac resistance per phase
b. Full-load armature resistance and synchronous reactance voltage drops
c. Voltage regulation at 0.8 PF leading and lagging.

6-14. Repeat Problem 6-13 using the assumption that the alternator is wye connected.

6-15. a. Tabulate results of Problem 6-13, using delta assumption and Problem 6-14, using wye assumption, and compute ratio of delta to wye for the following quantities: synchronous impedance per phase, armature resistance ac per phase, armature reactance per phase, armature current per phase, full-load armature-resistance voltage drop, full-load armature-reactance voltage drop, voltage regulation at 0.8 PF lagging, voltage regulation at 0.8 PF leading.
b. Using the ratios tabulated, explain why the voltage regulation and kVA ratings for either assumption must have a ratio of unity.

From the data given for Problems 6-16 through 6-19, assuming an effective to dc resistance ratio of 1.3, calculate voltage regulation at (a) unity power factor and at (b) power factors of 0.8 leading and lagging.

	Connection	RATED kVA	RATED Line volts	DC RESISTANCE TEST Line volts (dc)	DC RESISTANCE TEST Line current (dc)	OPEN-CIRCUIT TEST Line volts	OPEN-CIRCUIT TEST Field current (dc)	SHORT-CIRCUIT TEST Line current (rms)
6-16.	Y	2000	2300 rms	1.5 V	10 A	950 rms	10 A	rated
6-17.	Δ	1000	600	10.0	rated	275	25	rated
6-18.	Y	25	220	25.0	75	200	5	rated
6-19.	Y	500	2300	20.0	40	800	10	rated

6-20. a. Tabulate voltage regulation at power factors of unity, 0.8 lagging, and 0.8 leading as calculated in Problems 6-16 through 6-19
b. From your tabulation, draw inferences as to the magnitude of leading and lagging power factors from the magnitude of voltage regulation at unity power factor

ANSWERS

6-1(a) 10 (b) 1500 rpm (c) $66\frac{2}{3}$ Hz 6-2(a) 0.555 E rated (b) same (c) no effect 6-3(a) 216 V (b) 1600 A (c) 600 kVA 6-4(a) 254 V (b) 2270 A (c) 1000 kVA 6-5(a) 157.5 A (b) 272 A, 450 kW 6-6(a) 4830 W (b) 14,490 W (c) 3:1 6-7(a) 1500 lamps (b) 500 lamps 6-8(a) 7580.3 V (b) 8575 V (c) 7275 V (d) 0.798, 14.03, 0.732 per cent 6-9(a) 7543 V (b) 7800 V (c) 7275 V (d) 0.439, 3.86, −3.13 per cent 6-10. 1450 V 6-11(a) 515 V (b) 393.2 V 6-12(a) 1 Ω, 0.992 Ω (b) −8.18, 18 per cent 6-13(a) $\sqrt{3}$ × 174 A, 174 A (b) 82.2 V, 661 V (c) −10.85, 22.2 per cent 6-14(a) 301 A, (b) 47.5 V, 384 V (c) −10.9, 22.2 per cent 6-16 11.42, −14.28, 31.4 per cent 6-17 13, −15.5, 34.8 per cent 6-18 43.6 −4.48, 82 per cent 6-19 8.82, −13, 26.2 per cent

SEVEN

parallel operation

7-1. ADVANTAGES OF PARALLEL OPERATION

The old proverb of "not putting all one's eggs in one basket" is the fundamental principle governing parallel operation. A utility system usually consists of several generating stations, all operating in parallel. At each of the stations, there may be several ac alternators and/or dc generators, operating in parallel. There are numerous advantages to the subdivision of a generating system into several smaller stations, from an economic as well as from a military point of view. These advantages also apply to the use of several smaller generating units rather than a single large dynamo, although the latter is more efficient when loaded to its capacity. The chief advantages of parallel operation on both a system and station basis are:

1. If a single large unit is disabled, for whatever reason, the station is no longer functional; whereas, if one of several smaller units is in need of repair, the other smaller units are still available to provide service as needed.

2. A single large unit, for maximum efficiency, must be loaded to capacity. It is uneconomical to operate a large unit at light loads. Several smaller units operated in parallel may be removed or added as the service demands fluctuate; and each unit may be operated at near its rated capacity, thus providing maximum station and system efficiency.
3. In the event of repair, or shut-down for purposes of maintenance, smaller units facilitate maintenance, from the point of view of spares, replacements, and servicing of units.
4. As the average system and station demand is increased, additional units may be purchased and added to keep pace with this demand. The capital outlay is less initially, and growth corresponds to increasing average demand.
5. There is a physical and economic limit to the possible capacity of any single unit. For example, at a given generating station, the load may be 10 million kVA. Although units of up to several hundred thousand kVA have been operating, no single unit is constructed of sufficient capacity to supply so high a station or system demand.

For the reasons cited, parallel operation is advisable. It is customary at any given station to employ one large unit to handle the minimum demand, and to add other smaller units as the demand rises over a 24 hour period. When the growth of the station requires that several smaller units are continuously employed to handle the minimum demand, a second larger unit of equivalent capacity is added; and so on.

7-2. VOLTAGE AND CURRENT RELATIONS FOR SOURCES OF EMF IN PARALLEL

A parallel circuit is defined as one in which the *same* terminal voltage exists across all units (in parallel). When multiple sources of emf are connected in parallel, as shown in Fig. 7-1, the *same* voltage, V_L, at the bus bars, parallels the various sources as well as the load, Z_L (neglecting any drop in the lines interconnecting the various generators). The following relations hold true regardless of whether the sources of emf are batteries, dc generators, alternators, solar cells, power supplies, etc.*

$$V_L = I_L Z_L = \dot{E}_{g_1} - I_1 \dot{Z}_1 = \dot{E}_{g_2} - I_2 \dot{Z}_2 = \dot{E}_{g_3} - I_3 \dot{Z}_3 \qquad (7\text{-}1)$$

where E_g is the voltage generated by the source(s)
V_L is the terminal voltage at the bus
I_L is the total current delivered by the various sources to the load

* Equation 7-1 emerges from Millman's theorem in which the solution for V_L is obtained using Kirchhoff's current law

$$V_L = \frac{\dfrac{E_{g_1}}{Z_1} + \dfrac{E_{g_2}}{Z_2} + \dfrac{E_{g_3}}{Z_3}}{\dfrac{1}{Z_1} + \dfrac{1}{Z_2} + \dfrac{1}{Z_3}} = \frac{E_{g_1} Y_1 + E_{g_2} Y_2 + E_{g_3} Y_3}{Y_1 + Y_2 + Y_3}$$

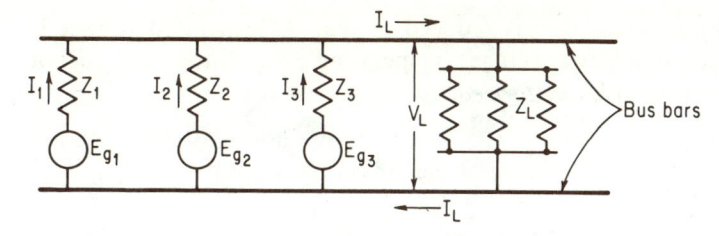

Figure 7-1

Voltage and current relations for sources of emf in parallel.

Z_L is the equivalent impedance (or resistance) of the load
Z_1, Z_2, Z_3 are the equivalent internal impedances (or resistances) of the source of emf
I_1, I_2, I_3 are the respective currents delivered by each of the sources of emf

In the above equation, it is not necessary that each of the sources produce the same generated emf or deliver the same current to supply the load. But in order for any unit to serve as a *source* of emf, as indicated by Eq. (7-1), it *is* necessary that the emf generated by that unit *exceed* the bus voltage V_L, in order to supply current *to* the bus.

If a source of emf produces a voltage E_g which is exactly equal to the terminal bus voltage, V_L, such a source is said to be *floating* across the line, i.e., it neither delivers nor draws current from the bus. For example, if $E_{g_3} = V_L$ in Fig. 7-1, I_3 is equal to zero; and there is no internal volt drop created in the source E_{g_3} because no current is delivered to or drawn from the bus.

If a source of emf produces a voltage E_g that is *less* than the bus voltage, current will be delivered by the bus, i.e., the other sources in parallel, *to the source*. Since current is flowing *into* the source when the bus voltage exceeds the source voltage, the relation between them is expressed by the equation

$$\dot{V}_L = \dot{E}_g + I_g \dot{Z}_g \quad \text{or} \quad \dot{E}_g = \dot{V}_L - I_g \dot{Z}_g \quad (1\text{-}9)$$

where I_g is the current in the generating source and Z_g is the internal impedance (or resistance) of the generating source.

When the source of emf is a rotating dynamo whose generated emf, E_g, *exceeds the terminal bus voltage*, the dynamo operation is called generator action [Eq. (1-10)] and the dynamo is operating in the *generator mode*.

When the generated emf of a dynamo is less than the voltage impressed across its armature, and the dynamo *receives* current from the bus [Eq. (1-9)], the dynamo operation is called motor action and the dynamo is operating in the *motor mode*. Consequently, any generator (in parallel with a bus) whose excitation is reduced so that the generated voltage is *less* than the terminal bus voltage, is operating as a motor; and such a "generator" is said to be *motorized*.

> When operating either in the generator or motor mode, the total power *generated* by a dynamo on a per phase basis is
>
> $$P_{g_1} = E_{g_1} I_{a_1} \cos \theta_1 \tag{7-2a}$$
>
> where E_{g_1} is the generated voltage per phase
> I_{a_1} is the phase current
> θ_1 is the phase angle between E_{g_1} and I_{a_1}
>
> While the power *delivered* to or *received* from the bus, on a per phase basis, is
>
> $$P_{L_1} = V_{L_1} I_{L_1} \cos \theta_1 \tag{7-2b}$$
>
> where V_{L_1} is the terminal (phase) voltage at the bus or lines
> I_{L_1} is the (phase) current entering or leaving the bus or lines

In the *generator* mode, when a source is delivering power to the bus, P_g exceeds P_L by that power consumed within the source itself.

In the *motor* mode, when a "source" is receiving power from the bus and is "motorized", P_L exceeds P_g by the power consumed within the source itself. P_g is the *internal power developed* by the dynamo armature in the direction of rotation of the armature, as noted in Eq. (4-7) for the dc motor.

Example 7-1 below treats the distinctions between Eqs. (7-2a) and (7-2b) for 3 dynamos in parallel in which one is in the generator mode, one is floating, and one is in the motor mode. Example 7-1 uses dc dynamos to simplify calculations.

EXAMPLE 7-1: Three dc shunt dynamos driven by a prime mover each having a field resistance of 120 ohms and an armature resistance of 0.1 ohm are connected to a 120 V bus. Dynamos *A*, *B*, and *C* have generated voltages of 125 V, 120 V, and 114 V, respectively. Calculate for each dynamo
a. The line current drawn from or delivered to the bus, and the armature current.
b. The power drawn from or delivered to the bus, and the power generated.

Solution:

a. $I_{gA} = \dfrac{E_g - V_L}{R_a} = \dfrac{125 - 120}{0.1} = 50 \text{ A}$ \hfill (7-1)

$I_f = \dfrac{120 \text{ V}}{120 \text{ }\Omega} = 1 \text{ A}$

Dynamo *A* delivers 50 A to the bus, and has an armature current of 50 A + 1 A = 51 A.

For dynamo B, $I_{gB} = \dfrac{120 - 120}{0.1} = 0$

Dynamo B is floating, and has an armature and field current of **1 A**.

For dynamo C, $I_{gC} = \dfrac{V_L - E_g}{R_a} = \dfrac{120 - 114}{0.1} = 60$ A

Dynamo C receives **61 A** from the bus, and has an armature current of **60 A**.

b. 1. Power delivered to the bus by dynamo A is

$$P_{LA} = V_L I_L = 120 \times 50 = 6000 \text{ W} \tag{7-2b}$$

power generated by dynamo A is

$$P_{gA} = E_g I_a = 125 \times 51 = 6375 \text{ W} \tag{7-2a}$$

2. Since dynamo B neither delivers power to nor receives power from the bus,

$$P_B = 0$$

Power generated by dynamo B, to excite its field, is

$$P_{gB} = E_g I_a = 120 \times 1 = 120 \text{ W}$$

3. Power delivered by the bus to dynamo C is

$$P_{LC} = V_L I_L = 120 \times 61 = 7320 \text{ W}$$

while the internal power delivered in the direction of rotation of its prime mover to aid rotation is

$$P_{gC} = E_g I_a = 114 \times 60 = 6840 \text{ W}$$

7-3. PARALLEL OPERATION OF SHUNT GENERATORS

Shunt generators are well suited for parallel operation, particularly because of their inherently drooping voltage characteristic with load. This may be seen from an examination of Fig. 7-2, which shows two shunt generators (of unequal capacity or external characteristic) in parallel delivering current to a bus and an external load. Since they are in parallel, the same voltage V_L exists at the bus. Generator 2 delivers a current I_2, and generator 1 delivers a currrent I_1 to the bus and load.

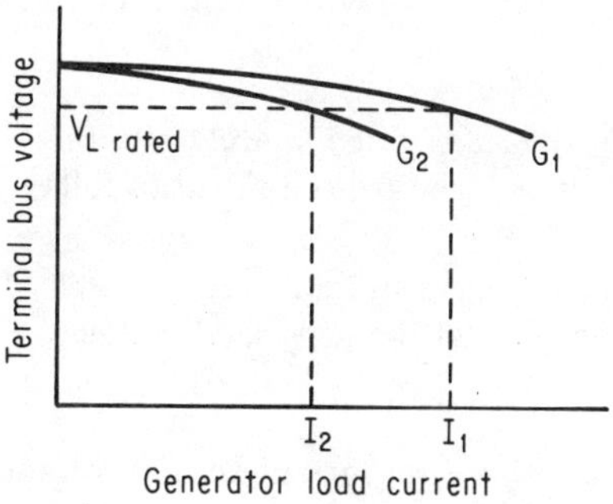

Assume now that the prime mover of generator 1 increases in speed momentarily. This will cause an increase in the generated voltage of generator 1, and also an increase over the terminal bus voltage, causing it to deliver more current to the load. Assuming that the load resistance remains constant, two conditions are operating: (1) the remaining generator(s) now carries less load and its terminal voltage rises, causing it to take back some share of the load in return; and (2) the sudden increase of load on generator 1 causes its voltage to drop, as indicated by the characteristic of Fig. 7-2.

Figure 7-2

Two dc shunt generators dividing load in parallel.

SEC. 7-3. / Parallel Operation of Shunt Generators

Thus, the two conditions described above tend to *oppose* and *reduce* the tendency of any paralleled shunt generator delivering current to a bus to take more than its share of the load. For this reason, *any* voltage source which possesses a *drooping* voltage characteristic will exist in stable equilibrium in parallel with other sources of *similar* characteristics.

Assuming that the load on *both* generators shown in Fig. 7-2 increases or decreases, the terminal bus or load voltage V_L will drop or rise, respectively, and each generator will carry a proportionate share of the load.

EXAMPLE 7-2: Generator 1 in Fig. 7-2 has a rating of 300 kW, and generator 2 has a rating of 600 kW, at a rated voltage (for both) of 220 V dc. If the no-load voltage for both generators is 250 V, assuming linear characteristics, calculate:

The total load and kW output of each generator when the terminal voltage is a. 230 V and b. 240 V. c. The per cent of rated kW carried by each generator at each voltage above.

Solution:

a. At 230 V:

Generator 1 carries $\dfrac{250-230}{250-220} \times 300 \text{ kW} = \dfrac{2}{3} \times 300 \text{ kW} = \mathbf{200\,kW}$

Generator 2 carries $\frac{2}{3} \times 600 \text{ kW} = \mathbf{400\,kW}$

b. At 240 V:

Generator 1 carries $\dfrac{250-240}{250-220} \times 300 \text{ kW} = \dfrac{1}{3} \times 300 \text{ kW} = \mathbf{100\,kW}$

Generator 2 carries $\frac{1}{3} \times 600 \text{ kW} = \mathbf{200\,kW}$

c. Both generators carry no load at 250 V, $\frac{1}{3}$ rated load at 240 V, $\frac{2}{3}$ rated load at 230 V, and rated load at 220 V.

7-4. CONDITIONS NECESSARY FOR PARALLEL OPERATION OF SHUNT GENERATORS*

It is evident from Example 7-2 that, if two shunt generators are to share and divide the total load equally *in proportion* to their *kilowatt* output *ratings*, the following conditions are necessary:

1. Each generator must have the *same* rated *voltage* and the *same* voltage *regulation* (voltage drop from no-load to full-load).

2. The polarities of all parallel connected generators must *oppose* each other (i.e., plus vs. plus, minus vs. minus), and the generated voltages must be higher than the bus voltage.

As to the first requirement, there are three factors affecting voltage

---

* Throughout this discussion, it is assumed that the speed characteristics of the prime movers driving the generators are either constant or drooping. Since we are concerned with electric machines, no mention is made of prime-mover characteristics or of the various types which may be employed.

regulation of a self-excited shunt generator (Section 3-13), namely, the internal $I_a R_a$ drop, the armature reaction, and the decrease in field current produced by the drop in armature voltage. It is not essential that each of these effects be similar for two machines, but the resultant external characteristics of each should be similar in form.

Regarding the second requirement above, it should be noted in Fig. 7-3 that, without a load across the bus bars, the two generators are connected to the bus so that their **emfs** tend to oppose and drive current to the bus (and to each other).

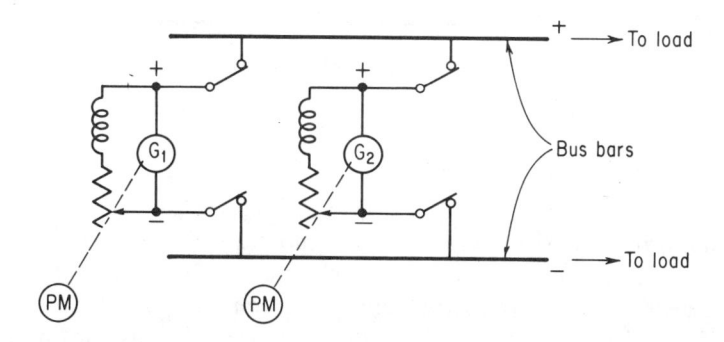

Figure 7-3

Two shunt generators in parallel.

7-5. PARALLEL OPERATION OF COMPOUND GENERATORS

It would appear, at first, that little difficulty should be involved in paralleling compound generators as easily as shunt generators. This is true, of course, providing the compound generators are differentially compounded, or cumulatively undercompounded. But cumulative flat or overcompound generators will *not* operate successfully in parallel without additional compensation. The *rising* external voltage characteristic of the flat and overcompound generators is one which leads to *instability*, as shown in Fig. 7-4a. As in the case of the shunt generators discussed previously, if the prime mover of generator 1 increases in speed, momentarily, the effect on the generator performance is:

1. Generator 1 develops a higher generated emf, causing it to take more load. But a greater load on generator 1 causes its voltage to rise. Generator 1 takes more load, rises in voltage, takes still more load, etc., until it carries *all* the load.

2. Generator 2 at the same time loses load, causing its voltage to drop, resulting in its taking less load with a resultant further drop in voltage, etc., until it carries no load and is even driven as a motor because of the difference in generated voltage between generators 2 and 1.

Figure 7-4

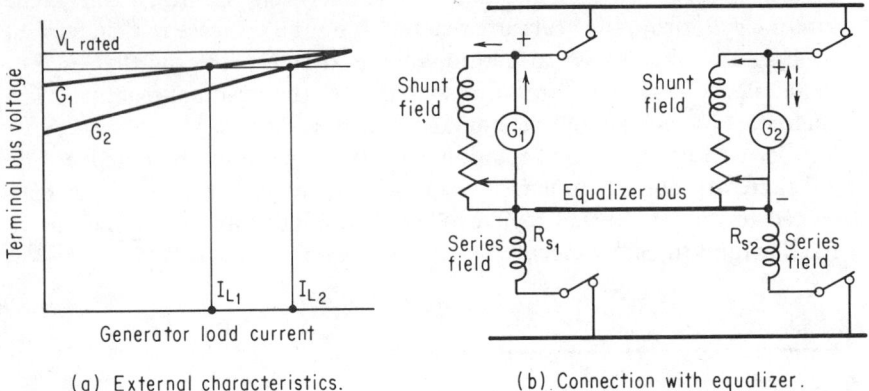

Two overcompound generators dividing load in parallel.

(a) External characteristics. (b) Connection with equalizer.

Stable equilibrium between cumulative compound generators is accomplished by means of an *equalizer*, a low-resistance cable or bus, connected to the *armature side* of the series field of the *same polarity* on *each* machine. In effect, the equalizer parallels *all* the series fields, as shown in Fig. 7-4b, of all the compound generators connected in parallel.

7-6. CONDITIONS NECESSARY FOR PARALLEL OPERATION OF COMPOUND GENERATORS

By paralleling the series fields of all compound generators in parallel, the equalizer maintains the *same* voltage across *all* series fields. This electrical relation makes for stable equilibrium and has the same effect as that produced by a drooping voltage characteristic in equalizing the load. The action of the equalizer is as follows:

1. Assume that generator 1 increases in voltage because of an increase in its prime mover speed, as before. The increase in voltage, due to a rising voltage characteristic, produces an increase in load. The increased armature and series field current of generator 1 causes an increase in the voltage of all the series fields (of all the generators) paralleled by the equalizer.

2. Generator 2 (and all other generators across the bus) produces a higher generated voltage, owing to the increased current in its series field, causing it to take a greater share of the load, which in turn causes its voltage to rise correspondingly.

Thus, the rise in voltage of the other generators tends to oppose and reduce the tendency of generator 1 to carry all the load in much the same way as the rise in voltage (owing to decreased load) of shunt generators opposes the tendency of one generator to carry all the load.

Even if the excitation of generator 2 were deliberately reduced to cause it to motorize and reverse its armature current, the action of the two generators is still stablized by the equalizer. As shown in Fig. 7-4b, only the current in the armature reverses; the current in both the series and the shunt fields remains the same. Since the path through the series fields is of higher resistance than the equalizer, most of the armature current for generator 2 is returned through the equalizer, and generator 2 operates as a shunt motor.

In addition to Conditions 1 and 2 of Section 7-4 for shunt generators, the following *additional* requirements are necessary for *compound* generators:

3. An equalizer must be connected on the *armature* side of the series field on the side of *same polarity for each* machine.
4. The resistance of all series fields must be approximately *inversely* proportional to the *capacities* (kW ratings) of the generators paralleled.

As to the third requirement above, Fig. 7-5 shows the effect of connecting an equalizer on the armature side of each machine but on sides of opposite polarity. This happens quite frequently in the laboratory when students fail to connect the series field of their compound generators to the side of same polarity. Although each generator, *independently*, operates as a cumulative compound generator, it is impossible to connect an equalizer between them successfully without producing a short circuit across the armatures of both machines *when paralleled*. As shown in Fig. 7-5, the drops across the series fields are small and the equalizer is shorting the positive and negative sides of the line.

Regarding the fourth requirement above, since it is the same voltage across the paralleled series fields which produces the equalizing action, the full-load $I_s R_s$ drop must be the same for each series field of each compound generator. For machines of different armature (and correspondingly series field) current values, the higher the rating, the lower its series field resistance.

It is precisely for this reason that, when a diverter is used to adjust the cumulative compound characteristic of a generator, it can no longer be placed in parallel with the series field of a given machine, since it diverts *all* the generators equally because of the paralleling action of the equalizer connection. Compound generators in parallel are diverted by means of *series* diverters, as shown in Fig. 7-6.

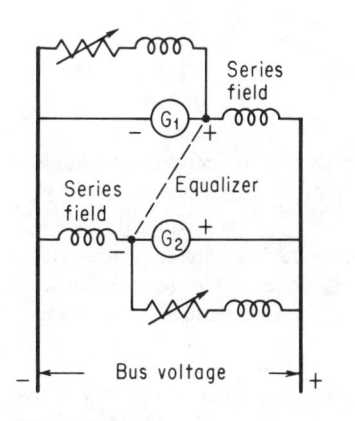

Figure 7-5

Possible erroneous connection of series fields and equalizer in laboratory producing a short circuit.

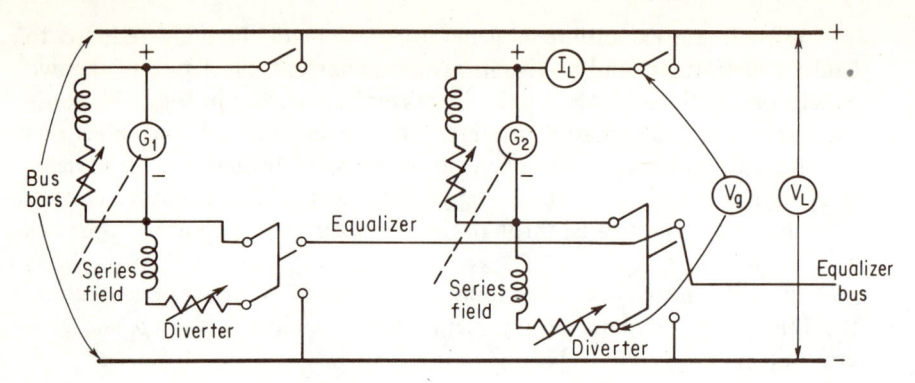

Figure 7-6

Use of diverter and switches for paralleling series fields of compound generators.

7-7.
PROCEDURE FOR PARALLELING GENERATORS

The procedure used to parallel incoming generators to a bus (or to parallel one generator to another) is basically the same, whether shunt or compound generators are used. It consists of the following steps (with reference to Fig. 7-6):

1. Assume that generator 1 is already on the line and delivering current to a load. For purposes of the laboratory, generator 1 is the bus.
2. Generator 2 is brought up to rated speed and adjusted for rated (bus) voltage. (The degree of compounding is assumed to have been previously checked by placing a small load across the generator and shorting the series field. If the voltage drops, it is cumulatively compounded; if the voltage rises, it is differentially compounded.) The polarity of the switch connections and of the series field is also assumed to have been checked with a voltmeter.
3. The double-pole switch of generator 2 is closed to parallel the series fields. The field rheostat should be adjusted to give rated bus voltage, $V_g = V_L$. The single-pole switch may now be closed, and generator 2 is floating on the line.
4. The field rheostat resistance should be reduced to increase the field current of generator 2 so that it may take some of the load carried by the bus.
5. To remove a generator from the bus, decrease the field current until that generator is floating. Open the switches, disconnecting it from the bus. Reduce the speed of the prime mover until the generator no longer rotates.

7-8.
CONDITIONS NECESSARY FOR PARALLELING ALTERNATORS

The conditions stated for the shunt generator in Section 7-4 represent, in essence, the basic requirements for parallel operation of all voltage sources, namely: (1) that the load voltage characteristics of the sources should be the same or nearly similar; and (2) that the polarities of the sources must be equal and opposite with respect to each other, at all times. These basic requirements must be expanded somewhat in order to apply to ac machines; namely,

1. The effective (ac) values of voltage are the same, i.e., all machines must have the *same effective* rated voltage.
2. The voltages of all alternators connected in parallel must have the *same waveform*.
3. The voltages must be exactly opposite in phase (with respect to two alternators or to a given alternator and the bus).
4. The frequencies of all alternators to be paralleled must be the same (that is, the product of their poles and speeds must be the same).
5. The combined over-all alternator voltage and prime-mover speed characteristics should be drooping with application of load.
6. For polyphase machines only, the phase sequence of the polyphase voltages of the incoming alternator must be the same as the bus.

In effect, *all* of the requirements listed (with the exception of the fifth) are met by the single statement that the "*polarities of the sources must be equal and opposite at all times.*" The fifth requirement is met by the statement that "the load-voltage characteristics of the sources should be the same or similar." The additional qualifications are introduced because we are dealing with an alternating voltage possessing a particular waveform, usually assumed to be sinusoidal, as a result of using distributed armature windings see Fig. 2-17c.

We shall deal first with the synchronization or parallel operation of single-phase alternators before proceeding to the synchronization of polyphase alternators because the former is less complex and involves only the first five conditions stated above.

7-9. SYNCHRONIZING SINGLE-PHASE ALTERNATORS

The polarity conditions stated in Section 7-8 are shown in Fig. 7-7a for two single-phase alternators operating in parallel, connected as shown in Fig. 7-7b. The instantaneous polarities generated by each alternator oppose each other at each instant, as shown by the waveforms of Fig. 7-7a and the current directions of Fig. 7-7b. If alternator 1 is taken as a reference, voltage E_1 is opposed at each end every instant by voltage E_2, as shown in the figures. A phasor diagram of this "internal or local" opposition created between the two alternators is shown in Fig. 7-7c, where they are equal and opposite. But if they are equal and opposite, the bus voltage, E_{gp}, should be zero. We are quite aware that it is *not* zero, however. It is fairly obvious then that, in considering phasor diagrams for parallel operation, we shall have to distinguish between those phasor diagrams used to represent voltages in a local circuit, i.e., those generated between two machines or within a single machine, and in the external circuit, i.e., the relation between the bus voltage and the load current it supplies.

(a) Local circuit showing opposing waveforms.

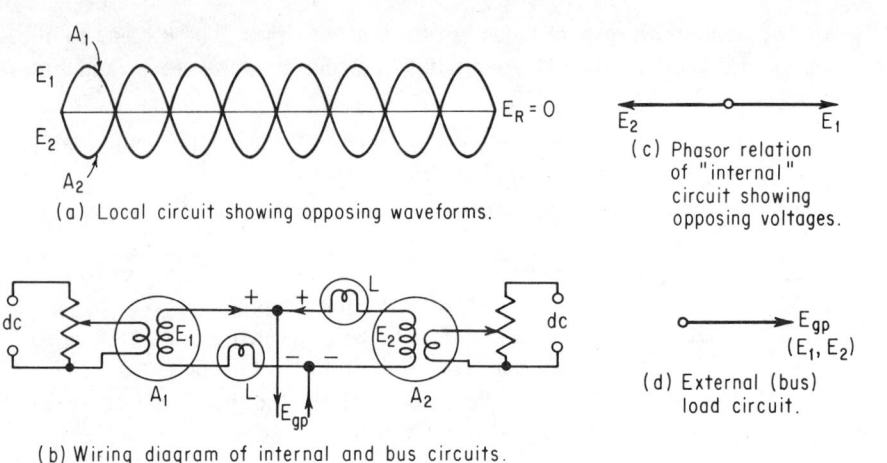

(b) Wiring diagram of internal and bus circuits.

(c) Phasor relation of "internal" circuit showing opposing voltages.

(d) External (bus) load circuit.

Figure 7-7

Synchronization of single-phase alternators.

The diagram of Fig. 7-7a represents perfect synchronization since both single-phase alternator voltages are equal and opposite at each instant, having the same frequency, waveform, and effective value of ac voltage. Let us consider what might happen as a result of a slight change in frequency of E_2 with respect to E_1.

Let us assume that the prime mover of E_2 suffers a reduction in speed, causing its frequency to drop. Previously, the resultant voltage in Fig. 7-7a was zero, since the graphical sum of the equal and opposite voltages was zero. In Fig. 7-8a, however, the voltages have the same effective values but differ in frequency. The resultant waveform is shown in Fig. 7-8b. Observe that, when E_1 and E_2 are exactly out of phase with each

(a) Local circuit waveforms.

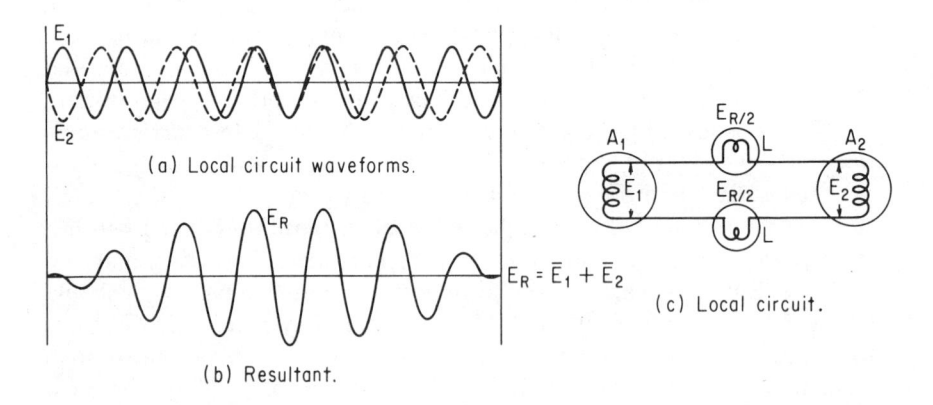

(b) Resultant.

$E_R = \bar{E}_1 + \bar{E}_2$

(c) Local circuit.

Figure 7-8

Effect of frequency difference between two alternators.

other and alternating, the resultant emf is zero; but, when they are in phase and alternating, the resultant waveform is a maximum alternating wave.

The lamps in the local circuit of Fig. 7-7b will always remain dark if the two waveforms are of the same frequency, form, and effective value of voltage, since the resultant voltage produced in the local circuit, E_r, is zero. In the case of Fig. 7-8c, however, the lamps in the local circuit will flicker whenever the effective value of $E_r/2$ is sufficient to cause them to light, i.e., whenever the resultant voltage is a maximum. Careful examination of the waveforms will reveal that the frequency difference between E_1 and E_2 is one cycle. The lamps in Fig. 7-8c will go from a dark period to a bright, and back to a dark period again, completing one cycle of pulsation and representing the frequency difference between the two waves.

Since it is possible to use lamps as a means of detecting *frequency differences* between two alternators, let us consider lamp synchronization of two single-phase alternators, as shown in Fig. 7-9a. In this method of synchronization, the lamps are connected as they were in Fig. 7-8c, known as the *dark-lamp* method.

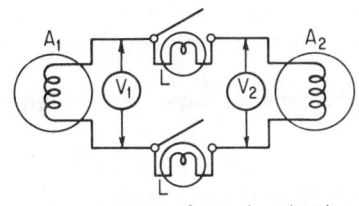

(a) Dark lamp method of synchronization.   (b) Bright lamp method of synchronization.

Figure 7-9

Synchronization of single-phase alternators by bright and dark lamp methods.

Synchronization is accomplished when voltages V_1 and V_2 are equal, and the lamps are dark. At this instant, the switch may be closed, paralleling the alternators. There are two disadvantages to the dark-lamp method of synchronization, however: (1) there could be an appreciable voltage difference between the two alternators but still insufficient difference to cause the lamps, in series, to glow; and (2) the operator who must close the switch has no way of knowing at what exact instant, during successive lamp pulsations, the voltage difference is actually zero.

The disadvantages stated above are overcome by the *bright-lamp* method shown in Fig. 7-9b, where the lamp connections are reversed. Now maximum brightness occurs when the waveforms are exactly equal and opposite (resultant voltage zero), and are dark when the resultant voltage is a maximum. This is illustrated by the following examples.

EXAMPLE 7-3: Alternator 1 in Fig. 7-9a has a terminal voltage of 220 V and a frequency of 60 Hz, whereas alternator 2 has a terminal voltage of 222 V and a frequency of $59\frac{1}{2}$ Hz. With the switch open, calculate:
a. The maximum and the minimum effective voltage across each lamp
b. The frequency of the voltage across the lamps
c. The peak value of the voltage across each lamp
d. The number of maximum light pulsations per minute.

Solution:

a. $E_{max}/\text{lamp} = E_1 + E_2 = \dfrac{220 + 222}{2} = 221$ V(rms)

$E_{min}/\text{lamp} = \dfrac{E_2 - E_1}{2} = \dfrac{222 - 220}{2} = 1$ V

b. $f = 60 - 59\frac{1}{2} = \frac{1}{2}$ Hz

c. $E_{peak} = \dfrac{221 \text{ V}}{0.707} = 313$ V

d. $n = \frac{1}{2}$ cycle/sec $\times$ 60 secs/min = **30 pulsations/min**

EXAMPLE 7-4: Each alternator of Fig. 7-9b generates a voltage of 220 V ac. Alternator 1 has a frequency of 60 Hz, and alternator 2 a frequency of 58 Hz. With the switch open, calculate
a. The maximum effective voltage across each lamp and its frequency
b. The phase relation at the instant maximum voltage occurs
c. The minimum effective voltage across each lamp and its frequency
d. The phase relation at the instant that minimum voltage occurs.

Solution:

a. $E_{max}/\text{lamp} = \dfrac{220 + 220}{2} = 220$ V

$f = 60 - 58 = 2$ Hz

b. The voltages are equal and opposite in the local circuit.

c. $E_{min}/\text{lamp} = \dfrac{220 - 220}{2} =$ **0 at zero frequency**

d. The voltages are in phase in the local circuit.

Examples 7-4 and 7-5 indicate the obvious advantages of the bright-lamp method of synchronization, since the difference in light intensity at a peak value is immediately discernible to the eye, which is quite sensitive to intensity differences. As a result, the paralleling switch may be closed at the precise instant that the voltages are equal and opposite. Assuming that the single-phase alternators have been connected in parallel, let us consider the conditions under which they will divide the load.

7-10. EFFECTS OF SYNCHRONIZING (CIRCULATING) CURRENT BETWEEN SINGLE-PHASE ALTERNATORS

Assume that the alternators of Figs. 7-7b or 7-9b are connected to a load bus feeding a lagging load. Assume also that, at the instant of paralleling these alternators, the frequencies and excitations were identical as shown in Fig. 7-7a. If the alternators are identical in design, i.e., if they have the same armature resistance and synchronous reactance, the phasor diagram of Fig. 7-10 will show the relation between the two alternators under conditions of load. Using alternator 1 from Fig. 7-7c as reference, the relation of all voltage drops in both alternators may be represented as shown in Fig. 7-10. Note that all the voltages and currents of the machines are in opposition with respect to each other. This is in complete agreement with the representations of Figs. 7-7a, b, and c where the voltages and currents are in opposition in the closed loop created by the two machines. Note that in this "ideal" phasor diagram, the terminal voltages are equal and opposite, and the generated voltages are equal and opposite, i.e., a full 180° apart.

The phasor diagram is also represented with respect to the load circuit, using alternator 1 as a reference. In this common-load circuit, the sum of I_{a_1} and I_{a_2} produces the load current I_L, while the load voltage, V_L, is the same as the terminal voltage or phase voltage, V_p, of the alternator. Since the two machines are identical, all internal voltage drops may be superimposed, and the generated voltages are equal and bear the same phase relation to the terminal voltage and the load current.

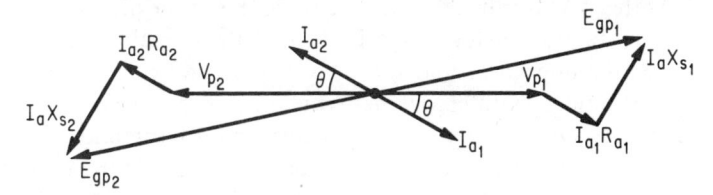

Figure 7-10
Internal "ideal" phasor diagram showing opposing voltages in parallel.

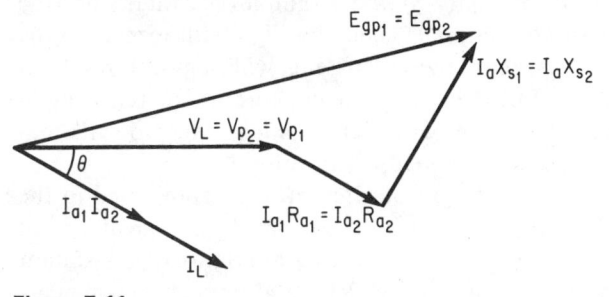

Figure 7-11
External phasor diagram with respect to load circuit (ideal) condition.

The alternators of Figs. 7-10 and 7-11 are perfectly synchronized because there is no voltage difference between them at the instant of synchronization or as a result of a change in induced emf because of internal voltage drops due to load. There is no synchronizing and circulating current between the alternators, therefore, as shown by Example 7-5.

EXAMPLE 7-5: If the alternators of Examples 7-3 and 7-4 each have an effective armature resistance of 0.1 ohm and a reactance of 0.9 ohm, calculate the synchronizing current in the armatures of both alternators if the switch between them is closed at the proper instant for paralleling.

Solution:

In Example 7-3,
$$E_r = 222 - 220 = 2 \text{ V}$$
$$I_s = \frac{E_r}{Z_1 + Z_2} = \frac{2 \text{ V}}{0.2 + j1.8} = \frac{2\angle 0° \text{ V}}{1.81 \angle 83.65° \text{ ohms}} = 1.105 \angle -83.65° \text{ A}$$

In Example 7-4, $E_r = 220 - 220 = 0 \text{ V}$
$$I_s = 0$$

It should be noted that the difference in circulating synchronizing current is not due to the lamp methods, but rather to the difference in generated emf per phase of the two alternators.

Example 7-5 and the diagram of Fig. 7-10 permit us to write an equation for the synchronizing current, circulating between two alternator armatures or a given alternator and its bus, namely:

$$I_s = \frac{\dot{E}_{gp_1} - \dot{E}_{gp_2}}{\dot{Z}_{p_1} + \dot{Z}_{p_2}} = \frac{E_r}{(R_{a_1} + R_{a_2}) + j(X_{s_1} + X_{s_2})} \quad (7\text{-}3)$$

where I_s is the synchronizing current, circulating in the alternator armature of the local circuit between the two alternators (or the alternator and the bus), per phase

E_r is the phasor difference between the generated voltages of the two alternators (or a given alternator and the equivalent generated voltage produced by remaining alternators feeding the bus), per phase $E_{gp_1} = E_{gp_2}$

R_a and X_s are the armature resistance and synchronous reactance, respectively, for each alternator, per phase

It should be noted, from Eq. (7-3) and Example 7-5, that, since the synchronous reactance of alternators is generally high with respect to the armature resistance, the synchronizing current will lag the resultant voltage, E_r, by almost 90°. But the synchronizing current is circulating in the armatures of *both* machines. What effect does it produce on the machines, and what is its effect on the load distribution?

Let us assume that the excitation of alternator 2, represented in the phasor diagram of Fig. 7-10, was increased so that E_{gp_2} is greater than E_{gp_1}. In the local or internal circuit of the two alternators, a resultant emf, E_r, will be produced, as shown in Fig. 7-12, and a synchronizing current, I_s, will flow. Since the induced voltages are exactly opposite in phase, the resultant voltage, E_r, is in phase with the larger induced voltage,

E_{gp_2}. The synchronizing current I_s lags E_{gp_2} and E_r by almost 90°, and leads E_{gp_1} by more than 90°, $(180 - \theta)$, as shown in Fig. 7-12.

The synchronizing current, as defined in Eq. (7-3) above, circulates only in the local circuit, and is limited only by the synchronous impedance of the two alternators (neglecting leads, bus bars, etc.) in parallel.

The synchronizing power generated by alternator 2 as a result of its increased excitation and armature current flow, I_s, is

$$P_2 = E_{gp_2} I_s \cos \theta \tag{7-4}$$

where θ is the angle between E_r (or E_{gp_2}) and I_s.

The synchronizing power produced by the synchronizing current in generator 1 is

$$P_1 = E_{gp_1} I_s \cos(180 - \theta) = -E_{gp_1} I_s \cos \theta \tag{7-5}$$

where $(180° - \theta)$ is the angle between $E_{gp_1} I_s$ and I_s.

Since the expression $\cos(180° - \theta)$ in Eq. (7-5) is greater than 90°, the synchronizing power, P_1 in alternator 1 is *negative*, whereas the power generated by alternator 2 is *positive*. Thus, *generator* action is produced in alternator 2, but motor action is produced in alternator 1. In the case of the latter, negative power (or power received by a generator) is indicative of motor action.

The numerator of Eq. (7-3), however, indicated that E_r is the difference of the generated voltages and [since $I_s \cos \theta$ is the same in Eqs. (7-4) and (7-5)] we may therefore write an equation for the true power loss (transformed into heat) which must be supplied mechanically by the prime mover of alternator 2.

$$E_r I_s \cos \theta = P_2 - P_1 = I_s^2 (R_{a_1} + R_{a_2}) \tag{7-6}$$

The power relations of the three equations above may be summarized in a single equation quantitatively as

$$\overset{(7\text{-}4)}{E_{gp_2} I_s \cos \theta} = \overset{(7\text{-}5)}{E_{gp_1} I_s \cos(180 - \theta)} + \overset{(7\text{-}6)}{E_r I_s \cos \theta} \tag{7-7}$$

or, qualitatively

Generated Power = Motor Power + Power Losses

Total power delivered by alternator 2 = Synchronizing power to alternator 1 + power losses
(tending to pull alternator 2 back)   (tending to push alternator 1 ahead)

Ignoring the external load on the two alternators, then, we may consider the effects of synchronizing current and the distribution of synchronizing power in the following illustrative example.

EXAMPLE 7-6: Two single-phase alternators are synchronized so that their **emf**s are opposed exactly by 180°, as shown in Fig. 7-12. But the **emf** of alternator 1 is adjusted to 200 V and the **emf** of alternator 2 is adjusted to 220 V. At the instant of closing the synchronizing switch, paralleling the two machines, if each alternator has an armature resistance of 0.2 Ω and a synchronous reactance of 2 Ω, calculate:

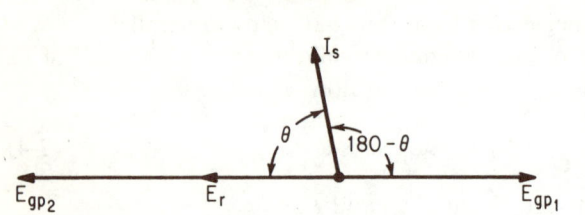

a. The generator action developed by alternator 2
b. The motor action or synchronizing power delivered to alternator 1
c. The power loss in both armatures, and the terminal voltage of both alternators
d. Draw a phasor diagram showing the voltage relations and all voltage drops.

Figure 7-12
Internal synchronizing current produced as a result of increased excitation of alternator 2.

Solution:

a. $E_r = E_2 - E_1 = 220 - 200 = 20$ V

From Eq. (7-3), $I_s = \dfrac{E_s}{Z_{p_1} + Z_{p_2}} = \dfrac{20 \text{ V}}{0.2 + 0.2 + j(2.0 + 2.0)}$ (7-3)

$= \dfrac{20 \angle 0° \text{ V}}{4.02 \angle 84.3° \text{ Ω}} = 4.98 \angle -84.3°$ A

From Eq. (7-4), $P_2 = E_{gp_2} I_s \cos \theta$
$= 220 \times 4.98 \cos 84.3° = \mathbf{108.9 \text{ W}}$ (the total power delivered by alternator 2)

b. From Eq. (7-5)

$P_1 = E_{gp_1} I_s \cos(180° - \theta) = -200 \times 4.98 \cos 84.3°$
$= \mathbf{-99 \text{ W}}$ (the synchronizing power received by alternator 1)

c. Power loss $= P_2 - P_1 = 108.9 - 99.0 = \mathbf{9.9 \text{ W}}$ (7-6)

Checked by $E_r I_s \cos \theta = 20 \times 4.98 \cos 84.3° = 9.9$ W or $I_s^2(R_{a_1} + R_{a_2}) = (4.98)^2 \times 0.4 = 9.9$ W, as given in Eq. (7-6)

From Fig. 7-13, V_{p_2}, the terminal phase voltage of alternator 2, is (from Eq. 7-1),

$V_{p_2} = E_{gp_2} - I_s Z_{p_1} = 220 - [4.98 \times 2.01 \angle 84.3°]$ (1-10)
$= 220 - 10 = \mathbf{210 \text{ V}}$ (generator action)

From Eq. (7-2)

$V_{p_1} = E_{gp_1} + I_s Z_{p_1} = 200 + [4.98 \times 2.01 \angle 84.3°]$ (1-9)
$= 200 + 10 = \mathbf{210 \text{ V}}$ (motor action)

d. The phasor diagram (see Fig. 7-13)

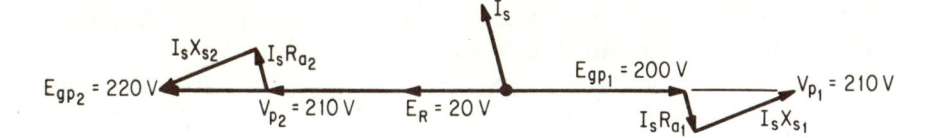

Figure 7-13

Phasor diagram for Example 7-6, part d.

Note that the terminal bus voltage of the generated voltages in Ex. 7-6 is

$$\frac{E_{gp_1} + E_{gp_2}}{2} = \frac{220 + 200}{2} = 210 \text{ V}$$

This is true, however, only when the two generators have the *same* internal armature resistance and synchronous reactance.

Example 7-6 points up once again the important relations concerning the differences between generator action and motion action, first presented in Section 1-20, as well as some new concepts, namely:

1. In the *motor* mode, electric energy from some outside source is delivered to the dynamo. The voltage generated by the motor is less than the voltage applied across its terminals by its internal impedance drop.
2. In the *generator* mode, the generated voltage, produced as a result of mechanical power supplied by a prime mover, exceeds the terminal voltage by its internal impedance drop.
3. The only power *loss* sustained [Eq. (7-6) neglecting friction, windage, etc.] is the armature copper loss in both armatures. The *synchronizing* power is a power which has been transferred (99 W in Example 7-5) from alternator 2 to alternator 1, causing the latter to tend to run as a motor in the same direction as it is being driven by its prime mover. Since power can *only* be dissipated by resistance, this power is available at the bus for use by the load in addition to that generated by both alternators, driven by their own prime movers.
4. Any time a motor is connected across a supply, it is in parallel with that supply. Current delivered to (and drawn by) the motor from the supply is indicative of motor action. If the motor, by some means, is driven in the same direction by its coupled load such that it sends current *to* the supply, it is operating in the generator mode when its generated voltage exceeds the bus voltage.

In addition to the production of a small power loss, *the primary effect of synchronizing current is the production of synchronizing power.* The phasor diagrams represented in Figs. 7-12 and 7-13 are only *instantaneous* diagrams of an instantaneous change produced in the generated emf of one or more alternators in parallel with a bus. Immediately on the production of this difference, a synchronizing power is produced which causes (1) the alternator generating the synchronizing power to *drop back* in phase as a result of the increased load and electromagnetic counter torque; and (2) the alternator receiving the synchronizing power to *advance* in phase because of the motor action produced in the same direction as its prime mover.

The effect of synchronizing current when the emf of one alternator is increased (by increasing its field current) is shown in Fig. 7-14, based on the above two instantaneous changes produced by synchronizing power. Recall that all phasors rotate in a counterclockwise direction. Note that alternator 2 has *dropped* back from its original position, while alternator 1 has *advanced* from its original position, bringing the induced **emf**s *closer* together. Since the resultant emf, E_r, is the phasor sum of the two induced **emf**s, it is no longer in phase with E_{gp_2} but lags it by some small angle. The synchronizing current, in accordance with Eq. (7-3), still lags this resultant by the angle θ. The angle by which alternator 2 will drop back in phase, and alternator 1 will advance in phase, depends on (1) the relative magnitudes of E_{gp_1} and E_{gp_2} (which determine the magnitude of E_r), and (2) the magnitude of the resultant synchronizing current circulating in both armatures.

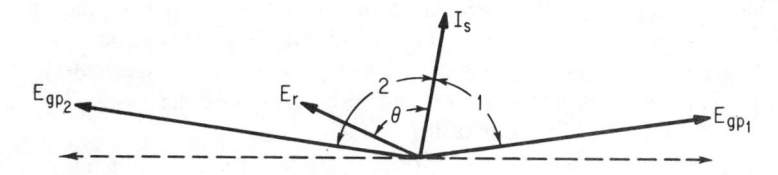

Figure 7-14

Adjusted phase positions as a result of synchronizing power transfer between two alternators.

An equilibrium is established where no synchronizing power or motor action is produced by either alternator, since angle 1 is less than 90° (for alternator 1) and angle 2 is less than 90° (for alternator 2) in accordance with the relation $E_{gp}I_s$ cos (the angle between them). The only power produced, then, is $E_r I_s \cos \theta$, and this armature copper power loss is made up by the relation

$$E_{gp_1}I_s \cos \text{(angle 2)} + E_{gp_2}I_s \cos \text{(angle 1)} = E_r I_s \cos \theta \qquad (7\text{-}8)$$

It should be noted that since the values of generated voltage of each alternator are the same in Fig. 7-14 as in Fig. 7-12, but are now closer in phase, the resultant voltage E_r and the synchronizing current I_s are *both* increased, and are not the same in Eq. (7-8) as in Eq. (7-7). This small additional increase in power loss, $I_s^2(R_{a_1} + R_{a_2})$, produced by the increase in synchronizing current, is supplied by both alternators in accordance with Eq. (7-8).

In the same way, if the field current of alternator 1 is suddenly increased so that its generated voltage exceeds that of alternator 2, a synchronizing power is supplied by alternator 2. In both instances, the terminal voltage of both machines is the *same* for the simple reason that they are in parallel. Since the synchronizing current is all *internal*, and since whatever occurs internally produces the same voltage with respect to the external circuit, changing the resistance of the field rheostat and field current of *any* alternator in parallel with other alternators *does not affect the division of load between them.* (Unlike the dc generators previously studied, a transfer of load *cannot* be accomplished by changing the generated voltage of the alternator.) The circulating current in the alternator armature, superimposed on the load current supplied to the external load, may change the power factor of the particular alternator and possibly reduce its capacity. This is shown in Fig. 7-15, which represents the combined effect of the load shown in Fig. 7-11 and the increased excitation of alternator 2 shown in Fig. 7-12.

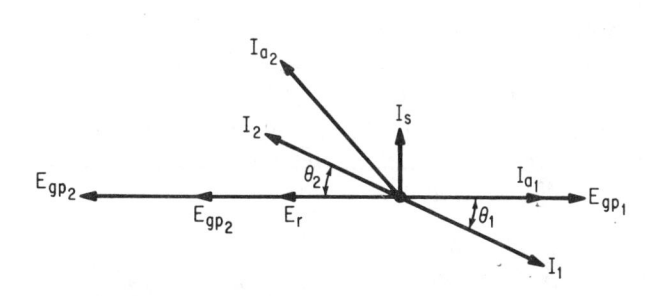

Figure 7-15

Change in load power factor of both alternators as a result of synchronizing current.

It should be noted that the effect of the adjusted phase position shown in Fig. 7-14 is not reflected in Fig. 7-15. The reason is that the angles are very small, as indicated by Eq. (7-8) and Example 7-6, where the total power loss is comparatively small. Originally, I_1 and I_2 are the equal load currents shown in Fig. 7-11, each lagging the generated voltage by the same angle, θ_1 and θ_2, respectively. When the field of alternator 2 is overexcited to produce a higher value of E_{gp_2} and E_r, respectively, the circulating synchronizing current, I_s, produced flows through the armatures of both alternators in addition to the load current. The resultant load current of generator 1 is reduced, and the power factor of generator 1 is improved. At the same time, the resultant armature current of generator 2 is increased, and its power factor is more lagging (worse). The poorer power factor of alternator 2 reduces its capacity to deliver (useful) current to the load.

Figure 7-15 also makes it possible to perceive the *stabilizing* effect of synchronizing current. Since the synchronizing current, I_s, lags the generated voltage in alternator 2 but leads the generated voltage in alternator 1, it produces a *demagnetizing* action in the *former* and a *magnetizing* action in the *latter* (Sections 5-9 and 6-7).

Thus, the synchronizing current produced as a result of overexciting a given alternator tends to: (1) cause that alternator to deliver more synchronizing power to the remaining alternators in parallel; (2) cause that alternator's voltage to operate at a poorer power factor with consequent demagnetization of its air-gap flux; and (3) cause the remaining alternators to improve their power factor with resultant magnetization of their air-gap flux.

Thus, any tendency of a given alternator to take on additional load as a result of increased generator voltage is automatically stabilized without any appreciable transfer of load.

This raises two questions: (1) How can the bus voltage be raised or lowered? (2) How is the transfer of load between alternators accomplished?

The only way to raise or lower the bus voltage without affecting the power and power factors of the individual alternators supplying load in parallel, is to increase or decrease the voltage of *all* parallel alternators *simultaneously.* This answers the first of the preceding questions; the second is discussed below.

7-11.
LOAD DIVISION
BETWEEN
ALTERNATORS

Let us assume that the two alternators, as shown in Figs. 7-10 and 7-11, are operating under ideal conditions in parallel, i.e., their generated **emf**s are equal, they are carrying equal loads, and they have identical internal synchronous impedance drops. Let us assume that the prime mover of alternator 1 tends to increase in speed, causing its generated emf E_{gp_1} to advance in phase as shown in Fig. 7-16. Operating originally as ideal paralleled generators, their induced **emf**s were at all times equal and opposite, and no resultant emf is produced. But now, prime mover 1 increases in speed, and a resultant emf E_r is produced by the frequency difference between the two machines and indicated by the advance of generated voltage E_{gp_1} in Fig. 7-16. As in the case of the preceding section, the resultant voltage, E_r, causes a circulating current I_s to be produced in the armature of both generators in accordance with Eq. (7-3). This synchronizing current causes a synchronizing power to be generated by generator 1, $P_1 = E_{gp_1} I_s \cos$ (angle 1). This synchronizing power con-

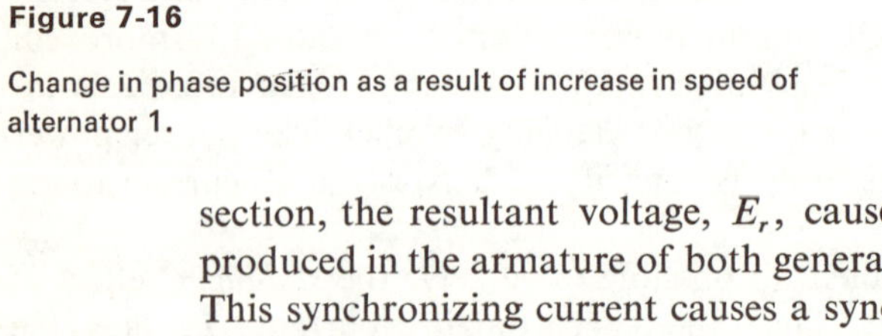

Figure 7-16
Change in phase position as a result of increase in speed of alternator 1.

tains an armature power loss component ($E_r I_s \cos \theta$) and a synchronizing power transfer component, which is the power transferred to alternator 2 to produce motor action.

Alternator 1 is delivering power by generator action, and alternator 2 is receiving power by motor action. The power received by alternator 2 is $E_{gp2} I_s \cos$ (angle 2), a negative power which is the difference between the power generated by alternator 1 and the armature copper losses of the two alternators [Eq. (7-8)]. Since the prime mover of alternator 1 is more heavily loaded as a result of the additional power generated, it will tend to drop back in phase and in speed. Alternator 2, on the other hand, as a result of receiving synchronizing power and motor action, will tend to advance in phase. Thus, once more, the synchronizing current acts in such a way as to keep the alternators continously in synchronism.*

It should be noted in Fig. 7-16 that the synchronizing power received by alternator 2 really depends on the angle θ. For a given angle of advance α of alternator 1 over its former position, angle 1 between the induced emf E_{gp_1} of alternator 1 and the synchronizing current, I_s, depends on θ. But θ depends on the internal synchronous impedance of the alternators. If θ is small, angle 1 is large, and the cosine of a large angle is a small value. In order to develop the same synchronizing power in alternator 1, more synchronizing current would be required for a small value of θ. An extremely high value of alternator synchronous impedance, however, might also reduce the synchronizing current faster than the decrease in angle 1 between the synchronizing current and the generated voltage. Obviously, then, a reasonably *high ratio* of synchronous reactance to armature resistance will produce a *rapid* and *sufficient* synchronizing power to assure successful parallel operation, although it may result in poorer regulation (Sections 6-7, 6-8, and 6-9). In general, therefore, it may be stated that those alternators which operate more successfully in parallel are those which tend to have poorer regulation.

Regarding changes in prime mover speed or sudden applications or removal of load, for maximum stability in parallel, alternators should have (1) a high ratio of synchronous reactance to armature resistance, and (2) a sufficiently low total impedance so that small changes in its phase advance angle (α) will produce large values of synchronizing current and power. This is illustrated in Examples 7-7 and 7-8.

EXAMPLE 7-7: The alternators of Example 7-6 each have a generated voltage of 230 V and an impedance of 2.01 $\angle$ 84.3° ohms. The prime mover of alternator 1 drives it 20° ahead of its correct position. Calculate:
a. The synchronizing current
b. The synchronizing power developed by alternator 1

---

* The prime movers, for this reason, must have flat or drooping speed characteristics, so that increased loading will tend to reduce speed and decreased loading will tend to increase speed, thereby aiding the above effect of the synchronizing current.

c. The synchronizing power received by alternator 2
d. The losses in the armature.

Solution:

$$E_2 = 230 \angle + 180° = -230 + j0$$
$$E_1 = 230 \angle 20° = 216 + j78.6$$
$$E_r = E_2 + E_1 = -14 + j78.6 = 79.8 \angle 100.1° \text{ V}$$

a. $I_s = \dfrac{E_r}{Z_1 + Z_2} = \dfrac{79.8 \angle 100.1° \text{ V}}{2(2.01 \angle 84.3° \, \Omega)} = \dfrac{79.8 \angle 100.1° \text{ V}}{4.02 \angle 84.3° \, \Omega}$ (7-3)
   $= \mathbf{19.85 \angle 15.8° \text{ A}}$

b. $P_1 = E_{gp_1} I_s \cos(E_{gp_1}, I_s) = 230 \times 19.85 \cos 4.2°$ (7-4)
   $= 4560 \times 0.9973 = \mathbf{4558 \text{ W}} = \text{power delivered to the bus}$

c. $P_2 = E_{gp_2} I_s \cos(E_{gp_2}, I_s) = 230 \times 19.85 \cos 164.2°$ (7-5)
   $= \mathbf{-4400 \text{ W}} \quad \text{(power received from bus)}$

d. Losses: $P_1 - P_2 = 4558 - 4400 = \mathbf{158 \text{ W}}$ (7-6)
   Check: $E_r I_s \cos \theta = 79.8 \times 19.85 \cos 84.3° = 158 \text{ W}$ (7-6)
   Double check: $I_s^2 R_{aT} = (19.85)^2 \times 0.4 = 158 \text{ W}$ (7-6)

EXAMPLE 7-8: Repeat Example 7-7 if each alternator has an impedance of $6 \angle 50° \, \Omega$.

Solution:

a. $I_s = \dfrac{E_r}{Z_1 + Z_2} = \dfrac{79.8 \angle 100.1° \text{ V}}{12 \angle 50° \, \Omega} = \mathbf{6.65 \angle 50.1° \text{ A}}$ (7-3)

b. $P_1 = E_{gp_1} I_s \cos(E_{gp_1}, I_s) = 230 \times 6.65 \cos 30.1° = \mathbf{1322 \text{ W}}$ (7-4)

c. $P_2 = E_{gp_2} I_s \cos(E_{gp_2}, I_s) = 230 \times 6.65 \cos 129.9° = \mathbf{-982 \text{ W}}$ (7-5)

d. Losses: $P_1 - P_2 = 1322 - 982 = \mathbf{340 \text{ W}}$ (7-6)
   Check: $E_r I_s \cos \theta = 79.8 \times 6.65 \cos 50° = \mathbf{340 \text{ W}}$ (7-6)
   Double check: $I_s^2 R_{aT} = (6.65)^2 (12 \cos 50°) = \mathbf{340 \text{ W}}$ (7-6)

Note that the use of a higher impedance at a lower Q, namely X_s/R_a, has resulted in (1) a *reduction* in synchronizing power, and (2) an *increase* in *losses* despite the reduction in synchronizing current. It is for this reason (as well as others) that a high ratio of armature reactance to resistance is preferred, *despite* its effect on regulation. Finally, as noted in Sec. 6-8, voltage regulation may be offset by use of voltage regulators which vary the field excitation to maintain constant voltage at the output regardless of changes in load.

7-12. HUNTING OR OSCILLATION OF ALTERNATORS

The preceding sections (Sections 7-10 and 7-11) indicated that an instantaneous synchronizing current is produced whenever the field current of an alternator is increased or its prime-mover speed is raised. The effect of the synchronizing current is to produce an instantaneous synchronizing power which causes the alternator *generating* the power to drop *back* into synchronism, and the alternators *receiving* the power to *advance* ahead of synchronism. It would appear that these tendencies, coupled with the drooping speed characteristics of the prime movers driving the alternator, should result in a situation of extreme stability and equilibrium. This would be true, indeed, if the speed of the prime mover were constant for a full cycle of revolution.

Unfortunately, a prime mover of a reciprocating nature (such as a steam, diesel, or gasoline engine) may have a constant average speed in rpm, but does not have a constant speed over one complete cycle of revolution. During the *power* stroke of a reciprocating steam engine, for example, the alternator is driven *ahead* of synchronism; while on the *return* stroke of the engine, the alternator falls *behind* synchronism. A 30-pole alternator operating at 60 Hz will have an average prime-mover speed of 240 rpm, or 4 rps. In one revolution, or one-quarter second, the alternator has gone through 15 cycles of alternation! The electrical response of the alternator to even slight changes produced in prime mover speed is almost instantaneous. Unfortunately, however, the rotors of the alternators are fairly heavy and have a great deal of inertia.

If, during the power stroke, one alternator is driven slightly ahead of synchronism, it will instantly deliver a synchronizing power to other alternators. The other alternators receive this instantaneous power but, because of their inertia, their response is slow. The advancing alternator now produces more and more synchronizing current in an attempt to bring the other alternators up to synchronism, at the same time falling back itself. The power transferred may be so great and the countertorque demand on the advancing alternator so high, that, instead of falling back to synchronism, it actually falls *behind* the other alternators. As it falls behind, its prime mover is on the *return* stroke (which doesn't help it at all); and in falling behind, it instantly receives power *from* the bus and acts like a *motor*.

The synchronizing power received plus the extra boost of the prime mover on its power stroke once more causes the original alternator to surge even farther ahead than it did the first time. The inertia of the other alternators, in not responding immediately to the power instantly received, causes the advancing alternator to develop an even larger synchronizing current than before. In effect, the alternator is "hunting" for the stable synchronous speed but cannot find it.

This repeated periodic oscillation or *hunting*, above and below the synchronous speed of the alternator driven by a reciprocating engine, continues to be amplified, each successive oscillation increasing over the previous one. If instruments such as ammeters and wattmeters are connected, the hunting may be observed as the ammeter current rises and falls and the wattmeter power reverses periodically, in response to the reception and generation of synchronizing power. Since this condition is one that will not cease of its own accord and is not self-limiting (for the reasons described above), it is necessary to take steps in the combined prime-mover and alternator design to eliminate it. The following are some of the techniques employed to reduce hunting.

1. *Amortisseur* (literally, "killer") or damper windings, consisting of squirrel-cage bars, are placed in the rotor pole faces to "kill" the hunting effect. When the field poles are rotating past the armature at synchronous speed, no voltage is induced in the short-circuit damper winding. If the speed of the rotor increases or decreases below synchronous speed, a voltage and high short-circuit current is produced in the damper winding. By Lenz's law this current sets up a flux which opposes the force that produced it, i.e., the change in speed.
2. The prime mover shaft may be equipped with a large and heavy flywheel. This increases the inertia of the prime mover and assists it in producing a more constant speed throughout a single revolution.
3. Dashpots or viscous fluid dampers are used on the fuel throttles or governors of the prime mover engines to prevent their immediate response to minor and sudden changes of demand for more or less power from the alternator.
4. Prime movers are employed which have a uniform power stroke over one complete revolution; steam or gas turbines, for example, have this characteristic.

7-13. SYNCHRONIZING POLYPHASE ALTERNATORS

All of the preceding discussion, while developed in terms of the single-phase alternator, applies equally to the polyphase alternator, since all considerations and calculations are done on a per phase basis for a three-phase or other polyphase alternator. The only differences occur in the method of synchronization and the criteria of phase sequence (the sixth requirement in Section 7-8). This requirement states that the *phase sequence* of the incoming alternator to be paralleled must be the same as that of the bus, i.e., the existing alternators already in parallel.

As stated in Section 2-13, there are only *two* phase sequences possible for a three-phase alternator, for the simple reason that there are only two possible directions of rotation in which the poles may sweep across the armature windings. Figure 7-17 shows an alternator on the left about to be paralleled to an alternator (or bus) on the right. The rotor of the left-hand alternator rotates counterclockwise, while that on the right rotates clockwise. Yet the phase sequence for both machines is the same (*ABC-ABCA*), as indicated by the phases encountered by a unit north pole in rotating about the armature and inducing voltages. The phase sequence may be checked rather simply by connecting a small induction motor across the bus and observing the direction of rotation. Then the induction

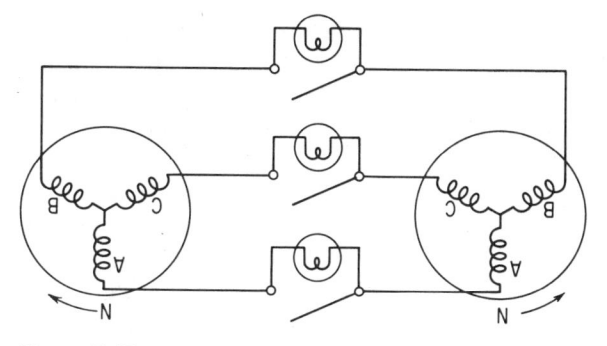

Figure 7-17

Phase sequence of alternators and dark lamp method of synchronization.

motor is connected across the incoming alternator, and, if the direction of rotation is the same, the phase sequence of the incoming machine is the same as the bus. If the induction motor rotates in the opposite direction, any two of the three incoming alternator leads (leads to the switches in Fig. 7-17) may be reversed, and the correct phase sequence is assured. The phase sequence may also be checked by a *phase-sequence indicator* (Section 7-15).

Synchronization may be accomplished by using lamp methods similar in principle to those used with the single-phase alternator shown in Fig. 7-9. The method employed in Fig. 7-17 is the *dark-lamp* method. Even if the effective phase and line voltages of the incoming and running machines are identical, and even if the frequencies of the alternators are identical, the lamps in Fig. 7-17 may not be dark. There is a rare possibility that the voltages will tend to "lock" in precise opposition, phase to phase. Thus, if the lamps remain steadily at a particular brightness, it indicates that both the incoming and the running machines have the same *frequency*, but that a voltage difference is produced by either (1) a fixed phase displacement between the induced **emf**s of the alternators, or (2) a difference in their effective phase voltages.

After ruling out the second possibility by means of a voltmeter, it will be necessary either to speed up or to slow down the incoming alternator slightly in order to find the precise moment to close the synchronizing switch (i.e., when the lamps are dark) as the lamps flicker in unison. If the lamps do *not* flicker in unison, the phases are not properly connected to the switches, or else the phase sequence is incorrect. Reversing any two leads will remedy the difficulty.

The disadvantage of using the dark-lamp method with polyphase alternators is the same as with single-phase alternators as discussed in Section 7-9, where it was found difficult to determine, even at a low rate of flicker, the middle of the dark period (when the alternators are exactly in synchronism and the **emf**s exactly 180° with respect to each other).

As in the case of single-phase alternators (Fig. 7-9b), the *bright-lamp method* may be used to indicate the instant of synchronization by maximum lamp brightness. Figure 7-18a shows the bright-lamp connec-

(a) Bright lamp method.

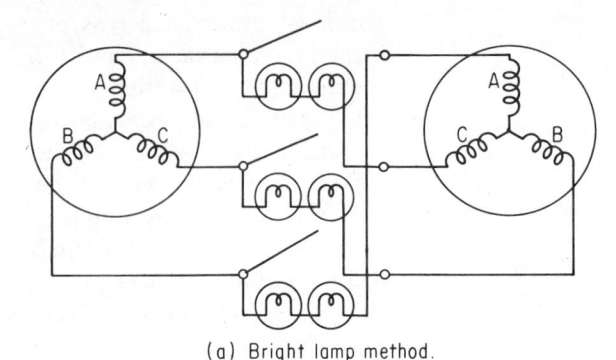

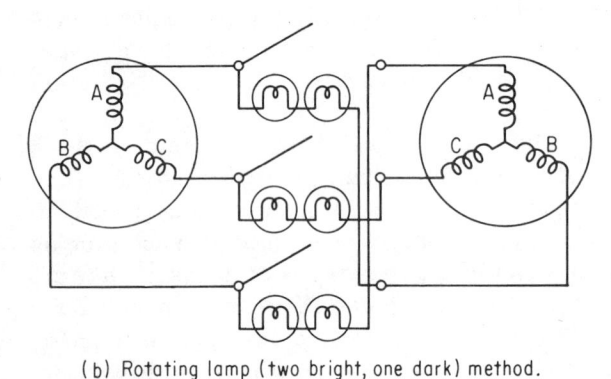

(b) Rotating lamp (two bright, one dark) method.

Figure 7-18

Lamp methods for synchronization.

tions for paralleling three-phase alternators in which all three lamp connections have been reversed from that shown in Fig. 7-17.

Figure 7-18b shows a third method, called the *rotating-lamp method,* in which the lamps will flicker two bright, one dark, and two dark, one bright, successively. The synchronizing switch is closed when the two outer lamps in Fig. 7-18b are bright and the center lamp dark. The advantage of this method is that it permits synchronization in terms of both maximum and minimum brightness. The figure also shows the use of lamps in series to prevent burnout of lamps at peak voltages. With high-voltage alternators, potential transformers are used, with either the lamps or the *synchroscope,* described below.

7-14.
SYNCHROSCOPES

Under commercial operating conditions, it may be difficult at times, using the lamps, to tell whether the incoming alternator is slow or fast. In the laboratory, it is an easy matter to increase the prime-mover speed (usually a variable-speed motor is used) and observe the lamp flicker. If the flicker *slows* down with *increasing* speed, the frequency of the incoming alternator is below that of the running alternator or bus.

An instrument called a *synchroscope* has been devised, with a rotating pointer (which indicates whether the incoming machine is slow or fast) and a fixed index to show the precise instant of synchronization when the paralleling switch should be closed. Synchroscopes are manufactured in a variety of designs, namely, the *polarized-vane* type, the *moving-iron* type, and the *crossed-coil* type. The synchroscope is designed for operation on single-phase circuits and therefore may be used to synchronize single-phase as well as polyphase alternators. Because it is basically a single-phase device, it *cannot* detect phase sequence; this must be checked by either an induction motor or a phase-sequence indicator. Nor can it detect voltage differences; this must be done by a voltmeter.

The *polarized-vane type* of synchroscope circuit and dial are shown in Fig. 7-19. The vane or pointer is polarized to the frequency of the running machine by means of a rotor coil. The stator winding consists of two coils (phases) distributed about the instrument's circumference in the same manner as a single-phase, split-field, induction motor, and is connected to the incoming machine. The rotating field of the stator rotates at the frequency of the incoming machine, while the iron vane is polarized (magnetized) at the frequency of the bus or running machine. When the frequencies are exactly in synchronism, the pointer will align itself in a fixed vertical position, as shown in Figs. 7-19b and c. In the latter figure, note that, when the rotating field is at right angles to the pointer, the pointer is unmagnetized. If the pointer were slightly magnetized owing to a difference in frequency it would move slowly to the left or right. If the field rotates at 61 Hz and the vane is magnetized at 60 Hz the vane will rotate clockwise at a speed of 1 rps. Conversely, if the field rotates at 58 Hz the vane will rotate counterclockwise at a speed of 2 rps. When the frequencies are identical, the pointer "locks" at a fixed position which indicates the *phase difference* between the two voltages of the alternators.

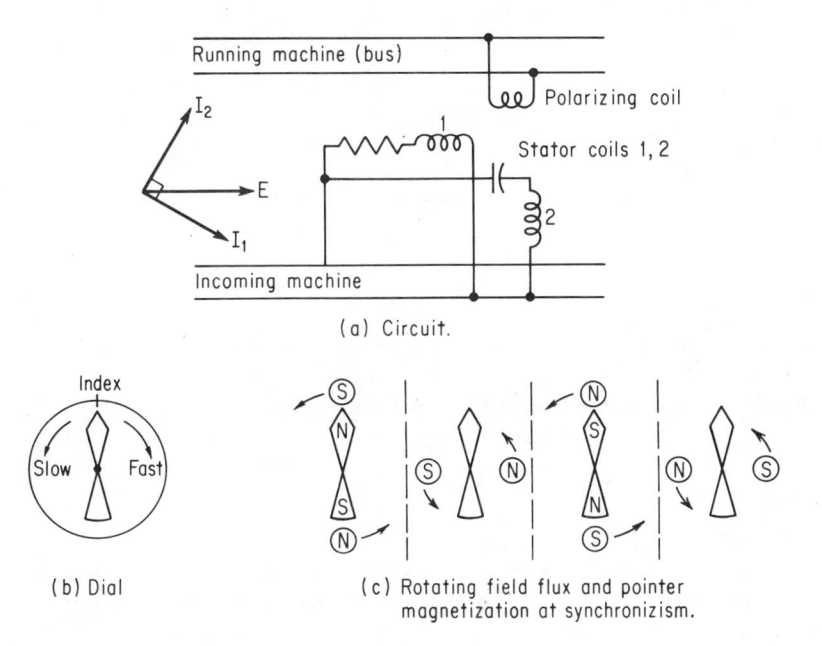

Figure 7-19

Synchroscope circuit dial and principle of operation (polarized-vane type).

SEC. 7-14. / *Synchroscopes*

7-15. PHASE-SEQUENCE INDICATOR

Although the synchroscope quite accurately can provide better indication than lamp methods of the instant when synchronization should occur, or whether the incoming alternator is slow or fast, it cannot indicate phase sequence (which the lamps provide quite nicely). It was stated earlier that polyphase induction motors may be used, but these may not be available nor as portable or convenient to use as a *phase-sequence indicator*. The basic circuit of a phase-sequence indicator is shown in Fig. 7-20a, and a simplified version in Fig. 7-20b. It consists of two identical neon lamps and a capacitor connected in wye, as an unbalanced three-phase load. The circuit resistances are designed to create a potential above and below the ignition potential of the neon lamps such that, for one phase rotation (1-3-2), the unbalanced load causes a greater voltage drop from the neutral N across lamp 1, E_{N1}. Note that, because of the unbalance, the phase voltages to neutral are as great as the line voltages, E_{1-2}, E_{2-3}, etc. A reversal of the phase sequence (1-2-3) will illuminate lamp N_3, similarly, and extinguish lamp N_1.

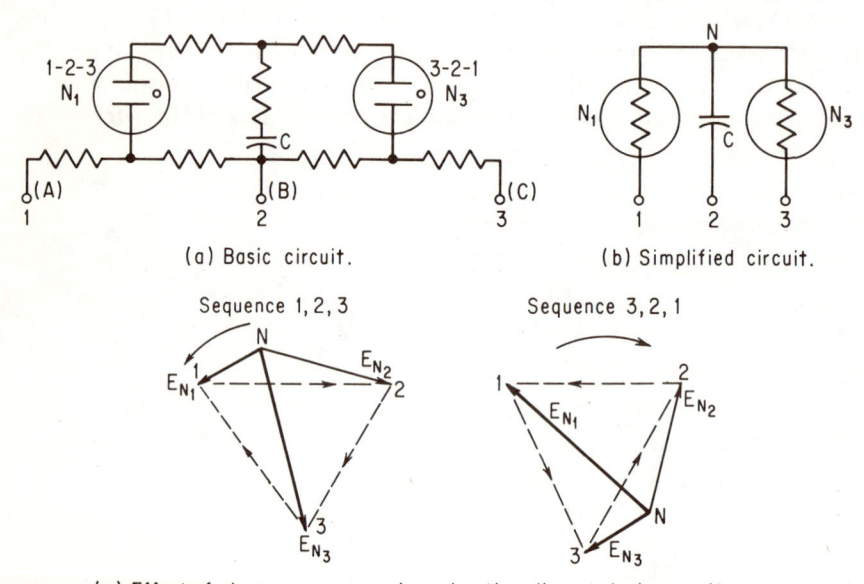

(a) Basic circuit.

(b) Simplified circuit.

(c) Effect of phase sequence on lamp ignition, line, and phase voltages.

Figure 7-20

Phase sequence indicator.

7-16. SUMMARY OF PROCEDURE FOR PARALLELING POLYPHASE ALTERNATORS

The following may serve as a summary of the steps necessary in placing a polyphase alternator in parallel with other alternators across a bus.

1. The alternator is brought up to rated speed, and its effective line voltages are adjusted to the bus voltage by means of a voltmeter.

2. The phase sequence is checked by means of a phase-sequence indicator or synchronizing lamps.
3. The frequency of the incoming alternator is compared to the bus frequency by means of a synchroscope or a lamp method. If the incoming machine frequency is low, the prime-mover speed is increased; if high, the speed is decreased.
4. The paralleling switch is closed at the instant the lamps or synchroscope indicate that the phase-to-phase voltages are exactly equal and opposite. The incoming alternator is now floating on the line.
5. The alternator is made to take on load by increasing the speed of its prime mover.
6. The power factor at which the alternator carries reactive power is adjusted by means of its field rheostat.
7. The bus voltage is adjusted by adjusting all field rheostats simultaneously.

BIBLIOGRAPHY

Bewley, L. V. *Alternating Current Machinery*, New York: The Macmillan Company, 1949.

Bewley, L. V. *Tensor Analysis of Electrical Circuits and Machines*, New York: Ronald Press, 1961.

Carr, C. C. *Electrical Machinery*, New York: John Wiley & Sons, Inc., 1958.

Crosno, C. D. *Fundamentals of Electromechanical Conversion*, New York: Harcourt, Brace, Jovanovich, Inc, 1968.

Daniels. *The Performance of Electrical Machines*, New York: McGraw-Hill, Inc., 1968.

Fitzgerald, A. E. and Kingsley, C. *The Dynamics and Statics of Electromechanical Energy Conversion*, 2nd Ed., New York: McGraw-Hill, 1961.

Fitzgerald, A. E., Kingsley, Jr. C., and Kusko, A. *Electric Machinery, 3rd Ed.* New York: McGraw-Hill Inc. 1971.

Gemlich, D. K. and Hammond, S. B. *Electromechanical Systems*, New York: McGraw-Hill, 1967.

Hindmarsh, J. *Electrical Machines*, Elmsford, N.Y.: Pergamon Press, 1965.

Jones, C. V. *The Unified Theory of Electrical Machines*, New York: Plenum Publishing Corp., 1968.

Majmudar, H. *Introduction to Electrical Machines*, Boston: Allyn and Bacon, 1969.

McFarland, T. E. *Alternating Current Machines*, Princeton, N. J.: D. Van Nostrand Co., Inc., 1948.

Puchstein, A. F., Lloyd, R., and Consad, A. G. *Alternating Current Machines*, 3rd Ed., New York: Wiley/Interscience, 1954.

Robertson, B. L. and Black. L. J. *Electric Circuits and Machines*, 2nd Ed., Princeton, N. J.: D. Van Nostrand Co., Inc., 1957.

Selmon. *Magnetoelectric Devices: Transducers, Transformers and Machines*, New York: Wiley/Interscience, 1966.

Siskind, C. S. *Direct-Current Machinery*, New York: McGraw-Hill, 1952.

Skilling, H. H. *Electromechanics: A First Course in Electromechanical Energy Conversion*, New York: Wiley/Interscience, 1962.

Thaler, G. J., and Wilcox, M. L. *Electric Machines: Dynamics and Steady State*, New York: Wiley/Interscience, 1966.

White, D. C., and Woodson, H. H. *Electromechanical Energy Conversion*, New York: Wiley/Interscience, 1959.

QUESTIONS

7-1. Give 5 advantages of using dc generators or alternators in parallel rather than a single unit to supply electrical system load.

7-2. Four single-phase alternators are connected in parallel to a bus. Alternators 1 and 2 are delivering energy to the bus. Alternator 3 is receiving energy from the the bus and alternator 4 is "floating" on the line. Give the equation expressing the relation between generated voltage and terminal voltage for:
   a. alternator 1 and alternator 2
   b. alternator 3
   c. alternator 4.

7-3. Repeat Quest. 7-2 for four dc generators similarly coupled to a bus this time giving equations for:
   a. power generated by dynamos 1 and 2 and power delivered to the bus
   b. power generated by dynamo 3 and power delivered to dynamo 3 by the bus
   c. Repeat (b) for dynamo 4.

7-4. For dc shunt generators operating in parallel, explain:
   a. why a condition of stable equilibrium exists between these generators for sudden increases or decreases in load
   b. two conditions which will cause proportionate load division between them
   c. why the prime movers driving the generators must also have constant or drooping speed characteristics.

7-5. For dc compound generators operating in parallel, explain:
   a. two additional conditions for maintaining proportionate load division with changes in load
   b. why a diverter for a given compound generator cannot be connected in parallel with the series field of that particular machine
   c. the manner in which the characteristics of the compound generators are adjusted by means of diverters

d. why the equalizer must be connected on the armature side of the series field and also on the side of same polarity for each machine.

7-6. Give the steps for bringing a dc generator in parallel with a group of other dc generators connected to a bus.

7-7. Assuming a given generator G_1 is operating in parallel with other generators, what is the effect of:
a. reducing the generated voltage of G_1 to the bus voltage
b. reducing the generated voltage of G_1 below bus voltage
c. uncoupling generator G_1 from its prime mover.

7-8. State:
a. The conditions necessary for successful parallel operation of alternators.
b. Summarize these conditions in a single statement.

7-9. Draw phasor diagrams representing:
a. voltage relations between two single-phase alternators in parallel with respect to the local circuit consisting of the two alternators
b. voltage relations between the bus voltage and load current supplied by these alternators.

7-10. Repeat Quest. 7-10 representing the waveforms graphically.

7-11. Draw a diagram showing the dark lamp method of synchronization of two single-phase alternators and explain the conditions under which:
a. the lamps will always remain dark
b. the lamps will always remain bright
c. the lamps will flicker in unison.

7-12. a. Given two alternators having rated voltages of 120 V each, what must be the rated voltage of the lamp used in Fig. 7-9(a) (see Ex. 7-3)?
b. State two disadvantages of the dark lamp method of alternator synchronization.
c. Explain why the bright lamp method [Fig. 7-9(b)] overcomes the above disadvantages.

7-13. With reference to synchronizing current:
a. define it by comparing it with load current, I_L, and armature current, I_a
b. write an equation for determining its magnitude for two alternators in parallel, per phase
c. what is its effect on load distribution?
d. what is its effect on the alternator which has a phase angle of less than 90° between its generated voltage and the synchronizing current?
e. Repeat (d) for angles of greater than 90°
f. Explain the primary effect of synchronizing current.

7-14. What is the significance of:
a. a positive synchronizing power and how does it affect an alternator in parallel?
b. Repeat (a) for a negative synchronizing power
c. the fact that the positive synchronizing power does not equal the negative synchronizing power?

7-15. If the primary effect of synchronizing current is the production of synchronizing power, draw
  a. the instantaneous phasor diagram for two alternators having unequal generated voltages 180° out of phase and show the resultant generated voltage and synchronizing current produced. Indicate positive and negative powers
  b. the instantaneous phasor diagram after synchronizing power has been effective in advancing the alternator receiving the power and retarding the alternator generating the power.

7-16. a. Give three effects of overexciting a particular alternator operating in parallel which results in a stabilizing of all the alternators in synchronism.
  b. What is the effect of overexciting a particular alternator on the load distribution of alternators in parallel?

7-17. With regard to alternators in parallel, explain how:
  a. the bus voltage is raised or lowered
  b. the load may be removed from a particular alternator
  c. the load is added to a particular alternator.

7-18. Alternators are designed to have a low overall impedance but a high ratio of synchronous reactance to armature resistance.
  a. What is the advantage of a low impedance with respect to synchronization?
  b. What is the advantage of a high ratio of X_s/R_a?

7-19. a. What is the cause of hunting of an alternator in parallel above and below the synchronous speed?
  b. Give four techniques which tend to eliminate hunting.

7-20. a. What differences exist in the criteria for paralleling polyphase alternators as compared to single-phase alternators?
  b. Give two methods of checking phase sequence of polyphase alternators.
  c. Does the synchroscope give an indication of phase sequence? Explain.

7-21. Summarize the 7 steps necessary to place an incoming alternator in parallel with other alternators to carry a share of the load.

7-22. Assuming a given alternator is operating in parallel with other alternators, and carrying a proportionate share of load, explain the effect of:
  a. reducing its excitation so that its emf is below bus voltage
  b. reducing its prime mover speed so that its emf is below bus voltage
  c. uncoupling the alternator from its prime mover.

7-23. a. What name is given to the synchronous dynamo when operating under conditions described in (c) or (a) of Quest. 7-22?
  b. At what speed will the dynamo operate under the conditions described in (a), (b) and (c) of Quest. 7-22?

PROBLEMS

7-1. Two generators are connected in parallel with a load of $\frac{6}{7}$ Ω. Generator A is adjusted to a generated voltage of 124 V and has an armature resis-

tance of 0.1 Ω. Generator B is adjusted to a generated voltage of 125 V and has an armature resistance of 0.05 Ω. Neglecting the field current drawn by each generator, the brush voltage drop and the voltage drop due to armature reaction, calculate
a. Current delivered by each generator to the bus
b. Current delivered by the bus to the load
c. Terminal bus voltage.

7-2. Repeat Problem 7-1, with the excitation of generator A lowered so that its generated voltage is 120 V and generator B raised so that its generated voltage is 127 V.

7-3. Repeat Problem 7-1 with the excitation of generator A lowered so that its generated voltage is 118 V and generator B raised so that its generated voltage is 128 V.

7-4. a. Referring to Problem 7-2, what is the effect on generator B if generator A is disconnected from the bus? Why?
b. Referring to Problem 7-2, what is the effect on generator B if the internal armature resistance of generator A is increased? Why?

7-5. a. Referring to Problem 7-3, what is the nature of the operation of generator A? Why?
b. What is the term usually given to the generated voltage of dynamo A?
c. What is the effect of decreasing the excitation of dynamo A on
1. The current drawn by dynamo A from the bus
2. The torque produced by dynamo A
3. Direction of torque produced by dynamo A compared to torque of its prime mover.
d. What is the terminal voltage of dynamo A? Give the equation stating operation of dynamo A
e. Give the equation stating operation of dynamo B.

7-6. Two 50 kW, 250 V, 1200 rpm identical shunt generators have a 10 per cent voltage regulation. One generator is delivering half its rated load at a terminal voltage of 262.5 V when the second generator is floated on the line in parallel with it. Assuming constant speed and excitation of both generators, calculate
a. Total power delivered at a bus voltage of 262.5 V
b. Maximum power delivered without exceeding the rating of the original generator
c. Power delivered by each generator in (b).

7-7. A 10 kW, 125 V, 1800 rpm shunt generator is operated in parallel with a 5 kW, 125 V, 1200 rpm shunt generator. The 10 kW dynamo has an 8 per cent while the 5 kW dynamo has a 10 per cent voltage regulation, respectively. Assuming that both dynamos are paralleled at a no-load voltage of 135 V and the speed is constant with application of load, for a total load current of 100 A delivered to a load, calculate
a. The load current delivered by each generator
b. The kilowatts delivered to the load
c. The kilowatts delivered by each generator.

7-8. Repeat Problem 7-7 when the total load current is 60 A.

7-9. Repeat Problem 7-7 when the total load current is 120 A.

7-10. a. Assuming that each of the generators of Problem 7-7 must be loaded to rated load when the load current is 120 A, calculate the no-load excitation voltage for each machine before paralleling.
b. If the two generators remain in parallel with the external load removed, which dynamo will operate as a motor and which as a generator?

7-11. Two 250 V, dc compound generators of unequal rating are connected in parallel. Generator A has a rating of 50 kW and B a rating of 100 kW. To enable the generators to have the same degree of compounding, the series field of generator B is shunted with a resistance of 0.05 Ω. Generators B and A each have a series field resistance of 0.05 Ω If both generators are connected short shunt, determine
a. Whether the generators will operate satisfactorily in parallel. If not, why not?
b. The necessary resistance which will provide satisfactory parallel operation and show the circuit which will accomplish this
c. The voltage drop across each series field in parallel when both generators are carrying rated load and half-rated load, respectively.

7-12. Two single-phase alternators are to be synchronized using the dark lamp method of synchronization shown in Fig. 7-9(a). Alternator A has a terminal voltage of 220 V and a frequency of 60 Hz. Alternator B has a terminal voltage of 222 V and a frequency of 61 Hz. Calculate:
a. Maximum effective voltage across *each* lamp
b. The frequency of lamp flicker
c. Minimum effective voltage across *each* lamp.

7-13. Each of the alternators in Problem 7-12 has a frequency of 60 Hz. Calculate the effective voltage across each lamp when
a. The phase difference between the alternators is 0°
b. The phase difference between the alternators is 180°
c. Calculate the frequency of lamp flicker in (a) and (b).

7-14. In the previous problems the two single-phase alternators, A and B, each have an effective resistance of 0.1 Ω and a synchronous reactance of 1.0 Ω, respectively. They are properly synchronized (i.e., their emf is 180° with respect to each other in the local circuit) but at the instant of synchronizing, the emf of alternator A is 220 V and alternator B is 210 V (rms). At the instant that the synchronizing switch is closed, and before the alternators adjust their phase positions, calculate
a. Resultant voltage between alternators
b. Synchronizing current
c. Power factor angle between (a) and (b)
d. Power developed by alternator A and mode of operation
e. Power developed by alternator B and mode of operation
f. Power loss in both armatures and synchronizing power
g. The terminal voltage of each alternator (bus voltage).

7-15. Two wye-connected, 150 kVA, 3980 V, three-phase alternators are synchronized without load so that alternator A is 20 electrical degrees ahead of its proper 180° local circuit phase position with respect to

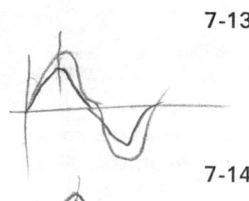

alternator B. Each alternator has a resistance of 0.12 Ω/phase and a reactance of 1.275 Ω/phase, respectively. At the instant that the synchronizing switch closes, and before the alternators adjust their phase positions, calculate
   a. The phase difference between the alternators in electrical degrees and volts rms
   b. The synchronizing phase current and the power factor angle
   c. The power developed per phase by alternator A and the mode of operation
   d. The power developed per phase by alternator B and the mode of operation
   e. Power loss in both armatures and the synchronizing power transferred through the bus bars, assuming equal power losses in each alternator
   f. Terminal voltage of each alternator per phase and line.

7-16. Two three-phase, wye-connected alternators, A and B, are to be paralleled to a set of common bus bars. The armature resistances and synchronous reactances per phase are 0.1 Ω and 1.0 Ω for alternators A and B, respectively. The line voltage of A is adjusted to 2500 V and B is adjusted to 2300 V and are exactly in phase opposition at the instant paralleled. Calculate:
   a. Resultant voltage between alternators per phase
   b. Synchronizing current per phase
   c. Power factor angle between (a) and (b) above
   d. Power developed by alternator A and mode of operation
   e. Power developed by alternator B and mode of operation
   f. Power loss per phase
   g. Terminal bus voltage per phase and line.

ANSWERS

7-1(a) 40 A, 100 A (b) 140 A (c) 120 V   7-2(a) 140 A, 0 (b) 140 A (c) 120 V
7-3(a) −20 A, 160 A (b) 140 A (c) 120 V   7-6(a) 26.25 kW (b) 75 kW (c) 50 kW, 75 kW   7-7(a) 28.6 A, 71.4 A (b) 12.62 kW   (c) 3.61 kW, 9.01 kW
7-8(a) 17.15 A, 42.85 A (b) 7.78 kW (c) 5.56 kW, 2.22 kW   7-9(a) 34.3 A, 85.7 A (b) 149.2 kW (c) 42.7, 106.5 kW   7-10(a) 137.5 V, 135 V (b) Conductor A, Motor B   7-12(a) 221 V (b) 1 Hz (c) 1 V   7-13(a) 221 V (b) 1 V (c) zero
7-14(a) 10 V (b) 4.98 A (c) 0.09932 (d) 108.8 W (e) −103.8 W (f) 103.8 W (g) 215 V   7-15(a) 160°   (b) 312.5 A   (c) 7.15 kW   (d) 693 kW motor (e) 69 kW, 704.5 kW/phase   (f) 2300 V   7-16(a) 118 V   (b) 58.75 A (c) 0.09932 (d) 8440 W (e) 7750 W (f) 690 W (g) 2400 V

EIGHT

ac dynamo torque relations— synchronous motors

8-1. GENERAL

It bears repeating that all electric motors, dc and ac, act as generators while the motor action is taking place. In the previous chapter on parallel operation, it was stated that, when a dynamo (ac or dc) is connected in parallel with a bus or another source of emf, it may act as (1) a generator, if its induced emf exceeds the bus voltage (and it generates power to the bus); or (2) a motor, if its induced emf is less than the bus voltage (in which case it receives power from the bus). It was also stated that a motor armature connected across a bus may be considered as being in parallel with that bus. In considering the parallel operation of the single-phase and polyphase alternator, it was demonstrated that two factors would cause an alternator to "motorize" and receive synchronizing power from the bus (or other alternators in parallel); namely (1) a decrease of field current and generated emf (below the bus voltage), and (2) a decrease in the *instantaneous speed* of the ac dynamo. When either of these conditions occur, the ac dynamo is operating as an ac synchronous motor.

It should be noted that the term "instantaneous speed" was used. The speed of an ac synchronous motor is determined [Eq. (2-16)] by the number of poles and the frequency, i.e., $S = 120f/P$. Since the frequency of the bus supplying the motor is constant, and since the number of poles is also constant, it is evident that the ac synchronous motor is a *constant-speed motor*. The manner in which its speed changes *instantaneously* (from moment to moment), as a result of the load and the magnitude of current which its stator receives from the bus, will be discussed in detail in this chapter.

Not only does a synchronous motor armature require and receive ac current from the bus, but, like any (double-excited) ac synchronous dynamo, it requires a dc excitation for its field. On large synchronous motors, the *exciter* (a dc shunt generator) is placed on the same shaft as the motor, and a small portion of the motor torque is required to generate the dc required for field excitation. Because of the possibility of variation of field excitation, the ac synchronous motor possesses one characteristic possessed by no other ac motor type—the power factor at which it operates may be *varied* at will.

A second, and somewhat unusual, characteristic of the polyphase (and single-phase) synchronous motor is that (like some single-phase motors) it is *not inherently self-starting*. Like the ac alternator, it must be brought up to speed by some auxiliary means, and then connected across the line.

Another peculiarity of synchronous motors is their susceptibility to hunting, (Sec. 7-12), particularly when the loads are subject to sudden changes or are not uniform over one cycle of revolution, as in the case of punch presses, shears, compressors, or pumps. The use of *damper windings* in the rotor construction has ended that problem and, at the same time, has made it possible for the synchronous motor to become self-starting as an induction motor.

Today, the synchronous motor is widely used, and its popularity has never been greater. In certain horsepower sizes and speed ranges it outsells the polyphase induction motor.*

Synchronous motors have the following specific advantages over induction motors: (1) synchronous motors can be used for power-factor correction in addition to supplying torque to drive loads; (2) they are more efficient (when operated at unity power factor) than induction motors of corresponding horsepower and voltage rating; (3) the field pole rotors of synchronous motors can permit the use of wider air gaps

---

* Differences in cost between induction motors and synchronous motors of the same horsepower, speed, and voltage rating vary because of relative stator and rotor manufacturing techniques for both types of motors. In moderate sizes, ranging from 50 to about 500 hp at low speeds, the synchronous motor is less expensive. In large sizes at high speeds, the synchronous motor is again less expensive. As new techniques are developed, these differences in initial cost may be eliminated or conversely accentuated.

than the squirrel-cage designs used on induction motors, requiring less bearing tolerance and permitting greater bearing wear; and (4) they may be less costly for the same horsepower, speed, and voltage ratings.

8-2. CONSTRUCTION

Basically, the construction of an ac synchronous motor is the same as the alternator (Section 6-2). The stator has a single-phase or polyphase winding identical to that of the alternator. The rotor is generally a salient-pole rotor, except in types of exceedingly high speed. In order to eliminate hunting and to develop the necessary starting torque when an ac voltage is applied to the stator, the rotor poles contain pole-face conductors which are short-circuited at their ends, as shown in Fig. 8-1. This *amortisseur* or *damper winding* consists of solid copper bars embedded at the surface of the pole face and short-circuited at each end by means of a shorting strip, as shown in Fig. 8-1b.

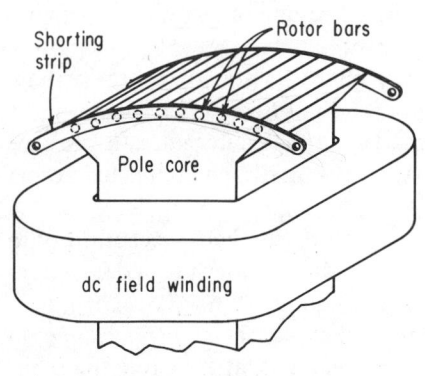

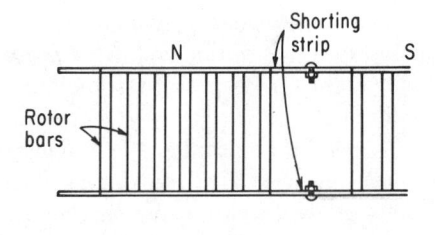

(a) Pole of an ac synchronous dynamo

(b) Amortisseer, damper, squirrel cage or starter tisseur.

Figure 8-1

Pole of an ac synchronous dynamo showing damper winding.

8-3. OPERATION OF THE SYNCHRONOUS MOTOR

As stated in Section 8-1, the synchronous motor is *not* inherently self-starting, i.e., it will not start by itself *without* a damper winding. This is shown in Fig. 8-2, where an alternating current is applied to the stator winding, and where the instantaneous direction of current in the coil sides of a given armature coil, *A* and *B*, is shown. Both the north and south poles will be subjected to an electromagnetic torque (the left-hand motor rule) moving the poles to the left (conductors to the right). The next instant, $\frac{1}{120}$ of a second later, the frequency reverses the direction of current in the coil, and the poles receive a torque in the opposite direction, as shown in the figure. Because

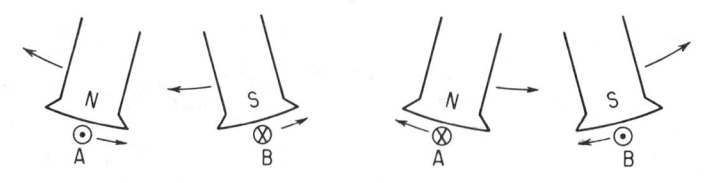

Figure 8-2

Zero resultant torque developed by stator conductors of a synchronous motor when rotor is at a standstill.

of the high inertia of the rotor, the resultant torque produced in one second is zero, since the rotor has, in effect, been *pushed alternately* clockwise and counterclockwise 60 times in that second, assuming a frequency of 60 Hz.

However, if by some means the rotor is moving clockwise at some speed near or at synchronous speed, as shown in Fig. 8-3, torque *will* be developed by coil-sides A and B to cause the rotor to continue to move clockwise. The *space movement* of the pole in electrical degrees at synchronous speed *corresponds* to the 180° reversal of direction of current in the armature coil, and the resultant torque produced is in the same direction.

The armature winding consists of many coils in series in each phase of a polyphase ac synchronous dynamo. The three-phase current in the armature conductors of the stator produces a uniform rotating magnetic field (Section 9-3, Fig. 9-1), rotating at a speed, $S = 120f/P$. The relation between the rotating field of the stator and the rotor poles is shown in Fig. 8-4a. The north and south poles, respectively, of the rotor, rotating at a synchronous speed, are *locked in synchronism* with the resultant

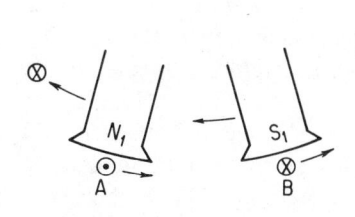

(a) Instantaneous torque.

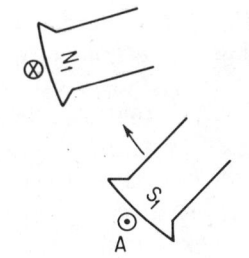

(b) Torque produced by reversal of current.

Figure 8-3

Torque in same direction when rotor rotates at synchronous speed.

SEC. 8-3. / Operation of the Synchronous Motor

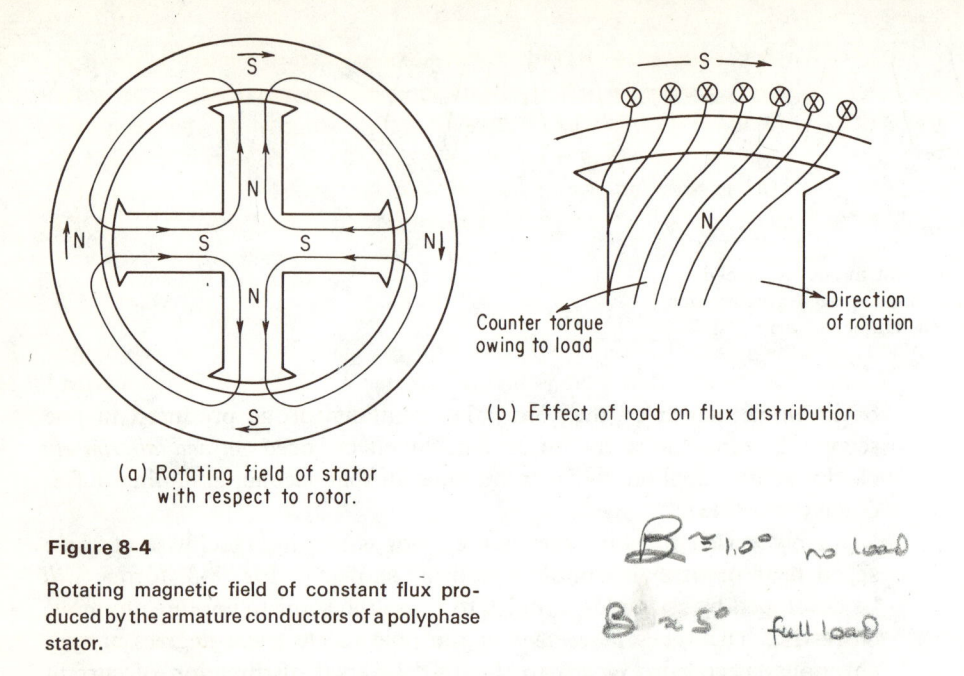

Counter torque
owing to load

Direction
of rotation

(b) Effect of load on flux distribution

(a) Rotating field of stator
with respect to rotor.

Figure 8-4

Rotating magnetic field of constant flux produced by the armature conductors of a polyphase stator.

armature synchronous rotating field of the stator. Thus, a rotor N pole is locked in synchronism with a stator S pole and vice versa, both rotating clockwise in synchronism at the synchronous speed. If a load is placed on the shaft of a synchronous motor, the countertorque created by the load will cause the rotor to drop back *momentarily* but it will continue to rotate at the same speed with respect to the rotating stator field.* The rotor speed is still at synchronous speed, however, with respect to the rotating field, but the rotor flux or mutual air-gap flux is reduced slightly, as shown in Fig. 8-4b, because of the increased air-gap reluctance.

If the countertorque is so great that it exceeds the maximum torque developed, and if the rotor "slips" out of synchronism, the synchronous motor will stop. Thus, <u>a synchronous motor will either run at synchronous speed or not at all</u>. Indeed, as the rotor is slowing down, the rotating fields of the stator slips by the rotor field poles so rapidly that it is unable to lock synchronously or "mesh" with the rotating stator field. This is why a rotor at standstill is also unable to start. At one instant, a unit N pole of a rotor is attracted to an approaching stator S pole, producing torque in a counterclockwise direction in Fig. 8-4b, and the next instant the same N pole is attracted in the opposite direction by a passing rotor S pole, producing torque in a clockwise direction, or a net torque of zero.

* This is not the same as slip speed in an induction motor, where the stator rotates at the synchronous speed but the rotor speed must always be less than the synchronous speed.

8-4. STARTING SYNCHRONOUS MOTORS

It is evident, then, that the synchronous motor must be brought up to a speed sufficiently close to synchronous speed, in order to lock into synchronism with the rotating field. The means by which it is brought up to speed are: (1) a dc motor coupled to the synchronous motor shaft; (2) using the field exciter generator as a dc motor; (3) a small induction motor of at least one pair of poles less than the synchronous motor; and (4) using the damper windings as a squirrel-cage induction motor.

The first method is sometimes used in laboratories with synchronous motors *not* equipped with damper windings. Generally, the synchronous motor is intended as the constant-speed prime-mover for the dc generator. But in order to bring the motor up to synchronism, the dc generator is operated as a motor, and the ac synchronous dynamo is synchronized to the ac supply (Section 7-16) as an alternator. Once in parallel with the supply, the synchronous dynamo operates as a motor. The dc "motor" will not act as a generator if its field current is increased so that its generated emf exceeds the dc bus.

The second method is actually the same as the first, except that the exciter (a dc shunt generator) is operated as a motor, and the ac synchronous dynamo is synchronized to the ac supply (Section 7-16).

The third method, using an auxiliary induction motor with fewer poles, involves the same synchronizing procedure for the ac synchronous motor as an alternator. At least one pair of poles fewer is required on the induction motor to compensate for the loss in induction motor speed due to slip.

In all of the three methods discussed above, it is necessary (1) that there is little or no load on the synchronous motor, and (2) that the capacity of the starting motor (dc or ac) is between 5 and 10 per cent of the rating of the synchronous motor coupled to it.

By far the most common method of starting a synchronous motor, however, is as an induction motor using the damper windings. This method is the simplest and requires no special auxiliary machines (see Sec. 8-5, below).

8-5. STARTING A SYNCHRONOUS MOTOR AS AN INDUCTION MOTOR BY MEANS OF ITS DAMPER WINDINGS

The *amortisseur* or damper winding is shown in Figs. 8-1a and b. It should be noted that the shorting strip, which short-circuits the rotor bars, contains holes for bolting to the next set of damper windings on the next pole. In this way, a complete squirrel-cage winding is formed; and although the bars are not of the capacity to carry the rated synchronous motor load, they are sufficient to start the synchronous motor as an induction motor. Where extremely large synchronous motors are started as induction motors, various methods are employed to reduce the starting current drawn from the bus. Since these methods are the same as those required in starting large induction motors, they will be covered in Chapter 9 (see Section 9-14 *et seq.*). The methods discussed there include, specifically, wye-delta starting line-resistance starting, reduced-voltage transformer starting, etc.

It is practically impossible to start a synchronous motor with its dc field energized. Even when left de-energized, the rapidly rotating magnetic field of the stator will induce extremely high voltages in the many turns of the field winding. It is customary, therefore, to short-circuit the dc field winding during the starting period; whatever voltage and current are induced in it may then aid the damper winding in producing induction motor action. In very large synchronous motors, field-sectionalizing or field-splitting switches are used which short-circuit individual field windings to prevent cumulative addition of induced voltages from pole to pole. (Such induced high voltages may puncture field insulation.)

Among the advantages of synchronous motors over induction motors is the fact that the air gap of a synchronous motor is greater (Section 8-1). The induction winding of the rotor therefore develops on starting, a fairly high ratio of rotor reactance to resistance. Although this may result in higher starting currents and lower power factors to develop the same torque or even less torque, it does result in improving the no-load slip-speed of the synchronous motor. Thus, when the short-circuit is removed from the field and dc is applied to the rotor field winding, at or near synchronous speed, the rotor easily pulls into synchronism with the rotating stator field. Occasionally, when a stroboscopic lamp is used to measure synchronous motor speed, students in the laboratory have observed no instantaneous change in speed when direct current is applied, indicating that the rotor has already pulled into synchronism "by itself." This is the principle by which the hysteresis motor and the synchronous-induction motor (Section 8-29) operate; namely, the rotor iron becomes *magnetized* by the stator flux, and the rotor pulls into synchronism without requiring dc field excitation, using *reluctance* torque, Secs. 1-2 and 8-17.

Another phenomenon occasionally encountered in the laboratory, when the field circuit is energized, is a sudden inrush of alternating current and a consequent "thump" indicating a sudden line disturbance. It occurs when the field **mmf** produces poles on the rotor which are directly under the rotating field of the *same* polarity, as shown in Fig. 8-5. This causes a sudden reduction in the air-gap flux, which reduces the generated **emf** and *suddenly* increases the armature current. The consequent reduction in torque causes the motor to slip back

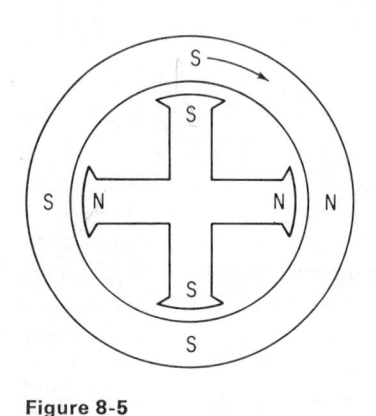

Figure 8-5

Magnetization of field that results in "slipping a pole."

240

CHAP. EIGHT / *AC Dynamo Torque Relations—Synchronous Motors*

a pole, 180 electrical degrees, where it now operates in synchronism, and the line current is restored to normal.

> In summary, when starting a synchronous motor on its damper windings:
> (1) the dc field winding is shorted and ac is applied to the stator, bringing the motor up to no-load speed as an induction motor
> (2) dc is applied to the field winding and the field current is adjusted to provide minimum ac line current.

8-6. STARTING A SYNCHRONOUS MOTOR UNDER LOAD

The synchronous motor starts and runs at or near synchronous speed as an induction motor on its induction type damper windings. In the discussion of Section 8-5, the assumption was made that, (1) the motor is lightly loaded and (2) in order to develop a speed close to synchronous speed, the induction motor cage winding required low resistance and high reactance. But these latter characteristics in an induction motor produce low starting torque for the same armature current (Section 9-10). In general, these torques are about 30–50 per cent of the full-load torque. For certain types of loads, such as fans or air compressors, whose loads are a function of speed, such low applied torques may be tolerated, and the synchronous motor will come up to speed and lock into synchronism with such a load on its shaft. But the growing popularity of the synchronous motor created a demand for a synchronous motor which could develop higher starting torques, ranging from full load up to 300 per cent of full load, and capable of starting under heavy load.

The starting torque of synchronous squirrel-cage rotor bars may be improved by the use of higher resistivity alloys in the bars. This, however, does not bring the rotor as close to synchronous speed as low-resistance rotors, because the slip is increased due to the high resistance. Opening the field for an instant, and shorting it just before applying the direct current, will cause the rotor to surge forward sufficiently to lock it into synchronism.

The better technique, however, is to use a wound rotor winding rather than squirrel-cage winding in the pole face, the so-called phase-wound damper winding. Such a rotor is recognized immediately because it employs five slip rings: two for the dc field winding and three for the ac wye-connected wound rotor winding. The starting performance of the *phase-wound damper* (or Simplex rotor) winding is similar to that of a wound rotor induction motor (Section 9-10), since external starting resistance is employed to improve the starting torque. The motor is started with full external resistance per phase, as shown in Fig. 8-6, and the dc field circuit shorted. The motor approaches synchronous speed as starting resistance is reduced and, when the dc field voltage is applied, the motor

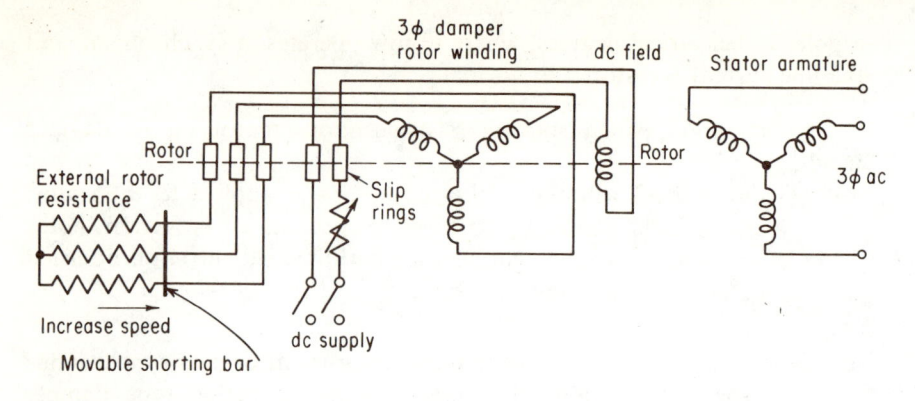

Figure 8-6

Schematic diagram of phase-wound damper synchronous motor (simplex rotor).

pulls into synchronism. By combining the high starting torque of the wound rotor induction motor (up to three times the normal full-load torque) with the constant speed and power factor correction running characteristics of the synchronous motor, the Simplex-type rotor synchronous motor has found many applications where starting under load is required in addition to constant speed.

8-7. SYNCHRONOUS MOTOR OPERATION

As stated in Section 8-1, all motors produce a generated voltage while motor action is taking place. During the transient period when a synchronous motor is being brought up to speed, as an induction motor, an armature current is flowing in its stator winding. This current is being limited essentially by the voltage induced and the current flowing in the rotor bars of the damper winding by transformer action (to be discussed in detail in the next chapter). Once the dc field is energized (and the rotor pulls into synchronism), however, the rotor flux induces an ac voltage in the stator conductors in accordance with Eq. (2-15). Since the synchronous motor is in parallel with the bus, the current which the motor draws as a result of motor action is a synchronizing current, and synchronizing power is required to maintain its rotor in synchronism with the frequency of rotation of the stator flux. Indeed, it was shown in Section 8-4 that, in some methods of starting without damper windings, the motor must be synchronized to the bus using the identical technique required for the parallel operation of alternators.

The phasor relation for an ac synchronous alternator delivering synchronizing current and power to another ac synchronous "dynamo" was discussed in Section 7-11 and shown in Figs. 7-14 and 7-16. In both these diagrams, it was noted that the phase positions of the generated

Figure 8-7

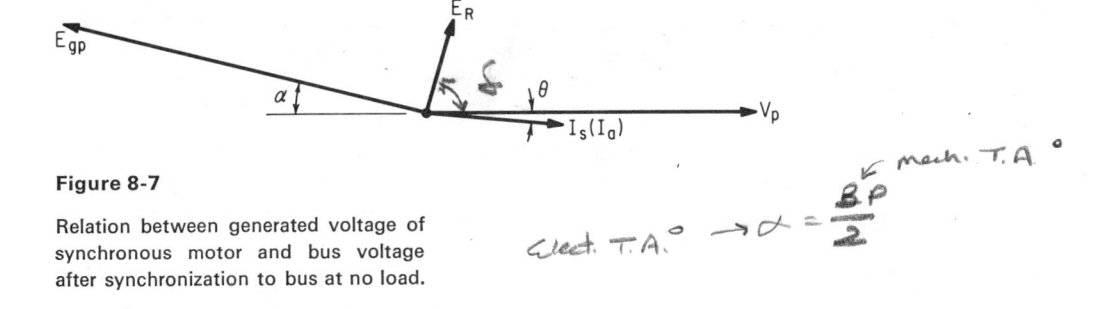

Relation between generated voltage of synchronous motor and bus voltage after synchronization to bus at no load.

voltages are less than 180° as a result of the synchronizing power. These figures are summarized in Fig. 8-7 with one minor change, namely, the terminal bus voltage supplying the synchronizing current is taken as the reference voltage and is represented as the terminal voltage per phase, V_p instead of E_{gp1}. It should be noted in the figure that the generated voltage, E_{gp} [Eq. (2-15)], is shown equal to the bus voltage (as it would be at the instant of synchronizing to the bus), but displaced from the 180° position by an angle α as a result of the synchronizing power *received*. Despite the fact that the generated voltage E_{gp} is equal to the bus voltage, it should be noted that the motor armature and synchronizing current I_s or I_a is a result of the resultant voltage, E_r. Assuming that the ac bus supply, consisting of one or more alternators in parallel, has negligible internal impedance (a justifiable assumption in most cases), Eq. (7-3), which expresses the value of the armature synchronizing current, may be simplified to:

$$I_a = \frac{\dot{V}_p - \dot{E}_{gp}}{R_a + jX_{sa}} = \frac{E_r}{Z_p} \tag{8-1}$$

where I_a is the armature current per phase drawn by the synchronous motor from the ac bus

V_p is the phase voltage applied to the armature of the synchronous motor stator

E_{gp} is the generated voltage per phase, generated in the armature conductors in accordance with Eq. (2-15)

E_r is the phasor difference between the armature applied voltage and generated voltage [see Eq. (8-3)] per phase

Z_p is the impedance per phase of the synchronous motor, consisting of R_a and X_{sa}

R_a is the effective armature resistance per phase

X_{sa} is the synchronous armature reactance per phase

The similarity between Eq. (8-1) above and the generic motor equation, Eq. (1-9), should be noted. It may be said, therefore, that the armature current drawn by an ac synchronous motor is limited by its impedance and its generated (counter) emf in much the same way that the armature current drawn by a dc motor is limited by its resistance and its counter emf.

There is one important difference, however, between an ac synchronous motor and a dc shunt motor. In the case of the dc shunt motor, as load is applied, the countertorque produces a drop in speed; the reduction in speed has the effect of reducing in turn the generated counter emf, thus permitting more armature current to flow [Eq. (1-9)]. The increased armature current supplies more driving torque [Eq. (4-4)] and develops more power in the armature [Eq. (4-7)]. In order to produce motor action and to enable the motor to receive current from the bus, the generated counter emf can never equal the bus voltage in a dc motor.

In an ac synchronous motor, however, the speed is *constant*, and the generated emf per phase, E_{gp}, is a function ($E_g = K\phi S$) of the dc field excitation only. The synchronous motor, therefore, is unable to draw more current from the supply as a result of decreased counter emf E_{gp}. If the motor is overexcited by a high dc field current, the generated emf per phase can *exceed* the terminal voltage. Since the field flux of a synchronous motor is *independent* of speed, how then does the synchronous motor adjust its armature current in Eq. (8-1) to develop increase power when load is applied to its shaft?

It was stated that, when the synchronous motor was synchronized to the bus, its generated phase voltage was equal and opposite to the voltage per phase of the bus. At this instant, the paralleling switch was closed. If the synchronous motor continues to be driven by its prime mover, its generated phase voltage will equal the bus voltage, as shown in Fig. 8-8a; and, as a dynamo, it is floating on the line (Section 7-2). Assume that the prime mover is suddenly uncoupled from the synchronous motor, which is precisely the situation when an unloaded synchronous motor pulls into synchronism upon excitation of its dc field. Since it is no longer being driven, it pulls back in phase, by angle α, developing a resultant voltage, E_r, in accordance with Eq. (8-1), and shown in Fig. 8-8b. But the angle α may be insufficient to maintain rotation of the synchronous motor, because very little synchronizing power is developed

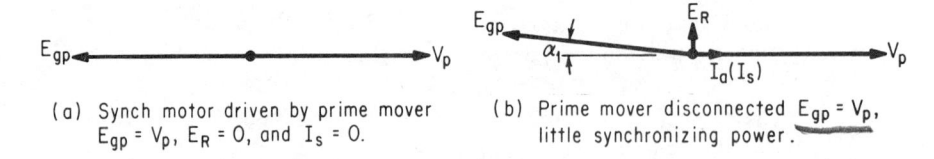

(a) Synch motor driven by prime mover $E_{gp} = V_p$, $E_R = 0$, and $I_S = 0$.

(b) Prime mover disconnected $E_{gp} = V_p$, little synchronizing power.

Figure 8-8

Steps in development of torque owing to synchronizing power drawn from bus by a synchronous motor (no load).

when the armature current is very small. It is necessary, therefore, for the rotating rotor field pole to drop back a few more electrical degrees (with respect to the rotating field produced in the stator armature) to maintain rotation.

As the rotor pole drops back in phase: (1) the resultant voltage difference E_r increases; and (2) the armature current, I_a, also increases [since Z_p in Eq. (8-1) may be assumed constant]. Figure 8-7 shows this situation, in which positive synchronizing power of magnitude $V_p I_a \cos\theta$ per phase, is delivered to the synchronous motor. Since the motor is no longer driven by a prime mover, this positive synchronizing power (as in the case of parallel alternators) does *not* pull the synchronous motor ahead so that its generated emf is displaced 180° with respect to the bus. Instead, the positive power produced is sufficient to overcome the synchronous motor countertorque produced by friction, windage, and other rotational losses of the motor. The rotor rotates therefore, at a constant synchronous speed with a fixed phase position lag, α, between the center of a rotor N pole and the center of an opposite S pole on the rotating field of the stator.

It should be noted that the no-load angle of lag, α, between the rotating field and the rotor poles, represents but a few electrical degrees. The number of *mechanical* degrees, β, is the same or even less than α, and the relation between them is

$$\beta = \frac{2\alpha}{P} \tag{8-2}$$

where P is the number of poles, and α is the number of electrical degrees.

It has been observed that, in the multipolar synchronous motor, the difference between the synchronizing position shown in Fig. 8-8a and the position in Fig. 8-7 or Fig. 8-8b is barely discernible by stroboscopic means (note displacement in Example 8-1). It should also be noted that, when E_{gp} is equal to V_p, the voltage and current are affected as follows:

1. The resultant voltage, E_r, (however small the angle α may be) leads V_p by less than 90°, since it is the phasor difference of two equal phasors as shown in Figs. 8-7 through 8-9.
2. The armature current, I_a, at no load, may be either slightly leading

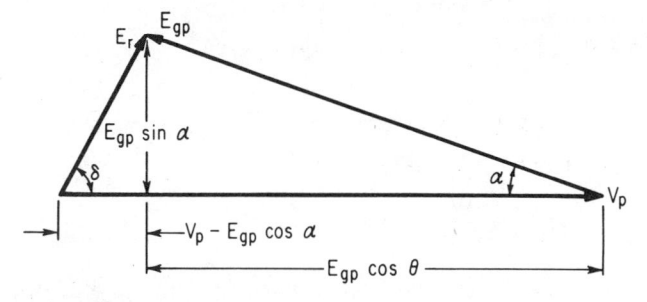

Figure 8-9

Determination of resultant armature voltage per phase.

SEC. 8-7. / Synchronous Motor Operation

(see Example 8-1), in phase with V_p, or even slightly lagging (as shown in Fig. 8-7), depending on the field excitation.

The computation of E_r for any values of the angle α, of the excitation voltage, and of the applied phase voltage of a given ac synchronous dynamo, may be determined from Fig. 8-9 and the following equation:

$$E_r = (V_p - E_{gp} \cos \alpha) + j(E_{gp} \sin \alpha) \tag{8-3}$$

where α is the torque angle, and all other terms have been defined in Eq. (8-1).

EXAMPLE 8-1: A 20 pole, 40 hp, 660 V, 60 Hz, three-phase, wye-connected, synchronous motor is operating at no-load with its generated voltage per phase exactly equal to the phase voltage applied to its armature. At no-load, the rotor is retarded 0.5 mechanical degree from its synchronous position. The synchronous reactance is 10 ohms, and the effective armature resistance is 1 ohm per phase. Calculate:
a. The rotor shift from the synchronous position, in electrical degrees
b. The resultant emf across the armature, per phase
c. The armature current, per phase
d. The power per phase, and the total power drawn by the motor from the bus
e. The armature power loss, and the developed horsepower.

Solution:

a. $\alpha = P\left(\dfrac{\beta}{2}\right) = 20\left(\dfrac{0.5}{2}\right) = 5°$ [from Eq. (8-2)]

b. $V_p = \dfrac{V_L}{\sqrt{3}} = \dfrac{660}{1.73} = 381$ V; $E_{gp} = 381$ V also, as given

$E_r = (V_p - E_{gp} \cos \alpha) + j(E_{gp} \sin \alpha)$
$= (381 - 381 \cos 5°) + j(381 \sin 5°)$
$= 1.54 + j33.2 = 33.2 \angle 87.3°$ V/phase

c. $Z_s = R_a + jX_s = 1.0 + j10 = 10 \angle 84.3°$ Ω/phase

$I_a = \dfrac{E_r}{Z_p} = \dfrac{33.2 \angle 87.3°}{10 \angle 84.3°} = 3.32 \angle 3.0°$ A/phase

d. $P_p = V_p I_a \cos \theta = 381 \times 3.32 \cos 3° = 381 \times 3.32 \times 0.999$
$= $ **1265 W/phase**

$P_t = 3P_p = 3 \times 1265$ W $=$ **3795 W**

e. $3 \times I_a^2 R_a = 3 \times (3.32)^2 \times 1.0 = 33$ W

Horsepower $= \dfrac{3795 - 33 \text{ W}}{746 \text{ W/hp}} =$ **5.3 hp**

In Example 8-1, the generated excitation voltage, E_{gp}, is *equal* to the bus voltage per phase, and the armature current is practically *in phase* with the armature applied voltage per phase, V_p. These two conditions

**8-8.
EFFECT OF
INCREASED LOAD
AT NORMAL
EXCITATION OF
SYNCHRONOUS
MOTOR ($E_{gp} = V_p$)**

constitute, for a synchronous motor, a state known as *normal* excitation. This state will be used as a *reference* for the effect of (1) the application of load to the shaft of the motor, and (2) changing the excitation at any given load. (See Fig. 8-11.)

Assuming that the field excitation of a synchronous motor is unchanged ($E_{gp} = V_p$), what is the effect on the armature current and the power factor of applying a load (or countertorque) to the motor shaft? The speed of an ac synchronous motor cannot decrease as a result of increase load; but the torque angle, α, Eq. (8-2) can and does increase, as shown in Fig. 8-10, parts a and b. It should be noted that the angle shown in the figure is β, the mechanical displacement of the rotor with respect to the stator. Thus, at no-load, there is a slight displacement angle between the center of a stator conductor (where the rotating field is a maximum) and the center of the pole core. At full load, the *mechanical* displacement is never more than 4 or 5° in polyphase multipolar synchronous motors. The effect on the power factor, the armature current, and the power drawn from the supply is shown in Example 8-2.

(a) No load.

(b) Increased load.

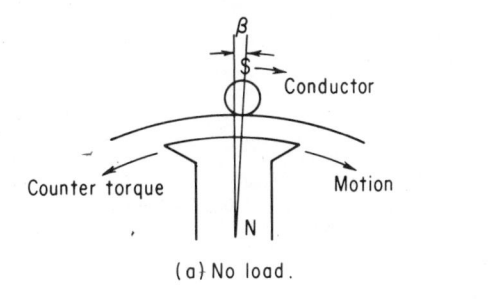

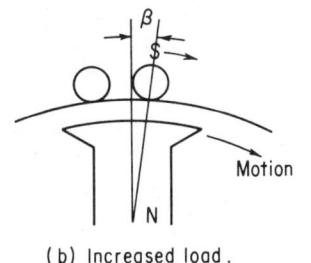

Figure 8-10

Effect of load on rotor position.

EXAMPLE 8-2: Repeat Example 8-1 with a mechanical displacement of 5° between rotor and synchronous position.

Solution:

a. $\alpha = \dfrac{P\beta}{2} = \dfrac{20 \times 5}{2} = 50°$ (electrical degrees)

b. $E_r = (V_p - E_{gp} \cos \alpha) + j(E_{gp} \sin \alpha)$
$= 381 - 381 \cos 50° + j381 \sin 50°$
$= 141 + j292 = 334\angle 64.2°$ V/phase

c. $I_a = \dfrac{E_r}{Z_p} = \dfrac{324 \angle 64.2° \text{ V}}{10 \angle 84.3° \Omega} = 32.4\angle -20.1°$ A

d. $P_p = V_p I_a \cos \theta = 381 \times 32.4 \cos 20.1° = \mathbf{11{,}600\ W}$

$P_t = 3P_p = 3 \times 11{,}600 = \mathbf{34{,}800\ W}$

e. $3I_a^2 R_a = 3 \times (32.4)^2\ 1.0 = \mathbf{3150\ W}$

$$\text{Horsepower} = \frac{34{,}800 - 3150\ W}{746\ W/hp} = \mathbf{42.5\ hp}$$

Example 8-2 shows quantitatively the increase in *both* the resultant voltage, E_r, and the armature current, I_a, as well as the increase in the

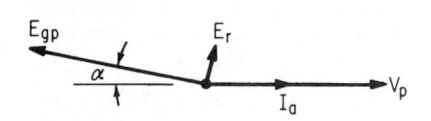

(a) No load, normal excitation $\theta = 0$.

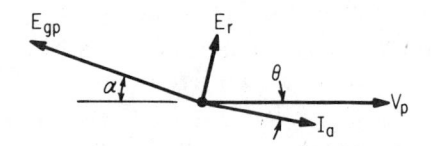

(b) Increased load, normal excitation E_R increases; I_a increases, lags by θ slightly.

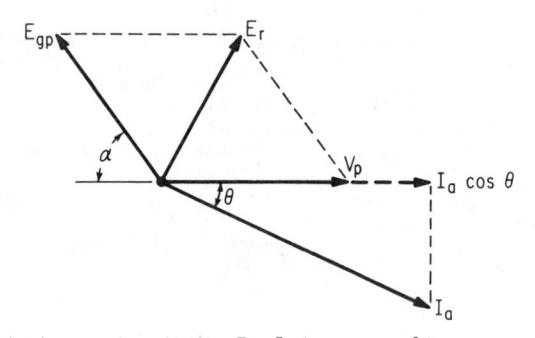

(c) Overload, normal excitation E_R, I_a increases; θ increases.

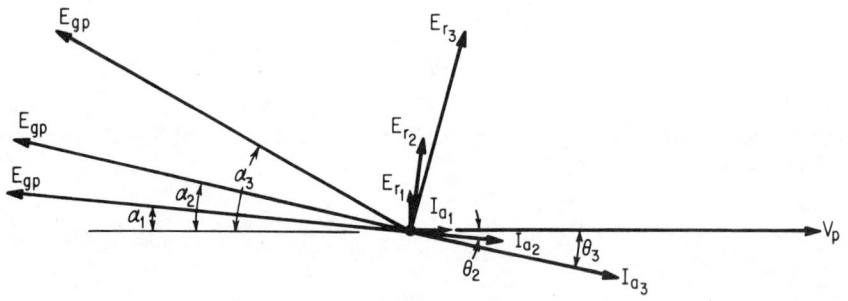

(d) Composite diagram for effect of load at normal excitation, $E_{gp} = V_p$.

Figure 8-11

Effect of load increases at normal excitation ($E_{gp} = V_p$).

power produced as a result of increased load. Note that the armature current increase has also produced *increased* armature copper *losses*.

A phasor summary of the effect of increased load at normal excitation is shown in Fig. 8-11d. Note that, as the load and α are increased, the resultant voltage difference E_r between the generated and the applied voltage to the armature, $(V_p - E_{gp})$, tends to increase rapidly, and the power factor angle, θ, increases slowly. Even under conditions of overload, shown in Fig. 8-11c, the power factor angle θ does not lag appreciably. This is seen rather clearly in the composite diagram in Fig. 8-11d. The armature current increases, with application of load, *at a faster rate* than the power factor angle, thereby automatically adjusting to load increases *essentially by current increase*. In Fig. 8-11c, the difference between I_a and $I_a \cos \theta$ (under conditions of overload) is not too great.

8-9. EFFECT OF INCREASED LOAD AT CONDITIONS OF UNDEREXCITATION ($E_{gp} < V_p$)

One cannot generalize as to the effect of load on a synchronous motor unless loading is studied under conditions of under, normal, and overexcitation. A composite diagram is shown in Fig. 8-12 for the synchronous motor under three load conditions when it is deliberately *underexcited*, $E_{gp} < V_p$. With a small load, and a torque angle of α_1, the armature current I_{a_1} almost lags the applied armature voltage per phase by 90°.

Unlike normal excitation, a fairly substantial armature current must flow to develop the same power $V_p I_{a_1}$ shown in Fig. 8-11d, because of the poor power factor. As the load is increased, however, the power factor improves. Because of the increased resultant voltage, more current flows; and since the power factor is increased the total power generated by the armature increases to meet the load. At very heavy loads, the effect of underexcitation produces a poorer power factor than normal

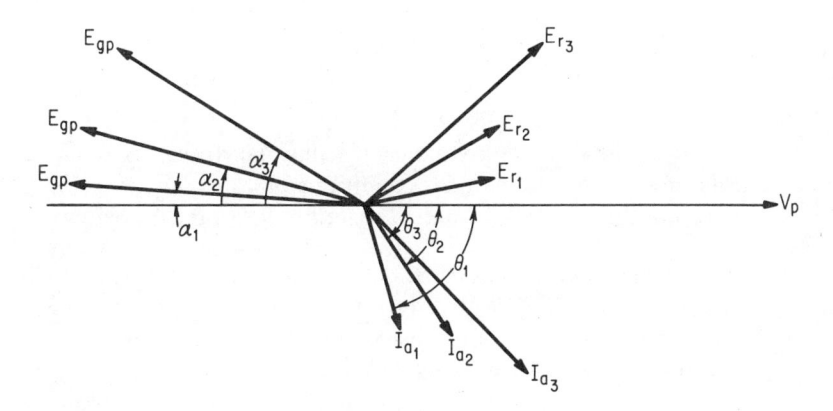

Figure 8-12

Effects of increased load at conditions of underexcitation ($E_{gp} < V_p$).

excitation, and a *much higher armature current* must flow to develop the same power with the same load compared to normal excitation.

**8-10.
EFFECT OF
INCREASED LOAD
AT CONDITIONS OF
OVEREXCITATION
($E_{gp} > V_p$)**

Unlike the shunt motor, it is possible for a synchronous motor to generate a voltage higher than its bus voltage and still draw current and power from the bus. This occurs because the generated emf and the bus voltage are not 180° out of phase, and E_r, the resultant voltage, still represents the phasor difference of the two voltages. When the load is small, as shown in Fig. 8-13, the resultant voltage E_{r_1} is almost in phase with E_{gp}, because the latter exceeds the bus voltage per phase. The quadrature armature current which flows, I_{a_1}, leads the applied voltage by almost 90°. Like the underexcited synchronous motor, as load is applied, the power factor improves (approaches unity power factor). The power factor angle *decreases at a faster rate* than the current increases thereby producing the necessary increased power to meet the increased applied load that is causing increases in torque angle α.

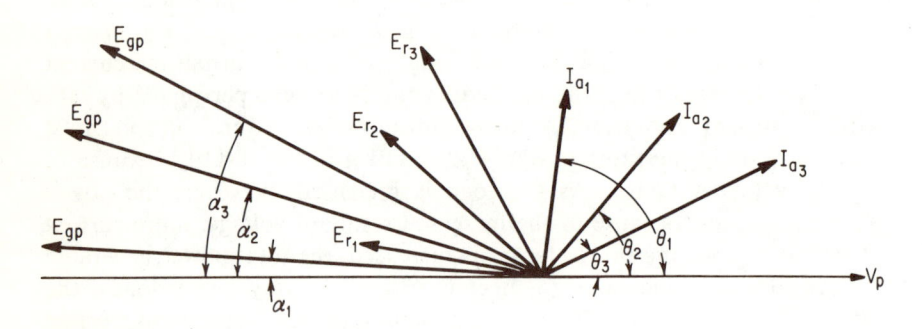

Figure 8-13

Effect of increased load at conditions of overexcitation ($E_{gp} > V_p$).

**8-11.
SUMMARY OF THE
EFFECT OF
INCREASED LOAD
(NEGLECTING
EFFECTS OF
ARMATURE
REACTION) UNDER
CONSTANT
EXCITATION**

In general, the following conclusions may be drawn as to the effects of increased load under conditions of constant excitation (neglecting the effects of armature reaction, Section 8-12 below):

1. As the mechanical *load increases*, the *armature current I_a increases, regardless of the excitation.*
2. If the synchronous motor is over- or under-excited, its power factor tends to approach unity *with increased load.*
3. When the motor is either over- or under-excited, the change in power factor is *greater* than the change in current with the application of load.
4. When the motor is normally excited, the change in current is greater

than the change in power factor *as load* is *increased*, and the power factor tends to become increasingly lagging.

**8-12.
EFFECT OF
ARMATURE
REACTION**

It was shown in Section 7-10 and Fig. 7-14 that a factor adding to the stability of synchronous alternators operating in parallel was that an overexcited synchronous alternator would tend to draw a more lagging current, which, by its demagnetizing action, would reduce the generated emf of the alternator. We have seen from the above discussion, however, that an over-excited synchronous motor tends to draw a leading current, which produces an increased demagnetizing action.

Whether acting as an alternator or as a motor, the synchronous dynamo is in parallel with and synchronized to the bus; and therefore it may be said:

1. An *overexcited* synchronous alternator will supply a lagging current to the bus, producing a *demagnetizing* effect as a result of armature reaction.
2. Similarly, an *overexcited* synchronous motor will draw a *leading current* from the bus, producing a *demagnetizing* effect as a result of armature reaction (Fig. 7-15).
3. Conversely, an *underexcited* synchronous alternator will supply a leading and *magnetizing* current to the bus.
4. Similarly, an *underexcited* synchronous motor will draw a *lagging magnetizing* current from the bus.

Thus, a synchronous ac dynamo, whether operated as a generator or as a motor, will have a *demagnetizing* armature reaction effect when *overexcited*, and a *magnetizing* armature reaction effect when *underexcited*. This effect obviously works toward normal excitation and extreme motor stability (see Fig. 8-17), as in the ac synchronous alternator. We have seen (Chapter 5) that armature reaction is an effect produced by and directly proportional to the armature current. The stabilizing effect is one additional reason for *not* compensating the armature reaction in ac synchronous dynamos as in dc dynamos. (see Fig. 8-17 and Sec. 8-13.)

The effect of armature reaction on the synchronous motor (whose excitation is maintained constant) as load is increased is summarized for various conditions of excitation below.

**8-12.1
Normal Excitation
(Fig. 8-11d)**

Increased load will tend to increase E_{gp}, the generated voltage per phase, as a result of the *slightly magnetizing* action produced by slightly *lagging* loads. The net effect is to maintain the armature current in phase with the bus voltage from no-load, at normal excitation.

8-12.2
Underexcitation
(Figs. 8-12 and 8-14)

Underexcitation produces a *magnetizing* armature reaction effect. The effect of increased load and consequent increased magnetization is to improve the power factor with increases in load. As shown in Fig. 8-14, the component of armature current which produces useful mechanical power is always $I_a \cos \theta$. At a small load, α_1, only a small portion of the total current drawn from the bus, I_{a_1}, is actually being used to drive the mechanical load. At the high load, α_3, a greater portion of I_{a_3} is in phase with V_p, developing the necessary power per phase $V_p I_{a_3} \cos \theta_3$ to derive the mechanical load applied to its shaft.

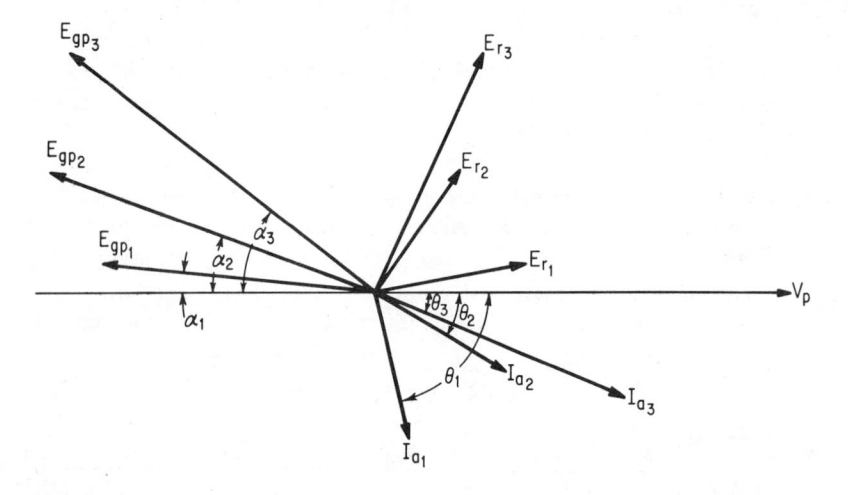

Figure 8-14

Effect of increased load and armature reaction at conditions of underexcitation.

8-12.3
Overexcitation
(Figs. 8-13 and 8-15)

Overexcitation produces a *demagnetizing* effect because of increased armature reaction with increased load. The net effect of such demagnetization is also to *improve* the power factor with increases of load. As shown in Fig. 8-15, only a small component of the total current I_{a_1} is producing useful mechanical

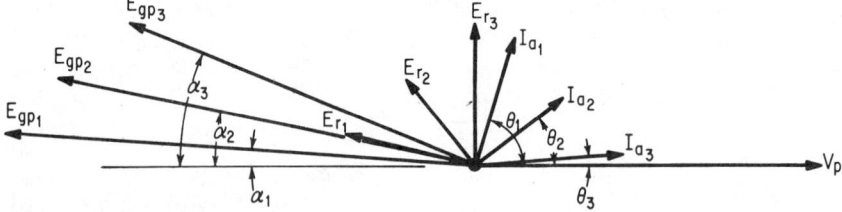

Figure 8-15

Effect of increased load and armature reaction at conditions of overexcitation.

power in driving the load at torque angle α_1; whereas, with a heavy load, practically all of the current produced by I_{a_3} is producing useful mechanical power.

> It may be concluded, therefore, that *armature reaction* produces a tendency to bring the armature current *toward a unity power factor* relation (in phase with the bus phase voltage) *as load is applied*, regardless of the state of excitation of the dynamo.

8-13. POWER FACTOR ADJUSTMENT OF SYNCHRONOUS MOTOR AT CONSTANT LOAD

Assume that a synchronous motor is operating at normal excitation (unity power factor) with a *given* mechanical load. Such a situation is shown in Fig. 8-16a, in which the dc field excitation has been adjusted for minimum current (or unity power factor) for the given load applied to the motor shaft. The phasor difference between the applied phase voltage V_p and the generated emf per phase E_{gp_1} is the resultant voltage E_r, producing the in-phase armature current I_a in accordance with Eq. (8-1), $I_a = E_r/Z_p$. The impedance per phase is assumed to be constant and to be lagging the voltage E_r by 90° throughout this discussion. The armature current, I_a, depends only on the magnitude and phase angle of E_r in developing the necessary mechanical power $V_p I_a$ required by the load on its shaft. What is the effect of either *decreasing* or *increasing* the excitation at a *given* load?

8-13.1 Effect of Decreased Excitation or Underexcitation

Figure 8-16b shows the effect of *decreasing* the excitation from E_{gp_1} to E'_{gp_1} at the same load angle α_1. The resultant voltage E_{r_1} causes a lagging current I_{a_1} to flow. Although I_{a_1} in Fig. 8-16b is larger than I_a in Fig. 8-16a, it is still insufficient to produce the necessary mechanical power $V_p I_a$ required by the load on its shaft. Note that in Fig. 8-16b, $I_{a_1} \cos \theta$ is equal to I_1, which is *less* than I_a. It is necessary, therefore, that the rotor be retarded from α_1 to α_2 in order to develop the necessary power. (Actually, since the effect of armature reaction at lagging loads tends to increase the generated voltage, angle α_2 should be less than that shown, because of an increase of E'_{gp_2} over E'_{gp_1}).

At angle α_2, generated voltage E'_{gp_2} produces a larger resultant voltage, E_{r_2}, which produces, in turn, current I_{a_2}. The in-phase component of I_{a_2} is just sufficient to produce the necessary mechanical power $V_p I_a$ required by the load on its shaft, as shown in Fig. 8-16b.*

---

* The construction of Fig. 8-16 does not take into account (1) the effect of armature reaction in changing the generated voltage, or (2) the effect of increased armature copper losses produced by the increased armature current, requiring an additional power input from the bus to the motor to supply the additional losses, representing an additional in-phase component. In the case of a lagging power factor, the additional magnetizing component produced by armature reaction would be offset by the increased losses, and the torque angle would still increase. In the case of a leading power factor, the change in torque angle would be slight because a current such as I_{a_1} would be required to maintain the additional loss (Fig. 8-16c).

(a) Normal excitation constant load.

(b) Decreased excitation constant load.

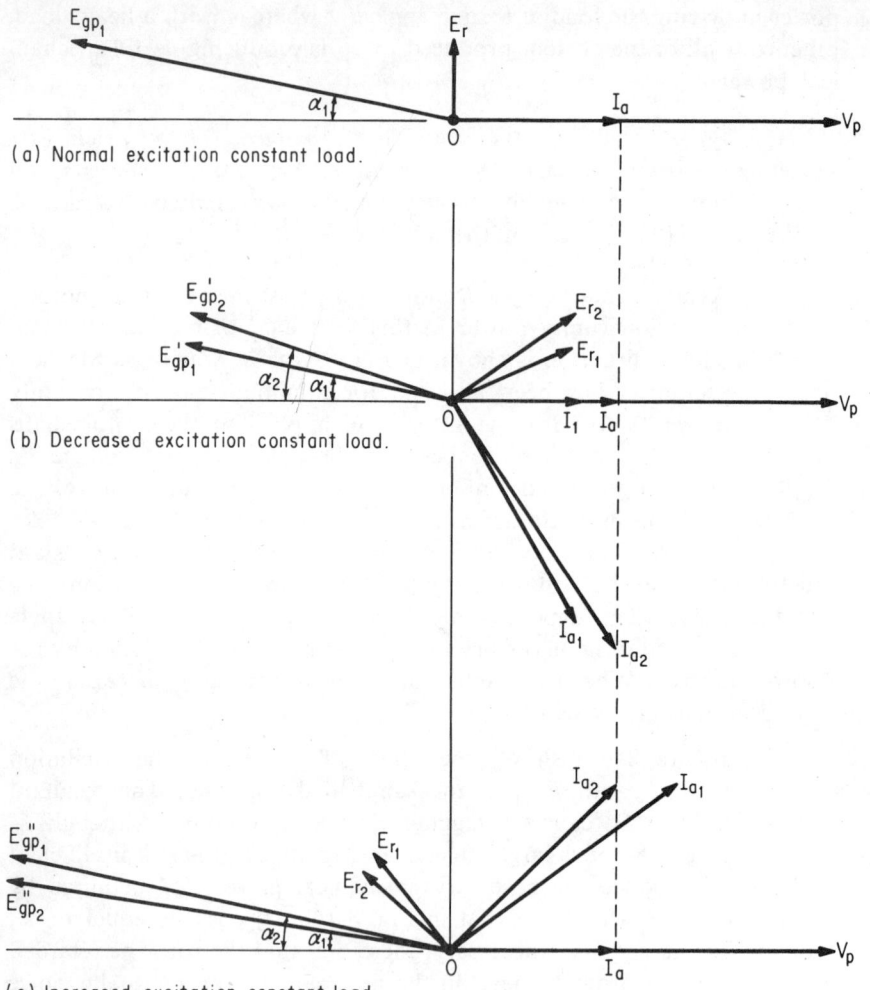

(c) Increased excitation constant load.

Figure 8-16

Effect of change of power factor (excitation) at constant load.

8-13.2
Effect of Increased Excitation or Overexcitation

Figure 8-16c shows the effect of increasing the excitation from E_{gp_1} in Fig. 8-16a to E''_{gp_1}, using the same load angle α. The resultant voltage E_{r_1} causes a leading current I_{a_1} to flow. I_{a_1} in Fig. 8-16c is *more* than is required to produce the necessary in-phase current, I_a. The load angle decreases, therefore, from α_1 to α_2, as shown in Fig. 8-16c. This decrease in load angle decreases E_{r_2} and I_{a_2} proportionately, to a value required to meet the necessary mechanical power $V_p I_a$ per phase.

Summary

The total power supplied by the bus to the motor is three times the power per phase, or $3V_p I_{a_2} \cos\theta$. As seen from Fig. 8-16, this power is relatively constant, ignoring the effects of armature reaction and losses due to change in armature current. The only appreciable change produced by a change in excitation over normal excitation has been (1) a change in *armature current*, and (2) a corresponding *change* in the *power factor* of the motor. Figure 8-16 also indicates that at any given load, *normal excitation* (i.e., the unity power factor condition where E_{gp} practically equals V_p and where the armature current I_a is in phase with the bus voltage per phase) may be obtained by simply adjusting the dc field rheostat to bring the line current to a *minimum*. Furthermore, neglecting the small increase due to armature copper power loss, at leading armature currents, the quadrature component of current which leads the applied voltage by 90° may be used for *system power factor correction* and for improvement of the alternator system operation supplying other (lagging) loads. The synchronous motor, therefore, is an *extremely versatile* motor, primarily because of its ability to maintain a constant excitation regardless of the power factor when operated from a constant potential bus. This last statement requires some explanation which will also aid in the understanding of the above power-factor characteristics.

The synchronous motor (like any doubly-excited dynamo) has two sources of excitation: (1) the ac stator excitation from the bus supplying its armature; and (2) the dc field exciting its rotor pole windings. As long as both dc and ac line voltages exciting a given synchronous motor (running at a given load) are constant, its excitation requirements are constant (by the law of the conservation of energy). Let us now assume, however, that the dc field is reduced, tending to demagnetize the air-gap flux. The motor draws a lagging component of armature current from the bus to magnetize the air gap and restore the mutual flux. Even at no load or at light loads, if the field circuit is suddenly opened, the synchronous motor would continue to operate in synchronism, because now its armature current lags the bus voltage by almost 90°, producing very great magnetization because of armature reaction, as shown in Fig. 8-17a. Assuming that the motor were operating at unity power factor in Fig. 8-17a, it would draw a current I_{a_1}. When the field is suddenly opened, however, the armature current increases rapidly, producing the very large armature reaction flux, $\phi_{a\text{-}r}$, in phase with the armature current. This produces a quadrature armature reaction voltage, $E_{a\text{-}r}$, whose magnetizing component E_m is in phase with E_{gp} and a small cross-magnetizing voltage E_c, shown in Figure 8-17a.

At unity power factor, with the armature phase current in phase with the bus voltage per phase, the dc excitation is normal and the synchronous motor requires no additional excitation from either the ac or

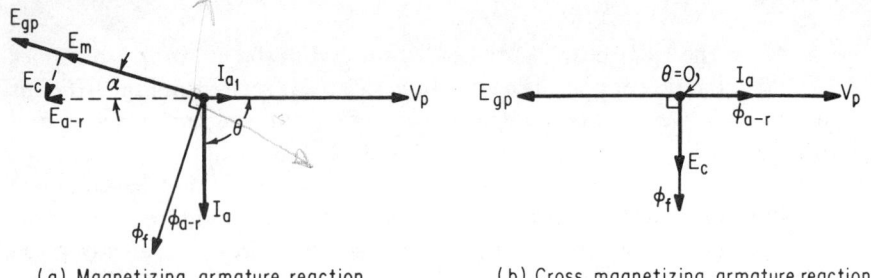

(a) Magnetizing armature reaction at lagging loads.

(b) Cross magnetizing armature reaction at unity power factor.

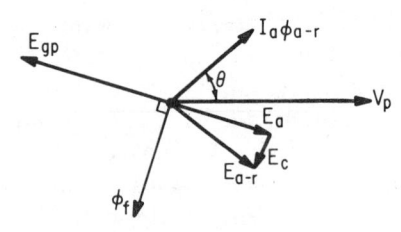

(c) Demagnetizing armature reaction at leading loads.

Figure 8-17

Tendency of synchronous motor to maintain constant excitation regardless of power factor because of effect of armature reaction.

the dc bus. The voltage E_c, resulting from armature reaction, is a cross-magnetizing voltage that does not possess a component which may affect the generated voltage, E_{gp}.

If the synchronous motor is now over-excited by increasing the dc field current or voltage, the synchronous motor draws a leading, demagnetizing current which attempts to produce an armature reaction flux, $\phi_{a\text{-}r}$, almost completely out of phase with the air-gap flux ϕ_f. The armature reaction flux $\phi_{a\text{-}r}$ produces, in the armature conductors, an armature reaction voltage $E_{a\text{-}r}$ having a large demagnetizing voltage component, E_d, as shown in Fig. 8-17c.

In summary, given mains of constant potential, the ability of the synchronous motor to maintain both its mutual air-gap flux (ϕ_f in Fig. 8-17) and its induced emf constant over a wide range of load and power factors, puts it in the same class as the transformer (Ch. 13) and the induction motor (Ch. 9), which possess similar characteristics in this respect.

8-14.
V-CURVE OF A SYNCHRONOUS MOTOR

The preceding phasor relations may be summarized *graphically* (as shown in Fig. 8-18) and determined experimentally in the laboratory, using the set-up shown in Fig. 8-19. It has been stated above that, when the field current of a synchronous motor is reduced, a lagging armature current I_{a_1} (Fig. 8-16)

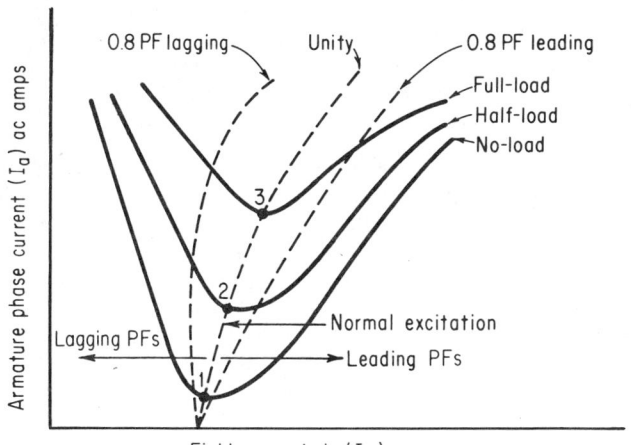

Figure 8-18

Families of V-curves for a synchronous motor.

(a) Relation between armature current and field current for various loads.

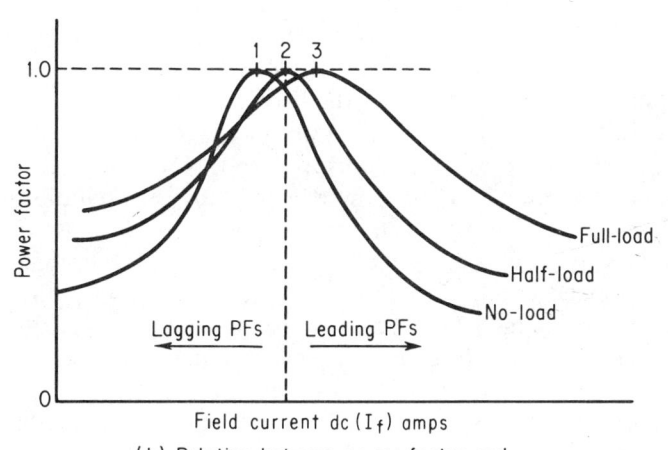

(b) Relation between power factor and field current for various loads.

is produced which exceeds the minimum current at unity power or at normal excitation. Similarly, when the motor is overexcited, the armature current also rises (Fig. 8-16) and exceeds the current required at normal excitation to develop the necessary torque, at any given load. By applying a given constant load to the shaft of a synchronous motor and varying the field current from under-excitation to over-excitation, recording the armature current at each step, the curves of Fig. 8-18a are obtained. The ac armature current is plotted against the dc field current for no-load, half-load, and full-load values, respectively.

SEC. 8-14. / V-Curve of a Synchronous Motor

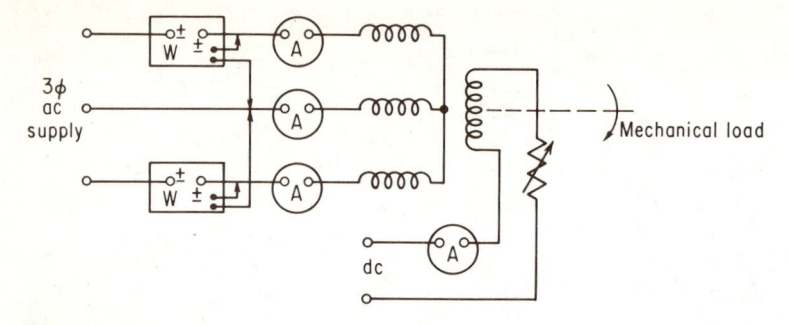

Figure 8-19

Laboratory connections for obtaining V-curves.

The wattmeter connections shown in Fig. 8-19 will also yield the power factor for each value of armature and field current at any given load condition.* Thus, as shown in Fig. 8-18b, the power factor (as determined from the wattmeter readings) is plotted against the field current for the various given loads. Note that both curves show that a slightly increased field current is required to produce normal excitation as the load is increased (points 1, 2 and 3, respectively). Note also that at no load, the armature current at unity power factor (normal excitation) is not zero (Fig. 8-11a) but some small value of ac armature current per phase, necessary to produce torque to counterbalance rotational losses. As load is applied (neglecting armature reaction) not only does the armature current rise (Fig. 8-11d) but it is also necessary to increase the excitation to bring the armature current back in phase with the bus phase voltage V_p.

Each of the curves in the family, therefore, will have a shift to the *right* as the *load* is *increased*, as shown in Figs. 8-18a and b, to provide the excitation required to obtain the same power factor (0.8 lagging, unity, or 0.8 leading) at an increased load. Thus, the V-curves represent the phasor diagrams, and vice versa for various conditions of load and power factor.

The V-curves also verify a basic point shown in Fig. 8-16, in which it was demonstrated that, if the excitation is varied at any given value of applied mechanical load, the torque angle α_1 must change in order that the developed power per phase, $V_p I_a \cos\theta$, may remain the same. Thus, at point 2 on Fig. 8-18a, if the excitation is increased the load increases

---

* The method employed in Fig. 8-19 is the two-wattmeter method. Since the synchronous motor is a balanced three-phase load, the one-wattmeter method, three-wattmeter method, industrial analyzer, or polyphase wattmeter would do as well in providing the power factor of the load. For a description of these methods, cf. H. W. Jackson, *Introduction to Electric Circuits*. 3rd. ed., Englewood Cliffs, N. J.: Prentice-Hall, Inc., Secs. 23-8, 9. 1970.

and more power is developed; Fig. 8-16c verifies this also. Similarly, if the excitation is decreased from point 2 in Fig. 8-18a, the load decreases and less power is developed; Fig. 8-16a verifies this also. The following examples illustrate that, if the torque angle is assumed constant, the increase in excitation results in an increased developed horsepower at a more leading power factor.

EXAMPLE 8-3: A six pole, 50 hp, 440 V, 60 Hz, three-phase, wye-connected motor has an effective armature resistance of 0.1 ohm and a synchronous reactance of 2.4 ohms/phase. When the motor is operating at a torque angle, α, of 20 electrical degrees, and the motor is under-excited, producing a generated phase voltage of 240 V, calculate:
a. the armature current, the power factor, and the horsepower developed by the armature
b. the same items as in (a) when the generated phase voltage is 265 V
c. the same items as in (a) when the generated phase voltage is 290 V.

Solution:

a. $V_p = \dfrac{V_l}{\sqrt{3}} = \dfrac{440 \text{ V}}{1.73} = 254 \angle 0° = 254 + j0 \text{ V}$

$E_g = 240 \angle 160° = 240(-\cos 20° + j \sin 20°) = -225.5 + j82.2 \text{ V}$

$\dot{E}_r = \dot{E}_g + \dot{V}_p = 28.5 + j82.2 = 86.8 \angle 70.85° \text{ V}$

$I_a = \dfrac{E_r}{Z_s} = \dfrac{86.8 \angle 70.85°}{0.1 + j2.4} = \dfrac{86.8 \angle 70.85° \text{ V}}{2.41 \angle 87.6° \, \Omega} = 36 \angle -16.75° \text{ A}$

Power factor $= \cos \theta = \cos 16.75° = $ **0.9575 lagging**

$P_d = 3 E_g I_a \cos (E_g, I_a) = 3 \times 240 \times 36 \cos 176.75°$
$= 3 \times 240 \times 36(-\cos 3.25°)$
$= $ **25,800 W** (drawn from the bus)

Horsepower developed $= \dfrac{P_d}{746} = \dfrac{25,800 \text{ W}}{746 \text{ W/hp}} = $ **34.6 hp**

b. $V_p = 254 + j0;$ $E_g = 265 \angle 160° = 265(-\cos 20° + j \sin 20°)$
$= -249 + j90.8$

$\dot{E}_r = \dot{V}_p + \dot{E}_g = 5 + j90.8 = 90.8 \angle 86.85° \text{ V}$

$I_a = \dfrac{E_r}{Z_s} = \dfrac{90.8 \angle 86.85° \text{ V}}{2.41 \angle 87.6° \, \Omega} = 37.7 \angle -0.75°$ or $37.7 \angle 0°$ A(practically)

Power factor $= \cos \theta = \cos 0° = $ **1.0 or unity power factor**

$P_d = 3 E_g I_a \cos (E_g, I_a) = 3 \times 265 \times 37.7 \cos 160°$
$= 3 \times 265 \times 37.7(-\cos 20°)$
$= $ **28,200 W** (drawn from the bus)

Horsepower developed $= \dfrac{P_d}{746} = \dfrac{28,200 \text{ W}}{746 \text{ W/hp}} = $ **37.8 hp**

c. $V_p = 254 + j0;$ $E_g = 290 \angle 160° = 290(\cos -20° + j \sin 20°)$
$= -272 + j99.2$

SEC. 8-14. / V-Curve of a Synchronous Motor

$$\dot{E}_r = \dot{V}_p + \dot{E}_g = -18 + j99.2 = 100.5 \angle 100.3° \text{ V}$$

$$I_a = \frac{E_r}{Z_s} = \frac{100.5 \angle 100.3° \text{ V}}{2.41 \angle 87.6° \text{ }\Omega} = 417. \angle 12.7° \text{ A}$$

Power factor $= \cos \theta = \cos 12.7° = $ **0.9757 leading**

$$P_d = 3 E_g I_a \cos(E_g, I_a) = 3 \times 290 \times 41.7 \cos 147.3°$$
$$= 3 \times 290 \times 41.7(-\cos 32.7°)$$
$$= 30{,}600 \text{ W (drawn from the bus)}$$

$$\text{Horsepower developed} = \frac{P_d}{746} = \frac{30{,}600 \text{ W}}{746 \text{ W/hp}} = 40.9 \text{ hp}$$

The solutions to Example 8-3 have been summarized in the following table:

PART	GENERATED VOLTAGE (volts)	POWER DRAWN (watts)	HORSEPOWER DEVELOPED (hp)	POWER FACTOR
(a)	240	25,800	34.6	0.9575 lagging
(b)	265	28,200	37.8	unity
(c)	290	30,600	40.9	0.9757 leading

Example 8-3 proves that if the applied load on the shaft remains constant (34.6 hp), and if the excitation is increased, the torque angle must necessarily *decrease* to develop the *same* power at the shaft. Thus if excitation is increased to unity power factor, since 37.8 hp is developed, the increased developed torque would reduce the torque angle to less than 20°. Similarly, at the leading power factor, the torque angle would be reduced still more.

8-15. COMPUTATION OF TORQUE ANGLE AND GENERATED VOLTAGE PER PHASE FOR A POLYPHASE SYNCHRONOUS MOTOR

In the case of the dc shunt motor, as load is increased the computation of decreased speed and increased torque is determined from the change in the counter emf. In the case of the synchronous motor, as the load is increased, the speed and the counter emf are relatively constant, and the increase in torque angle α produces the resultant voltage difference, E_r, whose magnitude and phase determine the armature current drawn by the motor from the bus. We have seen that the induced emf varies with excitation, and that its phase relation varies with the torque angle. But neither of these (E_g and α) are known quantities at any given value of load current. Is it possible to determine the torque angle and the counter emf from the normally measured power, voltage, and current drawn from the bus?

As with any ac synchronous dynamo, effective armature resistance per phase, R_a, and the quadrature synchronous reactance per phase,

X_s, may be determined from the dc stator resistance and the open-circuit and short-circuit test measurements (Section 6-10). For any given value of armature current, therefore, the synchronous impedance drop $I_a Z_p$ and its phase angle are known. But this voltage drop is the same as the resultant voltage E_r, which is the difference between the phase voltage applied to the armature and the generated armature voltage per phase, as shown in Fig. 8-9 and expressed in Eq. (8-1). Thus, if the terminal voltage per phase $V_p \angle 0°$ and $E_r \angle \delta$ (or $I_a Z_p$ and the angle it makes with the reference phase voltage) are known, then E_{gp} for any power factor may be readily obtained by means of the cosine law relationship, as demonstrated below.

It was shown in Section 8-7 that in Fig. 8-9

$$\dot{E}_{gp} = \dot{V}_p - \dot{E}_r = \dot{V}_p - I_a \dot{Z}_p \qquad (8\text{-}1)$$

This equation may be rewritten, using the cosine law as

$$E_{gp}^2 = E_r^2 + V_p^2 - 2E_r V_p \cos \delta \qquad (8\text{-}4)$$

where δ is the angular difference, at any power factor [see Eqs. (8-5), (8-6), and (8-7) below] between V_p and E_r
V_p is the applied bus voltage per phase
E_r is the synchronous impedance voltage drop at any given armature current $I_a Z_p$

As shown in Fig. 8-20, the angular difference, δ, may be computed for various power factor conditions from

At unity power factor	$\delta = \beta$	(8-5)
At leading power factor	$\delta = \beta + \theta$	(8-6)
At lagging power factor	$\delta = \beta - \theta$	(8-7)

where β is $\tan^{-1}(X_s/R_a)$, obtained from the synchronous impedance measurement (Sec. 6-10) and θ is the power factor angle between V_p and I_a, from wattmeter measurements.

Once the generated voltage per phase, E_{gp}, has been determined [using Eq. (8-4) above], it is possible to determine the torque angle, α, at any power factor, since it represents the angle opposite E_r, shown in the various constructions of Fig. 8-20, using the cosine law relationship*

$$\alpha = \cos^{-1} \frac{E_{gp}^2 + V_p^2 - E_r^2}{2 E_{gp} V_p} \qquad (8\text{-}8)$$

where all the quantities have been defined previously.

* As shown in Fig. 8-20, α may be determined as $\arctan I_a X_s / (V_p - I_a R_a)$. Perhaps the simplest determination of α emerges from the sine law where $\alpha = \arcsin E_r / E_{gp} \sin \delta$, as shown in Fig. 8-9.

$E_{gp} = (V_p - I_a R_a) + jI_a X_s$

$\alpha = \arctan \dfrac{I_a X_s}{V_p - I_a R_a}$

$\cos \alpha = \dfrac{E_{gp}^2 + V_p^2 - E_r^2}{2 E_{gp} V_p}$

(a) Unity PF relations.

$\delta = \beta + \theta$

$E_{gp} = V_p + I_a Z_p \cos(180 - \delta) + jI_a Z_p \sin(180 - \delta)$

$\alpha = \arctan \dfrac{I_a Z_p \sin(180 - \delta)}{V_p + I_a Z_p \cos(180 - \delta)}$

$\cos \alpha = \dfrac{E_{gp}^2 + V_p^2 - E_r^2}{2 E_{gp} V_p}$

(b) Leading PF relations.

$\delta = \beta - \theta$

$E_{gp} = (V_p - I_a Z_p \cos \delta) + jI_a Z_p \sin \delta$

$\alpha = \arctan \dfrac{I_a Z_p \sin \delta}{V_p - I_a Z_p \cos \delta}$

$\cos \alpha = \dfrac{E_{gp}^2 + V_p^2 - E_r^2}{2 E_{gp} V_p}$

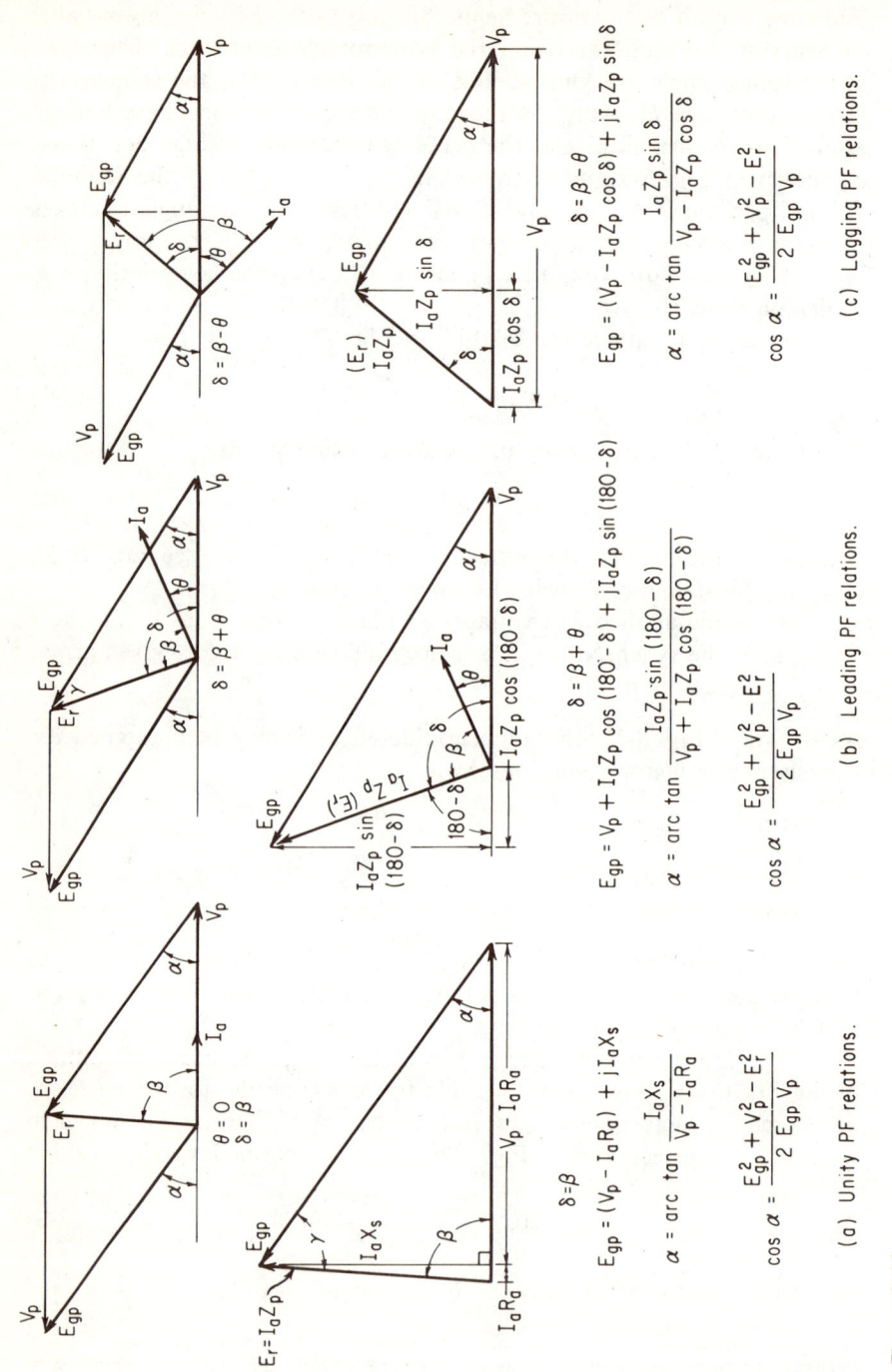

(c) Lagging PF relations.

Figure 8-20

Phasor relations between applied and generated voltage per phase for computation of α at any power factor.

It is also possible to represent the induced voltage per phase, E_{gp}, in terms of its horizontal and vertical components, using V_p as a reference as shown in Fig. 8-20. Using the horizontal and vertical components shown in the various constructions, we may express the generated voltage per phase at various power factors as

At unity power factor

$$\dot{E}_{gp} = (\dot{V}_p - I_a \dot{R}_a) + jI_a \dot{X}_s \qquad (8\text{-}9)$$

At leading power factor

$$\dot{E}_{gp} = [\dot{V}_p + I_a \dot{Z}_p \cos(180 - \delta)] + j[I_a \dot{Z}_p \sin(180 - \delta)] \qquad (8\text{-}10)$$

At lagging power factor

$$\dot{E}_{gp} = (\dot{V}_p - I_a \dot{Z}_p \cos \delta) + j(I_a \dot{Z}_p \sin \delta) \qquad (8\text{-}11)$$

where all the terms have been defined above.

The corresponding torque angles derived from the relations above are

At unity power factor

$$\alpha = \tan^{-1}\left(\frac{I_a X_s}{V_p - I_a R_a}\right) \qquad (8\text{-}12)$$

At leading power factor

$$\alpha = \tan^{-1}\left[\frac{I_a Z_p \sin(180 - \delta)}{V_p + I_a Z_p \cos(180 - \delta)}\right] \qquad (8\text{-}13)$$

At lagging power factor

$$\alpha = \tan^{-1}\left(\frac{I_a Z_p \sin \delta}{V_p - I_a Z_p \cos \delta}\right) \qquad (8\text{-}14)$$

where all the terms have been defined previously.

As in the case of the ac synchronous alternator, a simplification of the above diagrams results if all the phasors are projected to the armature current (I_a) as a reference (see Section 6-7, Fig. 6-4). The three conditions of unity, leading, and lagging power factors, shown in Fig. 8-20, are reproduced once more in Fig. 8-21, using the current as a reference.

The resulting relations are not only simpler to visualize and comprehend but, in addition, the following equations are somewhat simplified and, as will be shown, unified as well, compared to Eqs. (8-9) through (8-11). In addition, they are in the same form as that used for the regulation of an ac synchronous alternator, and they permit comparison as well as unified understanding from the point of view of an ac synchronous dynamo.

General equation: $E_{gp} = (V_p \cos \theta - I_a R_a) + j(V_p \sin \theta \pm I_a X_s)$ $+ =$ leading
                                                                                                                                        $- =$ lagging

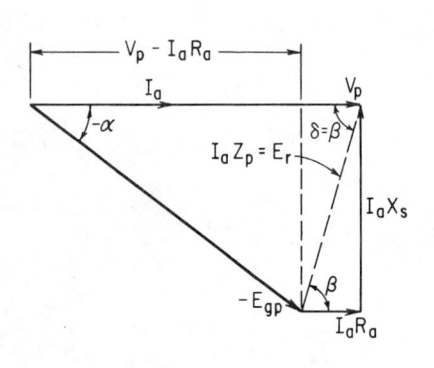

$E_{gp} = (V_p - I_a R_a) + j I_a X_s$

$\alpha = \tan^{-1} \dfrac{I_a X_s}{V_p - I_a R_a} = \sin^{-1} \dfrac{E_r}{E_{gp}} \sin \delta$

(a) Unity PF relations.

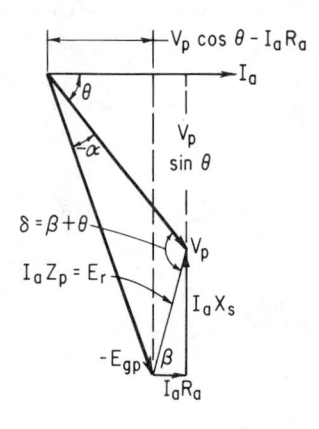

$E_{gp} = (V_p \cos \theta - I_a R_a) + j(V_p \sin \theta + I_a X_s)$

$\alpha = \tan^{-1} \dfrac{V_p \sin \theta + I_a X_s}{V_p \cos \theta - I_a R_a} - \theta = \sin^{-1} \dfrac{E_r}{E_{gp}} \sin \delta$

(b) Leading PF relations.

$E_{gp} = (V_p \cos \theta - I_a R_a) + j(V_p \sin \theta - I_a X_s)$

$\alpha = \theta - \dfrac{V_p \sin \theta - I_a X_s}{V_p \cos \theta - I_a R_a} = \sin^{-1} \dfrac{E_r}{E_{gp}} \sin \delta$

(c) Lagging PF relations.

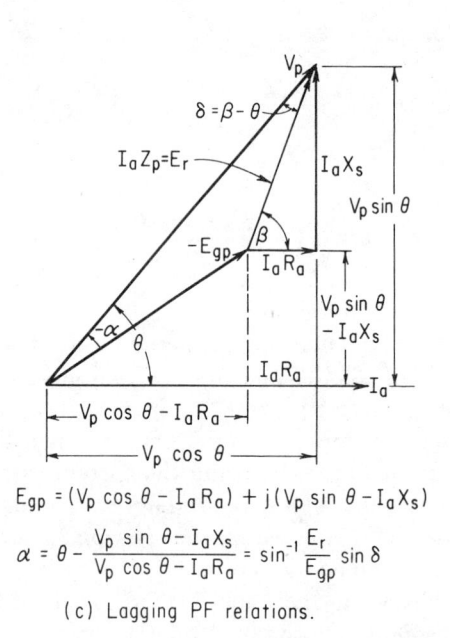

Figure 8-21

Vector relations of Figure 8-20 redrawn with current as a reference.

The results of the phasor representations shown in Fig. 8-21 permit us to express the generated voltage per phase at the various power factors as

At unity power factor

$$E_{gp} = (V_p - I_a R_a) + jI_a X_s \qquad (8\text{-}15)$$

At leading power factor

$$E_{gp} = V_p \cos\theta - I_a R_a + j(V_p \sin\theta + I_a X_s) \qquad (8\text{-}16)$$

At lagging power factor

$$E_{gp} = V_p \cos\theta - I_a R_a + j(V_p \sin\theta - I_a X_s) \qquad (8\text{-}17)$$

where all the terms have been defined above.

It is now possible to combine the equations above into a single general equation for the computation of the generated voltage per phase of a synchronous motor, for any and all conditions of power factor, in the form of*

$$E_{gp} = (V_p \cos\theta - I_a R_a) + j(V_p \sin\theta \pm I_a X_s) \qquad (8\text{-}18)$$

where, in the quadrature expression, $+$ is used for leading power factor and $-$ is used for lagging power factor.

Note that, at unity power factor, Eq. (8-18) is identical to Eq. (8-15) since $\sin\theta$ is zero and $\cos\theta$ is unity.

Figure 8-21 shows the manner in which the torque angle may be determined from the real and quadrature components of the generated voltage per phase. The corresponding torque angles at the various power factors are

At unity power factor

$$\alpha = \tan^{-1}\left(\frac{I_a X_s}{V_p - I_a R_a}\right) \qquad (8\text{-}19)$$

---

* A comparison of Eq. 8-18 with Eq. 6-8 will reveal the similarity in form for the determination of the generated emf in an ac synchronous dynamo and in a transformer.

As an alternator, $\qquad E_{gp} = (V_p \cos\theta + I_a R_a) + j(V_p \sin\theta \pm I_a X_s)$

As a synchronous motor, $\quad E_{gp} = (V_p \cos\theta - I_a R_a) + j(V_p \sin\theta \pm I_a X_s)$

In the real term, the voltage drop across the armature resistance is added to the real component of the terminal voltage for generator action, and the voltage drop is subtracted from the real component of the terminal voltage for motor action.

In the quadrature term, for generator action, the quadrature voltage drop across the reactance is added to the quadrature component of the applied voltage for lagging loads, and is subtracted for leading loads, to compensate for the demagnetizing armature reaction of lagging loads and for the magnetizing armature reaction of leading loads. For motor action, on the other hand, the reverse is true; in that case, lagging loads are magnetizing and leading loads are demagnetizing.

At leading power factor

$$\alpha = \tan^{-1}\left(\frac{V_p \sin\theta + I_a X_s}{V_p \cos\theta - I_a R_a}\right) - \theta \qquad (8\text{-}20)$$

At lagging power factor

$$\alpha = \theta - \tan^{-1}\left(\frac{V_p \sin\theta - I_a X_s}{V_p \cos\theta - I_a R_a}\right) \qquad (8\text{-}21)$$

As in the case of the generated voltage per phase, it is possible to unify the three equations for the torque angle into a single general expression which takes all power factors into account,

$$\alpha = \theta - \tan^{-1}\left(\frac{V_p \sin\theta \pm I_a X_s}{V_p \cos\theta - I_a R_a}\right) \begin{cases} \text{in the numerator:} \\ + \text{ (leading power factor)} \\ - \text{ (lagging power factor)} \end{cases} \qquad (8\text{-}22)$$

where α is always the difference between the two angles, regardless of which is greater.

In view of the relative simplicity of Eqs. (8-18) and (8-21), and the remarks in the footnote to Eq. (8-18), the reader may wonder, at this point, why it was necessary to use the presentation of Fig. 8-20 and all of the equations related to it. An examination of the constructions of Fig. 8-20 will show that both E_{gp} and α are represented in Fig. 8-21 in the reverse direction (hence the negative sign). It is relatively difficult, furthermore, to obtain a value (for the angle between the resultant voltage E_r and the applied phase voltage) from the diagrams of Fig. 8-21 in terms of the angles β and θ. Finally, since the bus voltage per phase, V_p, is the reference in both ac alternator and synchronous motor operation, the diagrams of Fig. 8-20 are more meaningful when empirical computations are made.

Depending on personal preference, any of the three methods shown above may be used. The following example illustrates all three methods and is a verification of all the equations presented above.

EXAMPLE 8-4: A two-pole, 1000 hp, 6000 V, 60 Hz, three-phase, wye-connected synchronous motor has an effective armature resistance of 0.52 ohm and a synchronous reactance of 4.2 ohms/phase. The efficiency of the motor at the rated load, 0.8 PF leading, is 92 per cent, neglecting field losses due to dc excitation. Calculate
   a. The generated emf per phase, E_{gp}. In solving this part, use the following methods: (1) the cosine law solution, Eq. (8-4); (2) the voltage reference solution, Eq. (8-10); and (3) the universal equation solution, Eq. (8-18)
   b. The torque angle, α
   c. The mechanical power developed by the armature at the rated load in watts and in horsepower
   d. The internal torque developed.

Solution:

$$\text{Input watts} = \frac{1000 \text{ hp} \times 746 \text{ W/hp}}{0.92 \text{ efficiency}} = 811{,}000 \text{ W}$$

Line and phase armature current

$$I_a = \frac{\text{input power}}{\sqrt{3}\, E_L \cos\theta} = \frac{811{,}000 \text{ W}}{1.73 \times 6000 \text{ V} \times 0.8} = 97.6 \text{ A}$$

[Note: at unity power factor, I_a would be $97.6 \text{ A} \times 0.8 = 78.0$ A]

$$V_p = \frac{V_L}{\sqrt{3}} = \frac{6000 \text{ V}}{1.73} = 3460 \text{ V}$$

$Z_p = R_a + jX_s = 0.52 + j4.2 = 4.22 \angle 82.93°\ \Omega; \quad \beta = 82.93°$

$E_r = I_a Z_p = 97.6 \times 4.22 = 412$ V at 0.8 PF leading

$\cos\theta = 0.8$, and $\theta = 36.8°$

At 0.8 PF leading, $\delta = \beta + \theta = 82.93° + 36.8° = 119.73°$

a. By the cosine law method, using Eq. (8-4)

$$\begin{aligned}E_{gp} &= \sqrt{E_r^2 + V_p^2 - 2E_r V_p \cos\delta}\\ &= \sqrt{412^2 + 3460^2 - 2(412)(3460)(-\cos 60.27°)}\\ &= \sqrt{170{,}000 + 12{,}000{,}000 + 2{,}850{,}000 \cos 60.27°}\\ &= \sqrt{13{,}580{,}000} = \mathbf{3683 \text{ V}}\end{aligned}$$

By the voltage reference method, using Eq. (8-10)

$$\begin{aligned}E_{gp} &= [V_p + I_a Z_p \cos(180 - \delta)] + j[I_a Z_p \sin(180 - \delta)]\\ &= 3460 + 412 \cos 60.27° + j412 \sin 60.27°\\ &= 3664 + j358 = \mathbf{3683 \text{ V (check)}}\end{aligned}$$

By the universal equation method, using Eq. (8-18)

$$\begin{aligned}E_{gp} &= (V_p \cos\theta - I_a R_a) + j(V_p \sin\theta - I_a X_s)\\ &= [(3460 \times 0.8) - (97.6 \times 0.52)]\\ &\quad + j[(3460 \times 0.6) - (97.6 \times 4.2)]\\ &= 2714.2 + j1665 = \mathbf{3683 \text{ V (check)}}\end{aligned}$$

b. solving for α, using Eq. (8-13)

$$\alpha = \tan^{-1}\left[\frac{I_a Z_p \sin(180 - \delta)}{V_p + I_a Z_p \cos(180 - \delta)}\right]$$

$$= \left[\frac{412 \sin 60.27°}{3460 + 412(-\cos 60.27°)}\right] = \frac{358}{3664} = \mathbf{5.5°}$$

Solving for α, using Eq. (8-22)

$$\alpha = \theta - \tan^{-1}\left(\frac{V_p \sin\theta + I_a X_s}{V_p \cos\theta - I_a R_a}\right)$$

$$= 36.8° - \tan^{-1}\left[\frac{(3460 \times 0.6) + (97.6 \times 4.2)}{(3460 \times 0.8) - (97.6 \times 0.52)}\right]$$

$$= 36.8° - \tan^{-1}\left(\frac{2485}{2714}\right) = 36.8° - 42.3° = \mathbf{5.5° \text{ (check)}}$$

Solving for α using sine law

$$\delta = \beta + \theta = 82.93° + 36.8° = 119.73°$$

$$\alpha = \sin^{-1} \frac{E_r}{E_{gp}} \sin \delta = \sin^{-1} \frac{412}{3683} \sin 119.73° = 5.5° \text{ (check)}$$

c. Mechanical power developed per phase is the product of the induced emf per phase, the armature current, and the cosine of the angle between them [see Figs. 8-20(b) and 8-21(b)].
Total mechanical power

$$P_d = 3E_{gp}I_a \cos(\alpha + \theta) = 3 \times 3683 \times 97.6 \cos(5.5 + 36.8°)$$
$$= 796{,}000 \text{ W}$$

$$\text{Internal horsepower developed} = \frac{P_d}{746} = \frac{796{,}000 \text{ W}}{746 \text{ W/hp}} = \mathbf{1065 \text{ hp}}$$

d. Torque $T = \dfrac{\text{hp} \times 5252}{S} = \dfrac{1065 \times 5252}{S} = \mathbf{1552 \text{ lb-ft}}$ [from Eq. (4-15)]

8-16. USE OF THE SYNCHRONOUS MOTOR AS A CORRECTOR OF THE POWER FACTOR

Example 8-4 showed that a synchronous motor is capable of delivering its rated mechanical power and simultaneously drawing a leading current from the bus. This leading current may be used to raise the over-all power factor of energy supplied by the bus or the system to other loads, in parallel with the synchronous motor. It is desirable to raise the over-all power factor of the system for several reasons. The alternators and/or transformers supplying the system are rated in terms of their current-carrying capacity. Thus, the total useful power they can deliver to the various connected loads on a system depends directly on the nature and power factor of the loads which they supply. In a system where fluorescent lamps are used almost universally for lighting, and where induction motors are used almost universally for power, the over-all system power factor may be as low as 0.6 PF. The system generation and supplying apparatus only will deliver 60 per cent of the total power that could be delivered at unity power factor. Furthermore, the lower the system power factor, the *greater* is the current required to supply the *same* kilowatt load. The increased current required from the mains to supply a low power-factor load results in *greater* transmission line voltage drop and greater transmission line power *losses*, as well as greater losses and a resulting decreased efficiency in the generating and system equipment supplying the load.

An improvement in the over-all system power factor at any given load results, therefore, in (1) released supply capacity for additional load; (2) reduced line voltage drop and better over-all voltage regulation; (3) increased over-all system (lines and equipment) efficiency; and (4) consequent lower operating costs. It is precisely for this last reason that in some areas an industrial occupancy is *penalized* for a *low power factor* and will be charged for energy at a higher rate than another occupancy which consumes energy at a higher power factor. In these areas, a consumer requiring additional motors for expansion or replacement in his plant will seriously consider the purchase of synchronous motors which may be over-excited to operate at 0.8 PF leading and, at the same time, deliver rated mechanical power to drive a load (see Sec. 8-18.)

Figure 8-22 shows the extent of improvement (in the power factor of a lagging load) produced by a synchronous motor. Assume a lagging load I_L which lags the bus voltage V_p by some phase angle θ_L. If a synchronous motor is added in parallel with the load across the same bus, and if its field is over-excited, the current I_m drawn by the synchronous motor in maintaining constant field excitation (Section 8-13) will be a leading current which leads the bus voltage by θ_m. The resultant current drawn from the bus, I_T, is the phasor sum of I_L and I_m, shown in Fig. 8-22.

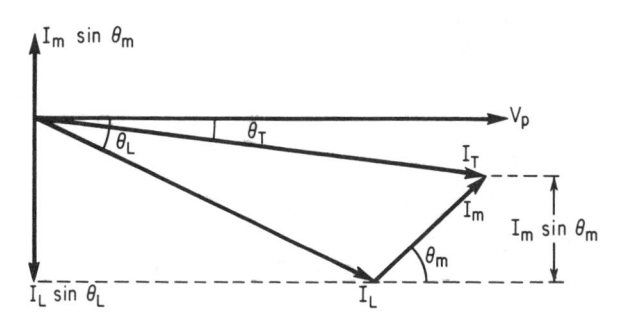

Figure 8-22

Power factor improvement through the use of an overexcited synchronous motor.

The original quadrature (*lagging*) component of the load current $I_L \sin \theta_L$ is, in effect, neutralized in part by the *leading* quadrature component of motor current $I_m \sin \theta_m$. The final power factor angle θ_T has thus been brought closer in phase with the bus voltage, with the consequent advantages of improved power factor. The power consumed by the motor per phase is represented by $V_p I_m \cos \theta_m$, which includes the useful mechanical power to drive a load (in addition to a small proportion of motor losses).

The benefit derived from synchronous motor power-factor correction in reducing lagging load current is shown in Example 8-5.

EXAMPLE 8-5: A factory consumes a total power of 2000 kW at 0.6 PF lagging from a transformer vault whose primary line voltage is 6000 V. In adding a new building wing to be devoted exclusively to electroplating, a dc motor generator set is to be purchased, to deliver approximately 750 kW. The choice of motor lies between a 1000 hp, 6000 V synchronous motor at 0.8 PF leading (Example 8-4), and a 1000 hp, 6000 V induction motor whose full-load power factor is 0.8 lagging. Assuming an efficiency of 92 per cent for each motor, calculate
  a. The total load current and the power factor using the induction motor
  b. The total load current and the power factor using the synchronous motor
  c. The per cent reduction in load current produced by using (b) as a percentage of (a)
  d. The overall power factor improvement.

Solution:

a. The induction motor load $= \dfrac{\text{hp} \times 746 \text{ W/hp}}{\text{efficiency}}$

$= \dfrac{1000 \times 746}{0.92} = 810{,}000$ W

Lagging current drawn by the induction motor

$$I_l = \dfrac{\text{watts}}{\sqrt{3} E_L \cos\theta} = \dfrac{810{,}000 \text{ W}}{1.73 \times 6000 \times 0.8} = 97.6 \angle -36.9° \text{ A}$$

Lagging factory load current

$$I'_l = \dfrac{\text{watts}}{\sqrt{3} E_L \cos\theta} = \dfrac{2{,}000{,}000}{1.73 \times 6000 \times 0.6} = 321 \angle -53.1° \text{ A}$$

Motor load current $= 97.6 \angle -36.9°$ A $= 78.0 - j58.5$
Factory load current $= 321 \angle -53.1°$ A $= 192.5 - j256.5$
Total load current, $I_l + I'_l$ $\qquad\qquad = 270.5 - j315.0$
$\qquad = 416 \angle -49.3°$ A (at a power factor of **0.651** lagging)

b. The synchronous motor load $= 97.6 \angle +36.9°$ A
$\qquad\qquad\qquad\qquad$ (at a power factor of 0.8 leading)

Motor load current, $I_t = 97.6 \angle 36.9°$ A $= 78.0 + j52.5$
Factory load current, $I'_l = 321 \angle -53.1°$ A $= 192.5 - j256.5$
Total load current, $I_l + I'_l$ $\qquad\qquad = 270.5 - j204.0$
$\qquad = 340 \angle -36.9°$ A (at a power factor of **0.8** lagging)

c. Per cent reduction in load

$\dfrac{\text{Original load} - \text{final load}}{\text{Original load}} = \dfrac{416 - 340}{416} \times 100 = \mathbf{18.25}$ **per cent**

d. Power factor improvement:
Using the synchronous motor in lieu of the induction motor will raise the total system power factor from **0.651** lagging to **0.8** lagging.

8-17. DEVELOPED ELECTROMAGNETIC TORQUE PER PHASE OF A SYNCHRONOUS MOTOR

The relation between external or internal horsepower, torque, and speed for any motor was given as

$$\text{hp} = \dfrac{TS}{5252} \quad \text{or} \quad T = \dfrac{5252 \text{ hp}}{S} \qquad (4\text{-}15)$$

The power drawn from the bus by the synchronous motor armature per phase is

$$P_p = V_p I_a \cos\theta \qquad (8\text{-}23)$$

and the power developed by the synchronous motor armature per phase has been given as the product of its generated emf per phase, E_{gp}, the armature current, I_a, and the cosine of the angle between them [based on Eq. (4-7) for the dc dynamo], or

Power developed per phase: $\qquad P_d = E_{gp} I_a \cos(E_{gp}, I_a) \qquad (8\text{-}24)$

If the horsepower developed in Eq. (4-15) is expressed as developed power in watts, the developed torque per phase is

$$T = \frac{5252 P_d}{S \times 746} = \frac{7.04 P_d}{S} = \left(\frac{7.04}{S}\right) E_{gp} I_a \cos(E_{gp}, I_a) \qquad (8\text{-}25)$$

From Eq. (2-16), in which $S = 120f/P$, the developed electromagnetic torque per phase of any ac synchronous dynamo, in lb-ft, can be derived from

$$T = \left(\frac{7.04 P}{120 f}\right) E_{gp} I_a \cos(E_{gp}, I_a) \qquad (8\text{-}26)$$

where P is the number of poles
E_{gp} is the generated emf per phase of the ac synchronous dynamo from Eq. (6-8) or (8-18)
I_a is the armature current per phase
$\cos(E_{gp}, I_a)$ is the cosine of the angle between the two quantities as determined by a phasor diagram
f is the frequency in Hz

Thus, regardless of the mode of operation, Eq. (8-26) will express the electromagnetic torque developed by the armature conductors of an ac synchronous dynamo either (1) as a motor, to produce rotation of the rotor; or (2) as a generator, to produce countertorque as a result of the flow of current generated in the armature conductors.

It has been stated that, as the mechanical load applied to the shaft of a synchronous motor increases, the torque angle α increases, producing an increase in the resultant voltage difference E_r (or $\dot{E}_{gp} - \dot{V}_p$), which in turn causes the bus to supply more armature current to the motor. In Eqs. (8-24) and (8-26), the angle between E_{gp} and I_a depends not only on the magnitude of the torque angle α, as shown in Figs. 8-20 and 8-21, but also on the power factor angle θ. Is there a *maximum* power that a synchronous motor can develop and is there a maximum torque angle beyond which the motor will pull out of synchronism?

Apparently there is—but what is it? If we could develop an expression similar to Eq. (8-24), which yields the power drawn (from the bus by a synchronous motor) in terms of torque angle α only, that expression could be substituted in Eq. (8-26) to yield the maximum or pull-out torque for the synchronous motor. The algebraic derivation for this equation is relatively complex.

It is possible to "derive" it from a phasor construction, however, as in Fig. 8-23, which shows the power per phase delivered to a load by an ac synchronous motor or a bus. Let θ represent the angle by which current I_a lags bus voltage V_p, with α as the defined angle between E_{gp} and V_p. The difference $E_{gp} - V_p$ is E_r, which is the same as $I_a Z_p$. Neglecting the

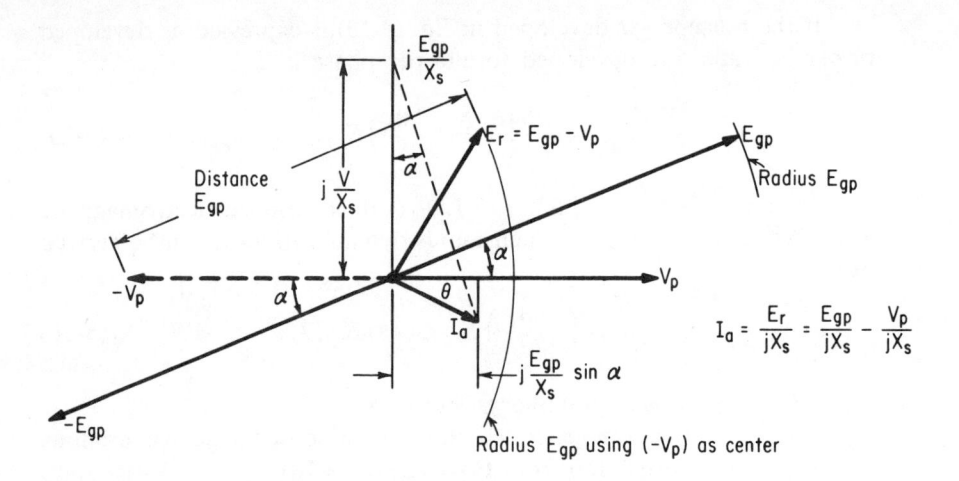

Figure 8-23

Construction for the derivation of maximum power and torque of an ac synchronous dynamo.

resistance of the armature, I_a lags E_r by 90° and is the same as $I_a jX_s$. The armature current may then be represented by

$$I_a = \frac{E_r}{jX_s} = \frac{E_{gp} - V_p}{jX_s} = \frac{E_{gp}}{jX_s} - \frac{V_p}{jX_s}$$

The locus of the armature current, I_a, is therefore a circle, whose center as shown in the construction is jV_p/X_s and whose radius is jE_{gp}/X_s. This information permits construction of jE_{gp}/X_s at angle α with the vertical axis, from which the current jV_p/X_s is known. The power delivered by the bus per phase is $V_p I_a \cos\theta$, or V_p times the projection of I_a on V_p. But this projection is also equal to $j(E_{gp}/X_s)\sin\alpha$. Therefore, the electric power developed per phase, P_p, by a rotating ac synchronous dynamo is

$$P_p = \left(\frac{V_p E_{gp}}{X_s}\right) \sin\alpha \tag{8-27}$$

which, when substituted in Eq. (8-26), yields the torque per phase or

$$T = \left(\frac{7.04P}{120f}\right)\left(\frac{V_p E_{gp}}{X_s}\right) \sin\alpha \tag{8-28}$$

where all the terms have been previously defined.

An examination of Eq. (8-28) indicates that, for a given number of poles, frequency, bus voltage, synchronous reactance, and excitation (producing E_{gp}), the maximum torque can occur only when α equals 90 electrical degrees. Moreover, since all the other terms in Eq. (8-28) are

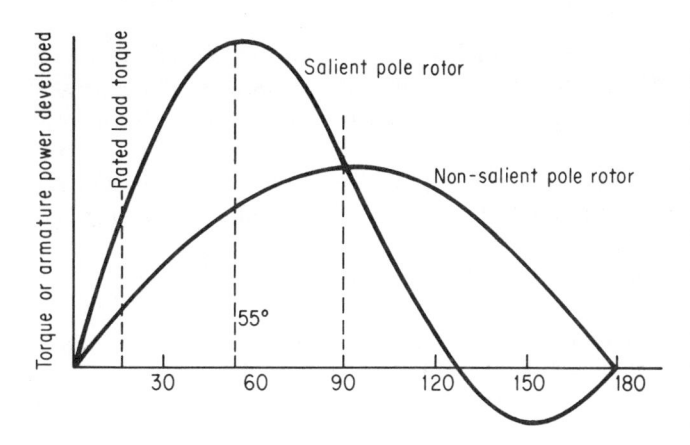

Figure 8-24

Power angle curves for two similar synchronous motors (same stator) showing maximum torque angles for salient and nonsalient rotor poles.

relatively constant for a given motor, the shape of the torque curve should be *sinusoidal*. The torque and the power developed by both salient and nonsalient (cylindrical) rotors is shown in Fig. 8-24. Maximum rotor torque *is* obtained at 90 electrical degrees in the case of the *cylindrical* (distributed winding) nonsalient rotor, which produces *no* quadrature armature reaction.* Maximum rotor torque is greater in the *salient*-pole machine, but occurs sooner, however, because of the so-called *reluctance torque* (Sec. 1-2) due to variation of the reluctance in the air gap produced by armature reaction (Section 6-11); the reluctance torque varies with sin 2α.

Reluctance torque is a maximum at small torque angles and, therefore, is of assistance in pulling a salient-pole synchronous motor into step during synchronizing (Sec. 8-5). Reluctance torque is a phenomenon which may also occur in an induction motor because of a variation in permeability between the stator and rotor teeth; reluctance torque may, in such cases, result in a synchronizing force called "sub-synchronous locking."

For the *salient*-pole machine shown in Fig. 8-24, maximum torque occurs at about 55 electrical degrees, and full-load torque at about 22 electrical degrees. The maximum torque for both salient-and nonsalient-pole machines is from 250 to 300 per cent of the full-load torque. Equations (8-27) and (8-28) are also of value in verifying differences between synchronous motors in terms of excitation or impedance. Since E_{gp} in Eq. (8-27) varies with the dc field excitation, and since V_p is the bus voltage (normally constant), if the excitation is reduced, as shown in Fig. 8-16b, the torque angle α must increase to develop the same mechanical power. Conversely, ignoring the effects of armature reaction, if the excitation E_{gp} is increased, as shown in Fig. 8-16c and by Eq. (8-27) and Example 8-3, the torque angle is *reduced*.

---

* For a fairly comprehensive treatment of the power and torque equations for salient and cylindrical rotor synchronous dynamos see L.V. Bewley, *Alternating Current Machinery*, New York: The Macmillan Company, 1949, Chapter 7, pp. 288–304.

Increasing the excitation, furthermore, increases the pull-out power and "stiffness" of the synchronous motor, not only because the generated voltage per phase E_{gp} is increased in Eq. (8-28), but also because the synchronous reactance X_s decreases with increased saturation. In general, moreover, a low-impedance (and low-X_s) synchronous motor will not require as great a torque angle to drive a given load [as shown in Eq. (8-28)] as a high-impedance machine, i.e., a low-impedance dynamo is stiffer in its opposition to change of torque angle with load.

EXAMPLE 8-6: Calculate the internal torque per phase and the developed total horsepower for the original excitation given in Example 8-3, using Eq. (8-28).

Solution:

$$T/\text{phase} = \left(\frac{7.04P}{120f}\right)\left(\frac{V_p E_{gp}}{X_s}\right) \sin \alpha \qquad (8\text{-}28)$$

$$= \left(\frac{7.04 \times 6}{120 \times 60}\right) \times \left(\frac{254 \times 240}{2.4}\right) \sin 20° = 50.8 \text{ lb-ft/phase}$$

$$\text{Horsepower} = \frac{TS}{5252} = \left(\frac{3 \times 50.8}{5252}\right) \times \left(\frac{120 \times 60}{6}\right) = 34.6 \text{ hp}$$

Note that the *same* result is obtained as in Example 8-3a. This is a verification of Eq. 8-28.

8-18. SYNCHRONOUS MOTOR RATINGS

Synchronous motors may be purchased in three *standard* ratings, namely: unity power factor, 90 per cent leading power factor, and 80 per cent leading power factor. Other ratings are obtainable from motor manufacturers by special quotation. A 100 hp unity PF synchronous motor has 80 per cent of the current capacity and armature current rating of a 100 hp 0.8 PF leading motor (the latter is a *larger frame size*). Both motors will deliver 100 hp at their pulleys; but the latter may be used also for power-factor correction at a leading power factor of 0.8, whereas the former is adjusted for minimum (unity power factor) current at full load. It is possible, of course, to operate any unity power factor (or other power factor rating) motor as an overexcited synchronous motor, but it must be realized that the horsepower rating of the motor can no longer be the same if the armature current is to be maintained within the rated value. (see "*V*" curves, Fig. 8-18a.)

8-19. SYNCHRONOUS CAPACITORS

A number of synchronous motors are manufactured without any shaft extensions whatever, ostensibly designed solely for power-factor correction and to be operated *without* any mechanical load. Any overexcited synchronous motor operated without load, however, may be classed as a *synchronous condenser* or *synchronous capacitor*. Although there is no mechanical load to contribute to armature current, the *V*-curves of Fig. 8-18 will show that, even without load, the armature current is high. This is not a disadvantage, however, because the synchronous capacitor is over-excited to that point where its armature current is equal to or exceeds its rated current.

As shown in Fig. 8-25, when a motor is over-excited without load, the resultant phase impedance voltage E_r is quite high, despite the very small torque angle, α, producing a relatively large leading armature current I_a which is practically at 90° with respect to the bus phase voltage. Alternating-current synchronous dynamos can be constructed less expensively in extremely high ratings and voltages than commercial fixed capacitors with the result that synchronous capacitors driving no mechanical load, whatever, are permitted to "float" on the line for purposes of power factor improvement. While it is usually commercially impractical to correct a given system power factor of any generating station to unity power factor (see Section 8-20), the following simple example will first demonstrate the technique in making power factor correction computations. A more direct method is shown in Sec. 8-21.

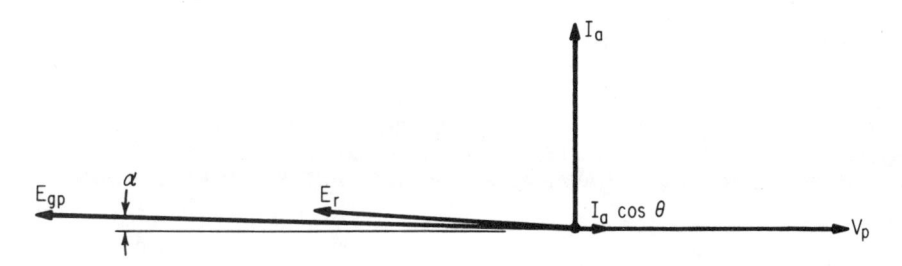

Figure 8-25

Synchronous capacitor phasor relationships.

EXAMPLE 8-7: A factory draws a lagging load of 2000 kW at a power factor of 0.6 from 6000 V mains. A synchronous capacitor is purchased to raise the overall power factor to unity. Assuming that the synchronous capacitor losses are 275 kW, calculate:
a. Original kilovars of lagging load
b. kvars of correction needed to bring the power factor to unity
c. kVA rating of synchronous capacitor and its power factor.

Solution:

a. $\text{kVA} = \dfrac{\text{kW}}{\cos \theta} = \dfrac{2000}{0.6} = \mathbf{3333 \text{ kVA}}$

Lagging kvars $= \text{kVA} \sin \theta = 3333 \times 0.8 = \mathbf{2667 \text{ kvars}}$

b. **2667 kvars of correction** are required to bring the power factor to unity

c. $\tan \theta = \dfrac{2667 \text{ kvars}}{275 \text{ kW}} = 9.68; \quad \theta = \text{arc tan } 9.68 = \mathbf{84.09° \text{ leading}}$

$\cos \theta = \mathbf{0.103 \text{ leading}}; \quad \text{kVA} = \dfrac{\text{kW}}{\cos \theta} = \dfrac{275}{0.103} = \mathbf{2755 \text{ kVA}}$

A synchronous capacitor at a power factor of 0.103 leading and a kVA rating of 2755 is required.

**8-20.
ECONOMIC
LIMIT TO
IMPROVEMENT OF
POWER FACTOR**

It is customary not to attempt correction of the power factor of a system all the way to unity power factor. There is an economic reason for this, despite the fact that large, high capacity synchronous capacitors are available and (for the same kVA rating) are usually less expensive than synchronous motors because (1) they are started and run without load, requiring no special heavy cage windings, and (2) they require smaller shaft diameters and lighter bearings, although their field windings are somewhat heavier.

The economic reason placing a limit on maximum power-factor correction can be inferred from the data in Table 8-1 below for a 10,000-kVA system.

**TABLE 8-1.
TOTAL REACTIVE KILOVOLT-AMPERES OF CORRECTION REQUIRED AT VARIOUS POWER FACTORS**

SYSTEM POWER FACTOR	OUTPUT IN kW	kvars AVAILABLE	kvars TO CORRECT FROM NEXT LOWER POWER FACTOR	CUMULATIVE TOTAL kvars REQUIRED IN CORRECTION
0.60	6000	8000	—	—
0.65	6500	7600	400	400
0.70	7000	7140	460	860
0.75	7500	6610	530	1390
0.80	8000	6000	610	2000
0.85	8500	5270	730	2730
0.90	9000	4360	910	3640
0.95	9500	3120	1240	4880
1.00	10,000	0	3120	8000

Table 8-1 shows that a 10,000 kVA system operating at 0.6 PF is capable of delivering only 6000 kW; whereas, at unity power factor, it could deliver 10,000 kW at the same current and the same line drop. Any increase in output, however, is at the expense of reactive kilovolt-amperes. In improving the power factor from 0.65 to 0.70, for example, there is an increase in output of 500 kW at a correction cost of 460 kvars. In improving the power factor from 0.80 to 0.85, the increase of 500 kW is made at a higher correction cost of 730 kvars.

At each successively higher power factor level, the kvar cost is greater for a further improvement of 0.05 in the power factor. In fact, in improving the power factor from 0.95 to unity, the 500 kW increase in output entails a correction cost of 3120 kvars. The monetary cost of the synchronous capacitors required for the increased capacities renders it generally economically prohibitive to raise the power factor much beyond the 0.85 level. Example 8-8 illustrates this point.

EXAMPLE 8-8: A 10,000 kVA system is operating at a power factor of 0.65, and the cost of a synchronous capacitor to improve the power factor is $60 per kVA. Neglecting the losses of the synchronous capacitor, calculate the cost of raising the power factor to:
a. Unity power factor
b. 0.85 power factor lagging.

Solution:

a. At the original load

$$kW = kVA \cos \theta = 10{,}000 \text{ kVA} \times 0.65 = 6500 \text{ kW at } \theta = 49.5°$$
$$kvars = kVA \sin \theta = 10{,}000 \text{ kVA} \sin 49.5° = 7600 \text{ kvars}$$

For unity power factor

kVA of synch capacitor = **7600 kVA** (neglecting losses)
Cost of synch capacitor = 7600 kVA × $60/kVA = **$456,000**

b. For 0.85 PF = cos 31.8°.
The total power, 6500 kW, remains the same and, therefore,

$$\text{Final system kVA is reduced to } \frac{6500}{0.85} \text{ kW} = 7650 \text{ kVA}$$

Final system kvars are reduced to 7650 kVA × sin 31.8° = 4030 kvars
Therefore, the kvars of correction added = 7600 − 4030 = 3570 kvars

For 0.85 PF, kVA of synch capacitor = **3570 kVA** (neglecting losses).

Cost of synch capacitor = 3570 kVA × $60/kVA = **$214,200**,

or less than half the cost in (a) above.

8-21. COMPUTATION OF SYNCHRONOUS MOTOR POWER FACTOR IMPROVEMENT USING THE KW-KVAR METHOD

The computation and prediction of the power factor improvement produced by adding a synchronous motor is greatly simplified by the kW-kvar method of representing the individual and combined loads in terms of the in-phase and quadrature components. Not only is this method significant for determination of the final power factor of a system when a synchronous motor is added. It also permits determination of the kVA rating and the power factor of a synchronous motor required to drive a given load and to produce the correction necessary to raise the over-all system power factor to a predetermined value. The following two examples illustrate two applications of the method.

EXAMPLE 8-9: The load on a three-phase system consists of 40,000 kW at a power factor of 0.8 lagging. A worn-out 7500 hp induction motor, which operated at a lagging power factor of 0.75 and an efficiency of 91 per cent, is to be replaced by a synchronous motor of the same hp rating, operating at either unity or a leading power factor, at the same efficiency.

Calculate:

a. The over-all system power factor using a unity power factor synchronous motor
b. The over-all system power factor using a leading 0.8 PF synchronous motor
c. The difference in kVA ratings of the two synchronous motors.

Solution:

The synchronous motor operates at the same efficiency as the induction motor that has been replaced, and therefore the total power of the system is unchanged. The solution involves construction of a table that shows the original condition of the system, the change, and the final condition. In the table, lagging kvars are shown as negative quantities, and leading kvars as positive quantities. All computed values in the tables are underlined. The table for each part of the solution is shown immediately below the computations for that part.

a. Original kVA $= \dfrac{40{,}000 \text{ kW}}{0.8 \text{ PF}} =$ **50,000 kVA**

Original kvars $= 50{,}000 \text{ kVA} \times 0.6 =$ **30,000 kvars**

Induction motor kW $= \dfrac{7500 \text{ hp} \times 746 \text{ W/hp}}{1000 \text{ W/kW} \times 0.91} =$ **6150 kW**

Final tan $\theta = \dfrac{23{,}380 \text{ kvars}}{40{,}000 \text{ kW}} = 0.584$; $\theta = \tan^{-1} 0.584 = 30.3°$

Cos $\theta = \cos 30.3° =$ **0.8625 lagging**

Final kVA $= \dfrac{\text{kW}}{\cos \theta} = \dfrac{40{,}000}{0.8625} =$ **46,300 kVA**

SYSTEM STATUS	kW	kvars	kVA	$\cos \theta$ PF
Original	40,000	−30,000	50,000	0.8 lagging
Remove Induction Motor	−6150	−6620	8200	0.75 lagging
Add Synchronous Motor	+6150	0	6150	1.0
Final	40,000	−23,380	46,300	0.8625

b. Synchronous motor kVA $= \dfrac{6150 \text{ kW}}{0.8 \text{ PF}} =$ **7680 kVA**

Synch motor leading kvars $= 7680 \text{ kVA} \times 0.6 =$ **4620 kvars**

Final kvars of system $= -30{,}000 - (-6620) + 4620$
$\qquad\qquad\qquad\qquad\quad =$ **−18,760 kvars**

Final tan $\theta = \dfrac{18{,}760}{40{,}000} = 0.468$; $\theta = \tan^{-1} 0.468 = 25.1°$

Cos $\theta = \cos 25.1° =$ **0.905 lagging**

Final kVA $= \dfrac{\text{kW}}{\cos \theta} = \dfrac{40{,}000}{0.905} =$ **44,200 kVA**

SYSTEM STATUS	kW	kvars	kVA	cos θ PF
Original	40,000	−30,000	50,000	0.8 lagging
Remove Induction Motor	−6150	−6620	8200	0.75 lagging
Add Synchronous Motor	+6150	+4620	7680	0.8 leading
Final	40,000	−18,760	44,200	0.905 lagging

    c. The synchronous motor is rated at 6150 kVA when operating at unity power factor.
The synchronous motor is rated at 7680 kVA when operating at 0.8 PF leading.
The use of a synchronous motor at 0.8 PF leading requires (7680 − 6150) = **1530 kVA** additional.

EXAMPLE 8-10: A 500-kVA load operates at a power factor of 0.65 lagging. It is desired to add to the system a 200 hp synchronous motor with an efficiency of 88 per cent, and to bring the total final load of the system (including the added motor) to a power factor of 0.85 lagging. Calculate:
a. The kVA and the power factor of the system with the motor added
b. The kVA rating of the synchronous motor, and the power factor at which it operate.

Solution:

Original kW = 500 kVA × 0.65 = 325 kW

Original kvars = 500 kVA × 0.76 = 380 kvar

Synchronous motor kW = $\dfrac{200 \text{ hp} \times 746 \text{ W/hp}}{1000 \times 0.88}$ = 169.5 kW

The results of the solution are presented below in a table, in which the computed values are underlined; lagging kvars are shown in the table as *negative* quantities, and leading kvars are shown as *positive* quantities.

a. Final kW = 325 + 169.5 = 494.5 kW

    Final kVA of system = $\dfrac{494.5 \text{ kW}}{0.85}$ = **582 kVA**

    System PF = kW/kVA = 494.5/582 = **0.85 lagging**

    Final kvars of system = 582 kVA × 0.525 = −306 (lagging) kvars

b. Synchronous motor kvars = Final kvars − Original kvars
= −306 − (−380) = +74 (leading) kvars

Synchronous motor tan θ = $\dfrac{\text{kvars}}{\text{kW}}$ = $\dfrac{74}{169.5}$ = 0.436

θ = tan⁻¹ 0.436 = 23.6°

Motor power factor, cos θ_m = cos 23.6° = **0.916 leading**

Synchronous motor kVA rating = $\dfrac{\text{kW}}{\cos \theta}$ = $\dfrac{169.5}{0.916}$ = **185 kVA**

SYSTEM STATUS	kW	kvars	kVA	cos θ PF
Original	325	−380	500	0.65 lagging
Added Synchronous Motor	169.5	+74	185	0.916 leading
Final	494.5	−306	582	0.85 lagging

8-22.
USE OF THE SYNCHRONOUS CAPACITOR AS A SYNCHRONOUS REACTOR

Normally a synchronous capacitor is operated at a constant and maximum excitation across the line at the *receiving* end of a plant or factory. Since it is uneconomical to correct the full-load power factor to unity (Section 8-20), the capacitor rating in kVA is considerably less than the total kVA drawn by the occupancy at full load. If the system load drops off throughout the 24 hr period, the power factor will tend to improve because the kVA rating of the capacitor is greater in proportion to the total kVA drawn by the load.

The voltage at the receiving end of a transmission line, V_r, is represented in Fig. 8-26 as drawing a load, I_L, at a lagging power factor angle θ. The impedance voltage drop in the transmission lines, IZ, is primarily inductive when the lines are carrying load current, causing the voltage V_r at the receiving end to be considerably less than the voltage E_s at the sending end. On the other hand, if the load drops to zero because of the distributed capacitance of the line, a reactive current I_c is consumed by the line which, for the same sending voltage, E_s, will produce, at the receiving end, a voltage V_r which greatly *exceeds* the sending voltage E_s.

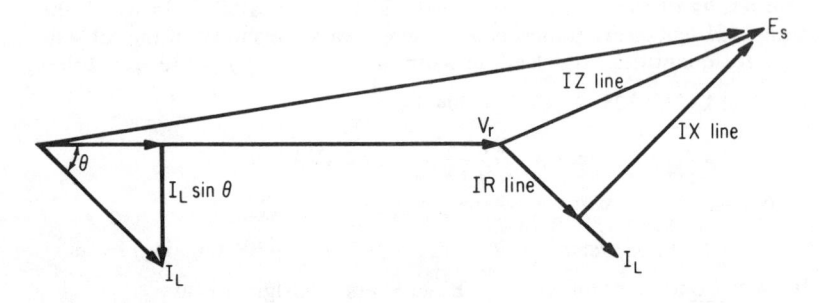

Figure 8-26

Original condition representing voltages at the sending and receiving end at full load.

Instead of maintaining the excitation of the synchronous capacitor constant, therefore, the synchronous capacitor dc excitation is controlled by a voltage regulator in a feedback system in which the dc excitation is *reduced* as the load decreases and the ac receiving voltage increases.

Since a decrease in load provides a rise in voltage at the receiving end, it is a simple matter to obtain fairly linear and accurate control over the range of load and of excitation of the synchronous capacitor. When the load drops to zero, as shown in Fig. 8-27, and the line drop is capacitive, it is necessary to counteract it by *under-exciting* the synchronous dynamo as a synchronous reactor, producing an *inductive* drop by virtue of the inductive current, I_r, drawn by the device from the supply system. Figure 8-28 shows the relation between E_s and V_r at full load and at no load. Note that the relation between the receiving two voltages is roughly the *same*. A synchronous capacitor used in this way maintains a constant voltage at the load regardless of variations in load current and power factor.

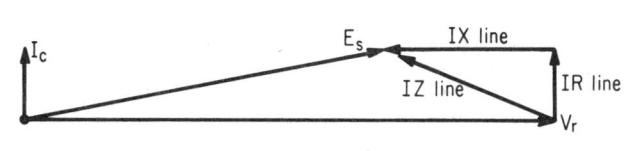

Figure 8-27

Original condition representing voltages at the sending and receiving end at no load.

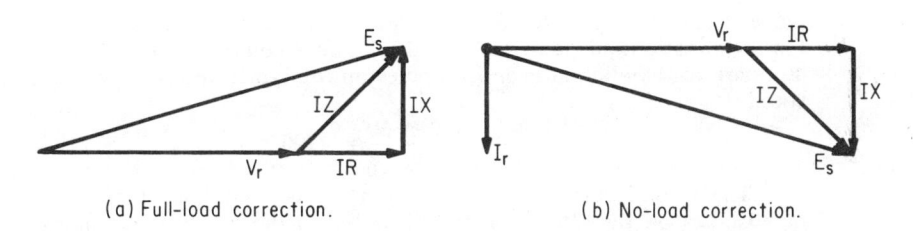

(a) Full-load correction.

(b) No-load correction.

Figure 8-28

Synchronous reactor effect on a transmission line.

8-23. USE OF THE SYNCHRONOUS MOTOR AS A FREQUENCY CHANGER

It was stated previously that, because of its *constant-speed* characteristic, the synchronous motor may be used to drive a dc shunt generator to maintain a relatively constant dc voltage from no load to full load. Because of its constant-speed characteristic, it may also be used to drive an ac alternator, either single-phase or polyphase, of various frequencies, maintaining a constant ac voltage and a constant frequency of the alternator.

There are specific advantages inherent in the use of higher-frequency equipment, although losses tend to increase with frequency. For the same kVA or hp rating, a higher-frequency dynamo or transformer may be smaller in size, requiring less iron to obtain the same degree of magnetic saturation. Furthermore, at 60 Hz the maximum speed obtainable with a bipolar synchronous motor is 3600 rpm ($S = 120f/P$). In order to obtain higher dynamo speeds, it is necessary to employ a *frequency changer*.

An ac to ac motor generator set in which a change of frequency occurs is called a *frequency changer*. Since the synchronous motor is coupled to the alternator, they are both operating at the same speed and, therefore

$$S = \frac{120 f_a}{P_a} = \frac{120 f_m}{P_m}$$

or

$$\frac{f_a}{P_a} = \frac{f_m}{P_m} \tag{8-29}$$

where f_a and f_m are the frequencies in Hz of the alternator and the motor, respectively, and P_a and P_m are the number of poles of the alternator and motor, respectively.

EXAMPLE 8-11: Determine the speeds and the number of poles for the alternator and motor, respectively, of three different frequency changers which will permit a frequency conversion from 60 to 400 Hz.

Solution:

Since $P_a/P_m = f_a/f_m = 400/60$ or 20/3, the ratio of f_a/f_m determines the combinations of poles and speed. The first combination must have 40 poles on the alternator and six poles on the synchronous motor at a speed

$$S = \frac{120 f}{P} = \frac{120 \times 60}{6} = 1200 \text{ rpm}$$

The second combination is an even-number multiple of $\frac{20}{3} = \frac{80}{12}$, or 80 poles on the alternator and 12 poles on the synchronous motor at a speed of 600 rpm.

The third combination is $\frac{20}{3}$ times $\frac{6}{6}$, or 120 poles on the alternator and 18 poles on the synchronous motor at a speed of 400 rpm.

8-24. THE SUPER-SYNCHRONOUS MOTOR

The term *synchronous* usually refers to speed. The supersynchronous motor does *not* operate at a supersynchronous speed, and its title is a misnomer. It would have been better to have called it a *supertorque* motor. The motor was developed by General Electric to provide a synchronous motor which was self-starting under heavy loads. In Section 8-6, the simplex-type rotor was described as a special construction having five slip-rings and employing a wound rotor in combination with the dc field winding. The simplex-rotor synchronous motor may develop torques between 250 and 300 per cent of full-load torque.

The "supersynchronous" motor is capable of developing *maximum* torque [Fig. 8-24, Eq. (8-28)] *on starting*. It, too, requires a special construction, however, and it is probably the most costly motor of its kind for a given horsepower rating. The rotor is the standard cage-type rotor with a dc field winding brought out to slip-rings on the rotor shaft. It is coupled directly to the mechanical load which it must drive. The entire stator, however, is free to rotate on trunnions, in the same manner as an ac dynamometer. But, whereas the latter is limited in its angular displacement, the stator of the supersynchronous motor is free to rotate on bearings at its synchronous speed. The stator armature winding, therefore, is also excited through slip-rings and is usually started at a reduced voltage by means of a three-phase variac or induction regulator. A large brake is provided around the outside of the stator frame to apply a braking action and to secure the stator in its running position.

Because the rotor is coupled to the load, when a reduced polyphase ac voltage is applied to the stator with brake released, the induction motor torque produced by the rotor poles reacts against the "stator" conductors; this reaction imparts to the stator a torque that is opposite in direction to the direction of rotation of the load (Fig. 8-3a). The stator picks up speed as the ac stator voltage is increased; and, as the stator reaches synchronous speed, full ac stator voltage is applied in addition to the dc field excitation. The stator pulls into synchronism with the rotor at a standstill, held by the inertia of the fixed heavy load coupled to its shaft. At this instant, the motor is operating as a synchronous motor without load, generating a counter-emf which limits its stator current.

The brake is now slowly applied to the rotating stator. Since the synchronous motor *must* run at synchronous speed, the reduction in stator speed must be made up by rotation of rotor speed in the opposite direction, i.e., for a synchronous speed of 1800 rpm, a stator speed of 1790 rpm counter-clockwise requires a rotor speed of 10 rpm clockwise. The torque angle α therefore increases to provide maximum torque in starting the heavy applied load. The armature current, although high, is limited by the generated emf in the stator. Reducing the speed of the stator by increased braking increases the speed of the rotor, until the stator is at a standstill and the rotor is rotating with the full applied load at synchronous speed. Supersynchronous motors, therefore, are maximum torque synchronous motors capable of *starting* and running at torques from 300 to 350 per cent of the full-load torque.

8-25.
SPECIAL TYPES OF SYNCHRONOUS MOTORS WHICH DO NOT EMPLOY DC FIELD EXCITATION

The discussion of the preceding sections has been concerned with polyphase synchronous motors in which the excitation of the rotor poles is provided and may be varied by direct current. There are types of synchronous motors, however, which do not employ dc field excitation, and these fall into two categories: (1) nonexcited synchronous motors, employing no field excitation whatever; and (2) ac field excitation motors or "synchro" motors. The "synchronous" motors falling into the latter category (employing ac field excitation) will be discussed in Chapter

11, along with other special dynamos. The various remaining non-excited synchronous motors in the first category, both polyphase and single-phase, will be covered in this chapter.

8-26.
THE SYNCHRONOUS-INDUCTION MOTOR

This motor was developed out of a demand for a self-starting polyphase synchronous motor in the smaller sizes (below 50 hp) requiring no dc field excitation and possessing the constant-speed characteristics of the synchronous motor. The rotor consists of a squirrel-cage winding (wound or cast) distributed uniformly around the rotor in the slots shown in Fig. 8-29a. When a

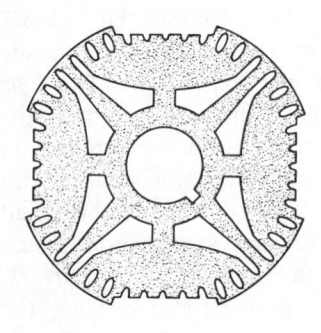

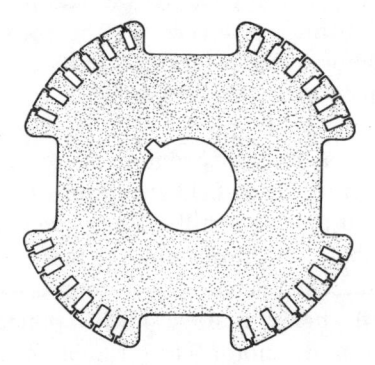

(a) Synchronous-induction motor.　　(b) Reluctance motor.

Figure 8-29

Differences between laminations and induction starting and running windings of nonexcited (hysteresis type) synchronous motors.

polyphase alternating current is applied to the standard polyphase stator armature, the motor starts as an induction motor. Because of the salient-pole rotor shown in Fig. 8-29a, the motor pulls into synchronism quite easily and develops the rapid and maximum torque of the *salient*-pole machine shown in Fig. 8-24 and discussed in the last few paragraphs of Section 8-17. Thus, the synchronous-induction motor develops reluctance torque (proportional to $\sin 2\alpha$) and is sometimes called a *polyphase reluctance* motor. But even this is a misnomer, because the synchronous-induction motor operates on the *combined* torque characteristics of the synchronous motor and the induction motor shown in Fig. 8-30. When designed with high-resistance rotor windings, fairly high starting torques of up to 400 per cent of full-load torque can be developed. On the other hand, the use of high-resistance rotor windings results in increased slip (Section 8-5), reduced efficiency, and less possibility of pulling into synchronism under load by means of reluctance torque. As shown in Fig.

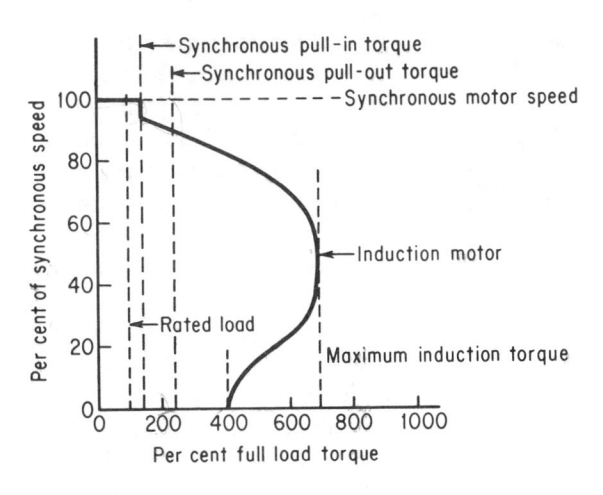

Figure 8-30

Synchronous induction motor speed-torque characteristic.

8-30, the synchronous pull in torque is developed at a little above the rated full-load torque.

As a synchronous motor, the synchronous-induction motor will operate at constant speed to a little beyond 200 per cent of full load. If the applied load exceeds 200 per cent of the full-load torque, it drops to its induction characteristic, where it will continue to operate as an induction motor up to almost 700 per cent of the full-load torque. Because the synchronous motor pull-out torque is about one-third the induction motor pull-out torque, the stator frame of a synchronous-induction motor is *three times as large* in size as an ordinary induction motor of the same output horsepower rating. Furthermore, since it operates from no load to full load as a nonexcited synchronous motor [Eq. (8-27)], an increased torque angle makes up for the lack of excitation, and the motor draws a high lagging current at a low power factor. This also results in reduced efficiency and necessitates the larger size of frame to dissipate heat.

In motors of relatively low horsepower, such as the synchronous-induction motor, the problems created by larger size and weight, low efficiency, and lagging current are unimportant in view of the advantages of constant speed, ruggedness, lack of dc excitation, high starting and running torque, and minimum maintenance that typify the synchronous-induction motor. Synchronous polyphase motors with dc excitation are rarely manufactured below 10 hp these days except by special order, because polyphase synchronous-induction motors are available up to 50 hp (and by special order from the manufacturer up to 100 hp).

8-27. RELUCTANCE MOTOR

Single-phase, *salient*-pole synchronous-induction motors are generally called *reluctance motors*. If the rotor of any uniformly distributed single-phase induction motor is altered so that the laminations tend to produce *salient* rotor poles, as shown in Fig. 8-29b, the reluctance of the air-gap flux path will be greater where there are no conductors embedded in slots. Such a motor, coming up to speed as an induction motor, will be pulled into synchronism with the pulsating ac single-phase field by the reluctance torque (Secs. 1-2 and 8-17) developed at the salient iron poles which have lower-reluctance air gaps.

The speed-torque characteristics of a typical single-phase reluctance motor are shown in Fig. 8-31. The motor starts at anywhere from 300 to 400 per cent of its full-load torque (depending on the rotor position of the unsymmetrical rotor with respect to the field windings) as a two-phase motor (Section 9-3) as a result of the rotating magnetic field created by a starting and running winding (displaced 90° in both space and time). At about three-quarters of synchronous speed, a centrifugal switch opens the starting winding, and the motor continues to develop a single-phase torque produced by its running winding only. As it approaches synchronous speed, the reluctance torque (developed as a synchronous motor) is sufficient to pull the rotor into synchronism with the pulsating single-phase field. The motor operates as a constant-speed, single-phase, non-excited synchronous motor up to a little over 200 per cent of its full-load torque. If it is loaded beyond the value of pull-out torque, it will continue to operate as a single-phase induction motor up to over 500 per cent of its rated torque.

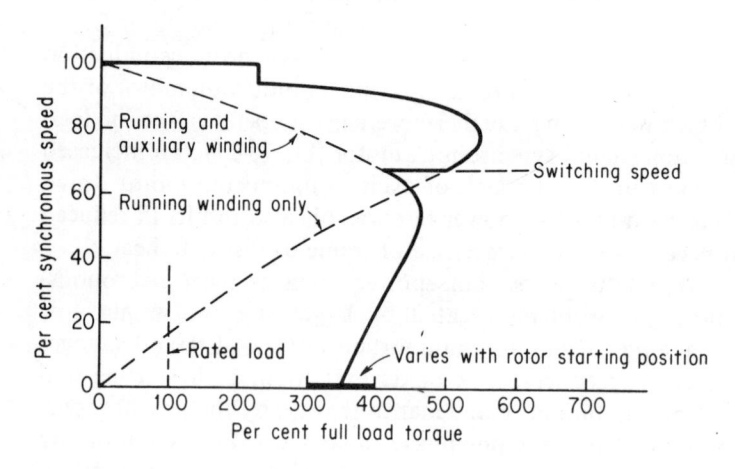

Figure 8-31

Reluctance motor speed-torque characteristic.

8-28. HYSTERESIS MOTOR

Single-phase cylindrical (*nonsalient*-pole) synchronous-induction or shaded-pole motors are classed as *hysteresis motors*. The difference between this motor and that of the preceding section is in (1) the *shape* of the rotor, and (2) the *nature* of the torque produced. Whereas the *reluctance* motor is pulled into synchronism and runs on *reluctance* torque, the *hysteresis* motor pulls into synchronism and runs on *hysteresis* torque. Hysteresis-type laminations, shown in Fig. 8-32, are usually made of hardened high-retentivity steel rather than commercial, low-retentivity dynamo steel.

(a) Hysteresis rotor. (b) Subsynchronous rotor.

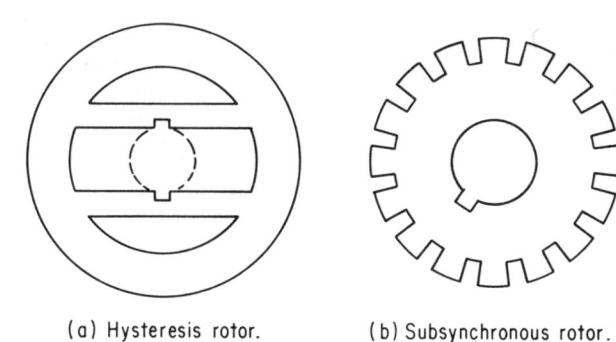

Figure 8-32

Two types of high retentivity laminations of nonexcited (hysteresis type) synchronous motors.

As a result of a rotating magnetic field produced by phase splitting or a shaded-pole stator (Chapter 10), eddy currents are induced in the steel of the rotor which travel across the two bar paths of the rotor shown in Fig. 8-32a. A high-retentivity steel produces a high hysteresis loss, and an appreciable amount of energy is consumed from the rotating field in reversing the current direction of the rotor. At the same time, the rotor magnetic field set up by the eddy currents causes the rotor to rotate. A high starting torque is produced as a result of the high rotor resistance (proportional to the hysteresis loss). As the rotor approaches synchronous speed, the frequency of current reversal in the cross-bars decreases, and the rotor becomes permanently magnetized in one direction as a result of the high retentivity of the steel rotor. With two field poles, the rotor shown in Fig. 8-32a will develop a speed of 3600 rpm at 60 Hz. The motor runs as a hysteresis motor on hysteresis torque because the rotor is permanently magnetized.

The amount of torque produced as a result of this magnetization is *not* as great as reluctance torque. But hysteresis torque is *extremely* steady, in both amplitude and phase, in spite of fluctuations of supply voltage, with the result that it is extremely popular in driving high-quality record players and tape recorders. Since reluctance torque can be produced more cheaply than hysteresis torque for the same fractional horse-power, high-torque hysteresis synchronous motors of good quality are more expensive than reluctance-torque synchronous motors of the same rating.

8-29. SUBSYNCHRONOUS MOTOR

Another type of hysteresis motor is the *subsynchronous motor* whose salient (yet cylindrical) pole lamination is shown in Fig. 8-32b. This motor starts in the same manner as the hysteresis motor described above. At synchronous speed, the rotor poles induced in a hysteresis rotor remain at fixed spots on the rotor surface as the rotor rotates in synchronism with the rotating magnetic field of the stator. It should also be noted, that hysteresis torque is effective when both of these rotors (Fig. 8-32) rotate at less than synchronous speed. For example, the subsynchronous rotor lamination shown in Fig. 8-32b has 16 poles and will rotate at 450 rpm. But hysteresis torque, unlike reluctance torque, is independent of rotor speed. If the rotor is turning at less than synchronous speed, the induced poles (which must go at synchronous speed since they are induced by the rotating stator field) move across the rotor surface at a "slip" speed, i.e., a speed equal to the difference between the synchronous speed and the rotor speed.

In the case of the subsynchronous motor, if the applied torque is *too great* at its normal synchronous speed based on the number of salient rotor poles, the motor will rotate at a subsynchronous speed determined by a whole-number multiple of the number of poles on the lamination (32 poles or 225 rpm in this case). <u>Since torque varies inversely as speed, as the speed drops, the subsynchronous motor develops more torque.</u> Geared down electrically, the subsynchronous motor can provide fairly high torques at a slow constant speed.

A final distinction between reluctance torque and hysteresis torque is that all reluctance-torque motors require induction-motor starting torque to come up fairly close to synchronous speed, where pull-in can occur as a result of reluctance torque (Sec. 1-2 and 8-17), and the motor operates as a reluctance motor at a constant synchronous speed. No reluctance motor, therefore, is self-starting.

<u>Hysteresis motors and subsynchronous motors, *are* selfstarting (as a result of phase splitting or shaded-pole techniques)</u> and come up to full synchronous speed, developing high rotor resistance because of hysteresis loss and, consequently, fairly high starting torque but poorer running hysteresis torque than reluctance-torque motors.

8-30. SOLID STATE DC FIELD SUPPLIES—STATIC SUPPLIES

Since dc is not readily available in most industrial occupancies, larger polyphase synchronous motors usually are equipped with an exciter (a dc self-excited shunt generator) mounted on the same shaft as the motor, as described in Sec. 8-4. The recent development of high current solid-state silicon rectifiers has encouraged the manufacture of synchronous motors without exciters and also the brushless synchronous motor.

In lieu of the exciter, a static dc power supply is readily obtained from the 3-phase supply which also supplies the stator of the synchronous motor. Such a supply is shown in Fig. 8-33, in which a Δ-Y transformation (Sec. 13-14) provides the desired reduced ac voltage to be rectified which, in turn, provides the desired dc voltage. Six silicon rectifiers are used to provide full wave rectification and 4 surge voltage suppressors are used to eliminate damage to the silicon rectifiers by voltage transients.

The capacitor, C, at the output, acts as a smoothing filter that provides almost pure dc to the field of the synchronous motor, via the brushes which are connected to the slip-rings of the synchronous motor rotor.

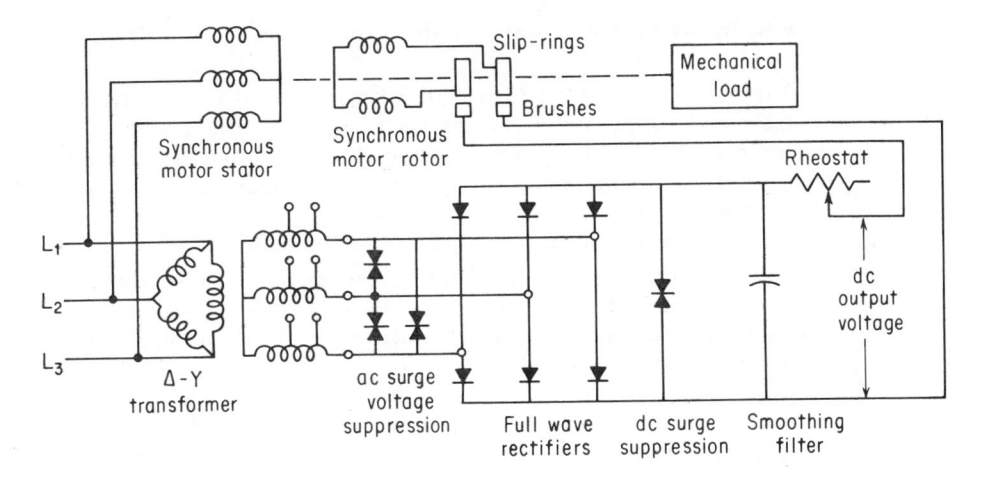

Figure 8-33

Solid-state dc static supply for synchronous motor.

8-31. **BRUSHLESS SYNCHRONOUS MOTORS**	The elimination of the exciter on the shaft of the synchronous motor eliminated the problems associated with commutation of a dc generator and sparking at the brushes connected to the commutator. But as shown in Fig. 8-33, it is still necessary to

supply the dc to the field via brushes and slip-rings. In order to completely eliminate slip-ring and brush maintenance, the *brushless synchronous motor* was developed.

A block diagram of one type of brushless synchronous motor is shown in Fig. 8-34. In effect, the system incorporates the rectification of Fig. 8-33 with the following modifications:

1. The silicon rectifiers of Fig. 8-33 are replaced by thyristors or SCRs.

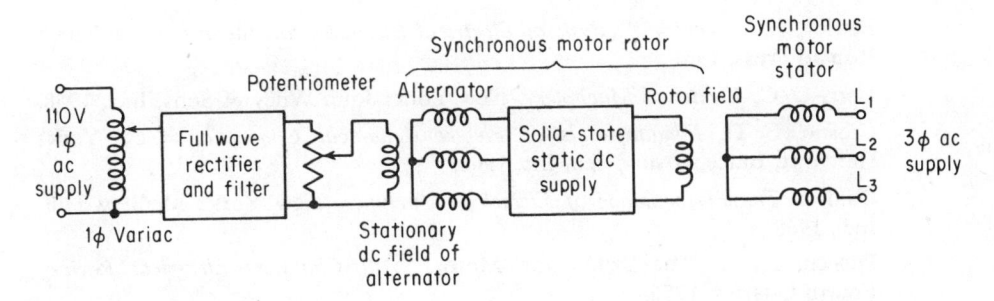

Figure 8-34

Brushless synchronous motor, block diagram form.

289

SEC. 8-31. / *Brushless Synchronous Motors*

2. The SCRs are fired by transistors which control the thyristor dc output.
3. The transformer of Fig. 8-33 is replaced by an ac alternator having a stationary dc field and a rotating polyphase armature in which ac voltages are generated. The dc excitation of the synchronous motor is controlled by the single-phase variac exciting the stationary dc field of the polyphase alternator which is on the same rotor shaft as the rotor field of the synchronous motor, as shown in Fig. 8-34.
4. The rotor of the synchronous motor, as shown in Fig. 8-34, carries the alternator armature, the static dc control and rectification system, consisting of transistors and thyristors described above and the rotor field of the synchronous motor.

The above described modifications shown in Fig. 8-34 provide a means of controlling the dc field excitation of a synchronous motor without the need for exciter or slip-rings or brushes of any kind.*

BIBLIOGRAPHY

Alger, P. L. *The Nature of Polyphase Induction Machines* New York: John Wiley & Sons, Inc., 1951.

———and Erdelyi, E. "Electromechanical Energy Conversion," *Electro-Technology*, September 1961.

Anderson, W. A. and Haegh, J. E. "Two-Speed Non-Exited Synchronous Motors," *Electro-Technology*, January 1959.

Bekey, A. "New High-Performance Designs for Small Synchronous Motors," *Electro-Technology*, November 1961.

Bewley, L. V. *Alternating Current Machinery*, New York: The Macmillan Company, 1949.

Bewley, L. V. *Tensor Analysis of Electrical Circuits and Machines*, New York: Ronald Press, 1961.

Carr, C. C. *Electrical Machinery*, New York: John Wiley & Sons, Inc., 1958.

Crosno, C. D. *Fundamentals of Electromechanical Conversion*, New York: Harcourt, Brace, Jovanovich, Inc, 1968.

Daniels. *The Performance of Electrical Machines*, New York: McGraw-Hill, Inc., 1968.

Dineen, R. J. "The Synduction Motor," *Allis-Chalmers Electrical Review*, Fourth Quarter, 1956.

Fitzgerald, A. E. and Kingsley, C. *The Dynamics and Statics of Electromechanical Energy Conversion*, 2nd Ed., New York: McGraw-Hill, 1961.

* Rosenberry, G. M. "The Brushless Synchronous Motor", *Electrical Manufacturing*, June 1960.

Fitzgerald, A. E., Kingsley, Jr. C. and Kusko, A. *Electric Machinery*, 3rd Ed. New York: McGraw-Hill Book Company, 1971.

Gemlich, D. K. and Hammond, S. B. *Electromechanical Systems*, New York: McGraw-Hill, 1967.

Hindmarsh, J. *Electrical Machines*, Elmsford, N.Y.: Pergamon Press, 1965.

"'Hybrid' Synchronous-Induction Motor." (Staff Report.) *Electrical Manufacturing*, May 1956.

Jones, C. V. *The Unified Theory of Electrical Machines*, New York: Plenum Publishing Corp., 1968.

Koenig, H. E. and Blackwell, W. A. *Electromechanical System Theory*, New York: McGraw-Hill, 1961.

Liwschitz, M. M., Garik, M. and Whipple, C. C. *Alternating Current Machines*, Princeton, N. J.: D. Van Nostrand Co., Inc., 1946.

Majmudar, H. *Introduction to Electrical Machines*, Boston: Allyn and Bacon, 1969.

McFarland, T. E. *Alternating Current Machines*, Princeton, N. J.: D. Van Nostrand Co., Inc., 1948.

Meisel, J. *Principles of Electromechanical Energy Conversion*, New York: McGraw-Hill, 1966.

Millermaster, R. A. *Harwood's Control of Electric Motors*, 4th Ed., New York: Wiley/Interscience, 1970.

Merrill, F. W. "Characteristics of Permanent-Magnet Synchronous Motors," *Electrical Manufacturing*, January 1956.

———. "The Permanent Field Synchronous Motor," *Electro-Technology*, December 1961.

Moore, R. C. "Synchronous Motor Torques," *Allis-Chalmers Electrical Review*, Third Quarter, 1959.

Nasar, S. A. *Electromagnetic Energy Conversion Devices and Systems*, Englewood Cliffs, N.J.: Prentice-Hall, Inc., 1970.

O'Kelly and Simmons. *An Introduction to Generalized Electrical Machine Theory*, New York: McGraw-Hill, 1968.

Puchstein, A. F., Lloyd, R., and Conrad, A. G. *Alternating Current Machines*, 3rd Ed., New York: Wiley/Interscience, 1954.

Robertson, B. L. and Black. L. J. *Electric Circuits and Machines*, 2nd Ed., Princeton, N. J.: D. Van Nostrand Co., Inc., 1957.

Schmitz, N. L., and Novotny, D. W. *Introductory Electromechanics*, New York: Ronald Press, 1965.

Seely, S. *Electromechanical Energy Conversion*, New York: McGraw-Hill, 1962.

Selmon. *Magnetoelectric Devices: Transducers, Transformers and Machines*, New York: Wiley/Interscience, 1966.

Skilling, H. H. *Electromechanics: A First Course in Electromechanical Energy Conversion*, New York: Wiley/Interscience, 1962

Synchronous Generators and Motors. (ASA C50.1.) New York: American Standards Association.

Thaler, G. J., and Wilcox, M. L. *Electric Machines: Dynamics and Steady State*, New York: Wiley/Interscience, 1966.

White, D. C., and Woodson, H. H. *Electromechanical Energy Conversion*, New York: Wiley/Interscience, 1959.

QUESTIONS

8-1. Name two factors which will cause an alternator to "motorize".

8-2. Give the equation which determines the average speed of a synchronous motor.

8-3. Why is the synchronous motor a special case of a doubly excited dynamo?

8-4. Because of the possibility of variation of field excitation, what three unusual characteristics are possessed by the synchronous motor which no other ac (or dc) motor exhibits?

8-5. Give four advantages possessed by synchronous motors over induction motors.

8-6. a. Explain why a synchronous motor is not inherently self-starting.
b. Explain why a synchronous motor will run at synchronous speed or not at all.

8-7. a. Give four methods used for starting synchronous motors.
b. Explain which of the above methods is most commonly used and why.

8-8. a. Why is it impossible to start a synchronous motor with its dc field energized?
b. What is meant by "slipping a pole" and under what conditions does it occur?

8-9. a. Describe the construction of a simplex rotor synchronous motor.
b. What advantages does it possess over the customary ac synchronous dynamo?
c. Compare the starting torque of the customary ac synchronous dynamo started on damper windings with the starting torque of a simplex rotor synchronous motor.

8-10. a. What reaction to increased load occurs in a shunt motor, enabling it to draw more line current?
b. Explain, using Eq. (8-1), why a synchronous motor rotor pole must drop back in phase with respect to the stator flux to develop synchronizing power. Compare this to shunt motor in (a) above.
c. Explain, using Eq. (8-2), why even a stroboscopic lamp will fail to detect the displacement of the rotor pole with application of load to the rotor shaft.

8-11. a. From the conditions given in Ex. 8-1, define "normal excitation."
b. Illustrate normal excitation by means of a phasor diagram showing V_p, I_a, E_r and E_{gp}.

c. What is the power factor and the power factor angle, θ, under conditions of *normal* excitation?

8-12. Under conditions of *normal* excitation, describe the effect of increased load on a synchronous motor's:
 a. torque angle, β or α
 b. power factor angle, θ
 c. armature current, I_a.

8-13. Repeat Quest. 8-12 a, b, c for the effect of increased load on a synchronous motor which is
 a. underexcited
 b. overexcited.

8-14. a. Regardless of the state of excitation of a synchronous motor, explain why armature reaction will have a tendency to improve the power factor of the motor.
 b. In view of your answer to (a), explain why armature reaction is not compensated in ac dynamos.

8-15. Given a synchronous motor operating with a constant load and normal excitation, explain the effect on torque angle, α, armature current, I_a, and power factor angle, θ, of a synchronous motor with:
 a. decreased excitation
 b. increased excitation.

8-16. a. Summarize the results of your answer to Quest. 8-15 graphically showing armature current vs. excitation field current and power factor vs. field current at a given load on the same curve. Use field current as abscissa.
 b. Using only an ac ammeter, explain how the unity PF condition of any loaded synchronous motor is easily obtained.

8-17. Draw the families of V-curves of a synchronous motor at no load, half load and full load showing:
 a. relation between armature and field current
 b. relation between power factor and field current.

8-18. Using the curves drawn above, explain:
 a. why it is necessary to increase the excitation to obtain minimum current with application of load
 b. why at any load the power factor decreases and the armature current increases if the field current is varied above and below normal excitation
 c. a synchronous motor may be "overloaded" with no load connected to its shaft.

8-19. Using the data and table developed from Ex. 8-3, explain why an increase in field excitation produces:
 a. an improvement in lagging power factor
 b. a reduction in torque angle for the same power developed at the output shaft.

8-20. State four reasons which compel utility companies to seek improvement in overall system power factor for a given kVA output.

8-21.   a. Give one inherent advantage of a synchronous motor over an induction motor as a source of mechanical power.
b. What is a synchronous capacitor and how may it be distinguished from a synchronous motor?
c. What three standard PF ratings are used for synchronous motors and why do they differ in size for the same hp rating?

8-22.   a. Is there an economic limit to improvement of PF and what is it?
b. In addition to correction of PF give an additional application of the use of a synchronous capacitor in transmission systems.
c. In addition to the correction of PF and a source of mechanical power, give an additional application of the synchronous motor.

8-23.   Compare salient and nonsalient synchronous motor rotors having the same stator with respect to:
a. magnitude of maximum torque
b. torque angle, α, in electrical degrees, at which maximum torque occurs
c. torque angle, α, in electrical degrees at which rated torque occurs.

8-24.   a. Describe the construction of a supersynchronous motor.
b. Compare the starting torque of a supersynchronous motor with that of a simplex rotor synchronous motor and a conventional synchronous motor.

8-25.   Construct a table showing the following nonexcited synchronous motors in the first column: synchronous-induction motor, reluctance motor, hysteresis motor and subsynchronous motor. Each of the following columns should be labelled: polyphase or single-phase, salient or nonsalient rotor, maximum starting torque, advantages, disadvantages, special features. Complete each row for each motor by filling in all blanks.

8-26.   List the advantages of brushless synchronous motors over conventional polyphase synchronous motors.

PROBLEMS

8-1.   Calculate:
a. The frequency of the voltage which must be applied to the stator of a 10 pole, three-phase, 220 V synchronous motor required to operate at 1200 rpm
b. The number of poles required for a 220 V, three-phase synchronous motor to operate at a speed of 500 rpm when 50 Hz is applied to the stator
c. The full load speed of a 36 pole, 60 Hz, 220 V synchronous motor.

8-2.   A 50 hp, unity PF 60 Hz, wye-connected, 220 V synchronous motor has a rated armature current of 108 A and runs at a speed of 450 rpm. The field excitation of the motor is adjusted to produce a _generated voltage_ equal to the _applied line voltage_ at no load, producing a torque angle of 1 mechanical degree. Calculate:

a. The number of poles
b. The number of electrical degrees by which the rotor lags the stator field, α
c. Resultant phase voltage, E_r, between applied phase and generated phase voltages
d. The angle, δ, between the resultant voltage, E_r, and the applied phase voltage, V_p
e. The phase current drawn by the motor if the phase impedance is $1.0 \angle 84.3°\ \Omega$
f. The power factor of the motor and θ
g. Total power drawn by the motor from the bus and power developed by the armature.

8-3. Repeat Problem 8-2 for a load producing a torque angle of 2.5 mechanical degrees.

8-4. Repeat Problem 8-2 for a load producing a torque angle of 3 mechanical degrees and a generated voltage per phase of 150 V

8-5. Repeat Problem 8-2 for a load producing a torque angle of 3 mechanical degrees and a generated voltage per phase of 100 V.

8-6. From your computations of Problems 8-2 through 8-5, inclusive, calculate for each load condition
a. Developed horsepower
b. Developed torque in lb-ft
c. Draw inferences as to the effect of increasing the load and maintaining the excitation constant (Problems 8-2 vs. 8-3)
d. Draw inferences as to effect of maintaining "constant" torque angle and decreasing the excitation (Problems 8-4 vs. 8-5).

8-7. Calculate the output horsepower and torque if the efficiencies of the synchronous motor at the various load and excitation conditions are
a. 70 per cent in Problem 8-2
b. 75 per cent in Problem 8-3
c. 85 per cent in Problem 8-4
d. 80 per cent in Problem 8-5
e. Account for differences between *developed* horsepower and torque as compared to *output* horsepower and torque.

8-8. The load on the motor of Problem 8-2 is increased until the load torque angle exactly equals the synchronous impedance angle of 84.3°; excitation is normal. At this breakdown point, calculate
a. Resultant phase voltage, E_r, between applied phase and generated phase voltages
b. The angle, δ, between resultant phase voltage, E_r, and applied voltage, V_p
c. The armature current and ratio of armature current to rated current (108 A)
d. Developed horsepower and torque
e. Ratio of developed horsepower and torque to rated output horsepower and torque.

8-9. A salient-pole, 12 pole, 220 V, three-phase, wye-connected, 100 hp synchronous motor has an armature resistance of 0.03 Ω and a synchronous reactance of 0.27 Ω/phase. Assume a generated voltage per phase at all times equal to the applied voltage per phase. For mechanical loads listed below, complete all the cells of the following table by making all necessary calculations.

β MECH. DEG.	α ELECT. DEG.	E_r (V)	I_p (A)	θ	$\cos\theta$	W_t (W total)	P_d (W)
5°							
10°							
	83.5°						
	83.65°						
	84.0°						
15°							
20°							

8-10. From the table in Problem 8-9, determine the point at which maximum or breakdown torque occurs. Explain
   a. Why maximum torque does *not* occur where α equals 90°
   b. What determines the point at which maximum torque will occur, ignoring effects of saturation and armature reaction
   c. Why the developed power does not continue to increase with increases in torque angle, E_r and I_a
   d. The effect of a reduced ratio of X_s/R_a (i.e., a stiffer rotor) on the point at which maximum torque will occur
   e. The change of power factor with increased load.

8-11. In Problem 8-9, the generated voltage per phase was assumed as equal to the applied voltage per phase; although this is rarely the case, even at a unity power factor. Assuming an efficiency of 88 per cent, for the motor of Problem 8-9, calculate
   a. Full-load armature current, full-load $I_a R_a$ and $I_a X_s$ volt drops, and full-load generated voltage per phase at a power factor of unity. Use any method or equation you desire.
   b. The number of electrical degrees, α, between stator and rotor
   c. The number of mechanical space degrees, β, between stator and rotor
   d. The resultant voltage, E_r, between applied voltage and generated voltage per phase, and its phase angle
   e. The power drawn by the armature from the bus and the electrical copper losses (armature only)
   f. The internal horsepower and internal torque developed at full load, unity power factor
   g. The maximum internal horsepower and internal torque developed (from Table, Problem 8-9)

h. Ratio of maximum torque and internal horsepower to rated torque and internal horsepower.

8-12. Assuming no change in any of the factors of Eq. (8-28), except α, calculate
a. The ratio of maximum to rated torque using the values of α obtained in Problems 8-9 and 8-11, respectively
b. Account for differences between this ratio and the ratio obtained in Problem 8-11(h).

8-13. A 100 hp, 550 V, eight-pole, wye-connected synchronous motor is connected to a load which constantly maintains a mechanical torque angle of 2.5 space degrees. The armature resistance and synchronous reactance per phase are, respectively, 0.2 Ω and 1.8 Ω. Assuming constant load regardless of excitation changes, in which the excitation is progressively increased from 225 V, 323 V, and 450 V generated voltage per phase, calculate and tabulate for ready reference and comparison the following
a. Resultant voltage per phase, E_r
b. Armature current per phase, I_p
c. Power factor
d. Total power input
e. Developed rotor torque.

8-14. For the three excitations given in Problem 8-13, calculate maximum torque developed by the rotor and tabulate in the same manner as in the previous problem.

8-15. From the tabulations of Problems 8-13 and 8-14, explain
a. Effect of increased excitation on developed torque
b. Why the power factor is always lagging in the tabulation of Problem 8-14.

8-16. A 100 hp, 2300 V, eight-pole, 60 Hz, wye-connected synchronous motor is designed to operate at a power factor of 0.9 leading. The efficiency at full load is 88 per cent and the armature resistance and reactance per phase are, respectively, 1 Ω and 20 Ω. When delivering rated load at rated power factor, calculate
a. Armature current per phase
b. Torque angle in electrical and mechanical space degrees
c. Generated voltage per phase
d. The torque angle required to produce an armature current at unity power factor, assuming that the excitation remains the same as in (c)
e. The armature current at unity power factor and per cent overload
f. The output horsepower at unity power factor load, assuming the efficiency is 85 per cent.

8-17. Assume that the excitation is reduced in Problem 8-16 so that the synchronous motor operates at unity power factor with rated load current flowing in its armature. To make this possible, calculate
a. Generated armature voltage per phase
b. Resultant voltage per phase
c. Torque angle in electrical and mechanical space degrees
d. The output hp at unity power factor load, assuming the efficiency is 90 per cent.

8-18. Compare the calculated and given data for Problems 8-16 and 8-17 with Appendix Table A-5 and account for differences in
a. Armature current
b. Horsepower.

8-19. Calculate the rating of a three-phase synchronous capacitor to raise the power factor of a 20,000 kW load operating at a power factor of 0.6 lagging up to
a. A power factor of 0.8 lagging (neglecting the losses of the capacitor)
b. Unity power factor (assuming a 10 per cent leading power factor synchronous capacitor)
c. In (b), how many more kilowatts may be added to the system to produce the original total of kilovolt amperes?

8-20. A 50,000 kVA, three-phase alternator is loaded to rated capacity at 0.7 PF lagging. A synchronous capacitor of 0.1 PF leading corrects the power factor of the system to unity. Calculate:
a. The additional kilowatts which may now be supplied by the alternator to the load at unity power factor
b. The kilovolt ampere rating of the synchronous capacitor required to perform this correction.

8-21. A 1300 hp, 0.8 PF, three-phase synchronous motor delivers 1300 hp at 0.8 PF leading in driving a mechanical load. The motor is connected across a line which is also supplying 1200 kVA at 0.6 PF to a load of several induction motors. Assuming a synchronous motor efficiency of 90 per cent, calculate
a. whether it is possible for the synchronous motor to bring the line power factor up to unity without exceeding the capacity of the synchronous motor
b. the final system power factor with the synchronous motor operated at rated capacity delivering 1300 hp and overexcited.

8-22. A three-phase synchronous motor raises the power factor of a system from 0.7 to 0.9 when connected across the line. If the input to the synchronous motor is 500 kVA at 0.8 PF leading, what was the original system load in kilovolt amperes before the motor was added?

8-23. A 20,000 kVA polyphase alternator is loaded to capacity with a 0.6 PF load consisting entirely of induction motors. If one 5000 hp induction motor operating at 0.65 PF is replaced with a 5000 hp synchronous motor rated 0.8 PF, having the same efficiency and full load current (i.e., 93 per cent), calculate
a. the increase in the system power factor
b. the number of additional kilowatts which may be used for incandescent lighting load.

8-24. An industrial area has a load of 4000 kVA at a lagging power factor of 0.6. An 800 hp synchronous motor having an efficiency of 88 per cent is added to supply rated motor load but primarily to improve the over-all power factor to 0.9 lagging. Calculate:
a. the power factor at which the synchronous motor operates
b. the kilovolt ampere rating of the synchronous motor

c. The useful hp output if the synchronous motor in (b) were operated to drive a load at unity power factor (assume same efficiency).

8-25. A frequency changer having two synchronous dynamos coupled together consists of a 10-pole, 50 Hz alternator driven by a 60 Hz motor. Calculate the number of poles required for the motor.

8-26. In using a 60 Hz synchronous motor to produce 400 Hz ac, specify the minimum number of poles required on each synchronous dynamo to accomplish frequency conversion.

8-27. Repeat Problem 8-26 for conversion from 60 Hz to 25 Hz.

8-28. A high speed centrifuge is designed to operate at a constant speed of 10,000 rpm in order to provide the necessary centrifugal force. A synchronous motor is selected for operation but only 60 Hz is available. Specify the frequency changer which will perform the conversion to operate the motor.

ANSWERS

8-1(a) 100 Hz (b) 12 (c) 200 rpm   8-2(a) 16 poles (b) 8° (c) 17.7 V (d) 86.8° (e) 17.7 A (f) 0.999 PF leading (g) 6750 W, 6650 W   8-3(a) 16 poles (b) 20° (c) 46.25 V (d) 79.85° (e) 46.25 A (f) 0.998 PF lagging (g) 17,550, 16,910 W 8-4(a) 16 poles (b) 24° (c) 61.8 V (d) 99.3° (e) 61.8 A (f) 0.996 PF leading (g) 22,700 W, 21,558 W   8-5(a) 16 poles (b) 24° (c) 54 V (d) 48.7° (e) 54 A (f) 0.813 PF (g) 16,750 W, 15,876 W   8-6(a) 8.85 hp, 22.65 hp, 29.0 hp, 21.3 hp (b) 103.3; 257; 338; 248 lb-ft   8-7(a) 6.33 hp, 73.8 lb-ft (b) 17.8 hp, 207.5 lb-ft (c) 25.85 hp, 302 lb-ft (d) 17.95 hp, 209.5 lb-ft   8-8(a) 170.3 V (b) 47.8° (c) 170 A 1.575 (d) 58.2 hp, 678 lb-ft (e) 1.162: 1   8-9. P_d only: 64,150; 109,400; 123,600, 124,000, 123,800, 123,500, 102,200   8-11(a) $E_g = 134.5$ V (b) 26.55° (c) 4.425° (d) 60.25 V, 83.65° (e) 84,800 W, 4500 W (f) 107.5 hp, 940 lb-ft (g) 166.3 hp, 1455 lb-ft   8-13. 179.5, 226.5, 232.5 lb-ft   8-14. 888 lb-ft, 1072 lb-ft, 1940 lb-ft   8-16(a) 23.65 A (b) 16°, 4° (c) 1580 V (d) 35.85° (e) 46.35 A, 196 per cent (f) 210 hp   8-17(a) 1387 V (b) 473 V (c) 19.95°, 4.99° (d) 113.5 hp   8-19(a) 16,667 kVA (b) 26,800 (c) 10,653 kW   8-20(a) 11,415 kW (b) 35,850 kVA   8-21(a) No (b) 0.9967   8-22 1320 kVA   8-23(a) 0.822 PF (b) 6200 kW   8-24(a) 0.37 PF (b) 1832 kVA (c) 2160 hp   8-25. 12 poles   8-26. Alternator: 40 poles, motor: six poles   8-27. 10 poles, 24 poles   8-28. motor: 72 poles, alternator: 10 poles

NINE

polyphase induction (asynchronous) dynamos

9-1. GENERAL

In the preceding chapters, we have considered the dc dynamo which operated with a dc voltage applied to its field winding and (as a result of rotation or commutation) an ac voltage applied to its armature winding. We also considered the ac synchronous dynamo which, like the dc dynamo, had dc voltage applied to its field winding and ac voltage applied directly to its armature winding. Both of these dynamos are considered to be doubly excited machines because two sources of excitation voltage are required for their normal operation. In Sections 8-25 through 8-29, special types of synchronous motors were discussed: those in which the field poles were permanently magnetized required only one source of excitation for their normal operation (no load to full load); but such machines are truly considered to be doubly-excited machines.

At first impression, the induction dynamo might also be thought of as a singly-excited dynamo, because only polyphase alternating current is applied to its stator. It will be shown, however, that an ac voltage of varying frequency is induced in its rotor in much the same way that an ac voltage is induced by transformer action, in the secondary of a transformer (Ch. 13). The induction dynamo, consequently is a doubly-excited dynamo which has an ac voltage impressed on *both* its stator (armature) winding and its rotor winding.* The voltage impressed on the stator armature is an excitation voltage of (normally) constant frequency and (normally) constant potential supplied from a polyphase or single-phase bus, identical to that of the synchronous dynamo. The voltage impressed on the rotor is an *induced* voltage of variable frequency and potential produced as a consequence of rotor speed with respect to synchronous speed.

Of all the types of motors studied thus far (dc and ac polyphase synchronous), the squirrel-cage induction motor (SCIM) is the simplest in construction. It has no commutator, no slip-rings, nor any moving contacts between the rotor and stator. This construction produces many advantages including maintenance-free operation, application to remote locations, and operation in severe environments where dust and other abrasive materials are a factor. For this reason, it is currently the most widely used ac polyphase motor.

While the induction motor is, perhaps, the simplest of all motors to run and operate, the theory of its operation is highly sophisticated.

9-2. CONSTRUCTION

The general construction of the induction dynamo was discussed in Section 2-5 and shown in Fig. 2-4. The stator armature is identical to that of the ac synchronous dynamo and requires no elaboration. The rotor core of an induction motor is a laminated steel cylinder in which the copper or cast aluminum conductors are *cast* or *wound* parallel (or approximately parallel) to the axial shaft in slots or holes in the core. The conductors need not be insulated from the core, because the induced rotor currents follow the path of lowest resistance, i.e., the copper or cast aluminum or copper alloy conductors of the rotor winding.

In the *squirrel-cage* rotor, the rotor conductors are short-circuited at each end by continuous end rings; hence the name, "squirrel-cage." In the larger rotors, the end rings are welded to the conductors rather than cast. Squirrel-cage rotor bars are not *always* parallel to the axial

---

* For this reason, in the design of a "universal" dynamo, the cylindrical rotor usually has a low dc voltage rating and a correspondingly higher ac voltage rating because of its higher impedance to alternating current. When direct current is applied to the wound rotor, it operates as an ac synchronous dynamo. As an induction generator or motor, however, the ac voltage induced is considerably higher. Furthermore, in various types of ac induction motor speed control systems this secondary voltage is applied (to the rotor of a line-excited induction motor) from another dynamo, called secondary "foreign" voltage control. See Kosow, *Control of Electric Machines*, Prentice-Hall, 1973, Secs. 7-5 through 7-9.

length of the rotor but may be skewed at an angle to the rotor axis to prevent cogging, to produce more uniform torque, and to reduce magnetic "hum" during motor operation.

Wound rotors are wound with copper conductors, usually insulated from the iron core, and are connected in wye in three-phase machines or in a star pattern in polyphase induction dynamos. Each phase-winding end is brought out to *slip rings* that are insulated from the rotor shaft. Normally, the rotor winding is not connected to an ac or dc supply, but either may be used [the former in concatenation, the latter in the universal dynamo].* Usually a balanced three-phase or polyphase variable resistor is connected to the slip ring brushes as a means of varying the total rotor resistance per phase. Because of their higher initial cost and increased maintenance costs, wound rotors are only used (1) where higher starting torques are required, (2) where speed control is desired and (3) where foreign voltages are introduced to the rotor.

9-3.
PRODUCTION OF A ROTATING MAGNETIC FIELD BY APPLICATION OF POLYPHASE ALTERNATING CURRENT TO STATOR ARMATURE

A rotating and constant resultant magnetic field rotating at a synchronous speed ($S = 120f/P$) may be produced by any polyphase group of windings displaced in space on an armature if the currents flowing through the windings are also displaced in time, For example, if a two-phase winding is also physically displaced on a stator by 90°, a constant rotating field will be produced (Section 10-5) since the phase currents are also displaced in time. All three-phase induction dynamos, therefore, in order to produce a constant rotating magnetic field, require three identical and individual windings, displaced on the stator by 120°, and carrying current also displaced 120° in phase or time. Figure 9-1a shows the phasor diagram of the currents flowing in a three-phase stator armature of phase sequence *ABCABCA*. The graphical relation and sinusoidal variation of each current for one cycle is shown in Fig. 9-1b. Figure 9-1c shows the space displacement of a typical concentrated three-phase winding, wye connected.

From our study of ac synchronous dynamo windings (Sections 2-14 through 2-17) we are aware that the conductors of each phase belt are distributed uniformly over the stator armature. The diagram of Fig. 9-1c, which employs concentrated (rather than distributed) coils, permits us to predict the resultant flux produced by all the coils of one phase. Each phase group in Fig. 9-1c consists of 12 conductors (or 6 coils) per phase, in which corresponding "finishing" ends of each phase, F_A, F_B, and F_C, are connected to a common point. The corresponding "start" of each phase, S_A, S_B, or S_C, is connected to the three-phase voltage supply. At time t_1, shown in Fig. 9-1b, the current in every coil in each winding (phase belt) is shown. At time, t_1, phase A is at a maximum in one

* See Footnote previous page.

Figure 9-1

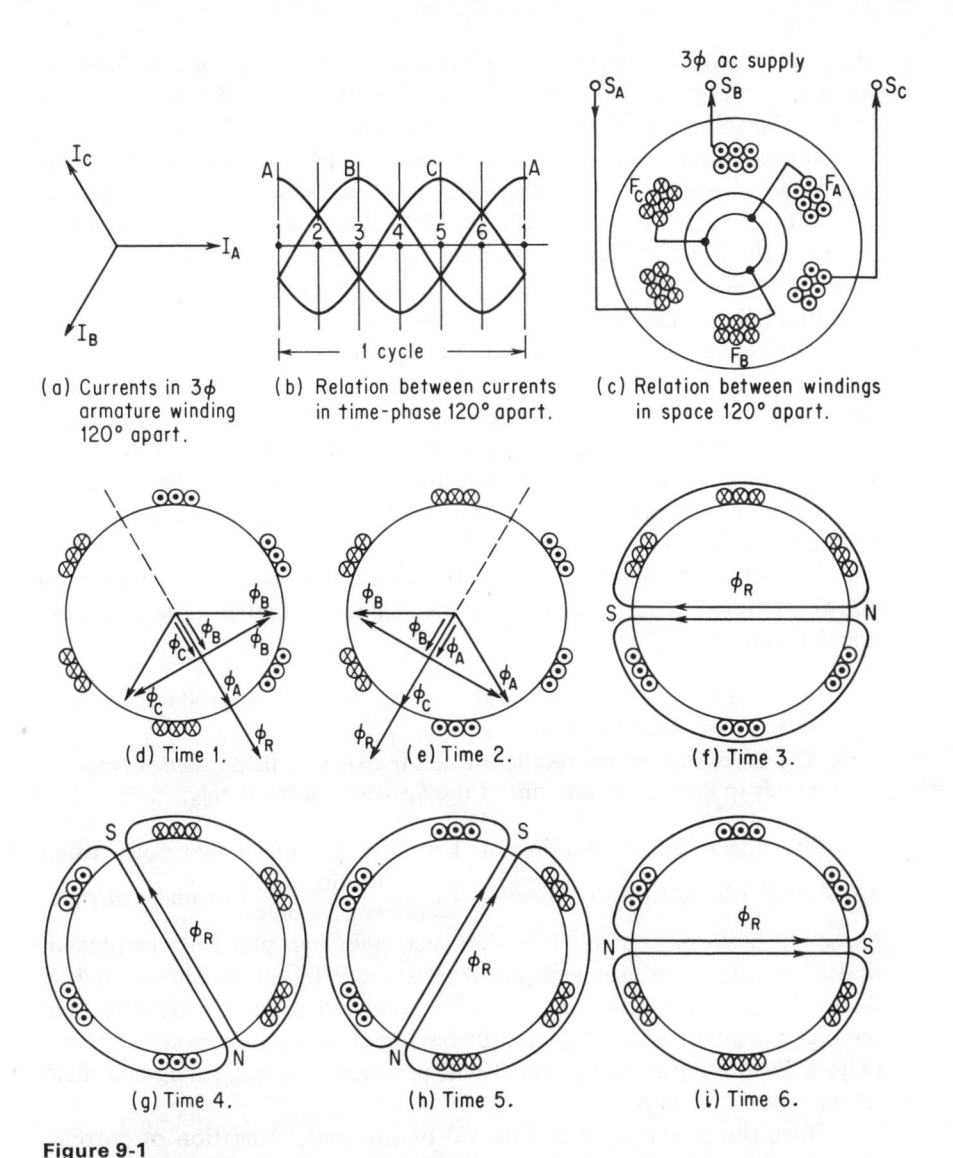

(a) Currents in 3ϕ armature winding 120° apart.
(b) Relation between currents in time-phase 120° apart.
(c) Relation between windings in space 120° apart.
(d) Time 1.
(e) Time 2.
(f) Time 3.
(g) Time 4.
(h) Time 5.
(i) Time 6.

Production of a constant rotating magnetic field at a synchronous speed.

direction, while the current in the coils of phase belts B and C is exactly 0.707 of its maximum value in the opposite direction. Note that, in both Figs. 9-1c and d, for time t_1, the current directions in B and C are opposite to that of A.

The fluxes produced by these phase belts are shown in Fig. 9-1d, using the right-hand rule. Note that ϕ_B and ϕ_C on ϕ_A produce a resultant flux, ϕ_R. The quadrature components of these projections cancel each other, since

SEC. 9-3. / *Production of a Rotating Magnetic Field*

they are equal and opposite in phase. The same procedure is followed at time t_2, where the current in phase B has reversed and is now equal to and in the same direction as A (which has now decreased to 0.707 of its maximum value). C is now at a maximum at time t_2. The resultant flux is plotted once again for each phase, and this time ϕ_A and ϕ_B produce quadrature components (which cancel each other) and components in phase with ϕ_C to produce a resultant ϕ_R of the *same* magnitude as at time t_1. Thus, at time t_2, 60 electrical degrees later than time t_1, the resultant flux has rotated 60°, but it is of *constant* magnitude.

Examining the axis of ϕ_R at times t_1 and t_2, it can be seen that the stator may be treated as a solenoid in which all of the conductors on one side of ϕ_R carry current *into* the stator, and all the conductors on the other side have current coming *out of* the stator. Thus, at time t_3, the resultant flux will be in a horizontal direction from right to left, entering the stator at a S pole on the left side and leaving the stator at a N pole on the left. The reader should check the directions of current and the time intervals shown in Fig. 9-1b against the specific time intervals shown at times t_3, t_4, t_5, and t_6, in the remaining figures. The following will be noted:

1. A *single constant rotating resultant* magnetic field is produced by a three-phase stator winding.
2. The space shift of the resultant rotating magnetic field *exactly* corresponds to the time-phase shift of the *frequency* of the supply.

The concentrated winding of Fig. 9-1c produced two poles using six slots for a three-phase winding $\left(\dfrac{6 \text{ slots}}{3 \text{ phase} \times 2 \text{ pole}}\right)$ or *one* slot/pole-phase. If a stator having *two* slots per pole per phase is employed, and if coils from the same phase are displaced by 90° as shown in Fig. 9-2a, the resultant coil currents will produce a rotating magnetic field having *four* poles. Similarly, a stator having *three* slots per pole per phase (Fig. 9-2b in diagrammatic form) will produce a rotating magnetic field of *six* poles, and so on.*

Since the period or time interval of sinusoidal variation of current, shown in Fig. 9-1b, is the same in the conductors, the speed of the rotating magnetic field varies directly with the frequency but inversely as the number of poles, verifying Eq. (2-16) once again ($S = 120f/P$). Thus, the speed of the rotating field in Fig. 9-1 at a frequency of 60 Hz is 3600 rpm. Similarly, the speed of the rotating field (with reference to a unit N pole) in Fig. 9-2a is 1800 rpm, and in Fig. 9-2b is 1200 rpm, and so on.

We may consider, then, that the nature of the stator winding, in terms of the frequency and the number of poles, will produce a rotating

---

* It can be shown that the number of poles, P, produced in the rotating magnetic field is $P = 2n$ when n is the number of slots/pole-phase (Sec. 2-16).

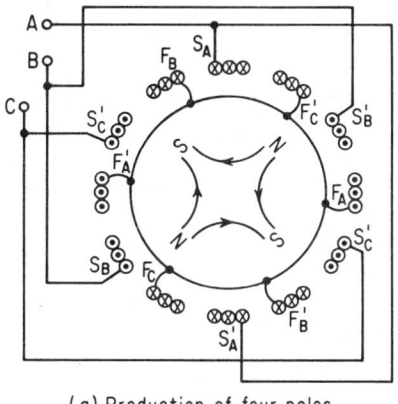

(a) Production of four poles
2 slots/pole phase.

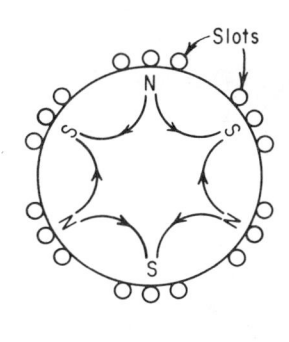

(b) Production of six poles
3 slots/pole phase.

Figure 9-2

Four- and six-pole rotating fields.

magnetic field of constant magnitude whose rotating or synchronous speed is expressed by Eq. (2-16), $S = 120f/P = 120f/2n$. Since the number of poles depends *only* on n (the winding employed), the synchronous speed of the rotating magnetic field of any *given* asynchronous dynamo is actually a function of *frequency*. (See Ex. 9-1.)

The rotating magnetic field produced and shown in Fig. 9-1 yielded a clockwise rotation for the phase sequence *ABCABC*, given in Fig. 9-1b. If *any two* of the line leads to the stator coils in Fig. 9-1c are interchanged, the reversed phase sequence will produce reversal of the direction of rotation of the magnetic field. As in the ac synchronous motor, it will be shown that the rotor rotates in the *same* direction as the rotating magnetic field, but now based on the *induction motor principle*. Consequently, the direction of rotation of *any* induction motor may be reversed (by simply reversing the phase sequence) by interchanging *any two* of the three line connections to the three-phase bus supply.

9-4. INDUCTION MOTOR PRINCIPLE

The principle of the induction motor may be illustrated simply using the apparatus shown in Fig. 9-3a. A permanent magnet is suspended from a string above a copper or aluminum turntable which pivots on a bearing set in a fixed iron plate. The field of the permanent magnet is thus completed through the iron plate. The pivot should be relatively frictionless and the permanent magnet must be of sufficient flux density. As the magnet rotates on the string, the disc below will be observed to rotate *with it*, regardless of the direction of rotation of the magnet. The disc follows the movement of the magnet, as shown in Fig. 9-3b, because of the induced eddy currents produced

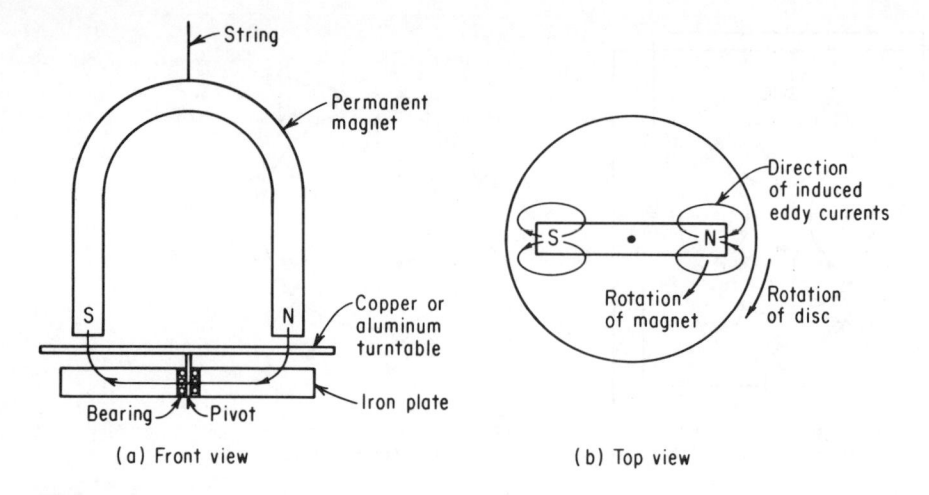

Figure 9-3

Induction motor principle.

by relative motion between a conductor (the disc) and a magnetic field. By Lenz's law (Section 1-6), the direction of the induced voltage (and the consequent eddy currents) produces a field tending to oppose the force or motion which produced the induced voltage.

In effect, as shown in Fig. 9-3b, the induced eddy currents tend to produce a unit S pole in the disc at a point under the rotating N pole of the magnet, and a unit N pole in the disc under the rotating S pole of the magnet. As long as the magnet continues to move, therefore, it will continue to produce eddy currents and poles of *opposite* polarity in the disc below it. The disc, therefore, rotates in the *same direction* as the magnet, but it must rotate at a speed *less than* that of the magnet. If the disc were to rotate at the same speed as the magnet, then there would be *no relative motion* between the conductor and the magnetic field, and no eddy currents would be produced in the disc.

It is because of the generator action that occurs, producing currents and a resultant opposing magnetic field, that the induction motor may be classed as a doubly-excited machine. Furthermore, as in all dynamos, while electromagnetic torque is the result of interaction between the magnetic fields produced by the two excitation currents, *generator action* is simultaneously occurring. In the ac *synchronous* motor, the motor action and the generator action occurred at the *synchronous* speed of the rotating magnetic field. In the ac induction motor, *neither* the motor action nor the generator action can possibly occur at the synchronous speed. For this reason, *induction principle* machines are classed as *asynchronous* or *nonsynchronous* induction dynamos.

As stated previously, the disc speed can never equal the speed of the magnet. If it did, the induced current would be zero, and no magnetic

CHAP. NINE / *Polyphase Induction Dynamos*

flux or torque would be produced. Thus, it must "slip" back in speed in order to produce torque. This results in a difference of speeds produced between (1) the synchronous speed of the rotating magnetic field, basically a function of frequency for any given induction dynamo; and (2) the "slip" speed at which the disc rotates as a result of torque produced by interaction between its field and the rotating magnetic field. This difference in speed between (1) and (2) is called the *slip speed* (or *revolutions slip*) and is usually expressed as a percentage of the synchronous speed (as per cent slip or just slip).

$$s = \frac{\text{revolutions slip}}{\text{synchronous speed}} = \frac{\text{synchronous speed} - \text{rotor speed}}{\text{synchronous speed}}$$

$$= \frac{(S - S_r) \times 100}{S} \tag{9-1}$$

or

$$S_r = S(1 - s) = 120\left(\frac{f}{P}\right)(1 - s) \tag{9-1a}$$

where *s* is the per cent slip (for purposes of computation, per cent is always converted to a decimal)
S is the synchronous speed ($120f/P$) in rpm of the rotating magnetic field produced by the stator
S_r is the speed of the rotor, in rpm

EXAMPLE 9-1: A three-phase induction motor stator has 3 slots per pole per phase. If the line frequency is 60 Hz, calculate
a. number of poles produced and total number of slots on the stator
b. speed, in rpm, of the rotating magnetic fields or poles
c. if the frequency is changed to 50 Hz, speed of the rotating magnetic field.

Solution:

a. $P = 2n = 2 \times 3 = $ **6 poles**

$$\text{total slots} = \frac{3 \text{ slots}}{\text{pole-phase}} \times 6 \text{ poles} \times 3 \text{ phases} = \textbf{54 slots}$$

b. $S = \dfrac{120f}{P} = \dfrac{120 \times 60}{6} = $ **1200 rpm** \hfill (2-16)

c. $S = \dfrac{120f}{P} = \dfrac{120 \times 50}{6} = $ **1000 rpm**

9-5. ROTOR CONDUCTORS, INDUCED EMF, AND TORQUE; ROTOR STALLED

As stated in Section 9-2, a commercial squirrel-cage induction-motor (SCIM) rotor is composed of copper or aluminum conductors or bars embedded in a laminated iron core. Figure 9-4a shows three of these rotor conductors (*A*, *B*, and *C*) and their relation to a multipolar magnetic field produced by a polyphase winding.

Assume that the rotor is stalled (at a standstill) and that

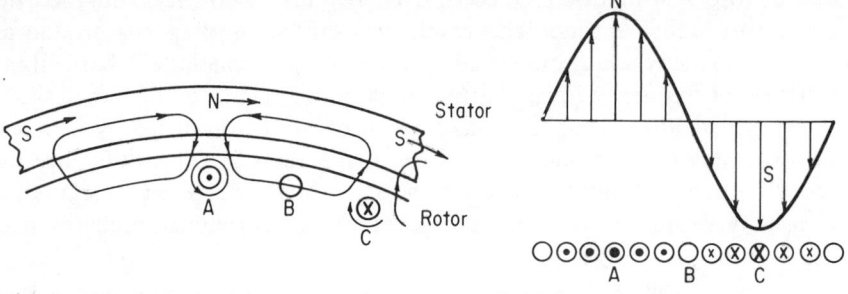

(a) Induced emfs produced in rotor conductors.

(b) Relation between field and rotor emfs.

Figure 9-4

Development of torque in rotor conductors.

the stator flux is rotating at synchronous speed in a clockwise direction with respect to (stationary) rotor conductor A, directly under a unit N pole. The direction of relative motion of conductor A is to the left, for purposes of determination of induced emf (right-hand rule, Section 1-5). This relative motion produces an emf toward the observer, and the emf current produces a flux which is counterclockwise around A, as shown. With respect to the field entering the iron core in the vicinity of conductor A, the force acting on conductor A as a result of the interaction between the magnetic fields produces repulsion on the left and attraction on the right-hand side of the conductor, or motion in the *same* direction as the magnetic field. By the hand rule (motor action requires the left-hand rule for the motion of the conductor) the conductor will be seen to develop electromagnetic torque tending to move the rotor also in the *same* direction as the magnetic field rotation.

Similarly, conductor C, directly under a unit S pole on the rotor in Fig. 9-4a, will produce an emf away from the observer, and a clockwise magnetic field resulting from an induced current producing motion in the *same* (clockwise) direction as the rotating magnetic field.

Conductor B, however, at the instant shown in Fig. 9-4a, is undergoing no change in flux linkage, and no emf is induced in it. All of the conductors lying between A and B on the rotor, therefore, will experience varying magnitudes of induced emf and rotor current (since they are short-circuited at each end), depending on whether they are directly under the unit pole or in the interpolar region, as shown in Fig. 9-4b. Thus, the *emf distribution* in the rotor conductors swept by the rotating magnetic field *resembles*, at any instant, the *flux distribution* as shown in Fig. 9-4b. Moreover, since the field is rotating at a synchronous speed, the direction of induced emf in any given conductor will vary sinusoidally in accordance with the magnitude of flux linking it, as shown in the figure. In effect, the *same* number of poles of *opposite* instantaneous polarity are produced in the rotor as there are on the stator. (See Ex. 9-1.)

At standstill, the frequency of the rotor induced emf is equal to the frequency of the rotating magnetic field. On the other hand, if the rotor were capable of rotating at the same speed as the rotating magnetic field, i.e., if conductor A moves at exactly the same speed as the unit N pole above, there would be no voltage induced in the rotor conductors, and the frequency of rotor alternation would be zero. The frequency of the voltages induced in the rotor, therefore, varies inversely with the rotor speed from a maximum (line frequency) at standstill to zero frequency at synchronous speed. But Eq. (9-1) shows that the slip *also* varies from a maximum at standstill to zero at synchronous speed. The rotor frequency may be expressed as a function of stator frequency and slip:

$$f_r = s \times f \qquad (9\text{-}2)$$

where f_r is the frequency of sinusoidal voltage and current induced in the rotor bars at any given slip, s, in hertz

s is the slip defined in Eq. (9-1), as the ratio of slip speed to synchronous speed or $(S - S_r)\,S$

f is the frequency of the stator (or line frequency) and of the rotating magnetic field, in hertz

EXAMPLE 9-2: A four-pole induction motor operating at a frequency of 60 Hz has a full-load rotor slip of 5 per cent. Calculate the rotor frequency
 a. At the instant of starting
 b. At full load.

Solution:

a. At the instant of starting, $s = (S - S_r)/S$, where S_r is the rotor speed. Since the rotor speed at that instant is zero, $s = [(S - 0)/S] = 1$ or unity slip The rotor frequency is

$$f_r = s \times f = 1.0 \times 60 \text{ Hz} = 60 \text{ Hz} \qquad (9\text{-}2)$$

b. At full load, the slip is 5 per cent (given above) and therefore

$$s = 0.05$$

$$f_r = s \times f = 0.05 \times 60 \text{ Hz} = 3 \text{ Hz} \qquad (9\text{-}2)$$

It should be noted that it is for precisely the reason illustrated in Ex. 9-2 that the wound-rotor induction motor may be used as a variable frequency changer when its rotor is driven at any speed and when the rotor-induced emf is picked off its slip-rings. (See Sec. 9-23.) When an induction motor is driven by a prime mover (and is operated in this manner), it is called an *induction generator*. Consequently, if it is standing still (unity slip), an induction generator will generate a (rotor) frequency, as shown in Ex. 9-2, of (1 × 60) or 60 Hz. If it is rotated at exactly synchronous speed in the same direction as the rotating magnetic field (zero slip), its generated (rotor) frequency is (0 × 60) or zero. If it is rotated at the same (synchronous speed) in the *opposite* direction (slip = 2), its generated frequency is (2 × 60) or 120 Hz. If it is rotated at twice synchronous speed in the opposite direction (slip = 3), its generated frequency is 180 Hz. Slips of greater than unity, as well as *negative* slips (rotation *exceeding synchronous* speed in the *same* direction), are possible, therefore, in an induction generator.* For the most part, in studying the characteristics of an induction *motor*, we shall be dealing with positive slips between unity (standstill) and zero (synchronous speed).

Since the rotor conductors have relatively low resistance (heavy bars of short length, short-circuited at each end) but are embedded in iron, they possess the property of inductance and, consequently, inductive reactance (Section 2-7). For a given induction dynamo rotor, the inductance (L_r) of the rotor bars is a fixed quantity (varying with the number of turns, the permeability, and the length and area of the magnetic circuit), but the rotor inductive reactance (X_r) will vary with the frequency of the rotor. The determination of rotor inductance directly and independent of frequency is a difficult matter, particularly for larger machines. It is customary, therefore, to determine the rotor reactance at standstill by means of a "locked rotor" test (used also in the determination of efficiency; see Section 12-13), and then use this reactance as a reference standard. Since rotor frequency increases with slip [Eq. (9-2)], and reactance varies with frequency, $(X_r = 2\pi f L_r)$, the rotor reactance at any rotor frequency is

$$X_r = sX_{lr} \qquad (9\text{-}3)$$

where s is the slip expressed as a decimal quantity, and X_{lr} is the locked rotor reactance at *standstill*.

It should be noted that the locked-rotor reactance should never be assumed as the maximum possible reactance because, as previously shown, the induction generator may develop rotor reactances exceeding the locked-rotor reactance at slips greater than unity. The locked-rotor reactance is merely a convenient standard or reference which simplifies computations.

If the frequency of the induced ac voltage in the rotor bars of an induction motor varies between zero at synchronous speed and the stator frequency at standstill, then by Eq. (2-14), $(E = k\phi f)$, the induced voltage in the rotor at any slip is also a function of the induced voltage

---

* The important point regarding the definition of slip is in terms of the direction of the magnetic field taken as a reference. If the rotating field has a synchronous speed of 1800 rpm clockwise, then a rotor speed of 1800 rpm counterclockwise produces a slip speed of $1800 - (-1800) = 3600$ rpm, and a slip of $\frac{3600}{1800} = 2$. Negative slips are also possible if the rotor is driven at a speed exceeding synchronous speed in the same direction, say 2000 rpm clockwise, producing a slip speed of $1800 - 2000$, or -200 rpm, and a slip of $-\frac{200}{1800}$, or -0.111.

at standstill with the rotor locked; or

$$E_r = sE_{lr} \tag{9-4}$$

where s is the slip expressed as a decimal quantity
E_{lr} is the voltage induced in the rotor at standstill with locked rotor
E_r is the induced emf in the rotor at any value of slip (positive, negative, greater or less than unity) and/or rotor frequency

Thus, rotor-induced voltage, rotor reactance, and rotor frequency, *all* vary as a function of *slip* from a "normal" maximum at standstill to zero when the rotor speed is equal to the synchronous speed (zero slip).

The torque developed at standstill by each individual conductor in the rotor* may be expressed in terms of the flux or current (producing the flux) in the stator and rotor, respectively, [from Eq. (4-4)] as

$$T = K_t \phi I_r \cos \theta_r \tag{9-5}$$

where K_t is a torque constant for the number of poles, the winding, the units employed, etc.
ϕ is the flux produced by each unit pole of the rotating magnetic field linking the rotor conductor
$I_r \cos \theta_r$ is the component of rotor current in phase with ϕ

The necessity for the term $I_r \cos \theta_r$ in Eq. (9-5), quite naturally emerges from the fact that while the induced voltages in the rotor conductors shown in Fig. 9-4b are in phase with the stator rotating magnetic field, the currents I_r in the rotor conductors are not in phase as shown in Fig. 9-5a. The rotor conductors have appreciable inductive rotor reactance due to slip [Eq. (9-3)], thus causing rotor current I_r to lag rotor voltage E_r by θ_r, as shown in Fig. 9-5a. Therefore, only that component of current in phase with the rotor flux will produce useful average torque. This is shown in Fig. 9-5b, where the product of ϕ and I_r is represented graphically.

It is now possible to derive an equation for the torque developed at standstill or under locked-rotor conditions (which is the starting torque) for the induction motor. Let R_r be the effective rotor resistance (at standstill) of all rotor conductors combined, and let X_{lr} be the locked-rotor

---

* As a unifying principle, the torque developed in each conductor of any doubly-excited dynamo is proportional to $\phi_1 \phi_2 \cos \alpha$, where ϕ_1 and ϕ_2 represent the resultant fluxes produced by the two excitation voltages, and α the angle between the fluxes. Thus, torque is produced by mutual repulsion or attraction of two magnetic fields. This principle, therefore, is equally true for electrodynamometer instruments and for dynamic speakers. Thus, Eqs. (4-4), (8-26), and (9-5) are all forms of Eq. (1-8) based on the electromagnetic force developed by a current-carrying conductor in a magnetic field.

Figure 9-5

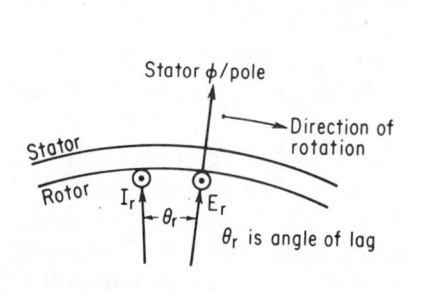

(a) Rotor current lags rotor voltage and flux.

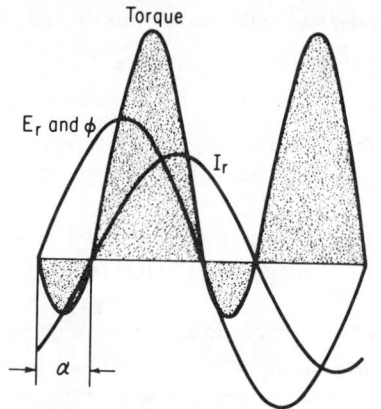

(b) Torque produced due to lag in rotor current.

Displacement of rotor current with respect to rotor voltage and consequent production of rotor torque.

reactance (at standstill) of all rotor conductors combined; then the rotor impedance at standstill, Z_{lr}, is

$$Z_{lr} = R_r + jX_{lr} = \sqrt{R_r^2 + X_{lr}^2}$$

and

$$\cos \theta_r = \frac{R_r}{Z_{lr}}$$

The current in the rotor at standstill is

$$I_{lr} = \frac{E_{lr}}{Z_{lr}} = \frac{E_{lr}}{R_r + jX_{lr}} = \frac{E_{lr}}{\sqrt{R_r^2 + X_{lr}^2}} \qquad (9\text{-}6)$$

where E_{lr} is the effective value of induced rotor voltage at standstill, and all other terms have been defined above.

Substituting into Eq. (9-5) I_{lr} from Eq. (9-6) and also the value of $\cos \theta_r = R_r/Z_{lr}$, the total starting torque developed by the induction motor rotor **at standstill** is

$$T_s = K_t \phi I_r \cos \theta_r = \frac{K_t \phi E_{lr}}{\sqrt{R_r^2 + X_{lr}^2}} \times \frac{R_r}{\sqrt{R_r^2 + X_{lr}^2}} = \frac{K_t \phi E_{lr} R_r}{R_r^2 + X_{lr}^2} \qquad (9\text{-}7)$$

where all the terms have been defined previously.

Note that the imaginary terms, as well as the square root, have been removed from the denominator of Eq. (9-7). Note also that, because the rotor is locked, and voltages are induced in it by transformer action, E_{lr} is proportional to ϕ, which in turn is proportional to the applied bus line or phase voltage on the stator winding V_p. Since ϕ is proportional to V_p, and since E_{lr} (by transformer action) is proportional to V_p, Eq. (9-7) may be simplified further to

$$T_s = \frac{K_t V_p^2 R_r}{R_r^2 + X_{lr}^2} \tag{9-7a}$$

But for a given squirrel cage induction motor (SCIM) since the effective rotor resistance R_r and the standstill rotor reactance X_{lr} are *constant* (for a given applied bus voltage at a constant frequency), they can be incorporated into a new torque constant, K'_t, and Eq. (9-7a) is finally simplified for starting torque (**at standstill**) to

$$T_s = K'_t V_p^2 \tag{9-8}$$

Equation (9-8) states that, for any given ac SCIM (particularly one of the squirrel-cage type which does not permit a variation in rotor resistance by some external means), the *starting torque* is solely a function of the *voltage applied to the stator winding*.* Reducing the applied rated phase voltage by one-half at starting, therefore, will produce one-quarter the starting torque produced with full voltage. Reducing the primary voltage also reduces the secondary and primary *current*, since the primary current reflects the current drawn by the secondary rotor resistance and reactance. The basic theory behind reduced-voltage starting of ac polyphase induction motors (Sec. 9-15) is to reduce primary (or stator) current.

EXAMPLE 9-3: A four-pole, 50 hp, 208 V, three-phase induction motor has a starting torque of 225 lb-ft and an instantaneous starting line current of 700 A (blocked rotor) at rated voltage. A reduced three-phase voltage of 120 V ac is applied to the line terminals. Calculate:
a. The starting torque
b. The starting current.

Solution:

a. $T_s = T_{orig}\left(\dfrac{V_s}{V_o}\right)^2 = 225 \text{ lb-ft} \left(\dfrac{120}{208}\right)^2 = 75 \text{ lb-ft}$ \hfill (9-8)

b. $I_s = I_{orig}\left(\dfrac{V_s}{V_o}\right) = 700 \text{ A} \left(\dfrac{120}{208}\right) = 403 \text{ A}$ \hfill (9-6)

---

* This equation emerges most naturally from the concept of the induction motor as a doubly-excited dynamo in which both windings are excited by alternating current. The excitation of the rotor depends on the magnitude of stator excitation by transformer action. The excitation of the stator rotating field also depends on the applied voltage; hence, both rotor and stator fluxes are a function of the applied voltage, as is the developed torque.

9-6. MAXIMUM TORQUE

We are aware that the standstill starting torque T_s of a SCIM may be as high as, or even higher than, its full-load value. We must also determine the maximum (or breakdown or pull-out) torque in terms of the running full-load torque.

Once the SCIM develops starting torque and rotates, its frequency, rotor reactance, and rotor-induced voltage are represented by sf, sX_{lr}, and sE_{lr}, respectively. As the rotor frequency and reactance decreases, the rotor-induced voltage also decreases as the motor speed rises. A decrease in rotor reactance increases the value of $\cos \theta_r$ in Eq. (9-5), but at the same time the decrease in rotor voltage tends to reduce the rotor current. For a given constant excitation, therefore, there must be a particular value of slip where the increase in $\cos \theta_r$ and the decrease in rotor current, I_r, produce a maximum value of torque in Eq. (9-5) ($T = K\phi I_r \cos \theta_r$). For any given slip, the rotor current is

$$I_r = \frac{sE_{lr}}{\sqrt{R_r^2 + (sX_{lr})^2}} \tag{9-9}$$

and, since

$$\cos \theta_r = \frac{R_r}{\sqrt{R_r^2 + (sX_{lr})^2}} \tag{9-10}$$

at any value of slip,* the running torque $T = K\phi I_r \cos \theta_r$ may be converted, using the same substitution techniques as in Eq. (9-7), to

$$T \text{ [for any slip]} = \frac{K_t \phi s E_{lr} R_r}{R_r^2 + (sX_{lr})^2}$$

But since the standstill, locked-rotor voltage, E_{lr}, is directly proportional to ϕ, the torque may be expressed by

$$T \text{ [for any slip]} = \frac{K_t \phi^2 s R_r}{R_r^2 + (sX_{lr})^2} \tag{9-11}$$

The *maximum* torque is obtained when Eq. (9-11) is differentiated with respect to the rotor resistance and set equal to zero (that is, when $dT/dR_r = 0$, which yields†

$$R_r = s_b X_{lr} \tag{9-12}$$

---

* The significance of θ_r should not be overlooked, even though it disappears in Eq. (9-11). Not only is it the angle by which the rotor current lags the induced motor voltage, but it is also the angle between the center of the stator pole and that rotor conductor which is carrying maximum instantaneous current, as shown in Fig. 9-5.

† The breakdown slip as a ratio R_r/X_{lr} is an approximation which is accurate for all practical purposes. A more precise value for s_b is $R_r/\sqrt{(R_p^2 + X_{lr}^2)}$, where R_p is the primary stator resistance per phase (usually neglected), and T_{max} is equal to $KV_p^2/[2(R_p + \sqrt{R_p^2 + X_{lr}^2})]$. See also Eq. (9-21) for another means of determining the maximum breakdown torque, as shown in Ex. 9-10.

CHAP. NINE / *Polyphase Induction Dynamos*

In other words, the *maximum* torque is obtained when the breakdown slip, s_b, is equal to R_r/X_{lr}. But, as noted above, ϕ^2 is proportional to V_p^2; and therefore the expression for the maximum torque that can be developed by any induction motor, substituting Eq. (9-12) in Eq. (9-11), is

$$T_{max} = \frac{KV_p^2}{2(s_b X_{lr})^2} \tag{9-13}$$

EXAMPLE 9-4: An eight-pole, 60 Hz SCIM is deliberately loaded to a point where pull-out or stalling will occur. The rotor resistance per phase is 0.3 ohm, and the motor stalls at 650 rpm. Calculate:
a. The breakdown slip, s_b
b. The locked-rotor reactance (the standstill reactance)
c. The rotor frequency at the maximum torque point.

Solution:

a. $S = \dfrac{120f}{P} = \dfrac{120 \times 60}{8} = 900$ rpm (2-16)

$s_b = \dfrac{S - S_r}{S} = \dfrac{900 - 650}{900} = \mathbf{0.278}$ (9-1)

b. $X_{lr} = \dfrac{R_r}{s_b} = \dfrac{0.3}{0.278} = \mathbf{1.08\ \Omega}$ (9-12)

c. $f_r = sf = 0.278 \times 60 = \mathbf{16.7\ Hz}$ (9-2)

9-7. OPERATING CHARACTERISTICS OF AN INDUCTION MOTOR

Assuming a SCIM is started with rated voltage across its stator line terminals (across-the-line starting), it will develop a starting torque in accordance with Eq. (9-7), which will cause its speed to rise. As its speed rises from standstill (100 per cent slip), its slip will decrease and its torque will increase up to that value of slip where the maximum torque is developed ($R_r = sX_{lr}$) in accordance with Eq. (9-12). This causes the speed to rise still further, reducing the slip and the developed induction motor torque simultaneously. The developed torque at starting, and at that value of slip producing maximum torque, both exceed (for the usual case) the applied torque of the load. The motor speed will rise, therefore, until the value of the slip is so small that the developed torque is reduced to a value *equal* to the applied torque. The motor will continue to run at this speed and *equilibrium* value of slip until the applied torque is either decreased or increased, in accordance with Eq. (9-11).

Figure 9-6 shows the relation between the starting, maximum, and rated full-load torques developed by the induction motor as a function of motor speed and slip. This figure is a graphical presentation of the developed motor rotor current and torque as a function of slip from the instant of starting (point *a*) to the steady-state running condition (usually between no load and full load—points *c* and *d*) where the developed and applied torques are equal. Note that at zero slip, the developed torque

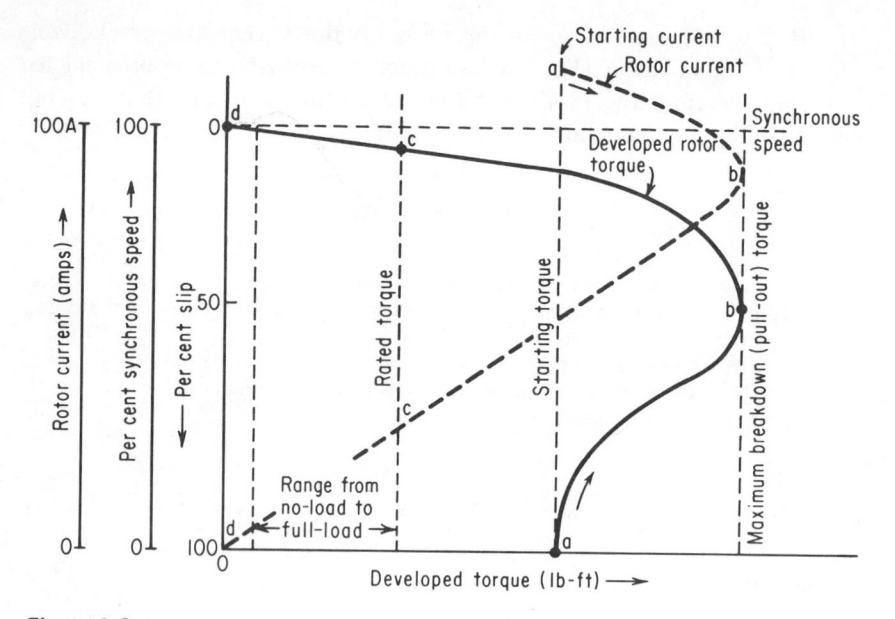

Figure 9-6

Effects of load on speed, developed torque, and rotor current.

and the rotor current (shown as a dotted line) are both zero because no induction motor action occurs at synchronous speed. Even at no load, it is necessary for the induction motor to have a small slip (usually a fraction of 1 per cent) in order to develop a small torque to overcome friction, windage, and other internal losses. The advantage of the presentation shown in Fig. 9-6 is that it is possible to visualize the acceleration of a given load from starting (point *a*) and the effects of a change in load on speed regulation, torque and rotor current.

The induction motor, as shown in Fig. 9-6, is a motor of essentially fairly constant speed from no load to full load (points *d* to *c* in the figure), having a speed characteristic resembling the shunt motor (Fig. 4-9).*
Let us consider the manner in which torque is developed from no-load to full-load, i.e., the normal running characteristics of the induction motor.

* An interesting parallel can be drawn between the shunt and the induction motors in that the decrease in speed of both these motors with load is just sufficient to yield an increase in current to produce the developed torque necessary to counterbalance the applied torque. In the shunt motor, this is accomplished by a decrease in counter emf. In the induction motor, the increase in slip (and also in rotor frequency, induced-rotor emf) produces the increase in rotor current and power factor necessary to counterbalance the applied torque.

316

CHAP. NINE / *Polyphase Induction Dynamos*

9-8. RUNNING CHARACTERISTICS OF AN INDUCTION MOTOR

The *normal running* characteristics of a squirrel-cage induction motor (SCIM) occur in the range from *no* load to *full* load (points *d* to *c* in Fig. 9-6). Let us consider the behavior of an induction motor rotor at a no-load speed slightly below the synchronous speed with increasing applications of load.

No-load condition: At no load, the slip is very small (a fraction of 1 per cent), and the rotor frequency, rotor reactance, and rotor-induced emf [Eqs. (9-2), (9-3), and (9-4)] are all very small. The rotor current is, therefore, small and only sufficient to produce the necessary no-load torque. Since the rotor current is small, the stator (primary) current is the phasor sum of its exciting current, I_e, and a primary load component, I_o, induced in the rotor by transformer action. Figure 9-7a shows the phasor sum of these currents at no load, where the open-circuited stator primary exciting current is I_e, i.e., the phasor sum of a combined hysteresis or power component, I_h, and a magnetizing component, I_m, required to produce the rotating stator flux. The power components, I_h and I_o, are both in phase with E_{gp}. The power factor at no load is represented by θ, the angle between I_{nl} and E_{gp}. Thus, $I_{nl} \cos \theta$ is the sum of I_o and I_h, i.e., the small stator current I_o produced by the rotor current and a primary loss component I_h due to hysteresis and eddy current in the stator and rotor iron. Note that because θ is large, the power factor is extremely small and lagging.

Half-load condition: As mechanical load is applied to the rotor, the speed decreases slightly. The small decrease in speed causes an increase in slip and in rotor frequency, rotor reactance, and rotor-induced emf [Eqs. (9-2), (9-3), and (9-4)]. The induced (secondary) rotor current increase is reflected as a primary stator current increase, I_{sr}, shown in

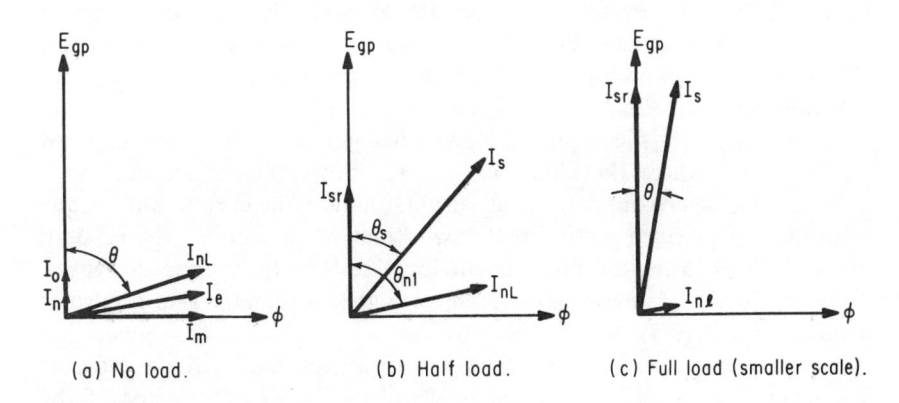

(a) No load.   (b) Half load.   (c) Full load (smaller scale).

Figure 9-7
Primary stator exciting and load current components, showing effect of increase in loads of power factor and stator current.

Fig. 9-7b. This primary stator current component I_{sr} is a power-producing component, like I_o, and is in phase with the primary induced voltage E_{gp} [Eq. (8-24)]. The phasor sum of the no-load current I_{nl} and the load component I_{sr} produces a stator current I_s at an improved power factor angle θ_s. Thus, the stator current has *increased* from I_{nl} to I_s, and the power factor angle has decreased from θ_{nl} to θ_s, both factors tending to produce more mechanical power in the armature and draw more power from the bus $(E_{gp}I_s \cos \theta_s)$.

Full-load condition: The SCIM will rotate at a value of slip which provides an *equilibrium* between the developed torque and the applied torque. As more load is applied, therefore, the slip increases because the applied torque exceeds the developed torque. When rated torque is applied to the induction motor shaft, the component of in-phase primary stator current drawn by the induction motor is large in comparison to the almost quadrature no-load current, as shown in Fig. 9-7c, and the power factor angle θ is fairly small. The full-load power factor varies from 0.8 (in small induction motors of approximately 1 hp) to about 0.9 or 0.95 (in large induction motors from 150 hp up).

Beyond full load: It might appear from Fig. 9-7 that further increases in load would produce improvement in the power factor, up to unity, and increased in-phase stator currents; but this is *not* the case. With increased load and slip, the rotor frequency continues to increase, and the increase in rotor reactance produces a *decrease* in the rotor power factor. Treating the induction motor as a transformer (Ch. 13), we might say that the transformer secondary has a lagging load which causes the primary power factor to lag as well. With loads beyond full load, therefore, the power factor approaches a *maximum* and then decreases rapidly. In order to produce the necessary increase in torque to meet the applied torque, the rotor and stator currents must increase to offset the decrease in power factor [Eq. (9-5)]. Figure 9-8a shows a large increase in overload current, I_{ol}, at a poorer power factor, θ_{ol}, than at the full-load stator current I_s shown in Fig. 9-7c.

The characteristics from no load to beyond full load are summarized in Fig. 9-8b, where the efficiency and the power factor are both at a maximum at approximately rated output, and the line current and torque continue to increase until breakdown (maximum torque). Note that, beyond breakdown, the line current increases but the torque decreases, because the *rate* of power factor decrease is greater than the rate of current *increase* in Eq. (9-5).

The subject of dynamo efficiency, generally, and induction motor efficiency, specifically, is covered in detail in Ch. 12. The shape of the efficiency curve shown in Fig. 9-8b may be explained briefly here as follows. At *light* loads the relatively large *fixed* losses in proportion to the small output produces a *low* efficiency. At *heavy* loads, the relatively large variable losses plus the fixed losses, again produce a *low* efficiency

(a) Phasor diagram.

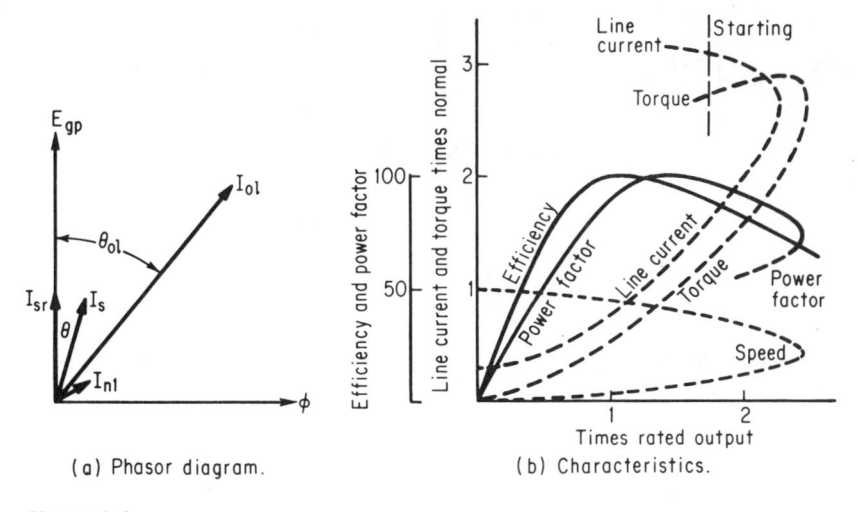

(b) Characteristics.

Figure 9-8

Effects of heavy loads on primary stator current and power factor.

despite the high output. Maximum efficiency occurs at moderate loads where the fixed and variable losses are equal (Sec. 12-6) and the output is approximately near rated values.

Note that maximum torque in Fig. 9-8b occurs at well beyond twice rated output, where the breakdown slip is that rotor frequency value at which the variable rotor reactance equals the rotor resistance. Since the effective rotor resistance of a squirrel-cage induction motor (SCIM) is practically constant, the maximum or breakdown torque depends on the rotor resistance (Eq. 9-12) in the final analysis.

**9-9.
EFFECT OF
CHANGE IN ROTOR
RESISTANCE**

The last statement of the preceding paragraph might imply a possibility of increasing the maximum or pull-out (breakdown) torque by increasing the rotor resistance. But Eqs. (9-11) and (9-13) indicate quite clearly that, if the rotor resistance and reactances are increased, the breakdown torque is decreased, since maximum torque varies inversely with these factors. What is the effect on the torque-slip characteristics of a change in rotor resistance?

In a squirrel-cage induction motor (SCIM), there is no way to introduce a change in rotor resistance once the rotor has been cast or manufactured. In a wound-rotor induction motor, (WRIM) however, it is a simple matter to introduce external resistance into the rotor circuit through the slip-rings, as shown in Fig. 9-9. If the shorting bar, shown in the figure, is moved to the extreme right, a maximum of rotor resistance is inserted in each phase of the wye-connected rotor. When moved to the extreme left, a minimum or zero *external* rotor resistance is introduced,

SEC. 9-9. / *Effect of Change in Rotor Resistance*

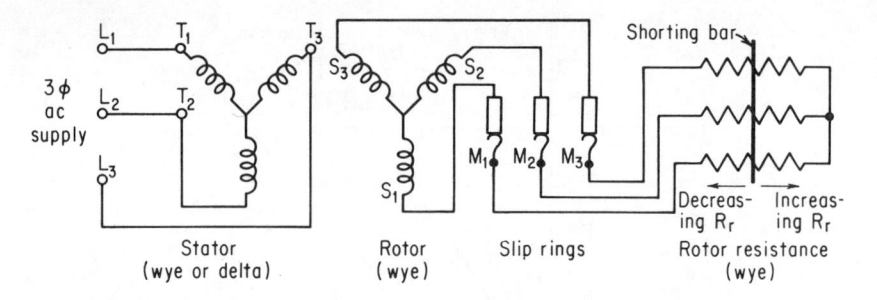

Figure 9-9

Wound-rotor induction motor (WRIM) external rotor resistance controller.

representing the equivalent resistance of a squirrel-cage rotor. The effect of a change in rotor resistance on both the starting and running characteristics may be determined by using a wound-rotor induction motor (WRIM). Let us consider each in turn.

9-10. STARTING CHARACTERISTICS WITH ADDED ROTOR RESISTANCE

At the instant of starting, the slip is *unity*, or one hundred per cent, since the rotor is at a *standstill*. The torque developed by the rotor is determined solely by the factors stated in Eq. (9-7); and, at the instant of starting, the torque is unaffected by the nature of the applied load. Assuming that the voltage applied to the stator is constant, the factors which determine the amount of starting torque developed are the rotor resistance and the standstill rotor reactance, i.e.,

$$T_{\text{starting}} = K_t'' \left(\frac{R_r}{R_r^2 + X_{lr}^2} \right) \qquad (9\text{-}7a)$$

If a variable resistance R_x is inserted in series with the rotor resistance R_r, of a WRIM, Eq. (9-7a) may be written for the starting torque as

$$T_s = K_t'' \left[\frac{R_r + R_x}{(R_r + R_x)^2 + X_{lr}^2} \right] \qquad (9\text{-}14)$$

where all the terms have been previously defined.

The power factor of the rotor may now be expressed as

$$\cos \theta_r = \frac{R_r + R_x}{\sqrt{(R_r + R_x)^2 + (X_{lr})^2}} \qquad (9\text{-}15)$$

It is fairly evident for a WRIM that changing the *total* rotor resistance in Eq. (9-15) above, will *change* the power factor at the instant of starting. Since the torque, as expressed in the fundamental torque equation [Eq. (9-5)], depends on the rotor power factor ($T = K_t \phi I_r \cos \theta_r$), for any given stator voltage, the excitation is constant, i.e., $\phi = k$. Raising the

320

CHAP. NINE / Polyphase Induction Dynamos

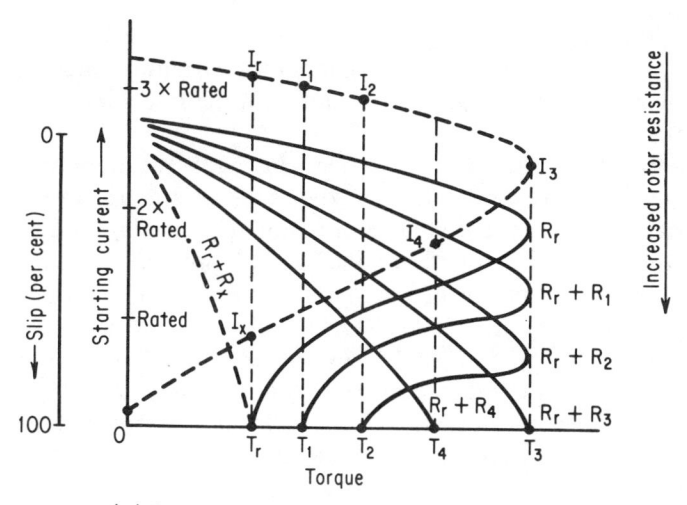

(a) Starting current vs torque.

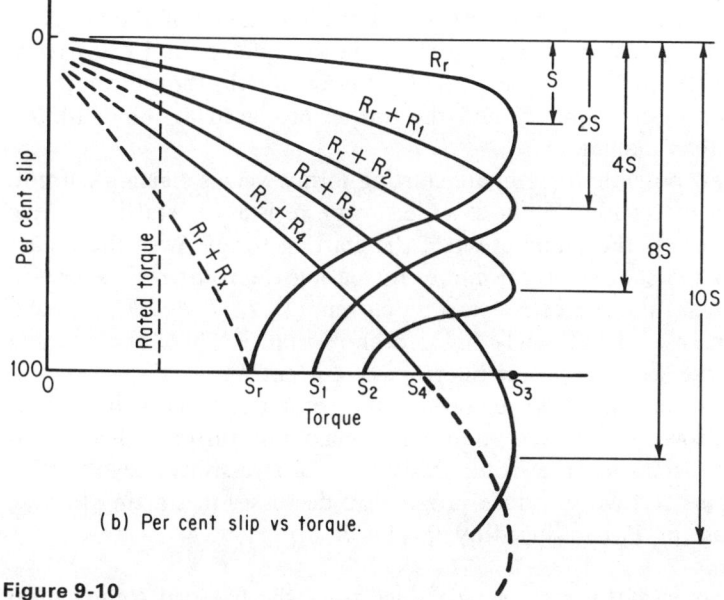

(b) Per cent slip vs torque.

Figure 9-10

Effect of change of rotor resistance on starting and running characteristics of a WRIM.

rotor resistance and the power factor at starting, therefore, will increase the total impedance, *reduce* the *starting current*, and, at the same time, *increase* the *starting torque*. The increase in power factor occurs at a *greater rate* than the decrease in rotor current, as shown in Fig. 9-10a. This figure represents a *family of curves* for various values of rotor resis-

321

SEC. 9-10. / *Starting Characteristics with Added Rotor Resistance*

tance similar to the single curve shown in Fig. 9-6. In Fig. 9-6, it was noted that, at unity slip or starting (point *a*), the rotor current is obtained by taking the intersection of the starting torque with the lower intercept of rotor current.

The family of curves shown as solid lines, in Fig. 9-10a, represent the slip-torque curves for various values of rotor resistance added to the basic rotor resistance, R_r, for the wound-rotor motor. The starting torque, T_r, which occurs with the basic rotor resistance R_r, will produce a starting current I_r on the higher intercept of the stator current curve shown in Fig. 9-10a. When resistance R_1 is added to each phase of the wye-connected wound rotor, shown in Fig. 9-9, by moving the shorting bar toward the right, a *new* torque-slip curve is created, as shown in Figs. 9-10a and b. It should be noted that, in Fig. 9-10b, as the rotor resistance is increased, the slip is increased in order to develop the same torque. Thus, for rotor resistance R_r, maximum torque occurs with slip *s*. For rotor resistance R_r plus R_1, maximum torque occurs with slip $2s$ (twice the original slip, *s*), and so on. Note that the total rotor resistance may be so high that maximum torque just occurs at about a slip of 100 per cent, i.e., $R_r + R_3$ equals sX_{l_r} when $s = 1$. Note that, if the total rotor resistance is higher, as in the case $R_r + R_4$, the maximum torque cannot be reached with a slip of unity and the torque produced at standstill, T_4, is thus less than maximum.

It is now possible to compare starting torque versus starting current, as shown in Fig. 9-10a. Without added rotor resistance, starting torque T_r is obtained by the intersection of the starting torque with the higher intercept of the stator current curve. Increasing the rotor resistance will always decrease the respective starting current (I_r, I_1, I_2 and I_3, I_4, etc.), even in the case of extremely high rotor resistance where the running torque T_4 intersects the lower intercept of current I_4.

As shown in Fig. 9-10a, increasing the rotor resistance will increase the starting torque up to the point where maximum torque is reached at stand-still. Increasing the rotor resistance progressively beyond this value (R_r plus R_3) will produce progressive decreases in starting torque, as in the case of T_4 (produced by R_r plus R_4).

EXAMPLE 9-5: The motor of Example 9-4 develops twice the full-load torque when started with its rotor short-circuited, and it runs at a full-load speed of 875 rpm. If 0.7 ohm per phase of added resistance is connected in series with the rotor, calculate
 a. The new full-load speed with added resistance
 b. The starting torque with added resistance.

Solution:

 a. The full-load slip, short-circuited, is

$$s = \frac{S - S_r}{S} = \frac{900 - 875}{900} = 0.0278 \qquad (9\text{-}1)$$

Since slip is proportional to rotor resistance and since the increased rotor resistance is $R_r = 0.7 + 0.3 = 1.0\,\Omega$, the new full-load slip with added resistance is

$$s_r = \frac{1.0\,\Omega}{0.3\,\Omega} \times 0.0278 = 0.0926$$

The new full-load speed is

$$S(1 - s) = 900(1 - 0.0926) = 900(0.9074) = \textbf{817 rpm} \tag{9-1a}$$

b. The original starting torque, T_o, was twice the full-load torque with a rotor resistance of $0.3\,\Omega$ and a rotor reactance of $1.08\,\Omega$ (Example 9-4). The new starting torque conditions may be summarized by the following table and computed from Eq. (9-14), where T_o is the original torque and T_f is the new torque.

CONDITION	R_r	X_{lr}	$T_{starting}$
Original	$0.3\,\Omega$	$1.08\,\Omega$	$2 \times T_{fl}$
New	$1.0\,\Omega$	$1.08\,\Omega$	?

$$T_o = K_t''\left(\frac{R_r}{R_r^2 + X_{lr}^2}\right) = K_t''\left[\frac{0.3}{(0.3)^2 + (1.08)^2}\right] = K_t''\left(\frac{0.3}{1.25}\right)$$
$$= K_t'' \times 0.24 \tag{7-9a}$$

$$T_f = K_t''\left[\frac{R_r + R_x}{(R_r + R_x)^2 + X_{lr}^2}\right] = K_t''\left[\frac{1.0}{(1.0)^2 + (1.08)^2}\right] = K_t''\left(\frac{1.0}{2.162}\right)$$
$$= K_t'' \times 0.463 \tag{9-14}$$

Then $\dfrac{T_f}{T_o} = \dfrac{K_t'' \times 0.463}{K_t'' \times 0.24} = 1.925$ and $T_f = \textbf{1.925}\,T_o$

Therefore, $T_f = 1.925(T_{fl} \times 2) = \textbf{3.95}\,T_{fl}$

The new starting torque with added rotor resistance has been almost doubled.

The data of Example 9-5, in which the combined rotor resistance is 1 ohm and the standstill reactance is 1.08 ohms, yielded a starting torque which is almost four times the full-load torque, compared to a starting torque which is twice the full-load torque without added rotor resistance. The effect of added rotor resistance in reducing the starting current is illustrated in Example 9-6.

EXAMPLE 9-6: The induced voltage per phase in the rotor of the induction motor in Examples 9-4 and 9-5 is 112 V. At the instant of starting the motor, calculate
a. The rotor current per phase, and the rotor power factor with the rotor short-circuited
b. Repeat (a) with added rotor resistance of $0.7\,\Omega$/phase.

Solution:

a. The locked-rotor impedance, per phase, is
$$Z_{lr} = R_r + jX_{lr} = 0.3 + j1.08 = 1.12 \angle 74.5° \, \Omega$$
$$I_r = \frac{E_{lr}}{Z_{lr}} = \frac{112}{1.12} = 100 \text{ A} \tag{9-6}$$
$$\cos \theta_r = \cos 74.5° = 0.267; \quad \text{or} \quad \cos \theta = \frac{R_r}{Z_{lr}} = \frac{0.3}{1.12} = 0.267$$

b. The locked-rotor impedance with added rotor resistance is:
$$Z_{lr} = 1.0 + j1.08 = 1.47 \angle 47.2° \, \Omega$$
$$I_r = \frac{E_{lr}}{Z_{lr}} = \frac{112}{1.47} = 76.3 \text{ A} \tag{9-6}$$
$$\cos \theta_r = \cos 47.2° = \frac{R_r}{Z_{lr}} = \frac{1.0}{1.47} = 0.68$$

As seen from Example 9-6, the rotor current at starting has dropped almost 25 per cent with added rotor resistance, while producing almost twice the starting torque as without added rotor resistance. At the same time, the rotor power factor increased from 0.267 to 0.68 as a result of adding rotor resistance. The sharp rise in the rotor power factor was more than enough, therefore, to compensate for the decrease in rotor current in Eq. (9-5) and to produce almost twice the original starting torque. It should also be noted that this value of starting torque occurs quite close to the rotor power factor which will *always* yield *maximum* starting torque at standstill, i.e., where R_r equals X_{lr} at a slip of unity, at cos 45°, which is a power factor of 0.707.

For a given wound-rotor induction motor, therefore, it is possible to *increase the starting torque* up to the *maximum* starting torque and, simultaneously, *decrease the starting current* by increasing the external rotor resistance per phase and the rotor power factor. Increasing the rotor power factor above 0.707 will *continue* to decrease the starting current, as shown in Fig. 9-10a, but the starting torque and maximum developed torque will tend to *decrease*, as well, as shown in Fig. 9-10b.

When large induction motors are started across the line, the starting current may be extremely high, causing an excessive voltage drop in the lines supplying current to a particular occupancy. In industrial occupancies, this reduction in the "house" line voltage may produce an objectionable reduction in the brightness of lamps as well as a disturbance in the operation of other motors and voltage-sensitive electronic equipment. The characteristics reduction in picture size of a television picture tube is a common consequence of this effect. It is not *always* necessary, furthermore, that an induction motor develop starting torques at or near the breakdown or pull-out torque. For most applications, a starting torque of approximately 1.5 to 2.0 times the rated torque is sufficient.

As shown in Figs. 9-10a and b, it *is* possible to develop the same starting torque as that which is produced without rotor resistance at a *much lower* value of starting current, using a combined rotor resistance of R_r plus R_x. Without rotor resistance, torque T_r in Fig. 9-10a will produce a high current I_r. With rotor resistance R_x, the same torque, T_r, is produced, and the *much reduced* starting current, I_x, results. This greatly reduced starting current is the result of operation at a higher power factor, raising the question posed in Example 9-7b below.

EXAMPLE 9-7: Using the data of the previous induction motor, determine
a. The added rotor resistance
b. The rotor power factor which will produce the same starting torque (twice the rated torque) as with the rotor short-circuited
c. The starting current.

Solution:

The new and the original conditions may be summarized in the following table.

CONDITION	R_r	X_{lr}	STARTING TORQUE
Original	0.3 Ω	1.08 Ω	$T_o = 2 \times T_{fl}$
New	$(0.3 + R_x)$ Ω	1.08 Ω	$T_n = 2 \times T_{fl}$

a. $T_o = K_t'' \left[\dfrac{0.3}{(0.3)^2 + (1.08)^2} \right] = K_t'' \left(\dfrac{0.3}{1.25} \right) = K_t'' \times 0.24$ (9-7a)

$T_n = T_o = K_t'' \left[\dfrac{0.3 + R_x}{(0.3 + R_x)^2 + (1.08)^2} \right] = K_t'' \times 0.24$ (9-14)

Simplifying, $0.3 + R_x = 0.24[(0.3 + R_x)^2 + (1.08)^2]$

Expanding and combining the terms yields $0.24R_x^2 - 0.856R_x = 0$

This is a quadratic equation having two roots, which may be factored as $R_x(0.24R_x - 0.856) = 0$, yielding

$$R_x = 0 \quad \text{and} \quad R_x = \dfrac{0.856}{0.24} = 3.57 \text{ Ω}$$

Note that this solution shows that the original torque is produced with an external resistance of either zero or twelve times the original rotor resistance. Therefore, $R_T = R_r + R_x = 0.3 + 3.57 = \mathbf{3.87 \text{ Ω}}$

b. $Z_T = R_T + jX_{lr} = 3.87 + j1.08 = 4.02 \angle 15.6° \text{ Ω}$

$$\cos \theta = \dfrac{R_T}{Z_T} = \dfrac{3.87}{4.02} = \cos 15.6° = \mathbf{0.963} \quad (9\text{-}15)$$

c. $I_r = \dfrac{E_{lr}}{Z_r} = \dfrac{112 \text{ V}}{4.02 \text{ Ω}} = \mathbf{28 \text{ A}}$ (9-6)

Note that the rotor current at starting is 28 per cent of the original starting current in (a) of Example 9-6.

Example 9-7 illustrates how it is possible to take advantage of the same torque that exists for both a high starting current and a low running current, shown in Figs. 9-6 and 9-10. The latter is used to reduce the current drawn from the mains and to reduce an objectionable voltage drop (see Secs. 9-14 and 9-21). It should be noted, however, that, in the case of added rotor resistance, as shown in Fig. 9-10b by the curve of R_r plus R_x, the starting torque is also the maximum torque.

9-11. RUNNING CHARACTERISTICS WITH ADDED ROTOR RESISTANCE

The running characteristics of the WRIM, or any induction motor in which resistance has been added to the rotor, is also shown in Fig. 9-10b. The no-load to full-load range of operation is represented between zero and rated torques. The curves indicate that slip is proportional to the magnitude of added rotor resistance. The *greater* the added rotor resistance, the *poorer* the speed regulation of the motor. It is possible, therefore, to vary the speed of a loaded wound-rotor motor to any speed *below* synchronous speed by adding or removing rotor resistance. As a result, the starting resistance which provides (1) high starting torques and (2) reduced starting currents, may also serve (3) as a means of speed control for speeds *below* synchronous speed. The effect of added rotor resistance on the running characteristic is illustrated in Example 9-8.

EXAMPLE 9-8: The motor of the previous problems (Examples 9-4 through 9-7) has a full-load speed of 875 rpm with the rotor short-circuited. The rotor resistance is 0.3 ohm. From Example 9-5, an added rotor resistance of 0.7 ohm provided a full-load speed of 817 rpm. Determine the full-load speeds for added rotor resistances of:
a. 1.7 ohms
b. 2.7 ohms
c. 3.7 ohms
d. 4.7 ohms.

Solution:

TOTAL ROTOR RESISTANCE $R_r + R_x$ (ohms)	SLIP	FULL-LOAD SPEED (rpm)
Given 0.3 0.3 + 0.7 = 1.0	Given 0.0278 0.0926	Given 875 817
Given (a) 0.3 + 1.7 = 2.0 (b) 0.3 + 2.7 = 3.0 (c) 0.3 + 3.7 = 4.0 (d) 0.3 + 4.7 = 5.0	Calculated 0.1855 0.278 0.371 0.463	Calculated 737.5 648 566 482

$$\text{Slip: } s_r = s_o\left(\frac{R_r + R_x}{R_r}\right) \qquad \text{Rotor speed: } S_r = S_o(1 - s_r)$$

a. $s_r = 0.0278\left(\frac{2}{0.3}\right) = 0.1855 \qquad S_r = 900(1 - 0.1855) = \mathbf{737.5 \text{ rpm}}$

b. $s_r = 0.0278\left(\frac{3}{0.3}\right) = 0.278 \qquad S_r = 900(1 - 0.278) = \mathbf{648 \text{ rpm}}$

c. $s_r = 0.0278\left(\frac{4}{0.3}\right) = 0.371 \qquad S_r = 900(1 - 0.371) = \mathbf{566 \text{ rpm}}$

d. $s_r = 0.0278\left(\frac{5}{0.3}\right) = 0.463 \qquad S_r = 900(1 - 0.463) = \mathbf{482 \text{ rpm}}$

Example 9-8 verifies, once again, that slip is proportional to rotor resistance.

9-12. INDUCTION MOTOR TORQUE AND DEVELOPED ROTOR POWER

The basic relation between external or internal horsepower, torque, and motor speed for any motor was given as

$$\text{hp} = \frac{TS}{5252} \tag{4-15}$$

As in the case of the synchronous motor (Section 8-17), it is necessary to evaluate the torque and/or the power developed by the induction motor rotor in terms of the voltage, current, and power factor of the induction motor stator. Thus, for any given slip or rotor speed, the running torque may be determined [Eq. (4-15)] if the power developed by the rotor is known. For any given slip, under running conditions, the rotor current is expressed by

$$I_r = \frac{sE_{lr}}{\sqrt{R_r^2 + (sX_{lr})^2}} \tag{9-9}$$

Dividing both the numerator and denominator by the slip, s, yields

$$I_r = \frac{E_{lr}}{\sqrt{(R_r/s)^2 + X_{lr}^2}} \tag{9-9a}$$

Equation (9-9a) implies that the rotor current under running conditions may be evaluated in terms of the standstill locked-rotor voltage and locked-rotor reactance per phase, together with a complex resistive term R_r/s. Thus, the rotor current and the power developed may be evaluated by considering the equivalent circuit of the rotor shown in Fig. 9-11a. At standstill, the slip is *unity*, and the circuit shown in the figure satisfies the standstill conditions given in Eq. (9-6). As the rotor rotates, the slip decreases, in turn increasing the "apparent" rotor resistance. The variable rotor resistance shown in Fig. 9-11a may be considered as consisting

(a) Based on Eq. 9-9(a). (b) Based on Eq. 9-16.

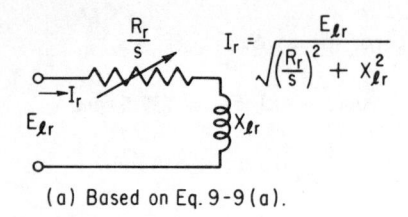

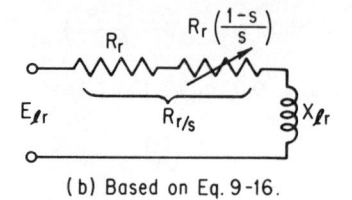

Figure 9-11

Equivalent circuits of rotor under running conditions.

of two terms, shown in Fig. 9-11b, namely,

$$\frac{R_r}{s} = R_r + R_r\left(\frac{1-s}{s}\right) \tag{9-16}$$

The first term in Eq. (9-16) represents the actual effective rotor resistance per phase, R_r, and the second represents an equivalent and fictitious load resistance term, $R_r[(1-s)/s]$, which varies directly with the load and with the slip under running conditions. Thus, the circuit of Fig. 9-11b truly represents the rotor under *running* conditions in terms of the induced locked-rotor voltage, the standstill locked-rotor reactance, and the fixed and variable rotor resistance components shown.

The fixed rotor resistance term R_r may be considered the *power loss component* of the rotor for any given rotor current, and the variable term, $R_r[(1-s)/s]$, may be considered the *rotor developed power term* (for power developed by the rotor conductors) in order to produce torque. Multiplying each of the above terms in Eq. (9-16) by the square of the rotor current, therefore, to obtain the power expressions, we get the basic expression

$$\frac{I_r^2 R_r}{s} = I_r^2 R_r + I_r^2 R_r\left(\frac{1-s}{s}\right) \tag{9-17}$$

or

$$\frac{\text{rotor power input}}{\text{per phase}} = \frac{\text{rotor copper loss}}{\text{per phase}} + \frac{\text{rotor power developed}}{\text{per phase}}*$$

Equation (9-17) is most significant, not only in the computation of the rotor power developed and the running torque of an induction motor, but also as the basis for efficiency determination from blocked-rotor

---

* The similarity between Eq. (9-17) and Eq. (4-7) should be noted. In the latter case, the power developed by a dc armature is the power supplied to the armature minus the armature copper loss. Equation (9-17) states that the power developed by an induction motor rotor is the power supplied to the rotor minus the rotor copper loss.

tests (cf. Chapter 12, Section 12-13). It is of interest to note that, by Eq. (9-17), the rotor power input at *any* given load or slip is the rotor copper loss at that load divided by the slip. Further, the rotor power developed is always the difference between the rotor input and the rotor copper loss.

It is now possible to express the developed rotor torque in terms of Eq. (4-15) as

$$T = \frac{5252 \text{ hp}}{S_r} = \frac{5252 P_d}{746 S_r} = \frac{7.04 P_d}{S_r} \quad \text{or} \quad 7.04\left(\frac{P_d}{S_r}\right) \quad (4\text{-}15a)$$

where S_r is the rotor speed at any value of slip.

But the rotor power developed by an induction motor, P_d, is from Eq. (9-17)

$$P_d = I^2 R_r \frac{(1-s)}{s} = P_{in}(1-s) \quad (9\text{-}18)$$

where P_{in} is the rotor power input to all phases.

The torque developed by the rotor, therefore, is

$$T = \frac{7.04 P_d}{S_r} = \frac{7.04 P_{in}(1-s)}{S_r} = \frac{7.04 P_{in}}{S_r/(1-s)}$$

Further, it was shown in Eq. (9-1a) that the synchronous speed, S, is equal to $S_r/(1-s)$. From this relationship, it follows that the torque can be expressed as

$$T = 7.04\left(\frac{P_{in}}{S}\right) \quad (9\text{-}19)$$

where P_{in} is the total rotor power input (to all phases) and S is the synchronous speed, $120f/P$, in rpm.

EXAMPLE 9-9: A four-pole, three phase, 60 Hz, 220-V, delta-connected, 1 hp WRIM has a wye-connected rotor having one-fourth the number of turns of the stator. The full-load speed is 1740 rpm. The rotor resistance is 0.3 ohm, and the locked-rotor reactance is 1 ohm. Calculate:
a. The locked-rotor voltage per phase
b. The rotor current per phase under running conditions
c. The rotor power input at full load (the total three-phase power input)
d. The rotor copper loss at full load
e. The rotor power developed, in watts and in horsepower
f. The rotor torque developed from (e), using Eq. (9-19).

Solution:

a. $E_{lr} = \dfrac{220 \text{ V}}{4} = 55$ **V/phase with the rotor blocked**

b. $s = \dfrac{S - S_r}{S} = \dfrac{1800 - 1740}{1800} = 0.0333$ (9-1)

$I_r = \dfrac{E_{lr}}{\sqrt{(R_r/s)^2 + X_{lr}^2}} = \dfrac{55 \text{ V}}{\sqrt{(0.3/0.0333)^2 + (1.0)^2}} = \dfrac{55}{\sqrt{82}}$ (9-6)
$= 6.075$ A/phase

c. $P_{in} = 3\left(\dfrac{I_r^2 R_r}{s}\right) = 3\left[\dfrac{(6.075)^2 \times 0.3}{0.0333}\right] = 993 \text{ W}$ (9-17)

d. The rotor copper loss $= I_r^2 R_r = P_{in} \times s$ (9-17)
$= 993 \text{ W} \times 0.0333 = 33.1 \text{ W}$

e. The rotor power developed $= P_{in} - P_{loss}$ (9-17)
$= 993 - 33.1 = 959.9 \text{ W}$

Check from Eq. (9-18): $P_d = P_{in}(1 - s)$ (9-18)
$= 993(1 - 0.0333) = 959.9 \text{ W}$

Horsepower developed $= \dfrac{P_d}{746} = \dfrac{959.9 \text{ W}}{746 \text{ W/hp}} = 1.286 \text{ hp}$

f. $T = \dfrac{\text{hp} \times 5252}{S_r} = \dfrac{1.286 \times 5252}{1740} = 3.88 \text{ lb-ft}$ (4-15a)

Check from Eq. (9-19):

$T = \dfrac{7.04 P_{in}}{S} = \dfrac{7.04 \times 993}{1800} = 3.88 \text{ lb-ft}$ (9-19)

It is also possible to compute the maximum or stalling torque from Eq. (9-19) if the rotor power input is known at the value of slip at which maximum or breakdown torque occurs, or

$$s_b = \dfrac{R_r}{X_{lr}}$$ (9-12)

The current in the rotor at the maximum torque point is

$$I_r = \dfrac{E_{lr}}{\sqrt{(R_r/s_b)^2 + (X_{lr})^2}}$$ (9-9a)

Substituting Eq. (9-12) in Eq. (9-9a) produces

$$I_r = \dfrac{E_{lr}}{\sqrt{2 X_{lr}^2}} = \dfrac{E_{lr}}{\sqrt{2} \times (X_{lr})} = 0.707\left(\dfrac{E_{lr}}{X_{lr}}\right)$$ (9-20)

But the rotor power input, P_{in}, at (any value and) the breakdown value of slip [Eq. (9-17)] is

$$P_{in} = \dfrac{I_r^2 R_r}{s_b} \quad \text{per phase}$$ (9-17)

Substituting Eqs. (9-20) and (9-12) in Eq. (9-17) yields, at maximum or breakdown,

$$P_{in} = \frac{E_{lr}^2}{2X_{lr}} \qquad \text{W/phase} \tag{9-21}$$

The value of P_{in} thus obtained may be substituted in Eq. (9-19) to yield the maximum or breakdown torque, as demonstrated by Example 9-10.

EXAMPLE 9-10: Calculate the maximum torque which can be developed by the 1 hp motor given in Example 9-9, and the per cent slip and the speed at which maximum torque is developed.

Solution:

$$P_{in} = \frac{E_{lr}^2}{2X_{lr}} = \frac{(55)^2}{2 \times 1} = 1513 \text{ W/phase} \times 3 \text{ phases} = 4539 \text{ W} \tag{9-21}$$

Substituting in Eq. (9-19),

$$T_{max} = \frac{7.04 P_{in}}{S} = \frac{7.04 \times 4539 \text{ W}}{1800} = \mathbf{17.72 \text{ lb-ft}} \tag{9-19}$$

Then

$$s_b = \frac{R_r}{X_{lr}} = \frac{0.3}{1.0} = \mathbf{0.3} \tag{9-12}$$

and

$$S_r = S(1 - s) = 1800(1 - 0.3) = \mathbf{1260 \text{ rpm}} \tag{9-1a}$$

If the running and maximum torques can be computed from the rotor power input, it should be possible to compute the starting torque in this manner as well. Since the slip is *unity* at the instant of starting, Eq. (9-9a) is simplified to

$$I_r = \frac{E_{lr}}{\sqrt{R_r^2 + X_{lr}^2}} \qquad \text{at the instant of starting} \tag{9-9b}$$

and the rotor power input per phase at a slip of unity is

$$P_{in} = I_r^2 R_r \tag{9-17a}$$

Substituting Eq. (9-9b) in Eq. (9-9b) in Eq. (9-17a) yields

$$P_{in}, \text{ per phase} = \frac{E_{lr}^2}{R_r^2 + X_{lr}^2} \times R_r \qquad \text{at the instant of starting} \tag{9-22}$$

which may be substituted in Eq. (9-19) to yield the starting torque.

EXAMPLE 9-11: Calculate the starting torque developed by the 1 hp motor given in Example 9-9.

Solution:

$$P_{in} = 3\left(\frac{E_{lr}^2}{R_r^2 + X_{lr}^2}\right) \times R_r = \left[\frac{3 \times (55)^2}{(0.3)^2 + (1.0)^2}\right]0.3 = 2495 \text{ W} \quad (9\text{-}22)$$

From Eq. (9-19), starting toque is

$$T_s = 7.04\left(\frac{P_{in}}{S}\right) = 7.04\left(\frac{2495}{1800}\right) = 9.76 \text{ lb-ft} \quad (9\text{-}19)$$

It will later be seen (Section 12-13) that, in performing the locked-rotor test (which is used to determine effective rotor resistance and locked-rotor reactance for an induction motor as part of the conventional method of determining induction motor efficiency), the rotor power input computed in Example 9-11 is most easily obtained, from which the starting torque is computed as in Example 9-11 above.

A convenient short cut that is sometimes employed in the solution of induction motor torque problems is an equation for the rotor torque at any value of slip s based on the maximum torque which has been computed at the breakdown or pull-out slip, s_b. This may be computed from the equation

$$T = T_b\left[\frac{2}{(s_b/s) + (s/s_b)}\right] \quad (9\text{-}23)$$

where T is the torque at any slip, s, and T_b is the maximum breakdown torque occurring at slip s_b.

EXAMPLE 9-12: The maximum torque in Example 9-10 was 17.72 lb-ft at a slip of 0.3. Calculate:
a. The full-load torque at a slip of 0.0333 (from Example 9-9)
b. The starting torque at a slip of 1.0 (from Example 9-11).

Solution:

From Eq. (9-23),

a. $T = 17.72 \text{ lb-ft}\left[\frac{2}{(0.3/0.0333) + (0.0333/0.3)}\right] = 17.72\left(\frac{2}{9.111}\right)$
$= 3.88 \text{ lb-ft}$

b. $T = 17.72 \text{ lb-ft}\left[\frac{2}{(0.3/1.0) + (1.0/0.3)}\right] = 17.72\left(\frac{2}{3.63}\right) = 9.76 \text{ lb-ft}$

Note that this method yields the same answers as those obtained in Examples 9-9 and 9-11.

9-13. MEASUREMENT OF SLIP BY VARIOUS METHODS

In testing induction motors and determining the slip at various conditions of load, it is essential that the value of slip obtained be accurate, since it appears in all of the calculations used in the above equations. For this reason, slip is almost never determined by the measurement of rotor speed directly, because a small measurement error in rotor speed will produce a large error in the value of slip. For example, the full-load speed in Example 9-9 is 1740 rpm. If the rotor speed is measured with a tachometer (either mechanical or electric) which has a two per cent error (not unusual for such a device) the reading may be 1740 ± 35 rpm, or as high as 1775 rpm or as low as 1705 rpm. The actual value of slip, as determined from Example 9-8, is 0.0333; whereas that from a measurement employing a tachometer with a two per cent error yields (using the low value) $s = (1800 - 1705)/1800$ or 0.05275. The error produced in the value of slip which might be used is, therefore, $(0.05275 - 0.0333)/0.0333 = 58.26$ per cent. A *small* two per cent *error* in the measurement of *rotor speed* has resulted in a *large* 58.3 per cent error in the computation of *slip*. This error in slip would be carried through all of the equations employed above, yielding spurious results. For this reason, rotor speed *per se* is rarely measured, and attempts are made to *measure slip directly*.

Perhaps the most simple and direct method of measuring slip, i.e., the difference between synchronous speed and rotor speed, is to compare the speed of the induction motor with a small synchronous motor's speed as shown in Fig. 9-12a. The synchronous motor employed must have the same number of poles as the induction motor. At the end of each shaft is a cylinder of phenolic or other suitable insulating material into which has been fitted a circular slip ring to which a small contactor is connected, as shown in the figure. The slip rings are connected to a voltage

(a) Electromechanical counter.

(b) Mechanical differential counter.

(c) Stroboscopic method (6 poles).

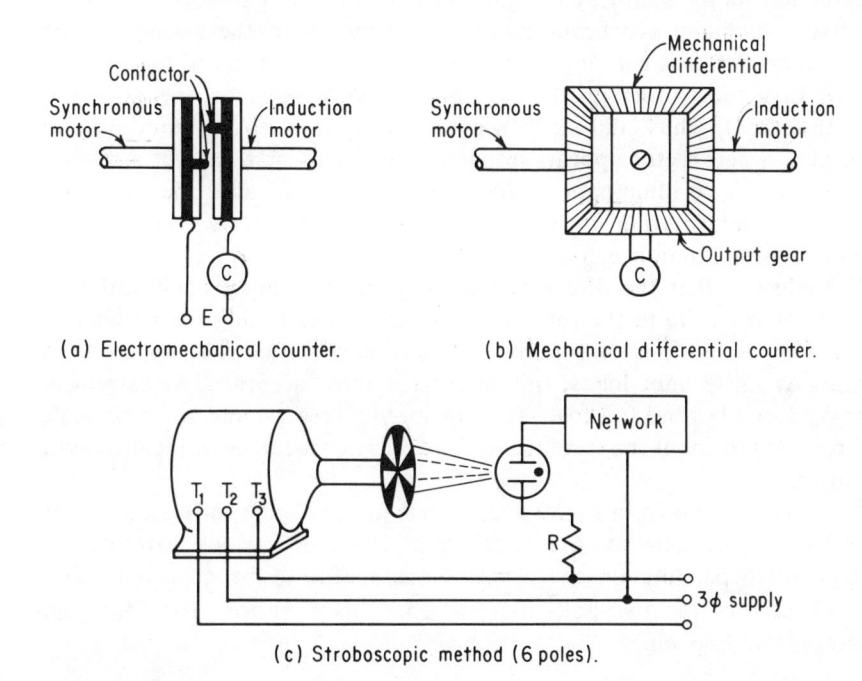

Figure 9-12

Various methods of measuring slip directly.

333

SEC. 9-13. / *Measurement of Slip by Various Methods*

source in series with an electric pulse counter (either of the electromechanical relay type or an electronic digital pulse counter). The synchronous motor always runs at synchronous speed, which is the speed of the rotating field of the induction motor. Each time the induction motor slips a revolution, the synchronous and induction motor contactors close the circuit and register a pulse. The number of pulses registered per minute (ppm) is the *slip speed* in rpm, from which slip is readily computed by Eq. (9-1).

A similar method is shown in Fig. 9-12b, illustrating the use of a *mechanical differential* whose output gear rotates at a speed equal to the difference of the two input gears to which the synchronous and induction motor shafts are connected. A mechanical counter or low-speed electric or mechanical tachometer is used to record the slip speed directly. This method has a disadvantage in loading the induction motor slightly because of the friction and drag in the gearing of the differential, and it should not be employed with smaller induction motors for the measurement of slip.*

An *optical* measurement that produces no loading effect whatever, and which may be used on the smallest of motors, is the *stroboscopic* method shown in Fig. 9-12c. This method utilizes a gas tube of sufficient size and brightness containing an inert gas: neon, argon, or xenon are most commonly used. A pulse-forming and deionizing network provides pulses which are synchronized to the frequency of the line supplying the induction motor stator. The light emitted from the glow tube occurs in bursts which are synchronized to the frequency of the supply. If the induction motor is rotating at synchronous speed, a burst of light occurs each time the sectors on the disc of the motor advance one pole (60° in this case). Thus, the stroboscopic disc appears to be standing still because a new sector appears in place of the old (each sixth of a revolution) to receive illumination. Whenever, because of slip, the induction motor is not rotating at synchronous speed, the sectors receive a full burst of illumination before they have completed 60 degrees of rotation. The effect is that the disc gives the illusion of rotating backward in a direction opposite to the rotation of the motor. Counting the number of apparent revolutions of rotation in one minute yields the slip speed in rpm. At fairly light loads, this method is most accurate. At extremely heavy loads beyond full load, the slip speed increases, and some practice is required to count the revolutions slip accurately because of the increased slip speed.

The disc shown in Fig. 9-12c is used for a six-pole induction motor whose synchronous speed is 1200 rpm. Discs of various patterns are required, depending on the synchronous speeds to be measured: two black sectors for two-pole motors; four black sectors for four-pole motors; and so on.

---

* An electric differential with less effective drag, Fig. 11-14 may be employed for small induction motors.

In factories employing fluorescent lighting, the stroboscopic effect described above may sometimes produce the hazardous illusion that rapidly rotating machinery is rotating slowly or even standing still. This danger is eliminated by connecting various adjacent luminaires to different phases of a three-phase supply, or by means of phase-shifting networks in the ballasts of certain luminaires. In this way, the illumination bursts are randomly distributed so that no periodic bursts of maximum to zero illumination are produced at or near the frequency of the rotation of the machine.

9-14. INDUCTION MOTOR STARTING

In most occupancies, residential as well as industrial, small SCIMs up to a few horsepower may be started across the line with little objectionable drop in the supply voltage and little or no delay in accelerating to their rated speed. Similarly, large SCIMs (even up to several thousand horsepower) may be started across the line without any damage or objectionable change in supply voltage, providing the supply mains are of sufficiently large capacity. For example, in the vicinity of a generating station or hydroelectric plant, it is customary to open and close the locks of the dams by means of induction motors of several thousand horsepower connected directly across-the-line, without resorting to the use of special starting equipment. Consequently, "across-the-line" starting need *not* be avoided if the mains are of *sufficient* capacity to provide the rated voltage and current required by the induction motor, since such starting does *not* in any way damage the induction motor.*

Although there are some exceptions among the various classifications of SCIMs (Section 9-21), an induction motor usually requires approximately six times its rated current when the rated voltage is applied to its stator. At the instant of starting, the rotor current (and the stator current, therefore) is determined by its locked-rotor impedance, $R_r + jX_{lr}$. Thus, if the stator voltage were reduced to one-half its value, the starting current

---

* Where the supply mains are of *limited* capacity in comparison to the starting current drawn by an induction motor, however, there is the possibility that, because of the heavy current line drop and the reduction in line voltage, the starting motor (and other motors as well) may not develop sufficient torque to accelerate the load and, as a result, may draw excessive rotor and stator currents. The motor and line protective equipment may after a short interval disconnect the motor, requiring that the motor be started once more, again with an accompanying line-voltage disturbance. The frequent voltage fluctuations may also affect electronic equipment and lighting to such a degree that some alternate method of induction motor starting would be required to limit the starting current. If the *lines* feeding the SCIM have unequal impedances, the stator voltages may become unbalanced, severely unbalancing line currents, also causing protective equipment to clear motor. In fact, a 1-2% unbalance in stator line voltages may cause a 20% unbalance in line currents producing localized motor heating and winding breakdown. The consumer is usually unaware of source voltage unbalance because he only measures the stator line voltages with the motor operating. He is unaware that the motor has balanced its stator voltages by unbalancing its currents, sometimes severely.

would also be reduced by the same amount, i.e., to approximately three times the rated current. But Eq. (9-8) indicates that, if the stator line voltage is reduced to one-half its value, the torque is reduced to one-quarter of its original value. Therefore, the desirable *reduction* in motor line *current* has been achieved at the cost of an *even greater* and undesirable *reduction* in *starting torque*. If the motor is started under heavy load, this is of some consequence, and the likelihood is that the motor will start with difficulty or may not start at all. On the other hand, if the motor is started without load, as in the case of a fan or some machine tools, the torque reduction may not be serious and the current reduction is advantageous.

The theory of various methods of starting will be discussed in this chapter.*

**9-15.
REDUCED
VOLTAGE—AUTO-
TRANSFORMER
STARTING**

Three-phase SCIMs may be started at reduced voltage by using a single three-phase autotransformer† (or *compensator*) or three single-phase autotransformers, as shown in Fig. 9-13a. The taps on the transformer vary from 50 to 80 per cent of the rated voltage. If the motor fails to accelerate the load at the lowest voltage, higher voltage taps may be tried until the proper and desired starting torque is obtained. Figure 9-13a is a diagrammatic representation of a commercial type; the drawing does not include the relays, undervoltage protection, or contacts usually associated with manual or automatic compensator starters. The triple-pole-double-throw

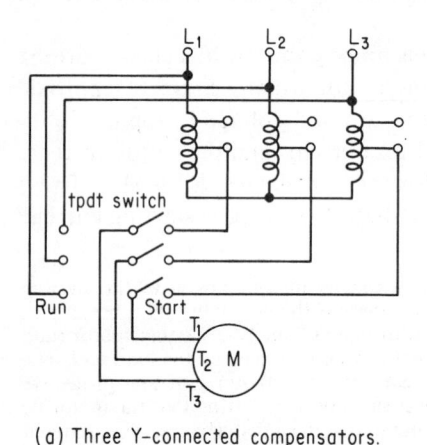

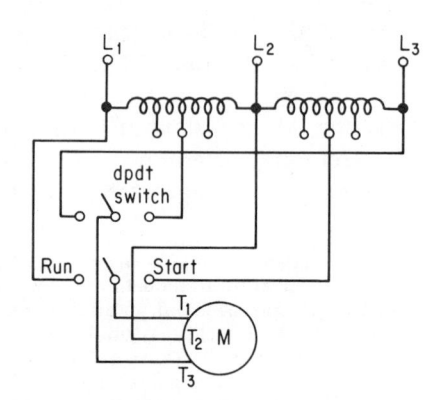

(a) Three Y-connected compensators. (b) Two open delta-connected compensators.

Figure 9-13

Reduced voltage autotransformer starting.

* For a discussion of commercial manual and automatic starters, see Kosow, *Control of Electric Machines* Prentice Hall, 1973, Chs. 5 and 7.
† The theory of autotransformers is covered in Sec. 13-12.

(tpdt) switch is thrown to the "start" position and left there until the motor has accelerated the load to almost full speed. It is then rapidly thrown to the "run" position, where the motor is connected directly across the line.

The compensator starter is used *only* during the period of starting, and its current rating, based on such *intermittent* duty, is somewhat *less* than that of a comparable transformer, which might be used to supply an induction motor continuously from some higher voltage source. The autotransformer acts in *two* ways to reduce the current drawn from the mains: (1) to reduce the motor starting current by a reduction in voltage, and (2) by the transformer turns ratio under which the *primary* line current is less than the *secondary* motor current. Since the turns ratio also represents the voltage ratio, the starting line current is reduced, therefore, by the square of the turns ratio, as illustrated in Example 9-13 below.

Because the compensator is used only intermittently, a saving (the elimination of one transformer) is effected by using two transformers in open delta (or V-V) as shown in Fig. 9-13b. This arrangement produces a small current unbalance in the center leg, L_2, of about 10 to 15 per cent of the starting current; but this unbalance is not excessive and does not affect the motor starting performance materially.

EXAMPLE 9-13: A three-phase, 208 V, 15 hp SCIM has a rated current of 42 A and a starting current of 252 A at the rated voltage. The full-voltage starting torque is 120 lb-ft. A compensator used in connection with the motor for starting employs 60 per cent taps. Calculate:
  a. The motor starting current at reduced voltage
  b. The motor line current, neglecting the transformer exciting current and losses
  c. The motor starting torque at reduced voltage
  d. The reduced voltage line current at starting as a percentage of current at full voltage
  e. The reduced voltage motor torque at starting as a percentage of full voltage torque.

Solution:

a. $I_{sm} = 0.6 \times 252$ A = **151.2 A to the motor**
b. $I_L = 0.6 \times 151.2$ A = **90.72 A drawn from the mains**
c. $T_s = (0.6)^2 \times 120$ lb-ft = **43.25 lb-ft**
d. Per cent line current at starting = $\dfrac{90.72 \text{ A}}{252 \text{ A}} \times 100 =$ **36 per cent of line current at full voltage**
e. Per cent motor starting torque = $\dfrac{43.25 \text{ lb-ft}}{120 \text{ lb-ft}} \times 100 =$ **36 per cent of starting torque at full voltage**

As shown by Example 9-13, a reduction of voltage to 60 per cent of the rated voltage results in a reduction in both the line current and the torque to 36 per cent of rated value. Similarly, a reduction in voltage to about 70 per cent would produce approximately one-half (49 per cent) of the starting torque and about one-half of the starting line current. The 70 per cent tap is a compromise value that is customarily used in practice.

**9-16.
REDUCED VOLTAGE,
PRIMARY RESISTOR
OR REACTOR
STARTING**

If a resistor or a reactor is inserted *in series* with each of the stator or primary line connections, the high starting current produces an immediate voltage reduction across the stator terminals. The motor torque is reduced in proportion to the square of the voltage applied to the stator terminals, but the line current is reduced *only in proportion to the reduction in voltage*. Figure 9-14a shows such a circuit, in which either resistances or reactance coils may be used to produce a sufficient reduction in stator voltage at the instant of starting.

This method of starting is sometimes called "primary impedance acceleration," and its effect is shown in Fig. 9-14b. The slip-torque curve for the motor at full voltage is shown. Using either primary resistance or primary reactance, a reduction in stator voltage at the instant of starting produces the torque reduction indicated. If this voltage (and the primary current) were constant, the motor torque curve would follow the dotted line shown in the figure. As the motor accelerates, however, the voltage across the stator increases because of the reduction in line current (less voltage drop across the series impedance), and the torque increases as the square of the increase in voltage.

Reduced-voltage starting by means of series stator resistance will *improve the starting power factor*, but somewhat *greater losses* are produced; and the maximum torque is not as great for the same series impedance using an equivalent reactor. But, as indicated by Fig. 9-14b, for the same impedance, the starting current and the torque at the instant of starting are the same for resistor and for reactor starting. The advantages of reactor starting in reduced losses, as well as increased maximum torque, are somewhat offset by the increased cost of the reactors. For this reason, reactor starting is reserved for induction motors of larger hp, generally.

**9-17.
WYE-DELTA
STARTING**

Most polyphase SCIMs are wound with their stator windings delta- (or mesh-) connected. A number of manufacturers furnish induction motors with the *start* and *finish* of each *phase belt* brought out for purposes of *external* connection. In the case of three-phase motors, these may be connected to the line either in wye or in delta. When wye-connected, the phase voltage impressed on the winding is $1/\sqrt{3}$, or 57.8 per cent of the line voltage. By means of switching, therefore, as shown in Fig. 9-14c, it is possible to start an induction motor in wye at a little more than half its rated voltage, and then run it in delta, with full phase and line voltage impressed. Since the torque varies as the square of the impressed stator voltage, the reduction

(a) Primary resistor or reactor starting.

(b) Effect of stator impedance.

(c) Wye-delta starting.

(d) Part-winding starting.

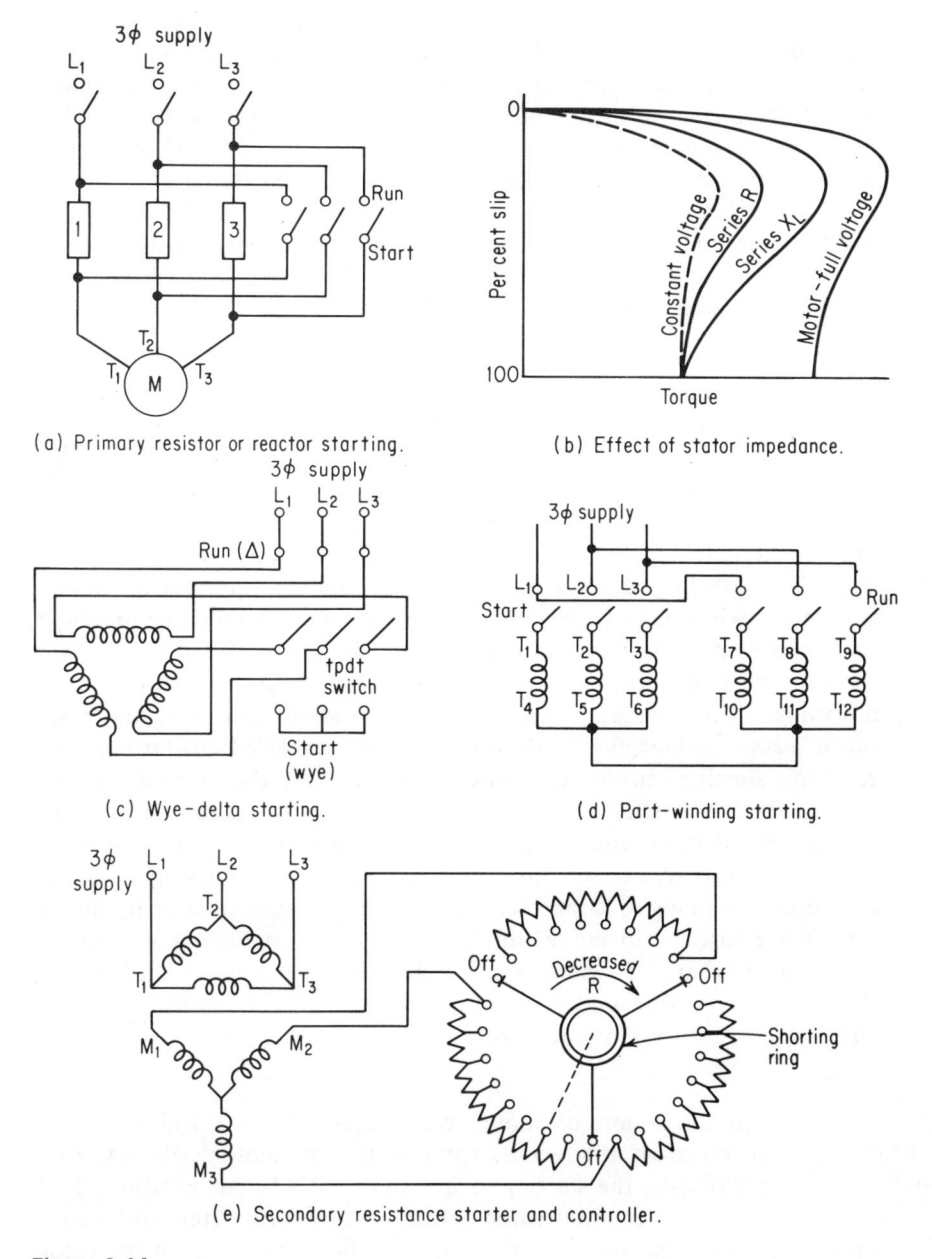

(e) Secondary resistance starter and controller.

Figure 9-14

Starting methods for induction motors.

in voltage when wye-connected will produce about one-third of the normal full-voltage starting torque.

Where such lower starting torques are permissible, with a starting current of approximately 58 per cent of the normal starting current, this

SEC. 9-17. / *Wye-Delta Starting*

relatively inexpensive method of starting is frequently employed. It goes without saying that such a motor (with six stator terminals in the case of a three-phase induction motor) is somewhat more costly than the conventional induction motor; but its cost is less than that of compensators or primary stator impedances and the associated starters.

The switching from wye to delta should be made as rapidly as possible to eliminate high transient currents due to momentary loss of power.*
For this reason, *spring-loaded* triple-pole double-throw (tpdt) *snap-action* switches are used, rather than knife-blade switches.

**9-18.
PART-WINDING
STARTING**

Frequently, polyphase SCIMs are designed with part-windings, i.e., two identical phase windings each of which will produce the same number of poles and the same rotating magnetic field.

The advantage of such windings is that they may be connected in series for higher-voltage systems or in parallel for lower-voltage systems; and such a motor, from the manufacturer's point of view, is more marketable. Figure 9-14d shows a 220/440 V part-winding squirrel-cage induction motor which is to be used on a 220 V line. On starting, only one wye section is employed. Two switches are shown in Fig. 9-14d to simulate the type of starter employed with part-winding induction motors. The advantage of part-winding starting is that the stator resistance and reactance is *twice as high* as if both windings were paralleled on starting. The resulting starting current is about 65 per cent of the normal starting current (with the windings paralleled), and the starting torque is about 45 per cent of the normal starting torque. The motor is thus started on half its winding, wye-connected; and when the motor comes up to speed, the second winding is connected in parallel. Because of a pronounced dip in the torque-slip curve during starting, the manufacturer usually recommends that part-winding starting should be accomplished where the motor is starting under light-load or no-load conditions, as in the case of fans, blowers or drill presses.

**9-19.
WOUND-ROTOR
STARTING**

No discussion of motor starting is complete without some mention of the wound-rotor induction motor. As indicated previously, the starting torque of a WRIM may be adjusted by means of external rotor resistance to provide starting torques up to the motor's maximum torque. By limiting the current in the rotor circuit [Eq. (9-6)], and by providing a higher power factor and torque at the instant of starting, the stator line current is reduced considerably. The diagram shown in Fig. 9-9 is represented once again in Fig. 9-14e without the slip-rings indicated and without the details of con-

---

* S. B. Toniolo, "Behavior of Induction Motors After Short Interruptions of Supply," *Elettrotecnica*, Vol. 30, 1943, pp. 181–184.

struction of the resistance controller. The shorting bar is now in the form of a ring, operated by an insulated handle.

In the "off" position, even with the stator energized, the motor will not turn because the rotor is open-circuited. The motor is started at the first contact with maximum rotor resistance (and maximum torque), and is accelerated by moving the handle clockwise, decreasing the rotor resistance. In the final position, the rotor is completely short-circuited. If the resistors used in a secondary resistance starter have a *continuous-duty* rating, the starter may serve as a speed-controller as well (Fig. 9-10b). The combination of the ability to develop high starting torque for starting under load, plus its speed-control advantages, have led to increased popularity of the WRIM for numerous applications.

9-20. DOUBLE-CAGE ROTOR LINE-STARTING INDUCTION MOTOR

It should be noted that, with the exception of the transformer and stator impedance methods of starting at reduced voltage, special constructions are required for wye-delta, part-winding, and wound-rotor induction motors (WRIMs). In response to a strong demand for a simple construction *line-starting* induction motor which would *not* produce objectionable line fluctuations, the double-cage rotor was developed. The slip-torque curves for the normal induction motor (class A, Fig. 9-16) indicate that the *running* characteristics (fairly constant speed with load) are *excellent*, but that the *starting* characteristics are *poor* (high current at low power factor). Conversely, omitting the advantage of speed control, the WRIM (described in the preceding section), has *excellent starting* characteristics but *poor running* characteristics (slip proportional to rotor resistance); in addition, the purchase of a WRIM entails the increased cost of the controller and its associated resistors. The WRIM has a high rotor resistance on starting, whereas the conventional induction motor has a low rotor resistance during its running period.

The "double-deck" or *double-squirrel-cage* rotor shown in Fig. 9-15 was designed to provide an improved line-start induction motor. Depending on the horsepower rating, several variations of rotor construction are used. Figure 9-15b shows a larger-capacity cast rotor, in which two sets of rotor bars of different *alloys* are employed, having the same or a different cross-sectional area. The *upper* bar is constructed of a *high-resistance* copper alloy, and the *lower* bar may be of *low-resistance* copper or cast aluminum. The *upper* bars are *close* to the rotating magnetic field and are embedded in iron so that, when the bars are carrying current, their self-inductance and leakage reactance are *small*. The *lower* bars are embedded *deeply* in the slots and are separated from the stator iron by a large magnetic gap, producing a *high* self-inductance and a large leakage reactance.

On starting, therefore, when the *rotor frequency* is *high* and the

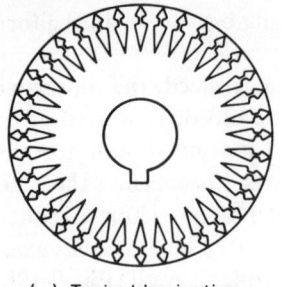

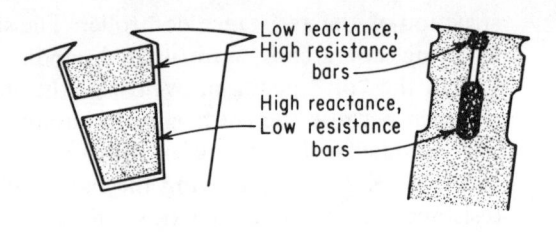

(a) Typical lamination.   (b) Different alloys.   (c) Die cast aluminum rotor.

Figure 9-15

Double-cage rotor construction.

same as that of the line, the *impedance* of the *lower* winding is much *higher* than that of the upper winding. Most of the rotor current is induced, therefore, in the *upper* winding, which is designed so that its high resistance *equals* its reactance *on starting*, developing maximum starting torque.

As the motor accelerates, however, the rotor frequency decreases, and the impedance of the lower or inner winding also decreases, causing more and more current to be induced in the inner winding as well. At small values of slip, therefore, when the motor is in its normal full-load range, most of the current flows in the inner winding of *low resistance*, yielding high efficiency (low rotor copper loss) and good speed regulation (slip proportional to resistance).

In smaller motors, as shown in Figs. 9-15(a) and (c), the laminations are designed in such a way that, by virtue of shape and cross-sectional area, a common casting alloy may be used. The lower bars have a larger area and low resistance with high reactance, and the upper bars of smaller area have a high resistance and low reactance.

9-21. COMMERCIAL INDUCTION MOTOR CLASSIFICATION

The development of the double-cage rotor created a versatility in induction motor rotor design which has led to a variety of slip-torque characteristics. By *properly proportioning* the double-cage winding, manufacturers have developed *numerous variations* of the single or *normal* cast rotor design. These variations result in starting torques greater or less than the normal design, and reduced starting currents, as well. In order to distinguish between the various available types, the National Electrical Manufacturer's Association (NEMA) has developed a code-letter system in which each SCIM type is manufactured in accordance with a particular design standard and is placed in a certain class, designated by a letter.* Since each letter classification specifies a difference in rotor construction, the following description will serve to clarify the selection of SCIMs for various types of service.

9-21.1
Class Letter A SCIM

As shown in Fig. 9-16, the Class A motor is a *normal* or standard SCIM manufactured for constant speed use. It has large slot areas (for good heat dissipation capacity) and fairly deep slot rotor bars. During the starting period, the current density is high near the rotor surface; during the running period, the current density is distributed fairly uniformly. This difference makes for some high resistance and low reactance at starting, resulting in a starting torque from 1.5 to 1.75 times the rated (full-load) torque. The fairly high starting torque and low rotor resistance produce a fairly rapid acceleration toward rated speed. As shown in Fig. 9-16, the Class A SCIM has the *best* speed

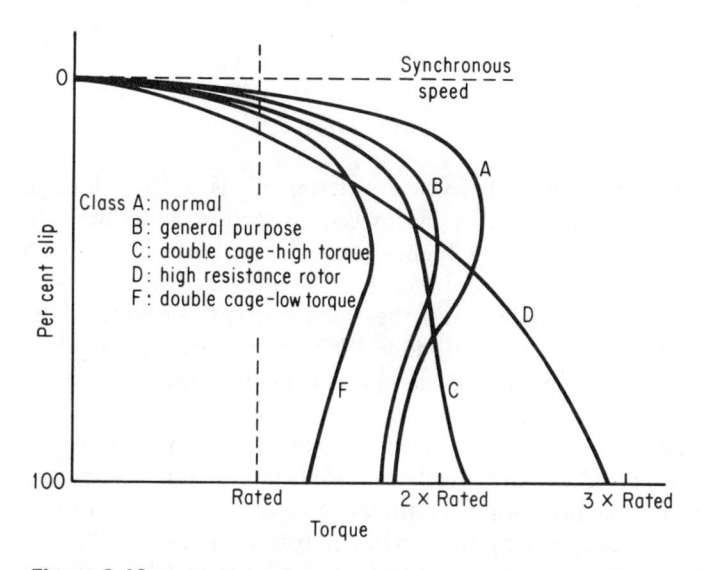

Figure 9-16

Slip-torque characteristics of commercial squirrel cage induction motors (NEMA standards).

---

* The NEMA class letters discussed in this section should not be confused with the Code Letters of the National Electric Code (NEC). The particular part of the NEC referred to here is given in the Appendix as Table A-6. Each letter represents a specific ratio of starting kVA input per unit rated horsepower. Table A-6 is used to determine the size of fuse required for a given motor. The Code letters in Table A-6 apply to all ac motors, single-phase or polyphase, whereas the NEMA class letters refer to induction motors only. The two groups of letters are unrelated: NEMA class letters have nothing whatever to do with NEC Code letters.

SEC. 9-21. / *Commercial Induction Motor Classification*

regulation (about 3 to 5 per cent). But its starting current unfortunately, varies from five to seven times the normal rated current, making it less desirable for line starting, particularly in large capacity sizes. In sizes below 5 horsepower, however, the Class A SCIM is frequently started across the line; and, because of its rapid acceleration, it does not produce extremely high current effects, which are undesirable.

9-21.2
Class Letter B SCIM

This letter designates a SCIM that is sometimes called the *general purpose* motor. As shown in Fig. 9-16, its slip-torque curve resembles the normal (Class A) motor quite closely. Its rotor is embedded somewhat deeper in slots than the normal Class A SCIM, and the greater depth tends to increase the starting and running rotor reactance. The increase in reactance at starting reduces the starting torque a little but *reduces* the *starting current*, as well. A slightly lower value of field excitation is also used on this motor to produce the reduced starting current and characteristic shown in the figure. Starting currents vary from 4.5 to 5 times the rated current; and in the larger sizes, above 5 horsepower, reduced-voltage starting is still employed with this class of motor. Because of their somewhat lower starting current and almost equivalent characteristics, Class B SCIMs are generally preferred over Class A motors in the larger sizes.

9-21.3
Class Letter C SCIM

The SCIM identified by class letter C is a double-cage rotor motor described in the preceding section and shown in Fig. 9-15. It develops a higher starting torque from 2 to 2.5 times the rated torque, compared to Classes A and B, and a (lower) starting current from 3.5 to 5 times the rated current. Because of its high starting torque, it accelerates rapidly. When used with heavy, high-inertia loads, however, the thermal dissipation of the motor is limited, since most of the current concentrates in the upper winding. Under conditions of frequent starting the rotor may have a tendency to overheat. It is better suited to sudden large loads but of a kind having low inertia. Figure 9-16 shows that this motor continues to develop increased torque with increased slip all the way to maximum torque at standstill. The Class C SCIM, however, has poorer speed regulation than Classes B and A, as shown in Fig. 9-16.

9-21.4
Class Letter D SCIM

The class letter D designates a SCIM known as the *high-torque*, *high-resistance rotor* motor. The rotor bars are constructed of a high-resistance alloy, and they are placed in slots close to the surface or are embedded in small-diameter slots. The ratio of rotor resistance to reactance at starting is higher than in motors of the preceding class letters. As shown in Fig. 9-16, the starting torque of these motors approaches 3 times the rated torque, with a starting current from 3 to 8 times the rated load, depending on design. This motor is designed for heavy starting duty; but again, like the Class C SCIM, it is not recommended for frequent starts because of small cross section and poor thermal dissipating ability. It finds its best application in loads such as shears or punch presses which call for higher torque with the application of a sudden load. The speed regulation of this SCIM is the poorest of all classes, as shown in Fig. 9-16.

9-21.5 Class Letter F SCIM

The SCIM identified by class letter F is known as the *low-torque, double-cage* rotor motor. It is designed primarily as a *low-starting-current* motor, since it requires the *lowest starting current* of all classes from A to D. The Class F SCIM has very high rotor resistance in both its starting and its running windings, tending to increase the starting and running impedance, and reducing the starting and running current. The Class F motor was designed to replace the Class B motor. The Class F SCIM produces starting torques of about 1.25 times the rated torque and low starting currents from two to four times the rated current. Class F motors are manufactured generally in sizes above 25 horsepower for across-the-line service. Because of the relatively high starting and running rotor resistance, these motors have poorer speed regulation than the Class B motors, low overload capacity, and usually low running efficiency. When started with light loads, however, the low starting currents eliminate the necessity for reduced-voltage equipment, even in large sizes.

Table A-9 in the Appendix lists the starting currents and torques for 220 V, three-phase induction motors in sizes from $\frac{1}{2}$ hp to 200 hp for the five classes discussed above.

9-21.6 Speed Control of Cast Rotor Motors

Unlike the WRIM, the SCIM does not lend itself easily to speed control. Further, the poor speed regulation of some classes, such as Class D, shown in Fig. 9-16, may create the necessity for some means of speed variation to obtain a desired speed at the given load.

There are 3 possible ways to vary the speed of SCIMs, namely:

1. by changing the applied frequency to the stator,
2. by changing the number of poles of the stator (and consequently the rotor),
3. by reducing the applied voltage to the stator.

The second method, pole-changing, requires special motors designed for changing poles by means of suitable switching in which the windings are connected in series and parallel combinations. Such motors are called *multispeed* SCIMs. Multispeed SCIMs are available in 2-speed or 4-speed switching combinations but such techniques are limited to smaller polyphase SCIMs.

The third possible method, reducing the applied voltage to the stator while running at a given load, as shown by Eq. (9-13), reduces the maximum and rated torque in proportion to the square of the voltage applied to the stator. While this method is used extensively in single-phase SCIMs, it does *not* lend itself to polyphase motors for two reasons: 1) the applied voltage must be reduced considerably in order to produce the necessary reduction in speed and torque. This makes the speed regulation of the motor very poor and even somewhat unstable with changes in load 2) the rotor and stator currents increase as the voltage is decreased, maintaining the torque fairly constant. This is why the stator voltage must be reduced considerably in order to effect a change in speed, again producing very poor speed regulation.

The most useful method, therefore, is the first, involving a change in frequency and voltage applied to the stator, since it does not require special motors and applies equally to all classes of SCIMs. Polyphase SCIMs are often rated 50/60 Hz, but this does not imply that there is no difference in operation. At 5/6 of the frequency applied to the stator, the motor runs at 5/6 of the speed, in accordance with Eq. (9-1a), over its entire speed-load characteristic. Further, in order to maintain the same stator flux (and flux density), the stator voltage must be reduced in the *same proportion*. Unfortunately, at the lower speed, the ability of the motor to self-ventilate is reduced. At the same time, due to the lower stator voltage, the input stator current and power has increased (in the ratio of 6/5) producing more internal heat. The efficiency is also reduced. It is obvious, therefore, that an attempt to obtain lower speeds by frequency and stator voltage reduction, is somewhat limited.

The reverse is the case with higher speeds and, in view of the poorer speed regulation of some SCIM classes, higher speeds may be desired. At higher frequencies and correspondingly higher stator voltages, the input current and power are reduced, the efficiency and speed increased, and because of the increase in self-ventilation, the rating is increased. It is precisely for this reason that 400 Hz motors of the same hp are much smaller in size than 60 Hz motors.

In general, it may be said that SCIMs designed for operation at 50/60 Hz may be operated over a range of 25 to 180 Hz with the understanding

1) that the stator voltage must be increased or decreased in the same proportion as the frequency,* and
2) that at the lower frequencies and speeds, the rating is reduced, as is the efficiency.

It is generally accepted that to obtain speed control of a polyphase SCIM by frequency variation, the power source must be a polyphase electronic cycloconvertor or alternator driven by a variable speed prime

---

* See Probs. 12-2 and 12-3, as well as Exs. 13-4 and 13-5, Sec. 13-2.

mover. As the alternator speed is increased, *both* the alternator voltage and frequency are increased, which simplifies matters greatly in the control of speed by changing the frequency. No special voltage adjustments are necessary *but* it should be noted that such a method of speed control, particularly in larger hp sizes, is extremely costly. For this reason, various mechanical gear and cone pulley arrangements are often used in lieu of the above electrical methods.

9-22. INDUCTION GENERATOR

The slip-torque curves of Fig. 9-16 indicate (up to the point of maximum torque) that the torque is proportional to slip, i.e., as the slip decreases, the torque decreases. When the slip is zero at synchronous speed, the torque is zero. This might imply that, if the ac induction dynamo is driven at a speed exceeding synchronous speed, i.e., at a *negative* slip, the torque would be negative (the motor receives rather than delivers mechanical power) and generator operation occurs. The transition from motor to generator mode is a function of slip.

Unlike the ac synchronous alternator, which is driven at a synchronous speed with respect to other alternators feeding a bus, the induction generator *must* be driven at a speed *above synchronous* speed in order to deliver power *to* the bus. The output of the induction generator depends, therefore, on the *magnitude of the negative* slip, or on how fast *above synchronous* speed the rotor is being driven in the *same direction* of rotation that occurred when it operated as an induction motor. Any induction motor, regardless of type, will operate as an induction generator, although the low-resistance types of rotors produce higher currents for the same induced voltage.*

The *induction generator* is *not* a self-excited generator. It is thus necessary to excite the stator with an external polyphase source at its rated voltage and frequency. It will generate only when it is connected to a source of fixed voltage and frequency and if it is then driven at a speed above the synchronous speed set by the supply frequency. Figure 9-17 shows the phasor conditions for an induction motor operating at a lagging current

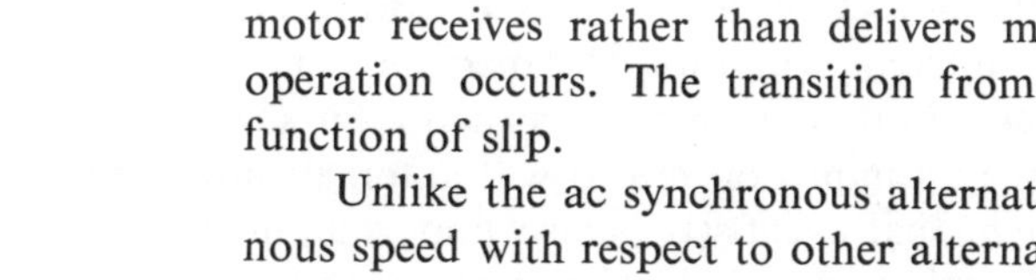

Figure 9-17
Approximate induction motor and induction generator phasor relations.

---

* Above synchronous speeds, induction generator action serves as a means of dynamic breaking automatically. Below synchronous speeds any polyphase SCIM is braked, dynamically, by (1) removing ac power from the stator and (2) replacing it with dc. The motor becomes a dc generator with a shorted armature producing heavy rotor currents and bringing the motor quickly to a stop. See Kosow, *Control of Electric Machines*, Prentice-Hall, 1973, Ch. 7.

at some power factor angle θ_m; the component of stator magnetizing current I_m is lagging the supply voltage by 90°. When the motor is driven by an external prime mover at a rate above synchronous speed, a generating voltage E_g produces a generating current I_g.

The generating current has as its component the exciting stator current, I_m, and the generating current I_g will always *lead* its generated voltage, E_g. As the negative slip is increased, the generated voltage, E_g, increases, increasing the generated leading current, I_g, to the bus. At about -0.05 slip, or -5 per cent, the induction generator is fully loaded (the rotor bars and stator windings carrying rated current). Because the induction generator *always* supplies a *leading current* to the line, it has possible application in power factor correction. It is not used for this purpose, however, since it requires a prime mover, whereas an overexcited synchronous capacitor does not.

The principle of the induction generator is important, however, in the *dynamic braking* of induction motors; the machine acts as an overspeed device and produces braking action when the motor speed exceeds synchronous speed, since an induction motor will automatically become an induction generator in that event. (It is also important in computing short-circuit system protection when induction motors are used.)

9-23. INDUCTION FREQUENCY CONVERTERS

The induced voltage in the rotor of a wound-rotor induction motor (WRIM) may be picked off the rotor slip-rings. If a WRIM is *driven* at any slip speed, the *induction generator* will act as a *frequency converter* merely by varying the rotor speed and rotor direction. For example, if the stator of an induction dynamo is excited by a 60 Hz, driving the dynamo in the same direction (as its motor rotation) will produce frequencies below 60 Hz supply, and at synchronous speed the frequency is zero. At standstill, the generated frequency is 60 Hz; and, when driven in the opposite direction at synchronous speed, the dynamo will generate 120 Hz. The general equation for the frequency of an induction converter is

$$f_{con} = f_{syn}\left(1 \pm \frac{S_{con}}{S_{syn}}\right) \qquad (9\text{-}24)$$

where f_{syn} is the synchronous stator frequency in hertz
S_{syn} is the synchronous speed ($120f/P$) in rpm
S_{con} is the speed of the converter in rpm

The *plus* sign in Eq. (9-24) is used when the direction of rotation is *opposite* to that of the stator rotating field, and the *minus* sign is used for a direction which is the *same* as that of the rotating field.

EXAMPLE 9-14: An eight-pole WRIM operating on a 60 Hz supply is driven by a variable speed prime mover as a frequency changer.

a. If it is operated first at 1800 rpm in the opposite direction, and then at 450 rpm in the same direction, what frequencies will result?
b. Calculate the speeds and directions required to obtain frequencies of:
  1. 25 Hz
  2. 400 Hz
  3. 120 Hz.

Solution: Using Eq. (9-24),

a. $f_{con} = f\left(1 + \dfrac{1800}{900}\right) = 60 \times 3 = $ **180 Hz** for 1800 rpm in the opposite direction as rotating magnetic field

$f_{con} = 60\left(1 - \dfrac{450}{900}\right) = $ **30 Hz** for 450 rpm in the same direction

b. 1. $25 = 60\left(1 + \dfrac{S_{con}}{900}\right)$

$S_{con} = \left(-1 + \dfrac{25}{60}\right)900 = \left(-\dfrac{35}{60}\right)900 = $ **−525 rpm**, or 525 rpm in the same direction as rotating magnetic field

2. $400 = 60\left(1 + \dfrac{S_{con}}{900}\right)$

$S_{con} = \left(-1 + \dfrac{400}{60}\right)900 = \left(\dfrac{340}{60}\right)900 = $ **5100 rpm** in the opposite direction

3. $120 = 60\left(1 + \dfrac{S_{con}}{900}\right)$

$S_{con} = \left(-1 + \dfrac{120}{60}\right)900 = (1)900 = $ **900 rpm** in the opposite direction compared to rotating magnetic field.

The magnitude of the induced rotor voltage of the frequency changer is purely a function of the relative rotor speed with respect to stator flux (as is the frequency produced). In Example 9-14(a), the voltage at 180 Hz is six times the voltage at 30 Hz, since a speed of 1800 rpm in the direction opposite to that of the synchronous speed (producing three times the voltage at standstill) bears a six-to-one ratio to a speed of 450 rpm in the same direction as that of the synchronous speed (producing one-half the voltage at standstill).

The use of WRIMs as frequency converters also finds application in the speed control of induction motors by concatenation and in various secondary "foreign voltage" techniques.*

BIBLIOGRAPHY

A-C Induction Motors. (ASA C50.2.) New York: American Standards Association.

* For a discussion of commercial methods of speed control of polyphase induction motors see Kosow, *Control of Electric Machines*, Prentice-Hall, 1973, Ch. 7.

Alger, P. L. *The Nature of Polyphase Induction Machines*, New York: John Wiley & Sons, Inc., 1951.

———— and Erdelyi, E. "Electromechanical Energy Conversion," *Electro-Technology*, September 1961.

———— and Ku, Y. H. "Speed Control of Induction Motors Using Saturable Reactors," *Electrical Engineering*, February 1957.

Best, I. W., Jr. "Applying Squirrel-Cage Induction Motors," *Electrical Manufacturing*, November 1956.

Bewley, L. V. *Alternating Current Machinery*, New York: The Macmillan Company, 1949.

Bewley, L. V. *Tensor Analysis of Electrical Circuits and Machines*, New York: Ronald Press, 1961.

Buchanan, C. H. "Duty-Cycle Calculations for Wound-Rotor Motors," *Electrical Manufacturing*, November 1959.

Carr, C. C. *Electrical Machinery*, New York: John Wiley & Sons, Inc., 1958.

Cook, J. W. "Squirrel-Cage Induction Motors Under Duty-Cycle Conditions," *Electrical Manufacturing* February 1956.

Cooney, J. D. "Applying ac Motors to Specific Types of Load," *Electrical Manufacturing*, July 1956.

Crosno, C. D. *Fundamentals of Electromechanical Conversion*, New York: Harcourt, Brace, Jovanovich, Inc, 1968.

Daniels. *The Performance of Electrical Machines*, New York: McGraw-Hill, Inc., 1968.

Fitzgerald, A. E. and Kingsley, C. *The Dynamics and Statics of Electromechanical Energy Conversion*, 2nd Ed., New York: McGraw-Hill, 1961.

Fitzgerald, A. E., Kingsley, Jr. C. and Kusko, A. *Electric Machinery*, 3rd Ed. New York: McGraw-Hill Book Company, 1971.

Fitzpatrick, D. "Reduced-Voltage Starting for Squirrel-Cage Motors," *Electrical Manufacturing*, March 1960.

Gemlich, D. K. and Hammond, S. B. *Electromechanical Systems*, New York: McGraw-Hill, 1967.

Hindmarsh, J. *Electrical Machines*, Elmsford, N.Y.: Pergamon Press, 1965.

Jones, C. V. *The Unified Theory of Electrical Machines*, New York: Plenum Publishing Corp., 1968.

Karr, F. R. "Squirrel-Cage Motor Characteristics Useful in Setting Protective Devices," AIEE Paper, 59–13.

Liwschitz, M. M., Garik M., and Whipple, C. C. *Alternating Current Machines*, Princeton, N. J.: D. Van Nostrand Co., Inc., 1946.

McFarland, T. E. *Alternating Current Machines*, Princeton, N. J.: D. Van Nostrand Co., Inc., 1948.

Majmudar, H. *Introduction to Electrical Machines*, Boston: Allyn and Bacon, 1969.

Meisel, J. *Principles of Electromechanical Energy Conversion*, New York: McGraw-Hill, 1966.

Nasar, S. A. *Electromagnetic Energy Conversion Devices and Systems*, Englewood Cliffs, N.J.: Prentice-Hall, Inc., 1970.

O'Kelly and Simmons. *An Introduction to Generalized Electrical Machine Theory*, New York: McGraw-Hill, 1968.

Puchstein, A. F., Lloyd, R., and Conrad, A. G. *Alternating Current Machines* 3rd Ed., New York: Wiley/Interscience, 1954.

Robertson, B. L. and Black, L. J. *Electric Circuits and Machines*, 2nd Ed., Princeton, N. J.: D. Van Nostrand Co., Inc., 1957.

Schmitz, N. L., and Novotny, D. W. *Introductory Electromechanics*, New York: Ronald Press, 1965.

Schohan, G. "Static Frequency Multipliers for Induction Motors," *Electrical Manufacturing*, April 1956.

Seely, S. *Electromechanical Energy Conversion*, New York: McGraw-Hill, 1962.

Selmon. *Magnetoelectric Devices: Transducers, Transformers and Machines*, New York: Wiley/Interscience, 1966.

Skilling, H. H. *Electromechanics: A First Course in Electromechanical Energy Conversion*, New York: Wiley/Interscience, 1962.

Thaler, G. J., and Wilcox, M. L. *Electric Machines: Dynamics and Steady State*, New York: Wiley/Interscience, 1966.

White, D. C., and Woodson, H. H. *Electromechanical Energy Conversion* New York: Wiley/Interscience, 1959.

Woll, R. F. "Applying the Wound-Rotor Motor," *Westinghouse Engineer*, March 1953.

QUESTIONS

9-1. Describe differences in construction between a synchronous motor and a squirrel cage induction motor (SCIM) with respect to
    a. stator construction
    b. rotor construction.

9-2. Give two types of induction motor rotor construction and describe each.

9-3. With respect to Quest. 9-2,
    a. which type of rotor permits introduction of a "foreign" voltage?
    b. how is this rotor made equivalent to a SCIM?

9-4. Reproduce Fig. 9-1 completely for a 12-slot armature having 2 slots/pole-phase producing a 4-pole magnetic field (Hint: see Fig. 9-2). Show:
    a. the resultant magnetic fields produced by each of the coil groups
    b. the rotation of the resultant fields for each instant in time
    c. that the phase shift of the resultant fields exactly corresponds to the time-phase shift of the frequency of the supply.

9-5. If n is the number of slots/pole-phase, show that the number of poles, P, corresponding to that produced in both the stator and the rotor of a three phase SCIM is $P = 2n$.

9-6. Explain how the nonsalient stator armature of a synchronous motor or an induction motor can produce
   a. the same number of poles on a SCIM rotor
   b. a rotating field whose synchronous speed is represented by $120f/P$ or $120f/2n$
   c. a stationary magnetic field (Hint: dc is a waveform of zero frequency).

9-7. a. Explain why the rotor of a SCIM is always forced to rotate in the same direction as the rotating magnetic field.
   b. Explain why reversing any two line leads reverses the direction of rotor rotation.
   c. Explain why the rotor speed is essentially a function of stator synchronous speed (which in turn is a function of stator frequency) and slip.

9-8. The induction motor is sometimes called a variable frequency transformer. In the light of this statement, explain:
   a. the conditions under which the rotor and stator frequencies are the *same*
   b. the conditions under which the rotor frequency is *smaller* than the stator frequency
   c. the conditions under which the rotor frequency is *greater* than the stator frequency
   d. the conditions under which the rotor (secondary) induced voltage and current are a maximum
   e. the conditions under which the rotor (secondary) induced voltage and current are a minimum.

9-9. a. Give 3 equations representing starting locked-rotor torque of a SCIM.
   b. Explain why starting torque of any *given* SCIM is essentially a function of voltage applied to the stator.
   c. If the stator voltage applied to a SCIM is half-rated voltage, what starting torque is produced?
   d. Repeat part (c) above for starting current produced.
   e. Is there any advantage in reducing the starting current of a SCIM?

9-10. a. State the equation representing the torque at any value of slip, defining all terms.
   b. State the equation representing the maximum torque developed in any SCIM, defining all terms. On what three factors does maximum torque depend?
   c. State, in equation form, under what conditions of slip maximum torque occurs.
   d. In part (c) which is the variable term and which is the fixed term for a given SCIM?
   e. Explain the expression "maximum torque occurs at that value of slip when the rotor reactance equals the effective rotor resistance", (see Eqs. 9-3, -12).

        f. Why must slip increase as load increases?

9-11. If the breakdown slip in Fig. 9-10b due to R_r is 0.2, find:
    a. the ratio of locked-rotor reactance to resistance at slip, s, and slip $4 \times s$
    b. the ratio of resistance to locked-rotor reactance at slip s and slip $4 \times s$
    c. which of the above is the s_b ratio?
    d. which curves in Fig. 9-10b produce an s_b ratio less than unity?
    e. which curves produce an s_b ratio greater than unity?
    f. at what slip is the rotor resistance always equal to the locked-rotor reactance?
    g. which curves in Fig. 9-10b have a higher locked-rotor reactance than resistance at the instant of starting (standstill)?
    h. which curves have a higher rotor resistance than locked-rotor reactance at standstill?

9-12. From the curves shown in Fig. 9-6, find:
    a. the approximate slip at which maximum torque occurs
    b. the slip at which starting torque occurs
    c. the approximate slip at which a value of running torque equals starting torque
    d. the approximate rotor current corresponding to the running torque in (c), above
    e. approximate slip at rated-load torque
    f. the approximate rotor current corresponding to the running torque in (e) above.

9-13. Explain why the PF (power factor) of a SCIM (squirrel-cage induction motor) is
    a. low at light loads
    b. low at heavy loads (reason not the same as in above)
    c. fairly high at approximately rated load
    d. important in determining the amount of rotor and stator current which must flow to develop the necessary torque to meet applied torque.

9-14. Explain why the efficiency of a SCIM is
    a. low at light loads
    b. low at heavy loads (reason not the same as in above)
    c. fairly high at approximately rated load
    d. a maximum, and the conditions at which the maximum occurs.

9-15. Explain why the torque of a SCIM is
    a. a maximum at loads well beyond rated output
    b. greater at starting than at rated loads
    c. essentially a function of effective rotor resistance.

9-16. Explain for a WRIM why
    a. it is possible to change effective rotor resistance
    b. a change in rotor resistance produces a change in starting torque
    c. increasing starting torque decreases starting current
    d. the maximum torque is the same for any value of rotor resistance up to the value of standstill rotor reactance
    e. starting torque is less than the maximum torque for values of rotor resistance greater than standstill (locked-rotor) reactance

f. current is reduced and PF increased under conditions of (e) above.

9-17. For a WRIM having a given rotor resistance, is it possible to have *two* values of:
a. slip producing the same torque? Explain.
b. torque producing the same rotor and stator current? Explain.
c. total rotor resistance which will still produce the same starting torque? Explain.
d. maximum torque? Explain.

9-18. Using the curves of Figs. 9-6 and 9-10 and related Eqs., explain why
a. the rotor current must always be a maximum at maximum torque rather than at the torque produced at instant of starting
b. the rotor current necessary to produce the developed torque varies inversely as the PF of the rotor
c. minimum starting current is obtained at a rotor PF of 0.707 at starting
d. increasing rotor PF above 0.707 decreases starting current but also decreases starting and maximum developed rotor torque.

9-19. a. Explain why the rotor resistance at standstill is smaller than the effective loaded rotor resistance under running conditions.
b. Express the relation between effective loaded rotor resistance under running conditions and the standstill rotor resistance in terms of an equation.
c. Using the terms of (b) above, define:
1. rotor power developed per phase
2. rotor copper loss, per phase
3. rotor power input per phase.
d. Express the terms of (c) above in a single equation.

9-20. The locked-rotor test for an induction motor is both simple and significant because it yields measurements of the locked-rotor voltage E_{lr} and locked-rotor reactance, X_{lr}, in addition to rotor resistance, R_r. Given the values of these 3, and the value of rated rotor current, I_r show, giving equations, how it is possible to compute:
a. maximum power input at breakdown
b. maximum (breakdown) torque
c. breakdown (maximum torque) slip
d. speed at which maximum torque is developed
e. rotor power input at any value of slip
f. torque at any value of slip
g. rotor power input at instant of starting
h. starting torque.

9-21. In the measurement of slip, explain why
a. measurement of rotor speed directly is never used to compute slip
b. an optical method is preferable to other methods of slip measurement
c. the stroboscopic method sometimes makes it difficult to produce accurate measurements of slip at heavier loads, beyond rated. What alternate method may be used?

d. why the number of poles produced by the stator is of significance in the proper choice of an optical disc used in the stroboscopic method.

9-22. In starting induction motors across the line, explain conditions under which a
a. high starting current is not objectionable
b. high starting current may be objectionable
c. WRIM may reduce its starting current
d. SCIM may reduce its starting current.

9-23. Compare the autotransformer, primary resistor, primary reactor and wye-delta methods of starting, using a 57.8% reduction in line voltage, as to:
a. starting current
b. starting torque
c. Which of the above methods produces the least primary (stator) starting current and why?

9-24. Compare part-winding starting with the methods described in Quest. 9-23 as to starting current and starting torque.

9-25. a. Explain, in terms of starting torque and starting current, why a line-starting double-cage rotor SCIM is preferable to a normal SCIM.
b. For the double-cage rotor, explain current distribution and power factor of the current, for both upper and lower rotor conductors on
1. starting conditions
2. running conditions.

9-26. For the 5 classes of SCIMs, make a table having the following *columns*: SCIM class, starting torque, starting current, maximum torque, speed regulation and application.
Compare each of the 5 classes (A, B, C, D, and F) by filling in all *rows* of the table.

9-27. a. List 3 possible ways that SCIMs having cast rotors may be speed controlled.
b. Which of these 3 methods is most universal in application to all SCIMs?
c. If the frequency applied to a SCIM stator is changed, why must the voltage be changed in the same proportion? (Hint. See Ex. 13-4.)
d. What are the disadvantages which result when speed is reduced by a reduction in frequency and voltage?
e. What are the advantages which accrue when speed is increased by an increase in frequency and voltage?
f. Why is a polyphase alternator driven by a variable speed prime mover ideally suited as the voltage source for speed control of a polyphase SCIM?

9-28. a. Why are WRIMs necessary?
b. Compare starting characteristics of WRIMs with SCIMs.
c. Compare running characteristics of WRIMs with SCIMs.
d. Give at least two applications of WRIMs not possible with SCIMs.

9-29. a. Under what conditions of slip does the asynchronous dynamo operate in the generator mode?

b. Why does an induction generator always supply a leading current to the bus? Explain, using phasor diagrams, how load of an induction generator is controlled.
c. Compare the use of an induction generator vs an overexcited synchronous motor as a power factor correction device, in light of (b) above.
d. Give one possible application for operating a SCIM in the generator mode. What are the limitations of such operation?

9-30. a. Under what conditions is a WRIM useful as a frequency converter?
b. How is it possible to obtain frequencies below the synchronous stator frequency?
c. Repeat (b) for frequencies above the synchronous stator frequency.
d. What is the advantage of using a WRIM as a frequency generator over a synchronous motor-generator set (See Sec. 8-23)?

PROBLEMS

9-1. The full load slip of a 60 Hz, 12-pole SCIM is 5.0 per cent. Calculate:
a. Full-load speed
b. The synchronous speed
c. Speed regulation.

9-2. A 60 Hz, six-pole SCIM has a full-load speed of 1140 rpm. Calculate:
a. Synchronous speed
b. Full-load slip
c. Speed regulation.

9-3. The rotor of an eight-pole, 3ϕ, 60 Hz, 208 V WRIM has 60 per cent of the number of turns per phase of the stator. The stator is delta connected and the rotor is wye connected and brought out to slip rings. Calculate rotor frequency and rotor voltage between slip rings under the following conditions
a. Rotor at standstill
b. Rotor at its full-load slip of 9 per cent
c. The rotor is driven by another motor in a direction opposite to the rotating magnetic field of the stator at a speed of 600 rpm
d. Repeat (c) when the speed is 900 rpm.

9-4. The rotor resistance and reactance of a SCIM rotor at standstill are 0.1 Ω/phase and 0.8 Ω/phase, respectively. Assuming a transformer ratio of unity, from the eight-pole stator having a phase voltage of 120 V at 60 Hz to the rotor secondary, calculate
a. Rotor starting current per phase
b. Rotor current at a full load slip of 6 per cent
c. Rotor current at that value of slip producing maximum torque.

9-5. A six-pole, 25 hp, 60 Hz, 440 V SCIM has a starting torque of 152 lb-ft and a full-load torque of 113 lb-ft. Calculate:
a. The starting torque when the stator line voltage is reduced to 300 V

b. The voltage which must be applied to the stator in order to develop a starting torque equal to the full load torque
c. The voltage which must be applied in order to operate motor at rated load from a 50 Hz source.

9-6. The starting current of the motor in Problem 9-5 is 128 A when rated voltage is impressed on the stator. Calculate:
a. Starting current when the voltage is reduced to 300 V
b. The voltage which must be applied to the stator in order not to exceed the rated line current of 32 A.

9-7. The locked rotor reactance of a six-pole, 60-Hz SCIM is three times the rotor resistance per phase. Calculate:
a. Slip
b. Speed
c. Rotor frequency at which maximum torque is developed.

9-8. From the data given for the SCIM of Problem 9-4, calculate
a. Rotor power input for maximum torque in watts [use Eq. (9-17)]. Check your answer using Eq. (9-21).
b. Maximum torque developed by the rotor in lb-ft
c. Torque at a slip of 5 per cent.

9-9. The rated speed of a six-pole, 60 Hz, three-phase induction WRIM is 1120 rpm with the wound rotor slip rings short circuited. Calculate the rated speed when
a. An external resistance per phase is added which equals the rotor resistance
b. Repeat (a) for an external resistance which is twice the rotor resistance
c. Repeat (a) for an external resistance which is three times the rotor resistance.

9-10. The full-load speed of a 12-pole, 60 Hz WRIM is 550 rpm. The standstill rotor reactance is 2 Ω and the rotor resistance per phase is 0.6 Ω, respectively. Calculate:
a. Slip and speed at which maximum torque is developed
b. The added rotor resistance to develop maximum torque at starting
c. The new full-load slip with added rotor resistance
d. Speed regulation with added rotor resistance
e. Ratio of full-load speeds with and without external rotor resistance.

9-11. In a six-pole, 60 Hz, polyphase SCIM, the full-load speed and torque is 1160 rpm and 7.2 lb-ft, respectively. The induced rotor emf is 50 V per phase. The standstill reactance is 0.8 Ω and rotor resistance 0.2 Ω/phase, respectively. For slips of 1.0, 0.75, 0.5, 0.25, 0.1, 0.05, 0.0333, 0.02 and 0.01, respectively, calculate and tabulate the following quantities: rotor emf, reactance, impedance, current and power factor, per phase.

9-12. From the tabulated data of Problem 9-11, calculate
a. Three-phase rotor power input using Eq. (9-3), $E_r I_r \cos\theta$, at the instant of starting
b. Three-phase rotor power input using Eq. (9-17) at the instant of starting
c. Starting torque using Eq. (9-19)

d. Maximum or breakdown torque using Eq. (9-23)
e. Full-load torque, from starting torque using Eq. (9-23). Compare with value given in Problem 9-11.

9-13. Using the value of full-load torque given in Problem 9-11
a. Calculate the torques at each value of slip tabulated in Problem 9-11
b. Plot a slip-torque curve using slip as ordinate and torque as abscissa.

9-14. Using the values of rotor current calculated in Problem 9-11 for each value of slip calculate
a. Rotor power input [using Eq. (9-17)]
b. Torque [using Eq. (9-19)]. Compare with values obtained in Problem 9-13.

9-15. The delta-connected stator of a three-phase, 60 Hz, six-pole 220 V WRIM has twice as many turns per phase as the rotor. The rotor resistance is 0.1 Ω/phase and the standstill reactance 0.5 Ω per phase. The full-load speed is 1140 rpm. Calculate:
a. Locked-rotor emf per phase, frequency, rotor current, and voltage between slip rings
b. Slip at which maximum torque will occur and corresponding rotor-phase current and total rotor power input from slip and from locked rotor values
c. Maximum breakdown torque
d. Starting and full-load torques

9-16. The total power supplied to a three-phase SCIM is 4000 W and corresponding stator losses are 150 W. Calculate:
a. Rotor power loss when the slip is 4 per cent
b. Total mechanical power developed
c. Output horsepower of motor if the friction and windage losses are 80 W
d. Over-all motor efficiency.

ANSWERS

9-1(a) 570 rpm (b) 600 rpm (c) 5.27 per cent   9-2(a) 1200 rpm (b) 5 per cent (c) 5.26 per cent   9-3(a) 215.5 V   60 Hz (b) 19.4 V, 5.4 Hz (c) 360 V, 100 Hz (d) 431 V, 120 Hz   9-4(a) 148.8 A (b) 64.8 A (c) 106 A   9-5(a) 70.7 lb-ft (b) 379 V (c) 367 V   9-6(a) 87.3 A (b) 110 V   9-7(a) $\frac{1}{3}$ (b) 800 rpm (c) 20 Hz   9-8(a) 9000 W (b) 70.4 lb-ft (c) 36.1 lb-ft   9-9(a) 1040 rpm (b) 960 rpm (c) 880 rpm   9-10(a) 0.3, 420 rpm (b) 1.4 Ω (c) 0.278 (d) 38.5 per cent (e) 0.807: 1   9-11 For a slip of 0.25, only: 12.5 V, 0.2 Ω, 0.283 Ω, 44.2 A, 0.707 PF   9-12(a) 2210 W (b) 2210 W (c) 12.95 lb-ft (d) 27.5 lb-ft (e) 7.22 lb-ft   9-13(a) Torques, starting with unity slip: 12.95 lb-ft, 16.5 lb-ft, 22.0 lb-ft, 27.5 lb-ft, 18.95 lb-ft, 10.56 lb-ft, 7.20 lb-ft, 4.375 lb-ft, 2.20 lb-ft   9-14(a) Rotor power inputs only: 2210 W, 2800 W, 3780 W, 4680 W, 3240 W, 1800 W, 1230 W. 745 W, 375 W   9-15(a) 110 V, 60 Hz, 215 A, 190.5 V (b) 0.2, 155.5 A, 36,300 W (c) 213 lb-ft (d) 81.9 lb-ft   9-16(a) 154 W (b) 4.95 hp (c) 4.85 hp (d) 0.904

TEN

single-phase motors

10-1.
GENERAL

The two preceding chapters were confined primarily to polyphase synchronous and asynchronous motors. There are numerous occupancies, industrial as well as residential, to which the electric utility has only brought a *single-phase* ac service. In all occupancies, furthermore, there is usually a need for small motors which will operate from a single-phase supply to drive various electric appliances such as sewing machines, drills, vacuum cleaners, air conditioners, etc. Generally, the term "small motor" means a motor of less than one horsepower, i.e., a *fractional horsepower* motor,* and most single-phase motors

---

* A small motor as defined by the American Standards Association (ASA) and the National Electrical Manufacturers Association (NEMA) is "a motor built in a frame smaller than that having a continuous rating of 1 hp, open type, at 1700 to 1800 rpm." Small motors are generally considered fractional hp motors, but since the determination is based on frame size, the following comparisons are of interest:
  1. A $\frac{3}{4}$ hp, 900 rpm, motor is not considered a fractional hp motor because its frame size, if used for an 1800 rpm motor, would yield a rating of more than 1 hp. Therefore, it is considered an *integral* hp motor of 0.75 hp $\times \frac{1800}{900} =$ 1.5 hp.
  2. A 1.5 hp, 3600 rpm, motor *is* a fractional hp motor because its frame size, if used for an 1800 rpm motor, would yield a rating of less than 1 hp of 1.5 hp $\times \frac{1800}{3600} = 0.75$ hp.

are indeed fractional horsepower motors. But single-phase motors are manufactured also in standard integral horsepower sizes: 1.5, 2, 3, 5, 7.5, and 10 hp for both 115 V and 230 V single-phase services and even for 440 V service in the 7.5 and 10 hp range (see Table A-4, Appendix). Special integral horsepower sizes range from several hundred up to a few thousand horsepower in locomotive service using ac series motors, single-phase.

The basic principles of single-phase ac motors are inherently those studied previously. The induction principle is extensively employed because of the simplicity of the rotor and because it obviates commutation difficulties. Various techniques are used to produce the rotating magnetic field required for starting with induction operation employing phase-splitting and shaded poles. It will be shown, however, that an induction rotor, once started by a rotating magnetic field, will continue to operate from a single-phase supply. Therefore, other starting techniques are employed, such as in reluctance start and repulsion start induction motors.

Single-phase synchronous, reluctance, and hysteresis motors were discussed in Sections 8-27 and 8-28, and subsynchronous motors in Section 8-29, primarily because their torque characteristics are so different from other classes of single-phase motors. The principle of the dc commutator motor is also used, mainly in fairly large integral horsepower sizes as the ac series motor and in fractional horsepower ratings as the *universal* motor. Small universal motors operate with voltages as low as 6 V or even lower (in the case of solar battery motors) at frequencies from direct current up to several hundred hertz, with speed ranges as high as 20,000 rpm.

The load and duty requirements for single-phase motors are easily as severe as those for polyphase machinery and perhaps even more so, because of a lack of routine maintenance procedures in domestic or residential occupancies. Single-phase ac series motors are designed for extremely rugged use in cranes, hoists, and traction service (electric locomotives) and these may range in size from a few horsepower up to several thousand. Single-phase synchronous alternators and motors are used quite commonly in railway service. Generally, the larger capacity machines are really unbalanced, three-phase, wye-connected, synchronous dynamos with one phase open and, as such, are used up to several thousand horsepower in M-G sets on locomotives.

Because a single-phase *induction* motor *is inherently not self-starting*, i.e., it does *not* have the true *revolving* magnetic field that is fundamental to the polyphase induction motor, various methods are employed to *initiate* rotation of the squirrel-cage rotor. As a consequence, a classification of single-phase induction motors based on the particular starting method has emerged (see Section 10-18), and these motors will be considered first, followed by a discussion of commutator-type single-phase motors.

**10-2.
CONSTRUCTION OF
THE SINGLE-PHASE
INDUCTION MOTOR**

The *rotor* of any single-phase induction motor is interchangeable with that of a polyphase SCIM (Class A or B), see Sec. 9-21. There is no *physical* connection between the rotor and the stator, and there is a uniform air gap between the stator and the rotor. The stator slots are distributed uniformly, and usually a single-phase double-layer lap "part-winding" is employed. A "simple" single-phase winding would produce no rotating magnetic field and starting torque for the reasons discussed in the next section. It is necessary, therefore, to modify or split the stator winding into two parts, each displaced in *space* and *time* on the stator.* Thus, there are *two* windings in *parallel*, both connected to the single-phase ac supply.

One of these stator windings, usually of appreciable impedance to keep the *running* current *low*, is called the main or *running winding*, and it is distributed in slots, uniformly spaced around the stator. The other winding, in parallel with the main winding, is the auxiliary or *starting winding*, which is also distributed uniformly about the stator but which begins in slots displaced 90° in *electrical space* from the main winding. The adjusted impedance and current of the auxiliary winding, with respect to the line voltage as a reference, usually is designed so that the current in this (starting) winding *leads* the current of the main winding, not necessarily by 90° but sufficiently to produce displacement in time as well as space.

In some designs, to be discussed later, the auxiliary winding is opened after the starting period, i.e., once rotor rotation begins. In other designs, the auxiliary winding remains in parallel with the main winding during starting as well as running. But the essential purpose of the auxiliary winding is to *produce rotation of the rotor*. This raises the question as to why a single-phase winding will *not* produce rotation in a SCIM by itself.

**10-3.
BALANCED
TORQUE OF
A SINGLE-PHASE
INDUCTION MOTOR
AT STANDSTILL**

The distributed single-phase winding (and, indeed, the winding of any phase) will tend to produce a net magnetic field, as shown in Fig. 9-1. The net magnetic field of the main winding distributed around the stator of a single-phase ac motor as shown in Fig. 10-1a, as having an instantaneous direction from left to right. Since the current in this field varies sinusoidally with the single-phase applied voltage, it will produce a concentrated field from right to left 180° later. The directions of current induced in

---

* The ASA (American Standards Association) defines a split-phase motor as "a single-phase induction motor equipped with auxiliary winding, displaced in magnetic position from, and connected in parallel with, the main winding."

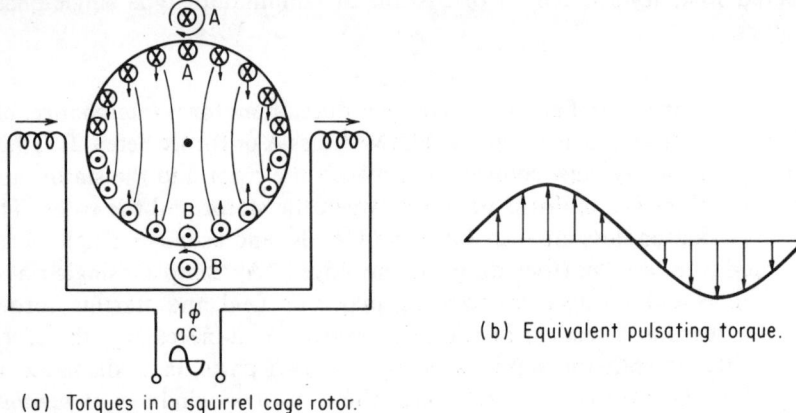

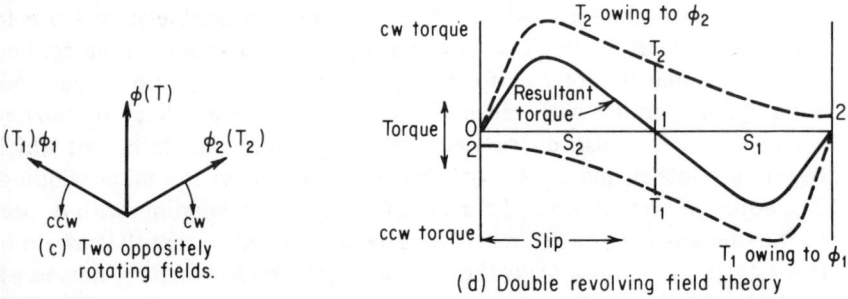

Figure 10-1

(a) Torques in a squirrel cage rotor.

(b) Equivalent pulsating torque.

(c) Two oppositely rotating fields.

(d) Double revolving field theory

Balanced torque at standstill in SCIM rotor excited by a single-phase winding.

the rotor by transformer action are shown in Fig. 10-1a. In accordance with Lenz's law, currents flow in these rotor conductors (conductors A and B, for example) in such a direction as to oppose the field (the right-hand corkscrew rule) producing them.

The direction of torque produced by these current-carrying conductors (left-hand rule) is shown in Fig. 10-1a, indicated by the arrows associated with each conductor. The clockwise torque produced by conductors in the right-hand half of the rotor is balanced by the counterclockwise torque conductors in the right-hand (same) half. The same is true of the left-hand half. Note that conductors A and B are incapable of producing a component of useful torque, even though they carry current, because the torque which they produce is directly at right angles to any motion of the rotor. The net torque is therefore zero. When the ac voltage reverses, the current directions and torques are reversed in the rotor conductors, but the net torque is still zero.

Since the magnitude of torque developed by each conductor depends on the magnitude of the resultant ac field, which in turn varies sinu-

362

CHAP. TEN / *Single-Phase Motors*

soidally with the ac supply, the resultant balanced pulsating torque may be represented by the diagram shown in Fig. 10-1b. The average value of pulsating torque for one full cycle is zero, as shown in Fig. 10-1b.

Any periodic oscillating or pulsating torque, represented in Fig. 10-1b, may also be represented as consisting of *two* oppositely rotating *torques* produced by two oppositely rotating *fields* of equal magnitude and angular velocity, shown in Fig. 10-1c. Since a (separate) *rotating* magnetic field will give rise to a resultant electromagnetic torque in a SCIM rotor, the torque produced by *each* of the opposite rotating fields is shown as dotted lines in Fig. 10-1d.

The dotted clockwise and counterclockwise torque curves are shown over the range from zero slip (synchronous speed) to 2.0 slip, i.e., synchronous speed in the reverse direction.* At a slip of unity (standstill) the starting torques, both in clockwise and counterclockwise directions, T_2 and T_1 respectively, are equal and opposite. Thus, no torque is produced by a single-phase pulsating field. The resultant torque of the two oppositely rotating fields is shown as a solid line in Fig. 10-1d, indicating that, if the slip of the rotor could be changed from unity at standstill to some other value, a net torque in either a **cw** or **ccw** (counterclockwise) direction might result.

> The resultant torque of the single-phase motor is zero *only* at a slip of unity or at synchronous speed in either direction. Once rotated in either direction, therefore, the single-phase motor will continue to rotate in that direction because a resultant net torque is produced to the left or the right of the standstill point shown in Fig. 10-1d.†

10-4. RESULTANT TORQUE OF A SINGLE-PHASE INDUCTION MOTOR AS A PRODUCT OF ROTOR ROTATION

Occasionally in the laboratory (or even in industry) one line lead to a three-phase induction motor is accidentally disconnected because of a poor mechanical connection or a blown fuse in that line. The motor will continue to rotate in the same direction as an induction motor, even under load, developing torque in accordance with and verifying Fig. 10-1d. The polyphase motor under these conditions is said to be "single-phasing." It can and will be shown, however, that once rotation is initiated, the single-phase motor is, in effect, a two-phase motor. Figure 10-2a shows the same SCIM rotor as in Fig. 10-1a which has been rotated by some external means in a clockwise (**cw**) direction. If the stator flux has an instantaneous direction, as shown in Fig. 10-2a, an emf is induced as a result of relative motion between the rotor conductors

---

* Section 9-5.

† In this respect, a single-phase motor without any auxiliary starting winding or starting device resembles an internal combustion outboard or lawn mower engine which must be started by a pull-string. Indeed, in the laboratory, the author has used a pull-string to illustrate Fig. 10-1d on single-phase motors in which the auxiliary winding has been disconnected.

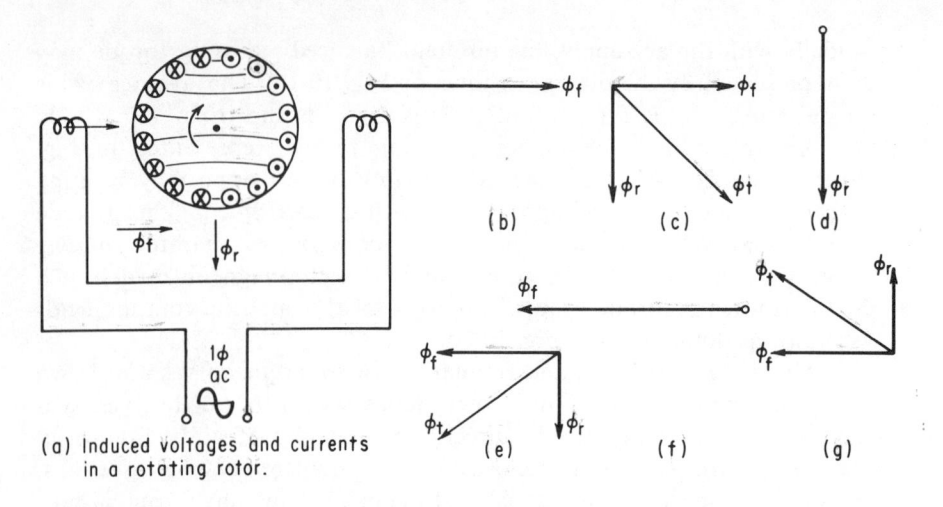

(a) Induced voltages and currents in a rotating rotor.

Figure 10-2

Rotor cross field produced by rotor rotation.

and the magnetic field (right-hand rule). This induced emf is sometimes called a "speed emf" Fig. 10-2a to distinguish it from the "transformer emf" Fig. 10-1a produced by transformer action. Both emf's are produced by a change in flux linkages: the *speed emf* as a result of relative motion between a conductor and its field; and the *transformer emf* as a result of the pulsating field.

The pulsating field is a relatively stationary field. Unlike the polyphase induction motor, therefore, the rotor frequency of the induced *speed emf* in conductors of a running motor is high (since it is proportional to speed) and the rotor reactance ($X_r \propto f$) is correspondingly high. The result is that, while a *speed emf* is produced in the rotor conductors at the instant shown in Fig. 10-2a, a rotor current will not flow until almost 90 electrical degrees later. When a current does flow in the rotor conductors, as shown in Fig. 10-2a, it will produce a rotor flux, ϕ_r, in the downward direction shown.

Figure 10-2b shows the pulsating field flux, ϕ_f, at a maximum, producing a maximum speed emf (but no rotor current or rotor flux) in the rotor bars in the direction shown in Fig. 10-2a. A brief instant later in time, however, a rotor current will flow producing a quadrature rotor flux, ϕ_r, as shown in Fig. 10-2c. Note that the pulsating field flux has decreased somewhat, and the rotor flux lags the field flux by 90 electrical degrees. An instant later, in Fig. 10-2d, the pulsating field flux is zero, but the rotor current is at a maximum producing maximum rotor flux, ϕ_r. An instant later, Fig. 10-2e, the field flux ϕ_f pulsates in the opposite direction and, at the same time, the rotor flux decreases. An instant later, at Fig. 10-2f, the stator flux is a maximum in the opposite direction (a two-pole machine is used for simplicity) and the rotor flux

Solution:

a. $I_s = 4\angle -15°$ A $= 3.86 - j1.035$ A
   $I_r = 6\angle -40°$ A $= 4.60 - j3.86$ A
   $I_{lr} = I_s + I_r = 8.46 - j4.895 = \mathbf{9.88\angle -30°}$ A
   Power factor $= \cos 30° = 0.866$ lagging
b. $I_s \cos \theta = \mathbf{3.86}$ A [from the above]
c. $I_r \sin \theta = \mathbf{-j3.86}$ A [from the above]
d. $\sin(40° - 75°) = \sin 25° = \mathbf{0.423}$

If the windings are displaced by 90° in space, and if their quadrature current components, which are displaced by 90° in time (Example 10-1), *are practically equal,* an equivalent two-phase rotating field is produced at starting which develops sufficient starting torque to accelerate the rotor, in the direction of the rotating field produced by the currents. (See Sec. 9-4.)

As the rotor accelerates, it generates its own speed emf (cross-field theory) and tends to produce a resultant torque by virtue of its own rotation in a particular direction (double-revolving field theory). The torque developed by the main pulsating field (produced by the running winding) exceeds that developed by both windings at a slip of about 15 per cent. It is evident, as well, from Example 10-1, that the running current alone would produce less loss since the losses of the starting winding would be eliminated. For *both* these reasons, as shown in Fig. 10-3a, a *centrifugal switch* (normally closed at standstill) is provided to open at a slip of about 25 per cent (maximum torque as a single-phase motor), and the motor pulls up to its rated slip (approximately 5 per cent or less, depending on the applied load) as a single-phase motor by virtue of its own cross-field.

In order to reverse the direction of rotation of any split-phase motor, it is necessary to reverse the terminal connections of the auxiliary starting winding with respect to the main running winding. This will produce a "two-phase" rotating field in the opposite direction. The reversal (plugging) can never be done under *running* conditions, obviously, as is sometimes done with polyphase induction motors. In the case of a single-phase resistance-start motor, nothing happens even if the centrifugal switch contacts are shorted and the starting winding is energized. Since the single-phase torque is greater than the "split-field" torque, the motor will continue to operate as a single-phase motor in the direction in which it was originally rotated. The resistance-start split-phase motor is classified therefore as a *nonreversible* motor.*

The rating of the starting winding is based on an *intermittent* rating only. If the centrifugal switch becomes defective and fails to open (usually due to welded contacts), the excessive heat produced by the high-resistance starting winding will so increase the stator temperature that both

---

* The split-phase capacitor-start motor, however, is a reversible motor. (See footnotes to Sections 10-5 and 10-7 for definitions and further distinctions.)

windings (starting and running) eventually burn out. Currently designed split-phase motors using improved insulations have extremely long life under normal operation, but many excellent motors have been destroyed as a result of a defective centrifugal switch. (If a motor tends to overheat, listening for the characteristic click of a centrifugal or magnetic-type relay switch will indicate whether the switch is functioning.) A clamp-on ammeter may also be used to measure the current decrease when the auxiliary winding is opened (see Example 10-1) indicating normal centrifugal switch operation. Then, the cause of overheating may be further investigated: lack of lubrication, defective bearings, excessive load, partially shorted winding, etc., until it is discovered and eliminated.

As stated above, the full-load slip of a split-phase resistance-start motor is about 5 per cent. The locked-rotor starting current varies from 5 to 7 times the rated current, and the starting torque from 1.5 to 2.0 times the rated torque.† The split-phase resistance-start motor is normally a *fractional* horsepower motor, and, since its rotor is small, it has a *low inertia* even when connected to a load. The result is that the relatively high starting current falls off *almost instantly* on starting, so that high starting current, *per se*, is not a major objection to this motor. The major objections to the motor are (1) its low starting torque; and (2) that, when heavily loaded, the slip exceeds 5 per cent, reducing the speed emf and producing an elliptical or pulsating torque which makes the motor somewhat annoyingly noisy. For this reason, the split-phase motor is used in appliances to drive loads which are themselves noisy: oil burners, machine tools, grinders, washing machines, dish-washers, fans, air-blowers, air compressors, and small water pumps.

Speed control of split-phase windings is a relatively difficult matter since the synchronous speed of the rotating stator flux is determined by the frequency and the number of poles developed in the running stator winding ($S = 120f/P$). The techniques involve using *part-windings* or *consequent* poles, as a means of changing the number of poles, as well as additional series stator impedance or variation in the impressed voltage to secure a modification of the torque characteristics, as given in Eq. (9-8) and Fig. 10-5. It must be pointed out, however, that all speed changes must be accomplished in a range *above* that at which the centrifugal switch operates ($s = 0.25$), and below synchronous speed. This results in a fairly limited range of speed control. Finally, since speed control is achieved by torque reduction, it cannot be used with heavy loads because breakdown results, as shown in Fig. 10-5.

† It can be shown that the *starting* torque of *any* split-phase motor, in accordance with Eq. (9-5) is $T_s = K_t \phi I_s \sin(\theta_s - \theta_r)$, where all terms are defined in Figs. 10-3b and 10-4b. Thus, for the same starting current the starting torque is proportional to the angle between the currents in the two windings. When the angle is 90°, the starting torque is maximized. [See Exs. 10-1d and 10-2c.]

**10-6.
SPLIT-PHASE
CAPACITOR-START
MOTOR**

The previously mentioned split-phase motor was termed a resistance-start motor because the impedance difference in the starting or auxiliary winding was due to the high resistance of its starting winding. As a means of improving the relatively low starting torque of the split-phase motor, a capacitor is added to the auxiliary winding to produce almost a true 90° current relation between the starting and running windings, rather than approximately 25°, as indicated by Exs. 10-1 and 10-2. Operating tests on fractional horsepower capacitor-start motors indicate that the current relation between the starting and running winding currents at the instant of starting are displaced by about 82°, whereas in the resistance-start motor the displacement is about 25°.

Empirically, it may be stated that the starting torque is proportional to the sine of the angle between the starting currents in the running and auxiliary starting windings. The ratio of starting torque of the capacitor-start to the resistance-start split-phase motor, as shown by Ex. 10-2 is of the order of sin 82°/sin 25°, or about 2.35 to 1, raising the starting torque of the capacitor-start motor to a range of 3.5 to 4.75 of the rated torque. The use of the capacitor also tends to reduce (to some extent) the initial total locked-rotor current, since it improves the power factor by providing a component of current which leads the applied voltage. Figure 10-4a shows a capacitor-start motor. Note that the *only* difference has been the addition of a capacitor in the auxiliary winding. The approximate capacitor values for capacitor-start motors of $\frac{1}{8}$ to $\frac{3}{4}$ hp are given in Table 10-1.

TABLE 10-1.
TYPICAL CAPACITOR VALUES FOR 60-HERTZ,
1725-RPM, SINGLE-PHASE
FRACTIONAL-HORSEPOWER CAPACITOR-START
SPLIT-PHASE MOTORS*

Horsepower (hp)	$\frac{1}{8}$	$\frac{1}{6}$	$\frac{1}{4}$	$\frac{1}{3}$	$\frac{1}{2}$	$\frac{3}{4}$
Capacitance microfarads (μF)	80	100	135	175	250	350

The values of the capacitors given in Table 10-1 are fairly large in comparison to the values of capacitors normally used in electronic amplifiers or even electronic power supplies. The development of fairly small, cylindrical, dry-type, electrolytic capacitors, rated for ac line voltages of 110 or 220 V (their average size is about $1\frac{1}{2}$ in. diameter by $3\frac{1}{2}$ in. long) is accompanied by rating the capacitor for an intermittent duty of a total of one minute of total use per hour, based on the number of

---

* Based on C. S. Siskind, *Electrical Machines, Direct and Alternating*, New York: McGraw-Hill Inc., 1959, Chapter 10, Table 9, p. 426.

starts. Capacitors are tested on this basis for twenty 3s periods (or forty $1\frac{1}{2}$s periods, etc.) equally distributed over an hour. Because the capacitor has an intermittent rating, (in addition to the intermittent rating of the starting winding), a defective starting switch, on a capacitor-start motor, may damage not only the windings but the capacitor as well. The same comments, therefore, regarding switch maintenance and replacement made in connection with the resistance-start motor apply with even more emphasis to the capacitor-start motor.

The capacitor-start motor, unlike the resistance-start split-phase motor, is a *reversible* motor. If temporarily disconnected from the supply, the speed of the motor drops to a slip of 20 per cent (about four times the rated slip of 5 per cent) and the centrifugal switch closes. If, at the same time, the line leads of the auxiliary starting winding are reversed with respect to the running winding, and reconnected to the supply, a rotating two-phase field will be set up opposite to the direction of rotor rotation. Unlike the resistance-start motor, whose main and auxiliary currents are displaced by only 25°, the capacitor start motor has a displacement of approximately 82° (See Ex. 10-2.) The torque proportional to this displacement (sin 82°) is 0.9903 of the torque at 90°.

In the capacitor-start motor, therefore, the "split-field" or "two-phase" torque *exceeds* the single-phase speed-emf torque produced by the rotor cross field. The reversed rotating field, moving opposite to the direction of rotor rotation, slows down the motor (reducing the speed emf and the cross-field torque still more), stops it, and reverses it in the opposite direction. The motor accelerates to 20 per cent slip in the reverse direction; and, when the centrifugal switch opens, the motor comes up to rated speed as a single-phase induction motor in the *opposite* direction.

Because of its higher starting torque, from 3.5 to 4.5 times the rated torque, and its reduced starting current for the same horsepower at the instant of starting (as shown in Example 10-2 below), the capacitor-start

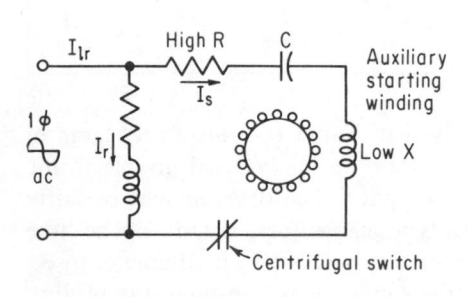

(a) Connection diagram.

(b) Phase relations.

Figure 10-4

Connection diagram and phase relations of a split-phase capacitor-start induction motor.

motor is currently manufactured in *integral* horsepower sizes up to 7.5 hp. (Resistance-start motor torques vary from 1.5 to 2.0 times the rated torque and never exceed $\frac{3}{4}$-hp size.) The integral horsepower capacitor-start motors are customarily dual-voltage types which may be used either on 115 V (windings in parallel) or 230 V (windings in series). The latter results in lower (approximately half) starting and running currents for the same horsepower rating. By virtue of their higher starting torque, capacitor-start split-phase motors are used for pumps, compressors, refrigeration units, air-conditioners, and larger washing machines, where a single-phase motor is required that will develop high starting torque under load and where a reversible motor is needed.

EXAMPLE 10-2: A capacitor is added to the auxiliary starting winding of the motor of Example 10-1 which causes the starting winding current to lead the supply voltage by 42°. The magnitude of currents in both the starting winding and the running winding remain the same, the latter lagging by the same amount as in Example 10-1. At the instant of starting, calculate
  a. The total locked-rotor current and the power factor
  b. The sine of the angle between the starting and running currents
  c. Compare the results with those of Example 10-1.

Solution:

a. $I_s = 4 \angle +42°$ A $= 2.98 + j3.15$ A
   $I_r = 6 \angle -40°$ A $= 4.60 - j3.86$ A
   $I_{lr} = I_s + I_r = 7.58 - j0.71 =$ **7.585 $\angle -0.7°$ A**
   $\cos (0.7°) =$ **0.9999**
b. $\sin [40° - (-42°)] = \sin 82° =$ **0.9903**
c. The locked-rotor (starting) current has been reduced from $9.88 \angle -30°$ A, to $7.585 \angle -0.7°$ A, and the power factor has risen from 0.866 lagging to unity (0.9999). The motor develops maximum starting torque ($T = KI_{lr}\phi \cos \theta$) with minimum starting current. The ratio of starting torques (capacitor to resistance-start) is

$$\frac{T_{cs}}{T_{rs}} = \frac{\sin 82°}{\sin 25°} = \frac{0.9903}{0.423} = 2.35$$

10-7. PERMANENT-SPLIT (SINGLE-VALUE) CAPACITOR MOTOR

Because of the reversible property of the split-phase capacitor-start motor, a single-phase motor has been developed having *two* permanent windings (usually wound with the same wire size and the same number of turns in both, i.e., the windings are *identical*). Because it runs continuously as a permanent split-phase motor, *no centrifugal switch* is required. The motor starts and runs by virtue of the quadrature phase-splitting produced by the two identical windings displaced in time and space. As a result, it does not possess the high running torque produced by either the resistance-start or the capacitor-start motor. Furthermore, the capacitor used in the permanent-split single-value capacitor motor is designed for continuous duty and is of an oil-filled type. The value of the capacitor is based on its optimum *running* rather than its starting characteristic. At the instant of starting, the current in the capacitive branch is very low. The result is that the permanent-split, singlevalue capacitor motor (unlike the capacitor-start motor) has a very poor starting torque, about 50 to 100 per cent of the rated torque.

As shown in Fig. 10-5a, a uniquely connected reversing switch permits the oil-filled capacitor to be shifted easily to either winding. It might appear that with poor starting torque and poor running torque, the prospects for this type of motor would be extremely dim. Yet its very weaknesses lead *directly* to its advantages, and the motor is very popular. It requires no centrifugal switch, and it is easily reversed because of its low running torque. Its low running torque makes it more sensitive, moreover, to voltage variations; and the permanent-split single-value capacitor motor is one of the few single-phase induction motors whose speed may be readily controlled by variations of line voltage, as shown in Figs. 10-5a and c.

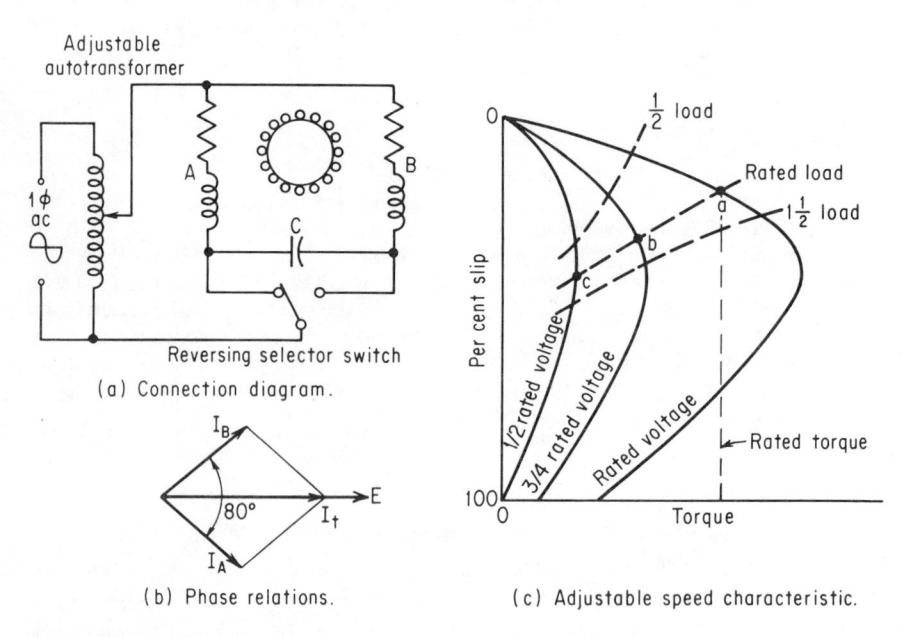

Figure 10-5

Connection diagram, phase relations, and voltage characteristics of permanent split-phase motor.

The phase relations of the motor under running conditions are shown in Fig. 10-5b for the given reversing selector switch position shown in Fig. 10-5a. Because of the fairly uniform rotating magnetic field created by equal windings whose equal currents are displaced by almost 90°, the torque is fairly uniform and the motor does *not* exhibit the characteristic pulsating hum developed by most single-phase motors when loaded. The value of the continuous-duty capacitor is selected so that the running currents are equal and displaced, as shown in Fig. 10-5b, between three-quarters load and full load. As previously stated, however, this value is too small for starting purposes, and results in starting torques between one-half to rated full-load torque see Fig. 10-5c.

In addition to its advantage as a reversing motor,* the permanent-split capacitor motor lends itself to speed control by voltage supply variation. Since the torque of any induction motor varies as the square of the voltage impressed on its stator, the torque-slip curve at three-quarters rated voltage is nine-sixteenths, or roughly one-half of that at full voltage, as shown in Fig. 10-5c. Similarly, at one-half rated voltage, the torque is roughly one-quarter of the rated torque. This is true, both at starting and running. If the motor is started at its rated voltage and is operating at its rated load (point *a*), a reduction in voltage to three-quarters of the rated voltage causes the speed to drop to point *b*. A further reduction to one-half the rated voltage causes the speed to drop to point *c*. In general, a heavy load tends to produce a greater speed drop with voltage change than a light load.

Various methods are used to adjust the applied ac stator voltage to produce the speed control desired including tapped transformers, variacs, potentiometers and tapped resistors or reactors. These methods are limited to speeds below synchronous speed.

Because of (1) its instant response as a reversing motor, (2) its quiet operation, and (3) the possibility of speed control, the permanent-split capacitor motor is used for exhaust and intake fans and blowers, office machines, and unit heaters.

10-8. TWO-VALUE CAPACITOR MOTOR

The permanent-split single-value capacitor motor has one particularly serious weakness, namely low starting torque.† Where starting conditions are not severe, this disadvantage is of no consequence. But where high starting torques are required, this disadvantage must be overcome. The *two-value capacitor motor* combines the quiet operation and limited speed control advantages of a running permanent-split capacitor motor with the high starting torque of the capacitor-start motor. Two capacitors are employed (as the name

---

* The ASA distinguishes between a *reversing* motor and a *reversible* motor. The permanent-split capacitor motor is a reversing motor because it can be reversed when running at rated load at rated speed. The capacitor-start split-phase motor is a reversible motor because it is capable of being reversed at well below rated speed, without requiring that the motor stop. The resistance-start split-phase motor is a non-reversible motor for reasons previously outlined.

† Since starting torque, $T_s = K_t \phi I_s \sin(\theta_r - \theta_s)$, the starting current I_s in the capacitive branch is low. A decrease in capacitive reactance will increase I_s. Hence, the need for a large starting capacitor.

implies) during the starting period. One of these, an electrolytic *starting* capacitor, similar to that used for the *intermittent* duty of the capacitor-start split-phase motor, is of fairly *high capacity* (about 10 to 15 times the value of the *running* capacitor) and is cut out of the circuit by a centrifugal switch when the slip reaches about 25 per cent.

Two methods are usually employed to obtain the necessary high capacitance at starting and the lower capacitance during running, both employing centrifugal switches to accomplish the crossover. The first method, similar to that described above, utilizes an electrolytic capacitor, as shown in Fig. 10-6a, in parallel with an oil-filled capacitor through a normally closed at starting centrifugal switch. The intermittently rated high-capacity electrolytic is disconnected at about 75 per cent synchronous speed, thus producing the necessary high starting torque. The motor then continues to accelerate as a capacitor motor, with the optimum capacitance value of the oil-filled running capacitor for running at or near rated load.

(a) Use of two capacitors and a centrifugal switch.

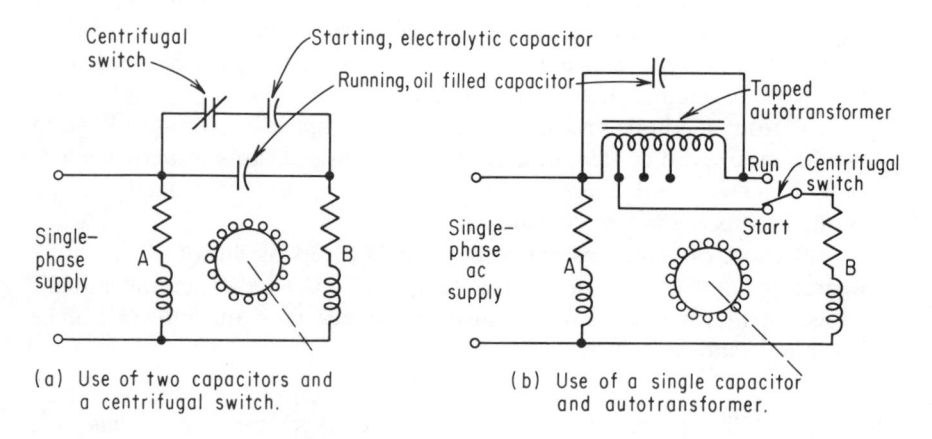

(b) Use of a single capacitor and autotransformer.

Figure 10-6

Connection diagrams for two types of two-value capacitor motors.

The second method (although it also uses a centrifugal switch) employs only *one* oil-filled high voltage capacitor in combination with an autotransformer, shown in Fig. 10-6b. This technique utilizes the transformer principle of reflected capacitive reactance from the secondary of the autotransformer back to the primary, in proportion to the square of the ratio of secondary to primary turns, i.e., $n^2 = (N_s/N_p)^2$. Thus, an autotransformer with 140 turns, tapped at the 20-turn point, shown in Fig. 10-6b, would reflect a 6 μF running capacitor to the primary as $(140/20)^2 \times 6$ μF (or $7^2 \times 6$ μF) or 294 μF (almost fifty times as much). Thus, a running oil-filled capacitor may be used for starting as well, provided that the voltage rating of the capacitor can withstand the step-up voltage produced by the transformation (in this case $n = 7/1$, and the capacitor voltage at starting would be 110×7, or 770 V, for a 110 V source).

As in the case of resistance-start and capacitor-start single-phase motors, a defective centrifugal switch may still cause serious damage. In the case of the type whose circuit is shown in Fig. 10-6a, if the switch fails to open and remove the large-capacity electrolytic, it may break down in time, since it only has an intermittent rating. In the case of the second type, shown in Fig. 10-6b, as the motor speed approaches synchronous speed, the secondary excitation voltage is extremely high, and the relatively expensive oil-filled capacitor (usually rated at about 1000 V ac) may be destroyed.

The primary advantage of the two-value capacitor motor is its high starting torque, coupled with quiet operation and good running torque. Still classed as a *reversing* motor, when the line leads of one winding are reversed, it is reversed in the usual manner. When the speed drops below 25 per cent slip during reversing, the centrifugal switch closes its starting contact, providing maximum torque as the motor slows down and reverses. The contacts open again when the motor is up to 75 per cent synchronous speed in the reverse direction. Frequent reversals, therefore, will reduce the life of the centrifugal switch. For this reason, where frequent reversals are accomplished, a permanent-split, single-value capacitor motor, using no centrifugal switch at all, is preferable.

The two-value capacitor motor has found recent application in a smaller home air-conditioning unit which uses this motor in its compressor and operates on a 15-ampere branch circuit.* The lower starting current and lower running current (7.5 A maximum) at an improved power factor over the capacitor-start motor, are obtained through the precise selection of starting and running capacitors for the fixed compressor load.

10-9.
SHADED-POLE
INDUCTION MOTOR

The single-phase motors considered in the preceding sections all employed stators having uniform air gaps with respect to their rotor and stator windings, which are uniformly distributed around the periphery of the stator. The starting methods employed thus far were generally based on a split-phase principle of producing a rotating magnetic field to initiate rotor rotation. Split-phase induction motors are manufactured in both fractional and integral horsepower motor sizes.

The shaded-pole motor is usually a small, fractional horsepower motor, not exceeding $\frac{1}{10}$ hp, but motors up to $\frac{1}{4}$ hp have been produced.

* The NEC permits a fixed appliance that draws a load no greater than 50 per cent of the rating of the branch circuit. On a 15-ampere branch circuit, the maximum current for an air conditioner is 7.5 amperes.

The great virtue of this motor lies in its utter simplicity: a single-phase rotor winding, a cast squirrel-cage rotor, and special pole pieces. No centrifugal switches, capacitors, special starting windings, or commutators are used. With but a single-phase winding, it is inherently self-starting. Both the double-revolving field theory and the cross-field theory, indicate that this is impossible. There must be some *auxiliary* means of producing the effect of a rotating magnetic field with a single-phase supply and only one stator winding.

Figure 10-7a shows the general construction of a *salient* two-pole shaded-pole motor. The special pole pieces are made up of laminations, and a short-circuited *shading* coil (or a single-turn solid copper ring) is wound around the smaller segment of the pole piece. The shading coil, separated from the main ac field winding, serves to provide a phase-splitting of the main field flux by delaying the change of flux in the smaller segment.

As shown in Fig. 10-7b, when the flux in the field poles tends to increase, a short-circuit current is induced in the shading coil, which by Lenz's law opposes the force and the flux producing it. Thus, as the flux increases in each field pole, there is a concentration of the flux in the main segment of each pole, while the shaded segment opposes the main field flux. At point c shown in Fig. 10-7e, the rate of change of flux and of current is zero, and there is no voltage induced in the shaded coil. Consequently, the flux is uniformly distributed across the poles. When the flux decreases, the current reverses in the shaded coil to maintain the flux in the same direction. The result is that the flux crowds in the shaded segment of the pole.

(a) General construction of 2-pole shaded pole motor.

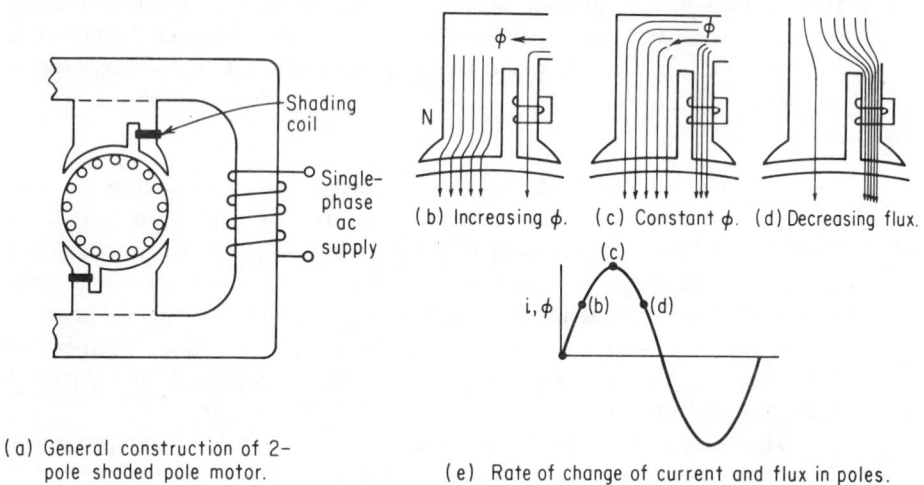

(b) Increasing ϕ.   (c) Constant ϕ.   (d) Decreasing flux.

(e) Rate of change of current and flux in poles.

Figure 10-7

General construction and principle of shaded-pole motor.

An examination of Figs. 10-7b, c, and d will reveal that at intervals b, c, and d, the net effect of the flux distribution in the pole has been to produce a sweeping motion of flux across the pole face representing a clockwise rotation. The flux in the shaded pole segment is always lagging the flux in the main segment in time as well as in physical space (although a true 90° relation does not exist between them). The result is that a rotating magnetic field is produced, sufficient to cause a small unbalance in rotor torques (double-revolving field theory) such that the clockwise torque exceeds the counter-clockwise torque (or vice versa), and the rotor always turns in the direction of the rotating field.

For the type of shaded-pole motor shown in Fig. 10-7, the rotation is clockwise since the flux in the shaded segment lags the main flux. In order to reverse the direction of rotation, it would be necessary to unbolt the pole structure and reverse it physically. To eliminate such a slow and complicated process, newer techniques have been devised for producing reversible shaded-pole motors.

The first of these methods is to connect the shading coils in series on corresponding shading segments, and short-circuit them through a switch. This is shown in Fig. 10-8a, where the shading coils on trailing salient-pole tips on one side are short-circuited for clockwise rotation, and those trailing pole tips on the opposite side of the pole are short-circuited for counterclockwise rotation. At no time, however, are both sets of trailing pole tips short-circuited.

The second method is generally used with *nonsalient*-pole stators. Two separate distributed windings, 90° in space with respect to the short-circuited shaded poles, are shown as windings A and B in Fig. 10-8b. When A (not shown as distributed) is energized, the flux pattern is in a clock-wise order: winding A, shaded pole A'; winding A (in the location of B), and shaded pole B'. When B is energized, the flux pattern is counterclockwise: winding B, shaded coil A', winding B distributed at A, and shaded coil B'.

The third method, shown in Fig. 10-8c, also employs a single continuous distributed winding with appropriate taps at the 90° points. When the taps of one set are energized by means of the dpdt switch, the rotor rotates clockwise. When the taps of a second set, displaced by 90° with respect to the shading coils, are energized, the motor rotates in the opposite direction.

The advantage of the distributed nonsalient stators over the salient-pole types shown in Figs. 10-7 and 10-8 is that, once rotation has been initiated, the rotor with the uniform flux gap tends to produce a more uniform rather than elliptical rotating magnetic field by virtue of its speed emf (cross-field theory).

The shaded-pole motor is rugged, inexpensive, small in size, and requires little maintenance. Its stalled locked-rotor current is only slightly higher than its normal rated current, so that it can remain stalled for short periods without harm. Unfortunately, it has a very low

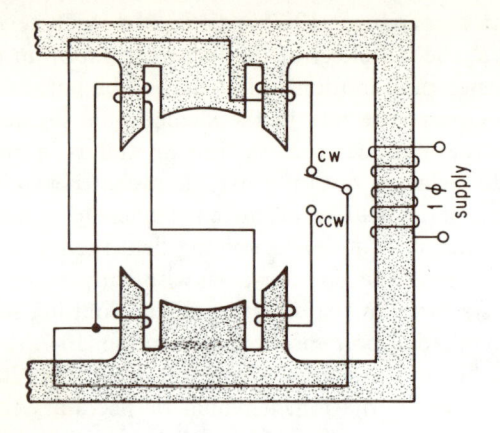

(a) Two pairs of shading coils short-circuited by switch. (Main winding not shown)

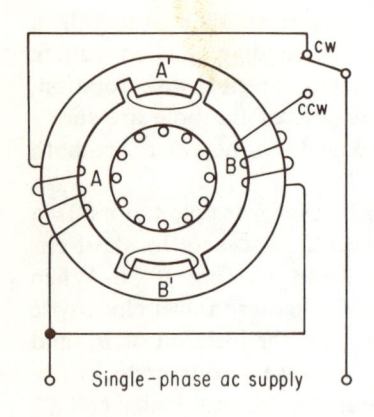

(b) Two individual distributed main fields at 90°.

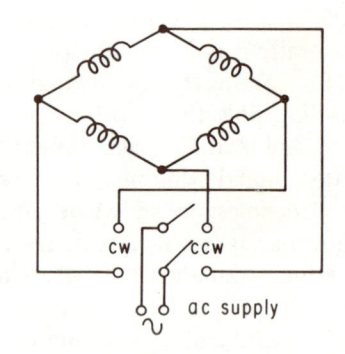

(c) Tapped distributed winding method.

Figure 10-8

Methods of reversing shaded-pole motors.

starting torque, low efficiency, and a low power factor. The last two considerations are not serious in a small motor. Its low starting torque limits its application to phono motors or turntables, motion picture projectors, electric rotisseries, small fans and blowers, vending machines, rotating store-window display tables, and other relatively light servomechanism loads.

In general, the basic methods of speed control employ a means of reducing the ac supply voltage to produce an increase in slip, similar to the capacitor motor, as shown in Fig. 10-5c.

**10-10.
RELUCTANCE-START
INDUCTION MOTOR**

Another induction motor employing a stator with a nonuniform air gap is the reluctance-start motor. Its rotor is the conventional squirrel-cage rotor, which develops torque once rotation has been initiated by the *reluctance principle*.* By virtue of unequal air gaps between the rotor and the nonuniform salient poles shown in Fig. 10-9, a sweeping effect is produced on the main field flux. The reluctance principle on which the motor operates is briefly that, where the air gap is small, the self-inductance in the field winding is great, causing the current in the field winding to lag the flux which produced it (in a highly inductive circuit, the current lags the voltage and the flux by almost 90°). Conversely, where the air gap is very great, self-inductance is reduced, and the current is more nearly in phase with the flux. The mutual air-gap flux is delayed, therefore, in the vicinity of the smaller gap, producing a sweeping effect similar to that produced in the shaded-pole motor. Since the fluxes are displaced somewhat in time and also in space, a rotating magnetic field is produced at all field poles, at instants t_1, t_2, and t_3, successively, as shown in Fig. 10-9.

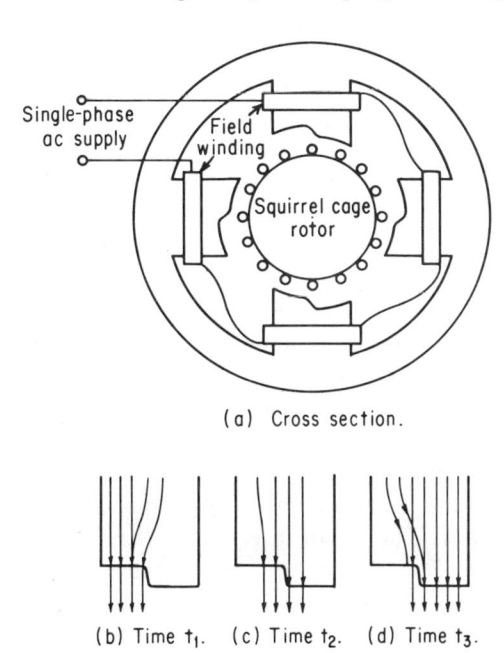

(a) Cross section.

(b) Time t_1. (c) Time t_2. (d) Time t_3.

Figure 10-9

Reluctance start-induction motor and development of rotating field.

The running torque characteristics of the salient pole reluctance-start induction motor are not as good as the nonsalient-pole shaded-pole motor. This is evident because, in order for the speed emf to develop a rotating magnetic field once rotation has been initiated, the air gap must be fairly uniform. Furthermore, the starting torque of the reluctance-start motor is also poorer than the shaded-pole motor (less than 50 per cent of the rated torque). Other than reversing the poles on the stator, there is *no* way of changing the direction of rotation of the reluctance-start induction motor. Operation is *always* in the direction from high to low air gap.

* The reluctance-*start* motor is an induction motor whose *starting* is initiated by the reluctance principle (Sec. 1-2). It is not the same as the reluctance motor (a nonexcited synchronous motor) discussed in Section 8-27. The single-phase reluctance motor, hysteresis motor, and subsynchronous motor are indeed, single-phase motors; but their operating similarity to the synchronous motor principle impelled the author to discuss them in Chapter 8 (see Sections 8-27, 8-28, and 8-29). This separation is intentional and avoids the customary confusion among the operating principles of these single-phase motors.

The shaded-pole motor is generally preferred to the reluctance-start induction motor since it costs no more to manufacture, has better starting and running torque, and is readily reversible. Speed control of the reluctance-start induction motor is the same as for shaded-pole motors.

10-11. SINGLE-PHASE COMMUTATOR MOTORS

The motors discussed so far in this chapter have all been single-phase SCIMs with squirrel-cage *cast rotors*, whose variations were primarily in the principle of starting. There is another group of motors called single-phase commutator motors, because the wound rotor of this kind of motor is equipped with a commutator and brushes. This group consists of two classes: (1) those operating on the repulsion principle (repulsion motors), in which energy is *inductively* transferred from the single-phase stator field winding to the rotor: and (2) those operating on the principle of the series motor, in which the energy is *conductively* carried both to the rotor armature and its series-connected single-phase stator field. (See Sec. 10-18.)

Because the repulsion principle is used as a method of starting for a type of induction motor known as the repulsion-start induction motor, we shall take up first the class of repulsion motors.

10-12. THE REPULSION PRINCIPLE

The stator of a repulsion motor is identical to that of an induction motor, in that the slots are distributed uniformly around the periphery of the stator. The winding is also a "standard" two-, four-, or six-pole winding, distributed around the stator to produce the necessary poles. No auxiliary or part windings are used. The stator structure of a repulsion motor is, therefore, identical to and interchangeable with the stator of any single-phase induction motor. The stator laminations are interchangeable with those of almost any induction motor (except the shaded-pole motor). In the drawings shown below a two-pole stator field will be represented for convenience.

The rotor of a repulsion motor is very similar to the "standard" dc armature, usually lap-wound (Sec. 2-11) with one or more pairs of brushes, depending on the pairs of poles for which the stator is wound. These brushes are *short-circuited*, however, and the brush rigging may be rotated to change the position of the brushes with respect to the polar axis. For the sake of simplicity, the armature of the repulsion rotor will be represented as a gramme-ring winding (Section 1-12), an early form of lap winding.

Figure 10-10a shows the instantaneous direction of field flux, ϕ_f, produced by a two-pole stator. In accordance with Lenz's law, an induced emf, which opposes the field flux, is set up in the gramme-ring armature winding of the rotor. The individually induced **emfs** combine vectorially in each of the two paths to produce a positive instantaneous polarity in the left-hand brush and a negative polarity in the right-hand brush.

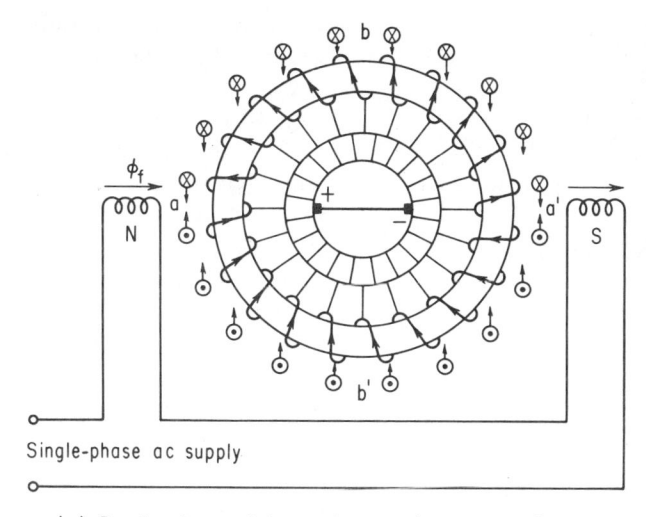

(a) Brush axis parallel to polar axis (hard neutral).

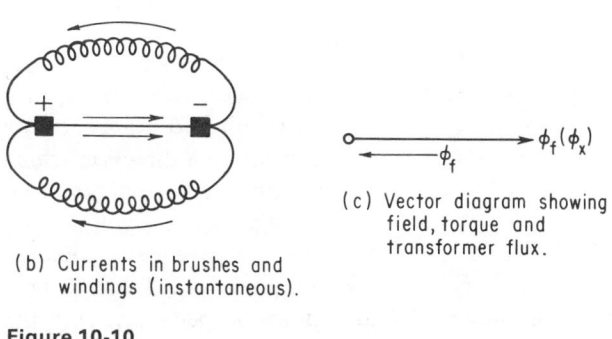

(b) Currents in brushes and windings (instantaneous).

(c) Vector diagram showing field, torque and transformer flux.

Figure 10-10

Hard neutral position showing maximum current, no torque.

Maximum current flows through the short-circuit conductor connecting the two brushes. When the field reverses, 180 electrical degrees later, the brush polarity also reverses because of the reverse induced **emfs**, and maximum current flows in the opposite direction. Figure 10-10b shows the instantaneous currents in the brushes and windings for the direction of field flux taken in Fig. 10-10a. No useful torque is produced as a result of this current flow, however; as may be noted from the perpendicular forces developed by each conductor orthogonal to the current direction and the magnetic field (left-hand motor rule). For each conductor under the N pole tending to produce clockwise torque (b' to a), there is a conductor tending to produce counterclockwise torque (b to a). The motor rotor is therefore stationary, with the rotor locked in a *hard* (maximum current) *neutral* position. Figure 10-10c also shows that all of the field flux ϕ_f is producing transformer flux ϕ_x (neglecting leakage), with the result that no orthogonal flux is produced by the rotor to react against this field. No rotation can be produced without such field interaction (see Sections 1-17 and 1-18).

If the brush axis is shifted 90° from the polar axis, as shown in Fig. 10-11, the same instantaneous **emfs** are produced in the armature as a result of the same instantaneous flux; but, owing to the brush position, no current flows in the external circuit. The sum of the **emfs** between brushes in each path is zero, as shown in Figs. 10-11a and b. Since no current flows, no perpendicular torque flux is produced by the conductors, as shown in Fig. 10-11c, and the rotor is at a standstill in a *soft* (zero

SEC. 10-12. / *The Repulsion Principle*

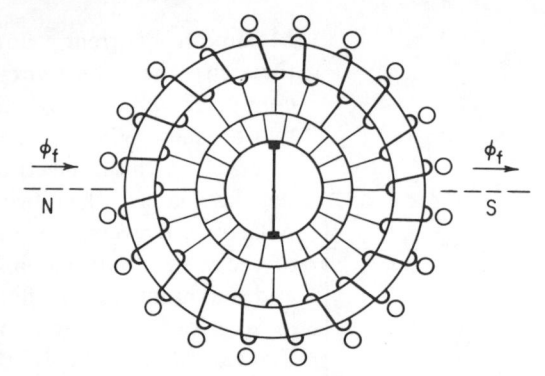

(a) Brush axis on mechanical neutral (soft neutral), zero current.

(b) Balanced emfs, no current in brushes and winding.

(c) Vector diagram showing field, flux and transformer flux.

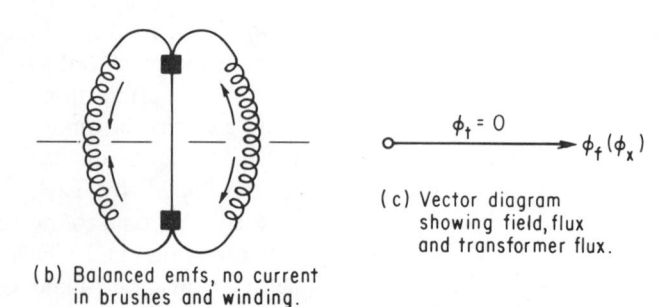

$\phi_t = 0$

$\phi_f(\phi_x)$

Figure 10-11

Soft neutral position showing zero current, no torque.

current) *neutral* position. Note that no current flow is shown in the conductors of Fig. 10-11a.

If the brush axis is shifted by some angle α from the polar axis, where α is less than 90°, rotation is produced. Although the **emfs** are still the same as in Figs. 10-10a and 10-11a, an unbalance of **emf** is produced in each path by the brush position, with the result that current flows in proportion to the voltage difference through the rotor conductors, as shown in Figs. 10-12a and b. The brush axis is *x-x'*, and the polar axis is still *a-a'*. Note that conductors from *a* to *x* and from *a'* to *x'* are carrying current in a direction which is opposite to their induced voltage, because voltage *a-x'* and voltage *x-a'* exceed voltage *a'-x'*. There is now an unbalance of conductors under N and S poles, respectively, carrying current in a given direction, and the rotor rotates in a clockwise direction as a result of these forces, as shown in Fig. 10-12a. Another way of indicating this, perhaps in a somewhat more sophisticated way, is shown in Fig. 10-12c. The net flux of the armature, ϕ_a, by Lenz's law must have been produced by a transformer flux ϕ_x which is a component of the field flux, ϕ_f. There is, therefore, a quadrature component of the field flux, ϕ_t, called the torque flux. The torque flux, ϕ_t, is a component of the field flux, ϕ_f, which is at right angles to the armature flux, ϕ_a, tending to produce rotation by the interaction (or repulsion) of mutually perpendicular fields. Thus, *repulsion* between the *rotor* and *stator fields* results in rotation of the rotor; hence, the name *repulsion motor*.*

It should be noted that the torque is zero at the zero degree (maximum current) hard neutral position, and that the torque is also zero at the 90°

---

* Most motor action is the result of similar repulsion and almost any motor could therefore be called a repulsion motor.

CHAP. TEN / *Single-Phase Motors*

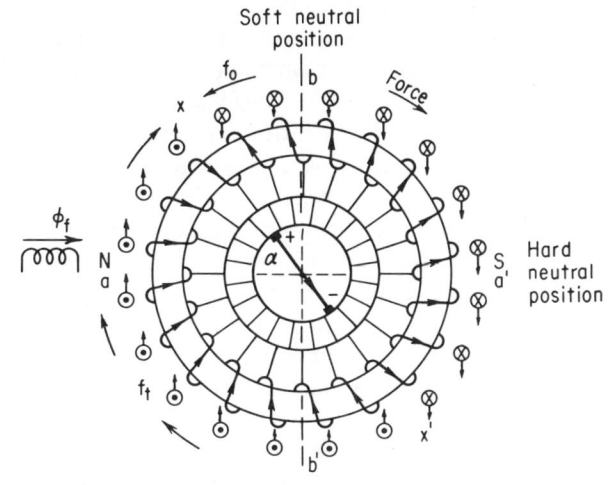

(a) Brush axis between hard and soft neutral.

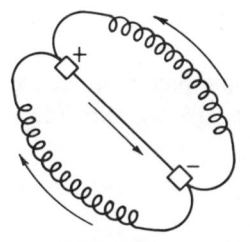

(b) Instantaneous currents in brushes and windings.

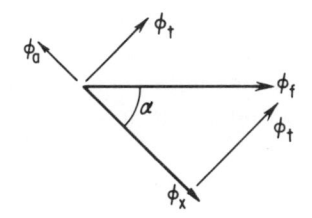

(c) Vector diagram showing main, transformer, and torque fluxes.

Figure 10-12

Brush axis between hard and soft neutral producing transformer and torque fluxes with accompanying motor torque.

(zero current) soft neutral position. Since the torque is a product of the armature flux (produced by the armature current) and the torque flux (produced by the field as a result of departure from the hard neutral position), it is evident that maximum torque will occur *closer to the hard neutral* position where the armature current is high. In commercial motors, maximum torque for a brush-shifting type of repulsion motor usually occurs where α is about 25 degrees from the hard neutral position.

It should be noted that *clockwise* rotation was produced when the brush axis was shifted from the hard neutral position in the *clockwise* or *same* direction. Shifting the brush axis counterclockwise from the hard neutral position causes counterclockwise rotation.

10-13. COMMERCIAL REPULSION MOTOR

Since repulsion motor rotors are usually lap-wound armatures, a four-pole ac stator winding would require two pairs of brushes; a six-pole stator, three pairs of brushes; etc. Each brush pair would be short-circuited and placed 90 mechanical degrees apart (for the four-pole) or 60 mechanical degrees apart (for the six-pole). Most commercial repulsion motors are either four- or six-pole motors, the poles produced by the type of ac stator windings employed. The stator windings, in addition, are split-field windings, consisting of two series-connected fields displaced 90 electrical degrees on the stator. The brushes are set along the axis of one of these windings, called the transformer field, ϕ_x, and 90 electrical degrees away from the other winding, called the torque field, ϕ_f. The relation of the armature to these windings is shown in Fig. 10-13a. Note that the resultant field, ϕ_r, produced by the two windings shown in Fig. 10-13b, has the same effect as that shown in the diagram of Fig. 10-12c *without the necessity for shifting the brushes.* The use of two separate windings results in a larger component of torque flux, reducing the armature current required to produce a given torque ($T = k\phi_t I_a \cos\theta$), as well as improving the power factor. In a commercial motor, the power factor is approximately unity at synchronous speed ($S = 120f/P$) and less than unity above and below synchronous speeds.

The repulsion motor, unlike the various single-phase induction motors, is capable of running at speeds *well above synchronous* speed

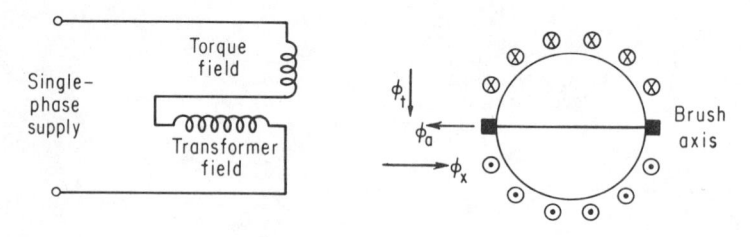

(a) Stator and armature fluxes.

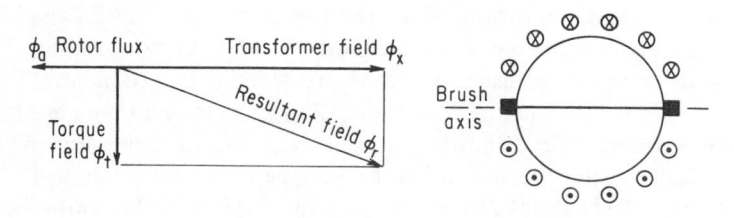

(b) Phase relation.

Figure 10-13

Commercial repulsion motor.

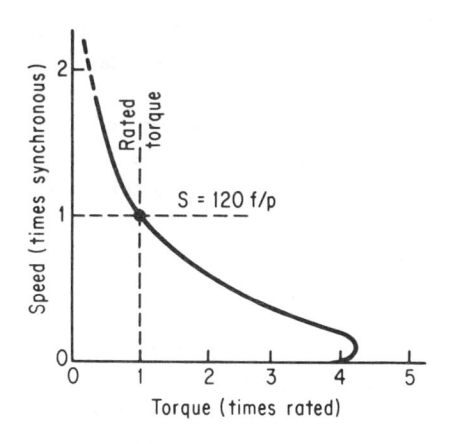

Figure 10-14

Speed-torque characteristic of a repulsion motor.

at light loads and well below synchronous speed when heavily loaded. Its speed-torque characteristic, shown in Fig. 10-14, most closely resembles a series motor. And this is scarcely odd, because it is, in effect, an inductively-coupled "series" motor in which the armature current is supplied by transformer coupling rather than by direct conductive coupling. Like the series motor, an increase in load on the repulsion motor produces an increase in armature (secondary) current and in its transformer coupled (primary) field current, proportionately. Thus, at heavy loads and at starting, the repulsion motor develops a starting current of about 1.5 to 2 times the rated load current.

The commercial repulsion motor shown in schematic form in Fig. 10-13a requires no brush shifting because the effect has been produced by the resultant flux of the stator windings. If it is desired to reverse the direction of rotation, reversing the connections of either field with respect to the other field will easily accomplish this. A few commercial repulsion motors, called "brush-shifting" motors, are produced with a lever arm connected to the brush rigging; these motors are capable of wide ranges of speed variation and smooth, stepless speed control, ranging as high as six times rated speed with the brushes at the maximum torque angle down to zero speed at the soft neutral (to avoid overheating and commutator arcing) for any given load.

It would appear that a single-phase motor of (1) such speed versatility, (2) excellent starting torque, and (3) low starting current, would be extremely popular. Furthermore, (4) the rotor, being independent of the stator, may be designed for low voltage, while the stator may be designed for several thousand volts. But, despite the advantages cited, the repulsion motor (1) is extremely noisy, (2) has poor speed regulation, and (3) requires periodic commutator maintenance. For these reasons, very few repulsion motors are manufactured at present. The repulsion principle, however, is more extensively applied to two other motors: (1) the repulsion-start induction motor; and (2) the repulsion-induction motor.

10-14. REPULSION-START INDUCTION MOTOR

If the commutator of a repulsion motor is completely short-circuited, it would produce a wound squirrel-cage-type rotor. It has already been shown that the stator of a repulsion motor is a distributed single-phase stator. Furthermore, the repulsion motor is capable of extremely high starting torque, whereas a number of other induction-principle single-phase motors are not. These facts led to the early development of the repulsion-start induction motor, one of the first types of single-phase motors to be commercially developed and sold in tremendous quantities.

The repulsion-start induction motor starts (as implied in its name) as a repulsion motor with its brushes set to the maximum torque position. When the load has been accelerated to about 75 per cent of synchronous speed, a built-in centrifugal device (similar to a governor) places a shorting ring in contact with the commutator bars, converting the armature to a squirrel-cage rotor. The motor then runs as an induction motor on its induction characteristic (Fig. 10-15).

Many types of governor-principle *centrifugal actuators* are used, and some of these lift the brushes from the commutator, simultaneously with the shorting action, to reduce brush wear and noise.

In fractional horsepower sizes, the repulsion-start induction motor has been largely replaced by capacitor-start and dual-capacitor induction motors. The reasons for this are (1) repulsion-start induction motors require more maintenance of commutator and mechanical devices; (2) they are not easily reversed (one has to shift the brush position or use separate field windings to do this); (3) they are more expensive in the same fractional horsepower rating; (4) they make quite a bit of noise on starting; and (5) the commutator arcing produces audible radio and visible television interference.

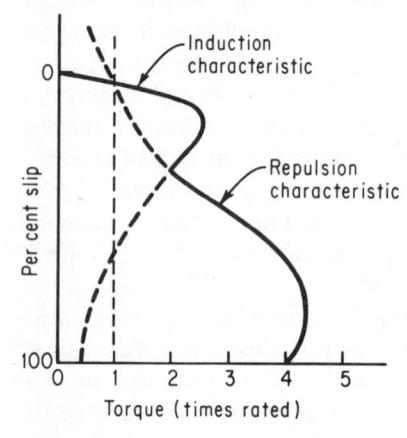

Figure 10-15

Speed-torque characteristic of repulsion-start motor.

In the integral horsepower sizes, however, the single-phase repulsion-start induction motor is still manufactured because of its (1) high starting torque, (2) low starting current, and (3) ability to accelerate a heavy load more rapidly than high-capacitance dual-capacitor motors.

10-15. REPULSION-INDUCTION MOTOR

As implied in its name, the repulsion-induction motor is a motor which combines the characteristics of the repulsion motor with those of the induction motor. Since the stator is the same for both a repulsion and an induction motor, the only modification required on the commutator-type rotor is the addition of a squirrel-cage winding. The repulsion-induction motor has a *double-cage* rotor in which the upper winding is connected to commutator bars and the lower (high-reactance) winding is an induction-type cage winding. The two rotor windings are insulated from each other, and the motor employs no governor or brush-lifting mechanism. With the squirrel-cage winding deeply embedded in the rotor slot, this cage winding, at starting,

386

CHAP. TEN / *Single-Phase Motors*

has an extremely high reactance and impedance. Little current flows in the cage winding, and the starting torque produced by this winding is negligible in comparison to the repulsion winding, as shown in Fig. 10-16.

Basically, the motor starts as a repulsion motor on its repulsion characteristic, producing about three to four times the rated torque. As the rotor accelerates, the rotor frequency and reactance of the low-resistance cage winding decreases, and more current is induced in the cage winding. For any given load on the rotor, the motor will operate as a combined repulsion and induction motor. If the load decreases, less slip is required and the motor speed rises, accelerated by its repulsion characteristic.

At the rated load, the motor operates at about synchronous speed (zero slip); and, since the cage-type winding on the rotor is cutting no flux, no current is induced in it. At loads less than rated, the repulsion motor tends to accelerate along the lines of its repulsion characteristic shown in Figs. 10-14. and 10-16. But now, at speeds above synchronous speed, the induction-type cage winding is being driven as an induction generator (see Section 9-22) whose prime mover is the repulsion motor. This tends to produce a counter torque which opposes the repulsion motor torque, with the result that the no load speed is only a little higher than the synchronous speed, as shown in Fig. 10-16, improving the speed regulation. The solid line in the figure represents the overall motor characteristic, a combination of the induction and the repulsion motor characteristics. In effect, this is the torque-slip characteristic which would be produced by a common induction motor stator having two rotors coupled on same shaft, one a repulsion type and the other an induction type.

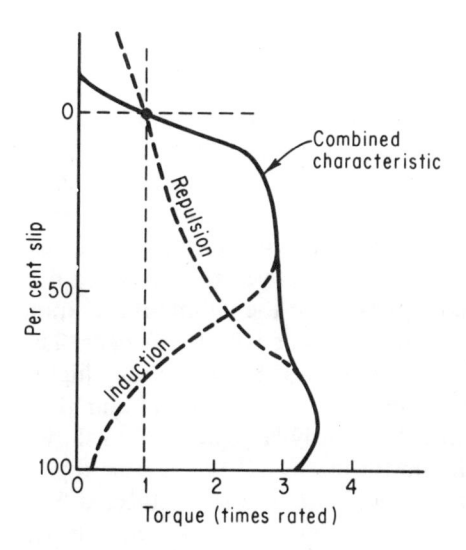

Figure 10-16

Speed-torque characteristic of repulsion-induction motor.

As shown in Fig. 10-16, the single-phase repulsion-induction motor has the advantages of (1) high starting torque, and (2) fairly good speed regulation. Its major virtue is (3) the ability to continue to develop torque under sudden, heavy applied loads without breaking down. It is manufactured in integral horsepower sizes and is used to drive fairly large reciprocating pumps and compressors where only single-phase power is available. Because of its fairly constant speed with the application of load, the repulsion-induction motor is also used in machine tools such as lathes and large boring milling machines where sudden stalling may be produced by heavy cutting. Single-phase repulsion-induction motors are also used in stokers, conveyors, compressors, and deep-well pumps.

Repulsion-induction motors are also reversed in the same manner as repulsion-start induction motors (by shifting the brushes with respect to the hard neutral or by reversing the transformer winding with respect to the torque field winding). Commercial sizes range from $\frac{1}{2}$ hp to about 15 hp.

**10-16.
UNIVERSAL
MOTOR**

There has always been a demand for a motor which might be used in small portable appliances and which would operate at any frequency from any of the sources of power available in various countries that might be visited by a tourist or traveller. An appliance containing such a motor, moreover, could be marketed internationally. Differences in voltage in various countries might be offset by using a tapped transformer or tapped resistor in conjunction with a motor of higher voltage rating.

At first impression, it would appear that the dc shunt motor is ideally suited as a universal motor. If the line leads from a dc supply feeding any dc shunt motor are reversed, the motor continues to run in the same direction. Application of alternating current as a voltage source to a shunt motor, however, results in very little starting or running torque. Since torque is the product of the interaction between the armature flux and the field flux ($T = k\phi_f I_a \cos \theta$), the reason for the failure of the shunt motor to produce as much torque as an ac motor is fairly evident. The shunt field is highly inductive, whereas the armature is essentially highly resistive. Thus, the armature and the field are not in phase, and their respective maximum fluxes are separate by a fairly large angle θ, as shown in Fig. 10-17a, producing practically no torque.

In the series motor, on the other hand, since a series-connected field and the armature always have the *same* current, their fields are always in phase, and the torque produced will be high ($\theta = 0$). Certain design modifications are necessary in the case of large series motors to insure good ac operation (Sec. 10-17). A small, fractional horsepower, dc series motor works equally well on ac for the reasons shown in Figs. 10-17b, c, and d. When the polarity of the supply reverses, both the polarity of the field and the direction of armature currents also reverse, continuing to produce torque in the same direction as shown.

The universal motor is designed for commercial frequencies from 60 hertz down to dc (zero frequency), and for voltages from 250 V to 1.5 V. A commercial universal motor may have a somewhat weaker series field and more armature conductors than a dc series motor of equivalent horsepower, for reasons which will be discussed in the next section. It is manufactured in ratings up to $\frac{3}{4}$ hp, particularly for vacuum cleaners and industrial sewing machines. In smaller sizes of $\frac{1}{4}$ hp or less, it is used in electric hand drills.

(a) Shunt motor.

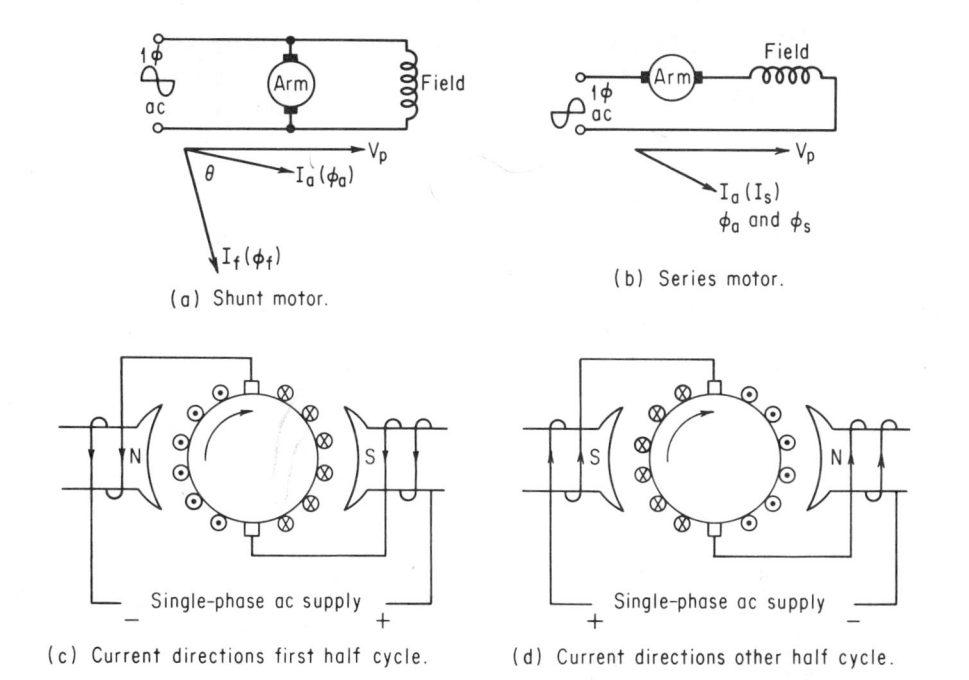

(b) Series motor.

(c) Current directions first half cycle.

(d) Current directions other half cycle.

Figure 10-17

Series or universal motor operation.

Like all series motors, the *no-load speed* of the universal motor is unusually *high*. Quite frequently, gear trains are built into the motor housing of some universal motors to provide exceedingly high torque at low speeds ($T = k\,\text{hp}/S$), and also to limit the high no-load speed by providing some loading through gear drag. In exceedingly small motors, $\frac{1}{20}$ hp or less, the full-load speed may be as high as 10,000 rpm, and the no-load speed considerably higher. When these motors are used in commercial appliances such as electric shavers, sewing machines, office machines, and small hand hair dryers or vacuum cleaners, they are always directly loaded with little danger of motor runaway.

10-17. AC SERIES MOTOR

Integral horsepower dc series motors, particularly in sizes above a few horsepower, operate poorly on alternating current. There is a great deal of sparking at the brushes, and the efficiency and power factor are low. Several design modifications are necessary in the series motor to improve its operation on alternating current. The modifications necessary and the reasons for each will be enumerated. Commercial ac series motors are designed:

1. With a more highly laminated field structure, to reduce the higher eddy current losses on alternating current.

2. With fewer series field turns, to reduce the reactive voltage drop across the series field and the losses due to eddy currents and hysteresis.
3. With more poles than corresponding dc machines, in order to restore the total torque [Eq. (4-3)].
4. With more armature conductors and more commutator segments, to compensate for the decrease in field flux [Eq. (4-3)].
5. With added resistance in series with the armature connections to the commutator, to reduce circulating currents, sparking at the brushes, and increased commutation difficulties caused by ac operation.
6. With special types of compensating windings, to reduce the increased armature reaction created by the increased number of armature conductors.
7. With special types of interpole windings, for the same reason as in item 6.
8. To reduce the series voltage drop across the interpole and compensating windings; these windings are often inductively connected rather than conductively connected, as shown in Fig. 10-18c. As the armature current increases because of increased load, the armature flux induces a higher current in these windings, and the effect of the interpole and compensating **mmf** is thus proportional to the load or the armature current. Since alternating current is produced in the armature conductors, even when connected to a dc supply, inductive coupling may be used for these windings even in the dc series motor. The voltage drop across these windings is usually small on direct current. There are fewer design problems with conductive coupling, however, shown in Fig. 10-18a.

With the modifications above, series motors operating on alternating current will perform in the same manner as their dc counterparts, producing the characteristic shown in Fig. 10-18d.

The phasor diagram for the conductively coupled ac motor is shown in Fig. 10-18b. The power developed by the armature of the ac series motor is $E_g I_a$ [Eq. (4-7)], and the armature current is thus limited by the generated armature counter emf, E_g, *plus* all of the impedance drops in series as shown by the equation in Fig. 10-18b.

The advantage of inductive coupling is an increase in the generated emf and the armature power, shown in Fig. 10-18c. With loose transformer coupling, the interpole and compensating windings are capacitively reflected to the armature, tending to reduce the armature impedance drop and to improve the power factor angle θ between V_p and I_a.

Large single-phase series ac motors have been supplanted in application by the simpler polyphase induction and synchronous motors, for the most part. They are still extensively used, however, in main line and interurban railway systems (traction service) for electric locomotives. These motors are designed for voltages below 300 V, taking their secondary power from 11,000 V, 25 Hz primary sources, the standard for railway service in the United States. The power ratings for railway motors vary from

Figure 10-18

AC series motor.

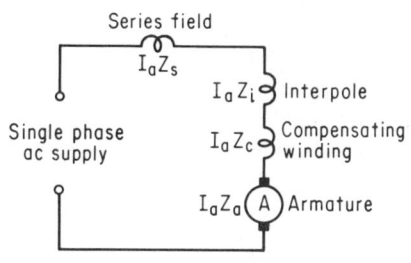

(a) Conductively coupled ac series motor.
(b) Phasor diagram for conductively coupled motor.
(c) Inductively coupled ac series motor.
(d) ac series motor characteristic.

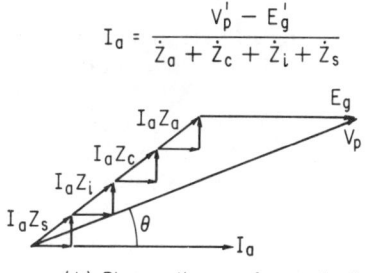

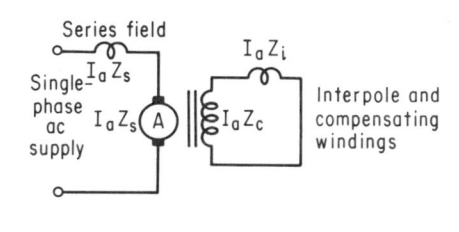

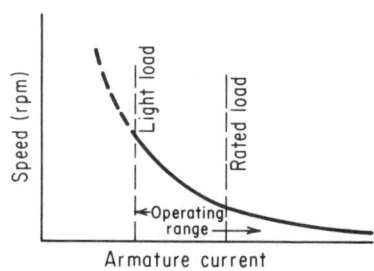

several hundred to slightly more than one thousand horsepower, with power factors of 0.95 and efficiencies of about 0.88 as a result of the modifications discussed above. European traction service uses a standard frequency of $16\frac{2}{3}$ Hz. The lower frequencies (below 60 Hz) are preferred because of the reduced series-impedance voltage drops at the lower frequency and the consequent higher developed power. This results in improved efficiency, as well, because of lessened eddy current and hysteresis loss.

Speed regulation is easily obtained through a change of supply voltage to the ac series motor by means of a tapped transformer or induction regulator. Since the efficiencies of transformers are extremely high, there is practically little loss in this method of armature or line voltage control as compared to the resistive method of speed control used with dc series motors. Reversal is also easily obtained by reversing the series field with respect to the armature through suitable switching.

10-18. SUMMARY OF TYPES OF SINGLE-PHASE MOTORS

Because of the many variations in principle of operation, and the fact that certain single-phase synchronous types were covered in Chapter 8, the following represents a summary of the principle of operation of all the various types of single-phase motors, with applicable section references, as described in this volume.

> I. Single-Phase Induction Motors
>   A. Split-phase motors
>   1. Resistance-start motor (Section 10-5)
>   2. Capacitor-start motor (Section 10-6)
>   3. Permanent-split (single-value) capacitor motor (Section 10-7)
>   4. Two-value capacitor motor (Section 10-8)
>   B. Reluctance-start induction motor (Section 10-10)
>   C. Shaded-pole induction motor (Section 10-9)
>   D. Repulsion-start induction motor (Section 10-14)
> II. Single-Phase Synchronous Motors
>   A. Reluctance motor (Section 8-27)
>   B. Hysteresis motor (Section 8-28)
>   C. Subsynchronous motor (Section 8-29)
> III. Commutator-Type Single-Phase Motors
>   A. Repulsion motor (Section 10-13)
>   B. Repulsion-induction motor (Section 10-15)
>   C. AC series motor (Section 10-17)
>   D. Universal motor (Section 10-16)

BIBLIOGRAPHY

Alger, P. L. and Erdelyi. E. "Electromechanical Energy Conversion," *Electro-Technology*, September 1961.

Bewley, L. V. *Alternating Current Machinery*, New York: The Macmillan Company, 1949.

Bewley, L. V. *Tensor Analysis of Electrical Circuits and Machines*, New York: Ronald Press, 1961.

Carr, C. C. *Electrical Machinery*, New York: John Wiley & Sons, Inc., 1958.

Crosno, C. D. *Fundamentals of Electromechanical Conversion*, New York: Harcourt, Brace, Jovanovich, Inc, 1968.

Crouse, C. H. "Capacitor Start/Run of 15 HP Single-Phase Motors," *Electrical Manufacturing*, March 1957.

Daniels. *The Performance of Electrical Machines*, New York: McGraw-Hill, Inc., 1968.

Douglas, J. F. H. "Reluctance Motors for High Torque Specifications," AIEE Paper CP 61-222.

Ellenberger, J. "Start-Run Protection of Split-Phase Motors" *Electrical Manufacturing* December 1959.

Fitzgerald, A. E. and Kingsley, C. *The Dynamics and Statics of Electromechanical Energy Conversion*, 2nd Ed., New York: McGraw-Hill, 1961.

Fitzgerald, A. E., Kingsley, Jr. and Kusko, A. C. *Electric Machinery*, 3rd Ed. New York: McGraw-Hill Inc., 1971.

Gemlich, D. K. and Hammond, S. B. *Electromechanical Systems*, New York: McGraw-Hill, 1967.

Hindmarsh, J. *Electrical Machines*, Elmsford, N.Y.: Pergamon Press, 1965.

Jones, C. V. *The Unified Theory of Electrical Machines*, New York: Plenum Publishing Corp., 1968.

Kasparian, C. "Shading Coils Energized for Motor Control," *Electro-Technology*, October 1961.

Liwschitz, M. M., Garik, M., and Whipple, C. C. *Alternating Current Machines*, Princeton, N. J.: D. Van Nostrand Co., Inc., 1946.

McFarland, T. E. *Alternating Current Machines*, Princeton, N. J.: D. Van Nostrand Co., Inc., 1948.

Majmudar, H. *Introduction to Electrical Machines*, Boston: Allyn and Bacon, 1969.

Millermaster, R. A. *Harwood's Control of Electric Motors*, 4th Ed., New York: Wiley/Interscience, 1970.

Puchstein, A. F., Lloyd, R., and Conrad, A. G. *Alternating Current Machines*, 3rd Ed., New York: Wiley/Interscience, 1954.

Robertson, B. L. and Black, L. J. *Electric Circuits and Machines*, 2nd Ed., Princeton, N. J.: D. Van Nostrand Co., Inc., 1957.

Selmon. *Magnetoelectric Devices: Transducers, Transformers and Machines*, New York: Wiley/Interscience, 1966.

Siskind, C. S. *Electrical Machines—Direct and Alternating Current*, 2nd Ed., New York: McGraw-Hill Inc. 1959.

Skilling, H. H. *Electromechanics: A First Course in Electromechanical Energy Conversion*, New York: Wiley/Interscience, 1962.

Smeaton, Motor *Applications and Maintenance Handbook*, New York: McGraw-Hill, 1969.

Thaler, G. J., and Wilcox, M. L. *Electric Machines: Dynamics and Steady State*, New York: Wiley/Interscience, 1966.

Veniott. *Fractional and Subfractional Horsepower Electric Motors*, 3rd Ed., New York: McGraw-Hill, 1970.

White, D. C., and Woodson, H. H. *Electromechanical Energy Conversion*, New York: Wiley/Interscience, 1959.

QUESTIONS

10-1. Define:
    a. integral hp motor
    b. fractional hp motor.

10-2. On the basis of the above definitions, explain why:

a. high speed motors (10,000 rpm and up) are usually fractional hp
b. low speed motors (below 50 rpm) are usually integral hp motors.

10-3. Explain
 a. why no single-phase SCIM is inherently self-starting
 b. why the various types of single-phase SCIMs are classified on the basis of starting method (see Sec. 10-18)
 c. whether there is a difference in *rotor* construction of a polyphase and single-phase motor
 d. why a single-phase, *single* stator winding produces no rotation of a SCIM rotor.

10-4. a. Explain why a single stator winding, as in Quest. 10-3d, does accelerate a SCIM rotor, if the rotor is first turned in a specific direction. Use the double-revolving field theory to illustrate your answer.
 b. Repeat (a) above, using the cross-field theory.

10-5. a. What is meant by elliptical torque?
 b. What conditions of load produce it?
 c. How is it possible to identify production of elliptical torque?
 d. Explain the mechanism by which the rotating magnetic field and torque become elliptical and the reasons for it.

10-6. In the design of the resistance-start single-phase induction motor, explain reasons for
 a. a starting winding consisting of few turns of smaller diameter copper wire as opposed to the use of many turns of heavier wire
 b. disconnecting the starting winding once the slip reaches 0.25 or less
 c. not requiring that the starting and running currents are equal to each other
 d. starting and running windings displaced in space by 90°.

10-7. a. Given a single-phase stator having two identical windings displaced in space by 90°, draw a wiring diagram using a switch and a resistor which will cause the motor to rotate in either direction.
 b. Illustrate the above diagram using a phasor diagram for each switch position.
 c. Is a high or low resistance resistor required? Explain.

10-8. a. Explain why the starting torque of a split-phase motor is relatively small.
 b. What is meant by plugging?
 c. On the basis of (a) and (b) above, explain why plugging a resistance-start motor while running will not reverse the direction of rotation.
 d. How is a resistance-start motor reversed at standstill?
 e. On the basis of the above, define a *nonreversible* motor and explain why the resistance-start motor is classified as a nonreversible motor.

10-9. a. Explain why the high starting current of a resistance-start motor is not an objectionable feature
 b. Give the major objections and three disadvantages of this motor
 c. Based on (b), give the major applications for this motor.

10-10. a. Is it possible to control the speed of a running split-phase motor with

the starting winding disconnected? Explain how, assuming no frequency change.
b. Within what range of slip is speed control possible by the method of (a)?
c. Is it possible to obtain speed control with heavy loads? Explain.

10-11. Define
a. a reversible motor
b. a reversing motor.
c. a nonreversible motor
d. a nonreversing motor

10-12. On the basis of the above definitions, explain
a. why a split-phase capacitor-start motor is classified as a reversible but not a reversing motor
b. why it is necessary to temporarily disconnect the motor from the supply in order to achieve reversal
c. Modify Fig. 10-4 using a dpdt switch to show how the reversal is accomplished. (Include a dpst line switch to the supply.)

10-13. a. Explain why the starting torque of a capacitor-start motor is more than twice that of a resistance-start motor.
b. Reducing the resistance of (using a heavier) starting winding to yield a high starting current increases the starting torque. Explain why.
c. Discuss pros and cons of using a low resistance, high starting current winding, in terms of starting current, economy, and size of stator.

10-14. a. Explain why capacitor-start motors are made in integral hp sizes and resistance-start motors are not.
b. Explain why capacitor-start motors are used in high starting torque applications while resistance-start motors are not.
c. Give several applications for capacitor-start motors.

10-15. a. The single-value capacitor motor is sometimes called a two-phase motor. Explain why.
b. Why does the above motor have low starting torque?
c. Why does the above motor have low running torque?
d. Give 3 advantages of this motor in comparison to split-phase capacitor-start motors.
e. Why are oil-filled rather than electrolytic capacitors used on this motor?

10-16. a. Distinguish between a reversible motor and a reversing motor. See Q. 10-11 above.
b. Explain, on the basis of the above definitions, why the permanent-split capacitor motor is a reversing motor.
c. Why is the above motor more easily adapted to speed control by voltage variation than other types of split-phase motors?
d. What applications are indicated for the above motor?

10-17. a. Why does the permanent-split, two-value capacitor motor produce high starting torque? What advantages does this motor have over other single-phase motors?
b. Why is an electrolytic capacitor used in the two-value capacitor motor?

c. Compare pros and cons of the use of an autotransformer and single oil-filled capacitor against the same oil-filled capacitor and an electrolytic starting capacitor, as shown in Fig. 10-6.

10-18. Explain why a defective (shorted) centrifugal switch constitutes a serious disadvantage to
a. the resistance-start motor
b. the capacitor-start motor
c. the two-value capacitor motor.

10-19. For the shaded-pole SCIM show
a. that the flux in the shaded part of the pole always lags the flux in the unshaded-pole segment, both in space and time, producing rotation of a SC rotor
b. why reversible shaded-pole motors use distributed winding techniques on nonsalient stators or wound shading coils on salient stators
c. advantages of nonsalient over salient stators
d. advantages and disadvantages of commercial types
e. methods of speed control.

10-20. For the reluctance-start SCIM show
a. that the flux in the low reluctance part of the pole always lags the flux in the high reluctance portion, both in time and space, producing rotation of a SC rotor
b. why the direction of rotation of a given motor is irreversible
c. why it is inferior to the shaded pole motor.

10-21. For the repulsion motor show
a. that the torque is zero at the hard neutral, zero degree, position
b. that the torque is zero at the soft neutral, 90° position
c. the position at which maximum torque is developed and why
d. how the direction of rotation is reversed.

10-22. For the commercial repulsion motor show
a. why 2 separate series-connected field windings are needed
b. why it might be considered an inductively coupled series motor
c. conditions under which speeds above synchronous speed may occur
d. the starting torque is very high and starting current is very low
e. why few repulsion motors are currently manufactured.

10-23. For the commercial repulsion-start SCIM show
a. principle of operation
b. special mechanical and switching devices employed
c. how the motor is reversed (give 2 methods)
d. advantages and disadvantages
e. why the motor is no longer made in fractional hp sizes.

10-24. For the commercial repulsion-induction motor, show
a. principle of operation and construction
b. special mechanical and switching devices employed
c. how to reverse the motor
d. advantages and disadvantages
e. applications and commercial sizes of manufacture.

10-25. a. Explain why a shunt (dc) motor will not work on ac satisfactorily, despite the fact that reversal of line polarity causes no change in direction of rotation.
b. Repeat for a series motor showing why it should operate intermittently on ac. (See Quest. 10-26a.)
c. Why is the speed of a universal motor independent of frequency?
d. What precautions are required in use of universal motors and how is this overcome?

10-26. a. What modifications are made to series (dc) motors to render them useful for commercial ac operation?
b. Why are single-phase series motor preferred for traction service rather than induction or synchronous motors?
c. What is the hp range of commercial ac series motors?
d. How are ac series motors speed controlled and reversed?

10-27. The following is suggested as a special assignment or term report. For *each* motor listed in Sec. 10-18, make a table of comparison with respect to the following categories:

A. Construction:
  1. stator.
  2. rotor.
B. Principle of operation, briefly described or named.
C. Starting conditions:
  1. Torque compared to rated.
  2. Current compared to rated.
D. Running conditions:
  1. Maximum torque compared to rated.
  2. Speed regulation.
E. Methods of speed control.
F. Reversal possibilities and how accomplished.
G. Advantages.
H. Disadvantages.
I. Commercial hp sizes.
J. Load applications.

PROBLEMS

10-1. In accordance with the definition given in Section 10-1, determine whether the following motors are integral or fractional horsepower motors
a. $\frac{3}{4}$ hp, 1200 rpm
b. $\frac{3}{4}$ hp, 1800 rpm
c. $1\frac{1}{2}$ hp, 6000 rpm
d. $1\frac{1}{4}$ hp, 3600 rpm.

10-2. A $\frac{1}{2}$ hp, split-phase, ac single phase motor draws a lagging current of

$3\angle-15°$ A in its starting winding and a current of $4.9\angle-40°$ A in its running winding with respect to the supply voltage of $230\angle 0°$ V. At the instant of starting, calculate
a. Total locked rotor current
b. Power factor and power consumed at starting
c. Power factor and power consumed under running conditions, assuming that the current to the running winding at full load is the same as that under starting conditions at the same power factor angle
d. Full-load efficiency.

10-3. a. Repeat Problem 10-2(a) and (b) if a capacitor is added to the starting winding causing a leading current in that winding of $2.38\angle 40°$ A at starting.
b. Compare the starting power factor and current of the capacitor-start motor with that of the conventional split-phase motor of Problem 10-2.
c. Calculate the value of the capacitor required to produce the leading current in the starting winding.
d. Compare the value of the capacitor computed in (c) with that given in Table 10-1 and account for difference.

10-4. A $\frac{1}{3}$ hp, single-phase, four-pole, ac split-phase motor draws 7.2 A from a 115 V, 60 Hz supply at a power factor of 75 per cent and runs at a speed of 1720 rpm when rated load is applied to its shaft. Calculate:
a. Full-load efficiency
b. Full-load slip
c. Rated output torque
d. Maximum torque if breakdown occurs at a slip of 30 per cent
e. Starting torque
f. Ratio of maximum and starting torques, respectively, to rated torque.

10-5. A two-pole, 60 Hz, 115 V reluctance-start, single-phase induction motor is rated at $\frac{1}{25}$ hp, 3300 rpm, and has a full-load efficiency of 60 per cent. Calculate:
a. Full-load power input
b. Full-load current if the full-load power factor is 0.65
c. Full-load slip and full-load torque
d. Maximum torque, if breakdown occurs at a slip of 20 per cent (in ounce-inches)
e. Starting torque
f. Ratio of maximum and starting torques, respectively, to full-load torque.

10-6. A dynamometer brake test on a fully loaded four-pole, 115 V, single-phase motor running at rated speed revealed the following data: power drawn by motor, 150 W; input current, 2.0 A; speed, 1750 rpm; length of brake arm, 12 in.; dynamometer scale reading, 6 oz at rated load. When operated at an overload to determine *maximum* torque, the following data was obtained: power drawn by motor, 550 W; input current, 10.0 A; speed 1400 rpm; dynamometer scale reading, 26.5 oz at maximum load. Calculate:
a. Efficiency at rated load

b. Power factor at rated load
c. Motor horsepower and rated torque
d. Maximum and starting torque
e. Efficiency and power factor at maximum torque
f. Rated torque from maximum torque.

10-7. A 110 V, two-value capacitor motor is started and run by means of a tapped autotransformer in combination with a single 5 μF capacitor, as shown in Fig. 10-6b. Calculate:
a. Effective values of capacitance in series with winding B in both the starting and running positions, respectively, and ac voltage rating of the capacitor using transformer taps having a ratio of 5:1 at starting and 1.2:1.0 in the running position
b. Repeat (a) using taps yielding starting and running ratios of 8:1 and 2:1, respectively.

10-8. The dual-value 60 Hz capacitor motor of Problem 10-7 has a main winding with a resistance of 30 Ω and an inductance of 0.1 H. The second (capacitor) winding has a resistance of 25 Ω and an inductance of 0.5 H. Calculate:
a. The relative values of starting current in each winding, the angle between these currents and the line current, power factor, and power input at the instant of starting using the autotransformer ratio of 5:1
b. Repeat (a) using a ratio of 8:1
c. Which of the two tapped autotransformer starting ratios should produce the higher starting torque, and why?

10-9. A loaded ac series motor draws a current of 6.5 A from a 115 V, 60 Hz source at a power factor of 0.85 lagging while delivering an output torque of 0.5 lb-ft at a speed of 5000 rpm. A series resistance used for speed control produces an ac voltage of 100 V across the motor, a current of 9 A at a power factor of 0.8 lagging, in order to produce the same torque at reduced speed. The combined armature and series field impedance drop of the motor is 0.5 Ω. Assuming that the impedance voltage drop is always in phase with the supply voltage and neglecting saturation, calculate
a. Reduced speed
b. Efficiency at the higher speed
c. Efficiency at the lower speed.

10-10. The design data for a $\frac{1}{8}$ hp, two-pole, 115 V universal motor gives the effective resistances of the armature and series field as 4 Ω ans 6 Ω, respectively. The output torque is 24 oz-in. when delivering rated current of 1.5 A (ac) at a power factor of 0.88 at rated speed. Calculate:
a. Full-load efficiency
b. Rated speed
c. Full-load copper losses
d. Combined windage, friction, and iron losses
e. The motor speed when the current is 0.5 A neglecting phase differences and saturation.

10-11. A 220 V, 7$\frac{1}{2}$ hp, three-phase, delta-connected induction motor draws a line current of 8 A and a total power of 546 W from the lines when running at

no load. If while running, a defective fuse causes one of the line wires to open, calculate
   a. The approximate value of line current
   b. The total losses
   c. Describe a simple test to distinguish the existing single-phase operation from a possible increase in current due to mechanical loading

10-12. The motor in Problem 10-11 is delivering a 5 hp load at an efficiency and power factor of 0.8 and 0.85, respectively, and the stator copper losses at this load are 4 per cent of the total input. If one of the lines is suddenly opened, calculate
   a. The new stator copper losses
   b. The approximate rating of this motor for single-phase operation, assuming that the total stator copper loss should not exceed the rated three-phase stator copper loss.

ANSWERS

10-1(a) integral (b) fractional (c) fractional (d) fractional   10-2(a) 7.73 A (b) 1530 W, 0.86 PF lagging (c) 862 W, 0.766 PF lagging (d) 0.433   10-3(a) 1278 W at 0.96 PF lagging (c) 32.4 μF   10-4(a) 0.4 (b) 0.0444 (c) 1.015 lb-ft (d) 3.44 lb-ft (e) 1.9 lb-ft (f) 3.39:1, 1.87:1   10-5(a) 49.7 W (b) 0.665 A (c) 0.0833 lb-ft, 0.0636 lb-ft (d) 14.9 oz-in. (e) 5.73 oz-in. (f) 1.22:1, 0.47:1   10-6(a) 0.622 (b) 0.653 PF (c) 72 oz-in. (d) 318 oz-in. 135 oz-in. (e) 0.598, 0.478 PF (f) 72.2 oz-in.   10-7(a) 125 μF, 7.2 μF, 550 V (b) 320 μF, 20 μF, 880 V   10-8(a) 2.31 A, 3.2 A, 94.2°, 3.8 A, 0.995 PF, 416 W (b) 1.012 A, 2.31 A, 25.7°, 3.25 A, 0.517 PF, 185 W (c) 5:1 ratio   10-9(a) 3090 rpm (b) 55.9 per cent (c) 26.5 per cent   10-10(a) 61.5 per cent (b) 5252 rpm (c) 22.5 W (d) 36.2 W (e) 17,350 rpm   10-11(a) 13.85 A (b) 1640 W   10-12(a) 558 W (b) 2.5 hp

ELEVEN

specialized dynamos

11-1.
GENERAL

The basic principles of ac and dc machines discussed in the preceding chapters have been utilized in the development of a variety of dynamos and combinations of dynamos. Such machines are used, in general, for the conversion of mechanical energy to electric energy, or vice versa. This chapter will deal with other kinds of dynamos and combinations of dynamos that, although they too accomplish similar energy conversions, are *more specialized* in nature and in application. The study of these specialized dynamos has been deferred until this point because, to understand and appreciate them, it is necessary to have a background knowledge of the machines discussed previously.

This chapter will be devoted to the following special types of electric machines: diverter-pole generators, third-brush generators, homopolar generators, dynamotors, synchronous converters, power selsyns and synchro-tie systems, three-wire generators, induction phase converters, self-synchronizing generators and motors (or selsyns), dc and ac servomotors, the Rosenberg generator, Amplidyne and multiple-field exciters such as the Rototrol and Regulex.

11-2. DIVERTER-POLE GENERATOR

When dc shunt generators are used for battery charging, the slightly drooping load-voltage characteristic with increasing load results in a longer charging period. As the battery voltage rises, the load on the shunt generator decreases, reducing the generator current and causing the voltage of the shunt generator to rise in order to maintain the charging current. The reduced current results in fewer ampere-hours of charge, requiring a longer charging time. It is possible, of course, to use a flat-compound dc generator, but even such a generator has a rising and drooping characteristic (see Fig. 3-13). It is also possible to use a shunt generator with a voltage regulator, in an attempt to maintain constant voltage; but such devices vary the field resistance of the shunt generator in incremental steps, and an absolutely flat characteristic with increased load is difficult to obtain.

The *diverter-pole* generator is an interesting modification of the shunt generator; its purpose is to produce an absolutely flat voltage characteristic from no-load to full-load. As shown in Fig. 11-1, it consists of a shunt-wound interpole generator in which a *magnetic shunt* is physically interposed between the interpole and the adjacent main pole of the same polarity. A diverter-pole winding, similar to that of a compound generator, is connected in series with the load as shown in Fig. 11-1b.

At no-load, there is no current, in the diverter-pole winding. A portion of the main field *flux*, therefore, is I_d, *diverted* through the magnetic shunt and into the diverter pole, as shown in Fig. 11-1a. This portion of shunted flux is diverted away from the armature mutual air-gap flux. As the load on a self-excited shunt generator increases, the terminal voltage will drop (for the three reasons noted in Section 3-13), all of which cause a decrease in the mutual air-gap flux. The turns are wound on the diverter-pole winding in such a way as to oppose the mutual flux created by its adjacent pole of like magnetic polarity. When the current taken by the load is light, the diverter pole still carries some portion of its no-load *flux*. As the load increases, the diverter-pole winding continues to create an increasing **mmf** in opposition to the main field flux, as shown in Fig. 11-1c. The net result is that, as the load increases, more of the diverted flux now enters the armature as mutual air-gap flux. This action counterbalances exactly the decrease of flux produced by (1) armature reaction, (2) increased $I_a R_a$ drop, and (3) decreased field current.

The smooth and flat terminal-voltage load characteristic is shown in Fig. 11-1d. When the load is greater than the rated load, the mmf from the diverter becomes excessive and tends to reduce the main field flux, causing the voltage to drop rapidly. This, too, is an added advantage, because on short circuit, the generator is self-protecting. The diverter-pole

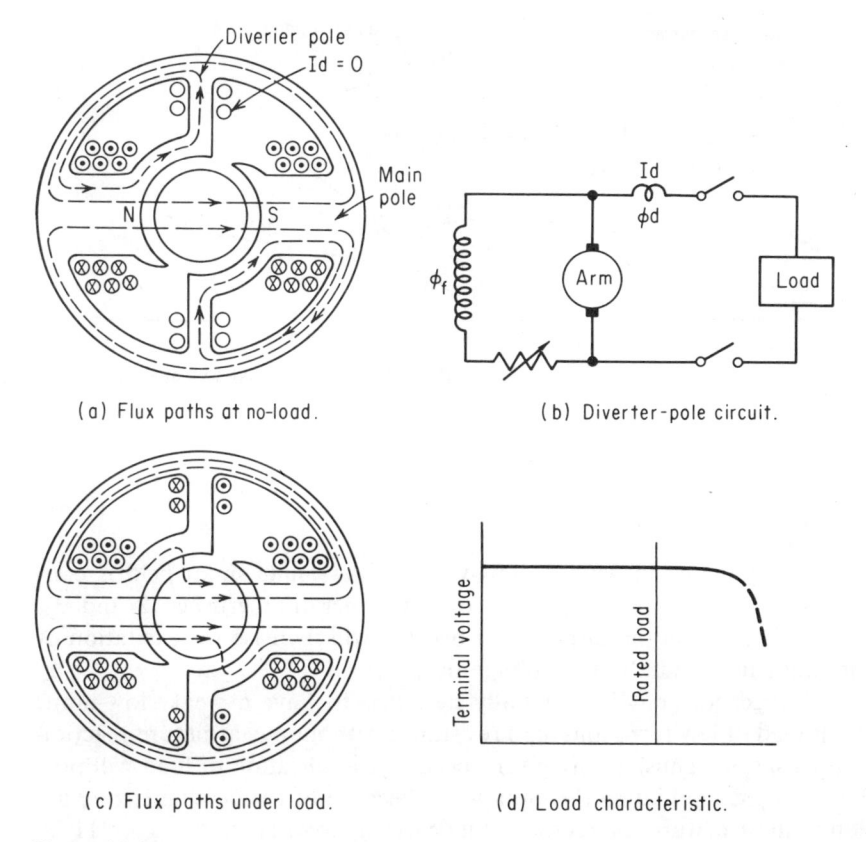

Figure 11-1

(a) Flux paths at no-load.

(b) Diverter-pole circuit.

(c) Flux paths under load.

(d) Load characteristic.

Diverter-pole generator operation and characteristic.

flux, ϕ_d, cancels the field flux, ϕ_f, almost completely if the generator is short-circuited, reducing the generated voltage sharply.

The diverter-pole generator finds its greatest application in battery-charging, where the current rate is high at starting and tapers off to a low rate for finishing, without requiring any adjustment by an attendant. It is also used in laboratories in lieu of an electronic power supply where absolutely constant high dc voltage and high current is required for experimental purposes. In these instances, the diverter-pole generator is driven by a synchronous motor.

11-3. THIRD-BRUSH GENERATOR

For many years, the third-brush generator was used for charging batteries on automobiles. It was developed in response to a demand for a generator capable of producing the proper characteristic for battery-charging over a *wide* prime mover *speed* range. As shown in Fig. 11-2a, the third-brush generator is essentially

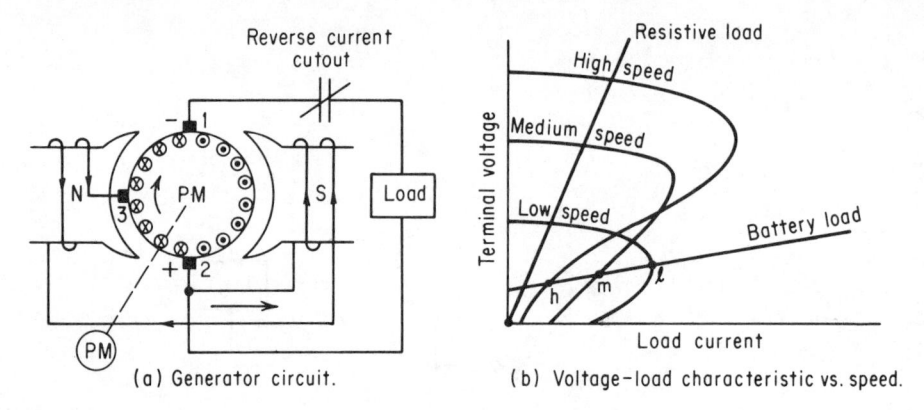

Figure 11-2

Third-brush generator.

(a) Generator circuit.

(b) Voltage-load characteristic vs. speed.

a shunt generator in which a low-voltage (low-**mmf**) shunt field is connected across the partial output of the generator (brushes 2 and 3). Moving brush 3 in the direction of brush 1 will increase the excitation of the shunt field and also the voltage output.

The generator is intentionally designed to have a weak, low-mmf, shunt field of few turns and high resistance without any armature-reaction compensation. Thus, at any given speed, the application of load will produce a rapid unbuilding (Section 3-12), because of the demagnetizing action of the armature conductors while carrying load current. Figure 11-2b shows three voltage-load characteristics at high, medium, and low speeds.

For a normal resistive generator load, the current delivered by the third-brush generator would produce a higher load current at a higher speed. But a storage battery has an extremely low internal resistance (in addition to an emf opposing the third-brush generator emf), and its characteristic intersects the voltage breakdown portions of the voltage-speed characteristic at points h, m and l. Thus, at *low* speeds, the generator is charging batteries at a *higher rate* than at high speeds. This is a desirable characteristic, since low-speed city driving (with frequent starts and stops) requires more frequent battery-charging than high-speed turnpike driving over long periods with few stops. To increase the charging rate, it is only necessary to adjust brush 3 in the direction of rotation, or toward brush 1.

The use of automatic voltage regulators, with cutouts to prevent motorizing of the generator and to adjust the voltage and current of a normal low voltage, high-capacity, shunt generator, eliminated the necessity for the low-capacity third-brush generator in automobile as well as marine power battery-charging. In recent years, moreover, the use of a polyphase alternator with suitable diode rectifiers has (1) completely eliminated the possibility of motorizing and (2) enabled dc output voltages at relatively low prime mover speeds.

**11-4.
HOMOPOLAR OR
ACYCLIC DYNAMO**

The *homopolar generator*, discovered by Faraday (see Section 1-11 and the footnote figure), is the answer to the usual student question as to whether a dc generator or motor could be constructed so that the conductors are always cutting flux in the same direction. Such a generator could never suffer from the usual eddy current or hysteresis losses which occur in all rotating machines, because it is basically an acyclic machine and requires no commutator. The acyclic homopolar generator in its commercial form is shown in Fig. 11-3a. The armature is a hollow brass or copper cylinder supported by a spider welded to the shaft. The brushes are heavy copper rings which make contact with the cylinder at each end. The fields are a pair of coils, each concentric to the shaft. The yoke is assembled in two halves to make construction possible. The polarity of the emf of the brushes is determined (right-hand rule) by the direction of rotation shown in the figure. Because of its *single* conductor, the homopolar generator produces a very low voltage (roughly 3 V at 1500 rpm), but the current is a function of physical size and flux density. Intermittent currents as high as 10,000 A have been developed, as well as continuous currents of approximately 6000 A on generators built by the Westinghouse Electric Company as early as 1896. If dc is supplied to the field and brushes, the dynamo will run as a *homopolar motor*.

(a) Commercial homopolar generator.

(b) Magnetohydrodynamic generator and electromagnetic pump.

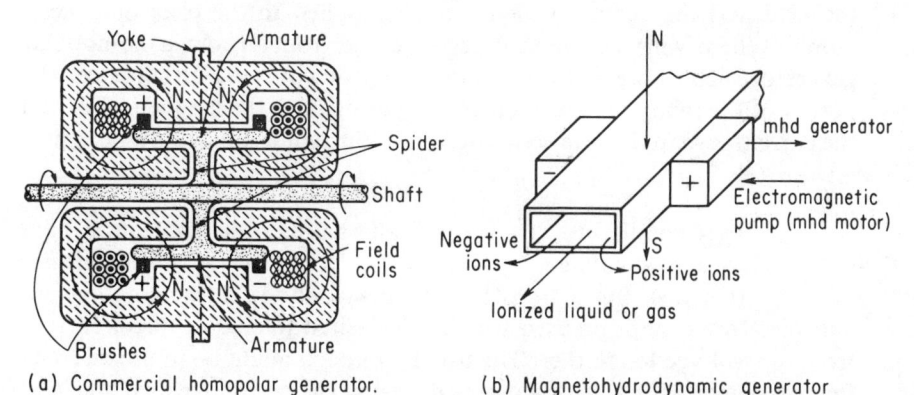

Figure 11-3

Homopolar (acyclic) dynamos.

Homopolar generators are used in railway service to provide high currents and strong magnetic fields in special railway cars containing apparatus for detecting concealed defects and cracks in railway tracks. During World War II, they were also used in marine service to excite large shipboard "degaussing" cables, effective against underwater magnetic mine devices.

The principle of the homopolar generator is currently undergoing extensive experimentation in MHD (magnetohydrodynamics) power generation.* An electrically conductive ionized liquid or gas (vaporized sodium or potassium) escaping from a nuclear fusion reactor at high speeds moves through an extremely strong magnetic field, as shown in Fig. 11-3b. The positive ions of the gas are driven toward one electrode (right-hand rule for the direction of the gaseous "conductor"), and the negative ions (electrons) toward the other electrode. The source of mechanical energy is the speed of the fast-moving gas escaping from the reactor. Thus, the current density (coulombs/second) at the electrodes is strictly a function of the quantity of matter passing the electrodes per second, while the potential is a function of the strength of the magnetic field and the speed of the ionized gas. By converting the direct magnetic field into an ac field, it is possible to generate alternating current at the electrodes as well.

The gas may be accelerated also by a homopolar motor principle, called the *electromagnetic pump*. This pump (an MHD motor) has no moving parts and, therefore, produces no leakage of radioactive gas or conductive cooling liquids used in the reactor. A voltage applied to the electrodes, shown in Fig. 11-3b, produces a force on the conductive gas tending to accelerate it. The force is proportional to the flux density, the current in the electrodes, and the mean length between them [Eq. (1-8)]. The pressure on the gas, therefore, is limited mainly by the flux density of the field and the current which can be supplied to the electromagnetic pump. Where will such high currents come from? From a homopolar generator such as that shown in Fig. 11-3a, of course. Furthermore, the output of the pump may be increased by using both the magnetic flux and the current several times in series by various ingenious liquid flow designs.†

11-5. DYNAMOTOR

It is possible to "transform" alternating voltages from a high to a low voltage, or vice versa, by means of a transformer (Ch. 13). But a transformer works only on alternating current and therefore cannot be used where it is desired to change direct current from one voltage to another. The usual approach might be to use an M-G (motor-generator) set, with a motor driven at the dc voltage available and the generator producing the desired dc voltage. The disadvantages of a *two*-unit M-G set are (1) high initial cost, and (2) lowered efficiency of conversion, because the over-all set efficiency is the *product* of the efficiencies of the two dynamos, respectively.

The conversion of dc voltages is more efficiently accomplished at lower unit cost by a dynamotor. The dynamotor is a single-frame structure having a single field and a single armature core with two separate windings.

---

* R. W. Porter, "Adventures in Energy Conversion," *Electrical Engineering*, October 1960, p. 801.

† See M. L. Vautrey, "L'emploi des pompes électromagnétiques," *Bulletin de la Société Française des Electriciens*, June 1960, p. 399; and M. B. Schwab, "Différents types de pompes électromagnétique," *Ibid.*, p. 404.

Each armature winding is insulated from the other in common armature slots and has its own commutator, one at each end, as shown in Fig. 11-4a. Because a common speed and a common field flux exist for both windings, the emf per conductor and per coil must be the same for each winding [Eq. (1-5)]. The total emf between brushes, therefore, is a function of the number of conductors in each armature path ($V_1/V_2 = Z_1/Z_2$, for a lap or wave winding), in accordance with Eq. (1-6). Neglecting the current taken by the field in Fig. 11-4a therefore, and assuming 100 per cent efficiency, the output power equals the input power, or

$$V_1 I_1 = V_2 I_2 \quad \text{or} \quad \frac{V_1}{V_2} = \frac{Z_1}{Z_2} = \frac{N_1}{N_2} = \frac{I_2}{I_1} \tag{11-1}$$

where N is the number of turns or $Z/2$ conductors per path.

(a) Dynamotor.

(b) Single-phase rotary synchronous converter.

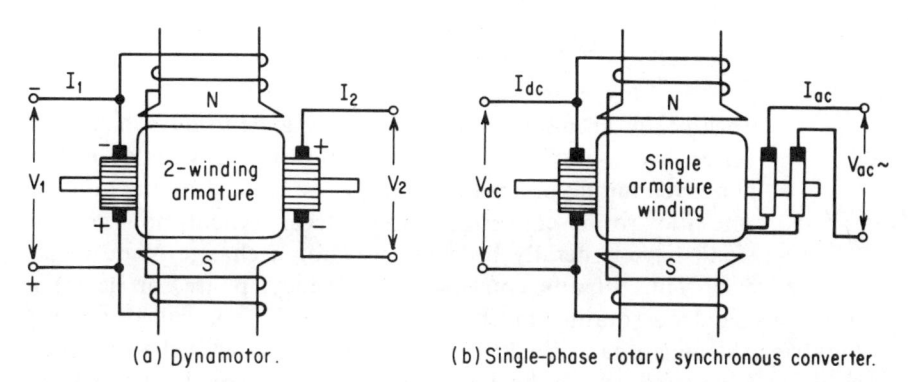

Figure 11-4

Dynamotor and single-phase rotary synchronous converter.

The similarity between Eq. (11-1) and the basic transformer Eq. (13-2b) should be noted. The advantages of the dynamotor, as implied above, are (1) lower initial costs compared to M-G sets of equivalent capacity, and (2) high operating efficiency, because the field, core, friction, and windage losses are the losses of a single machine (but the armature copper losses of two machines). The greatest disadvantage of the dynamotor in comparison to the M-G set it that the output voltage is a fixed function of the input voltage, in accordance with Eq. (11-1). Any attempt to increase output voltage by increasing the field flux results in a reduction in dynamotor speed. As a result, dynamotors are never equipped with field rheostats, and the field circuit is permanently connected internally.

Dynamotors are used in aircraft systems where weight is an important factor, and also on dc railway systems to provide 110 V for lighting from 550 V motor traction supplies. The commutation in dynamotors is excellent, and no armature-reaction compensation is necessary because the net (almost zero) armature flux is the result of load currents in the two windings traveling in opposite directions.† The net armature ampereturns, therefore, are just those required to provide sufficient torque to overcome rotational losses and those losses enumerated above, resulting in a small armature flux (and negligible armature reaction).

EXAMPLE* 11–1: An aircraft dynamotor operating on 28-V dc serves as a power supply for airborne communications equipment at 350 V dc and 500 mA. The high-voltage winding has 400 turns. Calculate:
a. The number of turns of the low-voltage winding
b. The input current if the efficiency is 90 per cent
c. The rating of the dynamotor.

Solution:

a. $N_l = N_h \left(\frac{V_l}{V_h}\right) = 400 \text{ turns} \left(\frac{28 \text{ V}}{350 \text{ V}}\right) = $ **32 turns**

b. $I_l = \left(\frac{I_h}{\text{eff}_y}\right)\left(\frac{N_h}{N_l}\right) = \left(\frac{0.5 \text{ A}}{0.9}\right)\left(\frac{400}{32}\right) = $ **6.994 A**

c. Output rating $= 0.5 \text{ A } (350 \text{ V}) = $ **175 W**

11-6. SINGLE-PHASE ROTARY CONVERTER

Technically speaking, a dynamotor is a dc-to-dc rotary converter, because it is a rotating dynamo that converts dc energy at a given input voltage to dc energy at a desired output voltage. The title "rotary converter" and the term "synchronous conversion" have generally been reserved for a change of dc to ac, or vice versa. Thus, if any conversion (frequency, phase, or dc/ac) is accomplished by a rotating machine, such a machine is called a rotary converter. A single-phase (synchronous) rotary converter is shown in Fig. 11-4b, and the similarity to the dynamotor should be noted. The dynamo shown may be considered a combination of either a dc motor and a synchronous single-phase alternator, or a synchronous motor in combination with a dc generator.

When ac is supplied to the slip-rings and dc is generated at the brushes, the dynamo is called a *direct* converter.

When dc is supplied to the brushes and ac is generated at the slip-rings, the dynamo is called an *inverted* converter.

An *essential* difference between the single-phase rotary synchronous converter and the dynamotor shown in Fig. 11-4 is that the former has only *one winding* brought out to slip-rings on one end and to a commutator on the other. Since alternating current is produced in the conductors *of any* dynamo (with the exception of the homopolar machine), only one winding is necessary if the dynamo is to be a dc-to-ac (or vice versa) converter Thus, as an ac/dc converter, alternating current is fed through the slip-rings

---

† Since the flux and the direction of rotation are the same, for a given direction of motor current (left-hand rule) there will be an opposite direction of generator current (right-hand rule). Also, by Eq. (11-1), ignoring losses, $I_1 N_1 = I_2 N_2$, and $I_1 N_1 - I_2 N_2 = 0$.

* These examples are theoretical, only, since in practice only polyphase converters are used. (See Sec. 11-7.)

to the armature conductors on the rotor, and the dynamo runs (if started in some manner) as a synchronous motor. The commutator converts the generated ac voltage to direct current. Similarly, as a dc/ac converter, the direct current in the external circuit of the dc motor is converted to alternating current in the armature winding to produce rotation, and the alternating current generated is picked off at the slip-rings as an ac output voltage.

The *single*-phase ac synchronous converter is never used in practice because: (1) it is not self-starting; (2) it has tendency to "hunt"; (3) it heats excessively; (4) it is inefficient compared to polyphase synchronous converters; and (5) for the same kW or kVA output, it would be physically larger than equivalent polyphase converters. But the principle of synchronous polyphase rotary converters is simplified through an analysis of the single-phase dc machine. Figure 11-5a shows the two-pole single armature winding of the converter shown in Fig. 11-4b with 12 coils and 12 commutator segments. Let us assume that the dynamo is operating as a dc motor at synchronous speed (3600 rpm at 60 Hz) in a clockwise direction.

Viewed as a dc motor, the applied voltage and the dc induced emf opposing this voltage are constant as long as the dc supply voltage and speed are constant. Thus, in each of Figs. 11-5a through d, the dc induced emf is the same because new conductors dynamically replace the old, continuing to induce emf with the rotation in accordance with Eq. (1-6). The same dc winding, however, is tapped at commutators 1 and 7, and brought out to slip-rings, as shown in Fig. 11-5. The ac single-phase induced voltage is picked off from brushes connected to these slip-rings.

Viewed as a single-phase alternator, the induced ac single-phase voltage is *not* constant. In Fig. 11-5a, with the taps aligned with the polar axis, the ac voltage is picked off as the sum of all the voltages between the N and S poles, i.e., the same as the dc voltage at any instant. In Fig. 11-5d, however, with the taps perpendicular to the polar axis, the ac voltage is picked off from conductor taps which are at the same potential points, and the ac voltage is zero. Thus the magnitude of the instantaneous ac voltage is not constant, but varies sinusoidally with the rotation as might be expected. Figure 11-5e summarizes the variation in instantaneous ac voltage at the slip-rings with respect to the average dc voltage between brushes for 90 degrees of rotation.

In predicting the output ac voltage and current, for a given input dc voltage and current, it must be remembered [Eqs. (1-5) and (1-6)] that the applied and induced dc voltage is an average value. Assuming that there is no ripple in the dc waveform (for a multiconductor machine), the average dc voltage value is the same as the maximum dc voltage value. Thus the effective value of the ac single-phase output voltage is

$$E_{ac} = \frac{E_{ac(max)}}{\sqrt{2}} = 0.707 E_{ac(max)} = 0.707 E_{dc(av)}$$

(a) $E_{ac} = E_{dc}(0°)$.

(b) $E_{ac} = 0.866 E_{dc}(30°)$.

(c) $E_{ac} = 0.5_{dc}(60°)$.

(d) $E_{ac} = 0(90°)$.

(e)

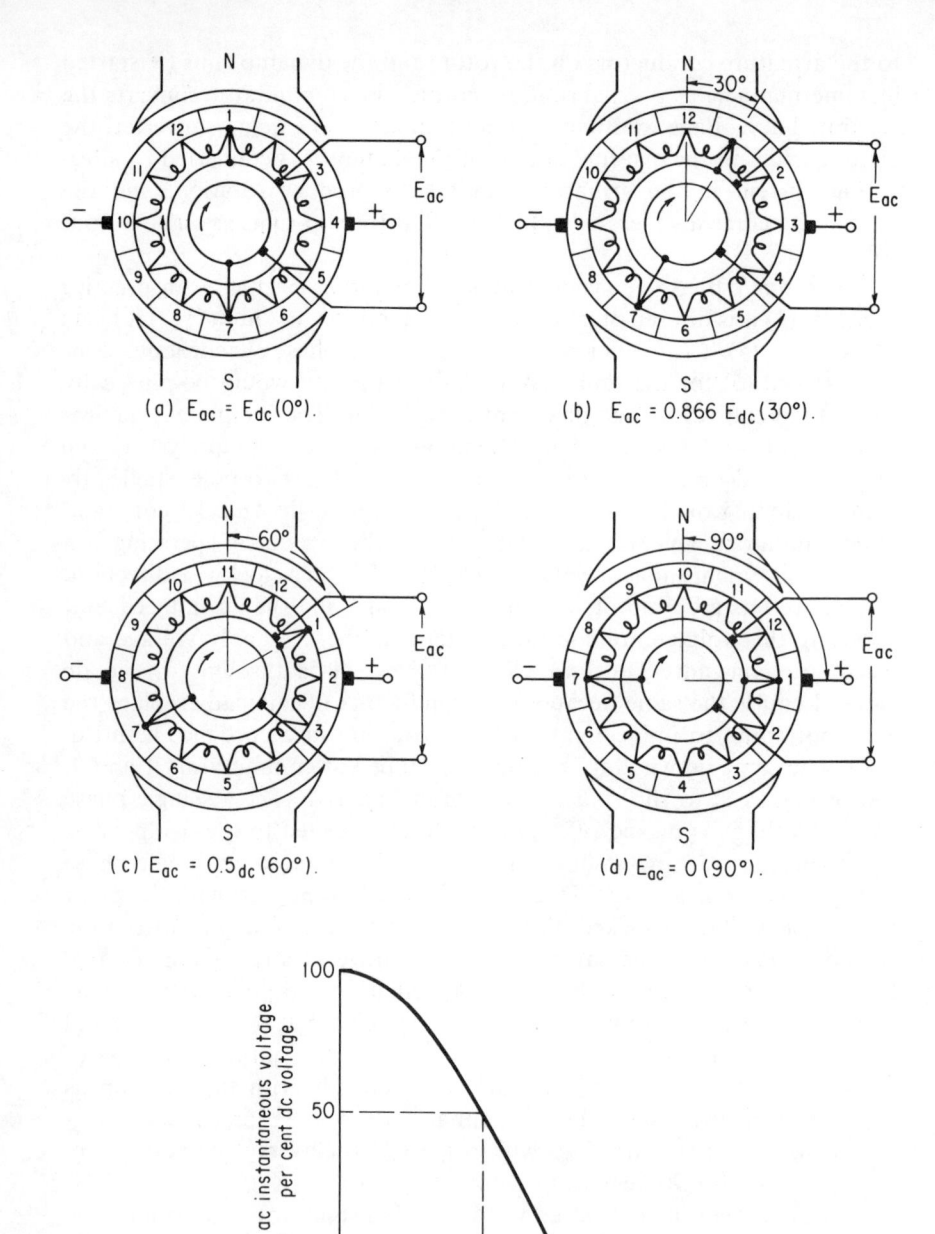

Figure 11-5

Variation in ac instantaneous voltage with respect to dc voltage in a single-phase synchronous converter.

or
$$E_{ac} = 0.707 E_{dc} \tag{11-2}$$

If, as in the case of the dynamotor, 100 per cent conversion efficiency is assumed, then

$$E_{ac}I_{ac} = E_{dc}I_{dc} = \frac{E_{ac}}{0.707} \times I_{dc}$$

and dividing both sides by E_{ac} yields

$$I_{ac} = \frac{I_{dc}}{0.707} = 1.414 I_{dc} \tag{11-3}$$

EXAMPLE 11-2: A 3-kW single-phase synchronous converter is operated as an inverted converter, i.e., it supplies alternating current from a dc source of 220 V. Assuming a 90 per cent conversion efficiency, calculate
a. The ac voltage and the ac current
b. The dc current supplied to the inverted converter.

Solution:

a. $E_{ac} = 0.707 E_{dc}$ [Eq. (11-2)]
$= 0.707 \times 220 \text{ V} = \mathbf{155.5 \text{ V}}$

$I_{ac} = \dfrac{\text{kW} \times 1000}{E_{ac}} = \dfrac{3 \times 1000 \text{ W}}{155.5 \text{ V}} = \mathbf{19.28 \text{ A}}$

b. $I_{dc} = \dfrac{I_{ac}}{1.414 \times \text{eff}} = \dfrac{19.28 \text{ A}}{1.414 \times 0.9} = \mathbf{15.15 \text{ A}}$

EXAMPLE 11-3: A 5-kW, single-phase synchronous converter operating at synchronous speed draws full-load current from its rated 220-V source at a power factor of 0.85. If the operating efficiency is 90 per cent, calculate:
a. The dc output voltage and current
b. The ac current supplied to the synchronous rotary converter.

Solution:

a. $E_{dc} = \dfrac{E_{ac}}{0.707} = \dfrac{220 \text{ V}}{0.707} = \mathbf{315 \text{ V}}$

$I_{dc} = \dfrac{\text{kW} \times 1000}{E_{dc}} = \dfrac{5 \times 1000 \text{ W}}{315 \text{ V}} = \mathbf{15.88 \text{ A}}$

b. $I_{ac} = \dfrac{\text{kW} \times 1000}{E_{ac} \times \text{PF} \times \text{eff}} = \dfrac{5000}{220 \times 0.85 \times 0.9} = \mathbf{29.7 \text{ A}}$

11-7. POLYPHASE ROTARY CONVERTER

In the single-phase converter previously considered, the armature winding of the two-pole dynamo was tapped at two points, 180 electrical degrees apart with respect to each other. If four poles were used, four taps would be required, each tap displaced by 180 electrical degrees.

* These examples are theoretical, only, since in practice only polyphase converters are used. (See Sec. 11-7.)

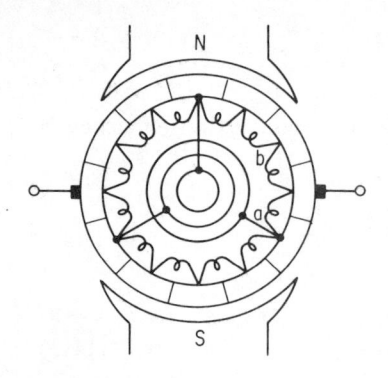

Figure 11-6

Two-pole three-phase synchronous converter showing taps and slip rings.

Similarly, as shown in Fig. 11-6, a two-pole, three-phase converter would require three taps, each 120 electrical degrees apart. If a four-pole dynamo were employed, it would require six taps (three for every pair of poles), each tap displaced by 120 electrical degrees. A polyphase three-phase converter requires three rings, a six-phase converter requires six rings, and a twelve-phase converter twelve rings, etc. The number of taps, T, for any polyphase armature is

$$T = nP' \qquad (11\text{-}4)$$

where n is the number of rings and P' is the number of pairs of poles in any converter.

Similarly, the number of electrical degrees between taps is

$$\delta = \frac{2\pi}{n} \text{ electrical radians}$$

$$= \frac{360}{n} \text{ electrical degrees} \qquad (11\text{-}5)$$

where n is the number of rings as in Eq. (11-4)

The induced voltage may be predicted for any dc armature of any given number of conductors, paths, poles, flux, and speed [Eq. (1-6)]. The effective ac voltage per phase obviously decreases as the number of phases, slip-rings, and taps increases, since the same dc armature must supply the ac voltage [Eq. (11-2)]. It can be shown, then, that a fixed relation exists for any polyphase synchronous converter between the ac voltage and the dc voltage (at average and at maximum voltage); namely

$$E_{ac} = \frac{E_{dc}}{\sqrt{2}} \sin \frac{\pi}{n} \qquad (11\text{-}6)$$

and similarly between the currents themselves, namely

$$I_{ac} = \frac{2\sqrt{2}}{n} I_{dc} \qquad (11\text{-}7)$$

where all the terms have been defined previously.

The validity of Eqs. (11-6) and (11-7) may be tested for the single-phase converter having $n = 2$ slip-rings in Eqs. (11-2) and (11-3).

Table 11-1 below uses Eqs. 11-4 through 11-7 to summarize the relations for single-phase, three-phase, six-phase, and twelve-phase converters.

TABLE 11-1.
POLYPHASE SYNCHRONOUS CONVERTER RELATIONS

QUANTITY	EQUATION	1-PHASE, 2-RING	3-PHASE, 3-RING	6-PHASE, 6-RING	12-PHASE, 12-RING
E_{ac} between rings	11-6	$0.707 E_{dc}$	$0.612 E_{dc}$	$0.354 E_{dc}$	$0.182 E_{dc}$
$I_{ac}{}^*$ in rings	11-7	$1.414 I_{dc}$	$0.943 I_{dc}$	$0.472 I_{dc}$	$0.236 I_{dc}$
δ, electrical degrees between taps	11-5	180	120	60	30
T taps per pole pair	11-4	2	3	6	12

* At unity power factor and 100 per cent efficiency.

EXAMPLE 11–4: A 500 kW, 600 V dc, 12 phase synchronous converter operates as a direct converter at a full efficiency of 92 per cent and a power factor of 0.93. Calculate:
a. The ac voltage between slip-rings
b. The dc output current
c. The ac current drawn from a 12-phase transformer-fed supply.

Solution:

a. $E_{ac} = 0.182 E_{dc} = 0.182 \times 600 \text{ V} =$ **109 V between slip-rings**

b. $I_{dc} = \dfrac{\text{kW} \times 1000}{E_{dc}} = \dfrac{500 \times 1000 \text{ W}}{600 \text{ V}} =$ **833.3 A**

c. $I_{ac} = \dfrac{0.236 I_{dc}}{\text{PF} \times \text{eff}} = \dfrac{0.236 \times 833.3}{0.93 \times 0.92} =$ **229.5 A**

As in the cases of the dynamotor and the single-phase synchronous converter, the polyphase converter is a dynamo in which the net armature **mmf** and the net flux are extremely small because the direct current flows in a direction opposite to the alternating current. The nature of the current waveforms in the various coils differs because the dc value is constant, whereas the ac value depends on the phase difference between coils as well as the power factor of the converter.

For example, in a coil lying midway between the taps (point *b* in Fig. 11-6), assuming that the power factor is unity, the ac and dc components of the current are exactly 180° out of phase, as shown in Fig. 11-7a. But in a coil *at* the taps, the dc and ac components can be only 90 electrical degrees with respect to each other, as shown by Figs. 11-5 and 11-6. Therefore, as shown in Fig. 11-5b and in Fig. 11-7b, the ac component is instantaneously zero when the dc component has its maximum or average value, and the waveforms are 90 electrical degrees apart. Thus, coils lying exactly at the taps have a resultant current waveform at unity PF

(a) Waveform at unity power factor in coils between taps.

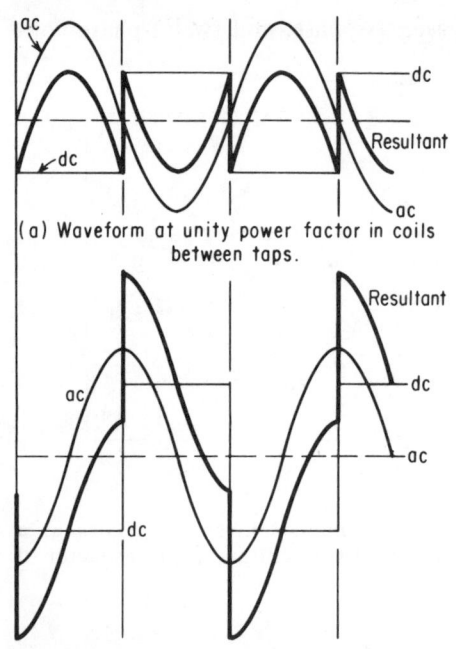

(b) Waveform at unity power factor in coils exactly at taps.

Figure 11-7

Comparison of resultant current waveforms between and at the taps of a synchronous converter.

similar to that shown in Fig. 11-7b, whereas coils lying midway between the taps have a resultant current wave-form similar to that shown in Fig. 11-7a. The heating effect produced by the effective value of the resultant current is greater, therefore, in coils lying close to the taps than in coils midway between the taps. A temperature gradient exists, therefore, between the hotter tapped coils and those coils midway between the taps.

It has been shown by Table 11-1, however, that increasing both the number of phases and the number of poles will increase the total number of taps on a synchronous converter. Increasing the number of taps reduces the temperature gradient and, in turn, increases the capacity of the synchronous converter, since capacity is dependent obviously on the maximum allowable current-carrying capacity of the hottest coil. Thus, as shown in Table 11-2, a three-ring, three-phase converter has a lower capacity than a six- or twelve-phase polyphase converter. Table 11-2 is based on 100 per cent conversion efficiency between the ac and dc inputs and outputs, respectively, and machines having the same number of poles.

TABLE 11-2.
EFFECT OF POWER FACTOR AND NUMBER OF PHASES ON RELATIVE OUTPUT AND HEATING OF A SYNCHRONOUS CONVERTER COMPARED TO OPERATION AS A DC GENERATOR*

POWER FACTOR	3-PHASE, 3-RING RELATIVE HEATING	3-PHASE, 3-RING RELATIVE OUTPUT	6-PHASE, 6-RING RELATIVE HEATING	6-PHASE, 6-RING RELATIVE OUTPUT	12-PHASE, 12-RING RELATIVE HEATING	12-PHASE, 12-RING RELATIVE OUTPUT	DC GENERATOR
1.0	0.565	1.33	0.268	1.93	0.209	2.19	1.0
0.95	0.693	1.20	0.364	1.66	0.299	1.83	1.0
0.90	0.843	1.09	0.476	1.45	0.404	1.57	1.0
0.85	1.02	0.99	0.609	1.28	0.528	1.38	1.0
0.80	1.23	0.90	0.768	1.14	0.676	1.22	1.0

* After C. S, Siskind, *Electrical Machines, Direct and Alternating*, 2nd Ed., New York: McGraw-Hill. Inc,. 1959.

Table 11-2 confirms the following conclusions, which have been indicated previously:

1. As the numbers of phases and rings are increased, the relative output of the synchronous converter increases, because the relative heating decreases.
2. As the number of poles increases, the relative output also increases (not shown in Table 11-2 but inferred from the above), because the relative heating decreases.
3. As the power factor decreases, the relative output also decreases, because the relative heating effect increases.
4. The relative output of any polyphase converter is generally greater than the same dynamo operating as a dc generator, because the temperature rise and heating effect of the former are generally less than those of the latter. (This is not true of the three-phase converter at power factors below 0.85, however.)

For the reasons cited above, large synchronous converters are designed for six- or twelve-phase ac inputs with many stator poles. The interconversion between the three-phase supply normally available, and the multiphase input required by the converter, is supplied through "interphase" multicoil transformers,* as shown in Fig. 11-8, connected in such a manner so as to provide the necessary transformation at the required input voltage. Balancing the increased transformer cost against the higher efficiency of 12- or 24-phase converters over six-phase, the latter is usually the most common. The three- to six-phase *star* transformer connection is the most common (See Sec. 13-20) method to provide the six-phase converter ac input.

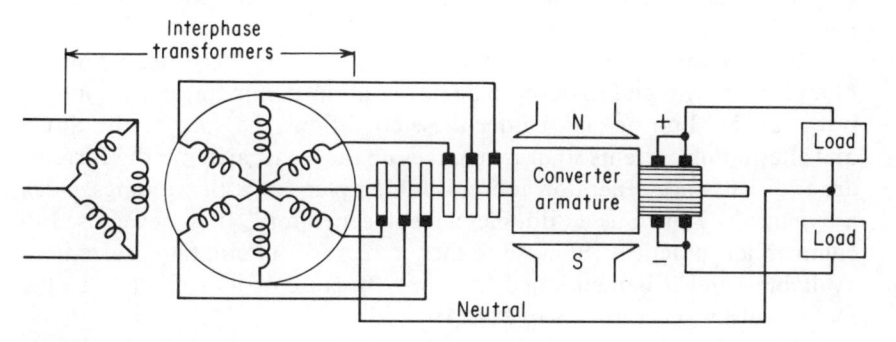

Figure 11-8

Three-to-six-phase converter showing transformation and production of a three-wire dc output.

*See Sec. 13-20.

SEC. 11-7. / Polyphase Rotary Converter

A voltage drop within the converter and in the system, with the application of increased load, requires some method of voltage control of the direct converter (alternating to direct current). One method is to use induction regulators or tap-changing transformers on the primary of the interphase transformers. Another method is to employ a booster (an ac polyphase alternator) on the same shaft as the converter, rated for a 15 per cent buck or boost in line voltage.

Synchronous converters must be started on their damper windings (as induction motors) and must be pulled into synchronism in the same manner as is used when starting synchronous motors. They may also be started as dc motors, where direct current is available, and synchronized to the bus; but this method is rarely employed.

The synchronous converter is particularly sensitive to sudden and large changes in load and to hunting, which may produce flashover between dc brush arms. For this reason, large converters are equipped with flash barriers (insulators) to quench any commutator arcing which may be initiated. Damper windings are essential to the stator field construction.

Direct synchronous converters usually supply from 250 V to about 750 V direct current (at 60 Hz), or to 1500 V direct current (at 25 Hz). The dc output voltage is limited by the maximum allowable voltage between commutator segments (about 15 V per segment). Where higher voltages are required, the dc outputs are series-connected.

Synchronous converters are supplied in capacities up to 4000 kW and small units are available as low as 100 kW. The large steel-tank mercury-arc rectifier, because of its higher efficiency at higher voltages, has, for the most part, displaced many of the applications of the synchronous converter as a source of direct current. Similarly, contemporary developments in the use of selenium and silicon rectifiers are now beginning to replace tank-type and gaseous thyratron and ignitron rectifiers, particularly in the lower capacity ranges (Sec. 13-21).

It is even possible to drive a synchronous rotary converter by means of a prime mover and to develop dc power at the commutator end and ac power at the slip-ring end. Under these conditions, the sum of the direct and alternating currents so drawn makes up the total capacity; the currents are additive, since generator action is taking place in both windings. Such a dynamo is known as a "double-current generator." It is never used in commercial practice (because either direct or alternating current is available), and it is mentioned here only because of its similarity to the Dobrowolsky generator (Sec. 11-8.4).

A synchronous converter may be used to supply a three-wire dc system. In such an application, the neutral wire is brought back to the point of neutral of the interphase transformer secondaries supplying the ac input, as shown in Fig. 11-8. As a result, the unbalanced current carried in the neutral of the dc system is returned to the converter armature through the transformer and the slip-rings. This is precisely the arrangement employed in the Dobrowolsky generator (Fig. 11-9d) to obtain a three-wire system.

11-8. THREE-WIRE SYSTEM GENERATORS

A three-wire distribution system (either an ac single-phase or a dc system) provides dual voltages to an occupancy. It provides the added advantage of the use of the higher voltage (at lower current) for high-power appliances (ranges and heaters, for example) and the lower voltage for lighting and other smaller loads. The great advantage of the three-wire system to the utility providing such service is that, despite the additional line conductor, a saving in copper results for the same power transmitted.* It also encourages the use of electrical energy and new appliances by providing voltages of greater versatility.

The three-wire ac system is supplied through the center-tapped secondary of one phase of a polyphase transformer. The three-wire dc system is usually supplied through a single generator in combination with auxiliary devices or combinations of generators. The following methods are employed to supply a three-wire dc system: (1) the two-generator method; (2) the balancer set; (3) the single-generator tapped-resistor method; (4) the Dobrowolsky generator; and (5) the synchronous converter. The first four methods are described and illustrated below. The synchronous converter was discussed in Section 11-7.

11-8.1 The Two-Generator Method

Two equal-voltage dc shunt or compound generators may be cascaded (to compensate for the line drop) to supply a three-wire distribution system, as shown in Fig. 11-9a, driven either by separate prime movers or by the same prime mover.

11-8.2 The Balancer Set

The two-generator method is undesirable when there is a heavy load unbalance in the line currents. The more heavily loaded generator has a large internal armature voltage drop, causing the voltage across the heavily loaded side to drop sharply. This disadvantage is overcome in the balancer set shown in Fig. 11-9b. This combination employs a main (230 V) generator, to furnish the balanced portion of the load, and two smaller (115 V) dynamos, to furnish the unbalanced portion of the load. When the load is balanced, the two dynamos are "floating" on the lines as unloaded motors. Whenever an unbalance occurs, the side carrying the heavier load suffers a drop in terminal voltage, and the dynamo on that side operates as a generator ($E_g > V_t$). At the same time, because of the voltage rise, the other dynamo (coupled to the same shaft) acts as a loaded motor and tends to reduce the voltage rise by

---

* It can be shown that doubling the voltage of transmission of a two-wire system results in a copper saving of 75 per cent. Adding the third or neutral wire to provide the original voltage, as well as the doubled voltage, results in a copper saving of 62 per cent compared to the original low-voltage, two-wire system.

(a) Two generator method.

(b) Balancer set method.

(c) Single generator tapped resistor method.

(d) Dobrowolsky generator method.

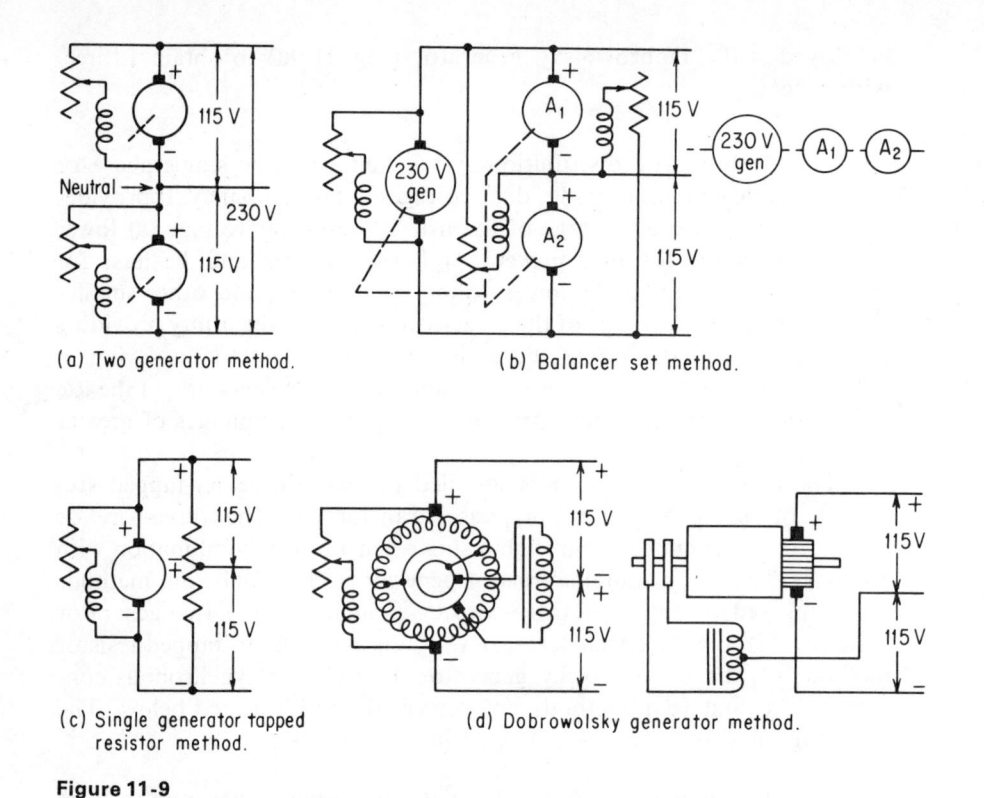

Figure 11-9

Methods of generation of a dc three-wire system.

driving its companion generator in the same direction, thereby supplying the energy required by the loaded smaller generator. As shown in Fig. 11-9b, a faster and improved response is also obtained by reversing the shunt-field line connections of the two dynamos. Assuming that A_1 is more heavily loaded and that its voltage drops below 115 V its field is now receiving a voltage which exceeds 115 V, assisting it to develop an increased voltage as a generator. At the same time, motor A_2 is receiving current from its increased voltage mains. A_2 tends to speed up as a result of the decreased voltage across A_1, which is also across the field of A_2. This action tends to maintain fairly good system regulation whenever any unbalance occurs. In commercial practice, regulation is also improved by adding series fields to dynamos A_1 and A_2.

11-8.3
Single-Generator
Tapped-Resistor
Method

The single 230 V generator, which supplies the balanced current for the balancer set, may also be used with a high-power center-tapped resistor, shown in Fig. 11-9c. This method is highly undesirable in comparison to the balancer set, because it provides no automatic method that tends to oppose the voltage unbalance as does the balancer set. As a result, in a manner similar to that of the two-generator method, the heavily loaded side engenders a cumulative increased voltage drop in the tapped resistor, as well as an increased voltage drop in the lines, both factors producing a decrease in the terminal voltage of the heavily loaded side. Furthermore, the power loss in the resistors results in a considerable reduction in the efficiency of operation. The method is shown here because it suggests the three-wire Dobrowolsky generator since the generator of Fig. 11-9c, like all generators, is producing alternating current internally. If a center-tapped inductance could be connected internally to this generator, a neutral would be provided.

11-8.4 Dobrowolsky Generator

As shown in Fig. 11-9d, the Dobrowolsky generator is indeed fundamentally a single-phase dc synchronous converter (Fig. 11-4b) driven as a double-current generator. The (only) ac load is the center-tapped reactance coil, either built-in as part of the armature winding or externally provided through slip-rings as shown in Fig. 11-9d. The former requires only one slip-ring, but the coil itself adds to the weight of the armature. The advantage of the tapped reactance coil (whether built-in or external) is that it has a high impedance to alternating current but a low dc resistance. Thus, the unbalanced direct current produces a small dc voltage drop in the reactance coil. This dc drop is small compared to the ac impedance drop across the sliprings, and the neutral voltage is maintained reasonably well under unbalanced conditions.

Since more direct current is flowing in one part of the coil than in the other, however, it is necessary that the reactor be rated in terms of the per cent unbalance it can withstand without excessive overheating. External coils are usually supplied to carry a maximum unbalance of 10 per cent, and special coils may be provided for higher unbalances if required. For this reason, it is poor practice to open one of the supply lines of a three-wire generator, since the heavy unbalance may burn out the reactance coil. This is another advantage of *external* reactance coils, since they may be replaced without disassembling the machine, thus facilitating maintenance.

11-9. EFFECT OF LINE RESISTANCE AND UNBALANCED LOADS IN THREE-WIRE SYSTEMS

If the loads connected from the line-to-neutral leads of a three-wire dc or ac system are unequal, unequal currents will flow in the lines and the difference current will flow in the neutral. Such a distribution system is said to be unbalanced, and the unequal currents will produce unequal line voltage drops. The solution of problems relating to these distribution systems involves the application of Kirchhoff's voltage and current relations. Three examples will be given below. Two of these will illustrate the effect of balanced versus unbalanced loads. The third will illustrate an alternative method of solution to the two previous examples.

EXAMPLE 11–5: A three-wire, 115/230 V dc system is used to supply two concentrated lighting loads, L_1 and L_2, at a distance of 2000 feet from the generating source, and a concentrated power load, L_3, 200 feet farther from the source. The power load, L_3, is 100 A, and each of the lighting loads is 200 A. The size of the line feeders is 500,000 circular mils, and the size of the neutral feeder is 250,000 circular mils. The resistivity of copper may be taken, for this example, as 11 ohms-circular mil/ft. Calculate:
a. the voltage across the loads $L_1, L_2,$ and L_3, respectively
b. voltage V_{1-2}.

Solution:

[See Fig. 11-10(a)]

a. Resistance $= \dfrac{\text{resistivity} \times \text{length}}{\text{area}} = \dfrac{\rho l}{A}$

$r_{1,2} = \dfrac{11\ \Omega}{500,000/1} \times 2000\ \text{ft} = 0.044\ \Omega$

$r_3 = \left(\dfrac{200}{2000}\right) 0.044 = 0.0044\ \Omega$

$r_N = \dfrac{11\ \Omega}{250,000/1} \times 2000\ \text{ft} = 0.088\ \Omega$

$I_1 = I_{L_1} + I_{L_3} = 200 + 100 = 300\ \text{A};$
$I_2 = I_{L_2} + I_{L_3} = 200 + 100 = 300\ \text{A}$
$I_N = I_{L_1} - I_{L_2} = 200 - 200 = 0$

Using Kirchhoff's voltage laws in which (1) the algebraic sum of all voltages in any closed-loop network is zero and (2) a voltage drop is encountered in a current-carrying resistance in the direction of the flow of current and a voltage rise is encountered in a direction opposite to the flow of current, we get:

Solving for E_{L_1}:
$-I_1 r_1 - E_{L_1} \pm I_N r_N + E_1 = 0$
$-(300 \times 0.044) - E_{L_1} + 0 + 115 = 0$
$E_{L_1} = \mathbf{101.8\ V}$

Solving for E_{L_2}:
$\pm I_N r_N - E_{L_2} - I_2 r_2 + E_2 = 0$
$0 - E_{L_2} - (300 \times 0.044) + 115 = 0$
$E_{L_2} = \mathbf{101.8\ V}$

Solving for E_{L_3}:
$-I_1 r_1 - I_3 r_3 - E_{L_3} - I_3 r_3 - I_2 r_2 + 230 = 0$
$-13.2 - (100 \times 0.0044) - E_{L_3} - (100 \times 0.0044) - (300 \times 0.044) + 230 = 0$
$E_{L_3} = 230 - 27.28 = \mathbf{202.72\ V}$

b. Voltage $V_{1-2} = E_{L_1} + E_{L_2} = 101.8 + 101.8 = \mathbf{203.6\ V}$

EXAMPLE Repeat Example 11-5 with a load L_1 of 100 A, L_2 of 150 A, and L_3 of
11–6: 200 A.

Solution:

See Fig. 11-10a

a. $I_1 = I_{L_1} + I_{L_3} = 100 + 200 = 300$ A;
$I_2 = I_{L_2} + I_3 = 150 + 200 = 350$ A
$I_N = I_{L_2} - I_{L_1} = 300 - 250 = 50$ A

Solving for E_{L_1}:
$-I_1 r_1 - E_{L_1} + I_N r_N + E_1 = 0$
$-(300 \times 0.044) - E_{L_1} + (50 \times 0.088) + 115 = 0$
$E_{L_1} = \mathbf{106.2\ V}$

Solving for E_{L_2}:
$-I_N r_N - E_{L_2} - I_2 r_2 + E_2 = 0$
$-(50 \times 0.088) - E_{L_2} - (350 \times 0.044) + 115 = 0$
$E_{L_2} = \mathbf{95.2\ V}$

Solving for E_{L_3}:
$-I_1 r_1 - I_3 R_3 - E_{L_3} - I_3 r_3 - I_2 r_2 + 230 = 0$
$-(300 \times 0.044) - (200 \times 0.0044) - E_{L_3} - (200 \times 0.0044)$
$\quad - (350 \times 0.044) + 230 = 0$
$E_{L_3} = \mathbf{199.64\ V}$

b. Voltage $V_{1-2} = E_{L_1} + E_{L_2} = 106.2 + 95.2 = \mathbf{201.4\ V}$

EXAMPLE Solve (a) Example 11-5 and (b) Example 11-6, using the *feeder-voltage-
11–7:* drop method* illustrated in Fig. 11-10b.

Solution:

This solution and Fig. 11-10b are both predicated on the convention that (1) a current flow from left to right [from the supply] represents a

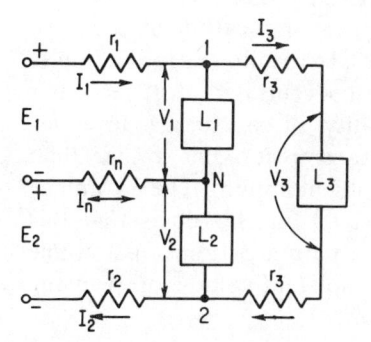

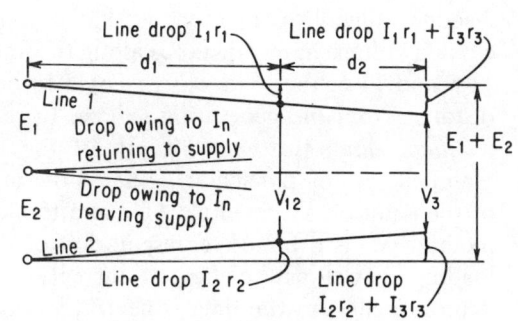

(a) Kirchoff's law solution.   (b) Feeder voltage–drop method of solution.

Figure 11-10

Three-wire systems with unbalanced loads illustrating Exs. 11-5, 11-6 and 11-7.

voltage drop in a *downward* direction and (2) a current flow from right to left [to the supply] represents a voltage rise in an upward direction. It is recommended that the reader draw a dimensioned diagram similar to Fig. 11-10b.

a. In Example 11-5, the line drop $I_1 r_1 = 300 \times 0.044 = 13.2 \text{ V} = I_2 r_2$ in magnitude, but *not* in direction.

$$\text{Line drop } I_3 r_3 = 100 \times 0.0044 = 0.44 \text{ V}$$

$$\text{Voltage } V_{1-2} = (E_1 + E_2) - \text{line drop } I_1 r_1 - \text{line drop } I_2 r_2$$
$$= 230 \text{ V} - 13.2 - 13.2 = \mathbf{203.6 \text{ V}}$$

and, since $I_N = 0$,

$$E_{L_1} = E_{L_2} = \frac{203.6}{2} = \mathbf{101.8 \text{ V}}$$

$$\text{Voltage } V_3 = E_{L_3} = (E_1 + E_2) - 2(I_1 r_1) - 2(I_3 r_3)$$
$$= 230 - (2 \times 13.2) - (2 \times 0.44) = \mathbf{202.72 \text{ V}}$$

b. In Example 11-6, the line drop $I_1 r_1 = 300 \times 0.044 = 13.2 \text{ V}$
Line drop $I_2 r_2 = 350 \times 0.044 = 15.4 \text{ V}$
Line drop $I_3 r_3 = 200 \times 0.0044 = 0.88 \text{ V}$
Line drop $I_N r_N = 50 \times 0.088 = 4.4 \text{ V}$ (leaving the supply)

From Fig. 11-10b

$$E_{L_1} = E_1 - I_1 r_1 + I_N r_N = 115 - 13.2 + 4.4 = \mathbf{106.2 \text{ V}}$$
$$E_{L_2} = E_2 - I_2 r_2 - I_N r_N = 115 - 15.4 - 4.4 = \mathbf{95.2 \text{ V}}$$
$$V_{1-2} = E_1 + E_2 - I_1 r_1 - I_2 r_2 = 230 - 13.2 - 15.4 = \mathbf{201.4 \text{ V}}$$

As a check:

$$V_{1-2} = E_{L_1} + E_{L_2} = 106.2 + 95.2 = \mathbf{201.4 \text{ V}}$$
$$E_{L_3} = E_1 + E_2 - (I_1 r_1 + I_2 r_2 + 2 I_3 r_3)$$
$$= 230 - (13.2 + 15.4 + 2 \times 0.88) = 230 - 30.36 = \mathbf{199.64 \text{ V}}$$

One advantage of the solution in Example 11-7, employing the feeder-voltage-drop method, is that it yields an indication of the anticipated voltage at *any* distance along the lines. In the illustrative Examples 11-5 and 11-6, the load was assumed to be concentrated at a particular distance from the generating source. In reality, of course, a load is *distributed* along the *entire* length of the transmission line rather than "lumped" at any particular point, as the problem implies. The advantage of the solution of Example 11-7 and the use of Fig. 11-10b is that this method yields a better representation of the nature of transmission-line loading, as well as the effect of current direction and voltage drop in the neutral feeder on the line to neutral voltages.

EXAMPLE 11–8: Using a dimensioned diagram similar to that shown in Fig. 11-10b, determine the voltages for the loads in Example 11-6 at the distances from the supply indicated in the table below. *Note:* The answers are given in the table, but the technique for solution is left to the reader.

Distance from Supply (ft)	Voltage From	Voltage To	Volts (V)
(a) 400	Line 1	Neutral	113.24
(b) 800	Line 1	Line 2	218.56
(c) 1200	Line 2	Neutral	104.32
(d) 1600	Line 1	Line 2	207.12
(e) 2100	Line 1	Line 2	202.72

11-10. INDUCTION PHASE CONVERTERS

When it is necessary to convert from one polyphase system to another, the primary and secondary windings of a number of single-phase transformers (or polyphase transformers) may be interconnected to accomplish this conversion.* Under *no* circumstances, however, is it possible for a given single-phase ac source to be converted into a polyphase supply through *transformer* systems (the reverse is a simple matter, obviously). This conversion is normally accomplished by a polyphase ac motor known as an *induction phase converter*.†

The induction phase converter is normally supplied from a single-phase high-voltage source through a step-down transformer whose secondary is divided into three (unequal voltage) series-connected windings, as shown in Fig. 11-11a. The stator winding of the converter is a two-phase winding in which the windings are displaced by 90 electrical degrees on the stator. One phase, consisting of two series-connected coils, numbers 1 and 2 in the figure, is connected in phase with the secondary of the step-down transformer. The other phase, which is displaced 90 electrical degrees on the stator, is *T*-connected with respect to the first phase winding, coils 3 and 4 in Figs. 11-11a and b. The rotor is a standard squirrel-cage rotor. The motor starts as a permanent split-phase two-phase motor without load and easily develops a speed emf to reach its no-load speed as a "two-phase" motor. The phasor diagram of transformer- and motor-induced emfs is shown in Fig. 11-11c. The induced emf in coils 1 and 2 is in phase with the transformer-induced emf E_{ac}. The induced emf in coils 3 and 4, because of their quadrature relation, exists 90 degrees out of phase with E_{ac}, producing voltage *OB*. The voltages, *AC*, *CB*, and *BA* are, therefore, a true three-phase voltage.

Induction phase converters are used to supply electric trolleys and buses from high-voltage overhead trolleys or double-track systems. Both the transformer and the converter are located in the vehicle. The single-phase voltage is converted to a three-phase supply, which is used to

---

* See Sections 13-19 and 13-20.

† The induction phase converter is discussed here rather than in Chapters 9 or 10 because of its similarity to the rotary converter and because of its specialized function. Unlike the single-phase rotary converter whose theory was covered in Sec. 11-6, the induction phase converter is self-starting as a 2-phase motor.

(a) Schematic of connections.

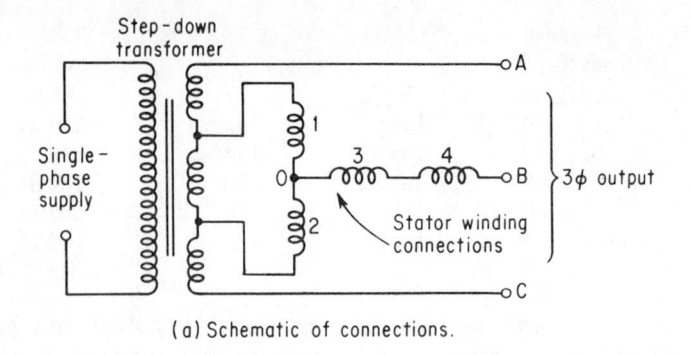

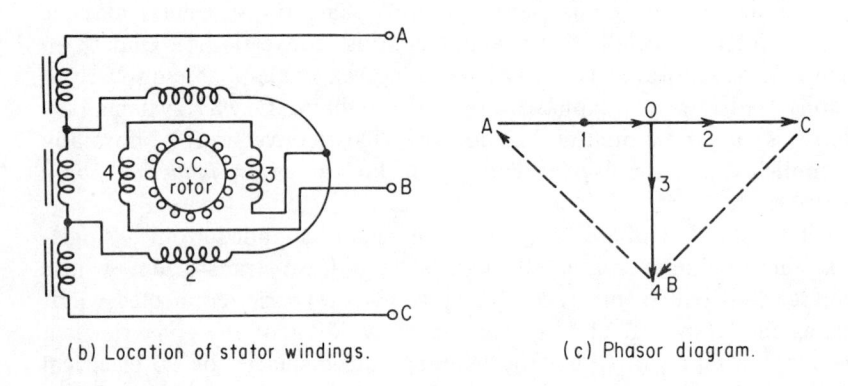

(b) Location of stator windings.

(c) Phasor diagram.

Figure 11-11

Induction phase converter.

operate a three-phase induction motor. The phase converter eliminates the need for two trolley wires or an extra "third" (track) rail. It is also more efficient than a 2-unit MG set in accomplishing this conversion.

As in the case of all converters, voltage output is controlled by voltage adjustment of the source using single-phase induction regulators, variacs or tapped transformers.

11-11. SYNCHRONIZING (SELSYN) DEVICES

The terms "selsyn" or "synchro," used widely in the literature and in industry, are but abbreviations of the term "self-synchronizing" to designate the same devices. There are essentially five basic self-synchronizing devices which fall into this category: the selsyn *transmitter*, the selsyn *receiver*, the selsyn *differential transmitter*, the selsyn *differential receiver*, and the selsyn *control transformer*. These devices are used in both open-loop and closed-loop control systems in a variety of ways to indicate, provide an offset, respond to, generate, or receive a signal indicating some angular position or linear displacement.

There are many types of selsyn devices; some operate on direct current and some are mechanical devices, hydraulic in nature. The discussion of

424

CHAP. ELEVEN / *Specialized Dynamos*

this section will be confined primarily to the five ac induction types cited above. These selsyn devices are essentially single phase in nature, the electric energy for their operation being supplied from a single-phase 60 Hz or 400 Hz supply. The *stators* of these devices are *identical* in construction (see Table 11-3 below) consisting of a distributed "three-phase" winding identical to that employed in a three-phase synchronous or asynchronous dynamo, i.e., synchronous motor, or a three-phase induction motor stator. The *rotors* of these devices are supplied through sliprings; they resemble either three-phase wound-rotor induction-motor rotors or single-phase salient- or nonsalient-pole rotors such as are used on synchronous alternators. Table 11-3 lists the constructional differences among the five types of ac selsyn devices.

As noted in Table 11-3, the only constructional difference between a transmitter and a receiver is a *damper*. The dampers used on selsyn *receivers* are essentially spring-loaded, friction, viscous, or magnetic damping devices designed to prevent oscillation of a receiver when actuated by a change in electric signal (see Fig. 11-23b) from a transmitter.

Unlike most rotating machines, selsyn devices do *not* rotate continu-

TABLE 11-3.
SELSYN CONSTRUCTION DIFFERENCES AND APPLICATIONS

DEVICE	STATOR WINDING	ROTOR WINDING	DAMPER	ROTOR FREE IN APPLICATION	APPLICATION
Transmitter*	3-phase	1-phase salient-pole	No	No	To convert a mechanical input to an electric output
Receiver*	3-phase	1-phase salient-pole	Yes	Yes	To convert electric shaft-position information to a mechanical output shaft position
Differential Transmitter*	3-phase	3-phase distributed	No	No	To compensate for errors in transmitter-receiver combinations by offsetting transmitter shaft
Differential Receiver*	3-phase	3-phase distributed	Yes	Yes	To indicate sum or difference of two transmitters
Control Transformer	3-phase	1-phase distributed	No	No	As an error detector to generate an electric rotor signal in proportion to the difference between the electric stator and the mechanical output of a servo system

* Any receiver may be used as a corresponding transmitter but not vice versa, since the transmitter does not possess a damper.

ously when energized. The only time any shaft movement occurs is when an unbalance of stator voltages is produced because of a change in the shaft position of the transmitter. Even then, the rotation is rarely, if ever, a complete revolution of the shaft of either the transmitter or the receiver in position control applications.

The simplest form of selsyn combination as a remote position-indicating device is shown in Fig. 11-12, in which the "three-phase" stators of a transmitter and a receiver are connected electrically, and both single-phase rotors are connected to a single-phase 60- to 400-Hz supply. The transmitter rotor is geared directly to the device (in this case a broadcasting antenna) whose directional position is to be transmitted and the transmitter rotor is, therefore, not free to turn. A total of five leads are required to transmit position information from the antenna transmitter (located possibly on a high mast or mountain top) to a remote indicator which might be in a studio or station. The function of the selsyn system, therefore, is to couple two shafts (electrically) so that when one (the transmitter shaft) is turned (mechanically) the other (the receiver shaft) is also turned (mechanically) to the same degree. Electric leads, even when there are five, are more flexible and practical for such coupling over distances, and even over relatively short runs, than mechanical shafting which is relatively rigid and impractical for distance transmission. For this reason, it is extensively used on ships and aircraft to provide control indications.

The principle upon which synchronization occurs is that **emf**s are induced from the rotors (of both the transmitter and the receiver) into the

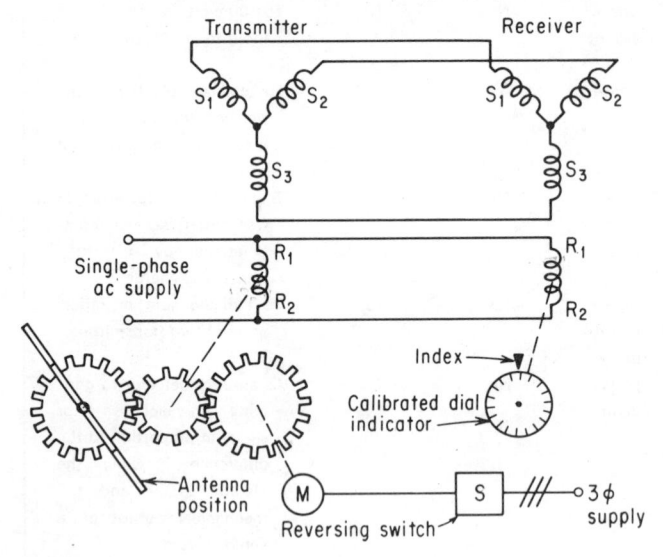

Figure 11-12

Simple selsyn transmitter—receiver system.

CHAP. ELEVEN / *Specialized Dynamos*

stator windings. Since these stator **emf**s are excited from a single-phase supply, they are all in phase; but they differ in magnitude with the position of their respective rotors. When the receiver rotor is in precisely the same position as the transmitter rotor, the **emf**s induced in each of the three "three-phase" windings are *equal* and *opposite*, and *no* stator currents are produced.

On the other hand, if the transmitter rotor is rotated mechanically to a new shaft position so that it is displaced with respect to its own stator winding, new **emf**s are produced in the transmitter stator with respect to the receiver stator. Stator currents flow, producing, in the receiver a *resultant stator field* that tends to rotate the receiver rotor (which is free to turn) to a position parallel to the stator field, i.e., such that no stator currents are produced. Since the rotors are always in phase (in parallel), in the new position by transformer action, the induced emf's in the receiver stator are equal and opposite to those in the transmitter, and the system is again in equilibrium. Any change in the transmitter rotor position, therefore, either clockwise or counterclockwise, will cause stator currents and a corresponding immediate change in the receiver rotor position until these currents are zero.

In some literature, the selsyn transmitter is called a "generator" and the receiver is called a "motor." Neither of these terms is technically correct, since a transmitter does not generate electric energy nor does a receiver convert electric energy into mechanical motion. Neglecting friction, the mechanical input to the trasmitter exactly equals the mechanical output of the receiver.

A single large transmitter is sometimes used in conjunction with two or more parallel-connected, remotely located receivers. The only limitation on such operation is that the transmitter unit must be of sufficient size and must have sufficiently low internal impedance to supply sufficient current and normal torque for each of the receivers whenever the transmitter rotor position is mechanically changed.

11-11.1 Differential Transmitter Shaft position information is sometimes required from two sources. If both sources are in correspondence, i.e., at the same position angle, then the difference in shaft positions is zero. For this purpose, a *differential transmitter* is used; one is shown in Fig. 11-13. If mechanical input 1 on the selsyn transmitter and input 2 on the differential transmitter are in correspondence, then the indicator rotor will not move. No currents are flowing in the stator windings of either the transmitter or the receiver. If, however, the rotor of input 1 is rotated, say 10° clockwise, the new voltage induced in the stator of transmitter 1 will cause currents in the stator of transmitter 2. The rotor of transmitter 2 cannot move because it, too, is connected to a load, and therefore, like any transformer, it transfers the energy to the stator of the receiver, causing receiver stator currents to flow and a net stator flux. This stator flux acts on the

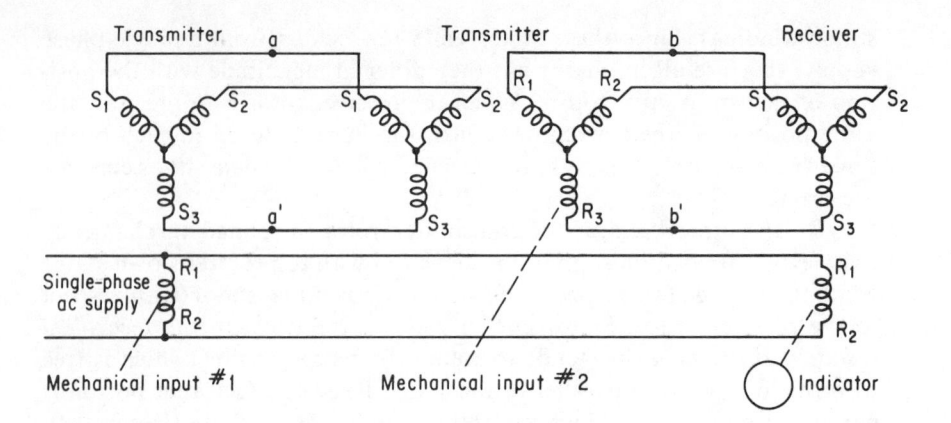

Figure 11-13

Use of a differential transmitter in a selsyn system.

rotor of the receiver, causing it to indicate the *difference* in rotor position between inputs 1 and 2.

If it is desired that the receiver should record the *sum* rather than the difference of the two mechanical inputs, it is only necessary to reverse the connections at a and a' and at b and b', i.e., S_1 and S_3 are reversed in both stator circuits. The indicator dial will now record the sum of inputs 1 and 2.

11-11.2 Differential Receiver

The use of a *differential receiver* is illustrated in Fig. 11-14. Two transmitters, geared to mechanical inputs, feed position information to the three-phase stator and rotor, respectively, of a differential receiver. The rotor of the receiver will rotate to that rotor shaft position at which no current flows in either its stator or its rotor wind-

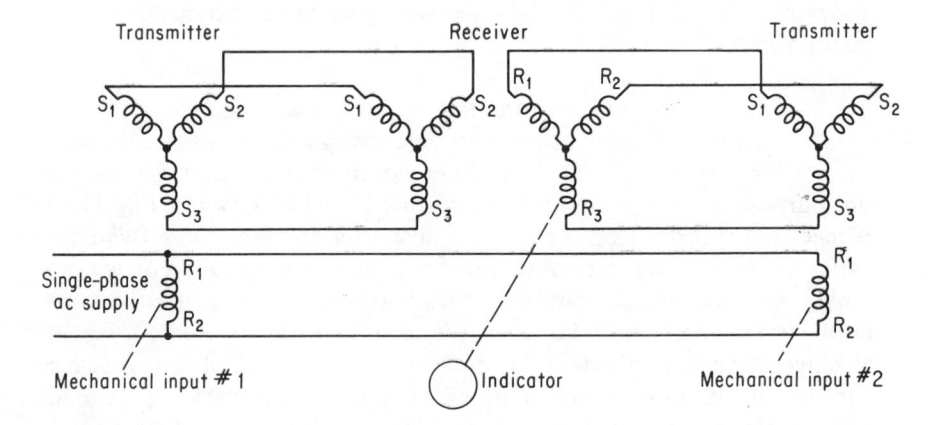

Figure 11-14

Use of a differential receiver in a selsyn system.

428

CHAP. ELEVEN / *Specialized Dynamos*

ings. This position represents the difference in rotor positions between each of the mechanical inputs, 1 and 2, respectively. Both of the circuits shown in Figs. 11-13 and 11-14 could be used in the measurement of slip (Section 9-13) in lieu of the mechanical differential employed in Fig. 9-12b. The indicator of Figs. 11-13 and 11-14 would rotate at a slip speed proportional to the difference in velocity inputs to the two transmitters.

11-11.3 Control Transformer

Figure 11-15 shows the use of a *control transformer* as an error detector in a closed-loop (servo) system. The desired mechanical position is converted to an electric signal brought to the stator of the control transformer by means of a selsyn transmitter. The actual mechanical position of the load is coupled mechanically to the rotor of the control transformer. Thus, the control transformer receives the desired position information in the form of an electric signal on its stator, and it receives the actual position in terms of the displacement of its rotor. If both of these are in correspondence, i.e., if the rotor of the control transformer is at 90° with respect to the transmitter rotor, the rotor of the control transformer produces no single-phase output. If there is an error between them, however, the voltage produced in the rotor by the resultant field of the stator is amplified and used to drive a two-phase motor in such a direction as to reduce the error. The amplified error signal drives the load (and the control transformer rotor) until a null is produced, i.e., until the load corresponds to the desired mechanical input position. Note that, in the system shown in Fig. 11-15, the transmitter could be located at some remote station, and only three leads are required to transmit the desired signal information over a distance to the closed-loop system located at the load.

The comparison of Figs. 11-12 through 11-15 with Table 11-3 will serve to indicate why dampers are required on only two of the selsyn devices: the differential receiver and the simple selsyn receiver. Their rotors are *never* connected to any load, whereas the rotors of the control

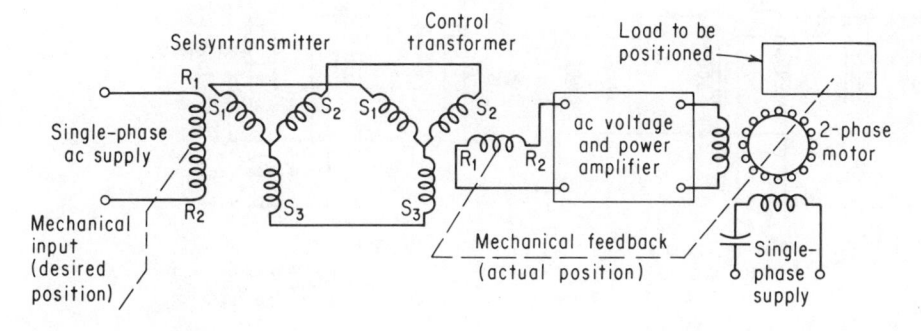

Figure 11-15

Use of a selsyn transmitter and control transformer (shown at null position).

transformer, the transmitter, and the differential transmitter are always mechanically coupled to a load.

It goes without saying that a selsyn system will *not* function properly if there is an open- or short-circuit in any of the stator or rotor connections. Overloads are particularly harmful to selsyn devices which do not rotate like motors nor possess rotor fans on their shaft for cooling. Any open circuit in any of the stator leads will cause continuous stator *overload* currents to be drawn from the rotor primaries.

11-11.4 Overload Indicators

As shown in Fig. 11-16a, *neon overload indicators* are connected through transformers in two of the stator line leads, S_2 and S_3. Should any stator lead become disconnected, the increased stator current produces sufficient potential to ignite the overload lamp. If any two stator leads are opened, there is no possibility of overload, so only two leads are metered.

(a) Overload indicators.

(b)

(c) Blown-fuse indicators in rotor circuit.

(d) Blown fuse lamp transformer and indicator lamp.

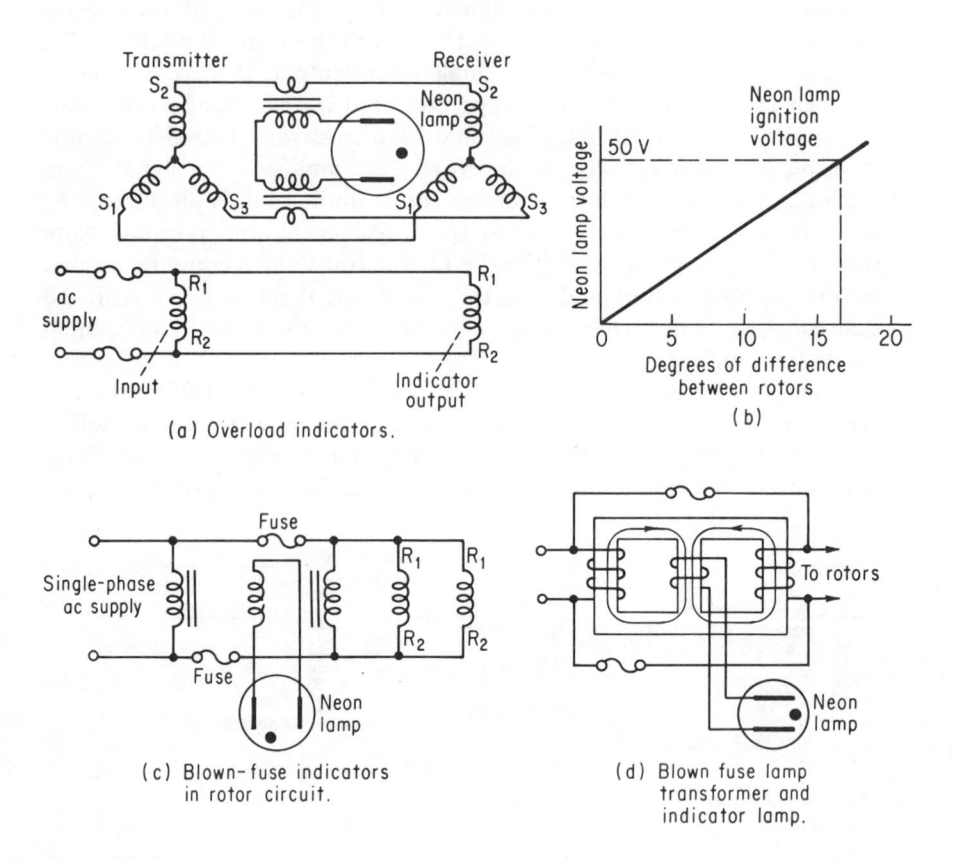

Figure 11-16

Overload and blown fuse indicators used on selsyn systems.

430

CHAP. ELEVEN / *Specialized Dynamos*

The overload indicators are also an indication of a *blocked receiver rotor*. Thus, if the rotor of the transmitter is mechanically turned and the receiver rotor is unable to turn mechanically in correspondence, the stator currents will increase in proportion to the degree of difference between the two rotors. As shown in Fig. 11-16b, the neon lamp will light if approximately 17° of difference are produced between rotors. The voltages developed across the neon lamps in the event of overload are also used to actuate audible alarm signals.

Any overload in the stator circuits is supplied by transmformer action from the rotors. The primary rotor circuits are fuse-protected on both sides of the ac line, as shown in Fig. 11-16a and c, to protect them from extreme overload currents drawn by the stators. The same neon lamp principle is used in the rotor circuit to provide indication of a blown fuse in the ac line supplying the rotors, but with one minor difference. The neon lamp is across the secondary of a transformer whose parallel primaries are connected in such a way as to produce opposing mmf's in the transformer iron, as shown in Fig. 11-16d. If either fuse opens, the left-hand primary shown in Figs. 11-16c and d is energized, but the right-hand primary winding is not. The secondary voltage across the neon indicating lamp is high (without the opposing mmf of both primaries) and the lamp ignites, simultaneously actuating an audible alarm signal.

11-11.5 Zeroing Selsyns

In order to calibrate the dials of selsyn devices, it is necessary to know the zero position. If these devices are not connected in any system, and if their rotor shafts are free to rotate, the simple transmitter, receiver, and control transformer may be zeroed as shown in Fig. 11-17a. The differential transmitter and the differential receiver may be zeroed as shown in Fig. 11-17b. *All* zero adjustments are normally performed at a *reduced* voltage to avoid damage to selsyn devices. When alternating current is applied, the rotor will lock in the zero position.

Once connected in a system, however, the shafts of the transmitter,

(a) Zeroing of selsyn transmitter, receiver or control transformer.

(b) Zeroing a differential transmitter or receiver.

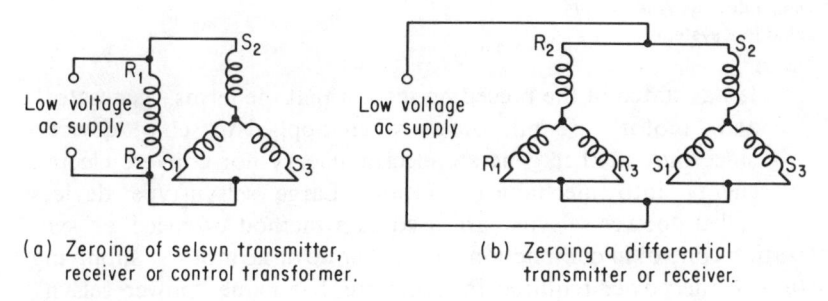

Figure 11-17

Method of zeroing selsyn devices when the rotors are free to rotate.

SEC. 11-11. / Synchronizing (Selsyn) Devices

the control transformer, and the differential transmitter are geared to a mechanical drive. It is possible to unclamp the stator and rotate it somewhat with respect to the rotor in order to zero it. The usual technique, however, is to rotate the mechanical drive mechanism slowly until the zero position is found by the null indication on the voltmeter shown in the various diagrams of Fig. 11-18. The control transformer and the differenrial transmitter are usually zeroed in two steps, a coarse zero and a fine zero procedure, using the circuits shown in Figs. 11-18b and c. The coarse zero step is required to avoid damage to rotor windings in the event of a large displacement of the rotor from its zero position, which is why a voltmeter is used in series with the rotors.*

(a) Transmitter.

(b) Control transformer.

(c) Differential transmitter.

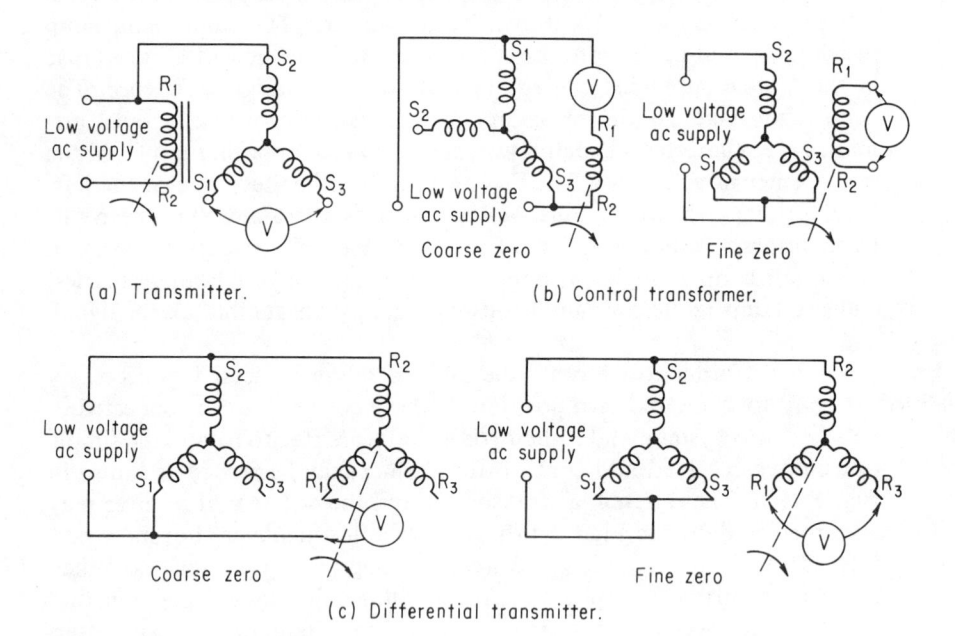

Figure 11-18

Methods of zeroing selsyn devices when rotors are mechanically coupled in a system.

11-12.
POWER SELSYNS AND SYNCHRO-TIE SYSTEMS

It was stated in the preceding section that the terms "generator" and "motor" are misnomers when applied to selsyn devices, since they neither generate electric energy nor convert electric energy into mechanical motion. Large selsyn-type devices called "power selsyns" are used as a method of speed or synchronization control in conjunction with either dc or ac motors, supplying a portion of the power required for rotation. The name "power selsyn"

* Space does not permit a complete discussion of synchro devices. For both additional theory and applications see E. Johnson, *Servomechanisms*, Englewood Cliffs, N.J.: Prentice-Hall, Inc., 1963, Chapter 6.

is a misnomer that appears quite frequently in the literature. These devices *are* actually small fractional or integral three-phase *wound-rotor induction motors* (Sec. 9-9), which *continuously* rotate at speeds below synchronous speed. Power "selsyns" *do* behave as synchronizing generators or motors.

A brief description of power synchronization produced by two such wound-rotor induction devices is shown in Fig. 11-19. The system shown is sometimes called the "synchro-tie" system. Each wound rotor is mechanically coupled to its respective load. Both the stators and the rotors of two wound-rotor motors are *electrically* tied in parallel as shown in Figs. 11-19a and b. When both prime movers (motors) are driving their respective mechanical loads at the same speed, the slip of each wound-rotor motor is the same, and the induced rotor voltages are equal in magnitude and opposite in phase with respect to each other. Thus, no currents flow in the rotor windings and no torque is produced by either rotor (actually the same as if the rotor were open-circuited).

(a) Synchro-tie system.

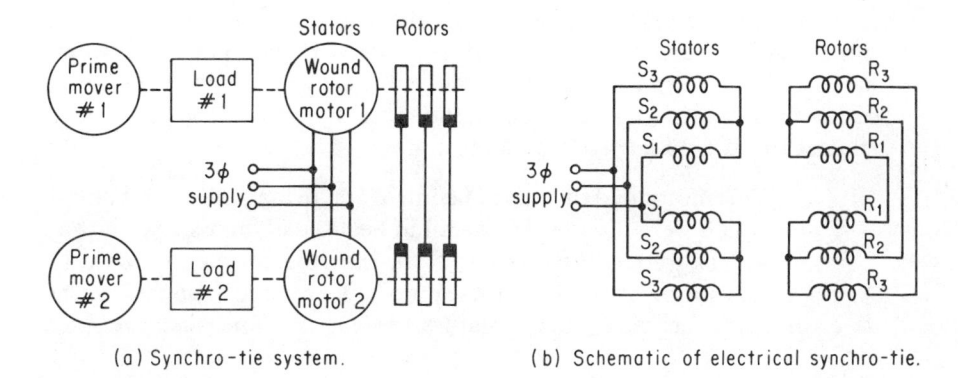

(b) Schematic of electrical synchro-tie.

Figure 11-19

Use of wound-rotor induction motors in a synchro-tie system.

When the loads are unbalanced, however, the prime mover with the lighter load will tend to run at a higher speed, causing its coupled wound-rotor motor to behave like an ac synchronous alternator in generating and delivering current to the rotor of the heavily loaded machine. The rotor current received by the heavily loaded motor tends to produce torque (motor action) in the direction of rotation and to aid its prime mover in increasing speed. Conversely, the wound-rotor machine acting as a generator requires torque for its generation in such a direction so as to oppose its own rotation (generator action), tending to reduce the speed of the lightly loaded prime mover.

The synchro-tie system thus serves to maintain a *constant* load on each prime mover, by *adding* load to the prime mover driving the *lightly loaded* shaft and by *removing* load from the prime mover driving the heavily loaded shaft. The amount of load that each wound-rotor induction dynamo can add or remove depends on the capacity of the wound-rotor dynamo in comparison to the prime movers employed. Usually they are about 10 per cent of the prime mover kVA or hp rating. Power "synchros" are extensively used for synchronizing motors of draw or lift bridges, in printing press machinery, in paper mills, and in steel mills.

11-13.
DC SERVOMOTORS

DC servomotors are dc motors driven by a current from dc electronic amplifiers or ac amplifiers with internal or external demodulators, saturable reactors, thyratron or silicon-controlled rectifier amplifiers, or any of various types of dc rotary amplifiers such as the Amplidyne, Rototrol, or Regulex amplifiers to be discussed later in this chapter. DC servomotors vary in size from 0.05 hp to 1000 hp.

The fundamental characteristics to be sought in any servomotor (dc or ac) are (1) that the motor output torque is roughly proportional to its applied control voltage (developed by the amplifier in response to an error signal), and (2) that the direction of the torque is determined by the (instantaneous) polarity of the control voltage. Four types of dc servomotors are used, and will be discussed in turn, namely: the shunt motor (either field or armature controlled), the series motor, and the permanent-magnet (fixed field excitation) shunt motor.

11-13.1
Field-Controlled Servomotor

This motor and this type of control are, in fact, the same as those which were discussed in detail in Section 4-7 and shown in Fig. 4-6, as well as Fig. 11-20a. Note that the torque produced by this motor is zero when there is no field excitation supplied by the dc error amplifier. Since the armature current is *always constant*, the

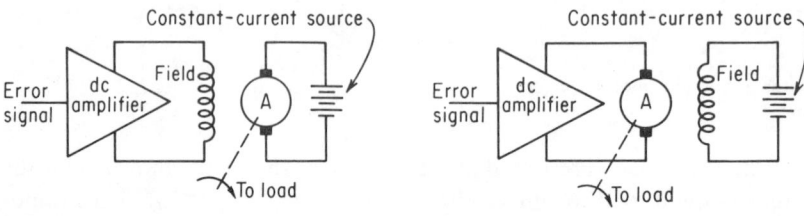

(a) Field controlled dc servomotor.

(b) Armature controlled dc servomotor.

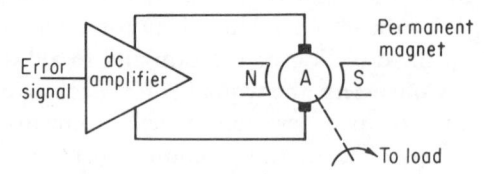

(c) Armature controlled permanent magnet dc servomotor.

Figure 11-20

DC servomotors, separately excited.

torque varies directly as the field flux and also as the field current up to saturation ($T = k\phi I_a$). If the polarity of the field is reversed, the motor direction reverses. The control of field current by this method is used only in small servomotors, however, because (1) it is undesirable to supply a large and fixed armature current as would be required for large dc servomotors, and (2) its dynamic response is slower than the armature-controlled motor because of the time constant of the highly inductive field circuit.

11-13.2 Armature-Controlled Servomotor

This servomotor employs a fixed dc field excitation furnished by a constant-current source, as shown in Fig. 11-20b. As stated, this type of control possesses certain dynamic advantages not possessed by the field-control method. A sudden large or small change in armature voltage produced by an error signal will cause an almost immediate response in torque, because the armature circuit is essentially resistive compared to the highly inductive field circuit.

The field of this motor is normally operated well beyond the knee of the saturation curve to keep the torque less sensitive to slight changes in voltage from the constant-current source. In addition, a high field flux increases the torque sensitivity of the motor ($T = k\phi I_a$) for the same small change in armature current. DC motors up to 1000 hp are driven by armature voltage control in this manner. If the error signal and the polarity of the armature voltage are reversed, the motor reverses its direction.

Large dc armature-controlled shunt motors are usually driven by amplidynes or multiple field exciters (Sections 11-16 and 11-17) called rotary amplifiers, where servomechanism requirements dictate the need for high-power drives.

11-13.3 Permanent-Magnet Armature-Controlled dc Servomotor

Small fractional hp low-torque "instrument" dc servomotors are employed, using permanent magnets for constant field excitation rather than a constant-current source, as shown in Fig. 11-20c. These devices are usually manufactured in 6 V and 28 V dc ratings. The field structure for these motors consists of Alnico VI alloy, cast in the form of a circular ring, about 1 in in diameter, completely surrounding the armature and providing a strong flux. The permanent-magnet motors are well compensated by means of commutating windings to avoid demagnetization of the field magnets whenever the dc armature voltage is suddenly reversed. Eddy currents and hysteresis effects in these motors are generally negligible, and the pole pieces are generally laminated to reduce arcing at brushes whenever a rapid change of signal voltage occurs. These devices are also controlled by armature voltage control in the same manner as the armature-controlled shunt motor.

11-13.4
Series Split-Field dc Motors

Small fractional-horsepower series split-field dc motors may be operated as separately excited field-controlled motors as shown in Fig. 11-21a. One winding is called the main winding and the other is called the auxiliary winding, although they are generally equal in **mmf** and are wound about the field poles in such a direction as to produce reversal of rotation with respect to each other. As shown in Fig. 11-21a, the motors may be separately excited and the armature may be supplied by a constant-current source. The advantages of the split field method of field control are that (1) the dynamic response of the armature is improved since the fields are always excited (there is no delay due to inductive time constant), and (2) a finer degree of control is obtained because the direction of rotation is more responsive to extremely small differences in current between the main and auxiliary windings.

(a) Separately excited.   (b) Directly excited.

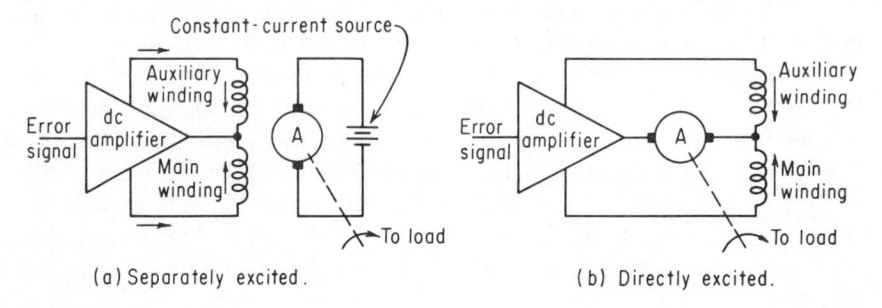

Figure 11-21

DC servomotors, split-field series type.

Larger series motors are operated using the configuration shown in Fig. 11-21b, because separate armature excitation using large constant currents is difficult to obtain. In this configuration, the armature current of the split-field series motor is the sum of the auxiliary and main winding currents. But when these series field currents are equal and opposite, no torque is produced. A slight decrease or increase in the auxiliary winding current will produce instantaneous torque and rotation in *either* direction. The series servomotor produces a high starting torque and a rapid response to slight error signals. The speed regulation is poor with this type of motor; but this drawback is generally not a major consideration in a servosystem since the load is usually fixed. The use of two windings in opposition reduces motor efficiency somewhat, although this is not much of a problem in smaller motors.

In general, dc "shunt" and series motors have greater rotor inertia than ac motors for the same horsepower rating because of the heavier winding of ac armatures. The added drag resulting from brush friction discourages the use of dc motors in extremely small and sensitive instrument servosystems. Small armatures are also skewed to reduce a phenomenon known as "slotlock" at low speeds. Commutation is also a problem with dc servomotors, although interpoles and compensating windings help the situation considerably. At high altitudes, however, because of a lack of oxygen, the oxide film may be abraded from the commutator, causing commutation difficulties.* Hermetically sealed small dc servomotors have been developed to overcome this particular problem. Even worse commutator problems arise from the fact that the motors operate most of the time from stalled or almost stalled (null) positions, and large currents flow to the slowly moving bars producing high arcing and pitting of the commutator. In addition, the arcing of any commutator motor produces radiation and radio interference. Finally, the brushes require periodic maintenance.

For all of the reasons given in the paragraph above, therefore, most of the smaller motors used in servomechanisms are of the ac two-phase or shaded-pole induction-motor type (see Section 10-9, Figs. 10-7 and 10-8).

11-14. AC SERVOMOTORS

The mechanical power of ac shaded-pole servomotors used in servomechanisms varies from $\frac{1}{1500}$ hp to $\frac{1}{8}$ hp. DC motors as described above are invariably used in higher ratings despite their stated disadvantages, because ac motors in the higher ratings are too inefficient; if they are constructed with the desired speed-torque characteristics required by servomechanism operation, ac motors are difficult to cool.

A schematic diagram of the *two-phase servomotor* is shown in Fig. 11-22a, and a diagram of the shaded-pole type is shown in Fig. 11-22b.

(a) Two-phase servomotor.

(b) Shaded pole servomotor.

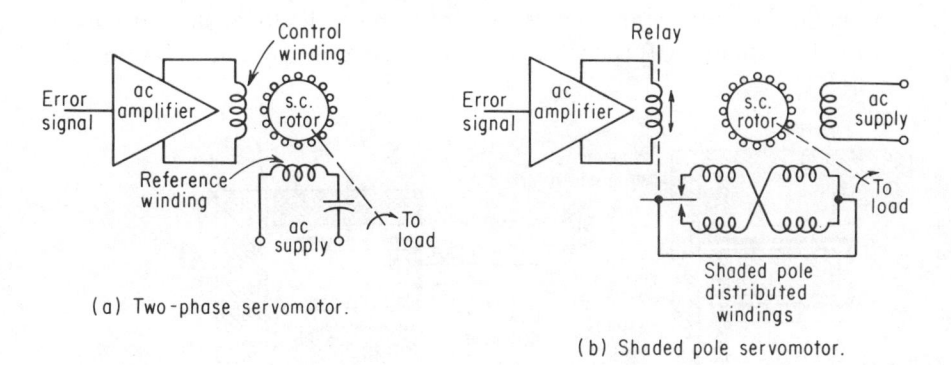

Figure 11-22
AC servomotors.

* The theory of interface films and their effect on resistance commutation is discussed in Kloeffler, Kerchner, and Brenneman, *Direct Current Machinery*, New York: The Macmillan Company, 1948, p. 322 ff.

The most common type employed is the four-terminal two-phase servomotor shown in Fig. 11-22a. This motor is a true two-phase motor, having two stator windings displaced 90° in space on the stator. The reference winding is usually excited through a capacitor by the fixed ac supply. With no error signal, the squirrel-cage rotor is at standstill. A small error signal of some particular instantaneous polarity with respect to the reference winding is amplified by the ac amplifier and fed to the control winding. Motor rotation is produced in such a direction so as to reduce the error signal, and the motor ceases to rotate when a null (zero error signal) is produced.

The *shaded-pole servomotor* shown in Fig. 11-22b employs a phase-sensitive relay to actuate those contacts that will produce a short-circuit of the shaded-pole winding to develop rotation in the desired direction. As with all shaded-pole windings, a single-phase ac field winding is connected to the ac supply. In the presence of an error signal sufficient to actuate the relay, one pair of shaded-pole windings is shorted; thereupon the servomotor rotates until the null is produced (at which point the relay drops out) and the motor stops. An error signal of opposite polarity will actuate the phase-sensitive relay to short-circuit a different pair of windings, causing rotation of the servomotor in the reverse direction (See Sec. 10-9).

It is fairly evident that the two-phase motor design of Fig. 11-22a is the better of the two types since it is capable of responding to small error signals. A shaded-pole servomotor will respond only when the amplified error signal is of sufficient magnitude to cause the relay to operate. The response of the two-phase servomotor to very small control signals is further improved by reducing the weight and inertia of the motor in a design known as a "drag-cup servomotor." These low-torque ac servomotors, shown in Fig. 11-23a, lend themselves extremely well to ac instrument servosystems. Since all of the iron for the magnetic circuit is stationary, the rotor consists only of a thin cylindrical shell of copper or brass, and its shaft is held in a single bearing. Because of its low inertia,

(a) Drag cup servomotor.

(b) Drag cup magnetic damper.

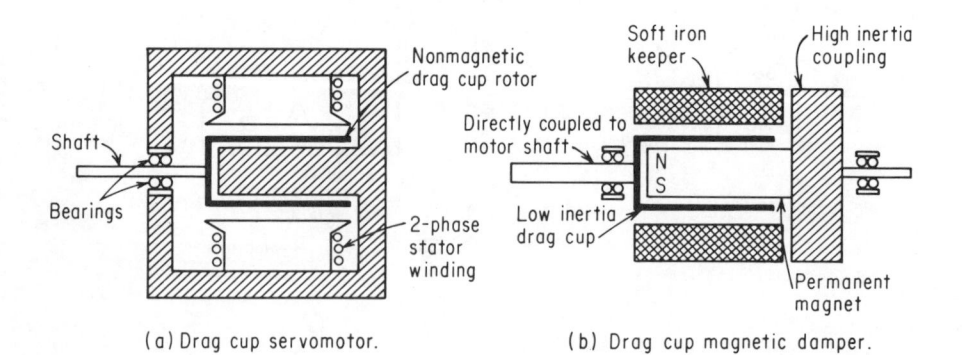

Figure 11-23

Drag cup ac servomotor and magnetic damper.

the drag-cup motor is thus capable of starting even when extremely small signals are applied to its control winding.

The drag-cup principle is also used to *damp* or slow down dc and ac servomotors so that they stop instantly when the error signal is at a null; in this way, they reduce hunting or overshooting whenever an error signal occurs. As shown in Fig. 11-23b, a low-weight, low-inertia drag-cup is coupled to the motor. The drag-cup surrounds a permanent magnet and is in turn surrounded by soft-iron keepers to preserve the retentivity of the permanent magnet. Any change in speed, i.e., starting, stopping, or reversal, will produce a damping action. The advantages of this method of damping are its long life and its resistance to wear.

11-15. ROSENBERG GENERATOR

Like the third-brush generator (Sec. 11-2), the Rosenberg generator was invented to deliver a constant current regardless of the speed of the prime mover. But, in addition it can also generate the same polarity regardless of the direction of rotation of the prime mover. No other generator possesses this unique property.

The third-brush generator (Section 11-2) is a generator with no armature-reaction compensation and with an essentially weak field to produce a rapid unbuilding with load. The Rosenberg generator is a second type of generator (and there are others) which utilizes armature reaction to good advantage in producing desirable operating characteristics. A simplified two-pole Rosenberg generator circuit is shown in Fig. 11-24a, connected to a storage battery which represents the generator load. The field poles of the generator have particularly *small* pole *cores* but *heavy* pole *shoes*. The cores are more than sufficient to carry the weak shunt field flux produced during operation, whereas the heavy shoes are

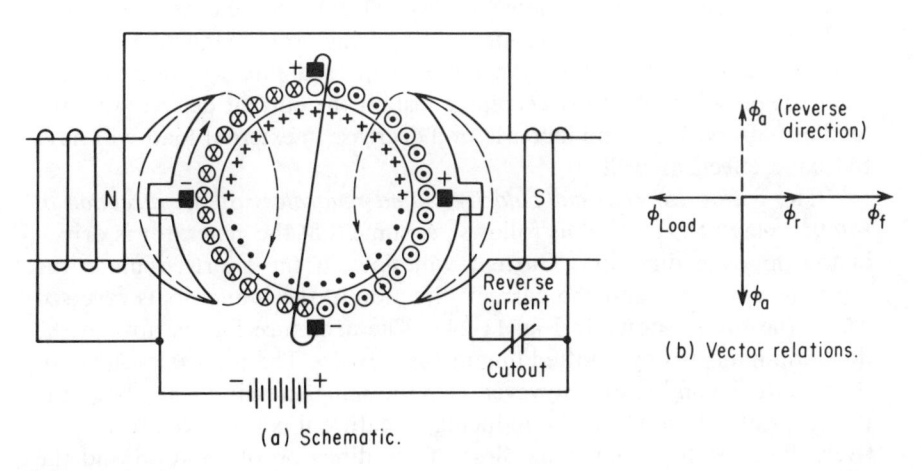

Figure 11-24

Rosenberg generator.

necessary to complete the magnetic circuit path required to carry the cross-magnetizing armature (reaction) flux. No armature-reaction compensation is provided; and, as will be shown, without armature reaction the generator is incapable of operation.

As the Rosenberg generator is rotated by its prime mover (usually the axle of a railway car) the dc voltage generated by its conductors is picked off its interpolar brushes and immediately short-circuited. The heavy short-circuit currents, aided by the heavy pole shoes, produce a high cross-magnetizing armature flux in the direction shown in Fig. 11-24a and (b). The rotating conductors driven by the prime mover of the generator are linked with this cross-magnetizing flux, and a new voltage is generated in all of the armature conductors. This voltage is indicated by the inner circle of dots and crosses shown in Fig. 11-24a. A second set of quadrature brushes on the commutator (not shown) of the generator, pick off this generated voltage and deliver it to the load, in this case a battery to be charged, through a normally closed cutout.

As shown in the phasor diagram of Fig. 11-24b, the original flux, ϕ_f, produces an armature voltage that is short-circuited by the external brushes, and a heavy armature flux, ϕ_a, is produced. The voltage resulting from ϕ_a is delivered to an external load, and the current in the armature conductors produces an armature load flux, ϕ_{load}, which directly opposes the field flux. The resultant main field flux, ϕ_r, is the difference between ϕ_f and ϕ_{load} necessary to establish and sustain the armature flux ϕ_a.

The Rosenberg generator, like the third-brush generator, was designed (1) to deliver a constant current regardless of speed, and (2) to generate the same polarity of emf regardless of the direction of rotation, as previously noted.

The feature of *constant current regardless of speed* follows directly from the *phasor* diagram shown in Fig. 11-24b. An increase of speed tends to increase the short-circuited voltage, the cross-magnetizing armature flux, and the load flux. An increase in load flux will decrease the resultant field flux, ϕ_r, thus keeping constant the current delivered to the load or battery. Either an increase or a decrease in external load will have the same effect, as well.

The feature of *constant polarity output regardless* of the *direction of rotation* may be explained as follows: Assume that the generator is driven in the opposite direction. Current will flow in the short-circuit in the opposite direction, and the current direction in the inductors is reversed within the circles shown in Fig. 11-24a. The armature flux is now in the direction indicated by a dotted line in Fig. 11-24b. The induced voltage in the armature conductors, however, is in the *same* direction since both the direction of rotation and the inducing armature flux are reversed respectively. The result is that, regardless of the direction of rotation and the magnitude of speed (above a certain minimum), the Rosenberg generator always produces a constant current in charging a battery. Its prime mover is a locomotive which runs forward and backward frequently.

The purpose of the cutout is to prevent the generator from being motorized when the car stops or falls below a certain minimum speed; in either of those situations, reverse current would be delivered by the battery to the generator.

The Rosenberg generator was introduced primarily because its theory of operation is so similar to the Amplidyne, a high-gain dc rotary power amplifier, discussed below.

11-16.
THE AMPLIDYNE

Regarded by some as the most important development in the field of rotating machinery in the first half of the twentieth century, the Amplidyne, along with the silicon-controlled rectifier, has been responsible for the continued use of and new developments in dc motors. Basically a dc generator, it is a *rotary power amplifier* capable of tremendous dc power gain and relatively rapid response. Like the simple single-phase induction motor (see Section 10-4, cross-field theory) the Rosenberg generator, the Rototrol and Regulex (to be considered subsequently) it falls into the class of *cross-field rotating machines*.*

The theory of the Amplidyne may be understood through the step-by-step development shown in Fig. 11-25. Consider a separately excited generator as shown in Fig. 11-25. In fairly comparable Amplidyne sizes, such a

(a) Separately excited generator.

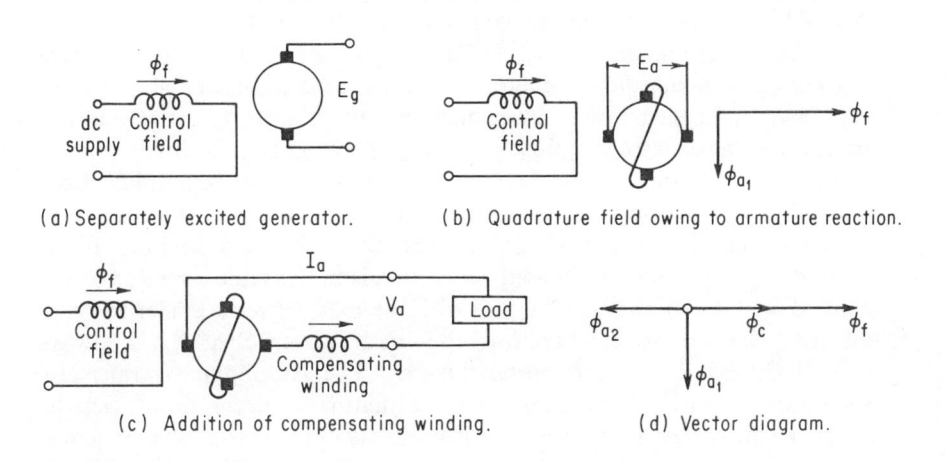

(b) Quadrature field owing to armature reaction.

(c) Addition of compensating winding.

(d) Vector diagram.

Figure 11-25

Amplidyne, step-by-step development.

* The author considered an organization of this volume based on synchronous, asynchronous, and cross-field dynamos, but abandoned it in favor of the present organization. All cross-field rotating dynamos exhibit (1) stator and rotor fluxes in quadrature; (2) generation of rotor currents which contribute to the excitation of quadrature fluxes; and (3) use of armature reaction to good advantage.

separately excited generator would require a control field excitation power which represents (at best) 1 per cent of its rated output power. Under these circumstances, then, we may consider the separately excited dc generator as a *rotary power amplifier* in which, for example, every ampere at 100 V (100 W) produces an output of 100 A (rated load) at 100 V. In effect, then, *any* generator may be considered a rotary power amplifier.

Figure 11-25b shows the same generator with its armature short-circuited. In order to develop the rated load without damaging the generator, it is necessary to reduce the field excitation to, say, one-hundredth of its original excitation. The rated short-circuit current in the armature conductors produces an armature reaction flux ϕ_{a_1}. Assuming that the armature flux is of the same magnitude as the field flux, the same voltage, E_a, is now generated in the armature at the same speed. Thus, a field current of $\frac{1}{100}$ of the original excitation, or $\frac{1}{100}$ A is now capable of generating the same voltage in a dynamo whose conductors can carry the same load current, i.e., 100 A at 100 V.

There are thus *two* stages of amplification produced within a *single* machine. The first stage of amplification occurred when, at a reduced voltage, sufficient field flux yielded the rated current in the short-circuit. The second stage of amplification occurred when the quadrature armature reaction flux produced a generated voltage in its armature conductors (since any generator is a rotary power amplifier). Since generation produces a power gain of 100, and since double generation has occurred in the Amplidyne, the overall power gain is 100 × 100, or 10,000.

The phasor diagram, Fig. 11-25d, shows the result of load current flowing in the armature conductors when a load is connected across E_a of Fig. 11-25b. The armature conductors produce a flux ϕ_{a_2} directly in opposition to the control field flux, ϕ_f. It would be highly undesirable to have quadrature armature reaction result from the second-stage amplification, because it would oppose the main field and act as a strong negative feedback limiting the output of the Amplidyne.* As shown in Fig. 11-25c, a compensating winding is added whose flux is in the same direction as the control field and whose **mmf** varies with the load current. Under any and all conditions of load, therefore, the load current flux ϕ_{a_2} is always neutralized by the compensating flux, ϕ_c. The Amplidyne is thus *two* generators in one, and by using its own armature excitation, it utilizes the same armature structure twice. The response is more nearly instantaneous, therefore, than if two machines were used, and the size is half of that required for two dynamos.

The commercial Amplidyne is designed to make the quadrature armature flux, ϕ_{a_1}, as large as possible. This in turn results in a large change in the output voltage, V_a, and the output current, I_a, with even

---

* In reality, it would be a Rosenberg generator whose output current is independent of speed as a result of strong internal feedback flux; See Section 11-14.

slight changes in the flux of the control field, ϕ_f. In addition, the commercial machine is equipped with several *pairs* of control windings to increase the versatility of the machine, each pair designed to produce normal saturation at its rated voltage. Thus, an Amplidyne might be equipped with a pair of 6 V dc control windings, a pair of 12 V, a pair of 24 V, etc. It is not necessary to operate the control windings in voltage pairs, however. The control windings are generally operated so that they oppose each other in a differential flux method (one winding being a reference and the other an error signal), although they may be operated aiding as well. Thus a 24 V *reference* winding may be "bucked" by a 110 V *feedback* winding.

Typical power gains vary with the capacity of the Amplidyne. A 400 W unit has a power gain of about 5000, and a 10 kW unit has a power gain of about 20,000. There is a limit to gain, however, and higher gain leads to slower (Amplidyne) response. The amount of amplification depends on the magnitude of the armature reaction flux resulting from the field flux. As the armature flux increases, a larger field flux is necessary to control it. But if the quadrature armature flux density is very high, the addition or subtraction of control field mmf requires a time lag before the armature flux will respond to changes in the control field.

The application of the Amplidyne in a closed loop as a dc power amplifier in alternator voltage control is shown in Fig. 11-26. (Further application of the Amplidyne and other rotary amplifiers in motor speed control are shown in Fig. 11-28). As shown in Fig. 11-26, a 2 kW Amplidyne is used to control the field voltage of a 200 kVA three-phase alternator. The alternator output voltage is converted to direct current by means of a step-down transformer and a rectifier, and is fed to one of the control

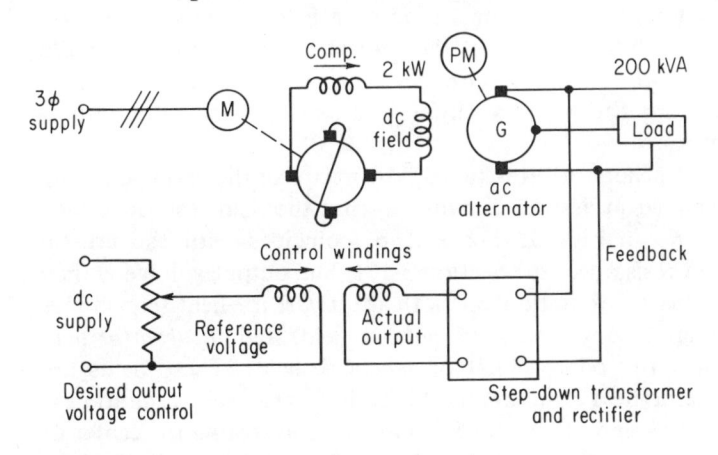

Figure 11-26

Alternator voltage control using Amplidyne as a dc amplifier in closed-loop system.

fields. The other control winding is fed from a separate dc (or rectified ac) source, and represents a means of setting the magnitude of the output voltage at a specific reference value. Assume that the voltage of the alternator rises; this in turn causes its dc control winding to "buck" the reference field and to reduce the excitation of the Amplidyne. The Amplidyne output to the field of the alternator thereupon decreases, and the alternator output consequently also decreases. Conversely, if the alternator voltage drops, the opposing action of the feedback winding is less, and the Amplidyne excitation is increased. The Amplidyne output to the alternator field increases, and the alternator voltage rises.

11-17. MULTIPLE-FIELD EXCITERS— ROTOTROL AND REGULEX

Various multifield dc rotary amplifiers have been developed since the emergence of the Amplidyne, arising from the fact that *any* generator whose excitation and output may be controlled is a rotary electromechanical power amplifier. Basically, there are three requirements for any rotary power amplifier: (1) a high degree of amplification or ratio of output armature power to control field input power; (2) relatively rapid and linear reponse to changes of voltage impressed on the control field windings; and (3) a minimum of at least two control fields for the purposes of detecting any discrepancy between the desired output (reference field) and the actual output (feedback control field).

Both the "Rototrol" (developed by Westinghouse) and the "Regulex" (developed by Allis Chalmers) employ *self-exciting* generators. The Rototrol is a self-excited dc *series* generator, and the Regulex is a self-excited dc *shunt* generator. In both these multifield exciters, the excitation of the self-energizing field is adjusted or "tuned," by means of a variable resistor, so that the generator is operating on a field resistance line coinciding exactly with the *linear* part of the saturation curve. Figure 11-27a shows the Rototrol circuit and its series generator saturation characteristic. Figure 11-27b shows the Regulex circuit and its shunt generator saturation characteristics.

Both of these generators are "tuned," by means of the variable tuning resistor in their field excitation circuits, so that the field resistance line, represented by R_f in Figs. 11-27a and b, coincides with the critical value of the field resistance (see Section 3-10). The output voltage is then adjusted by means of excitation on both the *reference* field (sometimes called the standard, comparison, or *pattern* field) and the *control* field (sometimes called the pilot, signal or *feedback* field). These fields are wound on the same field structure as the exciting field. The excitation is usually adjusted to some value between *a* and *b* representing the center of the linear portion of the critical field resistance line. Point *a* on each of the curves represents that value of excitation produced by the excitation field only, while all the values between *a* and *b* represent the additional excitation produced by the combined reference and control field fluxes. As

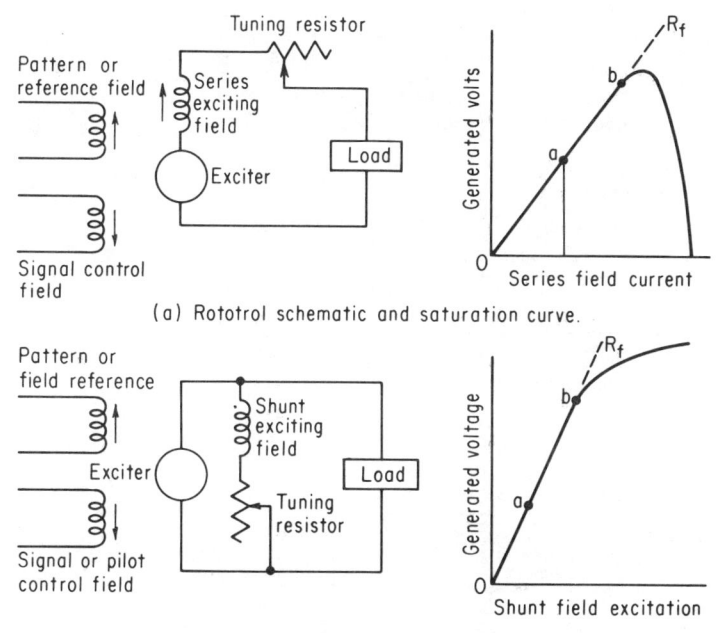

Figure 11-27

Multiple-field exciters.

shown in Figs. 11-27a and b, the reference and feedback control windings always produce fluxes which oppose each other (differential flux method), the reference field aiding the exciting field flux and the feedback field opposing both the reference field and the exciting field.

Applications of multiple-field exciters are to the same purposes and intentions as the Amplidyne. Figure 11-28 shows three basic circuits using the Rototrol for the control of voltage, current, and speed, respectively. Note that in two of these a "transducer" is necessary to convert to a dc voltage the quantity being controlled, i.e., current or speed in Figs. 11-28b and c.

A dc standard or reference source sets the desired pattern or reference flux. The opposing transduced output voltage permits the multifield exciter to increase or decrease its output, in order to maintain control of the voltage, current, or speed. To convert each of the applications shown for use with the Regulex, it is merely necessary to remove the series field from the exciter armature and to add a shunt field with its exciting resistor. Similarly, to convert each of the applications shown in Fig. 11-28 for use with the Amplidyne, it is merely necessary to remove the exciting resistor and to show a short circuit across the armature (see Fig. 11-28d with Fig. 11-26).

All dc generators exhibit *residual magnetism* in the field iron. Residual

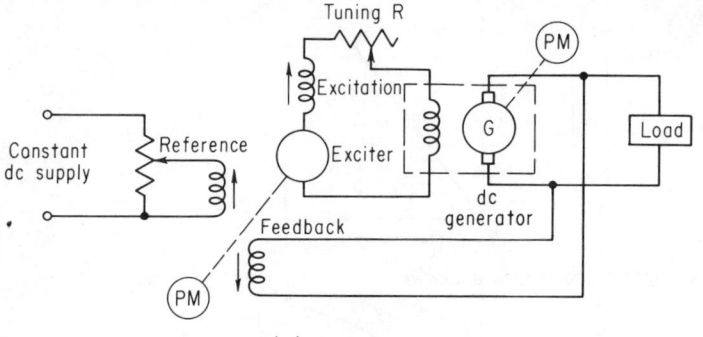

(a) Voltage control.

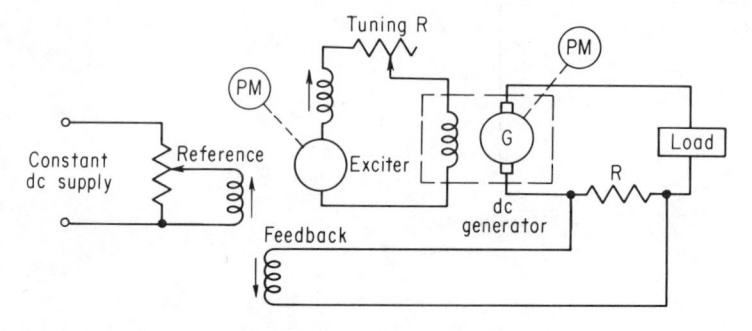

(b) Constant current control.

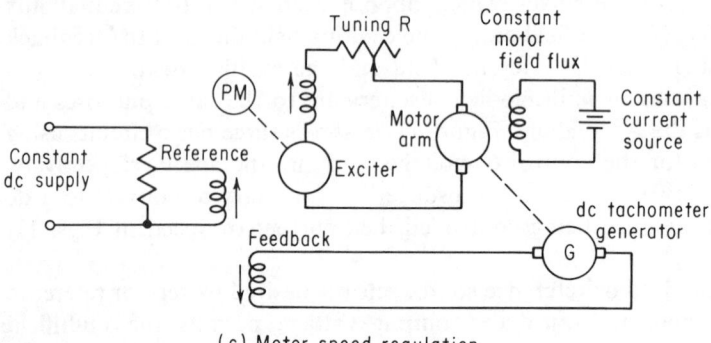

(c) Motor speed regulation.

Figure 11-28

Control of voltage, current, or speed by multifield (Rototrol) exciter.

magnetism along the curves of magnetization shown in Figs. 11-27a and b would destroy the linearity of the excitation and the output. All multiple-field exciters, including the Amplidyne, employ *demagnetization devices* to destroy residual magnetism. The most common method of de-

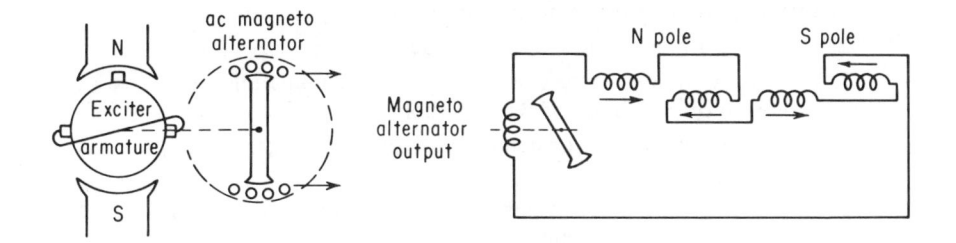

Figure 11-29

Multifield exciter demagnetization windings.

magnetization consists of a single-phase ac generator having a permanent-magnet (Alnico) rotor, as shown in Fig. 11-29. The output of the ac generator is fed to opposing ac windings located on the field poles. When the total control field excitation is reduced to zero, therefore, the ac windings completely neutralize the residual magnetism and the armature output voltage is zero.

It should be noted that neither the Rototrol nor the Regulex (single-stage amplifiers) is capable of the amplifications possible with the Amplidyne (a two-stage amplifier). They are confined specifically to low-power applications and are manufactured in the 500- to 2000-W range of output capacity. In the same small capacity, they are less expensive than the Amplidyne, and simpler to connect; in addition, they possess the advantage of less susceptibility to hunting because of slower response and smaller power amplification, both tradeoff disadvantages, as well.

11-18. BRUSHLESS DC MOTOR

The designation "brushless dc motor" cannot be used to describe one distinct motor design. In the past decade, a number of different designs have emerged in this category. At the present time the various designs may be classified under three general types:

a. Electronic commutation (brushless) dc motors;
b. The dc/ac inverter (brushless) dc motors, operating from a dc supply;
c. Limited rotation (brushless) dc motors.

Each of the three categories above contain motors using different designs and techniques. As of this writing, the state-of-the-art is so fluid and dynamic that no single design predominates in each category. This section is not intended to describe each type presently available. Rather, it introduces the reader to some of the more common types currently employed in spacecraft and more sophisticated control systems.

Brushless dc motors, while generally more expensive for the same hp rating, possess certain advantages over commutator and brush dc motors, namely:

1. they require little or no maintenance whatever;
2. they have a much longer operating life;
3. there is no arcing whatever in these motors eliminating the hazard of explosion or possibility of RF radiation;
4. they produce no brush or commutator particles or gases as byproducts of operation;
5. they are capable of operation submersed in fluids, combustible gases and may even be hermetically sealed;
6. they are generally more efficient than brush-type dc servomotors or conventional dc motors;
7. they provide a more rapid response (lower servotime constant) and a fairly constant output torque vs input current characteristic which lends itself to servomotor applications.

The disadvantages of the various types are:

1. greater overall total size because of additional space for associated electronics (although brushless motors themselves are usually smaller than conventional dc motors for the same hp);
2. higher initial cost (but reduced maintenance costs);
3. Somewhat limited choice (at present) in "stock" sizes necessitating "special" orders for particular applications.

11-18.1 Electronic Commutation (Brushless) dc Motor

All of the brushless dc motors under this heading have a wound stator and a permanent magnet (PM) rotor, as shown in Fig. 11-30a. Secured to the rotor shaft is some type of *rotor position sensor-transducer* which serves as an input to the solid-state switching system, eliminating the need for commutator and brushes.

The electronically commutated motor shown in Fig. 11-30a incorporates 3 solid-state switches (transistors) in series with its 3 stator windings, equivalent to a brush motor having 3 commutator bars. (Various commercial designs employ from 6 to 12 up to 100 or more stator windings and solid-state transistor or SCR switches.)

Secured to the motor shaft is a cam-shaped *light shield* which "senses" the rotor position and activates the transistor switch from its cutoff to a saturation condition, thus exciting the required stator torquer winding. The technique shown in Fig. 11-30a is a photoelectric sensing technique often used. Commercial electronic commutation motors use other sensing techniques such as magnetic transducers, Hall-crystal transducers, electrostatic sensors, electromagnetic induction coils, etc. The function of the transducer or sensor is to provide the signal to activate a particular transistor switch from its cutoff to saturation state. The transistor thus closes the circuit to its respective stator torquer coil. The motor shown in Fig. 11-30a operates as follows:

(a) Electronic commutation

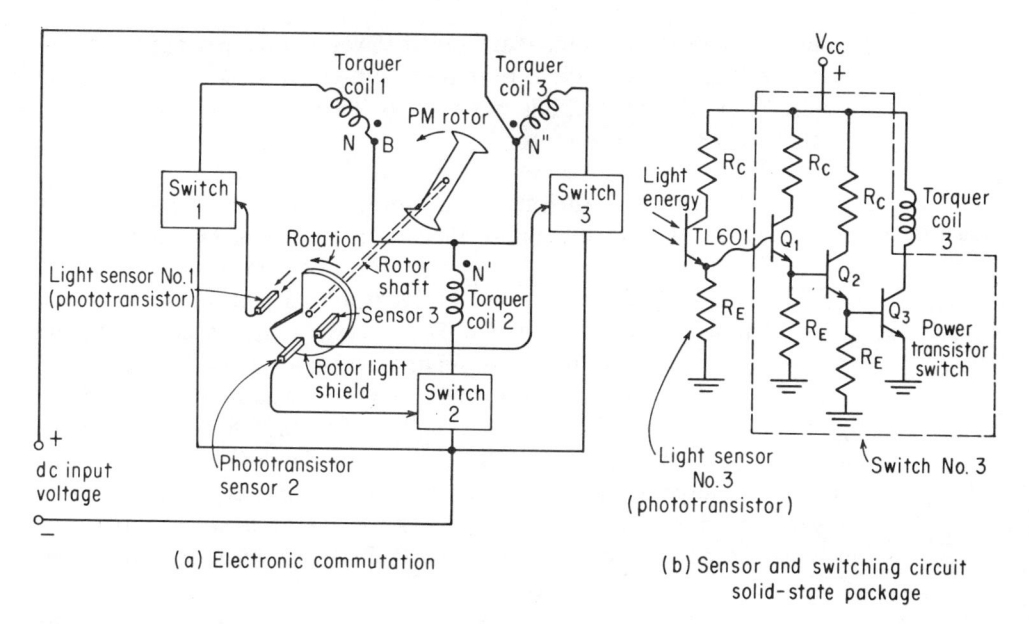

(b) Sensor and switching circuit solid-state package

Figure 11-30

Electronic commutation for brushless dc motor.

1. Phototransistor 1 activates transistor switch 1 in the position A shown. Phototransistors 2 and 3 are not activated since the light source (not shown) is blocked by the light shield.
2. Transistor switch 1 energizes coil 1, which is wound in such a direction to produce a pole of opposite polarity to the permanent magnet (PM) on the rotor. The rotor is attracted from position A to position B.
3. In position B, phototransistor 1 and corresponding coil 1 are deactivated and phototransistor 2 is activated. This in turn activates transistor 2 and energizes coil 2, which in turn attracts the rotor from position B to position C.
4. The action of the sensor and switch is to sequentially energize each of the torquing stator windings in turn to provide continuous rotation of the rotor shaft in the same direction (**ccw** in this case, as shown in Fig. 11-30a).

In commercial application, the phototransistor sensing and transistor switching circuitry employ stages of current amplification (Q_1 and Q_2) with a power transistor (Q_3) as the output switch in series with each respective stator torquer coil, as shown in Fig. 11-30b. The collector and emitter bias resistors are chosen to bias the phototransistor (TL 601) and transistors to full saturation when conducting and cutoff in the absence of a positive signal at the base of each *npn* transistor. Since the transistors are only conducting for a small fraction of one revolution, power losses in this amplifier are much lower than when operated continuously in the active region. In many commercial designs, the electronic transistor amplifier-switching package is located remotely from and external to the brushless motor.

11-18.2 The dc/ac inverter type (brushless) dc motor

A number of brushless dc motors employ an ac servomotor (Sec. 11-14) in conjunction with an electronic *inverter* for operation from a dc source. The electronic inverter package may be either separated or incorporated within the ac motor housing (integral type). The inverters are usually standard 12-V or 24- to 28-V dc inputs providing 50-, 60- or 400-Hz outputs to standard shaded pole or drag cup induction motors (Sec. 11-14) or hysteresis type, single-phase synchronous motors (Sec. 8-28). The inverter circuitry usually incorporates techniques for varying the output frequency and/or ac voltage to provide a variety of output speeds. Some electronic packages even include options for ac as well as dc *input*, resulting in a universal speed controlled package, with input frequencies from dc up to 400 Hz at a variety of input voltages.

While inverter type packages are not as efficient as electronic commutation packages, they have the advantages of constant speed and low inertia in the smaller sizes, typical of ac drag cup induction motor characteristics (Sec. 11-14).

11-18.3 Limited rotation (brushless) dc motors

The two previous classes of brushless dc motors are intended for continuous rotation operation. Limited rotation motors, however, only provide an output torque over a maximum of 180 degrees (plus or minus 90° in either CW or CCW direction). No commutation is required in such motors because no current reversal is necessary to produce continuous rotation.

Limited rotation motors have a PM rotor and a wound stator. When the stator is energized from a dc supply, the PM rotor is torqued either clockwise or counterclockwise, depending on the polarity of the supply energizing the stator winding, providing fairly constant torque over 90° of rotation in either direction.

Limited rotation motors find application as torquer motors for gyroscopic gimbals in stable elements of space platforms, in pen motors of chart recorders, as fine-adjustment position control torquers in servomechanisms, and as dc instrument servomotors such as dc tachometer indicators.

Some limited rotation brushless motors give the appearance of homopolar motors since they contain frameless "pancake" shaped discs without commutators which serve as torquers. It should be noted that the homopolar motor (Secs. 1-11 and 11-4) requires no commutator, but *cannot* be included as a "brushless" motor because brushes are required at each end of the rotor disc.

BIBLIOGRAPHY

Alexanderson, E. F. W., Edwards, M. A., and Bowman, K. K. "The Amplidyne Generator—A Dynamoelectric Amplifier for Power Control," *General Electric Review*, March 1940.

Alger, P. L. and Erdelyi, E. "Electromechanical Energy Conversion," *Electro-Technology*, September 1961.

Arrott, W. "The '*S*-Generator' Metadyne for High-Performance dc Drives," *Electrical Manufacturing*, May 1959.

Bacheler, A. T. "A Comparison of Adjustable Frequency ac and Synchronic Systems for Synchronized Drives," AIEE Paper CP 58-808

Bewley, L. V. *Tensor Analysis of Electrical Circuits and Machines*, New York: Ronald Press, 1961.

Campbell, S. J. "Integral Horsepower Synchro Systems," *Electrical Manufacturing*, November 1953.

Carr, C. C. *Electrical Machinery*, New York: John Wiley & Sons, Inc., 1958.

Crosno, C. D. *Fundamentals of Electromechanical Conversion*, New York: Harcourt, Brace, Jovanovich, Inc, 1968.

Daniels. *The Performance of Electrical Machines*, New York: McGraw-Hill, Inc., 1968.

Fitzgerald, A.E. and Kingsley, C. *The Dynamics and Statics of Electromechanical Energy Conversion*, 2nd Ed., New York: McGraw-Hill, 1961.

Fitzgerald, A. E., Kingsley, Jr. C. and Kusko, A. *Electric Machinery*, 3rd Ed. New York: McGraw-Hill Book Company, 1971.

Fink, R. A. "The Brushless DC Motor", *Control Engineering*, June 1970, pp. 75-80.

———. "The Brushless Motor: Types and Sources", *Control Engineering*, August 1970, pp. 42-45.

Gemlich, D. K. and Hammond, S. B. *Electromechanical Systems*, New York: McGraw-Hill, 1967.

Heumann, G. W. "Power Selsyn Operation," in *Magnetic Control of Industrial Motors*, New York: John Wiley & Sons, Inc., 1961, Vol. 2, p. 167.

Hindmarsh, J. *Electrical Machines*, Elmsford, N.Y.: Pergamon Press, 1965.

Howard, W. "Combining Hydraulic Torque Converters with ac Motors," *Electrical Manufacturing*, February 1957.

Jones, C. V. *The Unified Theory of Electrical Machines*, New York: Plenum Publishing Corp., 1968.

Levi, E. and Panzer, M. *Electromechanical Power Conversion*, New York: McGraw-Hill, 1966.

Littman, B. "An Analysis of Rotating Amplifiers," *Proc. AIEE*, Section P796, 1947.

Liwschitz, M. M. "The Multistage Rototrol," *Proc. AIEE*, Section P796, 1947.

Majmudar, H. *Introduction to Electrical Machines*, Boston: Allyn and Bacon, 1969.

Meisel, J. *Principles of Electromechanical Energy Conversion*, New York: McGraw-Hill, 1966.

Montgomery T. B. "Regulex—Instability in Harness," *Allis-Chalmers Electrical Review*, Second and Third Quarters, 1946.

Myles, A. H. "Fundamentals of Tie-Motor Control," *Control Engineering*, January 1959.

Nasar, S. A. *Electromagnetic Energy Conversion Devices and Systems*, Englewood Cliffs, N.J.: Prentice-Hall, Inc., 1970.

Newton, G. C., "Comparison of Hydraulic and Electric Servo Motors," *Proc. 5th Meeting, National Conference on Industrial Hydraulics*, Chicago: Armour Research Foundation, 1949.

O'Brien, D. G. "DC Torque Motors for Servo Applications," *Electrical Manufacturing*, July 1959.

O'Kelly and Simmons. *An Introduction to Generalized Electrical Machine Theory*, New York: McGraw-Hill, 1968.

Pestarini, J. M. *Metadyne Statics*, New York: John Wiley & Sons, Inc., 1952.

Puchstein, A. F. *The Design of Small Direct Current Motors*, New York: Wiley/Interscience, 1961.

Rosa. *Magnetohydrodynamic Energy Conversion*, New York: McGraw-Hill, 1968.

Saunders, R. M. "Dynamoelectric Amplifiers," *Electrical Engineering*, August 1950.

Schmitz, N. L., and Novotny, D. W. *Introductory Electromechanics*, New York: Ronald Press, 1965.

Seely, S. *Electromechanical Energy Conversion*, New York: McGraw-Hill, 1962.

Selmon. *Magnetoelectric Devices: Transducers, Transformers and Machines*, New York: Wiley/Interscience, 1966.

Siskind, C. S. *Electrical Machines—Direct and Alternating Current*, 2nd Ed., New York: McGraw-Hill Book Company, 1959.

Siskind, C. S. *Direct-Current Machinery*, New York: McGraw-Hill, 1952.

Skilling, H. H. *Electromechanics: A First Course in Electromechanical Energy Conversion*, New York: Wiley/Interscience, 1962.

Thaler, G. J., and Wilcox, M. L. *Electric Machines: Dynamics and Steady State*, New York: Wiley/Interscience, 1966.

Walsh, E. M. *Energy Conversion—Electromechanical, Direct, Nuclear*, New York: Ronald Press, 1967.

White, D. C., and Woodson, H. H. *Electromechanical Energy Conversion*, New York: Wliey/Interscience, 1959.

Williams, J. R. "The Amplidyne," *Electrical Engineering*, May 1946.

QUESTIONS

11-1. a. What disadvantages occur in using an ordinary shunt generator for dc battery charging?
b. Using the shunt generator voltage characteristic, explain why these disadvantages occur.
c. Using the diverter-pole generator characteristic, explain how the diverter-pole generator is ideally suited for battery charging.

11-2. Using Fig. 11-1, explain how the same mutual air-gap flux is maintained
a. at no load
b. at full load
c. if the **mmf** produced by the diverter causes too large an increase in output voltage. What modifications would be necessary? (Hint: see Sec. 3-19.)

11-3. What is the relative voltage output at high, medium and low speeds, of a third-brush generator having
a. a high resistance load?
b. a low resistance load?

11-4. For the third brush generator,
a. what specific design modifications are required compared to a shunt generator?
b. how is the charging rate increased, regardless of prime mover speed?
c. what is the function of the reverse-current cutout?
d. explain why it was supplanted in automotive battery charging by the
  1. shunt generator
  2. alternator.

11-5. a. Using the right-hand generator rule, determine the shaft polarity of the homopolar generator shown in the footnote figure to Sec. 1-11, if the top of the Faraday's disc is rotated toward the observer.
b. Why is the homopolar generator called the only true dc generator?
c. How can the polarity of the brushes of the commercial homopolar generator, shown in Fig. 11-3a be reversed? Give 2 methods.
d. How can the polarity of the MHD generator be reversed? Give 2 methods.
e. What modification is required to produce an ac MHD generator?
f. What are the possible advantages of MHD generation, using Faraday's principle, as compared to conventional steam-generation using an alternator?
g. What is the significance of the statement that "with nuclear fusion at high temperatures we are back, once again, to the first discovery by Faraday"? Explain.

11-6. a. The dynamotor has been called a "dc transformer." Explain.
b. Why is the dynamotor more efficient than a conventional dc MG set to convert dc from low to high voltages, and vice-versa?
c. List the advantages and disadvantages of the dynamotor vs the conventional MG set.
d. Why are dynamotors never equipped with field rheostats?

e. Why are dynamotors never compensated for armature reaction?
f. What is the function of the commutator at the output of a dynamotor?

11-7. a. Why is only one winding necessary in a synchronous converter but two are required on a dynamotor?
b. What is a direct converter?
c. What is an inverted converter?
d. Why are polyphase converters used extensively but single phase converters never used?

11-8. For a given dc generator kW rating, explain the effect on the kVA output of a polyphase synchronous converter of
a. increasing the number of phases and number of slip-rings
b. increasing the number of poles
c. increasing the PF of the load on the converter
d. using a winding which increases the number of taps on the rotor.

11-9. a. In view of your answers to Ques. 11-8a, explain why synchronous converters are usually 6-phase dynamos rather than 12- or 24-phase machines
b. What methods are used to control the dc output voltage of the converter shown in Fig. 11-8?
c. Why is a 6-phase star connection used in Fig. 11-8?
d. How is the direct converter of Fig. 11-8 started?

11-10. Under what conditions is it possible to obtain both dc and ac from a polyphase rotary converter?

11-11. a. List 4 methods of producing a 3-wire dc system.
b. Which of these methods is most preferable, in terms of overall efficiency and ease of maintenance?

11-12. From the current directions shown in Fig. 11-10a and the vertical direction of the voltage drops and rises shown in Fig. 11-1b, develop a rule for voltage drops
a. when the current flows from the source at left to load at right
b. when the current flows from the load at right to the source at left (Hint: See Exs. 11-6 and 11-7.)
c. Why is the above rule simpler to use than Kirchhoff's laws?

11-13. Why is the feeder voltage drop method using the above rule preferable to the lumped load method using Kirchhoff's laws? Give 3 reasons.

11-14. Using linear graph paper, 20 sq./inch, and a dimensioned diagram similar to Fig. 11-10b, verify the answers given in the table of Ex. 11-8.

11-15. What types of electrical devices or dynamos are best suited for converting large quantities of power
a. from dc to polyphase ac?
b. from polyphase ac to dc?
c. from 3ϕ ac to 12ϕ ac?
d. from 1ϕ ac to 3ϕ ac?
e. Explain the meaning of the phrase "no single-phase system can be transformed into a polyphase system."

11-16. For the induction phase converter shown in Fig. 11-1.
   a. compare its conversion efficiency against a 2 unit M-G set and explain why the latter is seldom used
   b. compare its method of starting and running with a single-phase to three-phase rotary converter and explain why the latter is never used
   c. explain how adjustment of the output voltage is achieved
   d. explain why only a single trolley wire and one track rail or a "double track" system is needed to supply it as against a three-phase system.

11-17. A farmer obtains from surplus sources a deep well jet pump powered by a 110 V, 1/2 hp, 3-phase motor. The electric utility refuses to supply 3-phase because of his remote location and small load requirements. The farmer's son, whose hobby is rewinding electric split-phase motors and transformers has just completed a reading of Sec. 11-10, and knows there is a solution in converting 110 V single-phase to 3-phase. But he does not know what the voltage ratings of the various components of Fig. 11-1c should be. Can you help him?
Hint: See Sec. 13-18 and Fig. 13-35c.

11-18. Give the purpose of the following selsyn devices:
   a. transmitter
   b. receiver
   c. control transformer
   d. differential transmitter
   e. differential receiver.

11-19. a. Which of the above 5 selsyn devices possess dampers and why? Can this be used to distinguish transmitters from receivers?
   b. On the basis of your answer to (a) above, explain why those which do not possess dampers are always mechanically coupled to a geared system.
   c. Why is it possible for a receiver to be used as a transmitter but impossible for a transmitter to be used as a receiver?
   d. Which of the 5 selsyn devices have salient-pole rotors and which have nonsalient poles?
   e. Explain why selsyn devices are not considered "rotating" machines.

11-20. Draw a diagram similar to Fig. 11-12 for the following shipboard applications:
   a. an indicator at the pilot's bridge showing position of his rudder
   b. an indicator in the engine room showing pilot's commands regarding engine speed and direction.

11-21. From a study of Fig. 11-12 explain
   a. how the operator can set his antenna to a particular bearing in degrees
   b. how the operator can determine whether there is mechanical blockage of gearing or antenna
   c. how it is possible for the operator to determine his navigational position from 2 land-based radio beacons of different known frequency and location. (Hint: Maximum radio signal means that the antenna is perpendicular to the source.)

11-22. a. Draw one aircraft application (correspondence of left and right wing flaps) using Fig. 11-13.
b. Repeat (a) using Fig. 11-14.

11-23. Using the principle shown in Fig. 11-15, design and draw a system for
a. lifting and lowering heavy cargos from ship to shore
b. turning the ship's rudder. (Include system of Quest. 11-20a.)

11-24. a. Explain why overloads are particularly harmful to (nonrotating) selsyn devices.
b. Give 3 possible conditions which will produce overloads in the primary rotor and secondary stator circuits of selsyn devices.
c. Explain why it is necessary to have both overload indicators and blown-fuse indicators on both the stator and rotor circuits respectively of selsyn systems

11-25. Explain the method of zeroing selsyn devices
a. when the rotors are free to rotate
b. when the rotors are mechanically coupled to a load.

11-26. a. Why is it necessary to zero the transmitters and receivers of all selsyns in a system?
b. How is the accuracy of the system affected by failure to adjust the dials to a fine zero position?

11-27. a. What are power selsyns or power synchros and why is this a misnomer?
b. Why is the term synchronizing generator or motor a better name for woundrotor induction motors used in a synchro-tie system?
c. Assuming that prime mover 2 in Fig. 11-19a increases in speed, explain the action of the synchro-tie system shown.
d. If prime movers 1 and 2 are 20 hp dc motors, what should the hp rating of the power synchros be to maintain the synchro tie?

11-28. Show how a synchro-tie system may be used to maintain a heavy scaffold in a horizontal position as it is raised by two motorized winches, one at each end. The lift motors are 50 hp induction motors. Draw a diagram showing all mechanical and electrical connections. Specify the rating of the power synchros.

11-29. a. What electronic sources are used to control dc servomotors?
b. Give 4 types of dc servomotors.
c. For each type, specifically indicate the hp range.

11-30. a. What is the advantage of the method of control shown in Fig. 11-20a and why is it limited to small motors? Why does the motor fail to run away in the absence of field current?
b. What is the advantage of the servomotor shown in Fig. 11-20b and why is this method particularly adapted to extremely large motors?
c. What is the advantage of the servomotor shown in Fig. 11-20c and why is its use confined primarily to dc instrument servomechanisms?

11-31. a. Give 2 reasons why series dc servomotors are generally of the split-field type.
b. What is the advantage of the split-field servomotor shown in Fig. 11-21a and why is it confined to fractional hp sizes?

11-32.
c. What is the advantage of the directly excited split-field series dc servomotor shown in Fig. 11-21b and why is it used in integral hp sizes?

11-32. a. For small and sensitive servosystems, give 4 reasons why ac servomotors are preferred to dc servomotors.
b. Why are ac split-phase, shaded-pole or drag-cup motors preferred?
c. Which types of servomotors lend themselves to large hp requirements?
d. Which types of servomotors generally lend themselves to small hp requirements?

11-33. a. Give one advantage of the two-phase servomotor over the shaded-pole type shown in Fig. 11-22.
b. Discuss the design of the drag-cup principle and show why it is extensively used in two-phase servomotor rotor designs.
c. How is the drag-cup principle used in magnetic damping of dc and ac servomotors?
d. What are the advantages of this method of damping compared to friction damping?
e. Why is damping necessary in a servomotor, specifically, and a servosystem, generally?

11-34. a. In what way is the Rosenberg generator unique from all other generators?
b. Describe the construction of the Rosenberg generator.
c. Describe the principle of operation of the Rosenberg generator.
d. How is constant current achieved regardless of speed?
e. How is constant polarity at the brushes maintained regardless of direction of rotation?

11-35. a. In what respects is the construction and theory of the amplidyne similar to the Rosenberg generator?
b. What three characteristics are exhibited by all cross-field dynamos?
c. Explain the statement, "Any separately excited generator is a rotary power amplifier."
d. Use your explanation in (c) to show how an alternator may be considered a dc-ac single-stage amplifier.

11-36. a. Explain how the amplidyne operates as a 2-stage dc-dc amplifier.
b. What is the typical power gain of an amplidyne?
c. Explain why higher power gains lead to slower response.

11-37. a. List three design requirements for any rotary power amplifier.
b. Using your answer to Quest. 11-35c, explain how a separately excited series generator may serve as a rotary amplifier, single-stage.
c. Repeat (b) for a separately excited shunt generator.
d. Over which portion of the magnetization curve are these amplifiers normally operated?
e. How is the voltage due to residual magnetism eliminated?

11-38. Draw a schematic diagram showing how a Regulex single-stage rotary feedback amplifier is used to maintain constant
a. output voltage of a dc generator
b. output current of a dc generator
c. speed of a dc motor.

11-39. Describe 3 general classes of brushless dc motors.

11-40. List 7 advantages and 3 disadvantages of brushless dc motors compared to conventional dc motors.

PROBLEMS

11-1. The third brush generator shown in Fig. 11-2a is customarily connected with its negative brush grounded to the automotive chassis frame. Explain what might occur if the generator is being driven at normal speed under each of the following conditions
   a. Accidental grounding of the third brush
   b. Accidental grounding of the positive brush
   c. Accidental connection of the field leads across the third brush and the wrong main brush
   d. Accidental connection of the field leads across the positive and negative brushes
   e. Reversal of direction of rotation of the prime mover (when tested in a laboratory before installation)

11-2. A homopolar generator consists of a copper disc 36 in in diameter rotated in a field of 80,000 lines/in^2 at a speed of 5000 rpm. The generator shaft is 2 in in diameter. Calculate:
   a. The voltage induced between the outer rim of the copper disc and the outer surface of the shaft
   b. If the disc in the above problem were held stationary and the field revolved about it at 5000 rpm in the *same* direction, explain the nature of the emf at the brushes
   c. If a dc supply of 67.7 V were connected to the brushes of the commerical homopolar generator shown in Fig. 11-3a observing the same brush polarity, what would be the direction of rotation of the homopolar motor? Explain.

11-3. A dc motor generator set having a generator output of 250 V at 320 mA and a $\frac{1}{6}$ hp driving motor whose full-load input is 12 V at 16 A, is replaced with a dynamotor of identical output having an input of 7.5 A from a 12-V dc source. Calculate:
   a. Over-all motor-generator set efficiency
   b. Efficiency of dc motor and generator, respectively
   c. Efficiency of dynamotor
   d. Saving per year if set operates 18 hr/day at a cost of $.08/kWhr.

11-4. A 25 kW rotary converter is powered from a 550 V, single-phase ac supply and operates at a power factor of 0.9 lagging and a full-load efficiency of 92 per cent. Neglecting voltage drops in the windings, calculate
   a. Output dc voltage and current
   b. ac current drawn from the supply.

11-5. The converter in Problem 11-4 is operated as an *inverted* converter from an 800 V, dc supply at the same full-load efficiency. Neglecting voltage drops in the windings calculate
   a. Output ac voltage and current
   b. dc current at full load.

11-6. A 1000 kW, 250 V, dc, synchronous six-phase converter operates as a direct converter at a full-load efficiency of 94 per cent and a power factor of 90 per cent lagging. Ignoring voltage drops, calculate
a. ac voltage between slip rings
b. dc output current at full load
c. ac current drawn from the secondary of a three-phase to six-phase transformer fed supply
d. Total kilovolt-ampere rating of three-phase to six-phase transformers.

11-7. The nameplate of a rotary converter bears the following information: 900 rpm, 550 V, dc, 500 kW, 60 Hz, 12 phase, 0.85 PF, 0.92 efficiency. Calculate:
a. Number of stationary field poles
b. dc output current at full load
c. ac voltage between slip rings
d. ac current
e. The kilovolt-ampere rating of each of three transformers required to perform the necessary phase transformation.

11-8. If a synchronous converter is required to deliver 1 kiloampere of dc, calculate the ac input current for a converter having an efficiency of 90 per cent at unity power factor equipped with
a. 12 slip-rings
b. 6 slip-rings
c. 3 slip-rings
d. 2 slip-rings, single phase.

11-9. Interpolating from data given in Table 11-2, calculate the following relative values
a. The power factor at which a 12-phase converter may operate and still have the same relative output as a three-phase converter operating at unity power factor.
b. Repeat (a) for the power factor of a six-phase converter
c. Per cent increase in relative output of a six-phase over a three-phase converter, each operating at 0.8 power factor
d. Per cent increase in relative output of a 12-phase over a three-phase converter operating at 0.9 power factor.

11-10. A 250 V/125 V, single-phase ac, three-wire service supplies a building whose load is essentially resistive. Load 1 (the western half of the building) is connected from line 1 to neutral and load 2 (the eastern half) is connected from line 2 to neutral. The line feeders and the neutral have resistances of 0.05 Ω each. Calculate the voltages across each load (loads 1 and 2, respectively) under the following load conditions and tabulate below:

	Load 1	Load 2	V_1	V_2
a.	300	300		
b.	300	200		
c.	100	300		
d.	300	0 (open fuse in line)		

11-11. A commercial electric range designed for operation on a 230 V/115 V, ac, single-phase three-wire system consists of four surface hotplate elements at 500 W, each connected across one side of the line to neutral, with a 3000 W oven element connected across the other side of the line and neutral. Each range installation is supplied with three No. 6 RH feeders having a resistance of 0.4 Ω/1000 ft. The average installation consists of a three-wire service run of approximately 600 ft to the occupancy. A number of consumers who have purchased the oven have complained to the manufacturer that the surface burners are hotter when the oven is on than when it is off. The manufacturer refers the consumers to the utility, claiming the condition a failure of the utility to regulate voltage. The utility claims that the design is poor and that it is the fault of the manufacturer.
  a. Show, by calculation, which claim is correct, computing voltage across surface elements with and without the oven element
  b. Calculate the per cent increase in power of the element
  c. Suggest a design modification which might eliminate the difficulty.

11-12. A competitive electric range is marketed having a 3000 W oven element rated at 230 V and four surface hotplate elements each rated at 500 W. One pair of surface elements is connected across each of the lines to neutral to form a balanced three-wire load. Calculate voltages across energized surface elements
  a. when oven element is off
  b. when one of each pair of elements is energized and oven is off
  c. when one surface element is on and oven is on
  All line and neutral feeders and services are the same as in Problem 11-11.

11-13. A 250 V/125 V, three-wire distribution system has various loads located as shown in the figure. Lines 1 and 2 are fed using No. 3 AWG (0.2 Ω/1000 ft) and the neutral is No. 6 AWG (0.4 Ω/1000 ft). For the loads indicated
  a. draw a diagram showing all currents and directions, including neutral currents

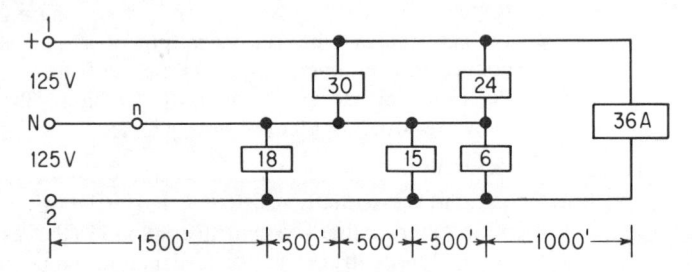

  b. calculate voltages across all loads
  c. calculate voltage from line 1 to neutral at a distance of 2500 ft from the source
  d. calculate the voltage from line 2 to neutral, 2000 ft from source
  e. which is the more heavily loaded side and why?
  f. what conclusion can you draw regarding the sizing of the supply feeders?
  g. why is the voltage higher from line 2 to neutral at a distance of 2000 ft from the supply than it is at 1500 ft from the supply?

11-14. The control field resistance of a Regulex multifield exciter is 250 Ω. A control field current of 80 mA produces an output voltage of 250 V across a 250 ohm alternator field winding whose voltage is regulated by the multifield exciter. Calculate:
a. The power amplification
b. The voltage amplification factor of the Regulex.

11-15. An Amplidyne used in the same applications with the same load as Problem 11-14 requires a current of 15 mA through a control field resistance of 125 Ω to produce the same output voltage. Calculate
a. Power amplification
b. Voltage amplification factor of the Amplidyne.

ANSWERS

11-2(a) 67.7 V   11-3(a) 41.67 per cent (b) 64.8 per cent, 64.3 per cent (c) 88.8 per cent (d) $53.60. 11-4(a) 777 V, 34.4 A (b) 54.9 A. 11-5(a) 565.5 V, 44.2 A (b) 34 A 11-6(a) 88.5 V (b) 4000 A (c) 2230 A (d) 1184 kVA. 11-7(a) 8 poles (b) 909 A (c) 100 V (d) 275 A (e) 213.3 kVA   11-8(a) 262 A (b) 524 A (c) 1048 A (d) 1573 A   11-9(a) 0.8344 PF (b) 0.865 PF (c) 29.5 per cent (d) 44 per cent   11-10(a) 110 V, 110 V (b) 105 V, 120 V (c) 130 V, 100 V (d) 95 V, 0 V   11-11(b) 12 per cent.   11-12(a) 110.33 V (b) 112.67 V (c) 109.66 V.   11-13(b) 147 V, 36 A load (c) 66.8 V (d) 109.9. V.   11-14(a) 156.5 (b) 12.5   11-15(a) 8889 (b) 133

TWELVE

power and energy relations; efficiency, ratings selection and maintenance of rotating electric machinery

12-1. GENERAL

Chapters 1 through 11 are concerned primarily with rotating electric *dynamos* of various types, both general and specialized. As such, they serve as *energy conversion* devices, converting either mechanical energy to electric energy or vice versa. In some instances, as in the case of the synchronous converter or dynamotor, electric energy is converted to mechanical energy which in turn produces electric energy once more. When and if this energy conversion occurs at a *uniform* rate, i.e., when the energy put into any given dynamo per unit time and delivered by the dynamo per unit time are both uniform and constant, we may consider that the dynamo is serving as a *power conversion* device.

A dynamo is, as implied in its name, a dynamic device. It will not accomplish a power (or energy) conversion when motionless or in a static state. It must be running or operating in order to convert energy.

For this reason, it is incapable of the property of energy storage. For this reason, also, in accordance with the law of conservation of energy, the total power received by a dynamo at any instant must equal the total power delivered by a dynamo at that instant. The total power received by a dynamo must equal its output (useful) power and its total power loss, in accordance with the law of conservation of power, or

$$P_{in} = P_{out} + P_{loss} \tag{12-1}$$

where P_{in} is the total power received by a dynamo
P_{out} is the useful power delivered by the dynamo to perform work
P_{loss} is the total loss produced within the dynamo as a result of energy conversion, i.e., $P_{in} - P_{out}$

It is evident from Eq. (12-1) that the power delivered to a dynamo must *always* be greater than the output power or power delivered *by* the dynamo to perform useful work. This means that a motor or a generator can never convert all of the power received into useful mechanical or electric output power. As stated also in Eq. (12-1), the difference between the dynamo input and output is its power loss, which does not perform useful work. Since this power loss does not produce either mechanical or electric energy (both of which are useful to the dynamo), it can only produce heat, light, or chemical energy. Almost all of the loss appears as heat energy or heat power.*

The greater the power loss in Eq. (12-1) as a percentage of the total input power, the greater the heat power and the hotter the dynamo, i.e., the temperature rise of the rotating machine.

The efficiency of a dynamo may be defined in terms of Eq. 12-1, therefore, as the ratio, η, where

$$\eta = \frac{P_{out}}{P_{in}}, \tag{12-2a}$$

$$= \frac{P_{in} - P_{loss}}{P_{in}}, \text{ for a motor} \tag{12-2b}$$

$$= \frac{P_{out}}{P_{out} + P_{loss}}, \text{ for a generator} \tag{12-2c}$$

As will be shown below, Eq. (12-2b) lends itself to motor efficiency while Eq. (12-2c) lends itself to generator efficiency.

The ratio of Eq. (12-2a), expressed as a percentage, is also a measure of the amount of heat power produced in proportion to the total input. A dynamo which operates at a high efficiency, or a high ratio of output to input power, produces comparatively little heat in proportion to its

---

* In commutator-type machines, a small portion of the loss produces visible light and other radiation (light energy) losses, but these are negligible in proportion to the heat loss.

input or output. Conversely, a dynamo which operates at a low efficiency produces a great deal of heat in proportion to its output.

Depending on the thermodynamic ability of the dynamo to dissipate the internally generated heat, the *temperature* of the machine will tend to *rise* until it finds that temperature at which its heat power dissipated *equals* the heat power loss generated internally. If this final equilibrium temperature is excessive; that is, if it exceeds the limit that the insulating materials used in the windings of the dynamo can withstand, then one of two alternatives is required: (1) external cooling devices must be employed in order that the dynamo rating (output power) remain the same; or (2) the output must be *reduced* (reducing the input and the losses) to a rating at which the losses and the temperature rise are *not* excessive. (See Secs. 12-16 & 12-19).

In the case of a motor, it is easier to measure the electrical input power than the mechanical output power, hence the form of Eq. (12-2b). In the case of a generator, it is easier to measure the electrical output power than the mechanical input power, hence the form of Eq. (12-2c). In both cases, therefore, the *losses* must be evaluated.

It is precisely for reasons of *output rating* (expressed either in horsepower for a motor or in kilowatts or kilovolt amperes for a generator) that an attempt is made to study those factors affecting the efficiency of a dynamo to insure that the *losses* and heat power are reduced and that the efficiency is high. We shall first consider factors affecting various types of heat *losses*, and then consider factors affecting dynamo ratings and the selection of dynamos in view of these *losses*.

12-2. DYNAMO POWER LOSSES

Dynamo power losses may be divided into two large classes: (1) those which are produced by the flow of current through the various parts of the dynamo windings called *electric* losses; and (2) those that are a direct function of the dynamic rotation of the dynamo, called *rotational* (or stray power) losses. The latter rotational (or *stray power*) losses are usually divided into two categories: (a) mechanical losses resulting from rotation, and (b) iron or core losses resulting from rotation.

The analysis of losses reveals that certain losses are a direct result of (and vary with) the *load*, whereas other losses are *independent* of the load. Table 12-1, therefore, is a breakdown of electric and rotational losses, listing *no-load* losses as well as *load* losses, and giving the formulas and equations containing those factors contributing to the loss. From this table, it is possible to generalize those losses which are a function of load and those which are independent of load.

Electric losses, as shown by Table 12-1, are those which result primarily from the flow of electric current. If, for example, the series field of a dc compound generator is short-circuited (all other conditions

> **TABLE 12-1.**
> **DISTRIBUTION OF DYNAMO POWER LOSSES**
>
> **A. ELECTRIC LOSSES**
>
> | Descriptions and formulas for loss components | Effects of the application of load |
> |---|---|
> | 1. dc field excitation loss
   Rheostat, $I_f^2 R_r$
   Field winding, $I_f^2 R_f$ $\Big\} V_f I_f$ | 1. Fairly constant with load, but may increase slightly, depending on the required regulation and the power factor—a function of I_f. |
> | 2. Armature winding loss, $I_a^2 R_a$ | 2. Increases with the square of the load. |
> | 3. ac stator excitation loss, $I_a^2 R_a$ | 3. Increases with the square of the load. |
> | 4. Rotor winding loss, $I_r^2 R_r$ | 4. Increases with the square of the load. |
> | 5. Brush and brush-contact resistance loss, $V_b I_a$ (or slip-ring loss). | 5. Increases with the square of the load. |
> | 6. Interpoles, compensating windings, series fields, control fields, etc. | 6. Increases with the square of the load. |
>
> **B. ROTATIONAL (STRAY POWER) LOSSES**
>
> | Descriptions and formulas for loss components | Effects of the application of load |
> |---|---|
> | *Mechanical Losses*
1. Bearing friction
2. Windage (air friction) on rotor
3. Brush friction
4. Ventilating fan windage
5. Bearing oil and/or cooling pump (if on the rotor shaft) losses | These losses are constant at a constant speed; they vary in direct proportion to changes in speed, only. |
> | *Core Losses (or iron losses)*
1. Hysteresis loss, $P_h = K_h B^x f V$
2. Eddy current loss, $P_e = K_1 B^2 f^2 t^2 V$ | These losses are constant at a constant speed and field flux; they vary in direct proportion to changes in flux and speed (frequency). |
>
> **C. STRAY LOAD LOSSES**
>
> Leakage fluxes in teeth, slot wedges, spider, pole faces, etc.
> Armature reaction fluxes in teeth, slot wedges, spider, pole faces.
> The stray load losses are usually assessed at 1 per cent of the output for generators above 150 kW and for motors above 200 hp; they are considered negligible for machines below these ratings.

remaining the same), the losses are reduced by the series field copper loss, and the efficiency is increased (although the voltage regulation might suffer as a result). Electric losses are sometimes referred to as "copper" losses, but neither the brushes nor the brush contact resistances are made of copper. Furthermore, armature and rotor windings are occasionally constructed of cast aluminum, and the term "winding" is more descriptive and technically more accurate than "copper." All of these electric losses tend to vary as the *square* of the load current except those, such as the field loss, which is independent of load, and the brush loss which varies directly with load.

Rotational losses are subdivided into those which are solely a function of speed (the so-called mechanical losses, which are essentially frictional losses) and those which are a function of *both* flux and speed (the so-called *core* losses). These losses occur when an iron armature or rotor structure rotates in a magnetic field, or when a change of flux linkages occurs in any iron structure. The hysteresis loss P_h is a measure of the electric energy required to overcome the retentivity of the iron in the magnetic flux path; using watts as the unit,

$$P_h = K_h B^x f V \qquad (12\text{-}3)$$

where V is the volume of iron in the dynamo subject to change of flux
K_h is a constant for the grade of iron employed
B is the flux density raised to the Steinmetz exponent. With modern values of dynamo iron, x is no longer 1.6 but closer to 2.0. (This is not to imply that, for a given volume, V, of iron the loss has increased, because K_h has been reduced considerably.)
and f is the frequency, in Hz, of reversal of flux

The *eddy current losses* occur not only in the dynamo iron but in *all* conductive materials within the flux path of the rotating or varying magnetic field of the dynamo. The eddy current loss, P_e, in watts, is:

$$P_e = K_1 t^2 B^2 f^2 V \qquad (12\text{-}4)$$

where K_1 is an eddy current constant for the conductive material
t is the thickness of the *conductive material*
B is the flux density
f is the frequency, in Hz, of reversal of flux
V is the volume of material subject to change of flux

For a dc dynamo the frequency, f, of reversal of flux varies with speed. Thus, the hysteresis loss varies directly with speed, whereas the eddy current loss varies as the square of the speed. Both hysteresis loss and eddy current loss vary approximately as the square of the flux density. For this reason, *core* losses are considered a function of *both* flux and speed. Core losses are essentially those which occur in the dynamo iron, hence the term iron losses.

Stray load losses represent additional losses due to the load, as described in Table 12-1C. These losses are larger in induction motors and other small gap machines. They represent:
(1) iron losses due to flux distortion (armature reaction) in dc machines and step-harmonics in ac machines
(2) skin effect losses in armature or stator conductors
(3) iron losses in the structural parts of machines.

12-3. POWER-FLOW DIAGRAMS

A clearer picture of the dynamo, operating as either a motor or a generator, is shown by the power flow diagrams in Fig. 12-1. On the left-hand side of the diagram is mechanical power, and on the right is electric power. Let us use this diagram as a means of analysis of generator and motor efficiency.

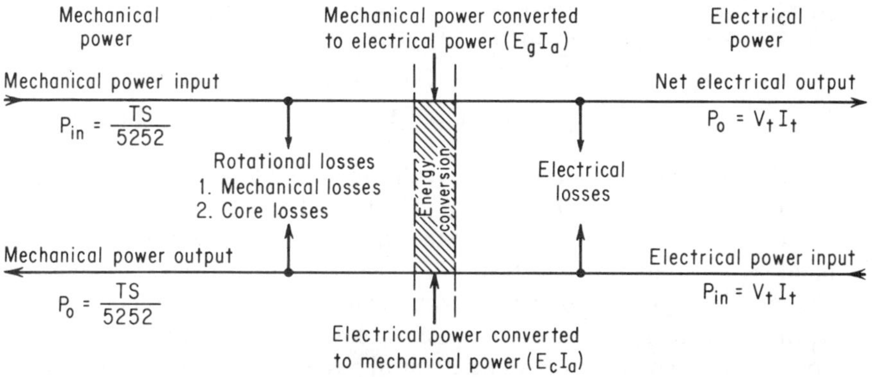

```
Mechanical                    Mechanical power converted          Electrical
  power                       to electrical power ($E_g I_a$)       power

Mechanical power input                                         Net electrical output
    $P_{in} = \dfrac{TS}{5252}$                                 $P_o = V_t I_t$
                    Rotational losses        Electrical
                    1. Mechanical losses     losses
                    2. Core losses

Mechanical power output                                       Electrical power input
    $P_o = \dfrac{TS}{5252}$                                   $P_{in} = V_t I_t$
                              Electrical power converted
                              to mechanical power ($E_c I_a$)
```

Figure 12-1

Combined power-flow diagram for motor or generator action.

12-3.1 Generator Power Flow

If mechanical power is applied to the shaft of a dynamo as the input, the power at the shaft is: $TS/5252$ horsepower. A dynamo driven mechanically as a generator sustains certain rotational losses. The difference between the rotational losses and the input mechanical power represents the net mechanical power which is converted to electric power by electromechanical conversion ($E_g I_a$). But the generator also sustains some internal electric losses, as well, which subtract from the electric power developed. The net electric power output, therefore, is $E_g I_a$ minus the electric losses, or the terminal voltage times the total current delivered to the load, $V_t I_t$, shown at the right in Fig. 12-1.

> In summary, for a dynamo operating as a *generator* or an *alternator*:
>
> $$\frac{\text{Electrical Output Power}}{} = \frac{\text{Mechanical}}{\text{Power Input}} - \left(\begin{array}{l}\text{Rotational losses}\\ + \text{ Electrical losses}\end{array}\right)$$
>
> $E_g I_a$ = Mechanical Power Input − Rotational Losses
>   = Electrical Output Power + Electrical Losses

12-3.2 Motor Power Flow

Electric power applied to the terminals (the right-hand side of Fig. 12-1) of a motor, $V_t I_t$, is immediately reduced by certain electric losses within the motor. The difference appears as electric power, $E_c I_a$ which is converted to mechanical power by electromechanical conversion. The mechanical power available, which has been produced by internal motor torque ($E_c I_a/746$) sustains some internal mechanical losses. The difference between these mechanical losses and the mechanical power produced as a result of electromechanical conversion is the mechanical output power.

In summary, for a dynamo operating as a *motor*:

$$\begin{matrix}\text{Mechanical} \\ \text{Output Power}\end{matrix} = \begin{matrix}\text{Electrical} \\ \text{Input Power}\end{matrix} - \begin{pmatrix}\text{Electrical Losses} \\ +\text{Rotational Losses}\end{pmatrix}$$

$$E_c I_a = \text{Electrical Power Input} - \text{Electrical Losses}$$
$$= \text{Mechanical Output Power} + \text{Mechanical Losses}$$

So the dynamo is, in reality, really very simple and straightforward, as shown in Fig. 12-1. The area of mechanical power is on the left-hand side of the dotted vertical line, and the area of electric power is on the right-hand side of the dotted vertical line. The center area is represented by the change of energy state, or *electromechanical conversion* (since energy can be neither created nor destroyed) where *no* loss occurs. Putting mechanical power into a dynamo involves mechanical power loss, change of state, electric power loss, and electric output. Putting electric power into a dynamo involves electric loss, change of energy state, mechanical power loss, and mechanical output. The reader should study Fig. 12-1 very carefully because it is fundamental to an understanding of the subject.

12-4. DETERMINATION OF LOSSES

It would be (and is) a relatively simple matter to (1) measure the mechanical input to a dynamo, and (2) use electric instruments to meter its electric output; and thus come up with the efficiency of the dynamo as a generator. In the case of smaller dynamos (below 1000 W), the efficiency is often determined *directly*, i.e., by direct measurement of the input and the output, using dynamometers or prony brakes, and sometimes even calibrated machines whose efficiencies are already known.

In the case of larger dynamos, however, it is neither economical, possible, or even convenient to determine efficiency by direct loading.* It *is* possible, nevertheless, to determine losses (now that we know what they are) or to simulate the loss conditions by certain *conventional* (running-light) methods or by *locked-rotor* tests, and to use this information in Eq. (12-2) to calculate efficiency. The value of the efficiency of all large electric rotating machinery is invariably a *computed value* based on specific no-load (*conventional*) measurements. We shall consider, first,

---

* A possible illustration of the difficulty entailed may be understood by considering the following hypothetical situation. Assume that a 1000 kVA alternator is built in a factory located in a suburban area. It has just been completed and is ready for a test of its relative efficiency. The plant manager asks, "Where can we get a 1 million W load?" His foreman says, "Let's ask the mayors of three or four of the nearest cities to lend us their cities for a few hours so that we can load our alternator."

determining the efficiency of the dc dynamo, followed by consideration of the ac synchronous dynamo and then the induction dynamo, using *conventional* methods in all cases.

12-5. DC DYNAMO EFFICIENCY

Regardless of whether a dc dynamo is to operate as a motor or as a generator, its *rotational* losses may be determined by running it as a motor without any mechanical load (no-load) at its rated speed with an impressed armature voltage (that corresponds to its generated or counter-emf under full load). The electric instrument connections for such a no-load, running-light test are shown in Fig. 12-2a. The dc terminal voltage, V_t, in this case is adjusted to (1) the full-load computed counter-emf, [E_c in Eq. (1-9)], if *motor* efficiency is to be determined; or (2) the full-load generated emf [E_g in Eq. (1-10)] if *generator* efficiency is to be determined.

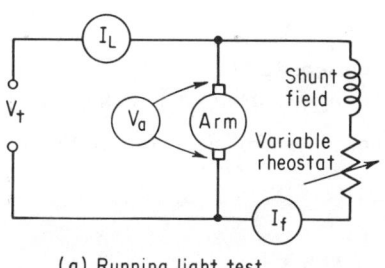

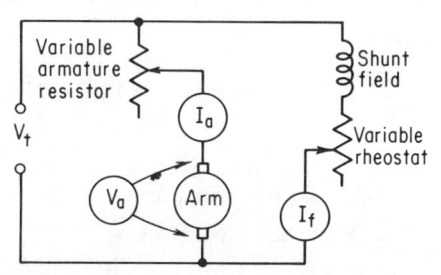

(a) Running light test.

(b) Duplication of load and flux speed.

Figure 12-2

Methods of determining rotational losses of dc dynamos.

Running a dynamo as a motor at no-load means that there is *no* mechanical output power. If the electric input is measured, therefore, and the electric losses computed, the difference between the total electric input and the computed electric loss *must* be the rotational losses of the motor at rated speed, as shown in Fig. 12-1. Stating this (for the dc dynamo as a dc motor) in terms of an equation:

Rotational Losses = Electric Power Input − Electric Losses
= Electric Power Input − (Field Circuit Loss + Combined Armature Circuit Loss)

$$\text{Rotational Losses} = V_a I_L - (V_a I_f + I_a^2 R_a)$$
$$= V_a I_L - V_a I_f - I_a^2 R_a$$
$$= V_a (I_L - I_f) - I_a^2 R_a$$

$$\text{Rotational (Stray Power) Losses} = V_a I_a - I_a^2 R_a \approx V_a I_a \qquad (12\text{-}5)$$

Equation (12-5) is a verification of Fig. 12-1, since it states that the rotational losses of a motor running at no-load (no mechanical output) equal the electric power input to the armature minus the *armature* electric losses ($I_a^2 R_a$). As will be shown in Example 12-1a, the armature electric losses at no-load are so small that they may be neglected, and the rotational losses may thus be assumed equal to $V_a I_a$ as stated in Eq. 12-5.

EXAMPLE 12-1: A 10 kW, 230 V, 1750 rpm shunt generator was run light as a motor to determine its rotational losses at its rated load. The applied voltage across the armature, V_a, computed for the test, was 245 V, and the armature current drawn was 2 A. The field resistance of the generator was 230 ohms and the armature circuit resistance measured 0.2 ohm. Calculate:
a. The rotational (stray power) losses at full load
b. The full-load armature circuit loss and the field loss
c. The generator efficiency at $\frac{1}{4}$, $\frac{1}{2}$, and $\frac{3}{4}$ of the rated load; at the rated load, and at $1\frac{1}{4}$ times the rated load.

Solution:

a. Rotational loss = $V_a I_a - I_a^2 R_a$ [from Eq. 12-5]
$$= (245 \times 2) - (2^2 \times 0.2) = 490 - 0.8 = \mathbf{489.2 \ W}$$

Note that 490 W may be used with neglible error because of negligible electric armature loss.

b. At the rated load,
$$I_L = \frac{W}{V_t} = \frac{10{,}000 \ W}{230 \ V} = \mathbf{43.5 \ A}$$

$$I_a = I_f + I_L = \frac{230 \ V}{230 \ \Omega} + 43.5 = \mathbf{44.5 \ A}$$

The full-load armature loss
$$I_a^2 R_a = (44.5)^2 \times 0.2 = \mathbf{376 \ W}$$

The field loss
$$V_f I_f = 230 \ V \times 1 \ A = \mathbf{230 \ W}$$

c. The efficiency at any load, for a generator, using Eq. (12-2c) is
$$\eta = \frac{\text{Output at that load}}{\text{Output at the load} + \text{Rotational loss} + \text{Electric loss at that load}}$$

Efficiency at $\frac{1}{4}$ load $= \frac{10{,}000/4}{(10{,}000/4) + 489.2 + [(376/16) + 230]} \times 100$
$= \mathbf{77 \ per \ cent}$

Efficiency at $\frac{1}{2}$ load $= \frac{10{,}000/2}{(10{,}000/2) + 489.2 + [(376/4) + 230]} \times 100$
$= \mathbf{86.2 \ per \ cent}$

Efficiency at $\frac{3}{4}$ load $= \frac{10{,}000 \times (3/4)}{[10{,}000(3/4)] + 489.2 + ([376(9/16)] + 230)} \times 100$
$= \mathbf{89 \ per \ cent}$

Efficiency at full load $= \dfrac{10{,}000}{10{,}000 + 489.2 + [376 + 230]} \times 100$

$= 90.1$ **per cent**

Efficiency at $1\frac{1}{4} \times$ the rated load (or $\frac{5}{4}$ rated load)

$= \dfrac{10{,}000 \times (5/4)}{[10{,}000(5/4)] + 489.2 + ([376(25/16)] + 230)} \times 100$

$= 90.6$ **per cent**

It should be noted that the efficiencies appear to increase with load in the above example. It should also be noted that there is a fixed (nonvarying) loss consisting of (1) the field loss of 230 W and (2) the rotational loss of 489.2 W, for a total of 719.2 W. This fixed loss exists even when the generator has an efficiency of zero, i.e., when it is not delivering any current to a load and its output is zero. There is also a variable loss factor, the armature copper loss $I_a^2 R_a$, which varies as the square of the armature current. Even at 125 per cent of the rated load, this variable copper loss component in Example 12-1 is 588 W, and still it is not as large as the fixed total loss of 719.2 W. At what point will maximum efficiency occur?

12-6. MAXIMUM EFFICIENCY

Analysis of Table 12-1 reveals that, for the dc dynamo, the sum of the field loss ($V_f I_f$) and the rotational loss (determined from the running-light test as $V_a I_a$) may be considered as a combined *fixed* loss which does not vary with the load current, I_a. (Example 12-1 ignored the brush and brush-contact loss $V_b I_a$ for the sake of simplification of the problem illustrating the determination of efficiency.) The *variable* loss, then, is the combined armature winding and associated armature current loss $I_a^2 R_a$ and $V_b I_a$, the former varying as the square of the armature current, the latter directly with the armature current.

Generator efficiency may be expressed at any load as*

$$\text{Efficiency, } \eta = \frac{\text{Output}}{\text{Output} + \text{Losses}} = \frac{VI_a}{VI_a + K + V_b I_a + R_a I_a^2}$$

where K is the field loss plus rotational loss or the *fixed* losses.

In order to determine the maximum efficiency, it is necessary to differentiate this expression with respect to I_a and set it equal to zero:

$$\frac{\delta \eta}{\delta I_a} = \frac{(VI_a + K + V_b I_a + R_a I_a^2)V - VI_a(V + V_b + 2I_a R_a)}{(VI_a + K + V_b I_a + R_a I_a^2)^2} = 0$$

* In this equation, the expression VI_a is used in lieu of VI_L in the numerator to create a common current term in both the numerator and the denominator. The difference is negligible, as we shall see.

which yields

$$VI_a + K + V_b I_a + R_a I_a^2 - VI_a - V_b I_a - 2R_a I_a^2 = 0$$

subtracting,
$$K - I_a^2 R_a = 0$$

or

$$K = I_a^2 R_a \tag{12-6}$$

Equation (12-6) thus states that maximum efficiency is obtained when the *fixed* losses, K, *equal* those losses which vary with the square of the load current. Since in most dynamos (whether direct or alternating current) those losses which vary directly with the load current are small (either electric slip-ring or brush losses), we may conclude that maximum efficiency occurs where the *fixed losses equal all the variable losses*. This relation applies equally well to *all* rotating machines, *regardless* of type; it applies to mechanical engines and turbines as well as to all electric dynamos covered in this volume and also to non-rotating devices such as transformers, power amplifiers, power supplies, etc.*

In Example 12-1 it was shown that, at half load, the efficiency was 86.2 per cent and that, at 125 per cent load, the efficiency was 90.6 per cent and still rising. Apparently, the increasing variable losses are not yet equal to the fixed losses even at this load of 125 per cent. Over the range of loads near maximum efficiency, the efficiency does not appear to vary a great deal, so that it is not important to attempt to obtain maximum efficiency at the rated load. Most commercial machines, in fact, exhibit maximum efficiency at somewhat slightly under the rated load. The method used to find that value of the rated load at which maximum efficiency occurs in the dynamo of Example 12-1 is illustrated in Example 12-2.

EXAMPLE 12-2: Using the data given for Example 12-1, calculate
a. the per cent of the rated load at which maximum efficiency will occur
b. the maximum efficiency in per cent
e. the efficiency at 1.5 times the rated load.

Solution:

a. $I_a^2 R_a = K = VI_f + V_a I_a$ [From Eq. 12-6; $K =$ field loss + rotational loss]

$I_a^2 R_a = 230 + (245 \times 2) = 720$ W

$I_a = \sqrt{\dfrac{720}{0.2}} = 60$ A ; $I_L = I_a - I_f = 60 - 1 = 59$ A

Per cent of rated load $= \dfrac{I_L}{I_L \text{ rated}} = \dfrac{59 \text{ A}}{43.5 \text{ A}} = $ **135.5 per cent**

* Maximum efficiency should *not* be confused with maximum electric power transfer, which occurs when the internal and external impedances are equal (at an efficiency of 50 per cent).

b. Maximum efficiency

$$= \frac{230 \times 59}{(230 \times 59) + 720 + 720} \times 100 = \textbf{90.75 per cent}$$

c. Efficiency at 1.5 times the rated load

$$= \frac{10,000 \times (3/2)}{[10,000(3/2)] + 489.2 + [376(9/4)] + 230} \times 100 = \textbf{90.55 per cent}$$

It should be noted that the above relation regarding maximum efficiency is true because the rotational losses were held constant, i.e., the generator or dynamo is assumed as being driven at a *constant* speed. In the case of a constant-speed motor, such as a synchronous motor, or a motor having good speed regulation, such as an induction motor or a shunt motor, the relation still may be used. In the case of *variable-speed* motors, however, it is necessary to plot each of the efficiencies computed versus the load current to determine *graphically* the value of load at which maximum efficiency will occur. The computation of the change in rotational loss is based on the assumption that the loss is a direct function of the change in speed. This is illustrated in Example 12-3.

EXAMPLE 12-3: A 600 V, 150 hp compound motor has a full-load nameplate rating of 205 A and a full-load speed of 1500 rpm. The resistance of the shunt field circuit is 300 ohms; the total armature circuit resistance is 0.05 ohm; and the series field resistance is 0.1 ohm. When it was run light as a motor at the rated speed with an applied voltage, V_a, of 570 V, the armature drew 6 A. The no-load speed of the motor was 1800 rpm. Calculate:
a. The rotational loss at full load and at $\frac{1}{4}, \frac{1}{2}, \frac{3}{4}$, and $1\frac{1}{4}$ times the rated load
b. The full-load variable electric losses, and also the variable electric losses at the loads given in part (a)
c. The motor efficiency at the loads given in part (a).

Solution:

a. Rotational loss $= V_a I_a = 570$ V $\times$ 6 A

$\qquad\qquad\qquad\quad = \textbf{3420 W at 1500 rpm (rated load)}$ \hfill (12-5)

Speed at $\frac{1}{4}$ load $= 1800 - \frac{300}{4} = 1800 - 75 = 1725$ rpm

Rotational loss at 1725 rpm $= \frac{1725}{1500} \times 3420$ W $= \textbf{3930 W}$

Speed at $\frac{1}{2}$ load $= 1800 - \frac{300}{2} = 1650$ rpm

Rotational loss at 1650 rpm $= \frac{1650}{1500} \times 3420$ W $= \textbf{3760 W}$

Speed at $\frac{3}{4}$ load $= 1800 - (\frac{3}{4} \times 300) = 1575$ rpm

Rotational loss at 1575 rpm $= \frac{1575}{1500} \times 3420$ W $= \textbf{3590 W}$

Speed at $\frac{5}{4}$ rated load $= 1800 - (\frac{5}{4} \times 300) = 1425$ rpm

Rotational loss at 1425 rpm $= \frac{1425}{1500} \times 3420$ W $= \textbf{3250 W}$

b. $I_a^2(R_a + R_s) = (203)^2(0.05 + 0.1) = \textbf{6150 W at full load}$
$\qquad\qquad\qquad\qquad\qquad\quad = $ **full load variable loss**

Variable losses at $\frac{1}{4}$ load = 6150 W × $(\frac{1}{4})^2$ = **384 W**

at $\frac{1}{2}$ load = 6150 W × $(\frac{1}{2})^2$ = **1535 W**

at $\frac{3}{4}$ load = 6150 W × $(\frac{3}{4})^2$ = **3450 W**

at $\frac{5}{4}$ load = 6150 W × $(\frac{5}{4})^2$ = **9600 W**

c. The efficiency of the motor = $\dfrac{\text{Input} - \text{Losses}}{\text{Input}}$ (12-2b)

where Input = volts × amperes × load fraction
Losses = field loss + rotational losses + variable electric losses

Input at $\frac{1}{4}$ load = 600 × 205 × $\frac{1}{4}$ = **30,750 W**   (rounded number)

at $\frac{1}{2}$ load = 600 × 205 × $\frac{1}{2}$ = **61,500 W**

at $\frac{3}{4}$ load = 600 × 205 × $\frac{3}{4}$ = **92,250 W**

at $\frac{4}{4}$ load = 600 × 205 × $\frac{4}{4}$ = **123,00 W**

at $\frac{5}{4}$ load = 600 × 205 × $\frac{5}{4}$ = **153,750 W**

Field loss for each of the conditions of load = 600 V × 2 A = 120 W
Rotational losses for each condition were calculated in part (a)
Variable electric losses for each condition were calculated in part (b)

Efficiency at $\dfrac{1}{4}$ load = $\dfrac{30{,}750 - (1200 + 3930 + 384)}{30{,}750}$

= **0.826 or 82.6 per cent**

at $\dfrac{1}{2}$ load = $\dfrac{61{,}500 - (1200 + 3760 + 1535)}{61{,}500}$

= **0.894 ro 89.4 per cent**

at $\dfrac{3}{4}$ load = $\dfrac{92{,}250 - (1200 + 3590 + 3450)}{92{,}250}$

= **0.912 or 91.2 per cent**

at $\dfrac{4}{4}$ load = $\dfrac{123{,}000 - (1200 + 3420 + 6150)}{123{,}000}$

= **0.9125 or 91.25 per cent**

at $\dfrac{5}{4}$ load = $\dfrac{153{,}750 - (1200 + 3250 + 9600)}{153{,}750}$

= **0.909 or 90.9 per cent**

These results are tabulated on page 475.

Example 12-3 indicates that, although the field loss is substantially constant, the rotational loss is decreasing in the same proportion as the speed. At the same time, the variable loss is increasing as the square of the armature current. In order to determine (with some exactness) the point at which maximum efficiency occurs, it would be necessary to choose *several* values of I_a immediately above and below the full-load point and to plot the resultant data of efficiency versus armature current graphically, at which maximum efficiency occurs. The value of I_a at which maximum efficiency occurs may then be evaluated graphically.

LOSSES, EXPRESSED IN WATTS					
Item	AT ¼ LOAD	AT ½ LOAD	AT ¾ LOAD	AT FULL LOAD	AT 5/4 LOAD
Input	30,750	61,500	92,250	123,000	153,750
Field loss	1200	1200	1200	1200	1200
Rotational losses, from part (a)	3930	3760	3590	3420	3250
Variable electric losses. from part (b)	384	1535	3450	6150	9600
Total of all losses	5514	6495	8240	10,770	14,050
η, Efficiency in per cent	82.6 per cent	89.4 per cent	91.2 per cent	91.25 per cent	90.9 per cent

An alternative method is a graphical plot of both rotational loss and variable electric losses as ordinates versus I_a as abscissa. The point at which these losses cross, exactly reveals the value of I_a graphically.

Examples 12-1 and 12-3 indicate that the data obtained from the running-light test (in which the dc dynamo is run light as a motor) may be used for the determination of *both* generator and motor efficiency. The calculations for the motor efficiency are a little more complex because of the change in speed.

12-7. DUPLICATION OF FLUX AND SPEED

An analysis of the losses shown in Table 12-1 will indicate that an assumption was made in Examples 12-1 and 12-3 that is not altogether correct with regard to rotational loss, namely, that the rotational loss varies only with speed if the machine excitation (field current) is held constant. With increased dynamo load, there is increased armature reaction, producing a change in flux density which affects the core losses. At the same time, there is also a change in the generated emf or counter emf (upon which the mechanical power developed by the armature depends) with increased load.

The generated emf or counter emf varies *directly* with the *flux* and the *speed*, and the rotational losses *also* vary directly with the *flux* and the *speed*. Therefore, for a dynamo whose speed varies (Example 12-3) or whose generated emf may vary (Example 12-1), it would be best to repeat the running-light test under several running conditions which duplicate the dynamo flux and speed conditions. As shown in Fig. 12-2b, a series resistance is connected in the armature circuit for the purpose of reducing the armature voltage of the dynamo to the calculated value of the generated emf or counter emf at the required (or reduced) speed.

Because the test is performed at no-load, the armature voltage drop is very small, and the applied voltage across the armature V_a may be taken as the generated emf or counter emf for any given condition. At any value of preset speed, therefore, the *rotational* loss is equal to the reading of $V_a \times I_a$ minus the small no-load copper loss which may also be neglected. Thus, by duplicating counter emf, the speed and flux conditions are also duplicated.

Example 12-4 illustrates the method for motor efficiency, but it may be used for generator efficiency as well.

EXAMPLE 12-4: A 600 V, 150 hp compound motor has a full-load speed of 1500 rpm and a full-load current of 205 A. The resistance of the shunt field is 300 ohms, the armature circuit resistance is 0.05 ohm, and the series field resistance is 0.1 ohm. At the rated load, calculate
a. the counter emf to be impressed across the armature when run light under the same conditions of flux and speed
b. the rotational loss if the armature current is 6 A when the appropriate armature voltage is impressed at 1500 rpm.

Solution:

At full load, $I_a = I_L - I_f = 205 \text{ A} - \dfrac{600 \text{ V}}{300 \text{ }\Omega} = 203 \text{ A}$

a. At full load, $E_c = V_t - I_a(R_a + R_s) = 600 - 203(0.15)$
$= 600 - 30.5 = \mathbf{569.5 \text{ V}}$

b. $P_{\text{loss}} = V_a I_a = 569.5 \times 6 = \mathbf{3410 \text{ W}} = \text{rotational loss}$

Note that the value of rotational loss obtained in (b) of Example 12-4 compares quite favorably with the value obtained at rated load in (a) of Example 12-3.

12-8. AC SYNCHRONOUS DYNAMO EFFICIENCY

Essentially, the only real difference between a synchronous alternator and a dc generator is the fact that, in the former, the armature is stationary and the field is rotating at a constant speed. The alternator armature effective (ac) resistance per phase is normally obtained in the same manner as was used in the synchronous impedance method for determining the alternator regulation (Section 6-10, Fig. 6-7a) from dc resistance measurements. The dc field circuit copper loss $V_f I_f$ is also determined by dc measurement.

As in the case of a dc dynamo, regardless of whether synchronous motor or alternator efficiency is to be determined, the ac synchronous dynamo is run as a synchronous motor without load at synchronous speed (the conventional running-light method). The field current is usually adjusted to the nameplate value corresponding to the power factor at which normal operation occurs or, in the case of a synchronous motor, to provide minimum current (unity power factor). Balanced three-phase armature currents at the rated line voltage and input power (using the method of either one, two, or three wattmeters) are recorded as shown in Fig. 12-3a. The rotational losses, as in the case of the dc

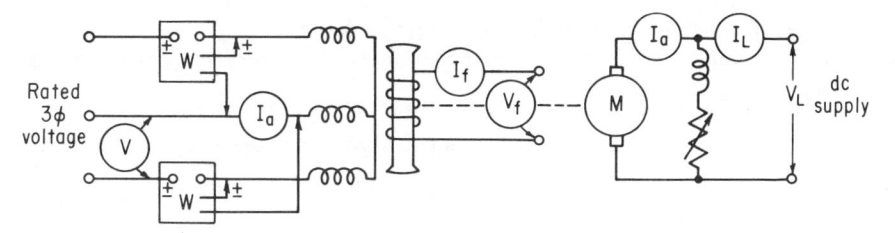

(a) Running light test of a synchronous dynamo.   (b) Calibrated dc dynamo method.

Figure 12-3

Methods of determining efficiencies of ac synchronous dynamos.

dynamo, are equal to the armature input minus the no-load copper losses, or

ac Synchronous Dynamo Rotational Loss (P_r)
= No-Load Armature Power Input — Armature Copper Loss

$$P_r = \sqrt{3} \times V_a I_a \cos \theta - 3I_a^2 R_a \qquad (12\text{-}7)$$

where I_a is the phase or line armature current and R_a is the effective armature resistance per phase.

Since both a synchronous motor and a synchronous alternator are operated at constant speed at a fixed frequency, the rotational loss may be considered constant. The full-load efficiency at unity or any other power factor is then determined by computation as in Example 12-5.

EXAMPLE 12-5: The three-phase, wye-connected, alternator tested by the synchronous impedance method in Example 6-4 is run light as a synchronous motor at its rated line voltage to determine its rotational losses. The no-load armature current is 8 A and the power input to the armature is 6 kW. An open-circuit line voltage of 1350 V is obtained from the alternator with a dc field excitation of 18 A at 125-V dc. Assuming that the core loss and dc excitation are unchanged from no-load to the rated load, calculate

a. rotational loss of the synchronous dynamo
b. field copper loss
c. electrical armature losses at $\frac{1}{4}, \frac{1}{2}, \frac{3}{4}$, and full load
d. efficiency at these loadings at a power factor of 0.9 lagging.

Solution:

From Example 6-4,

$$R_a = 0.45 \ \Omega/\text{phase}; \quad I_{a(f.l.)} = 52.5 \ \text{A}$$

a. From Eq. (12-7), $P_r = 6000 - (3 \times 8^2 \times 0.45)$
$= 6000 - 86.4 = $ **5914 W**   (rotational loss)

b. Field loss $= 125 \text{ V} \times 18 \text{ A} = $ **2250 W**

477

SEC. 12-8. / AC Synchronous Dynamo Efficiency

c. Full-load electric armature (copper) loss

$$= 3I_{a(\text{f.l.})}^2 R_a = 3 \times (52.5)^2 \times 0.45 = \textbf{3725 W}$$

Armature copper loss at $\frac{1}{4}$ load $= \frac{3725}{16} = \textbf{233 W}$

at $\frac{1}{2}$ load $= \frac{3725}{4} = \textbf{932 W}$

at $\frac{3}{4}$ load $= 3725 \times \frac{9}{16} = \textbf{2100 W}$

d. Efficiency in percentage

$$= \frac{(\text{Rated output times the load})}{(\text{Rated output times the load}) \text{ plus losses}} \times 100$$

Note: The output is given as 100 kVA, which is equal to 100,000 VA and, at a power factor of 0.9 lagging, the output becomes 90,000 watts. That quantity is used in the calculations below. The rotational loss was determined in part (a) to be 5914 W, and the field loss was determined in part (b) to be 2250 W. The variable electric armature (copper) losses were determined in part (c). In calculating the efficiency at various loads, the total of these losses appears in the denominator in each instance below. Per cent efficiency, η

at $\frac{1}{4}$ load $= \dfrac{(100{,}000 \times 0.9) \times (1/4)}{[(100{,}000 \times 0.9) \times (1/4)] + (5914 + 2250) + 233} \times 100$

$= \dfrac{90{,}000 \times (1/4)}{[90{,}000 \times (1/4)] + 8164 + 233} \times 100 = \dfrac{22{,}500}{30{,}897} \times 100$

$= \textbf{72.7 per cent}$

at $\frac{1}{2}$ load $= \dfrac{90{,}000 \times (1/2)}{[90{,}000 \times (1/2)] + 8164 + 932} \times 100 = \textbf{83.2 per cent}$

at $\frac{3}{4}$ load $= \dfrac{90{,}000 \times (3/4)}{[90{,}000 \times (3/4)] + 8164 + 2100} \times 100 = \textbf{86.8 per cent}$

at full load or $\frac{4}{4}$ load

$$\eta = \dfrac{90{,}000 \times (4/4)}{[90{,}000 \times (4/4)] + 8164 + 3725} \times 100 = \textbf{88.25 per cent}$$

12-9. VENTILATION OF ALTERNATORS

A portion of the rotational loss in Example 12-5 is the windage created by the rotating field and the ventilating fan located on the shaft of the alternator. Air cooling by means of internal fans is generally inadequate for alternators of larger sizes, however, and some method of sealed and forced ventilation is required to: (1) carry off the heat generated with a reasonable rise of temperature of both the alternator and the gaseous coolant employed; (2) provide a gaseous coolant with a lower windage loss and a possibly higher specific heat than air; (3) seal the alternator cooling system to keep out dirt and moisture (these alien elements would shorten the life of the alternator); (4) increase efficiency and (5) increase the alternator capacity.

In modern turbo alternators, which are driven at high speeds and which use sealed forced air-cooling, half or more of the total rotational losses at full load (see Example 12-5) are often the result of forcing air through the axial ducts provided in the rotor and the stator of the armature. Hydrogen has less viscosity than air, with about eight times the heat conductivity and about the same heat capacity per unit volume as air, so that cooling with a given flow of hydrogen gas compared to air (1) is more efficient in lowering the temperature of the alternator and (2) requires less windage loss in circulating the coolant. Two additional advantages of the use of hydrogen are (3) no oxidizing effect on insulation because of corona is produced with hydrogen and (4) a higher potential is required to produce corona in a hydrogen-air atmosphere than in an ordinary air atmosphere.

Some hydrogen-air mixtures are extremely explosive, however, and experiments show that explosions may be triggered within a range of 6 per cent hydrogen and 94 per cent air up to 71 per cent hydrogen and 29 per cent air. When there is more than 71 per cent hydrogen, the amount of oxygen in the gas is insufficient to support combustion. As a result, mixtures using 90 per cent hydrogen are employed, and there is no danger of explosion, even at high ignition temperatures.

A completely sealed system is required where hydrogen cooling is employed. The hydrogen is circulated by blowers and fans through the rotor and stator, and then passed over cooling coils inside the sealed casing; the coils contain a coolant—usually oil or water—to conduct the heat away from the circulating hydrogen. The gas is maintained at a pressure above atmospheric to prevent inward seepage of contaminating air, and the pressure is carefully metered to avoid and discover leaks.

Hydrogen cooling increases the over-all full-load efficiency by about 1 per cent, but it increases the alternator *capacity* by about 25 per cent. The later reason is the primary factor which justifies its use.

12-10. AC SYNCHRONOUS DYNAMO EFFICIENCY BY THE CALIBRATED DC MOTOR METHOD

A procedure recommended by the AIEE (now IEEE) for determining synchronous ac alternator (or motor) efficiency and alternator voltage regulation simultaneously is illustrated in Fig. 12-3b, in which a dc motor is coupled to the alternator, and described in the following steps. The motor is a *calibrated motor* whose efficiency is known over its entire no-load to full-load range.

Procedural Step	Purpose
1. Drive the alternator at synchronous speed without field excitation by means of the dc calibrated motor.	1. dc motor input times its (known) efficiency is the alternator input, which represents alternator friction and windage loss.

2. *Repeat step 1, but excite the alternator field at normal excitation, i.e., open-circuit excitation which will produce the rated voltage at the rated load.

2. The added motor input times its efficiency represents added alternator input or core loss (eddy current and hysteresis). The dc field copper loss of the alternator is also obtained (i.e., $V_f I_f$). The brush loss is also included. All these losses are fixed losses.

3. Reduce the field excitation to zero, short-circuit the armature of the alternator, and perform the synchronous impedance short-circuit test (Section 6-10), i.e., raise the field current until the rated armature current is produced at the rated speed.

3. The core loss is considered negligible because the excitation is so low. The motor input times its efficiency now represents the full-load armature copper loss plus friction and windage loss (step 1 above). Step 3 minus step 1 is the full-load armature copper loss.

4. Remove the short-circuit across the armature, and measure the open circuit armature voltage at this excitation (open-circuit test).

4. This yields the armature impedance and synchronous reactance per phase (the latter by computation in the usual manner).

The obvious advantage of the AIEE method, using the calibrated dc motor, is that ac synchronous alternator (or motor) efficiency and voltage regulation (by the synchronous impedance method) are simultaneously determined. Given the dc field loss, the full-load armature copper loss, and the rotational losses (friction, windage, and core loss) at the rated speed, the determination of efficiency is computed in the same way as illustrated in Example 12-5.

12-11. ASYNCHRONOUS INDUCTION DYNAMO EFFICIENCY

The induction dynamo, whether operated as an induction motor or as an induction generator, experiences a change in rotor speed with load as well as a change in the rotor frequency resulting from the speed change, as discussed in Chapter 9.

Unlike the ac synchronous machines, neither the rotational losses (which are functions of speed and frequency) nor the stator and rotor electrical losses (which are functions of load) are constant. Other methods must be employed, therefore, to determine the efficiency of ac synchronous induction machines. Since the efficiency of induction generators is *rarely* a matter of interest, the following dis-

---

* If it is desired to measure brush friction loss at the alternator's slip-rings and exciter brushes, these are then lowered, without excitation voltage, and the calibrated motor imput is again measured. The increase in imput over step 1 is the brush friction loss at synchronous speed.

cussion will be confined primarily to the ac asynchronous induction motor, both polyphase and single phase.

Two methods are generally employed. The first is a *conventional* method, in that it does not involve loading of the induction motor, called the *open-circuit* and *short-circuit* (blocked rotor) method. This method is generally employed on *extremely large* induction motors where loading is impractical, inconvenient, or uneconomical.

The second method is the AIEE load-slip equivalent-circuit method. This method is generally more accurate than the conventional blocked-rotor method, but it requires measurement of the slip at various loads from no-load to full- (rated) induction motor load. It is usually performed on smaller induction motors which may be loaded by prony brakes or electric generators. The latter statement suggests, of course, that, if loading is possible, a calibrated generator may be used to determine the efficiency. This is evidently a third procedure, and the efficiency may be electrically recorded as the ratio of the calibrated generator input (motor output) to electric motor input at any given load. As previously stated, the efficiency of small polyphase induction motors may also be determined (a fourth method) by means of a dynamometer or prony brake measurements to record their output at any given mechanical load, and the input is recorded electrically. The last three methods are only practicable with smaller induction motors. The first method is applicable to all induction motors, large or small.

We will consider only the conventional method (Sec. 12-13) and the AIEE method (Sec. 12-14) in detail.

12-12. EQUIVALENT RESISTANCE OF INDUCTION MOTOR

Both of the above two tests for the determination of induction motor efficiency require an expression of the equivalent resistance between lines of both the induction motor rotor and stator, referred to the stator, under blocked-rotor conditions. It is necessary, therefore, to derive this expression. Figure 12-4a shows a delta-connected induction motor stator and rotor at standstill. At the instant of starting, or with the rotor locked, the delta-connected stator may be considered the primary of a transformer, and the short-circuited rotor, the secondary. The total equivalent resistance between lines, R_{el}, referred to any two stator terminals in Fig. 12-4a, is, by parallel circuit theory:

$$R_{el} = \frac{R_a \times 2R_a}{R_a + 2R_a} = \frac{2}{3} R_a \tag{12-8}$$

and also

$$R_a = \frac{3}{2} R_{el} \tag{12-8}$$

If the rotor is locked, an excitation voltage of less than 10 per cent of

(a) Delta connected. (b) Wye connected.

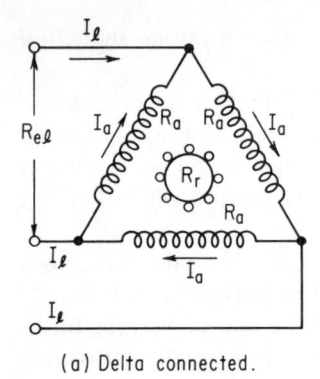

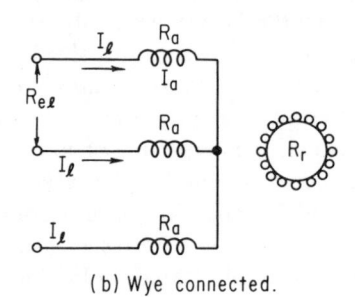

Figure 12-4

Determination of equivalent resistance of stator (and rotor) between lines of stator.

the rated voltage may be applied to the stator to develop the rated armature current, I_a, in the stator as a result of the rated load current in the rotor resistance R_r. At this reduced excitation, the core losses are negligible, $W_c \times B^2 \times (\frac{1}{10})^2$, as they are less than one one-hundredth of their rated voltage value.*

The stator power input represents, at standstill, only the equivalent copper loss of the stator and rotor combined, P_c, in Fig. 12-4a, or

$$P_c = 3I_a^2 R_a = 3\left(\frac{I_l}{\sqrt{3}}\right)^2 R_a = I_l^2 R_a$$

Substituting Eq. (12-8) for R_a in this equation yields

$$P_c = \tfrac{3}{2} I_l^2 R_{el} \tag{12-9}$$

where I_l is the current in the line of a three-phase induction motor and R_{el} is the total equivalent resistance between lines of a three-phase induction motor (representing the combined rotor and stator resistance), referred to the stator.

Similarly, assuming that the stator of the induction motor is wye-connected, as shown in Fig. 12-4b:

$$R_{el} = 2R_a \quad \text{and} \quad R_a = \frac{R_{el}}{2} \quad \text{(for a wye-connected stator)}$$

The input power or equivalent copper loss, for a wye-connected stator, at

---

* In some instances it may be necessary to use higher (over 10%) excitation voltages to obtain rated stator armature current. In such circumstances, the core losses may *not* be neglected. Under such conditions it is customary to use excitation voltages which produce half-rated current and copper loss and to modify Eqs. (12-9) and (12-9a) accordingly, in determining R_{el} and P_c.

reduced voltage is

$$P_c = 3I_a^2 R_a = 3I_l^2 R_a = \tfrac{3}{2} I_l^2 R_{el} \qquad (12\text{-}9a)$$

Note that this expression is the *same* as Eq. (12-9) for a delta-connected stator. It is therefore *completely unnecessary* to know whether the stator is wye or delta connected. The equivalent copper loss may be measured between lines, and the total equivalent resistance between lines may be determined from Eq. (12-9) for the combined total rotor and stator resistance, referred to the stator. This will be illustrated in Example 12-6.

12-13. INDUCTION MOTOR EFFICIENCY FROM OPEN-CIRCUIT AND SHORT-CIRCUIT (LOCKED-ROTOR) TESTS

As stated in Chapter 9, the circuit of an induction motor running under load (and at the instant of starting, as well) may be represented as a transformer [see Fig. 9-11 and Eqs. (9-9a) and (9-16)]. The conventional methods of determining transformer efficiency,* using the open-circuit (no-load) and short-circuit (locked-rotor) transformer tests, apply equally well to the induction motor. As in the transformer*, the efficiency determination is performed in two steps:

12-13.1 Open-Circuit, No-Load Test

The induction motor is connected to a line supply at its rated voltage and is run light without (any) load coupled to its shaft. Under these conditions, as in the case of previous "running-light" tests, the input to the induction motor stator armature represents (1) the rotational losses (core and mechanical losses), and (2) a small no-load equivalent stator and rotor copper loss. (The latter is not negligible as shown in Ex. 12-6b).

12-13.2 Short-Circuit, Locked-Rotor Test

The motor is disconnected, and its rotor is securely locked to prevent it from rotating. A low and gradually increasing three-phase voltage is applied (from either a three-phase variac or a polyphase induction regulator) to the stator until the rated nameplate line current flows. As in the transformer short-circuit test, and for reasons demonstrated in Section 12-11, the core (iron) losses are negligible, and there are no mechanical losses since the motor is stationary. The total power drawn by the motor thus represents the full-load electric copper losses of the stator and rotor. The equivalent total resistance (between lines) of the motor is computed from Eq. (12-9). This value is then used in the computation of rotational loss from the previous no-load test, as in the determination of efficiency.

Example 12-6 illustrates the treatment of data and the determination of efficiency by this method.

* See Sec. 13-8.

EXAMPLE 12-6: A three-phase, 60 Hz, 220 V, 5 hp, 0.9 power-factor, induction motor has a nameplate rated line current of 16 A and a speed of 1750 rpm. The data obtained by open- and short-circuit tests is:

	Open-Circuit Test	Short-Circuit Test
Line Current	6.5 A	16 A
Line Voltage	220 V	50 V
Polyphase Wattmeter	300 W	800 W (full-load equivalent copper losses)

Calculate:

a. The equivalent total resistance of the induction motor between lines.
b. The rotational losses.
c. The equivalent copper loss at $\frac{1}{4}$, $\frac{1}{2}$, $\frac{3}{4}$, and $1\frac{1}{4}$ rated load.
d. The efficiency at these load points.
e. The output hp at these load points.
f. The output torque at full load.

Solution:

a. $R_{el} = \dfrac{P_c}{(3/2)I_l^2} = \dfrac{800}{16^2} \times \dfrac{2}{3} = \mathbf{2.08\ \Omega}$

b. Rotational losses $= P_r - I_l^2 R_{el} = 300 - 6.5^2 \times 2.08 = 300 - 132$
$= \mathbf{168\ W}$

c. Equivalent copper loss at various load fractions

At $\frac{1}{4}$ load $= 800\ W \times (\frac{1}{4})^2 = \mathbf{50\ W}$

At $\frac{1}{2}$ load $= 800 \times (\frac{1}{2})^2 = \mathbf{200\ W}$

At $\frac{3}{4}$ load $= 800 \times (\frac{3}{4})^2 = \mathbf{450\ W}$

At full load $= \mathbf{800\ W}$ as given in short-circuit test data

At $1\frac{1}{4}$ load $= 800 \times (\frac{5}{4})^2 = \mathbf{1250\ W}$

d. Efficiency in percentage, for a motor is, by Eq. (12-2b)

$$\eta = \dfrac{\text{(Full-load input times fraction of load) minus losses}}{\text{(Full-load input times fraction of load)}} \times 100$$

The full-load input $= \sqrt{3} \times 220 \times 16 \times 0.9 = 5480$ W.

The rotational losses, from part (b) $= 168$ W.

The equivalent copper losses were computed in part (c).

Per cent efficiency

At $\dfrac{1}{4}$ load $= \dfrac{(5480/4) - (168 + 50)}{(5480/4)} \times 100 = \dfrac{1370 - 218}{1730} \times 100$
$= \mathbf{84.2\ per\ cent}$

At $\dfrac{1}{2}$ load $= \dfrac{(5480/2) - (168 + 200)}{(5480/2)} \times 100 = \mathbf{86.5\ per\ cent}$

At $\dfrac{3}{4}$ load $= \dfrac{[5480(3/4)] - (168 + 450)}{5480(3/4)} \times 100 = \mathbf{84.9\ per\ cent}$

$$\text{At } \tfrac{4}{4} \text{ load or full load} = \frac{[5480(4/4)] - (168 + 800)}{5480(4/4)} \times 100$$
$$= \mathbf{82.1 \text{ per cent}}$$

$$\text{At } \tfrac{5}{4} \text{ load} = \frac{[5480(5/4)] - (168 + 1250)}{5480(5/4)} \times 100 = \mathbf{79.3 \text{ per cent}}$$

e. Output in horsepower $= \dfrac{\text{Input} - \text{losses}}{746 \text{ W/hp}}$

Output in horsepower

$$\text{At } \tfrac{1}{4} \text{ load} = \frac{(5480/4) - 218}{746} = \mathbf{1.545 \text{ hp}}$$

$$\text{At } \tfrac{1}{2} \text{ load} = \frac{(5480/2) - 368}{746} = \mathbf{3.18 \text{ hp}}$$

$$\text{At } \tfrac{3}{4} \text{ load} = \frac{[5480 \times (3/4)] - 618}{764} = \mathbf{4.68 \text{ hp}}$$

$$\text{At } \tfrac{4}{4} \text{ load} = \frac{[5480 \times (4/4)] - 968}{746} = \mathbf{6.04 \text{ hp}}$$

$$\text{At } \tfrac{5}{4} \text{ load} = \frac{[5480 \times (5/4)] - 1418}{746} = \mathbf{7.28 \text{ hp}}$$

f. Output torque $= \dfrac{\text{Horsepower} \times 5252}{\text{Speed}}$

$$\text{At full load, output torque } T = \frac{6.04 \times 5252}{1750} = \mathbf{18.1 \text{ lb-ft}}$$

Several assumptions have been made in the determination of induction motor efficiency by the open- and short-circuit locked-rotor method shown in Example 12-6. As shown in the solution, a *constant* rotational loss was assumed at all load points. Table 12-1 indicates that mechanical losses such as bearing and windage are a function of speed. In addition, the iron core losses are a function of both speed and (in part) rotor frequency which increases with slip. The slightly increased frequency tends to counterbalance the decrease in speed, so that this assumption is justified. Furthermore, since the full-load slip rarely exceeds 5 per cent, and since the rotational loss is a small portion of the total loss as load increases, this error introduced in efficiency is not very significant.

The locked-rotor test, in addition, assumed negligible core loss. The core loss varies with the excitation voltage impressed on the stator under locked-rotor conditions. If the excitation voltage is a small per cent of the rated voltage, i.e., less than 10 per cent, the assumption is justified. But some induction motors, because of leakage and high standstill reactance, require as much as 33 per cent of the rated voltage in order to produce the rated line current to the stator. Under these circumstances, the locked-rotor core losses are not $\tfrac{1}{100}$ of the rated excitation loss, but $\tfrac{1}{9}$ of the rated excitation loss. The value is no longer negligible, and the data of the short-circuit locked-rotor test should be compensated accordingly as noted in the footnote to Eq. (12-8).

Finally, the locked-rotor test was performed at standstill at line frequency to determine the total equivalent ac effective resistance of the rotor and stator combined. This is not the frequency to which the rotor will be subjected at rated load and slip, and the rotor resistance component of the total equivalent resistance, R_{el}, may be somewhat high and the computed copper losses correspondingly high. This also yields a somewhat fallacious value of rotational loss from the no-load data. For extremely large induction motors, usually tested by this conventional no-load method, the efficiency thus computed is a *pessimistic* value, since *each* of the preceding assumptions yields a *lower* value of efficiency (from 2 to 3 per cent less) than really occurs under conditions of actual loading. The pessimistic value is preferred because the manufacturer can state that in actual use the motor will have a higher efficiency.

12-14. INDUCTION MOTOR EFFICIENCY FROM AIEE LOAD-SLIP EQUIVALENT-CIRCUIT METHOD

This method, recommended by the AIEE as the standard to be used in testing induction motors, yields a calculated value of efficiency which is somewhat closer to that under actual load conditions. The fundamental concept underlying this method stems from Eq. (9-9a) and Fig. 9-11 for the equivalent circuit of the rotor under running conditions and the discussion of Section 9-12. A fundamental relationship was derived, namely, that the rotor power developed, per phase, is equal to the rotor power input per phase less rotor copper loss per phase, or

$$\text{RPD} = \text{RPI} - \text{RCL}$$
$$(I_r^2 R_r)\frac{(1-s)}{s} = \frac{I_r^2 R_r}{s} - I_r^2 R_r \qquad (9\text{-}17)$$

From the above relationship, it may be seen that

$$\text{rotor power input (per phase and total)} = \frac{I_r^2 R_r}{s} = \frac{\text{rotor copper loss}}{\text{slip}}$$

from which

$$\text{rotor copper loss} = \text{slip} \times \text{rotor power input (per phase and total)}$$
$$(9\text{-}17a)$$

The significance of the relationship expressed in Eq. (9-17a) is that it is now possible to distinguish between and separate the rotor copper loss from the stator copper loss. The rotor copper loss is considered as varying with slip, whereas the stator copper loss varies with the stator current. The measurement of stator armature resistance, *per se*, is the same as that employed in the stator measurements of the ac synchronous alternator (Section 6-10). A measurement is made of the dc resistance between any two lines so as not to yield false results from induced effective resistance in the rotor, and a multiplying factor is used for the ac stator resistance between lines.

As in the case of the previous method (Section 12-12), a no-load, running-light test is made at the rated excitation; from the results of this test, the rotational loss is computed by subtracting from the input the stator copper loss. A load is then coupled to the shaft of the induction motor, and a complete load run is made under actual load conditions in which the input power, stator current, and slip are determined for each load step. The efficiency is then computed as indicated below, steps 1 to 5.

Thus, the AIEE method consists of three *tests*: (**1**) a total stator resistance measurement; (**2**) a running-light rotational loss measurement; and (**3**) direct loading to determine slip and the input power and stator current at each value of slip.

The load test data is treated in the following manner in order to determine efficiency:

1. The stator copper loss is calculated at each particular value of stator current and slip, using Eq. (12-9) (where R_{el} is now the value of stator resistance between lines determined by the stator resistance measurement). Note: $R_{E2} \cong 1.25 \ R_{dc}$ (between lines).
2. The stator copper loss is then subtracted from the metered input power at that particular value of slip to yield the rotor power input.
3. The rotor power input is multiplied by the value of slip in accordance with Eq. (9-17a) to yield the rotor copper loss at that particular load value; $RCL = s \times RPI$.
4. The total rotor loss at the particular load value is then computed as the sum of the rotational loss (test 2), and the rotor copper loss.(Step 3 above).
5. The rotor mechanical power output is then computed as the difference between the rotor power input (step 2) and the total loss (step 4) in accordance with Fig. 12-1, for any dynamo.

The AIEE method of efficiency determination may be illustrated for the motor of Example 12-6, using the additional data given in Example 12-7.

EXAMPLE 12-7: A dc resistance test of the motor of Example 12-6 yielded a resistance of 1 ohm between lines. When operated at the rated nameplate line current at rated load and rated speed, the power drawn from the 220-V supply was 5.5 kW, as recorded by the polyphase wattmeter of an industrial analyzer. Calculate:
a. Full-load efficiency.
b. The output horsepower and the torque at full load.
c. Compare the results obtained with those of Example 12-6.

Solution:

Preliminary calculations

$$R_{el} = 1.25 R_{dc} = 1.25 \times 1.0 \text{ ohm} = 1.25 \text{ ohms}$$

The rotational loss is equal to the no-load input less the no-load stator loss (cf. the open-circuit test, Example 12-6), and thus the rotational loss is equal to

$$300 \text{ W} - [\tfrac{3}{2} \times (6.5)^2 \times 1.25] = 300 - 79.1 = 220.9 \text{ W}$$

(Compare this rotational loss with that of Example 12-6.)
The full-load stator copper loss is $\tfrac{3}{2} \times 16^2 \times 1.25 = 480$ W
The rotor power input is equal to the total stator input at full load less the full-load stator copper loss, or $5500 - 480 = 5020$ W. (9-17)
The full-load slip is $(1800 - 1750)/1800 = 0.0278$.
The rotor copper loss [see Eq. (9-17a)] is equal to the slip times the rotor power input, namely, $0.0278 \times 5020 = 138.5$ W.
The total rotor loss, then, is the rotational loss plus the rotor copper loss, or $220.9 + 138.5 = 359.4$ W.
The rotor power output at full load is the rotor power input less the total rotor losses, or $5020 - 359.4 = 4660.6$ W.

a. Per cent efficiency $= \dfrac{\text{Output}}{\text{Input}} \times 100 = \dfrac{4661}{5500} \times 100$

$= $ **84.6 per cent**, at full load

b. Horsepower at full load $= \dfrac{4661}{745} = $ **6.25 hp**

Torque $T = \dfrac{\text{hp} \times 5252}{S} = \dfrac{6.25 \times 5252}{1750} = $ **18.7 lb-ft**

c. The method used in Example 12-6 gave a result of 6.04 hp at full load, compared with 6.25 hp by Example 12-7; and a torque of 18.1 lb-ft, compared with 18.7 lb-ft obtained in Example 12-7.

Note that the more realistic conditions of the AIEE method yield a higher value of efficiency (i.e., less pessimistic) and, for this reason, a correspondingly higher torque and output horsepower value, as well, at the rated load. It goes without saying that *if* the induction motor *can* be physically loaded, its efficiency should be determined using the more accurate and realistic AIEE method. On the other hand, the slightly pessimistic efficiencies obtained using the locked-rotor test, like the pessimistic regulation of an alternator by the synchronous impedance method, may be preferred in view of the ease of performance, the simplicity of calculation, and the confident knowledge that, under actual load conditions, the machine performance will surpass the calculated performance. Finally, the AIEE method requires direct loading and does not lend itself to testing of larger induction motors.

**12-15.
EFFICIENCY OF
SINGLE-PHASE
MOTORS**

The efficiency of *fractional* horsepower single-phase motors is usually determined by any of the following methods.

1. Small dynamometer test stands in which the motor is coupled to a resistance-loaded dc dynamometer generator whose stator is cradle mounted in trunions. A torque arm is welded to the dynamometer stator, restraining its rotation through a strain-gage load-cell or Chatillon-type scale for measuring the force or torque developed by the dynamometer. This method uses *direct* loading.
2. Calibrated generators (of known efficiency) may also be used to measure the relative efficiencies of fractional horsepower single-phase motors in the same manner as described in Section 12-10. This method also uses direct loading.
3. Integral horsepower single-phase motors may be tested by the conventional locked-rotor method. The technique is somewhat easier because of the relative simplicity of single-phase calculations, and no special equipment is needed.
4. The direct-loading AIEE method described may also be used if a more precise determination of efficiency is warranted (Section 12-14).
5. A prony brake is sometimes used in (1) above in lieu of the dynamometer generator with a torque arm and scale to read output hp by direct loading.

**12-16.
FACTORS
AFFECTING
RATINGS OF
MACHINES**

As stated previously, electric machines are rated in terms of their *output* capacities. Generators and alternators are rated in terms of their output kilowatt capacity (kW) or kilovolt amperes (kVA) at a given rated prime-mover speed and a rated terminal voltage. Motors (dc and ac) are rated in terms of output capacity in **shaft horsepower** at rated speed, full-load current, and applied voltage. When the electric machines are operated under these nameplate conditions, there is an implication that the temperature rise will not be excessive and that the machines will not overheat. While the manufacturer is aware that temporary overloads may be sustained, the rotating dynamos are not expected to carry sustained overloads for long periods. The consumer who, for reasons of economy, purchases a 10-hp motor to drive a 12- or 15-hp load continuously, runs the risk of *caveat emptor* in purchasing a product which (1) will deliver rated speed at rated load but not rated speed at an overload; (2) will overheat badly and (as a result) have a generally shorter life; and (3) will operate at a lower efficiency at the overload for the duration of its life. Thus, the lower initial cost is offset by poorer and more expensive running performance, coupled with the necessity for earlier replacement. For this reason, therefore, an indication is provided on all (electric rotating machinery) nameplates of the allowable temperature rise and duty cycle, as well as rated voltage, current, frequency, and speed.

12-17. TEMPERATURE RISE

The standard allowable temperature rise of currently manufactured electric machinery is 40°C above ambient temperature.

Thus, if the room temperature is 70°F or 21°C, a 40°C rise in temperature implies that the motor temperature may be as high as 61°C or 142°F. It goes without saying that this matter of "allowable" temperature rise cannot be carried to a compulsive absurdity, however. A motor located in a confined area next to a high-temperature device, such as a boiler or heater, may have an "ambient" temperature* of 140°F or 60°C. A 40°C rise over such an "ambient" temperature brings the motor temperature to 100°C, or the boiling point of water. Furthermore, even when the motor is not operating, the 140°F (60°C) "ambient" temperature of such a motor may be greater than its maximum limiting temperature based on the type of insulation employed. While any electric rotating machine will operate satisfactorily for some time at excessive temperatures, its life is shortened for precisely the *same* reason as if it were overloaded electrically.

Empirical studies show that for every 10°C increase in motor operating temperature over the recommended hottest-spot temperature limit (see Table 12-2), the winding life is cut in half. Conversely, for every 10°C reduction in motor operating temperature under the rated limit, the winding life is doubled.

Table 12-2 uses the standard maximum 40°C ambient temperature rise to assign a maximum allowable final temperature based on the limiting hottest-spot temperatures permitted for various classes of material.

The hottest-spot temperature allowed in a given dynamo, using a specific class of insulation, may be difficult to determine, since that spot may be buried in the stator or armature windings or may not be accessible for some other reason. Thermometers of the thermocouple or liquid-bulb type (mercury or alcohol) have to be located on the more accessible

---

* Ambient temperature, as defined by AIEE Standard No. 1, June 1947, is "the temperature of the medium used for cooling, either directly or indirectly, and is to be subtracted from the measured temperature of the machine to determine the temperature rise under specified test conditions." It it defined for particular cases as follows:

1. For self-ventilating apparatus, the ambient temperature is the average temperature of the air in the immediate neighborhood of the apparatus.
2. For air- or gas-cooled machines with forced ventilation or secondary water cooling, the ambient temperature is taken as that of the ingoing air or cooling gas.
3. For apparatus with oil or other liquid immersion of the heated parts where water cooling is employed, the ambient temperature is taken as that of the ingoing cooling water.

For the purpose of assigning a rating, 40°C is taken as the limiting ambient temperature of the cooling air or other gas.

TABLE 12-2.
LIMITING TEMPERATURES OF INSULATING MATERIALS

DESCRIPTION OF MATERIAL	INSULATION CLASS	ALLOWABLE TEMP. FOR 40°C AMBIENT TEMP. STD.	HOTTEST-SPOT MAXIMUM LIMITING TEMPERATURE
Cotton, silk, paper, or other organic materials neither impregnated nor immersed in liquid insulating materials.	O	50°C	90°C
1. Any of the above materials immersed or impregnated in liquid dielectrics. 2. Enamels and varnishes applied to conductors. 3. Films and sheets of cellulose acetate or other cellulose products. 4. Molded and laminated materials having cellulose filler or phenolic resins or other resins of similar property.	A	65°C	105°C
Mica, asbestos, fiberglass, or other inorganic minerals with a small proportion of Class A materials as binders and fillers.	B	90°C	130°C
1. Mica, asbestos, fiberglass, and similar inorganic materials with binding substances composed of silicone compounds. 2. Silicone compounds in rubbery or resinous form, or materials with equivalent dielectric and temperature properties.	H	140°C	180°C
Pure mica, porcelain, glass, quartz, and similar inorganic material in pure form (glass wool, spun-tapes, etc.)	C	No limit selected	

or outer portions of the dynamo, and such a temperature value depends on the thermodynamic temperature gradient created by the physical makeup of the machine. It is customary to add a correction of 15°C to the surface temperature to determine the hot-spot temperature. A higher and truer value of the internal maximum temperature is usually obtained by cold versus hot resistance measurements of the copper stator or rotor windings before and immediately after operation, using the temperature coefficient of copper.* A "hot-spot" correction of 10°C is generally added to the temperatures computed by resistance measurements.

Perhaps the best method of obtaining the hottest-spot temperature

* The equation
$$\frac{R_2}{R_1} = \frac{(1/\alpha)+t_2}{(1/\alpha)+t_1}$$
may also be used for materials other than copper. For copper, $1/\alpha$ equals 234.5; and the equation is solved for t_2 where t_1 is the ambient temperature.

is by means of several embedded temperature detectors. These are either thermocouples or temperature-sensitive resistive material which are permanently built into the machine and whose leads are brought out for temperature monitoring purposes. Well-placed detectors usually yield higher values of temperature than either winding-resistance or contact-thermometer techniques.

On the basis of Table 12-2, the reader might be inclined to conclude that, if even the poorest insulation, Class O, can withstand a maximum temperature of 90°C, there is nothing to be concerned about. Unfortunately, there is, since most temperature measurements are made at the surface or by winding-resistance techniques. It is precisely for this reason that the American Standards Association (ASA) sets limiting "observable" temperature designations.* This is in recognition of the fact that, under most conditions, surface temperature rather than internal temperature is recorded.

In addition to a temperature rating, other rating factors such as voltage, duty cycle, and speed are assigned to dynamos.

12-18. VOLTAGE RATING

The standard voltage ratings which have been adopted by NEMA are given in Table 12-3.

Note that, in Table 12-3, the voltage difference between

TABLE 12-3.
STANDARD VOLTAGE RATINGS FOR ELECTRIC
ROTATING MACHINERY

MACHINE	STANDARD VOLTAGE RATINGS
dc generators	125, 250, 275, 600 V
dc motors	120, 240, 550 V
ac single-phase motors	115, 230, 440 V
ac polyphase motors	110, 208, 220, 440, 550, 2300, 4000, 4600, 6600 V
ac alternators	120, 240, 480, 600, 2400, 2500, 4160, 4330, 6990, 11,500, 13,800, 23,000 V

dc generators and motors allows for a line voltage drop in the conductors supplying the motor. This is also true in the case of ac alternators and ac polyphase or single-phase motors. These voltage ratings also correspond to Tables A-3, A-4, and A-5 in the Appendix for dc and ac motors, although the tables do not include rated line current for the higher-voltage polyphase or synchronous motors shown above. As indicated in the appended tables, ratings are not available in the entire range of voltages.

* The allowable 40°C rating given general-purpose machines (rather than those values given in the first temperature column of Table 12-2) is a safety factor based on "observable" temperature rises for different types of machines. See *Rotating Electrical Machines*, ASA Standard C-50, 1943.

The higher-voltage dynamos usually are reserved for the higher-capacity ratings.

**12-19.
EFFECT OF DUTY
CYCLE AND
AMBIENT
TEMPERATURE ON
RATING**

In addition to temperature and voltage ratings, another rating factor is the *duty cycle*. The duty cycle of currently manufactured electric machinery is stated as either *continuous* duty, *intermittent* duty, *periodic* duty or *varying* duty.*

For the same horsepower or kVA rating capcity, the continuous-duty machine will be *larger* in size, physically, than the intermittent-duty machine. The larger size results from conductors of larger diameter and heavier insulation. Furthermore, a larger frame size presents a larger surface area from which heat may be dissipated; and this, too, results in a lower operating temperature for the same duration of operation. In general, a 10-hp *continuous*-duty motor may be considered a 12- or 13-hp *intermittent*-duty motor (although the rated speed may be somewhat less), since the temperature rise is not excessive if intermittently operated; see Example 12-8, Sec. 12-22.

The duty cycle is closely related to temperature, therefore, and is generally taken to include environmental factors also. A 100 kVA alternator (intermittent rating) might be converted to a 200 kVA alternator if it were continuously operated at the North Pole at an ambient temperature of $-80°C$, since all of the heat generated would still not be sufficient to overheat the alternator under such ambient conditions.

Just as the capacity rating and duty cycle are *reduced* by an *increase* in *ambient temperature*, so too are the capacity rating and duty cycle *increased* by an extreme *decrease* in ambient temperature.

In the same way, *totally enclosed* machines (without auxiliary forced ventilation, which do not permit ventilation and replacement of internal air) do not have as high a capacity rating as similar machines which are *not* totally enclosed and which are ventilated in such a manner that fresh air is drawn across the stator and rotor windings. (See Sec. 12-20.)

**12-20.
TYPES OF
ENCLOSURES**

The National Electric Manufacturers Association (NEMA) recognizes and defines the types of motor enclosures listed below. Both cost and physical size for totally enclosed motors is higher than open motors of the same hp rating, duty cycle and ambient temperature rise.

Waterproof Enclosure—a totally enclosed enclosure so constructed as to exclude water applied in the form of a stream from a hose, except that leakage may occur around the shaft providing it is prevented from entering the oil reservoir and provision is made for automatically draining such

* See Sec. 12-22 for definitions and calculations of rating based on duty cycle.

water. The means for the latter may be a check valve or a tapped hole at the lowest part of the frame for a drain pipe.

Dust-ignition-proof Enclosure—a totally enclosed enclosure so designed and constructed as to exclude ignitable amounts of dust or such amounts that might affect the performance rating.

Explosion-proof enclosure—a totally enclosed enclosure so designed and constructed as to withstand an explosion of a specified gas or vapor which may occur within it, and to also prevent ignition of specified gas or vapor surrounding it by sparks, flashes or explosions which may occur within the enclosure.

Totally Enclosed Enclosure—an enclosure which prevents the free exchange of air between the inside and outside of the enclosure but not sufficiently enclosed to be considered air-tight.

Weather-protected Enclosure—an open enclosure whose ventilating passages are so designed as to minimize entrance of rain, snow and airborne particles to the electric parts.

Guarded Enclosure—an open enclosure in which all openings giving direct access to live or rotating parts (except smooth motor shafts) are limited in size by the design of the structural parts or by screens, grills, expanded metal, etc. to prevent accidental contact with such parts. Such openings shall not permit passage of a cylindrical rod one-half inch in diameter.

Splash-proof Enclosure—an open enclosure in which the ventilating openings are so constructed that drops of liquid or solid particles, falling on it, at any angle not greater than 100° from the vertical, cannot enter either directly or by striking and running along a horizontal or inwardly inclined surface.

Drip-proof Enclosure—an open enclosure in which the ventilating openings are so constructed that drops of liquid or solid particles, falling on it, at any angle not greater than 15° from the vertical, cannot enter either directly or by striking and running along a horizontal or inwardly inclined surface.

Open Enclosure—an enclosure having ventilating openings which permit passage of external cooling air over and around the windings of the machine. When an internal fan is provided, such machines are called self-ventilating.

12-21. SPEED RATING; CLASSIFICATIONS OF SPEED AND REVERSIBILITY

Generators, converters, and alternators are all designed for a given constant speed whose value or rating is expressed on the nameplate. When driven by a prime mover at this rated speed, the generator, alternator, or converter should deliver constant (rated) voltage at the rated load.

Motors, however, are subject to speed change. A reduced speed will produce poor ventilation and overheating. Motors, therefore, are rated at the speed at which they will deliver their rated output horsepower at the rated voltage. When speed control is used on a motor, therefore, it cannot be expected that, for the same rated load current, a lower speed will produce the rated horsepower output. In general, as the speed *decreases*, the rating should be *decreased* proportionately.

A classification system based on the speed-load characteristics of motors has been developed by NEMA to define motors in terms of these characteristics, as follows:

Constant-Speed Motor—one whose speed varies a relatively small amount from no-load to full-load. While no definite limit has been set, it is usually considered that a speed regulation of approximately 20 per cent or better (less) is acceptable. This class includes shunt motors, squirrel-cage induction motors, synchronous motors, and various single-phase motors of the induction and synchronous types.

Varying-Speed Motor—one whose speed varies considerably from no-load to full-load, i.e., those whose speed regulation exceeds and is poorer than 20 per cent. Series motors, some compound motors, repulsion motors, and repulsion-induction motors fall into this category.

Adjustable-Speed Motor—one whose speed can be adjusted gradually over a considerable range, i.e., higher or lower than rated speed, but whose speed for any adjustment (speed regulation) will vary only a relatively small amount from no-load to full-load. The dc shunt motor is an excellent example of this type of motor.

Adjustable Varying-Speed Motor—one whose speed can be adjusted gradually over a considerable range, but whose speed for any adjustment will vary considerably from no-load to full-load, i.e., speed regulation poorer than 20 per cent. Series motors, some compound motors, repulsion-induction motors, and wound-rotor induction motors fall into this category.

Multispeed Motor—one whose speed can be adjusted for two or more definite values, but whose speed cannot be adjusted gradually and whose speed for any definite adjustment will vary only a relatively small amount from no-load to full-load. The induction motor, both polyphase and single-phase, having *consequent poles* is an excellent example of this category.

Non-Reversible Motor—a motor whose direction of rotation cannot be reversed, either while running or almost at standstill. A reluctance-start induction motor is a non-reversible motor.

Reversible Motors—a motor which may be reversed by changing certain *external* motor connections, even when the motor is running in one direction, without requiring that the motor stop. A capacitor-start motor is an example.

Reversing Motor—a motor which may be reversed at any time under *any* load condition, even when running at rated load and rated speed, by changing certain external motor connections.

All dc motors are reversing motors, using armature reversal plugging. Of the ac single-phase induction types, only the capacitor motor is a reversing motor. All polyphase induction motors are reversing motors, by plugging.

Table 12-4 below lists various groups of dc and ac motors, first, by their speed regulation characteristics and then by their speed variation characteristics.*

**TABLE 12-4.
MOTOR CLASSIFICATION BASED ON SPEED REGULATION AND SPEED VARIATION**

GROUP	MOTOR TYPE	SPEED REGULATION CHARACTERISTICS

1	Synchronous Motor
	a. Polyphase
	b. Single-phase
	  1. Reluctance Motor
	  2. Hysteresis motor
		Absolutely constant at synchronous speed,
		$S = 120f/P$

2	Asynchronous SCIM a. Polyphase b. Single-phase Shunt, Motor, dc	Relatively constant speed from no-load to full load, with somewhat higher no-load speed.
3	Polyphase SCIM, class D Compound Motor, dc	Moderate decrease in speed from no-load to full load.
4	Repulsion Motor Repulsion-Induction Motor Series Motor, dc and ac Universal Motor	Extremely large decrease in speed from no-load to full load. High speed at low or no-load. Very high starting torque, and low speed at high torque.

ADJUSTABLE SPEED VARIATION CHARACTERISTICS*

1	Polyphase SCIM or Synchronous Motor using adjustable frequency alternator	Speed variation range up to 6:1
2	dc motor using armature voltage and field rheostat control	Speed variation range up to 200:1
3	Single-phase and polyphase motors using mechanical speed adjustment systems or eddy-current clutches	Up to 25 hp, speed variation up to 16:1 Up to 100 hp, speed variation up to 100:1
4	dc motor using solid-state control of input waveform	Speed variation up to 200:1
5	Polyphase WRIM using a. secondary resistance control b. concatenation (foreign voltage control) c. Leblanc system d. Kramer control system e. Scherbius system f. solid-state foreign voltage control	Speed variation from 10:1 up to 200:1
6	Schrage Brush-shifting (BTA) motor	Speed variation up to 4:1
7	Brush-shifting repulsion motor	Speed variation up to 6:1
8	Multispeed SCIMs, polyphase and single-phase	Speed ratios of 2:1 or 4:1 but not adjustable in these ranges. Speed is definite with little change due to load.

* For a more detailed discussion of speed controls and speed control techniques, cf. Kosow, *Control of Electric Machines*, Englewood Cliffs, N.J.: Prentice-Hall, 1973.

12-22. FACTORS AFFECTING GENERATOR AND MOTOR SELECTION

In addition to some of the factors mentioned above, other factors are of importance in the selection of generators or motors for specific use.

In the case of a generator, synchronous converter, or alternator, such factors include: the type of prime mover; the method of mounting to be employed; direction of rotation; whether it is to be located in the open or in a totally enclosed building; the type of control which will be employed; the maintenance conditions in terms of accessibility; whether directly coupled, geared, or belted to the prime mover; and general humidity, atmospheric, or environmental conditions to which it will be subjected.

In the case of a motor, *duty service* (See Ex. 12-8) is perhaps the most important factor to be considered. The nature of load and overload frequency is a serious consideration; also the type of mounting, whether horizontal or vertical, and whether floor, ceiling, or wall-mounted; the type of speed control to be employed; the method of coupling to the load; and how frequently it will be stopped, started, and reversed; are factors affecting the type of motor to be selected and the horsepower rating capacity. Wherever possible, data based on tests with a temporary motor or by calculation should be used. Both the average and maximum *load* conditions should be considered, both in tests and in calculations. In some instances, the maximum load requirements may occur only at starting; whereas, in other cases, periodic overloads of short duration may exceed the starting requirements. Other factors include power source available, frequency, voltage fluctuations, reversing characteristics, speed range, method of mounting, space available, lubrication provisions, accessibility of brushes (if any), maintenance, coupling provisions, speed reduction techniques, type of enclosure (Sec. 12-20), cost per hp, starting and running torque, acceleration time, and breakdown torque.

Yet, as stated at the outset, duty cycle is perhaps most important. Four different types of duty cycles are classified by NEMA:

1. *continuous* duty—dynamo use requiring operation at fairly constant load for reasonably long periods of time.
2. *periodic* duty—load requirements recur regularly at periodic intervals over a reasonably long period of time.
3. *intermittent* duty—irregular occurrence of load requirements, including fairly long periods of rest at which no load occurs.
4. *varying* duty—both the loads and the periods of time at which the load requirements occur may be subject to a wide variation, without rest, over a reasonably long period of time without any regularity whatever.

Example 12-8 below shows the method of calculating hp rating, in terms of **rms hp**, for intermittent, varying duty and periodic duty motors.

As a general rule, for all dynamos the capacity selected should be such that the dynamo will be operating between three-quarters to full load most of the time. A dynamo which is larger than necessary will have a low running efficiency and higher operating cost in addition to higher initial cost. In the case of a generator, with a possible increased anticipated load, this may not be a problem. In the case of a motor driving a specific load, such as an induction motor, not only is the efficiency (of a larger than necessary motor) poor, but the power factor is also poor. Similarly, a dynamo which is too small has a lower operating efficiency, and is subject to overheating, shorter life, and increased maintenance and repair costs.

In a number of applications, it may be required to select a motor for service conditions to drive a load which varies widely over continuously repeated cycles. A drill, for example, may be used with various drill bits and may be driven into various thicknesses of various metals. The heating of the motor is determined not by the peak but by the **rms** values of current under various load conditions. Furthermore, the cooling period during standstill or idle time is *less* effective than when the motor is running, and, therefore, it is customary to divide idle periods by an empirical factor of approximately 3. The required horsepower capacity rating, therefore, is the rms "average" of the various instantaneous horsepower ratings throughout a given test cycle, as shown by Example 12-8.

EXAMPLE 12-8: EXAMPLE 12-8: A 200 hp test motor was employed to determine the best capacity rating for a varying load requirement duty cycle over a 30-min period. The test motor operated at 200 hp for 5 min, 20 hp for 5 min, and a rest period of 10 minutes followed by 100 hp for 10 min. Calculate the horsepower required for such an intermittent varying load.

Solution:

$$\text{rms hp} = \sqrt{\frac{[(200)^2 \times 5] + [(20)^2 \times 5] + [(100)^2 \times 10]}{5 + 5 + 10 + 10/3}} = 114 \text{ hp}$$

A 125-hp motor would be selected because that is the nearest larger commercial standard rating. This means that the motor would operate with a 160 per cent overload (at 200 hp) for 5 min, or one-sixth of its total duty cycle.

In general, most manufacturers of electric machinery employ applications engineers in their field service organizations to assist the consumer in the selection of the proper size and type of electric machine for a given load requirement. It is well to consult with one or more of these groups before purchasing a large piece of equipment which, if improperly selected, will result in high energy cost, inefficiency, poor service, overheating, breakdown, and increased maintenance costs.

12-23. MAINTENANCE

Preventive maintenance and routine inspection techniques conserve and prolong the life of electric machinery. Induction type machines require only periodic lubrication while some, equipped with self-lubricating "lifetime" bearings, require no lubrication

TABLE 12-5.
REFERENCE GUIDE TO PROBABLE CAUSES OF MOTOR TROUBLES

SYMPTOM OR TROUBLE / MOTOR TYPE	SPLIT-PHASE	AC SINGLE PHASE CAPACITOR START	AC SINGLE PHASE CAPACITOR START & RUN	SHADED-POLE	AC POLYPHASE (2 OR 3 PHASE)	BRUSH TYPE (UNIVERSAL, SERIES, SHUNT OR COMPOUND)
			*PROBABLE CAUSES			
Will not start	1, 2, 3, 5	1, 2, 3, 4, 5	1, 2, 4, 7, 17	1, 2, 7, 16, 17	1, 2, 9	1, 2, 12, 13
Will not always start, even with no load, but will run in either direction when started manually.	3, 5	3, 4, 5	4, 9		9	
Starts, but heats rapidly.	6, 8	6, 8	4, 8	8	8	8
Starts, but runs too hot.	8	8	4, 8	8	8	8
Will not start, but will run in either direction when started manually—over heats.	3, 5, 8	3, 4, 5, 8	4, 8, 9		8, 9	
Sluggish—sparks severely at the brushes.						10, 11, 12, 13, 14
Abnormally high speed—sparks severely at the brushes.						15
Reduction in power—motor gets too hot.	8, 16, 17	8, 16, 17	8, 16, 17	8, 16, 17	8, 16, 17	13, 16, 17
Motor blows fuse, or will not stop when switch is turned to off position.	8, 18	8, 18	8, 18	8, 18	8, 18	18, 19
Jerky operation—severe vibration.						10, 11, 12, 13, 19

*PROBABLE CAUSES
1. Open in connection to line.
2. Open circuit in motor winding.
3. Contacts of centrifugal switch not closed.
4. Defective capacitor.
5. Starting winding open.
6. Centrifugal starting switch not opening.
7. Motor over-loaded.
8. Winding short circuited or grounded.
9. One or more windings open.
10. High mica between commutator bars.
11. Dirty commutator or commutator is out of round.
12. Worn brushes and/or annealed brush springs.
13. Open circuit or short circuit in the armature winding.
14. Oil-soaked brushes.
15. Open circuit in the shunt winding.
16. Sticky or tight bearings.
17. Interference between stationary and rotating members
18. Grounded near switch end of winding.
19. Shorted or grounded armature winding.

SEC. 12-23. / Maintenance

whatever. Dynamos equipped with brushes require periodic brush, commutator or slip-ring maintenance in addition to lubrication. High-speed series-wound (dc, ac or universal) motors should not be selected for long and continuous duty cycles because the severe brush-sparking may require frequent commutator cleaning and brush replacement.

In lubricating electric machinery, excessive oiling is just as damaging as insufficient lubrication. Oil-gummed commutators and oil-soaked brushes may result in severe sparking of commutator machines. Oil leaking onto the stator may cause insulation breakdown of ac and dc stator windings.

Most types of electrical machinery require a minimum of maintenance confined only to minor lubrication. But many types of single-phase fractional hp motors of the split-phase and replusion types are equipped with centrifugal switches which may be a source of trouble that may damage the motor severely. If a centrifugal switch mechanism is "stuck in its running position", the motor fails to start. If "stuck in its starting position", the starting winding overheats and the motor fails to reach rated speed. The contacts of the switches may be gummed or oxidized or worn out, as well. Such mechanisms should be replaced rather than repaired.

Because maintenance is usually confined merely to routine lubrication, inspection becomes an important factor in prolonging life of machinery and should not be ignored. Four of the five senses are of extreme importance here: sight, sound, smell and touch. Visual inspection will reveal a number of troubles listed in Table 12-5. A noisy motor is an indication of worn bearings, overloading or single-phasing. A burnt odor, characteristic of burning insulation, is an indication of overload or breakdown. An overheated bearing or winding is detected by touch (the surface should not be so hot that one cannot hold one's hand on it).

Further, in troubleshooting, certain symptoms if identified (see Table 12-5) automatically eliminate others. If heating occurs and the temperature rises significantly, it automatically eliminates other probabilities such as blown fuse or failure to start.

The above list of 19 common motor troubles and their causes are not exclusive but should be useful in diagnosing most difficulties.* (See Table 12-5, previous page).

BIBLIOGRAPHY

Ahlquist, R. W. "Equations Depicting the Operation of the dc Motor," *Electrical Engineering*, April 1955.

Alger, P. L. *The Nature of Polyphase Induction Machines*, New York: John Wiley & Sons, Inc., 1951.

* Bodine Electric Company, *Fractional hp Motor and Control Handbook*, 3rd ed., 1968, page 99.

────── and Erdelyi. E. "Electromechanical Energy Conversion," *Electro-Technology*, September 1961.

Bewley, L. V. *Alternating Current Machinery*, New York: The Macmillan Company, 1949.

Bewley, L. V. *Tensor Analysis of Electrical Circuits and Machines*, New York: Ronald Press, 1961.

Buchanan, C. H. "Duty-Cycle Calculations for Wound-Rotor Motors," *Electrical Manufacturing*, November 1959.

Carr, C. C. *Electrical Machinery*, New York: John Wiley & Sons, Inc., 1958.

Cook, J. W. "Squirrel-Cage Induction Motors Under Duty-Cycle Conditions," *Electrical Manufacturing*, February 1956.

Crosno, C. D. *Fundamentals of Electromechanical Conversion*, New York Harcourt, Brace, Jovanovich, Inc, 1968.

Daniels. *The Performance of Electrical Machines*, New York: McGraw-Hill, Inc., 1968.

Fitzgerald, A. E. and Kingsley, C. *The Dynamics and Statics of Electromechanical Energy Conversion*, 2nd Ed., New York: McGraw-Hill, 1961.

General Principles upon Which Temperature Limits Are Based in the Rating of Electrical Equipment, New York: American Institute of Electrical Engineers. Publication No. 1.

Graphical Symbols for Electrical Diagrams. (IC1.), New York: National Electrical Manufacturers Association.

Heumann, G. W. "Motor Protection," in *Magnetic Control of Industrial Motors*, part 2, Chapter 6., New York: John Wiley & Sons, Inc., 1961.

Hindmarsh, J. *Electrical Machines*, Elmsford, N.Y.: Pergamon Press, 1965.

Industrial Control Equipment. (Group 25.) (ASA C42.25.) New York: American Standards Association.

Industrial Control Equipment. (UL508.) Chicago: Underwriters Laboratories, Inc. "Inherent Motor Overheat Protection Moves Inside the Field Coils," *Electrical Manufacturing*, November 1959.

Jones, C. V. *The Unified Theory of Electrical Machines*, New York: Plenum Publishing Corp., 1968.

Karr, F. R. "Squirrel-Cage Motor Characteristics Useful in Setting Protective Devices," AIEE Paper 59-13.

Lebens, J. C. "Positive Over-Temperature Protection—With Heat Limiters," *Electrical Manufacturing*, January 1958.

Levi, E. and Panzer, M. *Electromechanical Power Conversion*, New York: McGraw-Hill, 1966.

Libby, C. C. *Motor Selection and Application*, New York: McGraw-Hill Book Company, 1960.

Machine Tool Electrical Standards. Cleveland, Ohio: National Machine Tool Builders Association.

Majmudar, H. *Introduction to Electrical Machines*, Boston: Allyn and Bacon, 1969.

Morris G. C. "Duty-Cycle Motor Selection," *Electrical Manufacturing*, November 1958.

Motor and General Standards. (MG1.) New York: National Electrical Manufacturers Association.

Nasar, S. A. *Electromagnetic Energy Conversion Devices and Systems*, Englewood Cliffs, N.J.: Prentice-Hall, Inc., 1970.

O'Kelly and Simmons. *An Introduction to Generalized Electrical Machine Theory*, New York: McGraw-Hill, 1968.

Puchstein, A. F. *The Design of Small Direct Current Motors*, New York: Wiley/Interscience, 1961.

Schmitz, N. L., and Novotny, D. W. *Introductory Electromechanics*, New York: Ronald Press, 1965.

Selmon. *Magnetoelectric Devices: Transducers, Transformers and Machines*, New York: Wiley/Interscience, 1966.

Shoults, D. R., Rife C. J., and Johnson. T. C. *Electric Motors in Industry*, New York: John Wiley & Sons, Inc., 1942.

Siskind, C. S. *Direct-Current Machinery*, New York: McGraw-Hill, 1952.

Skilling, H. H. *Electromechanics: A First Course in Electromechanical Energy Conversion*, New York: Wiley/Interscience, 1962.

Slaymaker, R. R. *Bearing Lubrication Analysis*, New York: John Wiley & Sons, Inc., 1955.

Smeaton, Motor *Applications and Maintenance Handbook*, New York: McGraw-Hill, 1969.

Thaler, G. J., and Wilcox, M. L. *Electric Machines: Dynamics and Steady State*, New York: Wiley/Interscience, 1966.

Vaughan, V. G. and Glidden. R. M. "Built-In Overheat Protection for Three-Phase Motors," *Electrical Manufacturing*, August 1958.

—— and White. A. P. "New Hotter Motors Demand Thermal Protection," *Electrical Manufacturing*, February 1959.

Walsh, E. M. *Energy Conversion—Electromechanical, Direct, Nuclear*, New York: Ronald Press, 1967.

Veniott. *Fractional and Subfractional Horsepower Electric Motors*, 3rd Ed., New York: McGraw-Hill, 1970.

White, D. C., and Woodson, H. H. *Electromechanical Energy Conversion*, New York: Wiley/Interscience, 1959.

Wilcock, D. F. and Booser. E. R. *Bearing Design and Application*, New York: McGraw-Hill Book Company, 1957.

Wilt, H. J. "Circuit Factors in Motor Protection," *Electrical Manufacturing*, June 1959.

QUESTIONS

12-1. a. Give the relation between power and energy.
b. Under what conditions is an energy conversion device considered a power conversion device?
c. Is the dynamo capable of storing energy? Explain.

12-2. a. What is the relation between input and output power of a dynamo, in accordance with the law of conservation of power?
b. Why is the output power of a dynamo always less than the input power?
c. Why are the losses always the difference between input and output power?

12-3. a. Give the equation for dynamo efficiency in three forms.
b. State the application of each form.
c. Why does Eq. (12-2b) lend itself to a determination of motor efficiency?
d. Why does Eq. (12-2c) lend itself to a determination of generator efficiency?
e. On the basis of your answers to (c) and (d) above, why is it important to have some means of evaluating losses?

12-4. a. While dynamo losses may represent energy in the form of heat, light and chemical, why is it generally assumed that the major and only loss is the heat power loss?
b. Under what three major classes of loss is the heat power loss divided?
c. Define each of the major classes of loss listed in (b).
d. Is there justification for neglecting stray load loss below 150 kW and 200 hp.
e. How is the stray load loss evaluated above 150 kW and 200 hp?

12-5. a. List those electric losses which are fairly constant with variations in load.
b. List those electric losses which vary as a function of load in all dynamos.

12-6. a. Assuming that an alternator is driven at a constant speed with constant field excitation, which losses (both electric and rotational) are constant as a function of load?
b. Assuming that a synchronous motor with constant field excitation is loaded from no load to full load, list those losses (both rotational and electric) which are constant as a function of load.
c. Which of the losses under (a) and (b) above exist at no-load?

12-7. From your answers to Quests. 12-5 and 12-6, summarize for fairly constant speed dynamos
a. those losses which are reasonably constant, regardless of load
b. those losses which vary as a function of load

12-8. a. Explain why the hysteresis loss is confined only to magnetic materials while the eddy current loss extends to both magnetic and nonmagnetic conductive materials?
b. The core loss is sometimes called an iron loss. Why is this a misnomer?

12-9. Draw the power flow diagram for a dc shunt generator, only, showing
a. equation for mechanical input power
b. equation for power generated in the armature (electrical power developed).
c. core and mechanical losses, in terms of (a) and (b) above
d. equations for armature and field copper losses (electrical losses)
e. equation for output power
f. efficiency equation in terms of Eq. (12-2c).

12-10. Draw the power flow diagram for a dc shunt motor, only, showing
a. equation for electrical input
b. equation for armature and field copper loss
c. equation for electrical power converted to mechanical power
d. rotational and core losses
e. equation for output mechanical power
f. efficiency in terms of Eq. (12-2b).

12-11. a. Show that the change of state $E_g I_a/746$ or $E_c I_a/746$ involves no losses in converted watts to hp or vice-versa, in Fig. 12-1.
b. Convert 1 kW to units of heat power (Btu/min) and mechanical power (hp and ft·lb/min) respectively.

12-12. a. What is the advantage of the conventional methods of efficiency determination over direct loading?
b. Why is direct loading relatively impossible for larger dynamos?
c. Give one kind of conventional method which lends itself to dc dynamos.
d. Give other kinds of conventional methods which are used for synchronous and asynchronous dynamos.

12-13. a. Explain why the mechanical output of a dc dynamo is zero when run light as a motor, at some predetermined speed.
b. How is the rotational loss determined in (a), exactly, and approximately?
c. What specific losses are included in the rotational loss?

12-14. In the determination of the efficiency of a dc generator by conventional means, explain
a. why it is run light as a motor
b. how the electrical losses are evaluated at various load values
c. how the output is evaluated at various load values
d. how the efficiency is determined at any given load

12-15. a. From your answer to Quest. 12-7, explain, for constant speed dynamos, those dynamo losses which are constant regardless of load and those which vary as a function of load.
b. In terms of (a) above, explain under what conditions maximum efficiency may occur.
c. Given the fixed losses for a 3ϕ synchronous motor, W, and the resistance of the stator armature, R_a, show how it is possible to determine the value of armature current, I_a, at which maximum efficiency occurs.

12-16. a. Does the relation for maximum efficiency hold for variable speed dynamos?

- b. How is the value of load (at which maximum efficiency occurs) determined for variable speed dynamos?
- c. Give 2 graphical methods which may be used to determined the maximum efficiency load point in (b) above?

12-17.
- a. What assumptions are made in performing only one measurement in the running light method to determine rotational loss?
- b. Why is it more accurate to make a series of measurements duplicating the flux and speed under which the dynamo operates?
- c. What single factor must be duplicated which duplicates both flux and speed?
- d. Why is the counteremf-speed method more accurate in determining the efficiency of a variable speed dynamo than a single running light measurement?

12-18.
- a. What specific loss is evaluated when using the running light method for a synchronous dynamo, in the determination of conventional efficiency?
- b. What specific loss is evaluated when using the short-circuit test data for a synchronous dynamo, in the determination of conventional efficiency?
- c. What losses are considered fixed losses for an alternator?
- d. What losses are considered variable losses for an alternator?
- e. What factor in the efficiency equation is changed as a result of a change in PF of the alternator?

12-19.
- a. Give 4 advantages for the use of hydrogen over air in the cooling of alternators.
- b. Give 5 reasons why cooling is employed.
- c. Compare the increase in capacity by using hydrogen over air cooling.
- d. Compare the increase in efficiency by using hydrogen over air cooling.

12-20.
- a. Is the calibrated motor method of Sec. 12-10 a conventional or direct loading method for determining efficiency?
- b. What specific losses are evaluated in steps (1) and (2) above?
- c. What specific loss is evaluated in step (3)?
- d. What values are yielded in step (4) and are they necessary for a determination of efficiency? If not, what are they used for?

12-21.
- a. Give one conventional method used to determine efficiency of asynchronous dynamos.
- b. Give 3 methods which are used involving direct loading, and state why they are only applicable to smaller asynchronous machines.

12-22.
- a. In performing the open-circuit test, why is it not possible to ignore the rotor and stator copper losses of an induction motor?
- b. Under what circumstances may the core losses be ignored in performing the locked-rotor test of an induction dynamo?
- c. Under what circumstances must they be taken into account and how is this done in actual tests?
- d. In performing open- and short-circuit tests, what is the advantage of R_{el} expressing equivalent stator and rotor resistance between lines, in comparison to a measurement of resistance per phase?

12-23. a. Why is the locked-rotor test considered a pessimistic method?
b. Is there any advantage to a pessimistic efficiency determination?
c. Why does the AIEE method yield a more realistic value of asynchronous efficiency?
d. Contrast the advantages and disadvantages of the locked-rotor versus the AIEE methods.

12-24. a. Give 5 methods used to determine efficiency of single-phase motors.
b. Which of these methods requires special equipment?
c. For each of the 3 methods shown in (a) using direct loading, explain how the input and the output is measured?

12-25. a. What disadvantages occur from buying a dynamo which has a lower output rating than the average load for which it is intended?
b. What information is provided on the nameplates of most dynamos?

12-26. a. What is the meaning of the rating "40°C temperature rise" above ambient temperature"?
b. How is the temperature rating affected by the class of insulation employed on the dynamo windings?
c. What is the disadvantage of high ambient temperatures with respect to insulation, and dynamo life?
d. What is meant by hottest spot temperature?
e. Give 3 methods of measuring it practically.

12-27. a. Why is the voltage rating of a dc motor lower than for a dc generator?
b. Repeat (a) for the ac dynamo.
c. Why is the standard voltage range of available ac alternators higher than for ac motors?
d. Why is the voltage range of available ac polyphase motors higher than single-phase motors?

12-28. a. Why is the continuous duty larger than an intermittent duty dynamo?
b. How is the capacity of a given rated dynamo affected by 1) duty cycle, 2) ambient temperature, 3) enclosure and 4) forced ventilation?

12-29. a. Which types of enclosures are suitable for outdoor dynamos exposed to the weather?
b. In what locations are open enclosures permitted?
c. Under what actual physical circumstances (give examples) would a consumer desire a waterproof motor enclosure?

12-30. a. How is the rating of a dynamo affected by a decrease in speed?
b. Why are motors more subject to a change in rating due to speed than generators?
c. Distinguish between a varying speed motor and an adjustable speed motor.
d. Distinguish between an adjustable varying speed motor and the types defined in (c) above.

12-31. a. Explain why Table 12-4 distinguishes between the speed regulation characteristics and the adjustable speed variation characteristics.
b. What class of motors provides the best speed regulation with widest speed variation?

c. Can the speed of a hysteresis motor be varied? How?
d. What class of motors provides the lowest speed variation? Why?
e. Why is the polyphase alternator driven by a variable speed prime mover ideally suited for speed variation of synchronous and asynchronous motors? (See Sec. 9-21.)
f. What effect does increase in frequency and speed have on the rating in (e) above?

12-32. In addition to the factors of rated voltage, rated current, frequency, speed, duty cycle and temperature rise, list some other considerations affecting selection of
a. generators and alternators
b. motors.

12-33. a. Define 4 types of duty service.
b. What is meant by **rms** hp?
c. In the equation for calculating rms hp, why is the idle period divided by a factor of 3?
d. Assuming all other factors equal, on the basis of **rms** hp, arrange the following 1/4 hp motors in order of descending physical size: intermittent duty, varying duty, continuous duty and periodic duty.

12-34. a. Why is periodic inspection of dynamo operation an important factor in preventive maintenance and life of the dynamo?
b. Does a self-lubricating, explosion-proof SCIM require any periodic inspection? Explain.
c. Why is overlubrication just as harmful as under or no lubrication?
d. In making routine inspections, how are the human senses involved?
e. What are the limitations of human senses and what possible causes of troubles could only be revealed by instruments (See Table 12-5)?

12-35. With reference to Table 12-5, for each motor type listed in the columns,
a. total the number of *different* probable causes of failure
b. total the number of different symptoms of trouble
c. On the basis of (a) and (b) above, which single-phase motors are most trouble-free?
d. Compare your answer in (c) with polyphase motors and brush type motors. Draw conclusions.

PROBLEMS

12-1. An iron core choke coil subjected to a 60-Hz frequency has a core flux density of 60,000 lines/in² and produces a hysteresis loss of 2.5 W. Calculate the hysteresis loss if
a. The frequency is raised to 100 Hz
b. The flux density is reduced to 40,000 lines/in² (use $B^{1.6}$ for this grade of iron).

12-2. The eddy current loss of a 3 kVA, 208 V, 400 Hz transformer is 40 W and the hysteresis loss is 25 W at rated voltage and frequency, with a permis-

sible flux density of 52,000 lines/in². At full load, the copper loss is 65 W. Calculate the efficiency at
a. Rated load, unity power factor
b. $1\frac{1}{4}, \frac{3}{4}, \frac{1}{2}$, and $\frac{1}{4}$ rated load, unity power factor.

12-3. It is desired to use the transformer of Problem 12-2 at 60 Hz. Calculate:
a. The applied voltage which will maintain the same permissible flux density
b. The transformer rating in volt amperes
c. Hysteresis loss
d. Eddy current loss
e. Full-load efficiency
f. Conclusions regarding iron losses, rating, and efficiency, if both frequency and voltage of an electrical machine are reduced proportionately.

12-4. The efficiency of a 125 V, 5 hp motor is tested using a 250 V calibrated generator. The full-load efficiency of the calibrated generator is 0.86 when delivering 13 A at 247 V to a resistive load. The motor, running at rated speed of 1750 rpm, draws 36 A from the 125-V, dc supply. Calculate:
a. Motor output in watts and horsepower
b. Motor efficiency.

12-5. From data and results of Prob. 12-4 calculate full load losses of
a. The generator
b. The motor.

12-6. The rotational losses of a dynamo are separated into a frictional loss of 250 W and core losses of 180 W, when operated at rated speed of 1800 rpm and a flux density of 60,000 lines/in². Calculate the total rotational stray power loss when
a. The speed is increased to 2000 rpm
b. The flux density is increased to 70,000 lines/in² (use a factor of $B^{1.8}$) and speed is 1800 rpm
c. Both flux and speed are adjusted to raise the generated emf from 120 V to 125 V.

12-7. Preliminary calculations for determining the full-load stray power (rotational) loss of a 10 hp, 115 V, 1750 rpm shunt motor having armature and field circuit resistances, respectively, of 0.12 Ω and 52.5 Ω require
a. The voltage to be applied across the armature and
b. The speed at which the motor must be run at no load. Perform whatever calculations are necessary to obtain (a) and (b) (*cf.* Appendix Table A-3).

12-8. When operated under the conditions of voltage and speed determined in Problem 12-7, the 10 hp motor running at no load draws an armature current of 10 A. Calculate:
a. Full-load stray power loss
b. Full-load efficiency
c. Rotational and copper losses at half load ($I_L = 38$ A)
d. Half-load efficiency

e. Output horsepower when motor draws rated and half-rated line currents, respectively.

12-9. An 870 rpm, 120 V, 100 kW, long shunt compound generator has an armature resistance of 0.008 Ω, a series field resistance of 0.01 Ω, a brush volt drop of 1.2 V and a shunt field resistance of 30 Ω. Preliminary calculations for determining the full-load rotational losses by conventional methods require
   a. The voltage to be applied across the armature if the generator is to be run light as a motor
   b. The speed at which the generator is to be run as a motor at no load. Perform the necessary calculations.

12-10. When operated under conditions of voltage and speed determined in Problem 12-9, the armature current drawn from the mains is 32 A. Calculate:
   a. Full-load stray power loss
   b. Full-load efficiency
   c. Rotational and copper losses at half load
   d. Half-load efficiency.

12-11. For the calculations of Problem 12-10
   a. Tabulate the fixed and variable losses at both half and full load
   b. Explain why the efficiency is greater at half load than at full load
   c. Calculate the value of load at which maximum efficiency will occur
   d. Compute the maximum efficiency.

12-12. A 125 V shunt generator has a rated armature current of 80 A and the following losses at full load: friction and windage loss of 250 W, core loss 300 W, shunt field loss of 125 W, armature copper loss 500 W, brush contact loss 100 W. Assuming a maximum efficiency, which occurs in accordance with Eq. (12-6), where the approximately constant losses equal those losses that vary as the square of the load, calculate
   a. Armature current for maximum efficiency
   b. Full-load efficiency
   c. If the prime mover speed is reduced by 10 per cent and the flux (assume loss varies as $f^2 \times B^2$) is increased by 20 per cent to obtain the same rated voltage, calculate the full-load efficiency, assuming no change in rating due to inadequate cooling.

12-13. A 40 hp compound motor has an efficiency of 82 per cent when operated at rated load. The rotational stray power loss is 25 per cent of the full-load losses and may be assumed constant while all other losses vary as the square of the load current. Calculate the efficiency when the load is
   a. 20 hp
   b. 50 hp.

12-14. A 25 hp series motor has a theoretical maximum efficiency of 85 per cent when delivering three-quarters of its rated output. Assuming negligible brush contact loss, calculate
   a. Total losses at this load
   b. Rotational loss and variable copper losses at the maximum efficiency

c. Assuming a 20 per cent decrease in counter emf at full load, calculate the full-load rotational losses, copper losses and efficiency
d. Explain why full-load efficiency is higher than theoretical maximum.

12-15. A 1000 kVA, 2300 V, wye-connected, six-pole, 60 Hz alternator operates at a power factor of 0.8 lagging when delivering rated load. The dc armature resistance between terminals is 0.1 Ω and the dc field requires an input of 30 A at 125 V dc between slip rings from an exciter on the same shaft as the alternator. When driven at rated speed of 1200 rpm, using a calibrated motor and dynamometer, the following losses are determined: iron losses of 7.5 kW, friction and windage losses of 5.75 kW. Assuming an effective resistance factor of 1.25 times dc resistance, calculate
a. Full-load alternator efficiency at 0.8 PF
b. Per cent load at which maximum efficiency will occur
c. Maximum efficiency assuming no power factor or excitation changes.

12-16. A calibrated shunt motor is used to drive a 1000 kVA, 6990 V, three-phase alternator at rated speed. Data obtained is as follows:

Motor input (kW)	Motor efficiency	Conditions
1.0	0.56	1
15.2	0.85	2
30.0	0.88	3
60.5	0.90	4

Condition: (1) Motor uncoupled from alternator (2) Motor coupled to alternator, no alternator excitation. (3) Alternator field draws 60 A from 125-V, dc source to produce rated no-load voltage. (4) Armature of three-phase alternator is short-circuited and rated current flows in armature. From this experimental data, calculate
a. Alternator friction and windage losses when driven at rated speed
b. Core losses at rated speed
c. Field losses to produce rated voltage
d. Total full-load fixed losses
e. Full-load, half-load and maximum alternator efficiency at unity power factor
f. If the field excitation is increased to 75 A dc at 0.8 power factor lagging, calculate the full-load and half-load efficiency.

12-17. A 10 kVA, 250 V, wye-connected alternator, tested by open-circuit and short-circuit test procedure to determine regulation and effective armature resistance per phase, yields a value of 0.3 Ω/phase. When run light as a synchronous motor at rated voltage and speed, it draws a total power of 532 W and a no-load current of 7.75 A. Open-circuit test data indicates that the dc field requires an excitation of 2.0 A at 120 V dc to produce rated no-load voltage for unity power factor loading. Calculate:
a. Rotational loss
b. Field copper loss
c. Armature copper losses at $\frac{1}{4}$, $\frac{1}{2}$, $\frac{3}{4}$, and $1\frac{1}{4}$ rated load
d. Efficiency at the loadings listed in (c) at unity power factor.

12-18. A 7.5 hp, 220 V, three-phase, four-pole, 1750 rpm, 0.8 power factor, 22 A squirrel cage induction motor has the same number of stator as rotor turns. When operated at no load or rated voltage, the motor draws 9.5 A and a total power of 600 W. With the rotor blocked, the motor draws 22.0 A and 1100 W at a line voltage of 40 V. Calculate:
   a. Equivalent resistance between lines of the induction motor, referred to as the stator
   b. Rotational losses
   c. Equivalent copper loss at $\frac{1}{4}$, $\frac{1}{2}$, $\frac{3}{4}$, and $1\frac{1}{4}$ rated load
   d. Efficiency at each of these load points
   e. Load at which maximum efficiency occurs and the maximum efficiency
   f. Ratio of core losses at reduced voltage to those at rated voltage during blocked rotor test.

12-19. a. Recalculate the full-load efficiency in Problem 12-18 using the AIEE load-slip equivalent circuit method and the data given in the previous problem.
   b. Compare efficiency obtained in Problem 12-19 with Problem 12-18 and explain why the latter is pessimistic (i.e., the efficiency is lower).
   c. Tabulate for full load, $\frac{1}{4}$, $\frac{1}{2}$, $\frac{3}{4}$ and $1\frac{1}{4}$ rated load, the following values calculated by the AIEE load-slip method: stator copper loss, stator power input, rotor power input, slip, rotor copper loss, rotational losses, rotor total losses, rotor output and efficiency.
   d. Calculate the load at which maximum efficiency occurs and the maximum efficiency.
   e. Compare your answers to Problem 12-18e and Problem 12-19d and account for differences both in value and point at which maximum efficiency occurs.

12-20. Tests performed on a 10 hp, 220 V, four pole, three-phase induction motor yielded the following information

	Line Voltage (V)	Line Current (A)	Total Power (W)	Effective resistance of stator, per phase
No load	220	8.1	400	0.13 Ω
Rated load	220	27.0	8200 @ 1750 rpm	

Calculate:
a. Friction, iron and windage losses
b. Stator copper loss at full load
c. Full-load rotor power input
d. Full-load rotor copper loss
e. Full-load rotor output
f. Full-load output horsepower
g. Full-load torque
h. Full-load efficiency
i. Full-load power factor.

12-21. A dynamometer brake test on a 115 V, four-pole, single-phase capacitor-start induction motor yielded the following data

	Voltage (V)	Current (A)	Power (W)	Speed (rpm)	Dynamometer (Length of arm) (in)	Scale Reading (oz)
No load	115	4.8	203	1795	12	0
Rated load	115	5.8	493	1720	12	9.16
Max. break-down load	115	11.2	600	1440	12	20.65
Blocked rotor	60	5.8	150	0	12	7.93

Calculate:
a. Rated load efficiency, power factor, motor horsepower and rated torque
b. Maximum torque and starting torque, using computed maximum torque
c. Efficiency and power factor at maximum torque
d. Rated torque from maximum torque
e. Starting torque and power factor from blocked rotor data.

12-22. Using the AIEE load-slip method and pertinent data from Problem 12-21, calculate
a. Equivalent no-load rotor and stator resistance losses and effective resistance between lines referred to the stator
b. Rotational losses
c. Equivalent copper loss at rated load
d. Efficiency at rated load
e. Account for discrepancies in efficiency computed by this method.

12-23. A $7\frac{1}{2}$ hp, 115 V, dc motor has a full-load current of 58 A, an armature circuit resistance of 0.1 Ω and a full-load speed of 1750 rpm. When the shunt motor is connected to a 230-V source with suitable added field resistance to restore the field flux to its normal 115-V value, the armature current remains the same at an increased speed. Assuming that the increased speed and ventilation offsets the increase in rotational losses and that the electrical and mechanical design of the motor are suited to the new conditions, calculate
a. Counter emf and speed at the higher voltage
b. Horsepower rating at the higher voltage.

12-24. A 440 V, 30 hp induction motor used in a commercial laundry washer drives a load of 15 hp for 5 min followed by a 60 hp load for 5 min, followed by no load (drying period) for 15 min. If this cycle is repeated continuously over a 24 hr period, calculate
a. The factor by which the motor rating is exceeded
b. Determine if the rating is exceeded when the idle period is increased to 30 min.

ANSWERS

12-1(a) 4.167 W (b) 1.28 W   12-2(a) 0.958   (b) 0.956; 0.957; 0.948; 0.915
12-3(a) 31.2 V (b) 450 VA (c) 3.75 W (d) 0.9 W (e) 0.866   12-4(a) 5.01 hp
(b) 0.831   12-5(a) 525 W (b) 760 W   12-6(a) 478 W (b) 487.5 W (c) 447 W

12-7(a) 106.1 V (b) 1750 rpm   12-8(a) 1048 W (b) 0.778 (c) 1090 W, 164.5 W (d) 0.659 (e) 9.11 hp, 3.86 hp   12-9(a) 136.3 V (b) 870 rpm   12-10(a) 4310 W (b) 0.844 (c) 4070 W, 3150 W (d) 0.859   12-11(c) 473.5 A (d) 0.86   12-12(a) 93 A, 0.891 (b) 0.8875 (c) 0.885   12-13(a) 0.842 (b) 0.8.   12-14(a) 2470 W (b) 1235 W (c) 988 W, 2195 W, 0.853.   12-15(a) 0.965 (b) 120 per cent (c) 0.966   12-16(a) 12.92 kW (b) 13.48 kW (c) 7.5 kW (d) 33.9 kW (e) 0.94, 0.924, 0.9475 (f) 0.9275, 0.90.   12-17(a) 478 W (b) 240 W (c) 30 W, 120 W, 270 W, 480 W, 750 W (d) 0.77; 0.856; 0.883; 0.895; 0.896   12-18(a) 1.515 Ω (b) 395 W (c) 68.75 W, 275 W, 618 W, 1100 W, 1720 W (d) 0.723, 0.8, 0.798, 0.749, 0.743 (e) 0.802 @ 0.6 rated load (f) 0.033.   12-19(c) 0.677 W, 0.797 W, 0.819 W, 0.818 W, 0.808 W (d) 0.82I_L, 0.822   12-20 (a) 374.4 W (b) 284 W (c) 7541.6 W (d) 215.5 W (e) 7326.1 W (f) 9.825 hp (g) 29.5 lb-ft (h) 0.8925 (i) 0.796.   12-21(a) 0.452; 0.738; 0.572 lb-ft (b) 1.29 lb-ft; 0.496 lb-ft (c) 0.44; 0.465 (d) 0.572 lb-ft (e) 0.496 lb-ft, 0.431   12-22(a) 150 W, 4.46 Ω (b) 100 W (c) 150 W (d) 0.4925   12-23(a) 3590 rpm (b) 15.4 hp   12-24(a) 118.7 per cent (b) 103 per cent

THIRTEEN

transformers

13-1.
FUNDAMENTAL
DEFINITIONS

The transformer operates on the principle of *mutual induction* between two (or more) inductively coupled coils or circuits. A theoretical air core transformer is shown in Fig. 13-1 in which two circuits are coupled by magnetic induction. Note that the circuits are *not* physically connected (there is no conductive connection between them).

The circuit connected to the source of alternating voltage, V_1, is called the *primary* (circuit 1). The primary *receives* its *energy* from the alternating source. Depending on the degree of *magnetic coupling* between the two circuits (Eq. 13-1), energy is *transferred* from circuit 1 to circuit 2. If the two circuits are *loosely coupled*, as in the case of the air core transformer shown in Fig. 13-1, only a *small* amount of energy is transferred from the primary (circuit 1) to the secondary (circuit 2). If the two coils or circuits are wound on a common iron core, they are *closely coupled*. In such a case, almost *all* the energy received from the supply by the primary is *transferred by transformer* action to the secondary.

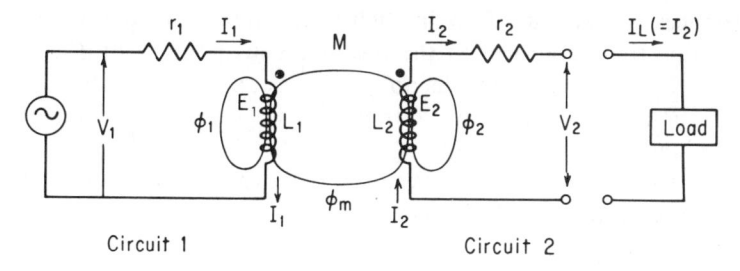

Figure 13-1

Air core transformer, inductively coupled, with defining symbols.

The following definitions apply to the transformer, as shown in Fig. 13-1, and are used throughout this chapter:

V_1 is the supply voltage applied to the primary, circuit 1, in volts
r_1 resistance of the primary circuit, in ohms
L_1 inductance of the primary circuit, in henries
X_{L_1} inductive reactance of the primary circuit, in ohms
Z_1 impedance of the primary circuit, in ohms
I_1 rms current drawn by the primary from the supply, in amperes
E_1 voltage induced in the primary coil (or circuit) by all the flux linking coil 1, in volts
E_2 voltage induced in the secondary coil (or circuit) by all the flux linking coil 2, in volts
I_2 rms current delivered by the secondary circuit to a load across its terminals
r_2 resistance of the secondary circuit (excluding load), in ohms
V_2 voltage which appears at the terminals of the secondary winding
L_2 inductance of the secondary circuit, in henries
X_{L_2} inductive reactance of the secondary circuit, in ohms
Z_2 impedance of the secondary circuit (excluding load), in ohms
ϕ_1 a leakage component of the flux linking coil 1 only,
ϕ_2 a leakage component of the flux linking coil 2 only,
ϕ_m the mutual flux, commonly shared by both circuits, linking coils 1 and 2
M the mutual inductance (a measure of the magnetic coupling) between the two coils (or circuits) produced by the mutual flux (ϕ_m) in henries.

Note the significance of the *dot convention*, used in Fig. 13-1, to show the instantaneous positive polarity of alternating voltage *induced* in both the primary and secondary windings as a result of transformer action. Thus, when V_1 is instantaneously positive, a voltage E_1 is induced in the primary winding of such polarity so as to oppose V_1, in accordance with Lenz's law, as shown in Fig. 13-1. Also note (from Fig. 13-1) that the current I_2 is in opposition to I_1. This is also in accordance with Lenz's law since I_1 produces ϕ_m. I_2 must flow in such a direction as to oppose I_1, and (at the same time) conform to the instantaneous polarity of E_2, as shown in Fig. 13-1. The instantaneous polarity of E_2 and I_2 establishes the instantaneous polarity of V_2 (upper terminal positive) and direction of current in the load.

The coefficient of coupling, k, between the two coils is a ratio of the mutual to the total flux defined as*

$$k = \frac{\phi_m}{\phi_m + \phi_1} = \frac{M}{\sqrt{L_1 \times L_2}} \qquad (13\text{-}1)$$

where all terms have been defined above.

If the two coils are *loosely* coupled, as in the air core transformer of Fig. 13-1, the terms ϕ_m and ϕ_2 are small in comparison to ϕ_1. As a consequence, the terms L_2 and M are small in comparison to L_1. Substitution in Eq. (13-1) yields a small value of coupling coefficient, k. This, in turn, yields a low value of E_2 and V_2 (in comparison to E_1 and V_1). For any given load, therefore, a small value of V_2 yields a small load current, I_2. Simply stated, then, for *loose* coupling, the power transferred to the secondary circuit, $E_2 I_2$, is relatively *small*.

Transformers having loose coupling are used primarily in high frequency (RF) communication and electronic circuits. Practically all of the transformers used in power and machinery applications, however, are tightly coupled iron core transformers.

If the coils or circuits are *tightly* coupled, and the leakage fluxes ϕ_1 and ϕ_2 are relatively small in comparison to ϕ_m, the mutual inductance M between the two coils is large as are the terms E_2, I_2 and V_2. In this case, the energy transformed $E_2 I_2 t$ is practically equal to $E_1 I_1 t$. Insofar as possible, in designing iron core power transformers, an attempt is made to achieve a coupling coefficient of unity, ($k = 1$) such that in Eq. (13-1) $M = \sqrt{L_1 L_2}$ as in the case of an *ideal* transformer.

The coupling between the two circuits is increased if portions of both coils are wound on the same coil form and the coil is placed on a relatively low reluctance magnetic core. Such designs tend to reduce leakage fluxes ϕ_1 and ϕ_2. But even under optimum designs, it is impossible to

---

*Equation 13-1 may be derived as follows. The mutual inductance, like any inductance is proportional to the induced voltage ($e_M = M\, di/dt$) where e_M is the voltage developed in coil 2 by that portion of the flux common to both coils 1 and 2, ϕ_m. In accordance with Faraday's law of electromagnetic induction (Eq. 1-5),

$$E_m \propto k N_1 N_2 \text{ but } N_1 \propto \sqrt{L_1} \text{ and } N_2 \propto \sqrt{L_2},$$

and therefore $E_m \propto M \propto k N_1 N_2 = k'\sqrt{L_1 L_2}$ where k' (or k) is the coefficient of coupling, by definition, and $\sqrt{L_1 L_2}$ is the geometric mean of the self inductances of the two coils.

achieve the ideal transformer—one having no primary or secondary leakage fluxes, and having unity coupling. The subsequent discussion begins with an ideal transformer, however, to simplify an understanding of the transformer relations which follow. The practical power transformer is then taken up, in turn.

13-2. IDEAL TRANSFORMER RELATIONS

Consider an ideal, iron core transformer as shown in Fig. 13-2 where leakage fluxes ϕ_1 and $\phi_2 = 0$ and $k = 1$. Such a transformer possesses only mutual flux ϕ_m, common to both the primary and secondary coils. When V_1 is instantaneously positive, as shown in Fig. 13-2, the direction of primary current I_1 produces the direction of mutual flux, ϕ_m, as shown. The primary induced voltage, E_1, in accordance with the dot convention and Lenz's law, produces a positive polarity

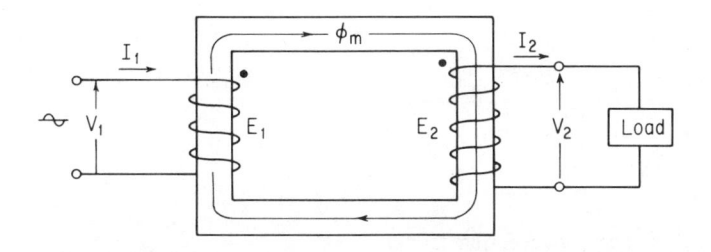

Figure 13-2

Iron-core transformer, ideal case.

at the top of the primary coil, which instantaneously opposes the applied voltage V_1. Similarly, at the secondary, for the direction of ϕ_m shown, the positive polarity of E_2 must be such as to create a demagnetizing flux opposing ϕ_m (Lenz's law). A load connected at the terminals of the secondary produces a secondary current I_2 which flows in response to the polarity of E_2 and produces a demagnetizing flux.

We are now in a position to comprehend qualitatively how a transformer develops secondary power and transfers power from primary to secondary, in the following way:

1. Assume an open circuit, infinite impedance or zero load on the secondary, and $I_2 = 0$.
2. As a result of the alternating mutual flux ϕ_m (created by the applied voltage), emfs E_1 and E_2 are produced having the instantaneous polarity shown with respect to ϕ_m (Fig. 13-2).
3. A small primary current, I_m, known as the magnetizing current, must flow even when the transformer is unloaded. The current is small because the primary induced voltage, E_1, opposes the applied voltage

V_1, at each instant. The magnitude of I_m is a function primarily of the reluctance of the magnetic circuit, $\mathscr{R}_m$, and the peak value of the mutual magnetizing flux, ϕ_{pm}, for a given number of primary turns.*

4. As shown in Fig. 13-3a, the small value of I_m lags the primary voltage by 90° producing ϕ_m.

(a) Primary relations, no-load

(b) Secondary relations, loaded transformer

(c) Primary relations, loaded transformer

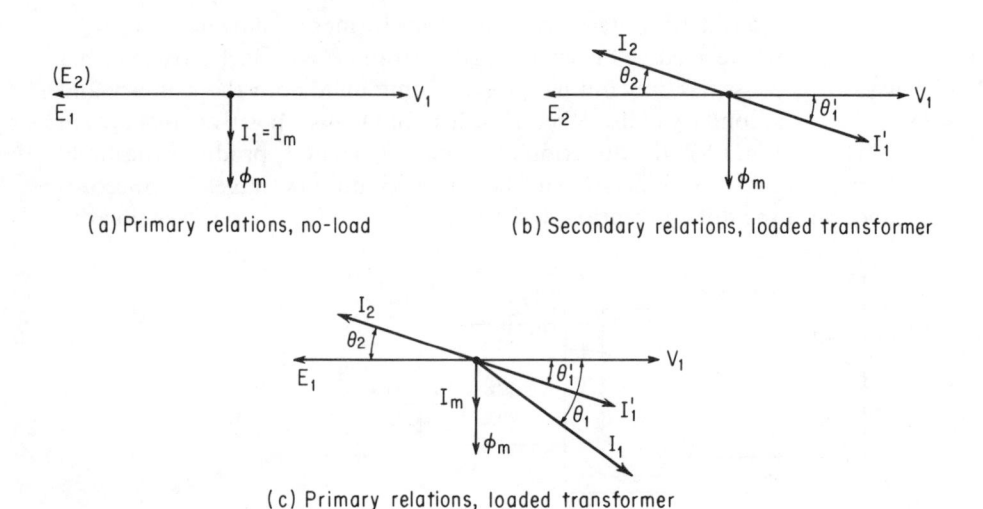

Figure 13-3

Phasor relations in an ideal transformer.

5. ϕ_m, in turn, requires 90° to produce the primary and secondary induced voltages, E_1 and E_2. These induced voltages are in phase with each other because they are both produced by ϕ_m. Note that E_1 in Fig. 13-3a opposes V_1 (Lenz's law). In the absence of load, Fig. 13-3a represents all current and voltage relations in the ideal transformer.

6. Assume a lagging (inductive) load connected across the secondary of the ideal transformer in Fig. 13-2. Such a load produces a current I_2 lagging E_2 by an angle θ_2 as shown in Fig. 13-3b.

7. The secondary ampere-turns, $I_2 N_2$, as shown in Fig. 13-2, tend to produce a demagnetizing flux which reduces both the mutual flux, ϕ_m, and the induced voltages, E_2 and E_1, instantaneously.

8. The reduction of E_1 causes a primary component of load current, I'_1, to flow in the primary circuit, such that $I'_1 N_1 = I_2 N_2$, restoring ϕ_m to its original value. Note in Fig. 13-3b that I'_1 lags V_1 by θ'_1 while I_2 lags E_2 by θ_2 such that $\theta'_1 = \theta_2$. The latter equality is necessary in order

---

* It can be shown that the peak value of magnetizing current, $I_{pm} = \phi_{pm}(\mathscr{R}_m/N_1)$ where $\mathscr{R}_m$ is the reluctance of the magnetic circuit, ϕ_{pm} is the peak value of the mutual magnetizing flux, and N_1 is the number of primary turns. cf. Jackson, *Introduction to Electric Circuits* (3rd ed.), Englewood Cliffs, N.J.: Prentice-Hall, Inc. 1970, p. 585.

that the restoring magnetizing primary turns $I'_1 N_1$ is equal and opposite to $I_2 N_2$, the load demagnetizing ampere-turns.

9. The effect of the primary component of load current I'_1 on the primary is shown in Fig. 13-3c where the primary current I_1 is the phasor sum of I_m and I'_1. Two points should be noted regarding the power factor relations of the primary circuit from this figure:
   a. the power factor angle of the primary has decreased from its unloaded original value of 90° to its loaded angle θ_1, and
   b. the power factor angle of the primary circuit is not exactly the same as the secondary circuit. (For a lagging load, $\theta_1 > \theta_2$.)

The above steps reveal the manner in which the primary circuit responds to a load on the secondary circuit. In a sense, the operation of the loaded transformer may be considered similar to the loading of a dc shunt motor* (Sec. 4-4).

The equality between the secondary demagnetizing mmf $I_2 N_2$ and the primary mmf load component which must flow to counteract the demagnetizing action, $I'_1 N_1$, as described in Step 8, above, may be summarized and rearranged as

$$I'_1 N_1 = I_2 N_2 \tag{13-2a}$$

or

$$\frac{I_2}{I'_1} = \frac{N_1}{N_2} = \alpha \tag{13-2b}$$

where α is the ratio of primary to secondary turns or the *transformation ratio*
I'_1 is load component of primary current
I_2 is the secondary or load current
N_1 and N_2 are the primary and secondary turns, respectively

The significance of the transformation ratio, α, in Eq. (13-2b), is that it is fixed (not constant) for any given (constructed) transformer depending on application. Consequently the load component of primary current may be calculated for any value of secondary load current, as shown in Ex. 13-1.

EXAMPLE 13-1: The high voltage side of a transformer has 500 turns while the low voltage side has 100 turns. When connected as a step-down transformer, the load current is 12 A. Calculate:

---

* As the load of a dc shunt motor is increased, an instantaneous decrease of speed and counter emf is accompanied by an increase in armature current from the supply to produce the necessary electromagnetic torque to meet the applied load. The reduction of counter emf in the dc motor in response to load increase is similar to the reduction in primary generated voltage E_1 of a transformer in response to load increase.

SEC. 13-2. / Ideal Transformer Relations

a. the transformation ratio, α
b. the load component of primary current.

Solution:

a. as a step-down transformer, the high voltage side is the primary and the low voltage side is the secondary. The transformation ratio, α, is

$$\alpha = \frac{N_1}{N_2}$$

$$= \frac{500 \text{ turns}}{100 \text{ turns}} = 5 \qquad (13\text{-}2b)$$

b. From Eq. (13-2b), $I'_1 = I_2/\alpha = 12 \text{ A}/5 = \mathbf{2.4 \text{ A}}$

The wording of Ex. 13-1 implies that either the low voltage or high voltage side of the transformer may be used as a primary (the side which is connected to the energy source). Thus, the transformation ratio for a given (constructed) transformer depends on its *application* as shown by Ex. 13-2.

EXAMPLE 13-2: Calculate the transformation ratio of the transformer in Ex. 13-1, when used as a step-up transformer.

Solution:

As a step-up transformer, the low voltage side is connected to the primary. The transformation ratio,

$$\alpha = \frac{N_1}{N_2} = \frac{100 \text{ turns}}{500 \text{ turns}} = 0.2 \qquad (13\text{-}2b)$$

Examples 13-1 and 13-2 show that the transformation ratio, α, is fixed for a given application but *not constant*. When used as a step-down transformer, $\alpha = 5$ but when used as a step-up transformer, $\alpha = 0.2$ (which is the reciprocal of 5). Since the terms step-up and step-down refer to voltages, as do high voltage and low voltage sides, the transformation ratio may be stated in terms of voltages using Neumann's quantification of Faraday's law (Eq. 1-1):

$$E_1 = N_1 \frac{d\phi_m}{dt} \qquad (13\text{-}3)$$

and

$$E_2 = N_2 \frac{d\phi_m}{dt} \qquad (13\text{-}4)$$

Since the rate of change of the mutual flux ($d\phi_m/dt$) linking the primary and secondary turns is the same, dividing Eq. (13-3) by Eq. (13-4) yields α in terms of voltages or

$$\alpha = \frac{N_1}{N_2} = \frac{E_1}{E_2} = \frac{V_1}{V_2} \qquad (13\text{-}5)$$

Equation (13-5) states that the primary to secondary voltage ratios are proportional to the primary to secondary turns ratios. It also verifies that the transformation ratio, α, is greater than unity for a step-down transformer but less than unity for a step-up transformer (see Exs. 13-1 and 13-2).

Equating Eqs. (13-2b) and (13-5) yields

$$\alpha = \frac{N_1}{N_2} = \frac{I_2}{I'_1} = \frac{E_1}{E_2} = \frac{V_1}{V_2} \qquad (13\text{-}6)$$

which may be transposed to yield the fundamental power relation between the primary and secondary,

$$E_1 I'_1 = E_2 I_2 \qquad (13\text{-}7)$$

and if the load component of primary current, I'_1 is much greater than the magnetizing current, I_m, we may write

$$E_1 I_1 = E_2 I_2 \quad \text{(where } I_m \text{ is negligible)} \qquad (13\text{-}8)$$

For an ideal, lossless transformer having neither primary nor secondary leakage fluxes (zero leakage reactances), we may say

$$V_1 I_1 = V_2 I_2 \quad \text{(for an ideal transformer)} \qquad (13\text{-}9)$$

Equation (13-9) verifies the fundamental definition of a transformer as a device which transfers energy from one circuit to another. For an ideal transformer, the volt amperes drawn from the alternating source, $V_1 I_1$, is equal to the volt amperes transferred to the secondary and delivered to the load, $V_2 I_2$, where all terms have been defined in Sec. 13-1.* Equation (13-9) also establishes a means of *rating* a transformer in voltamperes (VA) or kilovoltamperes (kVA) where V_1 and I_1 are the rated primary voltage and current, respectively, and V_2 and I_2 are the rated secondary voltage and current, respectively.

EXAMPLE 13-3: A 2300/115 V, 60 Hz, 4.6 kVA transformer is designed to have an induced emf of 2.5 volts per turn. Assuming an ideal transformer, calculate
    a. the number of high side turns, N_h
    b. the number of low side turns, N_l
    c. rated current of high voltage side, I_h
    d. rated current of low voltage side, I_l

---

* Note the significance between the letters V and E, denoting terminal and induced voltages, respectively. Note also, that like alternators, transformers are rated in kVA rather than kW. At extremely low PFs, a transformer may carry rated current and still transfer only a small amount of true power. The kVA rating insures that a transformer is never overloaded since it applies to all PFs from unity to zero.

e. transformation ratio when used as a step-up transformer
f. transformation ratio when used as a step-down transformer.

Solution:

a. $N_h = \dfrac{E_h}{2.5 \text{ V/t}} = \dfrac{2300 \text{ V}}{2.5 \text{ V/t}} =$ **920 turns**

b. $N_l = E_l \times \dfrac{1}{2.5}\dfrac{t}{V} = 115 \text{ V} \times \dfrac{1}{2.5}\dfrac{t}{V} =$ **46 turns**

c. $I_h = \dfrac{\text{kVA} \times 1000}{V_h} = \dfrac{4.6 \times 1000 \text{ VA}}{2.3 \times 10^3 \text{ V}} =$ **2 A** \hfill (13-9)

d. $I_l = \dfrac{\text{kVA} \times 10^3}{V_l} = \dfrac{4.6 \times 10^3 \text{ VA}}{1.15 \times 10^2 \text{ V}} =$ **40 A** \hfill (13-9)

e. $\alpha = \dfrac{N_1}{N_2} = \dfrac{N_l}{N_h} = \dfrac{46}{920} = \dfrac{1}{20} =$ **0.05**, as a step-up transformer

f. $\alpha = \dfrac{N_1}{N_2} = \dfrac{N_h}{N_l} = \dfrac{920}{46} =$ **20**, as a step-down transformer \hfill (13-2b)

In Ex. 13-3, the volts/turn ratio was given as 2.5 V/t, for both the high voltage and low voltage windings. It can be shown that this value is directly proportional to the peak value of the mutual flux, ϕ_{pm}, and the frequency, as expressed by the volts/turn ratio or*

$$\dfrac{E_2}{N_2} = \dfrac{E_h}{N_h} = 4.44 f \phi_{pm} \times 10^{-8} \dfrac{\text{volts}}{\text{turn}} = kf\phi_{pm} = kf(B_m A) \quad (13\text{-}10)$$

where B_m is the maximum permissible flux density and A is the area of the transformer core ($\phi_{pm} = B_m A$).

The significance of Eq. (13-10) cannot be disregarded because it establishes the maximum permissible mutual flux or maximum permissible flux density at a given frequency for a given rated voltage. Thus, transformers designed for operation at a given frequency may *not* be operated at another

---

* Neumann's quantification of Faraday's law states that the average emf induced in a coil of N turns is

$$E_{av} = \dfrac{N}{t}\phi_{pm} \times 10^{-8} \text{ V} \quad (1\text{-}1)$$

where t is the time it takes mutual flux to rise from zero to the peak value of mutual flux in maxwells. Assuming a sinusoidal input, having a frequency, f cycles per second, the flux rises to a maximum in a quarter cycle ($t = \tfrac{1}{4}f$) and

$$E_{av} = \dfrac{N\phi_{pm}}{\tfrac{1}{4}f} \times 10^{-8} = 4fN\phi_{pm} \times 10^{-8} \text{ V}$$

But since the form factor of a sine wave is the ratio of its effective to average value $(0.707/0.636 = 1.11)$, the effective value of the induced emf $= 1.11 E_{av}$ or

$$E = 1.11 E_{av} = 4.44 fN\phi_{pm} \times 10^{-8} \text{ V}$$

From which the volts/turn ratio is $E/N = 4.44 f\phi_{pm} \times 10^{-8}$ V. \hfill (13-10)
Note that Eq. (13-10) for the transformer is the same as Eq. (2-14) for the alternator.

frequency *without* corresponding changes in applied voltage, as shown by Ex. 13-4.

EXAMPLE 13-4: A 1 kVA, 220 V/110 V, 400 Hz transformer is desired to be used at a frequency of 60 Hz. Calculate:
a. the maximum rms value of voltage which may be applied to the high voltage side and maximum voltage output of the low voltage side
b. the kVA rating of the transformer under conditions of reduced frequency.

Solution:

a. to maintain the same permissible flux density in Eq. (13-10), both voltages of the high and low sides must change in the same proportion as the frequency

$$E_h = 220 \text{ V} \frac{60 \text{ Hz}}{400 \text{ Hz}} = 33 \text{ V}$$

and

$$E_l = \frac{E_1}{\alpha} = \frac{E_h}{2} = \frac{33.0 \text{ V}}{2} = 16.5 \text{ V} \quad (13\text{-}6)$$

b. the original current rating of the transformer is unchanged since the conductors still have the same current carrying capacity. Thus,

$$I_h = \frac{\text{kVA}}{V_h} = \frac{1 \times 10^3 \text{ VA}}{220 \text{ V}} = 4.545 \text{ A} \quad (13\text{-}9)$$

and the new kVA rating is

$$V_h I_h = V_1 I_1 = 33 \text{ V} \times 4.545 \text{ A} = 150 \text{ VA} \quad (13\text{-}9)$$

The significance of Ex. 13-4 is that it is possible to make frequency changes in transformer operation but only with corresponding changes in voltage. If the frequency and voltage are *both* reduced, the kVA capacity is correspondingly reduced. If the frequency and voltage are both increased, the kVA capacity is correspondingly increased (providing the maximum breakdown voltages of the transformer windings are not exceeded). Note that in either case, the maximum permissible flux density *must remain the same*. This is necessary so that the transformer does not overheat, as shown by Ex. 13-5.

EXAMPLE 13-5: Assuming that eddy current and hysteresis losses vary as the square of the flux density (see Eqs. 12-3 and 12-4), calculate the iron losses if the transformer of Ex. 13-4 is operated at rated voltage but at reduced frequency of 60 Hz. Assume the original iron losses of the transformer at 400 Hz to be 10 W.

Solution:

Since $E = kfB_m$, and the same primary voltage is applied to the transformer at reduced frequency, the **final** flux density, B_{mf} increases significantly above its **original** maximum permissible value B_{mo} to

$$B_{mf} = B_{mo} \frac{f_1}{f_f} = B_{mo}(400/60) = 6.67 B_{mo} \quad (13\text{-}10)$$

Since the iron losses vary approximately as the square of the flux density (see Eqs. 12-3, 12-4)

$$P_{iron} = (P_{orig})B^2 = 10 \text{ W } (6.67)^2 = 444 \text{ W}$$

Note from Ex. 13-5, even in the absence of load current, that the iron losses rise from 10 W to 440 W for a 1 kVA transformer which is operated at reduced frequency but at the same rated voltage. If the primary voltage had been reduced in the same proportion as the frequency, the iron losses would not have increased so dramatically (in fact, they would have decreased below 10 W because of the reduction in frequency).*

13-3.
REFLECTED
IMPEDANCE,
IMPEDANCE
TRANSFORMATION
AND PRACTICAL
TRANSFORMERS

The iron core transformer of Fig. 13-2 is shown again in Fig. 13-4a, with a load Z_L across its secondary output terminals. Note that if the secondary load is removed and the transformer is open-circuited, $I_2 = 0$; and the impedance, Z_L is infinite (since $Z_L = V_2/I_2$). For any value of load impedance, Z_L, the secondary impedance looking into the secondary terminals from the load as shown in Fig. 13-4b is

$$Z_2 = \frac{V_2}{I_2} \quad (13\text{-}11)$$

Similarly, the equivalent input impedance, looking into the primary terminals from the source, as shown in Fig. 13-4b is

$$Z_1 = \frac{V_1}{I'_1} \quad (13\text{-}12)$$

Since any change in load impedance and current at the secondary is reflected as a change in current at the primary, it is sometimes convenient to simplify the transformer to a single equivalent circuit. This involves reflecting the secondary impedance to the primary as

$$Z_1 = \frac{V_1}{I'_1}$$

but $V_1 = \alpha V_2$ as shown in Eq. (13-5) and $I'_1 = I_2/\alpha$ as shown in Eq. (13-2b), therefore

$$Z_1 = \frac{\alpha V_2}{I_2/\alpha} = \alpha^2 \frac{V_2}{I_2}$$

---

* It should be noted that the same applies to electric machines. It is possible to operate an alternator or motor at a reduced frequency but the rated voltage should also be reduced in the same proportion. The kVA rating is also correspondingly reduced, for reasons illustrated above. But the efficiency of the machine is *increased* because of lower core losses at the lower frequency, as shown by Eqs. (12-3) and (12-4).

Figure 13-4

(a) Ideal transformer with load

(b) Equivalent input and output impedance

$$Z_1 = \frac{V_1}{I_1'} \qquad Z_2 = \frac{V_2}{I_2}$$

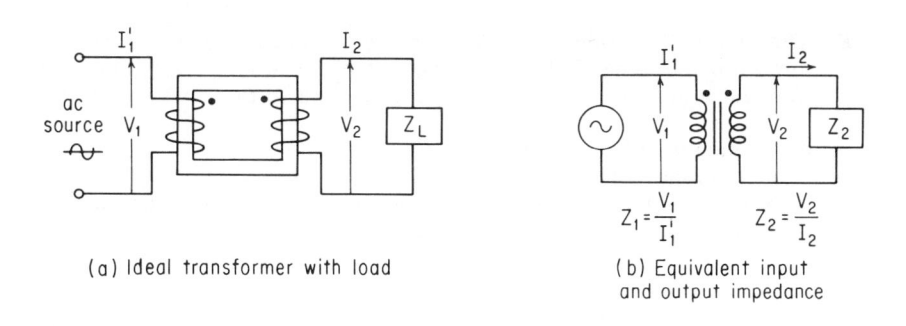

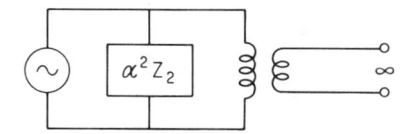

(c) Equivalent reflected impedance

Reflected impedance from secondary to primary.

but V_2/I_2 is the secondary impedance Z_2 as shown in Eq. (13-11) therefore

$$Z_1 = \alpha^2 Z_2 \quad \text{or} \quad \frac{Z_1}{Z_2} = \alpha^2 = \left(\frac{N_1}{N_2}\right)^2 \qquad (13\text{-}13)$$

Figure 13-4c shows the impedance looking into the input terminals from the source when the secondary impedance has been reflected back to the primary. It is now assumed that the secondary is open-circuited, as shown in Fig. 13-4c and that the impedance of the secondary transformer winding is negligible compared to the impedance of the load Z_L, which is equal to Z_2. Equation (13-13) states that the ratio of the input to output impedance is (equal to) the square of the transformation ratio. Since $Z_1 = \alpha Z_2$, this relation implies that transformers may serve as impedance matching devices so as to provide maximum power transfer from one circuit to another.* One common example is the case of an output transformer used to match the speaker load impedance to the output impedance of an audio amplifier. Example 13-6 illustrates the simplicity and elegance of Eq. (13-13) in the solution of input impedance calculations. Example 13-7 illustrates impedance matching of a servo-amplifier to a servomotor.

* In accordance with the maximum power transfer theorem, maximum power is delivered by a source to a load when the impedance of the load is equal to the internal impedance of the source. Since it is not always possible for the load to match the source impedance, transformers are used between source and load for such purposes.

EXAMPLE 13-6: The high voltage side of a step-down transformer has 800 turns and he low voltage side has 100 turns. A voltage of 240 V is applied to the high side and a load impedance of 3 Ω is connected to the low voltage side. Calculate:
a. secondary voltage and current
b. primary current
c. primary input impedance from ratio of primary voltage and current
d. primary input impedance from Eq. (13-13).

Solution:

a. $V_2 = \dfrac{V_1}{\alpha} = 240 \text{ V} \left(\dfrac{1}{800 \text{ turns}/100 \text{ turns}}\right) = \dfrac{240 \text{ V}}{8} = 30 \text{ V}$ (13-6)

$I_2 = \dfrac{V_2}{Z_2} = \dfrac{30 \text{ V}}{3 \text{ Ω}} = 10 \text{ A}$ (13-11)

b. $I_1 = \dfrac{I_2}{\alpha} = \dfrac{10 \text{ A}}{8} = 1.25 \text{ A}$ (13-2b)

c. $Z_1 = \dfrac{V_1}{I_1} = \dfrac{240 \text{ V}}{1.25 \text{ A}} = 192 \text{ Ω}$ (13-12)

d. $Z_1 = \alpha^2 Z_2 = 8^2 \times 3 \text{ Ω} = 192 \text{ Ω}$ (13-13)

EXAMPLE 13-7: An ac servoamplifier has an output impedance of 250 Ω and the ac servomotor which it must driven as an impedance of 2.5 Ω. Calculate:
a. the transformation ratio of the transformer to match the impedance of the servoamplifier to the servomotor impedance
b. the number of primary turns if the secondary has 10 turns.

Solution:

a. $\alpha^2 = \dfrac{Z_1}{Z_2}$ and $\alpha = \left(\dfrac{Z_1}{Z_2}\right)^{1/2} = \left(\dfrac{250}{2.5}\right)^{1/2} = (100)^{1/2} = 10$ (13-13)

b. $N_1 = \alpha N_2 = 10(10 \text{ turns}) = \mathbf{100 \text{ turns}}$ (13-2b)

13-3.1 Practical Transformer

A loaded, iron-core, practical transformer is represented in Fig. 13-5a. Although tightly coupled by the iron core, a small amount of leakage flux is produced at the primary and secondary windings, ϕ_1 and ϕ_2, respectively, apart from the mutual flux, ϕ_m, shown in Fig. 13-5a.

The primary leakage flux, ϕ_1, produces a primary inductive reactance, X_{L_1}. The secondary leakage flux, ϕ_2, produces a secondary inductive reactance, X_{L_2}. In addition, the primary and secondary windings are wound with copper having resistance. The internal resistance of the primary winding is r_1 and that of the secondary winding is r_2.

The winding resistances and reactances of the primary and secondary, respectively, produce voltage drops within the transformer as a result of the primary and secondary currents. Although these voltage drops are internal, it is convenient to represent them externally in series with an ideal transformer as shown in Fig. 13-5b. The ideal transformer shown in

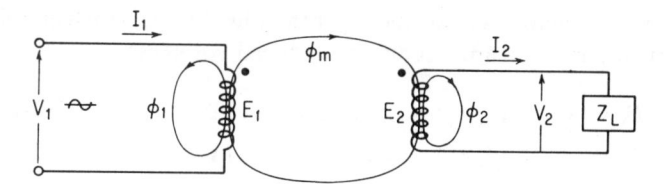

(a) Leakage fluxes in loaded practical transformer

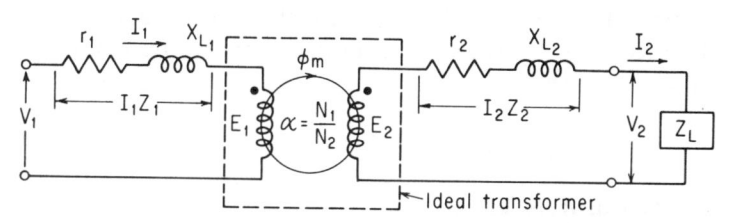

(b) Primary and secondary resistances and reactances causing voltage drops in practical transformer

Figure 13-5

Practical transformer.

Fig. 13-5b is assumed to have no resistance or reactive voltage drops in its windings. Leakage has been included by the primary voltage drop $I_1 Z_1$ and the secondary voltage drop, $I_2 Z_2$. Since these are inductive voltage drops, by simple ac theory, we can say that the primary internal transformer impedance is

$$Z_1 = r_1 + jX_{L_1} \quad \text{where all terms are defined in Sec. 13-1} \quad (13\text{-}14)$$

and the secondary internal transformer impedance is

$$Z_2 = r_2 + jX_{L_2} \quad \text{where all terms are defined in Sec. 13-1} \quad (13\text{-}15)$$

It is now possible to see the relation between the terminal and induced voltages of the primary and secondary, respectively. In accordance with Eq. (13-10), the induced primary and secondary voltages may be evaluated from the fundamental relation

$$E_1 = 4.44 f N_1 B_m A \times 10^{-8} \text{ V} \quad (13\text{-}16)$$

and

$$E_2 = 4.44 f N_2 B_m A \times 10^{-8} \text{ V} \quad (13\text{-}17)$$

where all terms have been defined above.

But since it is relatively difficult to evaluate B_m, the maximum permissible flux density of the transformer from voltage and current measurements, the following relations also emerge from Fig. 13-5b, which enable computation of the primary and secondary induced voltages:

$$\dot{E}_1 = \dot{V}_1 - (\dot{I}_1 \dot{Z}_1) = \dot{V}_1 - I_1(r_1 + jX_{L_1}) \tag{13-18}$$
$$\dot{E}_2 = \dot{V}_2 + (\dot{I}_2 \dot{Z}_2) = \dot{V}_2 + I_2(r_2 + jX_{L_2}) \tag{13-19}$$

Note from Fig. 13-5b and Eq. (13-18), that the voltage applied to the primary, V_1, is greater than the voltage induced in the primary winding E_1. Also note from Fig. 13-5b and Eq. (13-19) that the voltage induced in the secondary winding, E_2, is greater than the secondary terminal voltage, V_2. Thus, we may write

$$V_1 > E_1 \quad \text{and} \quad V_2 < E_2 \tag{13-20}$$

for a loaded, practical transformer.

EXAMPLE 13-8: A 2300/230 V, 500 kVA, 60 Hz step-down transformer has the following values: $r_1 = 0.1\ \Omega$, $X_{L_1} = 0.3\ \Omega$, $r_2 = 0.001\ \Omega$, $X_{L_2} = 0.003\ \Omega$. When the transformer is used as a step-down transformer and loaded to its rated capacity, calculate:
a. secondary and primary currents
b. secondary and primary internal impedances
c. secondary and primary internal voltage drops
d. secondary and primary induced voltages, assuming that the terminal and induced voltages are in phase
e. ratio of primary to secondary induced voltages, and primary to secondary terminal voltages.

Solution:

a. $I_2 = \dfrac{\text{kVA} \times 10^3}{V_2} = \dfrac{500 \times 10^3 \text{ VA}}{230 \text{ V}}$
$= 2.175 \times 10^3\ \text{A} = \mathbf{2175\ A}$ \hfill (13-9)

$I_1 = \dfrac{I_2}{\alpha} = \dfrac{2175\ \text{A}}{10} = \mathbf{217.5\ A}$ \hfill (13-2b)

b. $Z_2 = r_2 + jX_{L_2} = 0.001 + j0.003 = \mathbf{0.00316\ \Omega}$ \hfill (13-15)
$Z_1 = r_1 + jX_{L_1} = 0.1 + j0.3 = \mathbf{0.316\ \Omega}$ \hfill (13-14)

c. $I_2 Z_2 = 2175\ \text{A} \times 0.00316\ \Omega = \mathbf{6.88\ V}$
$I_1 Z_1 = 217.5 \times 0.316\ \Omega = \mathbf{68.8\ V}$

d. assuming that the terminal and induced voltages are in phase:
$E_2 = V_2 + I_2 Z_2 = 230 + 6.88 = \mathbf{236.88\ V}$ \hfill (13-19)
$E_1 = V_1 - I_1 Z_1 = 2300 - 68.8 = \mathbf{2231.2\ V}$ \hfill (13-18)

e. ratio of $\dfrac{E_1}{E_2} = \dfrac{2231.2}{236.88} = 9.43 = \alpha = \dfrac{N_1}{N_2}$ \hfill (13-5)

but $\dfrac{V_1}{V_2} = \dfrac{2300}{230} = 10$

Example 13-8 for a practical transformer shows that in order to obtain a given ratio of primary to secondary *terminal* voltages, the transformation ratio ($N_1/N_2 = E_1/E_2$) should be slightly *less* taking primary and secondary internal impedance drops into account. It also verifies the relations summarized in Eqs. (13-18) through (13-20).

EXAMPLE 13-9: From the primary and secondary terminal voltage and currents, respectively, in Ex. 13-8, calculate
 a. load impedance, Z_L
 b. primary input impedance, Z_P
 c. compare Z_L to Z_2 and Z_P to Z_1
 d. account for differences between these impedances in (c).

Solution:

a. $Z_L = \dfrac{V_2}{I_2} = \dfrac{230 \text{ V}}{2175 \text{ A}} = 0.1055 \; \Omega$

b. $Z_P = \dfrac{V_1}{I_1} = \dfrac{2300 \text{ V}}{217.5 \text{ A}} = 10.55 \; \Omega$

c. the impedance of the load, $Z_L = 0.1055 \; \Omega$ which is *much greater than* the internal secondary impedance, $Z_2 = 0.00316 \; \Omega$
The primary input impedance, $Z_P = 10.55 \; \Omega$ which is *much greater than* the internal primary impedance, $Z_1 = 0.316 \; \Omega$

d. It is essential for Z_L to be much greater than Z_2 so that the major part of the voltage produced by E_2 is dropped across the load impedance, Z_L. As Z_L is reduced in proportion to Z_2, the load current increases and more voltage is dropped internally across Z_2.

The primary input impedance, Z_P, represents the reflected load impedance, $Z_P = \alpha^2 Z_L = (10)^2 0.1055 = 10.55 \; \Omega$. Again, for a given value of primary load current drawn from the supply, the primary internal impedance, Z_1, should be small compared to the primary reflected impedance so that E_1 will be large relative to V_1.

13-4. EQUIVALENT CIRCUITS FOR A PRACTICAL POWER TRANSFORMER

The solutions and comparisons of Ex. 13-9 would imply that it might be possible to use impedance transformations to develop an equivalent circuit for the practical transformer. Such an equivalent is useful in solving problems related to the efficiency and voltage regulation of a transformer.

The circuit of Fig. 13-5b is shown in Fig. 13-6a with the load impedance and the secondary internal resistance and reactance reflected back to the primary. Note that the primary current, I_1, is the sum of a magnetizing current component, I_m, and a load component of current I'_1, in Fig. 13-6a. This is in accordance with the phasor relations of a loaded transformer shown in Fig. 13-3c. In addition, R_m represents the equivalent transformer power loss due to (hysteresis and eddy current) iron losses in the transformer core as a result of the magnetizing current, I_m. Note that

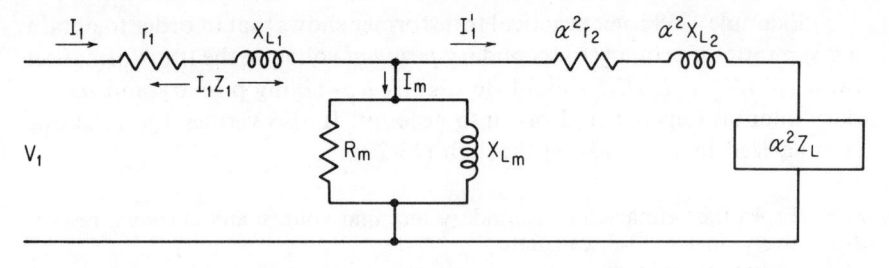

(a) Power transformer equivalent circuit

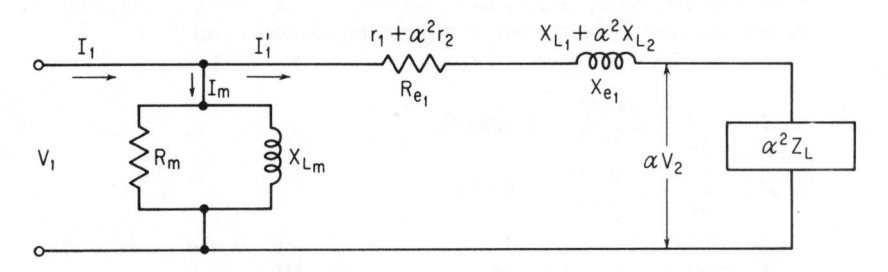

(b) Approximate equivalent containing lumped resistances and reactances

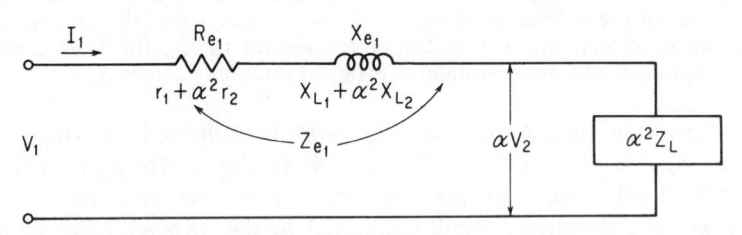

(c) Simplified equivalent circuit assuming negligible magnetizing current

Figure 13-6

Equivalent circuits for a power transformer.

R_m is shunted by X_{L_m} which represents the (open-circuit) transformer reactance component.

Figure 13-6a is a representation of the transformer that satisfies either the loaded or unloaded condition. If the secondary of the transformer shown is open-circuited, $I'_1 = 0$, and only I_m flows ($I_1 = I_m$) producing a small internal voltage drop due to primary impedance Z_1 (equal to $r_1 + jX_{L_1}$). Since the primary impedance Z_1 and the primary voltage drop $I_1 Z_1$ are relatively small, it is possible to obtain an approximate equivalent by *shifting* the parallel L-R branch directly across the supply, V_1. Doing this enables us to *lump* the internal resistances and reactances of the primary and secondary circuits, respectively, as shown in Fig. 13-6b, to produce the following equivalents:

530

CHAP. THIRTEEN / *Transformers*

$$R_{e_1} = r_1 + \alpha^2 r_2 = \text{equivalent resistance referred to the } \textit{primary} \quad (13\text{-}21)$$
$$X_{e_1} = X_{L_1} + \alpha^2 X_{L_2} = \text{equivalent reactance referred to the } \textit{primary} \quad (13\text{-}22)$$
$$Z_{e_1} = R_{e_1} + jX_{e_1} = \text{equivalent impedance referred to the } \textit{primary} \quad (13\text{-}23)$$

If the transformer is loaded to any extent, the load component of current, I'_1, exceeds the magnetizing current I_m, and the latter may be disregarded as negligible, as shown in the simplified equivalent of Fig. 13-6c. This figure permits a number of predictions regarding the transformer efficiency and voltage regulation (Secs. 13-5 to 8) as well as the evaluation of primary (and secondary) current. For a loaded transformer, the primary current, depending on the nature of the load, is

$$I_1 = \frac{V_1}{Z_{e_1} + \alpha^2 Z_L} = \frac{V_1}{(R_{e_1} + jX_{e_1}) + \alpha^2 (R_L \pm jX_L)}$$
$$= \frac{V_1}{[R_{e_1} + \alpha^2 R_L] + j[X_{e_1} \pm \alpha^2 jX_L]} \quad (13\text{-}24)$$

where $+jX_L$ represents the reactance of an inductive load and
$-jX_L$ represents the reactance of a capacitive load

EXAMPLE 13-10: For the transformer given in Ex. 13-7, calculate
a. Equivalent internal resistance referred to the primary.
b. Equivalent internal reactance referred to the primary.
c. Equivalent internal impedance referred to the primary.
d. Equivalent secondary load impedance of 0.1 Ω (resistive), referred to the primary.
e. Primary load current if the primary supply voltage is 2300 V.

Solution:

a. $R_{e_1} = r_1 + \alpha^2 r_2 = 0.1 + 10^2(0.001)$
$= 0.1 + 0.1 = \mathbf{0.2 \ \Omega}$ (13-21)

b. $X_{e_1} = X_{L_1} + \alpha^2(X_{L_2}) = 0.3 + 10^2(0.003)$
$= 0.3 + 0.3 = \mathbf{0.6 \ \Omega}$ (13-22)

c. $Z_{e_1} = R_{e_1} + jX_{e_1} = 0.2 + j0.6 = \mathbf{0.632 \ \Omega}$ (13-23)

d. $\alpha^2 Z_L = (10)^2 0.1 \ \Omega = 10 \ \Omega \text{ (resistive)} = \alpha^2 R_L$ (13-13)

e. $I_1 = \dfrac{V_1}{(R_{e_1} + \alpha^2 R_L) + j(X_{e_1} \pm \alpha^2 jX_L)}$ (13-24)
$= \dfrac{2300 \text{ V}}{(0.2 + 10.0) + j(0.6 + 0) \ \Omega}$
$= \dfrac{2300 \text{ V}}{10.2 + j0.6} \approx \dfrac{2300 \text{ V}}{10.21 \ \Omega} = \mathbf{225 \text{ A}}$

Note that the primary current value obtained in Ex. 13-10 of 225 A compares favorably with the primary current value of 217.5 A obtained in Ex. 13-8b.

The approximation of Fig. 13-6c neglects the magnetizing component, I_m, of the primary current, I_1. In effect, this means that the power factor angle of the secondary load is directly reflected to the primary without change. (Note that Fig. 13-3c shows a difference between the reflected angle θ'_1 and the primary power factor angle θ_1 due to the magnetizing component of current.) If the magnetizing current component of primary current is neglected, we obtain the simple equivalent phasor diagram for a loaded transformer under any condition of loading (leading, lagging or unity power factor), as shown in Fig. 13-7. The explanation of each of these phasor diagrams is given below:

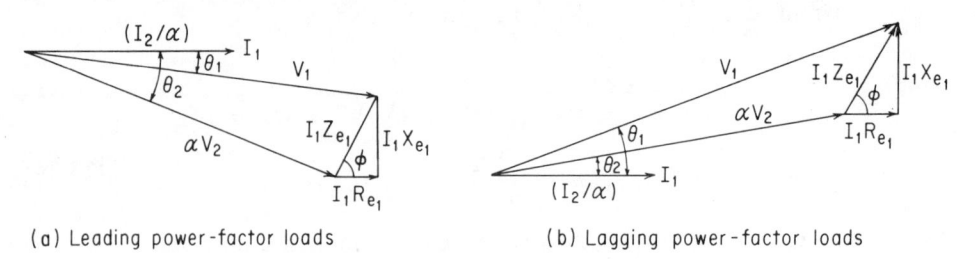

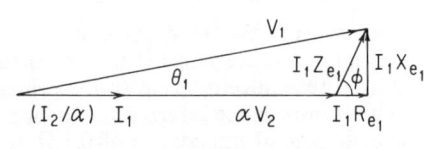

(a) Leading power-factor loads

(b) Lagging power-factor loads

(c) Unity power-factor loads

Figure 13-7

Power transformer under varying secondary load conditions and its effect on primary power factor angle.

      a. Leading power factor load—Fig. 13-7a. The reflected secondary load current, I_2/α, leads the reflected secondary load voltage, αV_2, by a leading power factor angle, θ_2. As shown in Fig. 13-6c, the phasor difference between αV_2 and V_1 is the equivalent impedance voltage drop $I_1 Z_{e_1}$. The resistive voltage drop $I_1 R_{e_1}$ is in phase with I_1. The reactive voltage drop, $I_1 X_{e_1}$, leads the current I_1 by 90°. Because of these equivalent voltage drops, V_1 still leads I_1 by angle θ_1. Leading angle θ_1, is of necessity, smaller than θ_2 because the transformer is internally inductive.

      b. Lagging power factor load—Fig. 13-7b. The reflected secondary load current, I_2/α, lags the reflected secondary load voltage, αV_2, by a lagging power factor angle, θ_2. The phasor difference between the reflected secondary voltage, αV_2, and the primary applied voltage, V_1, is the equivalent impedance voltage drop, $I_1 Z_{e_1}$. In this case, lagging power angle θ_1 is greater than lagging power factor angle θ_2 because the transformer is highly inductive internally and it tends to make the overall circuit more lagging.

c. Unity power factor load—Fig. 13-7c. The reflected secondary load current, I_2/α, is in phase with the reflected secondary load voltage, αV_2, at unity power factor, with a resistive load on the secondary of the transformer. The phasor difference between the reflected secondary voltage αV_2 and the primary applied voltage V_1 is the equivalent impedance voltage drop $I_1 Z_{e_1}$. The primary current I_1 lags V_1 by a small angle θ_1. With unity power loads on the secondary, the primary sees a small lag in primary current behind primary voltage due to the total internal equivalent inductance of the transformer.

13-5. VOLTAGE REGULATION OF A POWER TRANSFORMER

An examination of the three cases shown in Fig. 13-7 also reveals that the reflected secondary voltage, αV_2, differs from the primary applied voltage, V_1, by the equivalent impedance voltage drop referred to the primary, $I_1 Z_{e_1}$. (The phasor diagrams of Fig. 13-7 are similar to the phasor diagrams of Fig. 6-4 which are used to predict the voltages regulation of an alternator). This would imply that the relation between the reflected secondary terminal voltage under load, αV_2, and the primary voltage V_1, bears the same relation as that between the secondary terminal voltage and the secondary no-load voltage, respectively.

Figure 13-8 shows the three conditions of Fig. 13-7 with two modifications:

1. All values have been reflected to the *secondary* side of the transformer, and
2. secondary *voltage* is used as a reference phasor rather than the current.

The equation for equivalent resistance referred to the secondary is

$$R_{e_2} = r_2 + \frac{r_1}{\alpha^2} \qquad (13\text{-}25)$$

and the equation for equivalent reactance referred to the secondary is

$$X_{e_2} = X_{L_2} + \frac{X_{L_1}}{\alpha^2} \qquad (13\text{-}26)$$

From which the equivalent impedance referred to the secondary is

$$Z_{e_2} = R_{e_2} + jX_{e_2} \qquad (13\text{-}27)$$

For any given value of load current, I_2, therefore, it is possible to compute the induced voltage, E_2, and the voltage regulation of the transformer,

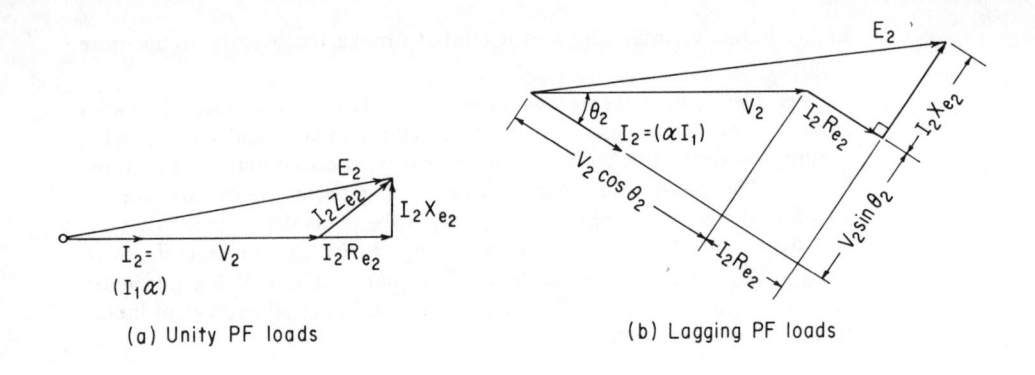

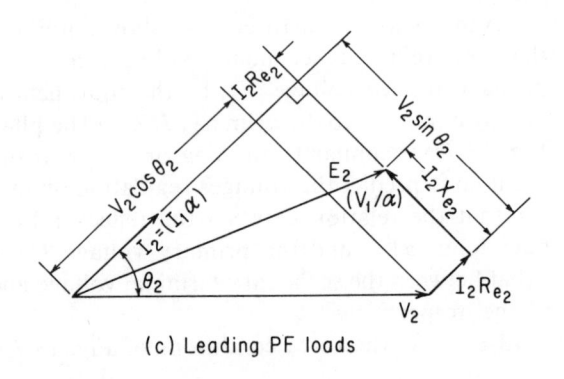

Figure 13-8

Power transformer secondary voltage regulation —all voltages and currents referred to secondary side—secondary voltage used as the reference phasor.

respectively, from the following relations emerging from Fig. 13-8.

$$E_2 = (V_2 \cos \theta_2 + I_2 R_{e_2}) + j(V_2 \sin \theta_2 \pm I_2 X_{e_2}) \qquad (13\text{-}28)$$

where the $+$ in the quadrature term is used for lagging and unity PF loads and the $-$ in the quadrature term is used for leading loads

Note the similarity between Eq. (13-28) for transformers and Eq. (6-8) for alternators. As in the case of dc generators and alternators, Eq. (3-9) may be rewritten for the voltage regulation, measured at the secondary terminals of a transformer:

$$\text{VR (percent voltage regulation)} = \frac{E_2 - V_2}{V_2} \times 100 \qquad (3\text{-}9a)$$

where E_2 is the no-load, induced secondary voltage, as determined from Eq. (13-28),

and V_2 is the full load (rated) secondary voltage of the transformer

EXAMPLE 13-11: Measurements made on a 2300/230 V, 500 kVA transformer provide the following values of equivalent reactance and resistance referred to the secondary (low-voltage side): $X_{e_2} = 0.006 \,\Omega$ and $R_{e_2} = 0.002 \,\Omega$.
Calculate:
a. induced emf, E_2, when the transformer is delivering rated secondary current to a unity PF load
b. repeat (a) for a load at 0.8 PF lagging
c. repeat (a) for a load at 0.6 PF leading
d. calculate the voltage regulation in (a), (b), and (c), respectively
e. account for differences in voltage regulation with reference to Fig. 13-8.

Solution:

Preliminary calculations:

Rated secondary current

$$I_2 = \frac{\text{kVA} \times 10^3}{V_2} = \frac{500 \times 10^3 \text{ VA}}{230 \text{ V}} = 2.175 \times 10^3 \text{ A} \qquad (13\text{-}9)$$

Full load equivalent resistance volt drop

$$I_2 R_{e_2} = (2.175 \times 10^3 \text{ A})(2 \times 10^{-3}\,\Omega) = \mathbf{4.35 \text{ V}}$$

Full load equivalent reactance volt drop

$$I_2 X_{e_2} = (2.175 \times 10^3 \text{ A})(6 \times 10^{-3}\,\Omega) = \mathbf{13.05 \text{ V}}$$

a. at unity PF, from Eq. (13-28)

$$\begin{aligned}E_2 &= (V_2 \cos\theta_2 + I_2 R_{e_2}) + j(V_2 \sin\theta_2 + I_2 X_{e_2})\\ &= (230 \times 1 + 4.35) + j(0 + 13.05) = 234.35 + j13.05 = \mathbf{234.5 \text{ V}}\end{aligned}$$

b. at 0.8 PF lagging

$$\begin{aligned}E_2 &= (V_2 \cos\theta_2 + I_2 R_{e_2}) + j(V_2 \sin\theta_2 + I_2 X_{e_2}) &(13\text{-}28)\\ &= (230 \times 0.8 + 4.35) + j(230 \times 0.6 + 13.05)\\ &= (184 + 4.35) + j(138 + 13.05) = 188.35 + j151.05 = \mathbf{241.8 \text{ V}}\end{aligned}$$

c. at 0.6 PF leading

$$E_2 = (V_2 \cos\theta_2 + I_2 R_{e_2}) + j(V_2 \sin\theta_2 - I_2 X_{e_2}) \qquad (13\text{-}28)$$
Note: minus sign for leading loads

$$\begin{aligned}&= (230 \times 0.6 + 4.35) + j(230 \times 0.8 - 13.05)\\ &= (138 + 4.35) + j(184 - 13.05) = 142.35 + j170.85 = \mathbf{222.2 \text{ V}}\end{aligned}$$

d. 1. VR at unity PF

$$VR = \frac{E_2 - V_2}{V_2} \times 100 = \frac{234.5 - 230}{230} \times 100 = \mathbf{1.956\%} \qquad (3\text{-}9a)$$

2. VR at 0.8 PF lagging

$$VR = \frac{E_2 - V_2}{V_2} \times 100 = \frac{241.8 - 230}{230} \times 100 = \mathbf{5.13\%} \qquad (3\text{-}9a)$$

3. VR at 0.6 PF leading

$$VR = \frac{E_2 - V_2}{V_2} \times 100 = \frac{222.2 - 230}{230} \times 100 = \mathbf{-3.39\%} \qquad (3\text{-}9a)$$

e. A transformer loaded to rated current by a unity PF load has a small (low) per cent regulation. E_2 is slightly larger than V_2 and leads V_2 by a small positive angle as shown by Ex. 13-11a and Fig. 13-8a.

A transformer loaded to rated current by a lagging load has a higher per cent regulation than when loaded at unity PF. E_2 is somewhat larger than V_2 than at unity PF, and leads V_2 by a relatively small positive angle, as shown in Fig. 13-8b.

A transformer loaded to rated current by a leading load may have a lower regulation, close to zero per cent or even negative as in the case of Ex. 13-11d3 above. The lower or negative regulation occurs because E_2 may be smaller than V_2, as shown in Fig. 13-8c.

Example 13-11 shows that the regulation of a transformer may be improved by a leading load current, I_2, with respect to the secondary terminal voltage, V_2. Such a (capacitive) leading current tends to counterbalance the internal inductive voltage drops due to the primary and secondary inductive reactance of the transformer, itself.

13-6. VOLTAGE REGULATION FROM THE SHORT-CIRCUIT TEST

The simplified equivalent circuit referred to the primary, derived in Fig. 13-6c, is shown again in Fig. 13-9a. If the low voltage secondary of a transformer is short-circuited,* both V_2, the secondary terminal voltage, and Z_L, the secondary load are zero. The equivalent circuit for such a transformer, with secondary short-circuited, is shown in Fig. 13-9b. The implication of Fig. 13-9b is clear. If a transformer secondary is short circuited, only the primary and secondary internal resistances and reactances are loading the transformer. Consequently, the current I_1 drawn from V_1 is determined solely by the equivalent internal impedance, Z_{e_1}, where $I_1 = V_1/Z_{e_1}$.

Figure 13-9 also makes possible another useful comparison. A

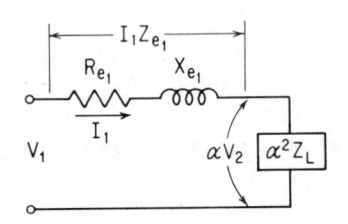

(a) Simplified equivalent circuit for loaded transformer

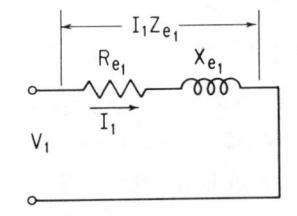

(b) Equivalent circuit of transformer with secondary shorted

Figure 13-9

Loaded and short-circuit transformer equivalent circuits, referred to the primary.

* The low voltage side is short-circuited because the winding has 1) a lower terminal voltage and 2) a higher current rating. While either side may be shorted, it is customary to short-circuit the low voltage side for these reasons.

normally loaded transformer, shown in Fig. 13-9a, has a small internal equivalent voltage drop, $I_1 Z_{e_1}$, in comparison to αV_2, the voltage dropped across the load. The major proportion of V_1 in Fig. 13-9a is dropped across αV_2. But with zero load impedance (secondary short-circuited) all of V_1 is dropped across the internal equivalent impedance, $I_1 Z_{e_1}$, as shown in Fig. 13-9b. Thus, when a transformer is short-circuited

$$V_1 = I_1 Z_{e_1} \tag{13-29}$$

from which

$$Z_{e_1} = \frac{V_1}{I_1} \tag{13-30}$$

and from simple ac circuit theory

$$R_{e_1} = \frac{V_1 \cos \theta}{I_1} = \frac{V_1 I_1 \cos \theta}{I_1^2} \tag{13-31}$$

and

$$X_{e_1} = \frac{V_1 \sin \theta}{I_1} = \frac{V_1 I_1 \sin \theta}{I_1^2} \tag{13-32}$$

Equations (13-30) through (13-32) clearly indicate that primary voltage, current and power measurements when the transformer is short-circuited will yield the parameters R_{e_1}, Z_{e_1} and X_{e_1}.

Figure 13-10 shows a typical arrangement of instruments and devices to obtain *short-circuit test* data of a given transformer. The procedure is as follows:

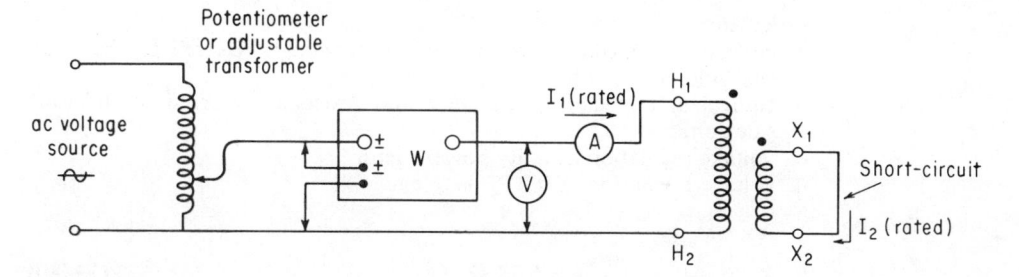

Figure 13-10

Typical instrument connections for short-circuit test to determine Ze_1, Xe_1, and Re_1.

1. With potentiometer or adjustable transformer set to provide zero output voltage, short-circuit the low voltage terminals, X_1-X_2, of the transformer.
2. Slowly and carefully increase the voltage using the adjustable transformer or potentiometer until *rated primary current* is recorded on the

SEC. 13-6. / Voltage Regulation from the Short-Circuit Test

ammeter. (Rated primary current is determined from the transformer rating in volt-amperes divided by the rated voltage of the high voltage side, VA/V_h.)

3. Record short-circuit power, P_{sc}; short-circuit voltage, V_{sc} and short-circuit primary current, $I_{sc} = I_1$ (rated).

4. Calculate Z_{e_1} from ratio of voltmeter to ammeter readings:

$$Z_{e_1} = \frac{V_{sc}}{I_{sc}} = \frac{\text{voltmeter reading}}{\text{ammeter reading}}$$

5. Calculate R_{e_1} from ratio of wattmeter reading to ammeter reading squared:

$$R_{e_1} = \frac{P_{sc}}{I_{sc}^2} = \frac{\text{wattmeter reading}}{(\text{ammeter reading})^2}$$

6. Calculate X_{e_1} from Z_{e_1} and R_{e_1} obtained in steps 4 and 5 above, using either:
   a. $X_{e_1} = \sqrt{Z_{e_1}^2 - R_{e_1}^2}$ or
   b. $\theta = \arccos(R_{e_1}/Z_{e_1})$
      $X_{e_1} = Z_{e_1} \sin \theta$

Short-circuit test data and calculations, as well as its application to transformer voltage regulation, is illustrated in Ex. 13-12.

EXAMPLE 13-12: A 2300/230 V, 20 kVA step-down transformer is connected as shown in Fig. 13-10, with the low voltage side short-circuited. Short-circuit data obtained for the high voltage side are:

wattmeter reading = 250 W

voltmeter reading = 50 V

ammeter reading = 8.7 A

Calculate:
a. equivalent impedance, reactance and resistance referred to the high side voltage
b. equivalent impedance, reactance and resistance referred to the low side voltage
c. voltage regulation at unity power factor
d. voltage regulation at 0.7 PF lagging.

Solution:

a. $Z_{e_1} = \dfrac{V_{sc}}{I_{sc}} = \dfrac{50 \text{ V}}{8.7 \text{ A}} = \mathbf{5.75 \ \Omega}$ \hfill (13-30)

$R_{e_1} = \dfrac{P_{sc}}{(I_{sc})^2} = \dfrac{250}{(8.7)^2} = \mathbf{3.3 \ \Omega}$ \hfill (13-31)

for X_{e_1}, $\theta = \arccos \dfrac{R_{e_1}}{Z_{e_1}} = \arccos \dfrac{3.3}{5.75} = \mathbf{55°}$

$X_{e_1} = Z_{e_1} \sin \theta = 5.75 \sin 55° = \mathbf{4.71 \ \Omega}$ \hfill (13-32)

b. $Z_{e_2} = \dfrac{Z_{e_1}}{\alpha^2} = \dfrac{5.75 \ \Omega}{10^2} = \mathbf{0.0575 \ \Omega}$ \hfill (13-13)

$$R_{e_2} = \frac{R_{e_1}}{\alpha^2} = \frac{3.3\,\Omega}{10^2} = 0.033\,\Omega$$

$$X_{e_2} = \frac{X_{e_1}}{\alpha^2} = \frac{4.71\,\Omega}{10^2} = 0.0471\,\Omega$$

c. rated secondary load current,

$$I_2 = \frac{\text{kVA} \times 1000}{V_2} = \frac{20 \times 10^3}{230\,\text{V}} = 87\,\text{A} \tag{13-9}$$

$$I_2 R_{e_2} = 87\,\text{A} \times 0.033\,\Omega = 2.87\,\text{V}$$
$$I_2 X_{e_2} = 87\,\text{A} \times 0.0471\,\Omega = 4.1\,\text{V}$$

Full load secondary induced emf at unity PF is

$$\begin{aligned}E_2 &= (V_2 \cos\theta_2 + I_2 R_{e_2}) + j(V_2 \sin\theta_2 + I_2 X_{e_2}) \\ &= (230 \times 1 + 2.87) + j(0 + 4.1) \\ &= 232.87 + j4.1 = 232.9\,\text{V}\end{aligned} \tag{13-28}$$

$$\begin{aligned}VR \text{ at unity PF} &= \frac{E_2 - V_2}{V_2} \times 100 \\ &= \frac{232.9 - 230}{230} \times 100 = \frac{2.9}{230} \times 100 = 1.26\%\end{aligned}$$

d. Full load secondary induced emf at 0.7 PF lagging

$$\begin{aligned}E_2 &= (V_2 \cos\theta_2 + I_2 R_{e_2}) + j(V_2 \sin\theta_2 + I_2 X_{e_2}) \\ &= (230 \times 0.7 + 2.87) + j(230 \times 0.713 + 4.1) \\ &= (161 + 2.87) + j(164 + 4.1) = 163.9 + j168.1 = 235\,\text{V}\end{aligned} \tag{13-28}$$

$$VR \text{ at 0.7 PF lagging} = \frac{E_2 - V_2}{V_2} = \frac{235 - 230}{230} = 2.175\% \tag{3-9a}$$

13-7. ASSUMPTIONS INHERENT IN THE SHORT-CIRCUIT TEST

The wattmeter of Fig. 13-10 essentially records the equivalent copper (resistance) losses of the primary and secondary, referred to the high voltage (primary) side, as implied in Eq. (13-31). It might appear that the wattmeter of Fig. 13-10 would provide a somewhat higher power reading since the ac input source must also supply the core or iron losses of the transformer. Such a higher reading conceivably yields a higher value of R_{e_1} from Eq. (13-31).

In actuality, however, as shown by Ex. 13-12, the voltage applied to the high side of the transformer during the short-circuit test is only a small fraction of the rated high-side voltage. The flux density, as shown by Eq. (13-10) varies directly as the voltage or $(\frac{50}{2300})$. The core losses, as shown by Ex. 13-5, vary approximately as the square of the flux density or as the square of the voltage applied to the primary. At rated voltage and frequency, the core loss of a transformer is hardly negligible. But in the case of the short-circuit test, the voltage applied to the primary is only a small fraction of the rated high-side voltage. Under these conditions, the core loss, which varies as the square of the voltage, may be considered negligible, as shown by Ex. 13-13.

EXAMPLE 13-13: Calculate the fraction of the core loss at rated voltage which is measured by the wattmeter in Ex. 13-12.

Solution:

Let P_h = rated core loss due to hysteresis and eddy currents at rated voltage. Since P_h is proportional to V_1^2, the core loss under short-circuit test conditions is

$$P_{h(sc)} = \left(\frac{V_{1sc}}{V_{1rated}}\right)^2 \times P_h = \left(\frac{50}{2300}\right)^2 \times P_h = (0.02175)^2 \times P_h$$
$$= 0.000474 P_h$$

Note from Ex. 13-13 that the core loss for a power transformer is indeed negligible, under extremely reduced high-side voltages during the short-circuit test. This is true for most power transformers with the possible exception of extremely small or high frequency iron core transformers. The core loss, therefore, may be neglected under most short-circuit testing conditions. (Note that core loss *cannot* be neglected in the short-circuit, locked rotor test of the induction motor as described in Sec. 12-13, last paragraph, precisely because the ratio of input to rated voltage is relatively high.)

13-8. TRANSFORMER EFFICIENCY FROM THE OPEN-CIRCUIT AND SHORT-CIRCUIT TEST

The rated or operating voltage core loss, as implied from the above discussion, may be obtained only by exciting any transformer winding at its rated voltage. Since most power transmission and distribution transformers have one or more very high voltage windings, it is customary and safer to perform the open-circuit test to determine core loss on the *lowest* voltage winding obtainable. The typical connections are shown in Fig. 13-11 for a two winding transformer, with rated voltage applied

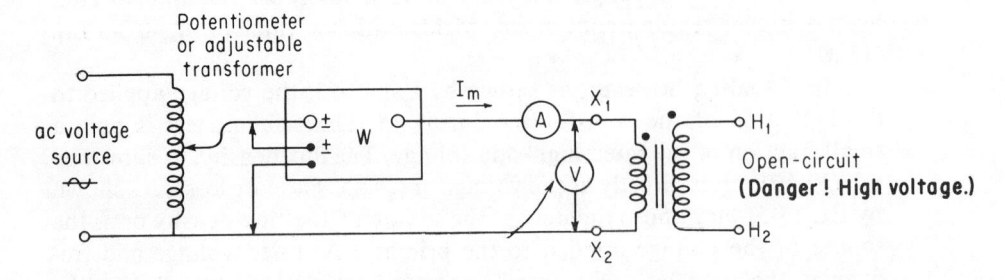

Figure 13-11
Typical instrument connections for open-circuit test to determine core loss.

540

CHAP. THIRTEEN / *Transformers*

to the low voltage terminals, $X_1 = X_2$, and the high voltage terminals, $H_1 = H_2$, open-circuited. Since rated voltage is applied to the low voltage side, rated voltage also appears at the terminals of the high voltage side. Care should be taken to see that these high voltage terminals are properly insulated from each other and from personnel (See Fig. 13-11).

The major purpose in performing the open-circuit test is to measure the core loss at rated voltage. The procedure for performing the open circuit test is as follows:

1. Bring potentiometer or adjustable transformer up from zero until rated voltage for the particular winding is recorded on the ac voltmeter.
2. Record open-circuit power, P_{oc}, rated voltage, V_r, and magnetizing current, I_m, as measured by the wattmeter, voltmeter and ammeter, respectively.
3. Calculate core loss from $P_h = P_{oc} - I_m^2 R_x$, where R_x is the resistance of the low voltage winding selected.

Because the transformer is open-circuited, the no load current is relatively small, as is the resistance of the particular low voltage winding on which the test is performed. In most instances, therefore, it is customary to use the wattmeter reading as the core loss without subtracting the small copper loss in the winding as a result of magnetizing current (see Ex. 13-14).

In addition to voltage regulation, it is possible to use the open-circuit *and* short-circuit test data to predict the efficiency of a transformer. It should be noted that both these tests employ *conventional* techniques rather than direct loading. (The advantages, as previously noted in Sec. 12-4, of conventional testing techniques are that small amounts of power are required to perform the tests, and often large loads are not available to test large transformers.) A transformer, whose secondary is open-circuited, only consumes power for its core losses, less than one per cent of its rated output. The power consumed during the short-circuit test, similarly, is very small, because the input power is essentially the rated copper loss of the transformer which, again, is less than one per cent of its rated output.

Equations (12-1) and (12-2) developed in the previous chapter for the dynamo apply equally to the transformer whose efficiency at any value of load is

$$\eta = \frac{P_{out}}{P_{out} + P_{losses}} = \frac{V_2 I_2 \cos\theta_2}{V_2 I_2 \cos\theta_2 + [\text{Core Loss} + I_2^2 R_{e_2}]} \quad (13\text{-}33)$$
$$\qquad\qquad\qquad\qquad\qquad\qquad\qquad\qquad\text{(fixed)}\quad\text{(variable)}$$

Note that the numerator of the above equation represents useful power transferred from the primary to the secondary and on to the load. The bracketed term in the denominator represents the losses which occur during this transfer. The losses are of two types 1) a *fixed* loss, the core loss and 2) a *variable* loss, the equivalent copper loss, referred to the secondary. It should also be noted that the only fixed term in Eq. (13-33) is the core loss. The useful output power term and the equivalent copper loss term *both* are functions of I_2, the secondary load current.

As stated for efficiency of dynamos in Eq. (12-6), maximum efficiency occurs when the fixed and variable losses are equal, or

$$I_2{}^2 R_{e_2} = P_h \quad \text{for maximum efficiency} \tag{13-34}$$

where P_h is the core loss; a fixed loss as determined from the open-circuit test.

The value of secondary load current at which maximum efficiency occurs is

$$I_2 = \sqrt{P_h/R_{e_2}} \quad \text{for maximum efficiency} \tag{13-35}$$

Further it should be noted that the power factor of the load, $\cos \theta_2$, determines the magnitude of the useful output power term in Eq. (13-33). For the same value of rated load current, I_2, a reduction in power factor is accompanied by a corresponding reduction in efficiency (see Fig. 13-12). Finally, as in the case of all machinery, electrical and otherwise, the efficiency curve of a transformer follows the same general form as dictated by Eq. 13-33. At relatively light loads, the fixed loss is high in proportion to the output, and the efficiency is low. Under heavy loads (output greater than rated), the variable (copper) loss is high in proportion to output and the efficiency is again low. Maximum efficiency, of course, occurs at that value of load where the fixed (core) loss equals the variable (copper) loss, as summarized in Eqs. (13-34) and (13-35). The efficiency curve, therefore, rises from zero (at zero output, no load) to a maximum at approximately half to full-rated load, and drops off again at heavy loads (beyond rated).

The following examples summarize the use of short-circuit and open-circuit test data to predict efficiency at various values of load, and the load at which maximum efficiency occurs for the particular transformer under test. Note in Ex. 13-14 that the conventional method employed requires but a small fraction of the kVA at which the transformer is rated (approximately 1.6% for the short-circuit test and even less for the open-circuit test) for performance prediction.

EXAMPLE 13-14: A 2300/208 V, 500 kVA, 60 Hz distribution transformer has just been constructed. It is tested by means of the open-circuit and short-circuit test, prior to being put into service as a step-down transformer, to predict its efficiency and regulation. The test data obtained is

Open-circuit test $V_l = 208$ V, $I_l = 85$ A, $P_{oc} = 1800$ W

Short-circuit test $V_h = 95$ V, $I_h = 217.5$ A, $P_{sc} = 8.2$ kW

From the above data, calculate
a. equivalent resistance referred to the low side
b. resistance of the low-side winding only
c. transformer copper loss of low-side winding during open-circuit test
d. transformer core loss when rated voltage is applied
e. can P_{oc} obtained in the open-circuit test be used as the core loss? Explain.

Solution:

a. From the short-circuit test,

$$R_{eh} = \frac{P_{sc}}{(I_h)^2} = \frac{8.2 \times 10^3}{(2.175 \times 10^2)^2} = 0.173 \ \Omega \qquad (13\text{-}31)$$

$$R_{el} = \frac{R_{eh}}{\alpha^2} = \frac{0.173 \ \Omega}{(2300/208)^2} = \frac{0.173 \ \Omega}{(11.05)^2} = \mathbf{0.001417 \ \Omega}$$

b. Resistance of the low-side winding only (cf. Ex. 13-10) $= \frac{R_{el}}{2} = \frac{0.001417 \ \Omega}{2} = \mathbf{7.1 \times 10^{-4} \ \Omega}$

c. $I_m^2 R_l = 85^2 \times 7.1 \times 10^{-4} = \mathbf{5.125 \ W}$

d. $P_{core} = P_{oc} - I_m^2 R_l = 1800 - 5.125 = \mathbf{1794.9 \ W}$

e. **Yes**, the power obtained in the open-circuit test may be used as the core loss. The error is approximately $5/1800 = 0.00278$ or 0.278%. This is within the error produced by the instruments used in the test. We may assume, therefore, that the core loss is **1800 W**.

EXAMPLE 13-15: Using the data of Ex. 13-14, calculate
a. efficiency of the transformer when the secondary is loaded by a purely resistive (unity PF) load to $\frac{1}{4}$, $\frac{1}{2}$, $\frac{3}{4}$, rated and $\frac{5}{4}$ rated load. Tabulate all losses, output and input as a function of load.
b. repeat (a) for same load conditions but with a load of 0.8 PF lagging
c. the load current at which maximum efficiency occurs regardless of PF
d. the load fraction at which maximum efficiency occurs
e. the maximum efficiency at unity PF.

Solution:

Preliminary data

Core loss = fixed loss = 1800 W, from open-circuit test data given in Ex. 13-14.

Copper loss at rated load = variable loss = 8200 W from short-circuit test data.

Full load output = kVA $\cos \theta_2$ = 500 kVA × 1 = 500 kW, from transformer rating.

a. Tabulation at **unity PF**

LOAD FRACTION (OF RATED LOAD)	CORE LOSS watts	COPPER LOSS watts	TOTAL LOSS watts	TOTAL OUTPUT watts	TOTAL INPUT (OUTPUT + LOSSES) watts	EFFI- CIENCY OUTPUT INPUT per cent
1/4	1800	512	2312	125,000	127,312	98.25
1/2	1800	2050	3850	250,000	253,850	98.47
3/4	1800	4610	6410	375,000	381,410	98.25
1	1800	8200	10,000	500,000	510,000	98.1
5/4	1800	12,800	14,600	625,000	639,600	97.8

b. Tabulation at **0.8 PF lagging**

1/4	1800	512	2312	100,000	102,312	97.7
1/2	1800	2050	3850	200,000	203,850	98.25
3/4	1800	4610	6410	300,000	306,410	97.9
1	1800	8200	10,000	400,000	410,000	97.6
5/4	1800	12,800	14,600	500,000	514,600	97.25

c. From Ex. 13-14a, $R_{el} = 0.001417 \, \Omega$

$$I_2 = \sqrt{\frac{P_h}{R_{e_2}}} = \sqrt{\frac{1800}{1.417 \times 10^{-3}}} = (1.27 \times 10^6)^{1/2} = \mathbf{1125 \text{ A}}$$

d. rated secondary current, $I_2 = \dfrac{500 \text{ kVA} \times 1000}{280 \text{ V}} = 2400 \text{ A}$

load fraction $= \dfrac{1125 \text{ A}}{2400 \text{ A}} = \mathbf{0.47}$

e. Efficiency $= \dfrac{V_2 I_2 \cos \theta_2}{V_2 I_2 \cos \theta_2 + (P_c) + (I_2^2 R_{e_2})}$

$$\eta = \frac{208 \times 1125 \times 1}{208 \times 1125 \times 1 + 2(1800)}$$

$$\eta = \frac{234{,}000}{234{,}000 + 3600} = \frac{234{,}000}{237{,}600} = \mathbf{98.48\%}$$

Several important conclusions may be drawn, from Ex. 13-14 and 13-15, regarding transformer efficiency. These are

1. At *no load* with rated voltage applied to one of the low voltage windings, the power drawn, P_{oc}, is essentially due to core loss. The small no load copper loss may be neglected.

544

CHAP. THIRTEEN / *Transformers*

2. Although the efficiency is zero at no load ($I_2 = 0$, and output $V_2 I_2 \cos \theta_2 = 0$), it rises quickly with small application of load to the secondary. As shown in Fig. 13-12, and in Exs. 13-15a and b, the efficiency is over 97% at 1/4 rated load.

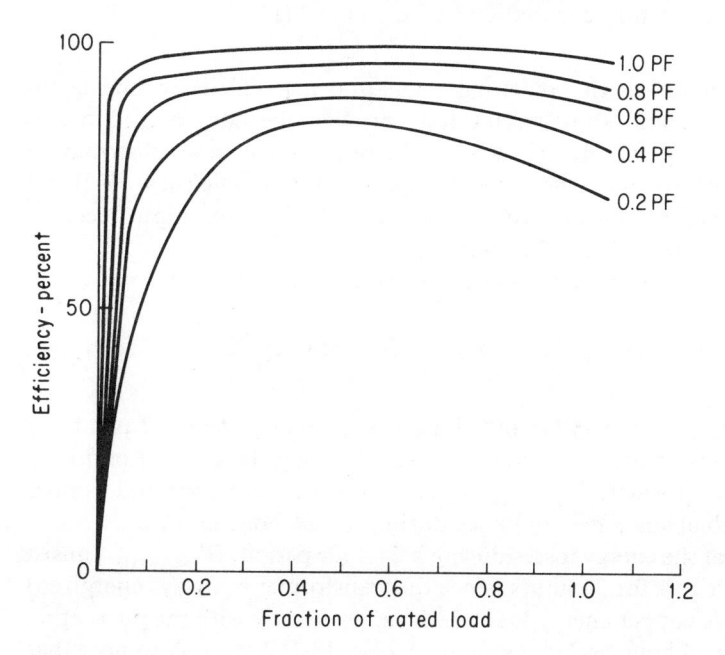

Figure 13-12

Transformer efficiency as affected by power factor of load.

3. Maximum efficiency occurs at approximately half load, as shown by Ex. 13-15d. Setting maximum efficiency at this value of load permits the transformer to maintain fairly high efficiency at light loads below this value and heavy loads above this value. The transformer, thus, maintains its high efficiency over a fairly broad spectrum of load values, as shown in Fig. 13-12.
4. For the same value of load current, the effect of decreasing values of power factor is a small reduction in efficiency, as shown in the family of curves of Fig. 13-12. Note from the curves and the tabulations (last column) of Ex. 13-15a and b, that at each load step the efficiency is higher at the higher power factor.
5. The load at which maximum efficiency occurs, however, for each curve of the family shown in Fig. 13-12, remains the *same* because the core losses and the variable copper loss ($I_2^2 R_{e_2}$) are *independent* of power factor. As in the case of (4) above, maximum efficiency is lower at this load point for lower power factors.

6. Efficiencies of transformers are somewhat higher than those of *rotating electrical machinery*, for the same kVA capacity, because the latter possess additional losses such as rotational losses and stray load losses. Thus, a well designed transformer is always more efficient than the rotating machinery connected as its secondary load.

13-9.
ALL-DAY
EFFICIENCY

In addition to predicting regulation and efficiency, the open-circuit and short-circuit tests provide the data useful in calculating *all-day efficiency* of transmission and distribution transformers, where by definition, **all-day efficiency** = (total *energy delivered* by a transformer to a load)/(total *energy* input received by the transformer), for a 24 hour period.

Stated in equation form, all day efficiency is expressed as

$$\text{all day efficiency} = \frac{W_{\text{out (total)}}}{W_{\text{in (total)}}} = \frac{W_{o1} + W_{o2} + W_{o3} \text{ etc.}}{W_{\text{out (total)}} + W_{\text{loss (total)}}} \quad (13\text{-}36)$$

where W_{o1}, W_{o2}, W_{o3} etc. are individual *energy* requirements drawn from the transformer by the load connected during the 24 hour period to the transformer. $W_{\text{loss (total)}}$ is the sum of (fixed) core and copper (variable) load *energy* losses during the 24 hour period.

Note that the energy losses during a 24 hour period, $W_{\text{loss (total)}}$, consist of a fixed core loss for 24 hours (since the transformer is always energized) plus a variable copper energy loss which varies directly with the fluctuating load over the 24 hour period, as shown by Ex. 13-16 below. Also note that Eq. (13-36) is a ratio of energies rather than powers, as shown in Ex. 13-16.

EXAMPLE 13-16: The 500 kVA distribution transformer of Ex. 13-14 is anticipated to have the following load requirements over a 24 hour period:

no load, 2 hours
20% rated load, 0.7 PF, for 4 hours
40% rated load, 0.8 PF, for 4 hours
80% rated load, 0.9 PF, for 6 hours
rated, load, unity PF, for 6 hours
125% rated load, 0.85 PF, for 2 hours

Assuming constant input voltage (and constant core loss), calculate
a. core loss over the 24 hour period
b. total energy loss over the 24 hour period
c. total energy output over the 24 hour period
d. all-day efficiency.

Solution:

a. $W_c = P_c t = \dfrac{1800 \text{ W} \times 24 \text{ hr}}{10^3 \text{ W/kW}} = $ **43.2 kWhr** = total energy core loss for 24 hr, including 2 hr at no load

b. From short-circuit test, equivalent copper loss at rated load = 8.2 kW, and the various energy losses during the 24 hour period are tabulated as,

% RATED LOAD	POWER LOSS kW	TIME PERIOD hr	ENERGY LOSS kWhr
20	$(0.2)^2 \times 8.2$	4	1.31
40	$(0.4)^2 \times 8.2$	4	5.25
80	$(0.8)^2 \times 8.2$	6	31.50
100	8.2	6	49.20
125	$(1.25)^2 \times 8.2$	2	25.62

Total energy load loss over 24 hour period = 112.88
(excluding 2 hr at no load)

c. total energy output over 24 hour period is tabulated as

% RATED LOAD	PF	kVA cos θ	kW	TIME PERIOD hr	W_{out}, ENERGY DELIVERED kWhr
20	0.7	$0.2 \times 500 \times 0.7$	140	4	560
40	0.8	$0.4 \times 500 \times 0.8$	160	4	640
80	0.9	$0.8 \times 500 \times 0.9$	360	6	2160
100	1.0	500×1	500	6	3000
125	0.85	$1.25 \times 500 \times 0.85$	531	2	1062

Total energy required by load for 24 hrs = 7422
(excluding 2 hr at no load)

d. all-day efficiency $= \dfrac{W_{out \text{ (total)}}}{W_{out \text{ (total)}} + W_{loss \text{ (total)}}}$

$= \dfrac{7422 \text{ kWhr}}{7422 + 43.2 + 112.88 \text{ kWhr}}$

$= \dfrac{7422 \text{ kWhr}}{7578.1 \text{ kWhr}} = 98\%$ (13-36)

Note that despite the varying load and PF conditions, the overall energy efficiency of the distribution transformer over a 24 hour period is relatively high. This is anticipated in view of the curves of Fig. 13-12 which show relatively high efficiency over a wide variation of load at relatively high power factors. Only complete lack of use or operation at extremely low power factors will result in low all-day efficiency for practical distribution transformers.

13-10. PHASING, IDENTIFICATION AND POLARITY OF TRANSFORMER WINDINGS

In addition to the open-circuit and short-circuit tests used to determine regulation, efficiency and all-day efficiency of commercial transformers, it is customary to perform a number of other tests before a transformer is put into service. Two such tests are concerned with the *phasing* and *polarity*, respectively, of the constructed transformer. *Phasing* is the process by which the individual terminals of the separate coil windings of a transformer are identified and corrected. The polarity test is performed so that the individual terminals of the separate coil windings of a transformer

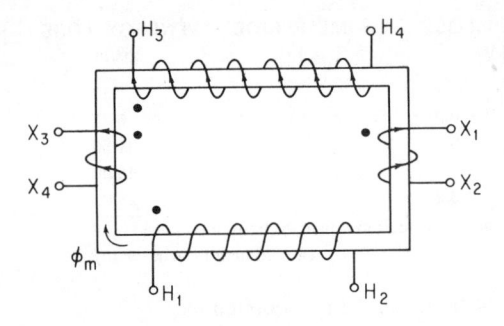

Figure 13-13

Determination of instantaneous polarity of transformers using dot convention.

may be marked or coded so that terminals having the same relative instantaneous polarity are identified. We will consider polarity, first, followed by phasing techniques.

Figure 13-13 shows a multicoil transformer having two high voltage windings and two low voltage windings. The high voltage coils (those having many turns) are coded using the letter "H" to designate their terminals. The low voltage terminals, as shown in Fig. 13-13 are designated by the letter "X".

As shown in Fig. 13-13, the instantaneous polarity is coded by a number subscript. The particular code shown in the figure uses an *odd*-number subscript to designate the instantaneous *positive* polarity of each winding. Note that the odd-numbered subscript also corresponds to the dot representing *positive induced emf* in each winding, shown in Fig. 13-13. Thus, in the event that coils are to be connected either in parallel or in series to obtain a variety of voltage ratios, the connection can be made properly with due regard for instantaneous polarity. The reader should verify for himself the manner in which a dot (or an odd number) is assigned to the windings in Fig. 13-13. Assume that the primary, H_1-H_2 is energized and that H_1 is instantaneously connected to the positive terminal of the supply. The mutual flux, ϕ_m, is set up instantaneously in the core in the clockwise direction shown. In accordance with Lenz's law, induced emfs are set up in the remaining windings in the direction shown. An alternative method to verify the dot convention of Fig. 13-13 is to compare the manner in which the coils are wound on the core. Coils H_1-H_2 and X_3-X_4 are wound in the same direction, therefore the dot is on the left terminal. Coils X_1-X_2 and H_3-H_4 are wound the same with respect to each other but opposite to H_1-H_2. These coils must have the dot on the right terminal to signify positive polarity and also a polarity opposite to H_1-H_2.

Unfortunately, it is impossible to examine a commercial transformer and deduce the direction in which the turns are wound, in order to determine either phasing or relative polarity of the coil terminals. A multiwinding transformer may have as few as 5 or as many as 50 leads brought out to a terminal block. If it is possible to examine the bare conductors of the coils, the diameter of the wire may provide some clue as to which leads or terminals are associated with high voltage or low voltage coils. Low voltage coils will have larger cross-sectional area conductors than high voltage coils. High voltage coils may also have heavier insulation than low voltage coils. However, this physical examination provides *no indication* whatever regarding polarity or phasing of coil taps or coil ends associated with individual coils which are insulated from each other.

**13-10.1
Transformer
Phasing**

Figure 13-14 shows a transformer whose coil ends have been brought out to a terminal block whose terminals have not (as yet) been identified as to phasing or polarity. A simple method for phasing the windings of the transformer is shown in Fig. 13-14. A series-connected 115 V lamp and a 115 V ac supply provides a

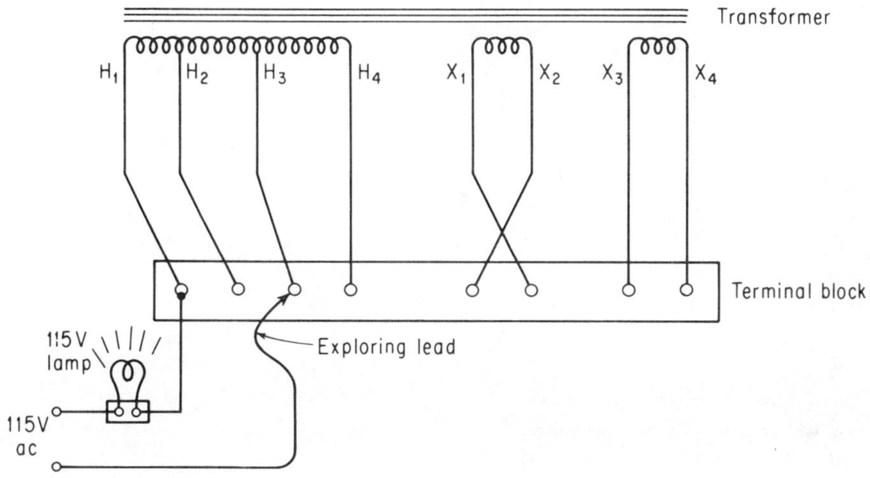

Figure 13-14

Test for phasing out terminals of transformer coils and coil taps using a lamp as a continuity tester. (An ac voltmeter may also be used in place of the lamp.)

means for effecting coil identification. If the load side of the lamp is connected to terminal H_1 as shown and the exploring lead is connected to terminal X_4, the lamp does not light. Moving the exploring lead from right to left along the terminal block produces no lamp indication until terminal H_4 is encountered. The lamp lights on terminals H_4, H_3 and H_2, indicating that only the four left hand terminals are part of a single coil. The relative brightness of the lamp may also provide some indication as to the taps. (The lamp is brightest when the leads are across H_1-H_2 and least bright across H_1-H_4). A more sensitive check of the phasing of coils and taps may be obtained using an ac voltmeter (1000 Ω/V) in place of the lamp with the voltmeter connected on its 150 V scale. The voltmeter will read the supply voltage for each tap of a common coil since its internal resistance (150 kΩ) is much higher than the resistance of the transformer winding. An electronic or battery operated ohmmeter

SEC. 13-10. / *Phasing, Identification and Polarity of Transformer Windings* may then be used to identify taps by resistance measurement and also verify coil windings by the continuity test.

13-10.2 Transformer Polarity

Having identified the coil ends by the above phasing tests, the relative instantaneous polarity is determined by the method shown in Fig. 13-15, using an ac voltmeter and a suitable ac voltage supply (either rated voltage or lower). The polarity test consists of the following steps:

(a) Polarity test

(b) Additive polarity ($V_t > V_r$)

(c) Subtractive polarity ($V_t < V_r$)

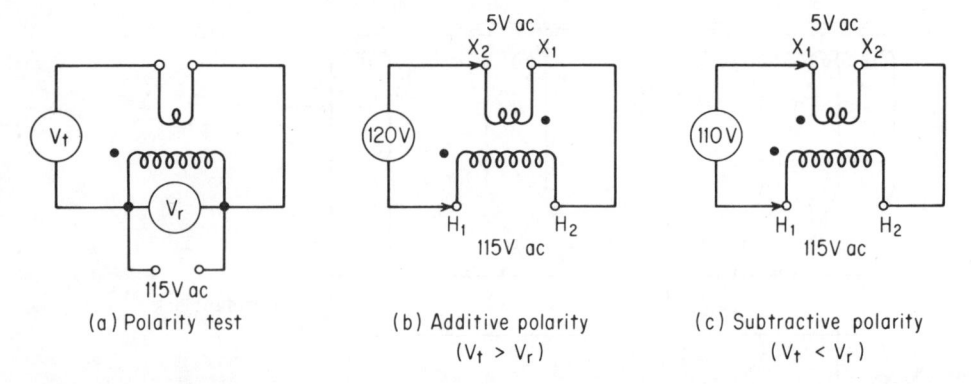

Figure 13-15

Test for polarity of windings of a transformer showing additive, subtractive polarity and terminal designations.

1. Select any high voltage winding and use it as a reference coil.
2. Connect a lead from one terminal of the reference coil to one terminal of any other winding of unknown polarity.
3. Designate the other terminal of the reference coil with a polarity dot (instantaneously positive).
4. Connect a voltmeter (ac) on highest range from dotted terminal of reference coil to other terminal of coil of unknown instantaneous polarity.
5. Apply rated (or lower) voltage to the reference coil.
6. Record voltage across reference coil V_r and test voltage between coils, V_t.
7. If the test voltage, V_t is *greater than* V_r, the polarity is *additive*, and the dot is designated on the test coil as shown in Fig. 13-15b.
8. If the test voltage, V_t is *less than* V_r, the polarity is *subtractive*, and the dot is designated on the test coil as shown in Fig. 13-15c.
9. Label the dotted terminal of reference coil H_1, and the dotted terminal of the test coil X_1 (or any suitable odd number designation).
10. Repeat steps 2 to 9 above for remaining windings of the transformer.

13-11. CONNECTING TRANSFORMER WINDINGS IN SERIES AND PARALLEL

The phasing and polarity tests described in Sec. 13-10 are fundamental when considering the ways that windings of a single multiwinding transformer or several individual transformers may be connected either in series or in parallel to obtain a variety of voltages. Let us first consider the multiwinding transformer shown in Fig. 13-13 having a rated voltage of 115 V for each high voltage winding and 10 V for each low voltage winding. Four possible combinations of voltage ratios are obtainable using this transformer, as shown in Fig. 13-16. These combinations are:

(a) High voltage coils in series, low voltage coils in series

(b) High voltage coils in series, low voltage coils in parallel

(c) High voltage coils in parallel, low voltage coils in series

(d) High voltage coils in parallel, low voltage coils in parallel

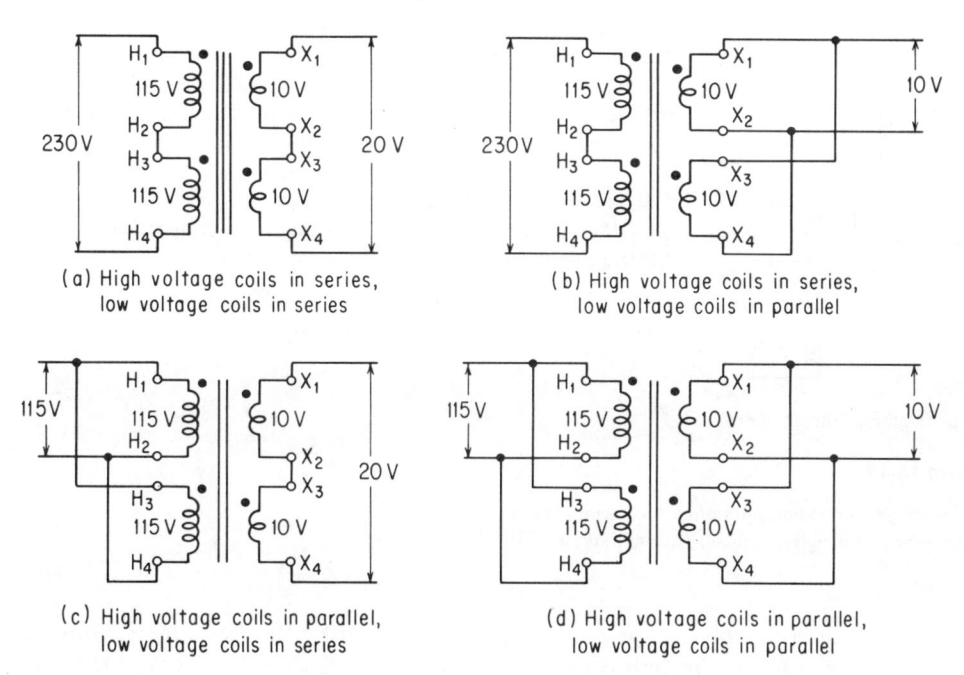

Figure 13-16

Connecting equal voltage transformer windings in series and parallel.

a. High voltage coils in series; low voltage coils in series (Fig. 13-16a).
b. High voltage coils in series; low voltage coils in parallel (Fig. 13-16b).
c. High voltage coils in parallel; low voltage coils in series (Fig. 13-16c).
d. High voltage coils in parallel; low voltage coils in parallel (Fig. 13-16d).

Note that when the coils are connected in *parallel*, coils having the SAME *voltage* and instantaneous polarity are paralleled (terminals having odd numbers are connected to one side of the line and even numbers to the other).

When connecting coils in *series*, coils of *opposite* instantaneous polarity are joined at one junction (an odd-numbered terminal is connected to an even-numbered terminal) so that the voltages are *series-aiding*. The induced voltages would oppose each other (yielding zero output voltage) if oppositely connected. (This last point, however, may be disregarded when connecting coils of unequal voltage ratings, as described below, Fig. 13-17).

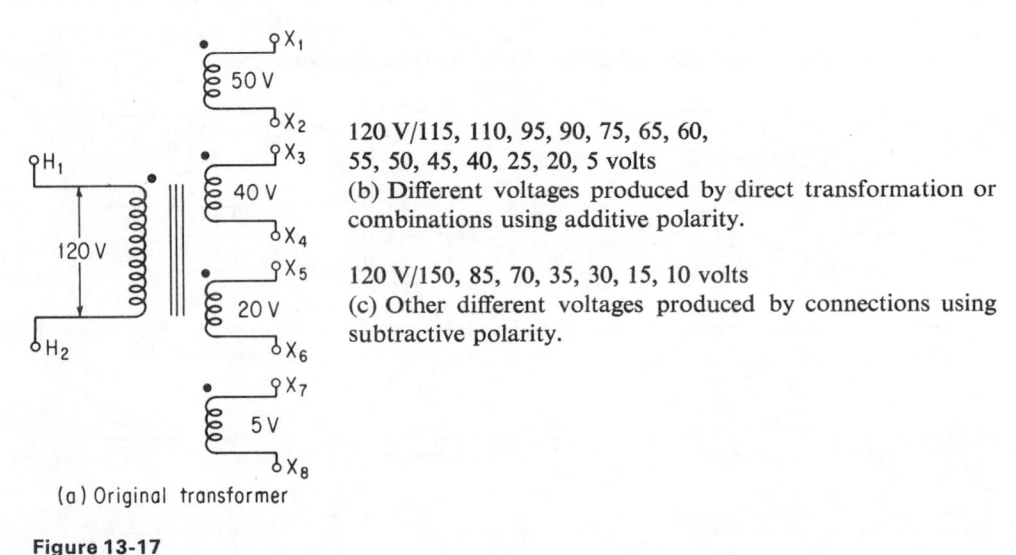

120 V/115, 110, 95, 90, 75, 65, 60, 55, 50, 45, 40, 25, 20, 5 volts
(b) Different voltages produced by direct transformation or combinations using additive polarity.

120 V/150, 85, 70, 35, 30, 15, 10 volts
(c) Other different voltages produced by connections using subtractive polarity.

(a) Original transformer

Figure 13-17

Connecting unequal voltage transformer windings in series-aiding and series opposing combinations.

Note that the voltage combinations produced by the four connections of Fig. 13-16a through d are, respectively: 230/20 V; 230/10 V; 115/20 V and 115/10 V. (While four voltage and current combinations are produced by these connections, only 3 ratios are produced, namely: 23/1 11.5/1 and 5.75/1).

Only coils of identical voltage ratings may be connected in parallel. The reason for this, as shown in Fig. 13-16d, is that when coils are paralleled, the induced voltages instaneously oppose each other. Thus, if two coils having unequal voltage ratings are paralleled, large circulating currents develop in both windings because the internal equivalent impedance of the windings is relatively small while the net difference between the (unequal) induced voltages is relatively large, as shown in Ex. 13-17.

EXAMPLE A 10 VA, 115 V primary filament transformer has two secondary windings,
13-17: 6.3 V and 5 V, respectively, having impedances of 0.2 Ω and 0.15 Ω, respectively. Calculate

a. rated secondary current when the low voltage secondaries are connected series-aiding
b. circulating current when the windings are paralleled and per cent overload produced.

Solution:

a. both coils must be series-connected and used to account for the full VA rating of the transformer. Hence,

$$I_2 = \frac{VA}{V_2} = \frac{10 \text{ VA}}{6.3 + 5 \text{ V}} = \frac{10 \text{ VA}}{11.3 \text{ V}} = \textbf{0.885 A}$$

rated current in 5 V and 6.3 V winding

b. when paralleled, the net circulating current is the net voltage applied across the total internal impedance of the windings or

$$I_c = \frac{6.3 \text{ V} - 5 \text{ V}}{0.2 + 0.15 \, \Omega} = \frac{1.3 \text{ V}}{0.35 \, \Omega} = \textbf{3.71 A}$$

The per cent overload is $(3.71 \text{ A}/0.885 \text{ A}) \times 100 = \textbf{419\%}$

Note that Ex. 13-17 shows that even when the voltage ratings appear to be almost the same they cannot be paralleled. *Only* windings having *identical* (same) voltage ratings may be connected in parallel with due regard for instantaneous polarity. If instantaneous polarity is disregarded, a short circuit immediately results, as shown in Ex. 13-18.

EXAMPLE 13-18: High-side short-circuit test data for the 20 kVA transformer shown in Fig. 13-16a is 4.5 V, 87 A, 250 W. Calculate:
a. equivalent impedance referred to the high side; coils series-connected
b. equivalent impedance referred to the low side; coils series-connected
c. rated secondary current; coils series-connected
d. secondary current if when the coils in Fig. 13-16a are short-circuited with rated voltage applied to the high voltage side, and percent overload produced.

Solution:

a. $Z_{e_h} = \dfrac{V_h}{I_h} = \dfrac{4.5 \text{ V}}{87 \text{ A}} = \textbf{0.05175} \, \boldsymbol{\Omega}$

b. $Z_{e_l} = Z_{e_h} \times \left(\dfrac{N_2}{N_1}\right)^2 = 0.05175 \, \Omega \left(\dfrac{20}{230}\right)^2 = \textbf{3.91} \times \textbf{10}^{-4} \, \boldsymbol{\Omega}$

c. I_2 rated $= \dfrac{20 \text{ kVA} \times 1000}{20 \text{ V}} = 1000 \text{ A} = \textbf{1} \times \textbf{10}^3 \textbf{ A}$

d. I_2 short-circuit $= \dfrac{20 \text{ V}}{3.91 \times 10^{-4} \text{ A}} = 5.1 \times 10^4 \text{ A} = \textbf{51} \times \textbf{10}^3 \textbf{ A}$

The per cent overload is $\dfrac{51 \times 10^3 \text{ A}}{1 \times 10^3 \text{ A}} \times 100 = \textbf{5100\%}$

Coils of unequal voltage ratings, however, may be *series*-connected either *aiding* or *opposing*. This gives rise to a number of interesting ratio possibilities in multiwinding transformers, as shown in Fig. 13-17. A total of 21 different voltage combinations, capable of delivering rated secondary current, are possible with the transformer shown in Fig. 13-17a (excluding autotransformer connections), using a 115 V primary. A total of 14 voltages emerge from the series-additive or direct transformation combinations, shown in Fig. 13-17b. In addition, 7 more voltages emerge from connections using subtractive combinations, shown in Fig. 13-17c. (The various ways of connecting the transformer to produce the above combinations is left as an exercise for the reader.)

It goes without saying that if the 120 V H_1-H_2 winding in Fig. 13-17 is not used as the primary, other combinations are possible such as using the 50 V winding (or the 40 V, etc,) as the primary. In these applications the transformer may be used as a step-up or step-down transformer with windings connected both series-aiding and opposing. Thus, many more combinations of voltage transformations are indicated than those shown as possibilities in Figs. 13-17b and c. Still other combinations are possible if winding H_1-H_2 is connected series aiding with winding X_7-X_8 (H_2 connected to X_7) and a 125 V primary is obtained. Similarly, connecting the high voltage winding to other low voltage windings would permit primary voltages as high as 230 V/5 V, and so on. Thus, the transformation possibilities of Fig. 13-17a are many. And still more are available if the transformer is connected as an autotransformer as described in the following section.

It should be noted, however, only when *all* the winding are used in aiding combinations is the full kVA capacity of the transformer generally realized, as indicated in the solution to Ex. 13-17a. Special transformers, however, are sometimes constructed to provide the full kVA capacity for any winding and/or ratio combination but such transformers are generally larger in size because more iron and heavier conductors are required for their construction. In using transformers in various series combinations, both aiding and opposing, therefore, consideration must be given to possible reduction in kVA capacity using windings in *isolation*, i.e., no conductive coupling between them.

13-12. THE AUTO-TRANSFORMER

All of the combinations discussed for the transformer of Fig. 13-17a in Sec. 13-11 assumed *isolation* between the primary and secondary. Transformations at higher efficiency and without severe reduction (indeed, even an increase) in kVA capacity are possible in an autotransformer providing we are willing to sacrifice isolation of the secondary from the primary circuit.

Theoretically, an *autotransformer* is defined as a transformer which has only one winding. Thus, a multiwinding transformer having isolated windings may be considered an autotransformer if all its windings are connected series-aiding (or opposing) to form a *single* winding. Such autotransformer connections are shown in Figs. 13-18a and b. At first

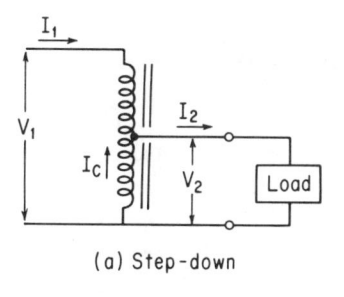

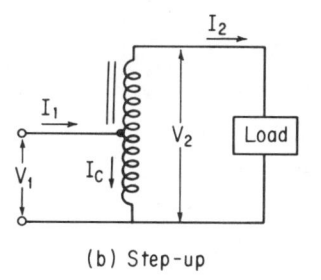

Figure 13-18

(a) Step-down (b) Step-up

Connections of autotransformer in step-down and step-up configurations.

glance, it might appear that the step-down transformer of Fig. 13-18a is nothing more than a voltage divider. But a glance at the current direction of that portion of the autotransformer *common* to both the primary and secondary circuits, I_c, shows that its current direction is reversed compared to an ordinary voltage divider. Further, in an ordinary voltage divider, I_1 is greater than I_2. But the autotransformer must obey Eq. (13-9) where $V_1 I_1 = V_2 I_2$. Hence if V_2 is smaller than V_1, I_2 must exceed I_1. Thus, for the circuit shown in Fig. 13-18a, as a step-down autotransformer

$$I_2 = I_1 + I_c \tag{13-37}$$

Figure 13-18b also proves beyond doubt that the autotransformer when used in the step-up configuration cannot possibly be a voltage divider. Here again, since $V_1 I_1 = V_2 I_2$ and $V_2 > V_1$, then $I_1 > I_2$. Thus, for the circuit shown in Fig. 13-18b, as a step-up autotransformer

$$I_1 = I_2 + I_c \tag{13-38}$$

Note the direction of I_c in Figs. 13-18a and b.

The autotransformer may also be made variable, however, in much the same way as the potentiometer is an adjustable voltage divider. *Variable autotransformers* consist of a single winding wound on a toroidal iron core, as shown in Fig. 13-19a. Such variable autotransformers, called *powerstats* or *variacs*, have a carbon wiper on a rotary shaft which makes contact with exposed turns of the transformer winding. Although the construction of Fig. 13-19a permits use as a step-down transformer only, the circuit of Fig. 13-19b permits both step-down and some step-up provision. (impossible in a potentiometer). Note that in both cases, however, only a *single* winding is employed. Variable autotransformers are extremely useful in laboratory or experimental situations requiring a wide range of voltage adjustment with little loss of power. The application of

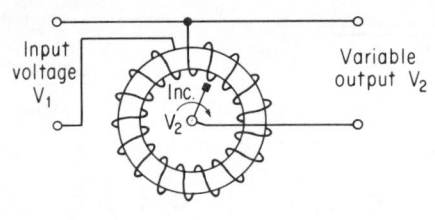

(a) Variable autotransformer

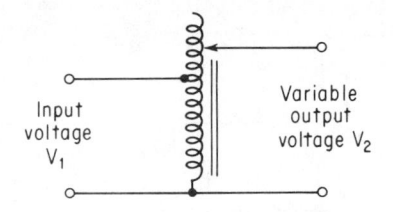

(b) Variable autotransformer with step-up and step-down provision

Figure 13-19

Variable autotransformer.

the variable autotransformer as a single phase motor speed control device is described in Sec. 10-7. The application of a *tapped* autotransformer as an impedance transfer device is discussed in Sec. 10-8.

It should be noted that the instantaneous current in the common portion of the autotransformer, I_c, shown in Figs. 13-18a and b, may occur in either direction, upward (away from) or downward (towards) with respect to the common connection, depending on whether the transformer is used as a step-down or step-up device. We shall see, also, that the direction of the instantaneous current is also a function of whether the common winding uses additive or subtractive polarity with respect to that part of the winding not common to both circuits (primary and secondary). Thus, the only way to determine the direction of current in the common winding is to draw instantaneous directions of primary current, I_1 and secondary current, I_2. The difference between these currents is and must be supplied by I_c.

Any ordinary two-winding isolation transformer may be converted into an autotransformer, as shown in Fig. 13-20. The original isolation transformer, with its polarity markings is shown in Fig. 13-20a. The transformer selected is a 10 kVA, 1200/120 V isolation transformer. It is desired to convert this transformer into an autotransformer, preserving *additive* polarity between the high voltage and low voltage sides. The connection for additive polarity is shown in Fig. 13-20b. This circuit is redrawn in Fig. 13-20c with the common terminal of the autotransformer at the top and redrawn once again in Fig. 13-20d with the common terminal at the bottom. Since the polarity is additive (as shown in Fig. 13-20d), the secondary voltage $V_2 = 1320$ V, while the primary voltage, V_1 is 1200 V. Although the original kVA of the isolation transformer is 10 kVA the arrangement shown in Fig. 13-20d results in a *marked increase* in kVA as noted from Ex. 13-18. Also note from Fig. 13-20d that the low voltage side has a higher current ($I_1 > I_2$) and I_c must flow *toward* the common terminal, in accordance with Eq. (13-38).

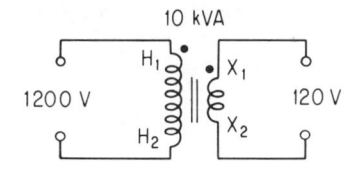

(a) Original 10-kVA isolation transformer

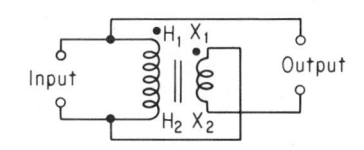

(b) Connection as a step-up transformer, using additive polarity

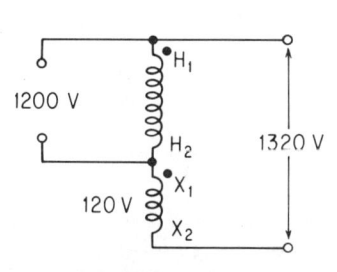

(c) Voltages produced by additive polarity

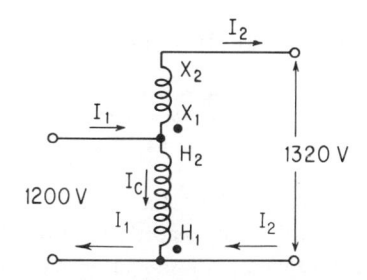

(d) Figure redrawn with common at bottom showing current relations

Figure 13-20

Isolation transformer connected as a step-up transformer using additive polarity.

EXAMPLE 13-19: For the 10 kVA, 1200/120 V isolation transformer shown in Fig. 13-20a connected as an autotransformer with additive polarity shown in Fig. 13-20d, calculate
a. original capacity of the 120 V winding in amperes
b. original capacity of the 1200 V winding in amperes
c. kVA rating of autotransformer of Fig. 13-20d using capacity of 120 V winding computed in (a) above
d. per cent increase in kVA capacity of autotransformer over isolation transformer
e. I_1 and I_c in Fig. 13-20d from value of I_2 used in part (c) above
f. per cent overload of 1200 V winding when used as an autotransformer
g. interpret and draw conclusions from the above calculations.

Solution:

a. $I_{\text{low side}} = \dfrac{10 \text{ kVA} \times 1000}{120 \text{ V}} = 83.3 \text{ A}$ (13-9)

b. $I_{\text{high side}} = \dfrac{10 \text{ kVA} \times 1000}{1200 \text{ V}} = 8.33 \text{ A}$ (13-9)

c. Since the 120 V winding is capable of carrying 83.3 A, the new kVA rating of the autotransformer is

$$V_2 I_2 = \dfrac{1320 \text{ V} \times 83.3 \text{ A}}{1000} = 110 \text{ kVA} \qquad (13\text{-}9)$$

SEC. 13-12. / The Autotransformer

d. per cent increase in kVA using isolation transformer as an autotransformer is

$$\frac{\text{kVA auto}}{\text{kVA isolation}} = \frac{110 \text{ kVA}}{10 \text{ kVA}} \times 100 = \mathbf{1100\%}$$

e. $I_1 = \dfrac{\text{kVA} \times 1000}{V_1} = \dfrac{110 \text{ kVA} \times 1000}{1200 \text{ V}} = \mathbf{91.75 \text{ A}}$

$I_c = I_1 - I_2$ (from Fig. 13-20d) \hfill (13-38)
$= 91.75 - 83.3 = \mathbf{8.42 \text{ A}}$

f. Percentage OL of 1200 V winding $= \dfrac{I_c}{I_{high}} = \dfrac{8.42 \text{ A}}{8.33 \text{ A}} \times 100 = \mathbf{101\%}$

g. As an autotransformer, the kVA has been increased to 1100% of its original value with the low voltage coil at its rated capacity and the high voltage coil at a negligible overload (1.01 × rated).

The dramatic increase in kVA capacity produced by connecting an isolation transformer as an autotransformer accounts for the smaller size in autotransformers of the same kVA capacity compared to ordinary isolation transformers. It should be pointed out, however, that only as the ratio of primary to secondary voltages approaches unity does such a marked increase in capacity occur. If there is a large ratio of primary to secondary voltages, the kVA capacity increase is not as marked. (For $\alpha > 10$, the kVA increase is less than 10%.)

The same isolation transformer using subtractive polarity connected as a step-down transformer is shown in Fig. 13-21. In order to produce a

(a) Connection as a step-up transformer using subtractive polarity

(b) Voltages produced by subtractive polarity

(c) Figure redraw with common at bottom showing current relations

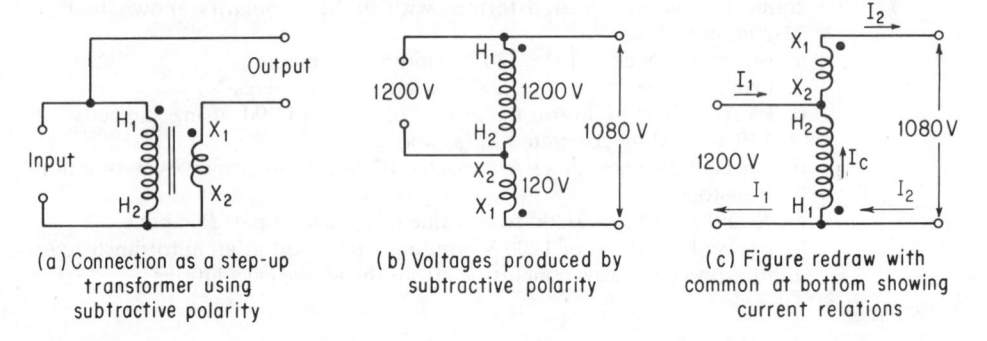

Figure 13-21

Isolation transformer connected as a step-down transformer using subtractive polarity.

single winding using *subtractive* polarity, it is necessary to connect X_2 to H_2, as shown in Fig. 13-21a. The voltages produced by this combination are shown in Fig. 13-21b, where it may also be seen that the transformer (despite its appearance) acts as a *step-down* transformer. This circuit is again redrawn in Fig. 13-21c and the instantaneous currents are added.

Note that current I_c is away from the junction in this case primarily because the autotransformer is a step-down transformer and $I_2 > I_1$. The current I_c must augment I_1 to equal I_2, as noted in Eq. (13-37).

As in the previous case of additive polarity, connecting the 10 kVA isolation transformer as a step-down autotransformer with subtractive polarity results in an increase in kVA rating, as shown by Ex. 13-19.

EXAMPLE 13-20: Repeat Ex. 13-19 for the 10 kVA, 1200/120 V isolation transformer connected as a step-down autotransformer with subtractive polarity shown in Fig. 13-21d.

Solution:

a. From Ex. 13-19a, $I_{\text{low side}} = 83.3$ A
b. From Ex. 13-19b, $I_{\text{high side}} = 8.33$ A
c. New kVA rating of the autotransformer is

$$\frac{V_2 I_2}{1000} = \frac{1080 \text{ V} \times 83.3 \text{ A}}{1000} = 90 \text{ kVA} \tag{13-9}$$

d. Per cent increase in kVA rating using isolation transformer as an autotransformer is

$$\frac{\text{kVA (auto)}}{\text{kVA (isol)}} = \frac{90 \text{ kVA}}{10 \text{ kVA}} = 900\%$$

e. $I_1 = \dfrac{\text{kVA} \times 1000}{V_1} = \dfrac{90 \text{ kVA} \times 1000}{1200 \text{ V}} = 75$ A (13-9)

$I_c = I_2 - I_1 = 83.33 - 75 = 8.33$ A (13-38)

f. Percentage OL of 1200 V winding $= \dfrac{I_c}{I_{\text{high}}} = \dfrac{8.33}{8.33} = 100\%$

g. As an autotransformer with subtractive polarity, the kVA has been increased to 900% of its original value as an isolation transformer with both the low voltage and high voltage coils connected at rated capacity.

Examples 13-19 and 13-20 prove that connecting an isolation transformer as an autotransformer results in an increase in kVA capacity. The increase in capacity varies with the connection (either subtractive or additive) and the ratio of transformation produced.

The question that now may be considered is "Why does the kVA rating of an isolation transformer increase when connected as an autotransformer?" Transformers, as we have seen, are fairly efficient devices. Practically all the energy received by the primary is transformed to the secondary by an isolation transformer. Furthermore, energy can neither be created nor destroyed. Then, why should the autotransformer be capable of greater energy transfer (than an isolation transformer) from primary to secondary?

The clue to the answer is found in the fact that there is *no* conductive link between the primary and secondary circuit of an *isolation* transformer.

In an isolation transformer, *all* the energy received by the primary must be *transformed* to reach the secondary. In an autotransformer, *part* of energy may be transferred *conductively* from the primary to the secondary and part of the energy may be transferred by transformer action. This difference accounts for the increase in kVA rating of the autotransformer. The autotransformer has the advantage of transferring energy conductively, as well as by transformer action, from the primary to the secondary, as we shall now see.

The circuits of Fig. 13-18 are shown in Fig. 13-22 with some added notation. The circuit of Fig. 13-22a shows a step-down auto transformer.

$V_P I_1$ transformed
$V_2 I_1$ transferred conductively

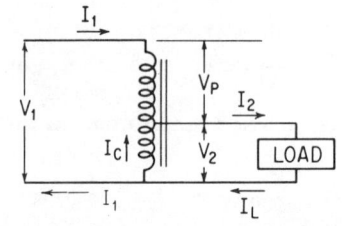

$V_S I_2$ transformed
$V_1 I_2$ transferred conductively

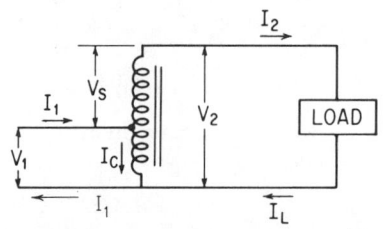

(a) Step-down currents and voltages

(b) Step-up currents and voltages

Figure 13-22

Notation for autotransformer in step-down and step-up configurations showing volt-amperes transferred conductively and by transformer action.

Since $I_2 = I_1 + I_c$ in this circuit, *all* of the current I_1 is *conducted* to I_2. The voltamperes transferred *conductively*, from primary to secondary, for a *step-down transformer*, is

$$\frac{\text{conductive voltamperes}}{\text{(transferred to secondary from primary)}} = V_2 I_1 \quad (13\text{-}39)$$

Since $V_2 + V_p = V_1$, the difference between V_1 and V_2 (or V_p) is a measure of the energy transformed. Thus, the voltamperes transferred from primary to secondary by transformer action, for a *step-down transformer*, is

$$\frac{\text{transformed voltamperes}}{\text{(transferred to secondary from primary)}} = V_p I_1 \quad (13\text{-}40)$$

For a step-up transformer, the same logic prevails. As shown in Fig. 13-22b, I_2 is that part of I_1 which is transferred conductively. Hence, the voltamperes conductively transferred from primary to secondary, for a *step-up* transformer, is

$$\frac{\text{conductive voltamperes}}{\text{(transferred to secondary from primary)}} = V_1 I_2 \quad (13\text{-}41)$$

Since $V_2 = V_s + V_1$, the difference between V_2 and V_1 (or V_s) is a measure of the energy transformed. Thus the voltamperes transferred from primary to secondary by transformer action, for a *step-up* transformer, is

$$\frac{\text{transformed voltamperes}}{\text{(transferred to secondary from primary)}} = V_s I_2 \quad (13\text{-}42)$$

For both the step-up and step-down transformer, the total amount of energy transferred from the primary to the secondary, measured in kVA is

$$\text{kVA (total)} = \text{kVA transferred conductively} + \text{kVA transformed} \quad (13\text{-}43)$$

Thus, for a *step-down* autotransformer

$$\text{kVA}_{total} = \frac{V_2 I_1}{1000} + \frac{V_p I_1}{1000} \quad (13\text{-}44)$$

While for a *step-up* autotransformer

$$\text{kVA}_{total} = \frac{V_1 I_2}{1000} + \frac{V_s I_2}{1000} \quad (13\text{-}45)$$

EXAMPLE 13-21: For the step-up autotransformer of Ex. 13-19 and Fig. 13-20d, calculate
a. kVA transferred conductively from primary to secondary
b. kVA transformed
c. Total kVA
d. Compare answer in (c) above to part (c) of Ex. 13-19.

Solution:

a. kVA conductively transferred $= \dfrac{V_1 I_2}{1000} = \dfrac{1200 \text{ V} \times 83.3 \text{ A}}{1000}$

$\qquad\qquad\qquad\qquad\qquad\qquad = 100 \text{ kVA} \qquad\qquad (13\text{-}41)$

b. kVA transformed $= \dfrac{V_s I_2}{1000} = \dfrac{120 \times 83.3}{1000} = 10 \text{ kVA} \qquad (13\text{-}42)$

c. kVA total $= 100 \text{ kVA} + 10 \text{ kVA} = 110 \text{ kVA} \qquad\qquad (13\text{-}43)$

d. The answer to (c) above is identical to that obtained in part (c) of Ex. 13-19.

Note that the original transformer of Ex. 13-19 had a rating of 10 kVA as an isolation transformer. In Ex. 13-21, connected as a step-up autotransformer, 100 kVA is additionally transferred conductively but 10 kVA is still transformed. The increase in kVA rating of the autotransformer is due to *conductive* (and not transformer) energy transfer.

EXAMPLE 13-22: For the step-down autotransformer of Ex. 13-20 and Fig. 13-21d (using subtractive polarity), calculate
a. kVA transferred conductively from primary to secondary
b. kVA transformed
c. Total kVA
d. Compare answer in (c) above with part (c) of Ex. 13-20

Solution:

a. kVA conductively transferred $= \dfrac{V_2 I_1}{1000} = \dfrac{1080 \text{ V} \times 75 \text{ A}}{1000} =$ **81 kVA**

b. kVA transformed $= \dfrac{V_p I_1}{1000} = \dfrac{120 \text{ V} \times 75 \text{ A}}{1000} =$ **9 kVA**

c. kVA total $= 81 \text{ kVA} + 9 \text{ kVA} =$ **90 kVA**

d. The answer to (c) above is identical to that obtained in part (c) of Ex. 13-20.

Note again in Ex. 13-22 that the 10 kVA isolation transformer reconnected as a step-down autotransformer using subtractive polarity has a much higher rating due to the kVA conductively transferred. The transformed kVA is still within the rating of the original isolation transformer. Again, the increase in kVA rating of the autotransformer is due to conductive (and not transformer) energy transfer.

13-13. AUTO-TRANSFORMER EFFICIENCY

The efficiency of a conventional isolation transformer, as shown in Fig. 13-12, is fairly high from relatively light loads to full load. As described in Sec. 13-8, only two classes of losses may be found in a conventional transformer: a fixed core loss and a variable copper loss in the primary and secondary windings. The latter loss increases as the square of the load current. Thus, the variable copper loss at 5/4 rated load is 25/16 (approximately 156%) of the loss at rated load.

It was also shown (Sec. 13-12) that the autotransformer transfers part of its kVA by conduction. Consequently, for the same kVA output, an autotransformer is somewhat smaller (less iron is used) than a conventional isolation transformer. Thus, the core losses are significantly less for the same output power in an autotransformer.

The autotransformer possesses only one winding, by definition, as compared to two in the conventional isolation transformer. Further, as shown in Fig. 13-23, the current carried by a portion of this winding is the difference between the secondary and primary currents. These two factors

Figure 13-23

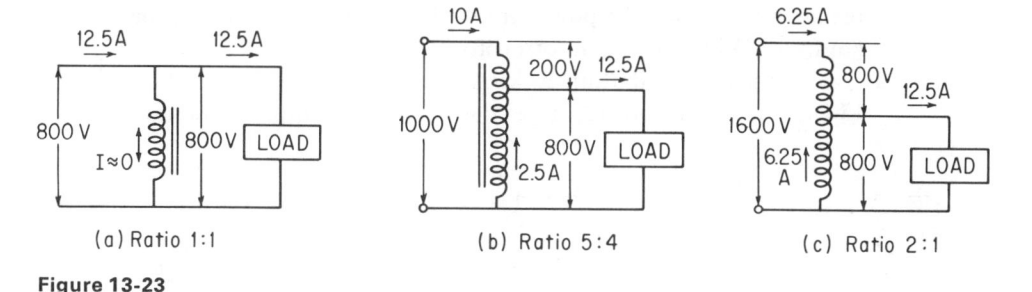

Effect of transformer ratio on autotransformer efficiency.

(a single winding and lower current) tend to reduce the variable load loss, as well.

The net effect is that autotransformers possess unusually high efficiency (99% and higher) approaching 100%. This efficiency, however, varies with the transformer ratio, as shown in Fig. 13-23. It is highest when the transformer ratio is closest to unity for reasons shown in Fig. 13-23a. Here all the energy is conductively transferred and the current in the transformer is extremely small (almost zero, except for a very small exciting current). The variable copper loss in the transformer winding in Fig. 13-23a is practically zero because of the relatively low resistance of the transformer winding and negligible exciting current.

When the transformer ratio $\alpha = 5/4$, as shown in Fig. 13-23b, only 1/5 of the entire transformer winding conducts a primary (not secondary) current of 10 A, while 4/5 of the winding conducts a current of 2.5 A. Again, this has the effect of reducing the variable copper loss and maintaining high efficiency, while delivering the same kVA to the load.

Even at a ratio of $\alpha = 2/1$, as shown in Fig. 13-23c, only half of the secondary load current appears in the single transformer winding, reducing the variable copper loss considerably in comparison to an isolation transformer delivering the same kVA to a load. Thus, we conclude (1) autotransformers are generally smaller in size and more efficient than conventional isolation transformers of the same kVA rating and (2) the efficiency of autotransformers increases as the transformation ratio approaches unity.

The reader may ask, "If autotransformers are so superior to conventional isolation transformers, why don't we use autotransformers exclusively?"

It has already been shown that conventional isolation transformers having several separate insulated windings may be used to provide a variety of voltage transformations, including possible connection as an autotransformer. This is not possible with a fixed, tapped autotransformer. But the reader may argue, "In power transmission and distribution, the voltages are fixed. Why not use autotransformers there?"

A typical distribution 23 kVA isolation transformer is shown in Fig. 13-24a with an autotransformer designed to accomplish the same purpose

(a) Isolation transformer

(b) Equivalent autotransformer

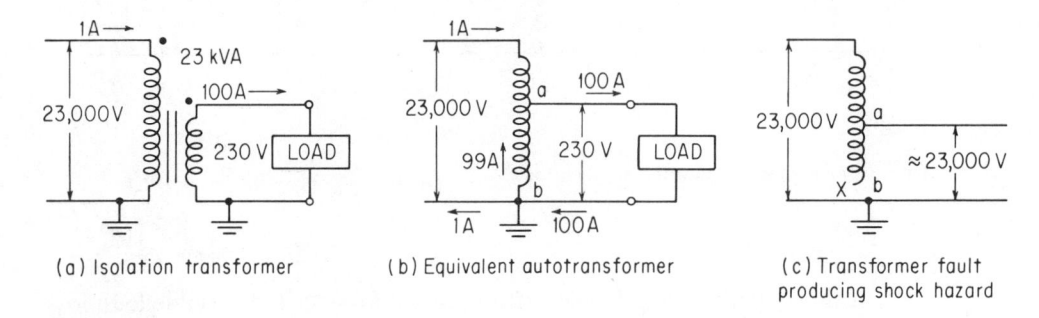

(c) Transformer fault producing shock hazard

Figure 13-24

Shock hazard possibility if autotransformer is used for power distribution.

in Fig. 13-24b. The function of a distribution transformer is to reduce the transmission voltage to some commercially safe value (230 V in this case). Assume that a fault (in this case, an open circuit) occurs in either the primary or secondary of the isolation transformer of Fig. 13-24a. In either case, no voltage appears at the load and the 23 kVA isolation transformer is replaced as soon as possible after the loss in voltage is reported.

The equivalent autotransformer is shown in Fig. 13-24b. Observe that junctions a and b carry the highest currents (100 A in this case). These junctions, therefore, develop hot spots which may result in opens. An open in the transformer winding at either points a or b, as shown in Fig. 13-24c, immediately places 23,000 V at the load! Of course, if the overcurrent protection devices (at either the distribution transformer or the load which it serves) are operating properly, the load is immediately disconnected. Nevertheless, during the short period that it takes the overcurrent protection devices to clear the circuit, some damage might occur. But even assuming that the load is removed, the autotransformer is now shown in Fig. 13-24c with a fault at point b. The danger to personnel is immediately evident since the entire transformer winding is at 23,000 V to ground. It is precisely for this reason that autotransformers are confined to relatively low voltages and restricted to machinery applications discussed in Secs. 9-15, 10-7, and 10-8. Here, their advantages of smaller size and weight, lower cost, and higher efficiency, dictate their use with minimum disadvantage.

13-14. THREE-PHASE TRANSFORMATION

To transform a three-phase source of voltage, either a bank of three identical single-phase transformers, shown in Fig. 13-25, are required or, alternatively, a single polyphase transformer having the six windings (as shown in Fig. 13-25) on a *common* iron core is used.*(Throughout this discussion we shall use individual single phase transformers but the same connections and results emerge from the use of identical individual windings on a polyphase transformer.) Note that the individual transformers of Fig. 13-25 have the same kVA rating, and high to low voltage rating. Also note that the transformers are individually phased and properly marked so that the odd-number subscript shows instantaneous positive polarity (Sec. 13-10) on both high and low voltage sides.

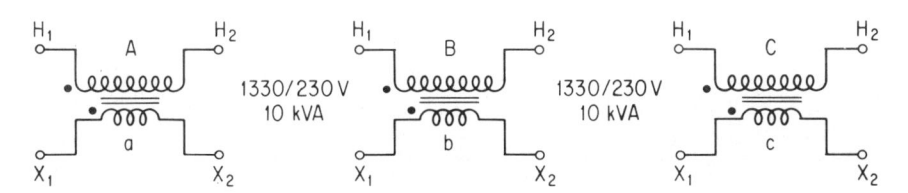

Figure 13-25

Three identical single-phase transformers (a, b, and c) showing polarity markings. Each transformer rated at 1330/230 V, 10 kVA.

Let us assume that the three phase line voltage available for excitation of the transformers is a 2300 V, 3 phase, 60 Hz supply, as shown in Fig. 13-26a. The three *line* voltages are displaced by 120°, as shown in Fig. 13-26a, and this relation represents the voltages between the three lines of the power source: V_{AB}, V_{BC} and V_{CA}, respectively, where each has a magnitude of 2300 V, somewhat in excess of the rated voltage of the high side of the individual transformers. This, of course, dictates that the individual transformers must be Y-connected, as shown in Fig. 13-26b. Note that in doing so, care is taken to insure that the instantaneous positive polarity (H_1 terminal) is connected to the supply while the H_2 terminal of each transformer is connected to a common junction (N). Note that the high voltage coils are designated A, B and C, in Fig. 13-26b, while the low voltage coils (as yet unconnected) are designated a, b and c, respectively.

The relation between the line voltages applied by the source and the phase voltages appearing across the individual transformer high voltage

---

* It will be shown in Sec. 13-17 that the use of individual transformers is preferred to a single polyphase transformer when continuity of service is required. A Δ–Δ bank may be operated in V–V with one transformer removed.

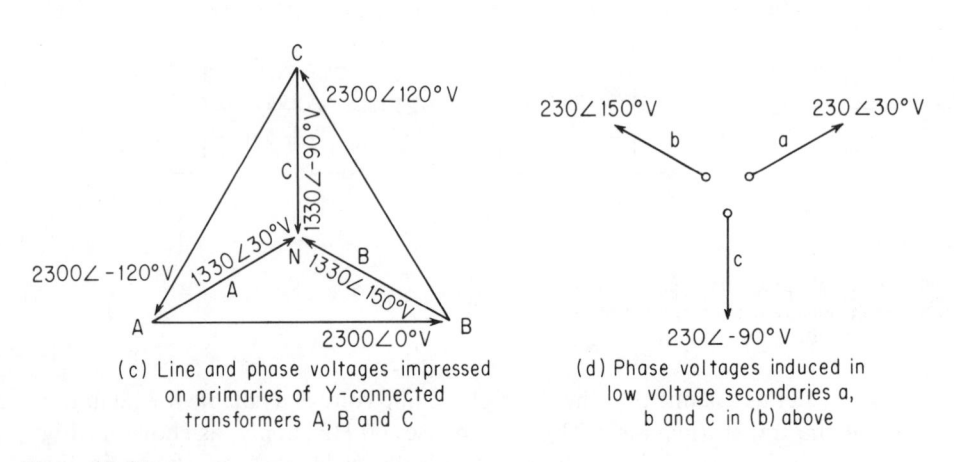

Figure 13-26

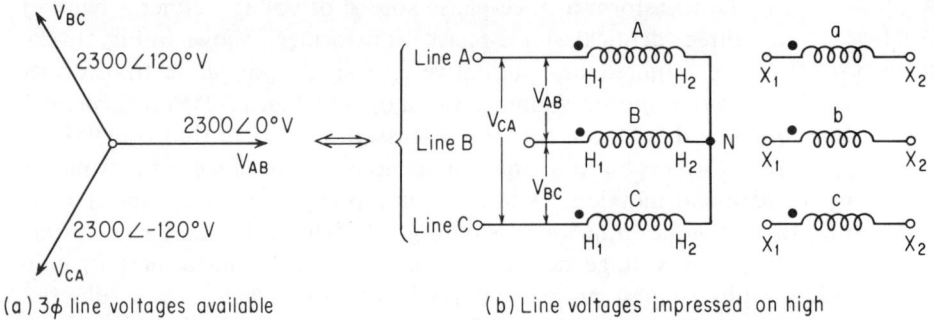

(a) 3φ line voltages available for excitation of transformers

(b) Line voltages impressed on high voltage sides of Y-connected transformers A, B and C

(c) Line and phase voltages impressed on primaries of Y-connected transformers A, B and C

(d) Phase voltages induced in low voltage secondaries a, b and c in (b) above

3-φ line and phase voltages impressed on high voltage sides of Y-connected transformers and phase voltages induced in low voltage sides.

coils is shown in Fig. 13-26c. (This figure is most important and should be studied carefully.) The (phase) voltage impressed on transformer primary A is $1330\angle 30°$ V. The phase voltage from terminal B to N, impressed on transformer B, is $1330\angle 150°$ V. The phase voltage from terminal C to N, impressed on transformer C, is $1330\angle -90°$ V.

In accordance with conventional three-phase circuit theory*, the phase voltage, V_p, is

$$V_p = \frac{V_L}{\sqrt{3}} = 0.577 V_L \tag{13-46}$$

where V_L is the line voltage

* For a complete discussion of three-phase systems, H. W. Jackson, *Introduction to Electric Circuits*, (3rd ed.), Ch. 23; Englewood Cliffs, Prentice-Hall, Inc. 1970.

Thus, the phase voltages shown in Fig. 13-26c are not only less than but are displaced from the line voltages by 30°. Note however that the phase voltages impressed on the 3 transformers, despite the phase shift of 30°, are still displaced 120° from each other in accordance with conventional three-phase theory.* Consequently, the *phasor sum of any two-phase voltages, is the line voltage* as shown in Fig. 13-26c. Thus, line voltage V_{AB}, is the phasor sum of the voltages across H_1-H_2 of coil A and H_2-H_1 of coil B (a reversal of phasor B), as shown in Figs. 13-26b and c.

As a result of transformer action, the voltages induced in the low voltage coils, *a, b,* and *c,* respectively, bear the *same* relation to each other as the phase voltages impressed on the primary, as shown in Fig. 13-26d. It is obvious that the 1330 V primaries must be Y-connected to the 2300 V supply so as not to exceed the voltage rating of the transformer high voltage windings. This, in turn, dictates the relation of the primary phase voltages shown in Fig. 13-26c and also the phase relations of the unconnected secondaries, shown in Fig. 13-26d. The relation developed in Fig. 13-26d, therefore, will be used as a reference for the various ways the secondaries may be connected.

Figure 13-27a shows the secondaries Y-connected in such a way that the X_2 terminals are brought to a common junction, n, and the instanta-

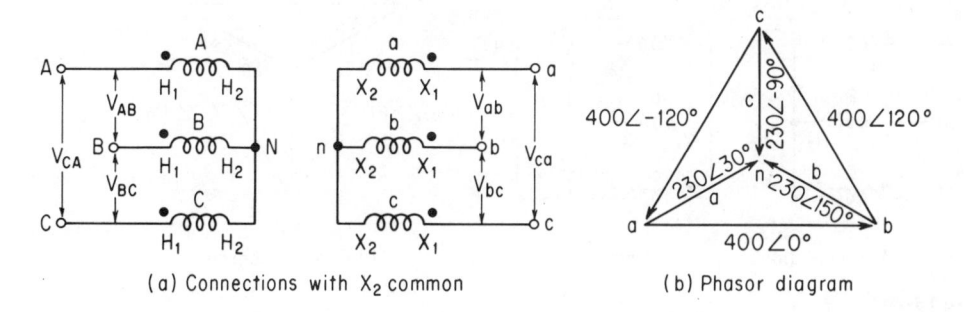

Figure 13-27

Secondaries (X_2 terminals) Y-connected and phasor diagram showing phase and line voltages.

neously positive (dotted) X_1 terminals are brought out to lines *a, b* and *c,* respectively. Using the phasor relations of Fig. 13-26d, the phase and line voltages of the Y-connected secondaries are shown in the phasor diagram of Fig. 13-27b. Note from Fig. 13-27b, in accordance with conventional three-phase circuit theory*, the phasor sum of any two phase voltages produces a line voltage. Thus, the voltage between lines *a* and *b,* V_{ab}, is the

---

* For a complete discussion of three-phase systems, H. W. Jackson, *Introduction to Electric Circuits,* (3rd ed.), Ch. 23; Englewood Cliffs, Prentice-Hall, Inc. 1970.

phasor sum of the voltages across X_1-X_2 of coil a and X_2-X_1 of coil b. Hence, V_{ab} as shown in Fig. 13-27b is the $\sqrt{3}$ times the phase voltages and displaced from it by 30°.

It is most important to compare Fig. 13-27b with Fig. 13-26c and note that for the particular connections made in Fig. 13-27a, there is no displacement between the line voltages of the secondaries and the line voltages of the primaries. Similarly, there is no displacement (phase shift) between the phase voltages of the primaries and the phase voltages of the secondaries. The differences between the two phasor diagrams are only differences in magnitude of voltages due to ratio of transformation. Thus, primary line voltage $V_{AB} = 2300 \angle 0°$ V, while secondary line voltage $V_{ab} = 400 \angle 0°$ V. Similarly, primary phase voltage $V_{AN} = 1330 \angle 30°$ while secondary phase voltage $V_{an} = 230 \angle 30°$ V.

The importance of making connections with *due regard for instantaneous polarity* cannot be overlooked both in the matter of paralleling secondaries and to obtain secondary line voltages of proper magnitude and phase relation.

Consider the simple matter of connecting the X_1 (instead of the X_2) terminals to a common junction, n, as shown in Fig. 13-28a. Since the in-

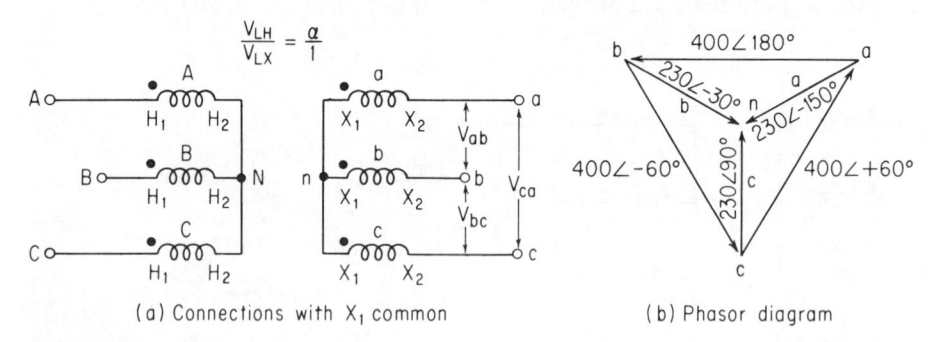

Figure 13-28

Secondaries (X_1 terminals) Y-connected and phasor diagram showing phase and line voltages. (Note phase and line reversals of phasors compared to Figure 13-27b.)

stantaneous polarities of the 3 secondaries are the same, the reader might reason (as does the student in the laboratory) that the three secondaries of Fig. 13-28 could be paralleled with the secondaries of Fig. 13-27. Yet this simple juxtaposition of connections produces a short-circuit when paralleled for the same reason that compels preservation of polarity in single phase transformers. Note that connecting the instantaneously positive (dotted) terminals to n rather than to the line produces a 180° reversal of both phase and line voltages, shown in the phasor diagram of Fig. 13-28b. Thus, while the magnitudes of all phase and line voltages are the same in Figs. 13-28b and 13-27b, they are 180° out of phase with respect to each other and can never be connected in parallel. Any attempt to do so produces an immediate short circuit. Thus, the secondaries of Figs. 13-28a and 13-27a *cannot* be paralleled.

An immediate short circuit also results on paralleling in the event that one of the secondary phase windings is accidentally reversed when making Y-connections. This is shown in Fig. 13-29a where coil b has its

(a) One winding accidentally reversed

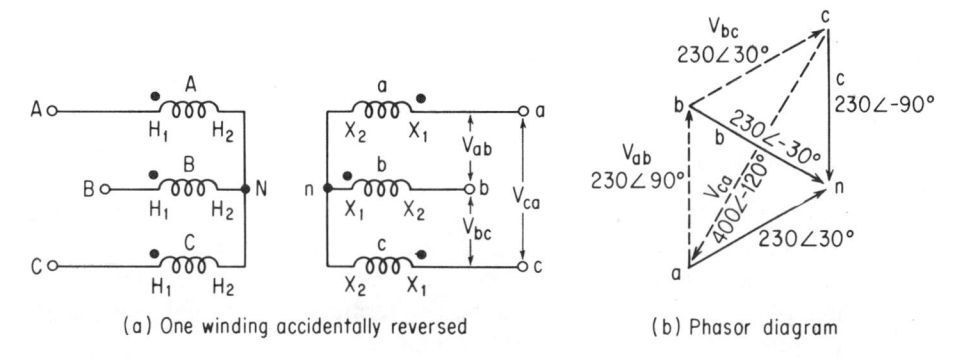

(b) Phasor diagram

Figure 13-29

Effect of accidentally reversing one phase winding (coil B of Y-connected secondaries) and unbalanced phase and line voltages produced.

dotted terminal X_1 connected to the common n junction rather than terminal X_2. Although the phase voltages are still 120° out of phase with respect to each other, as shown in Fig. 13-29b, the reversal of coil b produces reduced and phase displaced line voltages ($V_{ab} = 230\angle 90°$ and $V_{bc} = 230\angle 30°$) and also one line voltage ($V_{ca} = 400\angle -120°$ V) greater than the other two. Note that the 3 line voltages are no longer equal in magnitude nor displaced from each other by 120° and, moreover, no longer in phase with the line voltages of Fig. 13-27b. Thus, reversing one winding accidentally will not permit the Y-connected secondaries of Fig. 13-29a to be paralleled with the Y-connected secondaries of Fig. 13-27a.

Similarly, with all the primaries connected in Y, it is not ever possible to parallel Y-connected and Δ-connected secondaries. The low voltage coils of the transformers of Fig. 13-25 are shown connected in delta in Fig. 13-30a. Note that the series or *mesh* connection of the delta requires that coil ends of *opposite* instantaneous polarity are *mesh-connected* to form a closed loop. It is usually customary, as shown in Fig. 13-30a, to use a voltmeter to measure the resultant voltage, V_R, before *closing the delta* between terminal X_2 of coil c to terminal X_1 of coil a. Only when the voltmeter reads zero, is the voltmeter removed and the delta closed. Using the same impressed primary Y-connected voltages, the phase voltages of the secondary coils, a, b and c, respectively, are the same in Fig. 13-30b

SEC. 13-14. / Three-Phase Transformation

(a) Delta connection

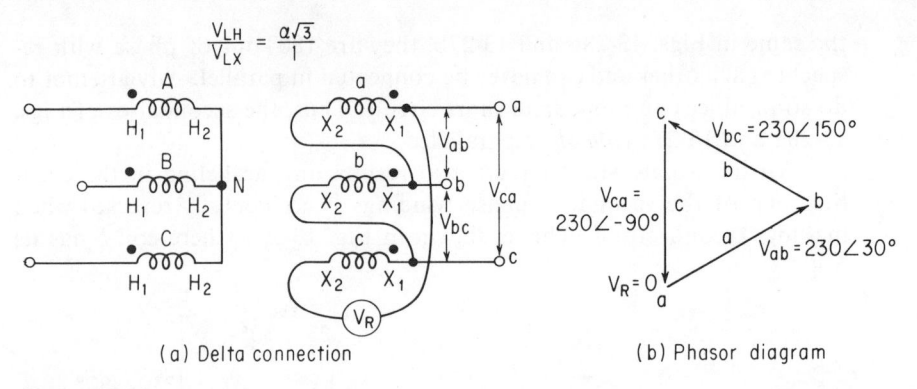

(b) Phasor diagram

Figure 13-30

Secondaries Δ-connected and phasor diagram.

as in Fig. 13-26d. When mesh-connected, therefore, the phasor sum of the secondary phase voltages should be zero. Note that the phase and line voltages of the delta-connected secondaries are the same, in accordance with three-phase circuit theory, since the line terminals are brought out across each phase of the mesh-connected winding.

The importance of a voltmeter reading before closing any mesh circuit, generally, and the delta of transformer secondaries, specifically, is shown in Fig. 13-31. One winding of the transformer (coil c) is accidentally

(a) Δ-connection with one winding accidentally reversed

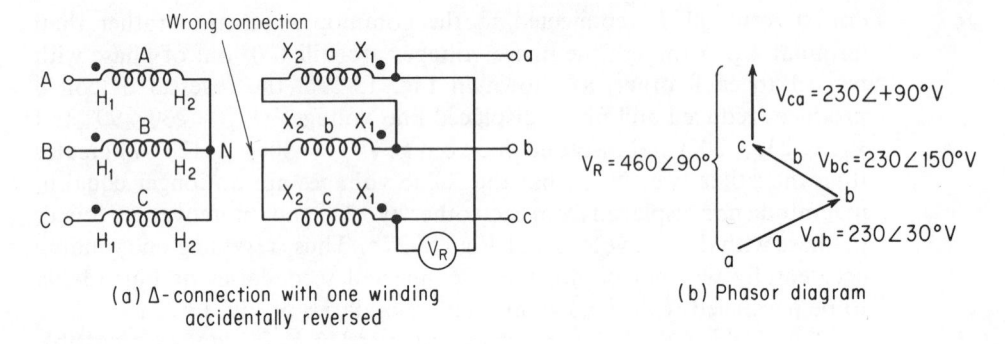

(b) Phasor diagram

Figure 13-31

Effect of accidentally reversing one phase winding (coil C) of Δ-connected secondaries.

reversed in the mesh. The instantaneous polarity of coil c is $230\angle +90°$ V instead of $230\angle -90°$ V, as represented in Fig. 13-26d. The phasor diagram of Fig. 13-31b shows that the voltage measured by voltmeter V_R is no longer zero but actually $460\angle 90°$ V (twice the phase voltage). If the delta is closed without using a voltmeter, a serious short circuit is produced causing high circulating (mesh) currents in the transformers.

It was noted above but bears repeating that the Δ-connected secondaries cannot be paralleled to the Y-connected secondaries. As shown in Fig. 13-27b, there is no phase shift between the secondary line voltages and the primary line voltages of Fig. 13-26a, in a Y-Y transformation. But a Y-Δ transformation, shown in Fig. 13-30b produces a 30° phase shift (and lower line voltages). Even if the voltage ratio of the Y-Δ transformers was predetermined to produce the same line voltages, the phase shift which ensues in a Y-Δ transformation prohibits parallel operation.

We may conclude therefore that with proper regard for primary and secondary line voltages, ratios, instantaneous polarity and proper connection of coils, the following parallel combinations discussed thus far are possible:*

a. Y-Y to Y-Y: no phase shift from primary to secondary line voltages
b. Y-Δ to Y-Δ: same 30° phase shift from primary to secondary in all transformers.
c. Δ-Y to Δ-Y: same 30° phase shift from primary to secondary in all transformers.
d. Δ-Δ to Δ-Δ: no phase shift from primary to secondary.
e. Δ-Δ to Y-Y: no phase shift from primary to secondary, but different voltage ratings.
f. Y-Y to Δ-Δ: no phase shift from primary to secondary but different voltage ratings are required.
g. Y-Δ to Δ-Y: same 30° phase shift from primary to secondary line voltages. Line to line voltage ratings must be the same, however.

The parallel combinations which are *not* possible, despite identical primary and secondary line voltages, involve those in which a phase shift is produced in one set and not in the other. Thus, a Y-Δ cannot be paralleled with a Δ-Δ. (The listing of combinations *not* possible is left as an exercise for the reader in Quest, 13-37).

EXAMPLE 13-23: An industrial plant draws 100 A at 0.7 lagging PF from the secondary of a 2300/230 V, 60 kVA, Y-Δ distribution transformer bank. Calculate:
a. power consumed by the plant in kW and apparent power in kVA
b. *rated* secondary phase and line currents of the transformer bank
c. per cent load on each transformer
d. primary phase and line currents drawn by each transformer
e. kVA rating of each transformer.

* This list is incomplete because it does not contain T-T and V-V transformers. See Table 13-1, Sec. 13-18.)

Solution:

a. $P_T = \dfrac{\sqrt{3}\, V_L I_L \cos\theta}{1000} = \dfrac{1.73 \times 230 \times 100 \times 0.7}{1000} = 28$ kW

$\text{kVA}_T = \dfrac{P_T}{\cos\theta} = \dfrac{28 \text{ kW}}{0.7} = 40$ kVA

b. rated $I_{P_2} = \dfrac{\text{kVA} \times 1000}{3 V_P} = \dfrac{60 \text{ kVA} \times 1000}{3 \times 230 \text{ V}} = 87$ A

rated $I_{L_2} = \sqrt{3}\, I_{P_2} = 1.732 \times 87$ A $= 150$ A

c. $\dfrac{\text{load current per line}}{\text{rated current per line}} = \dfrac{100 \text{ A}}{150 \text{ A}} = 0.67 \times 100 = 67\%$

d. $I_{P_1} = I_{L_1} = \dfrac{\text{kVA} \times 1000}{\sqrt{3}\, V_L} = \dfrac{40 \text{ kVA} \times 1000}{\sqrt{3} \times 2300 \text{ V}} = 10$ A

e. kVA/transformer $= \dfrac{\text{kVA}_T}{3} = \dfrac{60 \text{ kVA}}{3} = 20$ kVA

EXAMPLE 13-24: Repeat Ex. 13-23 using a Δ-Δ transformation, and compare primary line currents against Y-Δ transformation.

Solution:

a. $P_T = 28$ kW and $\text{kVA}_T = 40$ kVA from Ex. 13-23a.
b. rated $I_{P_2} = 87$ A; rated $I_{L_2} = 150$ A from Ex. 13-23b.
c. per cent load on each transformer $= 67\%$ from Ex. 13-23c.
d. $I_{P_1} = 10$ A from Ex. 13-23d.
   but $I_{L_1} = \sqrt{3}\, I_{P_1} = 1.73 \times 10$ A $= 17.3$ A
   The primary line current drawn by a Δ-Δ bank is $\sqrt{3}$ times the primary line current drawn by a Y-Δ bank.
e. kVA/transformer $= \dfrac{60 \text{ kVA}}{3} = 20$ kVA, same as in Ex. 13-23e.

It should be noted from Exs. 13-23 and 13-24 that the only difference between the Y-Δ and Δ-Δ distributions is that the current drawn from the primary mains is 1.73 greater for the latter. This clearly shows the advantage of using Y-Δ transformations where a step-down voltage distribution is required.

13-15. TRANSFORMER HARMONICS

Quite frequently, in laboratory or field tests, in using a voltmeter as a precautionary measure before closing the delta, as shown in Fig. 13-30, the voltmeter does not read zero nor does it read a voltage as high as twice the phase voltage, shown in Fig. 13-31b. Still, there is some reluctance to close the delta in the presence of a voltage which is sometimes as high as 50 V, particularly when the ac voltmeter is of a high impedance type (5 kΩ/V or more). The voltage measured in this case is due to the relatively large third harmonic present in the 3 transformers.*

All single phase transformers when excited at rated voltage produce a third harmonic. This results because the saturation curve of commercial transformer cores rise sharply and saturate quickly, causing the magnetization current to become distorted. Thus, a purely sinusoidal voltage (at a fundamental frequency) produces a magnetization current containing the fundamental frequency plus a large third harmonic component. In single phase transformers, the magnetization current is small compared to the load current and the resulting current waveform is only slightly distorted.

In three phase transformers, however, the three fundamental magnetization currents are displaced by 120° but the third harmonic currents are *in phase* (as are the 6th, 9th, 12th, etc. harmonics). The net result is that in the *absence of a closed* circuit (such as a *Y*-system or an open mesh), this tripled third harmonic component produces a secondary voltage waveform in each winding which contains a large third harmonic voltage. If a closed circuit is provided, such as in a mesh or delta connection, so that the third harmonics circulate, the harmonic is suppressed and no secondary voltage distortion is produced.

In a *Y-Y* transformation, however, there is no closed path for the third harmonic current in either the primary or the secondary, and the output voltage waveforms are distorted. This situation is overcome by providing a neutral line to ground at either the primary or the secondary (or both), thereby permitting a closed path for harmonic voltages and currents, as shown in Fig. 13-32. We may consider Fig. 13-32 as a generic diagram which meets almost every grounding situation. For example, in a *Y-Δ* transformation, the primary transformer neutral is connected to the

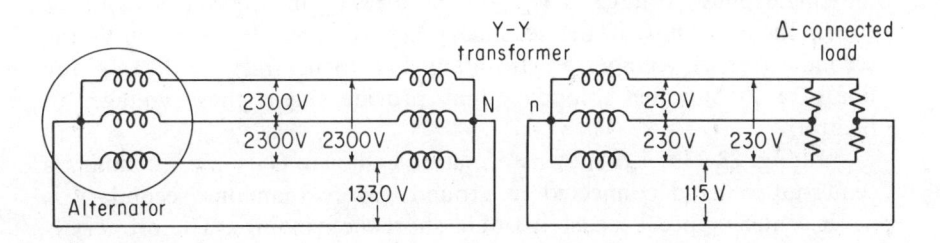

Figure 13-32

Alternator supplying Y-Y transformer with Δ-connected load.

* The writer has demonstrated in the laboratory that if the voltmeter is removed and an ac ammeter is connected in its place, the voltage disappears and the net current is zero! The third harmonic is immediately suppressed whenever a *closed* circuit is created for the delta or mesh.

source neutral, thereby suppressing harmonics. In a Δ-Y transformation, the secondary neutral (as shown in the right hand half of Fig. 13-32) is connected either to the neutral of one delta load or the neutral of a Y-connected load (Fig. 13-33a). When a Δ-Δ transformation is used, no neutral is necessary because the mesh connection provides a closed path for transformer harmonics.

13-16. IMPORTANCE OF NEUTRAL AND MEANS FOR PROVIDING IT

The neutral is fundamental to the suppression of harmonics in Y-Y systems. But in addition to this function, in Y-Y, Y-Δ, Δ-Y, or Δ-Δ systems, the neutral also provides the following advantages:

1. a path for unbalanced currents due to unbalanced loads,
2. a means by which *dual* electric service may be furnished (*both* higher 3ϕ voltage for higher power and motor loads and lower single phase voltage for lighting and domestic appliance loads),
3. a means by which the phase voltages (across Y-connected loads or Y-connected transformers) are balanced with respect to the line voltages.

In Fig. 13-33a, a Y-Y transformation is shown. The neutral from the source (an alternator or higher voltage transformer) is brought to the neutral of the transformer primaries and secondaries and also brought to the loads. A Y-connected 3ϕ load is connected to the neutral so that any phase current unbalance will not unbalance the phase voltages. Three phase Δ-connected loads are connected across the lines *a-b*, *b-c*, and *c-a*, respectively, as shown in Fig. 13-33a. Single phase loads are connected from one line to neutral, as shown. Note that for a Y-connected secondary, the single phase voltage is $V_L/\sqrt{3}$ or $0.577V_L$ in accordance with 3ϕ circuit theory. Thus, if the secondary line voltages, V_L, are 208 V, the secondary phase voltages V_p (from any line to neutral) are 120 V. *Any* line of a Y-connected secondary may provide single phase voltages to neutral.

In Fig. 13-33b, a Δ-Δ transformation is shown. Only *one* secondary is center-tapped and connected to ground. (More than one secondary is never center-tapped because it would short the windings. The primary is *never* grounded because it would short out the transformer at the source). The center-tapped secondary provides single phase voltages which are *half* of the line voltages ($V_L/2$). Note that both delta-connected and Y-connected three phase loads may be connected to the secondary of a delta transformer. If the Y-connected 3ϕ load is unbalanced, however, the unbalanced currents produce unbalanced phase voltages as shown in the phasor diagram of Fig. 13-33b. Furthermore, if the single phase loads are unbalanced, an unbalance of single phase voltages is produced due to line drop in the 3-wire transmission for reasons discussed in Sec. 11-9. Thus, the neutral in a Δ-Δ transmission will not prevent three phase or single phase voltage

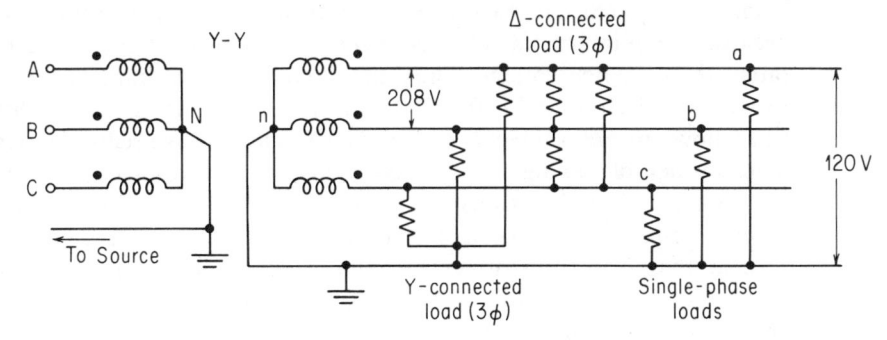

(a) Y-Y transformation with Y-connected and Δ-connected loads, and 1ϕ loads

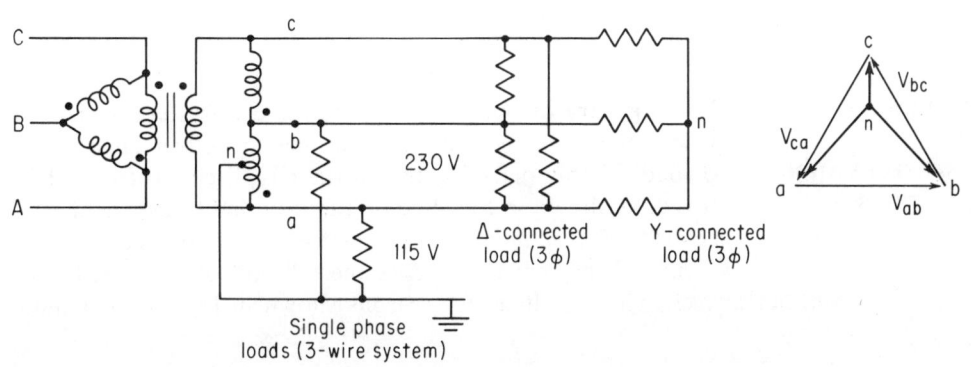

(b) Δ-Δ transformation with Y-connected and Δ-connected loads

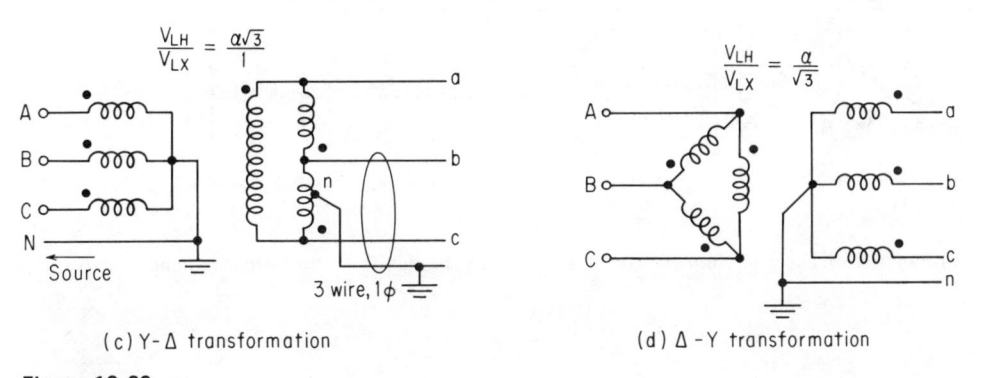

(c) Y-Δ transformation

(d) Δ-Y transformation

Figure 13-33

Single-phase and three-phase loads with provision of neutral in various transformations.

unbalance, although it does provide a path for unbalanced single phase currents, thereby tending to reduce voltage unbalance.

A Y-Δ transformation is shown in Fig. 13-33c. Note that the neutral of the primary is grounded to the source to suppress primary harmonics. A 3-wire single phase system is provided via a center-tapped neutral across lines b and c. Lines a, b and c may provide Y-connected and Δ-connected loads. Y-Δ systems are most often used in step-down distribution systems because the primary transformers need only be insulated for the phase rather than the line voltage. Thus, the rated primary voltage of the transformers required for a 23,000/230 V transformation, using Y-Δ configuration (shown in Fig. 13-33c) is only 13,300 V. This voltage reduction results in a considerable saving in construction costs of high voltage transformers.

Similarly, the Δ-Y transformation shown in Fig. 13-33d lends itself to *high voltage transmission* because it provides a higher secondary line voltage than the transformer secondary rating. Thus, a 230/23,000 V transmission may be provided by transformers whose secondary windings are rated for 13,300 V. Note, in Fig. 13-33d that a neutral is required at the secondary to suppress harmonics and provide the required neutral for Y-Y systems.

13-17.
V-V TRANSFORMATION RELATIONS—OPEN DELTA SYSTEM

If the primary of one transformer of a Δ-Δ system is accidentally opened, the system will continue to deliver energy to a 3ϕ load. If this defective transformer is disconnected and removed, as shown in Fig. 13-34a, the resulting transformer bank is called an *open delta* or *V-V* system. The system continues to supply 3ϕ power to the Δ-connected and Y-connected loads without any change in voltage for reasons shown in Figs. 13-34b and c.

(a) Removal of one transformer from a Δ-Δ system producing V-V transformer bank

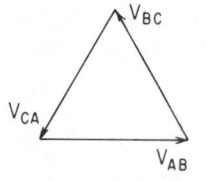

(b) Line voltages applied to V-V primary

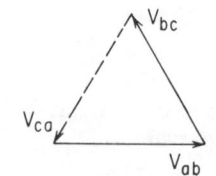

(c) Secondary line voltages produced by V-V transformer bank

Figure 13-34

Open delta or V-V transformer bank and phasor relations.

Figure 13-34b shows the phasor relations for the 3ϕ line voltages applied to the V-V primaries. The phasor relations for the phase and line voltages induced in the two secondaries are shown in Fig. 13-34c. Note that the phase and line voltages are the same. V_{ab} is the voltage induced in transformer secondary coil a. V_{bc} is the voltage induced in transformer coil b. The phasor sum of $V_{ab} + V_{bc}$ produces V_{ca} as shown in Fig. 13-34a and c. Consequently, 3 line voltages, 120° apart, are still produced by a V-V system.

The voltamperes supplied by each transformer in a V-V system is *not* half (50%) the total voltamperes but *rather* 57.7%. This may be proven as follows:

Since each transformer in V-V now delivers line (not phase) current, the voltamperes supplied by each transformer in open-delta compared to the total 3ϕ voltamperes is

$$\frac{\text{VA per transformer}}{\text{Total 3}\phi \text{ power}} = \frac{V_p I_p}{\sqrt{3} V_L I_L} = \frac{V_L I_L}{\sqrt{3} V_L I_L}$$

$$= \frac{1}{\sqrt{3}} = 0.577 \qquad (13\text{-}47)$$

Equation (13-47) also shows that if 3 transformers in Δ-Δ are delivering rated load and one transformer is removed, the overload on each remaining transformer is 173%, since the reciprocal of Eq. (13-47) is the ratio of the total load to the load per transformer. Finally, the above relation also implies that if 2 transformers are operated in V-V and loaded to rated capacity, the addition of a third transformer increases the total capacity by 173.2% (or by the $\sqrt{3}$). Thus, at an increase in cost of 50% for a third transformer, the capacity of the system is raised by 73.2%, in converting a V-V system to a Δ-Δ system.

EXAMPLE 13-25: Each of the Δ-Δ transformers, as shown in Ex. 13-24, is rated at 20 kVA, 2300/230 V and the bank supplies a 40 kVA load, at 0.7 PF lagging. If one defective transformer is removed for repair, calculate, for the V-V connection
a. kVA load carried by each transformer
b. per cent of rated load carried by each transformer
c. total kVA rating of the transformer bank in V-V
d. ratio of V-V bank to Δ-Δ bank transformer ratings
e. per cent increase in load on each transformer when one transformer is removed.

Solution:

a. $\frac{\text{load in kVA}}{\text{transformer}}$, in V-V $= \frac{40 \text{ kVA}}{1.73} = $ **23.1 kVA/transformer** (13-47)

b. per cent transformer load $= \frac{\text{load in kVA/transformer}}{\text{kVA rating/transformer}}$

$= \frac{23.1 \text{ kVA}}{20 \text{ kVA}} \times 100 = $ **115.5%**

c. kVA rating of V-V bank $= \dfrac{\sqrt{3} \times \text{kVA rating}}{\text{transformer}}$
$= 1.732 \times 20 \text{ kVA} = \mathbf{34.64 \text{ kVA}}$

d. ratio of ratings $= \dfrac{V\text{-}V \text{ bank}}{\Delta\text{-}\Delta \text{ bank}} = \dfrac{34.64 \text{ kVA}}{60 \text{ kVA}} = \mathbf{57.7\%}$ (13-47)

e. $\dfrac{\text{original load in } \Delta\text{-}\Delta}{\text{transformer}} = \dfrac{40 \text{ kVA}}{3} = \mathbf{13.33 \text{ kVA/transformer}}$

per cent increase in load $= \dfrac{\text{kVA/transformer in } V\text{-}V}{\text{kVA/transformer in } \Delta\text{-}\Delta}$

$= \dfrac{23.1 \text{ kVA}}{13.33 \text{ kVA}} \times 100 = \mathbf{173.2\%}$

It should be noted from Ex. 13-25 that while each transformer load has been increased by 173% as a result of removing one transformer from a Δ-Δ bank, the transformers are not seriously overloaded, as shown by Ex. 13-25b. Originally, the load on each 20 kVA transformer, in Δ-Δ supplying the 40 kVA load, was 13.33 kVA. Thus, the transformers in Δ-Δ, in Exs. 13-24 and 13-25 are not supplying their full rated capacity to the load.

As a result of the V-V connection, however, the load increases dramatically on each transformer, as shown in Ex. 13-25e, by 173.2%. The two transformers in V-V are thus each carrying a 15.5% overload. This overload may be sustained until such time as the third transformer is replaced.

Example 13-25 verifies the relations of Eq. (13-47) and the fact that each transformer carries 57.5% of the total load and not half as pointed out previously.

Utility companies often take advantage of the above relations by initiating a 3ϕ system using V-V transformer connections and adding a third transformer as the increasing load conditions require it. This simple expedient is easily justified since the return in added capacity (73%) more than meets the added cost (50% of 2 transformers).

Like the Δ-Δ (and Y-Y), the open delta or V-V produces no phase shift from primary to secondary line voltage as shown in Figs. 13-34b and c. It may be paralleled therefore with those transformers (of the same primary and secondary line voltages) having no phase shift (see Table 13-1).

**13-18.
T-T
TRANSFORMATION
RELATIONS**

Like the V-V transformation, it is possible to use only 2 transformers to provide 3-phase transformation, if they are connected T-T. Unlike the V-V, however, the T-T requires 2 special transformers, each of which is different from the other. Like the V-V, the T-T derives its name from the appearance of its connections, as shown in Fig. 13-35a.

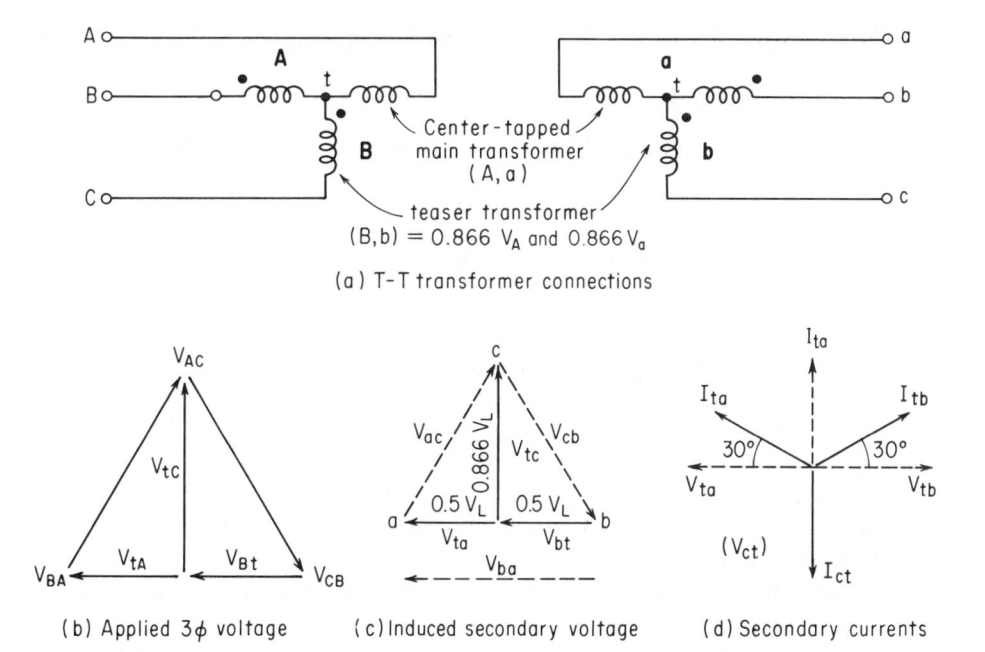

Figure 13-35

T-T transformer connections and phasor relations.

(a) T-T transformer connections

(b) Applied 3φ voltage

(c) Induced secondary voltage

(d) Secondary currents

The special transformers required for the *T-T* connection are a *teaser* transformer (*B, b* in Fig. 13-35a) whose rated primary and secondary voltages are 0.866 (or 86.6%) of the rated primary and secondary voltages of the *main* transformer (*A, a*). The main transformer is either a center-tapped transformer or a multiwinding transformer having two equal primary and secondary windings whose *total* (series-connected) rated voltage is 1.15 (or 115%) of the teaser transformer primary and secondary voltages, respectively.

The applied 3φ voltages to the primary of the *T*, V_{AC}, V_{CB} and V_{BA} are resolved into V_{Bt}, V_{tA} and V_{tC}, (for the instantaneous polarities shown) as shown in Fig. 13-35b, where *t* is the junction between the teaser transformer and the center-tap of the main in Fig. 13-35a. The induced secondary voltages are in phase with the components shown for the applied voltage, and are represented as the solid-line phasors of Fig. 13-35c. As may be seen from Fig. 13-35c, the line voltage V_{bc} is the phasor sum of $V_{bt} + V_{tc}$. Since V_{bt} is $0.5V_L$, where V_L is the secondary line voltage and V_{tc} is 0.866 V_L, which is the rated secondary voltage of the teaser transformer, then

$$V_{bc} = V_{bt} + V_{tc} \tag{13-48}$$
$$= 0.5V_L + j0.866V_L = V_L, \text{ the secondary line voltage} \tag{13-49}$$

579

SEC. 13-18. / *T-T Transformation Relations*

Similarly, as shown in Fig. 13-35c, line voltage V_{ac} is

$$V_{ac} = V_{at} + V_{tc}$$
$$= 0.5V_L + j0.866V_L = V_L, \text{ the secondary line voltage} \quad (13\text{-}50)$$

While the line voltage V_{ba} is

$$V_{ba} = V_{bt} + V_{ta}$$
$$= 0.5V_L + 0.5V_L = V_L \quad (13\text{-}51)$$

Since the three secondary line voltages shown in Fig. 13-35c are equal (to V_L) they are in an equilateral triangle 60° apart and their phasor relation is a true 120°, thus producing a true 3ϕ transformation of the original line voltages applied to the primary.

The phasor relations between the currents and the voltages for the main (center-tapped) and teaser transformer respectively are shown in Fig. 13-35d. The phasor sum of the currents is zero, as it is in any true 3ϕ system. Note, however, that the current in the teaser transformer I_{tc} is in phase (at unity PF) with its voltage V_{tc}. But also note that the currents in the main transformer lead or lag their component phase voltages by 30°. These current relations account for the *derating* of both transformers in the following way:

The total load supplied by the main transformer is

$$VA_{\text{main}} = V_{ta}I_{ta} \cos 30° + V_{tb}I_{tb} \cos 30°$$
$$= 0.5V_L I_L \cos 30° + 0.5V_L I_L \cos 30° = V_L I_L \cos 30°$$
$$VA_{\text{main}} = 0.866V_L I_L \quad \text{where } V_L \text{ and } I_L \text{ are the secondary} \quad (13\text{-}52)$$
$$\text{line voltages and currents, respectively}$$

Although the phase (and line) current of the teaser transformer is in phase with its respective line voltage, the teaser is also derated because its windings are rated for $0.866V_L$ and thus

$$VA_{\text{teaser}} = V_{ct}I_{ct} = (0.866V_L)I_L$$
$$= 0.866V_L I_L \quad (13\text{-}53)$$

As in the case of 2 transformers in V-V, the teaser and the main transformer each carry half the total 3ϕ load since

$$VA_{\text{total}} = VA_{\text{main}} + VA_{\text{teaser}}$$
$$\sqrt{3}V_L I_L = \frac{\sqrt{3}}{2}V_L I_L + \frac{\sqrt{3}}{2}V_L I_L = 0.866V_L I_L + 0.866V_L I_L$$
$$(13\text{-}54)$$

For 3ϕ transformations, the above relations show that there is no advantage whatever in using a *T-T* bank in comparison to a *V-V* transformer bank, for a variety of reasons. A *T-T* bank requires special (more costly) transformers and a special connection arrangement. (The *V-V* bank easily lends itself to addition of the third transformer and each transformer is a standard conventional single-phase transformer.) Thus, if transformers are used in *T-T*, the addition of a third transformer to produce a Δ-Δ bank is a difficult matter because the teaser transformer is only rated for $0.866V_L$.

It can also be show that the derating factor for both the *T-T* and *V-V* transformers is actually the same and no advantage is gained, therefore, through the use of *T-T* transformations. The ratio of the rated *V-V* bank load per transformer to the total load each transformer could supply)if both were used as transformers connected in single phase) is

$$\frac{V\text{-}V \text{ kVA}}{2 \times (\text{single phase kVA})} \text{ or } \frac{\sqrt{3} \text{ kVA}}{2 \text{ kVA}} = 0.866$$

this is exactly the same factor which emerges for each transformer in *T-T* as shown by Eqs. (13-53) and (13-54).

For the above reasons, therefore, the *T-T* transformer bank has little application except to serve as an introduction to the Scott connection, given in Sec. 13-19.

Like the *V-V* (*Y-Y* and Δ-Δ, as well), the *T-T* transformer as shown in Figs. 13-35b and c produces no phase shift from primary to secondary. It may therefore be paralleled with any of those transformers which exhibit no primary to secondary phase shift, providing the primary to secondary voltages are the same, and the same instantaneous polarity is preserved, as shown in Table 13-1, below.

TABLE 13-1.
TYPES OF 3-PHASE TRANSFORMERS WHICH MAY BE PARALLELED*

COLUMN A (0° PHASE SHIFT)	COLUMN B (30° PHASE SHIFT)
Y-Y	Δ-Y
Δ-Δ	Y-Δ
T-T	
V-V	

Table 13-1 implies that a *T-T* transformer may be paralleled with a *V-V*, a *T-T*, a *Y-Y* or a Δ-Δ transformer, but not with either a *Y*-Δ or Δ-*Y* transformer. Thus, any transformer in Column A may be paralleled

* Assuming same primary to secondary voltages, as well as proper regard for phase sequence, instantaneous polarity and phasing to produce same phase shift (see Sec. 13-14).

with itself or another transformer in Column A. Similarly, any transformer in Column B may be paralleled with an identical combination or another in Column B. Thus, a Δ-Y transformer may be paralleled with another Δ-Y combination or a Y-Δ transformer bank, since in either case a 30° phase shift is produced.

No transformer combinations in Column A may be paralleled with a transformer combination in Column B under any circumstances because the latter produces a 30° phase shift between the primary and secondary voltages.

13-19. THREE-PHASE TO TWO-PHASE TRANSFORMATIONS —SCOTT CONNECTION

Any *polyphase* system may be transformed (using suitable combinations of transformers) into another *polyphase* system. Given a 3ϕ supply, it is possible to obtain any polyphase system from 2 phase up to 24 phase (and higher).* It is desirable to use transformers because of their extremely high conversion efficiency.

The phasor diagram of Fig. 13-35c showing the phase voltages induced in the secondaries of the *T-T* transformers suggests a quadrature relationship between the two phasors. This is the same relation that exists in 2-and 4-phase systems. Both of these transformations are made using the so-called Scott connection.

Like the *T-T* connection, two special tapped transformers are required. The main transformer, shown in Fig. 13-36a, has a center-tapped primary (or two equal series-connected windings). The teaser transformer has a voltage rating which is $\sqrt{3}/2$ or 0.866 of the voltage rating of the main transformer. The secondaries of *both* transformers have equal voltage ratings and may be center-tapped (for 4-phase only as shown in Fig. 13-36b).

A 2-phase, 3-wire system is produced by connecting the ends of the transformers and bringing a neutral wire from this junction, as shown in Fig. 13-36a. If the 4 ends of the two transformers are brought out (*without* any junction) a 2-phase, 4-wire system is produced. The phasor diagram of Fig. 13-36a verifies the relation between the phase and line voltages of a 2-phase system, using the instantaneous polarity of the transformers as shown. The voltage between lines, V_L, is

$$V_L = \sqrt{2}\,V_P = \sqrt{2}\,V_a = \sqrt{2}\,V_b \tag{13-55}$$

---

* It is impossible to use transformers to convert a single phase to a 3ϕ system for large amounts of power. Of course a single-phase motor driving a polyphase alternator will produce a polyphase system but the efficiency of such conversion is the product of the efficiencies of the individual dynamos. A somewhat more efficient conversion is accomplished using a single dynamo called an induction phase converter (see Sec. 11-10). Very small amounts of power can be produced from a single-phase system using *R-C* phase shift networks to produce 2ϕ which in turn may be transformed to 3ϕ and higher.

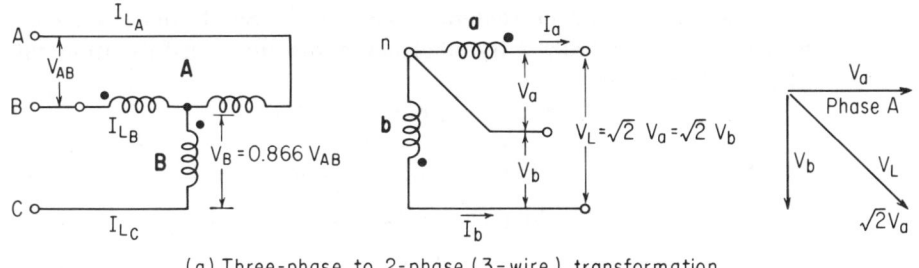

(a) Three-phase to 2-phase (3-wire) transformation

(b) Quarter-phase or 4-phase (5-wire) transformation

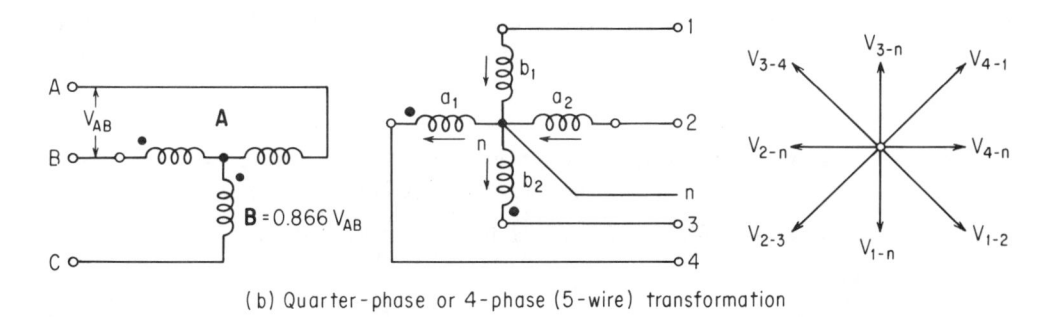

Figure 13-36

Scott connections for 2-ϕ and 4-ϕ transformations.

where V_p is the phase voltage or voltage induced in either transformer, V_a or V_b

A 2-phase, 5-wire system is produced by connecting the center taps of the transformer secondaries and bringing a neutral wire from this junction as shown in Fig. 13-36b. Such a system is sometimes called a quarter-phase (or 4-phase), 5-wire transformation. As shown in the phasor diagram, it produces 4-line voltages and 4-phase voltages (any line to neutral). The 4-line voltages are displaced from each other by 90° and the same is true of the 4-phase voltages. The line voltages are the phasor sum of any two-phase voltages and bear the same relation as that given in Eq. (13-55) or $V_L = \sqrt{2} V_p$.

The Scott connection is used to produce 3ϕ power from 2ϕ feeders (or vice-versa) in fairly large quantities to operate motors and other polyphase equipment at its rated voltage. Thus, an occupancy which has a 155/110 V, 2ϕ, 3-wire 2ϕ service may easily use a 440 V, 3ϕ, SCIM, with relatively little power loss, using a Scott 2ϕ to 3ϕ transformation, as shown in Ex. 13-26, below. As a choice between purchasing a special or new motor as opposed to one which is readily available plus the cost of transformers, the latter is usually the least expensive (transformers are less costly in the same kVA rating than rotating machines).

The voltage and current relations for the Scott connection emerge by first assuming both unity efficiency and unity transformation ratios. Thus, in Fig. 13-36a,

the output 2ϕ kVA = the input 3ϕ kVA or

$2V_a I_a = \sqrt{3} V_L I_L$ but for unity transformation $V_L/V_a = 1$, and $2I_a = \sqrt{3} I_L$, yielding each of the 2ϕ currents (I_a, I_b) at the output as

$$I_a = I_b = \frac{\sqrt{3}}{2} I_L \text{ for a unity transformation}$$

From Eq. (13-6), since the load ampere-turns of the secondary of any transformer must be equated to the primary ampere-turns (neglecting magnetizing current), we may write

$I_a N_2 = \frac{\sqrt{3}}{2} I_L N_1$, and since the transformation ratio, α is N_1/N_2 and the efficiency is η, we may write

$$I_a = \frac{\sqrt{3}}{2} I_L \alpha \eta = I_b, \text{ for } 3\phi \text{ to } 2\phi \text{ transformation} \tag{13-56a}$$

and

$$I_a = \frac{\sqrt{3}}{2} \frac{I_L}{\alpha \eta} = I_b, \text{ for } 2\phi \text{ to } 3\phi \text{ transformation} \tag{13-56b}$$

where I_a and I_b are the 2ϕ line currents
I_L is the 3ϕ (balanced) line current
α is the transformation ratio primary to secondary
η is the efficiency of the transformers

EXAMPLE 13-26: A 440 V, 100 hp, 3ϕ SCIM, having a PF of 0.8, is to be operated from a 155/110 V, 2ϕ supply. Assuming that the transformers selected have an efficiency of 98% at full load, calculate
a. motor line current (see Appendix A-5)
b. transformation ratio
c. current in the primary of the Scott transformers
d. kVA rating of the main and teaser transformers.

Solution:

a. $I_L = 123 \text{ A} \times 1.25 = \mathbf{154 \text{ A}}$

b. $\dfrac{N_1}{N_2} = \alpha \dfrac{V_a}{V_L} = \dfrac{110 \text{ V}}{440 \text{ V}} = \dfrac{1}{4}$

c. $I_a = \dfrac{\sqrt{3}}{2} \dfrac{I_L}{\alpha \eta} = \dfrac{\sqrt{3}}{2} \times \dfrac{154 \text{ A}}{0.25 \times 0.98} = \mathbf{544 \text{ A}}$ \hfill (13-56b)

d. $\text{kVA} = \dfrac{V_a I_a}{0.866 \times 1000} = \dfrac{110 \text{ V} \times 544 \text{ A}}{0.866 \times 1000} = \mathbf{69.1 \text{ kVA}}$ \hfill (13-53)

13-20.
THREE-PHASE TO SIX-PHASE TRANSFORMATIONS

Because of their relatively high efficiency, transformers serve as excellent polyphase transformation devices in providing higher polyphase systems (usually) from 3ϕ supplies. Such higher systems are particularly useful in half- and full-wave rectification because of the relatively lower ripple components (Sec. 13-21). Thus, where large quantities of dc power are required, it is not uncommon to convert 3ϕ to 6, 12 or even 24ϕ using transformers and suitable solid-state half- or full-wave rectifiers. While the theory of the higher order transformations is beyond the scope of this work, this section will provide some introduction to polyphase transformation theory by covering the 5 basic 3ϕ to 6ϕ transformations.

The type of transformer required to produce a true* 3ϕ to 6ϕ transformation is one having 2 separate but equal voltage secondaries. Three such individual single-phase transformers are required (although a single polyphase transformer with 6 separate secondaries may be used), as shown in Fig. 13-37a. (For a true 12ϕ conversion, transformers having 4 separate

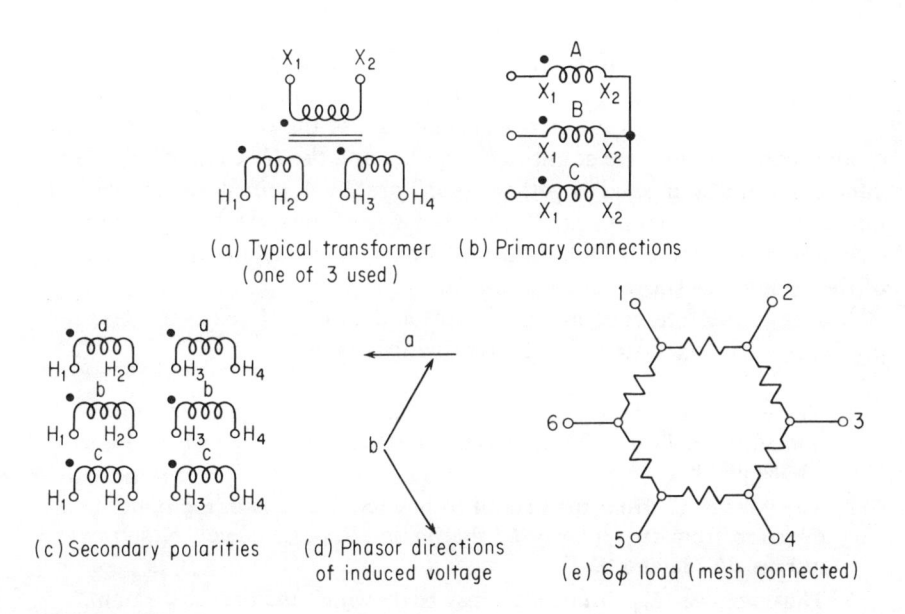

(a) Typical transformer (one of 3 used)
(b) Primary connections
(c) Secondary polarities
(d) Phasor directions of induced voltage
(e) 6ϕ load (mesh connected)

Figure 13-37
Typical transformer required for true 3- to 6-ϕ transformation with primary connection, phasor directions of secondary voltages and 6-ϕ load.

* A true transformation is one which produces a (desired) polyphase system at the secondaries without requiring interconnection of the secondary terminals to the load. Only the star and mesh connections produce *true* polyphase systems.

secondaries are required; for 24ϕ, 8 separate secondaries, and so on).

The three transformers are Y-connected (although Δ could be used) to a 3ϕ supply with proper regard for instantaneous polarity, as shown in Fig. 13-37b. For the application shown, it is desired to produce a high voltage 6ϕ system, hence the low voltage primaries and the high voltage secondaries. (The *same* primary connection will be used for *all* of the 5 types of 6ϕ transformations covered, and only the secondary connections are shown in Figs. 13-38 through 13-42, since each of these is different.) Note that the secondary instantaneous polarity and phasor direction of induced voltage in each of the secondaries is shown in Figs. 13-37c and d respectively. These voltages are 120° apart since they are produced by and from a 3ϕ supply, as expected. Thus, the instantaneous (dotted) terminals shown in Fig. 13-37c have the instantaneous phasor direction shown in the phasor diagram of Fig. 13-37d.

A typical 6ϕ load, mesh-connected, is shown in Fig. 13-37e. This load with its terminals (1 through 6) will be used to load all the types of 6ϕ transformations shown below.

13-20.1 6ϕ Star

The first transformation shown is the 6ϕ star. In the generic star connection, one end of all coils is connected to a common winding (thus the Y is a *special case* of the *star*). The H_2 ends of all secondary coils are connected together and the H_3 ends of all secondary coils are connected together; *both* are then joined to a *common* junction, n, as shown in Fig. 13-38a. The 6 free ends are brought out to terminals 1 through 6, which in turn is connected to terminals 1 through 6 of the 6ϕ load, as shown in Fig. 13-38a.

Even if the free ends are not connected to the 6ϕ load, the 6ϕ star produces a true 6ϕ system for the following reasons:

1. The voltage E_{n_1} from the neutral to line terminal 1 is in the same direction as the polarity of the phasor, H_2-H_1 of coil a, shown in Figs. 13-38a and b.
2. The voltage E_{n_2} from the neutral to line terminal 2 is in the opposite direction from the polarity of the phasor, H_4-H_3, of coil b, shown in Figs. 13-38a and b.
3. Thus, voltages E_{n1} through E_{n_6} may be drawn on the phasor diagram, shown in Fig. 13-38b.
4. The line voltages E_{61}, E_{12}, E_{23}, etc. are found using double subscript notation. Thus, $E_{61} = E_{6n} + E_{n1}$. This is the same as drawing a line from E_{n6} to E_{n1} on the phasor diagram of Fig. 13-38b.
5. Note that for the 6ϕ star connection, the magnitude of the line voltages (E_{12}, E_{23}, etc.) is the same as the phase voltages (measured from the neutral to any one line which is the same as the voltage across any coil).
6. Thus, even in the absence of the load, the relation shown in the phasor diagram of Fig. 13-38b exists, and the 6ϕ star is a true 6ϕ system.

When the terminals 1 through 6 of the 6ϕ supply is connected to terminals 1 through 6 of the 6ϕ load, the same line voltage is applied across each of the individual mesh-connected 6ϕ loads, as shown in Fig. 13-38a, ansuming no internal impedance drops due to load in the transformer secondary windings.

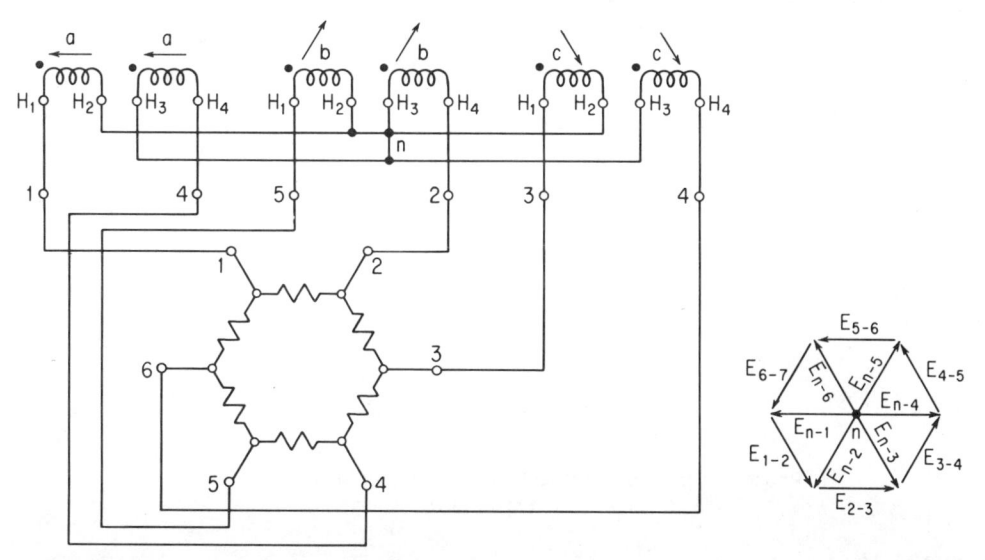

(a) Secondary connection for 6ϕ star and connections to 6ϕ load (b) Phasor diagram 6ϕ star

Figure 13-38

6-ϕ star secondary connections and phasor diagram.

13-20.2
6ϕ mesh

The secondary connections for the 6ϕ mesh are shown in Fig. 13-39a. Note that before closing the mesh, as in the case of a delta secondary (Sec. 13-14), a voltmeter is necessary to assure that the phasor sum of all series-connected mesh voltages is zero. A connection table is shown in Fig.13-39b to simplifythe connections and also to verify the phasor diagram shown in Fig. 13-39c. The coil ends have been lettered to simplify reference to end connections.

The phasor diagram of Fig. 13-39c is obtained as follows:

1. Begin with coil *a-b* which serves as a reference. With *b* connected to *n* (see table) voltage *n-m* must be in the opposite direction shown (since the arrow head corresponds to the dot).
2. With terminal *m* connected to *c*, voltage in coil *c-d* is in direction shown (arrow head corresponds to dot).
3. Voltage in coil *h-j* is as shown but we want voltage *j-h*, since *d* is connected to *j*. Hence the phasor is reversed as shown in Fig. 13-39c.
4. With *h* connected to *f*, phasor *f-g* is in the direction shown.

587

SEC. 13-20. / *Three-Phase to Six-Phase Transformations*

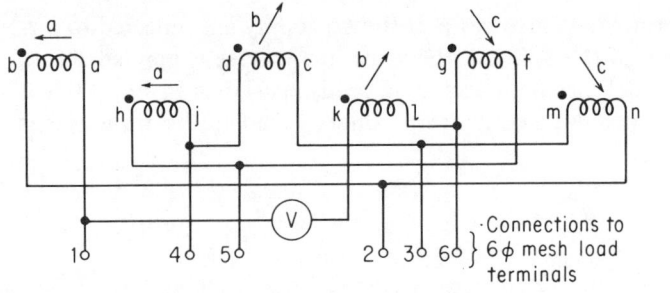

(a) Secondary connections for 6φ mesh

Connection table

b	to	n
m	to	c
d	to	j
h	to	f
g	to	l
k	to	Ⓥ
Ⓥ	to	a

(b) Connection table

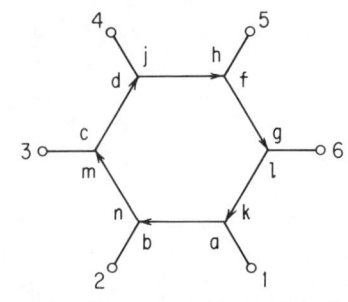

(c) Phasor diagram and connections to load terminals (load same as in Fig 13-38a)

Figure 13-39

6-φ mesh secondary connections (Load same as in Figure 13-38a).

5. With g connected to l, phasor k-l is in the direction shown. But we want phasor l-k which means a reversal of phase. This brings us back to a-b, the reference coil.

Note that the mesh phasor diagram shown in Fig. 13-39c produces a true 6φ system of voltages between the line terminals 1 through 6, respectively, regardless of connection to a 6φ load, and that the phase and line voltages of the mesh are the *same*.*

Occasionally when connecting the 6φ mesh, (Fig. 13-39a) a harmonic voltage appears on the voltmeter if the neutral of the Y-primary is not grounded. Grounding the neutral of the primary (or connecting the primaries in delta) will eliminate this harmonic voltage (Sec. 13-15).

13-20.3
6φ diametrical

The observant reader might discover in examining Fig. 13-38a and 13-39a that connections from the 6φ star and mesh, respectively, are brought out to the load in a particular pattern (1-4-5-

---

* By definition, this relation holds for *any* mesh system (the 3φ delta, the 6φ, 12φ mesh, and so on) since the phase voltage of a mesh is the same as its line voltage.

2-3-6). The opposite ends of the first star-connected transformer are brought out to terminals 1 and 4 of the load. Opposite ends of the second star-connected transformer are brought out to terminals 5 and 2 of the load. Opposite ends of the third star-connected transformer are brought out to terminals 3 and 6 of the load. This suggests the possibility of a 3ϕ to 6ϕ transformation using no special transformers or center-taps whatever!

This possibility is realized using the *diametrical* connection shown in Fig. 13-40a. If the (diametrically) opposite polarity ends of each trans-

(a) Secondary connections for 6ϕ diametrical (load same as in Fig 13-38a)

(b) Phasor diagram when interconnected via 6ϕ load

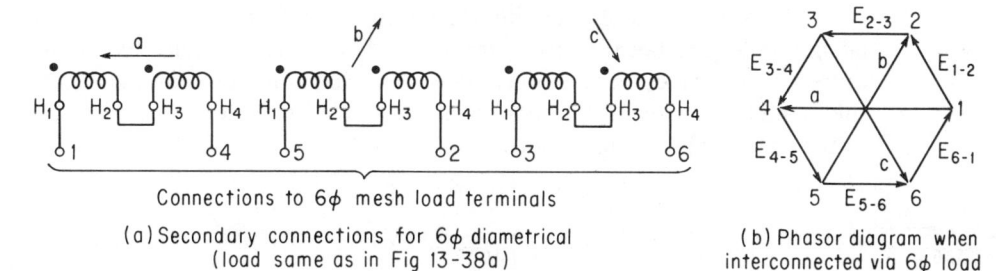

Figure 13-40

6-ϕ diametrical secondary connections and phasor diagram.

former secondary are connected to the load terminals in the sequence 1-4-5-2-3-6, as shown, a 6ϕ line voltage appears at the terminals of the load, as shown in Fig. 13-40b. Note that in the absence of a load connection, the induced voltages in the secondaries of transformers a, b and c are displaced by 120°, as shown in Figs. 13-40a and b. Connecting diametrically opposite ends in the proper sequence to the load terminals, as shown in the phasor diagram of Fig. 13-40b, automatically creates 6 line voltages which are properly displaced to form a 6ϕ system: E_{12}, E_{23} and so on, as shown.

As may be seen from Fig. 13-40b, the relation between the voltage induced in each (full) secondary winding, E_{1_4}, and a typical 6ϕ line voltage, $E_{12} = E_{6\phi}$ is

$$E_{12} = \frac{E_{14}}{2}$$

and hence

$$E_{6\phi} = \frac{E_{3\phi}}{2} = E_p \qquad (13\text{-}57)$$

where $E_{3\phi}$ is the full secondary voltage induced in each transformer and E_p is the "phase" voltage induced in each coil (H_1-H_2 or H_3-H_4) of a 2 winding transformer (if used).

SEE. 13-20. / *Three-Phase to Six-Phase Transformations*

The diametrical connection is perhaps the simplest of all 3ϕ to 6ϕ connections because no interconnection is required between secondaries and no special transformers are required. (Three identical single-phase isolation transformers may be used). It does *not* produce a true 6ϕ system, however, and if one of the lines to the 6ϕ load is opened, or an open appears in the 6ϕ mesh load, the phasor relation of Fig. 13-40b returns to a simple 3ϕ system.

13-20.4
6ϕ double-wye

Two other systems, which are also not "true" 6ϕ systems, are shown in Figs. 13-41 and 13-42. Both these systems (the double-Y and double-Δ) require interconnection to the load to produce 6ϕ line voltages. Because they are somewhat more complicated in required transformer connections, they are seldom used in comparison to the diametrical, except if a change in secondary line voltage is desired (as in the case of the double delta).

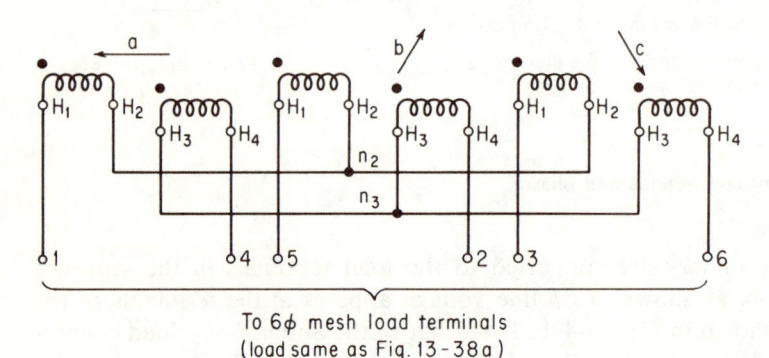

(a) Secondary connections for 6ϕ double-Y (neutrals not connected)

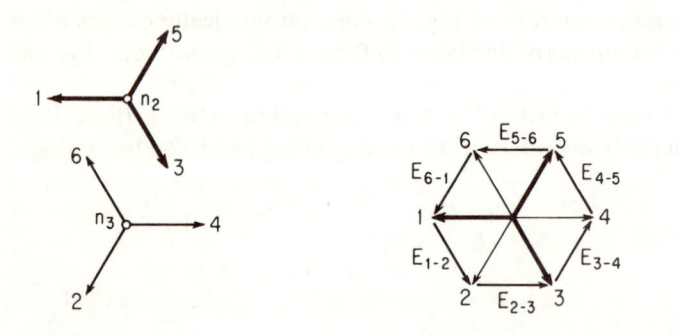

(b) Phasor diagram for double-Y

(c) Phasor diagram when interconnected via 6ϕ mesh load

Figure 13-41
6-ϕ double-Y secondary connections and phasor diagrams.

Figure 13-42

6-ϕ double-Δ secondary connection table and phasor diagram.

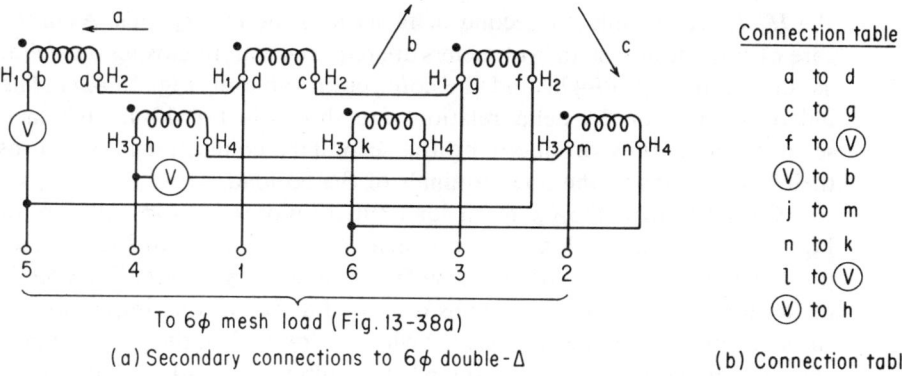

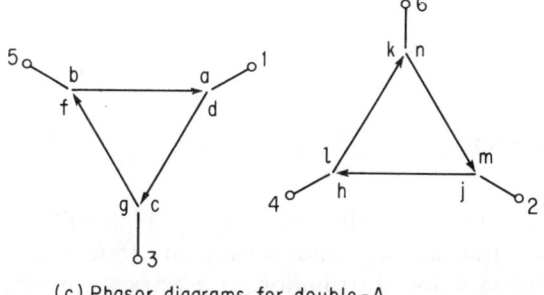

(a) Secondary connections to 6ϕ double-Δ

(b) Connection table

a	to d
c	to g
f	to Ⓥ
Ⓥ	to b
j	to m
n	to k
l	to Ⓥ
Ⓥ	to h

(c) Phasor diagrams for double-Δ

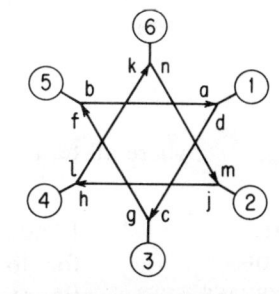

(d) Double-Δ phasor diagram, interconnected via 6ϕ mesh load

The *double-wye* connection is shown in Fig. 13-41a. Close examination reveals that this is almost the same as the 6ϕ star (Fig. 13-38a) with one minor exception—the two star terminals of the Y (n_2 and n_3) are *not* joined to a common junction. Thus, as shown in the phasor diagram of Fig. 13-41b, two separate 3ϕ Y systems are produced at terminals 1, 3 and 5, and also at terminals 2, 4 and 6.

Connection to the 6ϕ mesh load superimposes the (3ϕ) phasors of Fig. 13-41b on the 6ϕ load and on each other, as shown in Fig. 13-41c, producing line voltages displaced by 60°, as in any 6ϕ system. As in the case of the 6ϕ star, the magnitude of the 6ϕ line voltages is the same as the phase voltage (measured from a neutral to any one line) which is the same as the voltage across any secondary coil (H_1-H_2 or H_3-H_4).

13-20.5
6ϕ double-delta

The *double-delta* is the mesh analog of the double Y. Two separate 3ϕ delta connections are made having opposite instantaneous polarity, as shown in Fig. 13-42a. The first delta set uses

SEC. 13-20. / Three-Phase to Six-Phase Transformations

the H_1-H_2 coils while the second delta set uses the H_3-H_4 coils. As in the case of any mesh system, voltmeters are required before closing the delta, as shown in Fig. 13-42a and the connection table of Fig. 13-42b. The individual phasor 3ϕ delta relations are shown in Fig. 13-42c for each separate delta thus produced by the secondary connections, as well as their connections to the line terminals of the 6ϕ load.

Connection to the 6ϕ mesh load superimposes the (3ϕ) phasors of Fig. 13-42c on the 6ϕ load and on each other, as shown in Fig. 13-42d, producing line voltages displaced by 60°, as in any 6ϕ system. The double delta differs from all other systems discussed above in one important respect: it produces a line voltage (E_L) which is lower than the phase voltage (E_p). As shown in Fig. 13-42d the 6ϕ line voltage, E_{1_2} is less than the 3ϕ coil voltage $E_{cd} = E_{1_3} = E_p$, the phase voltage. The relation between them is

$$E_L = \frac{E_p}{\sqrt{3}} = 0.577 E_p \tag{13-58}$$

where all terms are defined above.

13-21. USE OF POLYPHASE TRANSFORMATIONS IN POWER CONVERSION

In addition to their uses in transforming 3ϕ ac to high voltages for long distance transmission and subsequent transformation to lower voltages for distribution of electrical energy, transformers are also used in the *conversion* of ac to dc. Polyphase transformations from 3ϕ to 6ϕ and even higher are a step in the rectification process. Several advantages may be given for polyphase rectification over single phase, namely:

1. lower ripple content of fundamental and higher order harmonics exist in the output waveform, requiring less complex smoothing filters,
2. transformers are used more efficiently in that the ratio of dc output power per kVA of transformer is higher for polyphase conversion,
3. a higher dc to ac (average to rms) voltage ratio emerges as the number of phases increases,
4. the overall efficiency of the conversion process increases. This is important when large amounts of ac power are to be converted to dc.

It is much more efficient, therefore, for a power company to supply ac to an industrial consumer requiring large quantities of dc, by transmission of energy over 3ϕ transmission lines at high voltages. Transformers are then used efficiently to (1) provide a secondary voltage suitable for rectification and (2) transform the 3ϕ supply into 6ϕ or higher to obtain the advantages cited above.

Table 13-2 on page 593 shows a comparison of various pertinent factors governing the choice of the number of phases which may be used for optimum rectification. The first row shows the ratio of output dc voltage to ac phase voltage, V_{dc}/V_{p_2}. This ratio increases as the number of phases increase to a theoretical maximum of 1.414. It represents, in effect, a measure of the useful dc to the ripple ac component present. It would thus appear that if 12ϕ rectification is used (or even 24ϕ), the theoretical maximum of the peak value of ac is practically reached within one percent (see Fig. 13-44b). Note that for any sine wave, $E_m = 1.414E_{p_2}$, where E_{p_2} is the secondary rms phase voltage of the transformer, and this value is the limiting dc value.

TABLE 13-2.
COMPARISONS BETWEEN POLYPHASE SYSTEMS OF RECTIFICATION

NUMBER OF PHASES, n	1 (FULL WAVE)	3	6	12	∞
$\dfrac{V_{dc}}{V_{p_2}}$	0.9	1.17	1.35	1.4	1.414
$\dfrac{E_h}{V_{dc}}$	0.667	0.25	0.057	0.014	0
$\dfrac{P_{dc}}{VA_2}$	0.54	0.675	0.551	0.400	0

The second row of Table 13-2 shows the ratio of E_h/V_{dc} (the amplitude of the major harmonic to the unfiltered dc output voltage) as the number of phases increase. Here again, as the number of phases increases, the harmonic content is reduced, thus reducing the ac ripple considerably. This factor, too, would tend to dictate use of higher polyphase transformer conversions and rectification.

The last row of Table 13-2 is a ratio of the rectified power to the volt-ampere rating of the secondary of the transformer winding. This ratio is sometimes called the *utilization* factor. A low utilization factor means a higher transformer cost for the amount of dc power produced. The utilization factor can be shown to be a theoretical maximum for 2.7 phases. Thus, it would appear that 3ϕ rectification provides the most efficient conversion in terms of transformer cost, despite its higher ripple content and lower ratio of dc to ac rms voltage.

Figure 13-43 shows the use of a Δ-Y high voltage primary to low voltage secondary conversion, using solid-state, half-wave rectification. The primary is delta to sup-

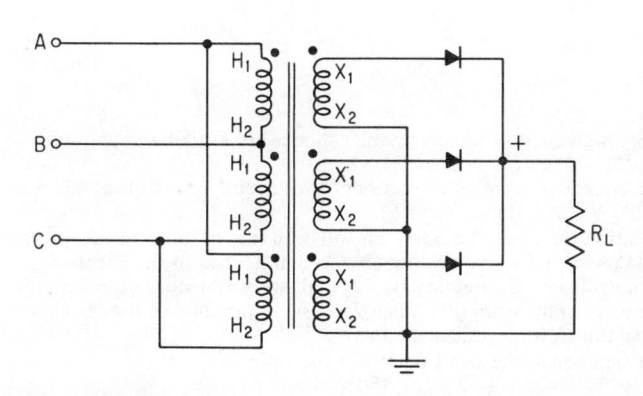

Figure 13-43
3-ϕ half-wave rectification.

press harmonics. The secondary neutral is grounded for the same reason. Yet for most commercial applications involving large quantities of dc power, the circuit of Fig. 13-43 is highly undesirable, despite its higher transformer utilization factor. The dc current component is always in the same direction in each secondary winding and this results in an excessively high excitation current due to dc magnetization of the iron core. The net effect results in overheating of the transformers. This, coupled with the disadvantages of higher ripple content and lower ratio of dc to ac rms voltage, dictates the use of 6ϕ rectification, despite a somewhat lower transformer utilization factor.

**13-21.1
6ϕ half wave rectification using diodes**

The 6ϕ star connection is preferred for most 3ϕ to 6ϕ transformations because it provides a 6ϕ neutral (or ground), as well as a *true* 6ϕ system. A total of 6 solid-state rectifiers are used of suitable peak reverse voltage (PRV) and current rating* to provide a half-wave rectified output of the 6ϕ secondary voltage.†

If it is desired to adjust the dc output voltage, depending on relative costs and amount of power involved, two methods are commonly employed. A three-phase variac may be used at the input to vary the supply voltage to the 3 delta-connected transformers, thus changing the secondary output phase voltage E_{p_2} and the output dc voltage, V_{dc} in Fig. 13-44a. Alternatively, in lieu of diodes D_1 through D_6, silicon controlled rectifiers (SCRs) in conjunction with a phase shifting circuit may be used to control the dc output voltage (Fig. 13-45).

The waveforms produced by the six diodes are shown in Fig. 13-44b, in the *absence* of filter capacitor C (shown in Fig. 13-44a). The effect of adding a filter capacitor on the output waveform is shown in Fig. 13-44c.

Note that the output waveform of Fig. 13-44b contains a relatively small ripple component so that it varies from E_m to E_x, where $E_m = 1.414 E_{p_2}$. The average value of the rectified dc for the output waveform of Fig. 13-44b is‡

$$V_{dc} = 1.35 E_{p_2} \tag{13-59}$$

---

* As of this writing, high current silicon rectifier diodes are available rated up to 500 A with PRVs up to 1200 V (G. E. type A295PN).
  Similarly, high current silicon controlled rectifiers are currently available at 470 A rms with PRVs up to 1200 V (G. E. type C290PB).
† Full wave rectification requires 12 diodes, one on each side of the 6 transformer windings shown in Fig. 13-44a. This involves possible loss of neutral in the event that one diode fails on the neutral side. Consequently, 12ϕ half-wave transformations may be preferred to 6ϕ full-wave transformations using the same number of diodes (12) despite a somewhat lower transformer utilization factor.
‡ It can be shown that for n phases (2 or more) the ratio of
$$\frac{V_{dc}}{E_{p_2}} = \frac{\sqrt{2}\,n}{\pi}\left(\sin\frac{180°}{n}\right)$$
Substituting $n = 6$ in this equation yields 1.35, as shown in Table 13-2 and Eq. (13-59) above.

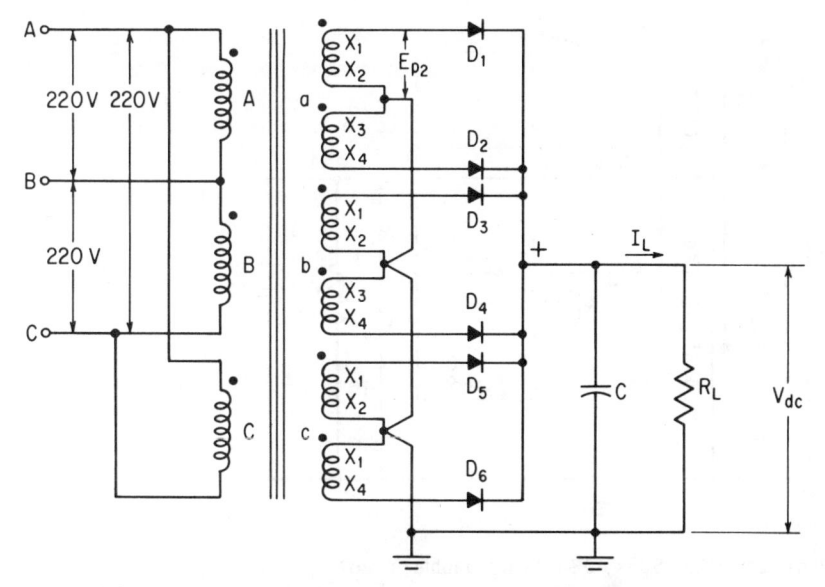

(a) Connections 3φ to 6φ half-wave rectification using 6φ star

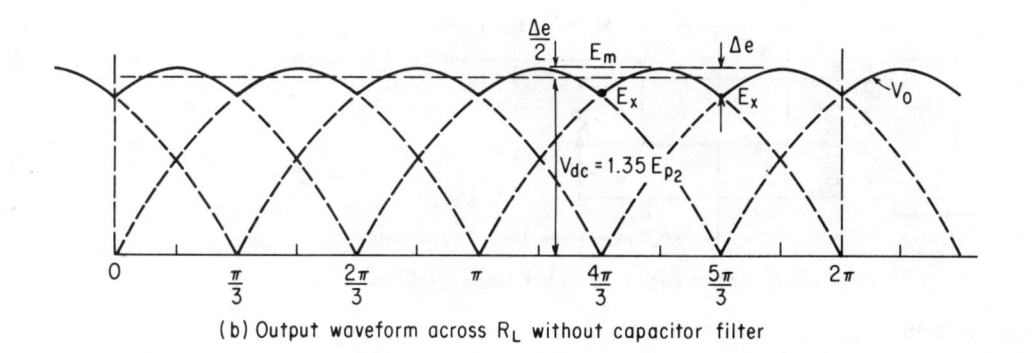

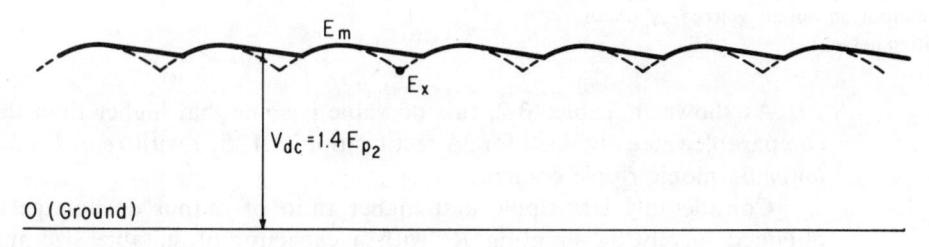

(b) Output waveform across R_L without capacitor filter

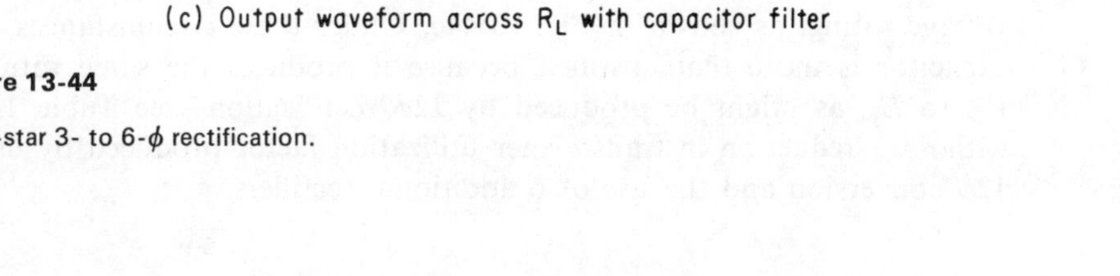

(c) Output waveform across R_L with capacitor filter

Figure 13-44

Delta-star 3- to 6-φ rectification.

SEC. 13-21. / Use of Polyphase Transformations in Power Conversion

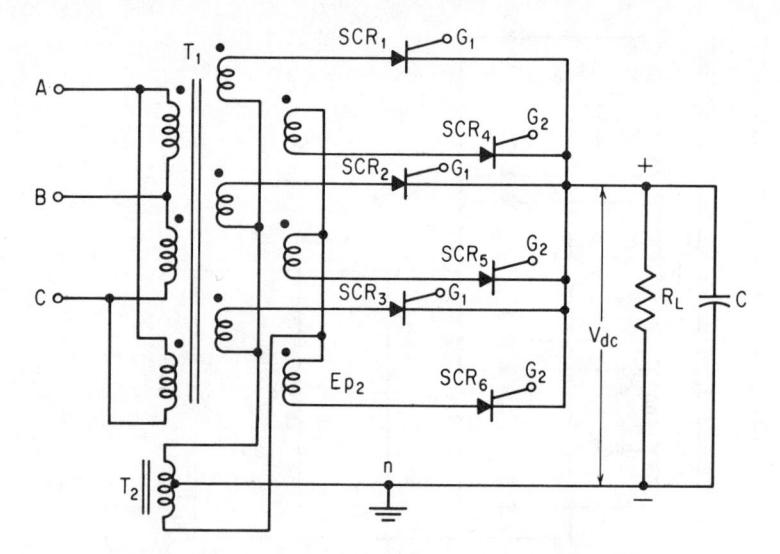

(a) Phase controlled SCRs, $3\phi\Delta$ to 6ϕ double-Y with interphase transformer, T_2

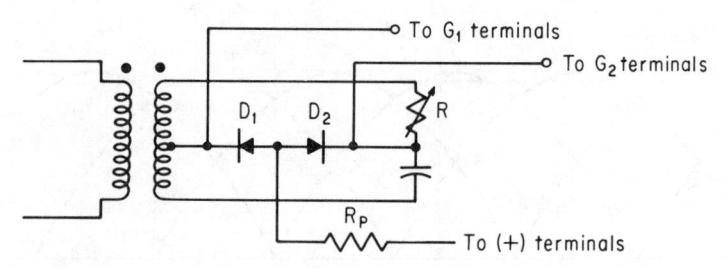

(b) Typical phase control circuit to adjust firing of SCRs

Figure 13-45

Double-Y 6-ϕ transformation with SCRs to control dc output voltage by phase shifting network.

As shown in Table 13-2, this dc value is somewhat higher than the comparable value obtained for 3ϕ rectification ($1.17E_{p_2}$) with *considerably lower* harmonic ripple content.

Considerably less ripple and higher ratio of output dc voltage is obtained merely by shunting R_L with a capacitor of suitable size and voltage rating, as shown in Fig. 13-44c. Under these circumstances, the capacitor is more than justified because it produces the same ratio of V_{dc} to E_{p_2} as might be produced by 12ϕ rectification (see Table 13-2) without a reduction in transformer utilization factor produced by using 12ϕ conversion and the use of 6 additional rectifiers.

**13-21.2
6φ double-Y
rectification
using SCRs**

A typical commercial technique used to provide an *adjustable* dc output involving use of SCRs is shown in Fig. 13-45a. Here, the conversion is a delta double-Y, 3φ to 6φ conversion. Note that an additional center-tapped reactance called an *interphase transformer*, T_2, is used between the neutrals of the two Ys. (If T_2 is omitted, a star would be produced.) The interphase transformer serves

1. to provide a center-tapped ground connection for the double-Y,
2. to act as a means of equalizing any dc voltage differences between the two halves by providing a high reactance to ac but a low resistance to dc,
3. to improve the voltage regulation of the circuit (i.e., its ability to maintain its output voltage from no load to full load).

The use of the double-wye connection provides all the advantages of 6φ conversion without the added disadvantage of lower transformer utilization factor. Since it inherently consists of two 3φ transformations, as shown in Table 13-2, it has a utilization factor of 0.675 rather than 0.551, which applies to the case of Fig. 13-44a. At the same time, when connected to the load, R_L, a 6φ system is produced and ripple is that of any 6φ rather than that of a 3φ system.

Control of the magnitude of the output voltage, V_{dc}, is achieved by means of a phase shift network shown in Fig. 13-45b. Variable resistance R controls the phase angle between the gate signal voltage at terminals G_1 and G_2 with respect to the ac input voltage. By varying R from 0 to a high value, the phase angle may be varied from almost 0 to 180°. The conduction of the SCR is thus varied from its maximum (at $\theta = 0$) to a minimum (when $\theta = 180°$).

The circuit of Fig. 13-45 is useful whenever variable voltage dc is required as in the control of speed of a dc motor by armature voltage control.

BIBLIOGRAPHY

Crosno, C. D. *Fundamentals of Electromechanical Conversion*, New York: Harcourt, Brace, Jovanovich, Inc, 1968.

Daniels. *The Performance of Electrical Machines*, New York: McGraw-Hill, Inc., 1968.

Hindmarsh, J. *Electrical Machines*, Elmsford, N.Y.: Pergamon Press, 1965.

Majmudar, H. *Introduction to Electrical Machines*, Boston: Allyn and Bacon, 1969.

Selmon. *Magnetoelectric Devices: Transducers, Transformers and Machines*, New York: Wiley/Interscience, 1966.

Skilling, H. H. *Electromechanics: A First Course in Electromechanical Energy Conversion*, New York: Wiley/Interscience, 1962.

Thaler, G. J., and Wilcox, M. L. *Electric Machines: Dynamics and Steady State*, New York: Wiley/Interscience, 1966.

QUESTIONS

13-1. For a transformer
    a. define primary and secondary windings
    b. high voltage side and low voltage side windings
    c. is it possible for any voltage winding to serve as the primary? Explain.

13-2. What is the significance of the dot convention for
    a. the primary?
    b. separate secondary windings?
    c. why are the induced emfs in all windings, including the primary, in phase?

13-3. a. What is meant by loose coupling and how are the values of M, k, V_2 and I_2 affected by it?
    b. What conditions produce loose coupling?
    c. Is it possible for the geometric mean of the primary and secondary inductances to be smaller than the mutual inductance? Explain.

13-4. a. Why is tight coupling desirable on power transformers?
    b. Is unity coupling ever possible in practice in commercial power transformers? Why not?
    c. Define an ideal transformer.

13-5. Explain, in detail, how a transformer transfers power from primary to secondary with a leading load on the secondary terminals. Use phasor diagrams to illustrate your explanation.

13-6. a. Define transformation ratio, α.
    b. Is the transformation ratio constant for a given transformer? Explain.
    c. Define transformation ratio in terms of 1) turns ratio, 2) current ratio and 3) voltage ratio.
    d. Why does the voltage ratio vary directly as the turns ratio?
    e. Why does the current ratio vary inversely as the turns ratio?

13-7. Why are transformers rated in kVA rather than watts, or kW?

13-8. Show that the volts/turn ratio of a transformer
    a. is proportional to the frequency and peak value of mutual flux
    b. follows the same (identical) relations as that for an alternator.

13-9. a. Is it possible for a 60 Hz transformer to operate on 400 Hz? Under what conditions?
    b. Why is it necessary to maintain the peak value of mutual flux and flux density at a constant value, regardless of frequency changes?

13-10. In operating a 400 Hz transformer at 60 Hz, explain
  a. why the voltage must be reduced in the same proportion as the frequency.
  b. why the kVA rating is reduced in the same proportion as the frequency.
  c. why the core losses are reduced considerably.
  d. why the efficiency of the transformer increases.

13-11. On the basis of Quest. 13-10 above, explain why a 1 kVA, 400 Hz transformer is smaller than a 1 kVA, 60 Hz transformer.

13-12. Is it possible to "increase" the kVA rating of 60 Hz transformers by appropriate changes in nameplate ratings of frequency and voltage? (See Prob. 13-9.) Explain in detail how this is done.

13-13. a. State the relation between secondary and primary impedance of a transformer.
  b. Explain, on the basis of (a) above, how the transformer serves as an impedance matching device.
  c. What is the purpose of impedance matching?
  d. Give two applications where impedance matching is necessary.

13-14. For a loaded practical transformer explain
  a. why the primary applied voltage V_1 is greater than the primary induced emf E_1
  b. why the secondary terminal voltage V_2 is smaller than the secondary induced emf, E_2
  c. why the transformation ratio is slightly less than the ratio of primary to secondary terminal voltages.

13-15. For a loaded practical power transformer, having load Z_L connected across its secondary terminals, explain why
  a. the load impedance is usually greater than the internal impedance of the secondary winding
  b. the primary input impedance is much greater than the impedance of the primary winding
  c. designers attempt to make the internal impedances as close to the internal resistances as possible, for both primary and secondary windings
  d. transformers must have primary and secondary internal resistance.

13-16. In developing the equivalent circuit for a power transformer fully loaded
  a. what is the advantage of such an equivalent circuit?
  b. why is the equivalent circuit referred usually to the primary?
  c. what assumptions are made in lumping primary and secondary resistance, and reactance, respectively?
  d. why is it possible to neglect magnetizing current?

13-17. If a practical transformer is loaded to capacity with a leading PF load, explain why the primary PF is higher than the secondary PF using
  a. phasor diagrams
  b. Eq. (13-24).

13-18. If a practical transformer is loaded to capacity with a lagging PF load, explain why the primary PF is lower than the secondary PF using
 a. phasor diagrams
 b. Eq. (13-24).

13-19. Taking secondary terminal voltage as the reference phasor, explain using both Eq. (13-28) and phasor diagrams
 a. why lagging loads produce poorer voltage regulation than unity PF loads
 b. why some leading loads produce better voltage regulation than unity PF loads
 c. Is it possible for a transformer to have a negative voltage regulation? Explain.

13-20. Compare Eq. (13-28) with Eq. (6-8) and discuss
 a. similarities
 b. differences
 c. explain why transformers have better voltage regulation than alternators for the same kVA rating.

13-21. In performing the short-circuit test of a transformer
 a. why is the low voltage side usually short-circuited? Give 2 reasons.
 b. using Fig. 13-9, explain why short-circuiting the secondary of a transformer provides only the equivalent impedance, resistance and reactance from primary measurements of voltage, current and power
 c. why are the core losses of the transformer considered negligible?
 d. under what circumstances must the core losses of the transformer be taken into account? Why?

13-22. a. Is the open-circuit test necessary to determine the voltage regulation of the transformer?
 b. What specific information is obtained from the open circuit test and where is it (only) used?
 c. Why is it customary to perform the open circuit test on the lowest voltage winding obtainable on the transformer?
 d. What precautions are necessary, however, in connection with (c) above?
 e. Is the core loss the same if a high voltage winding was used? Explain.

13-23. In determining *efficiency only*, using open- and short-circuit test data
 a. why is it unnecessary to calculate equivalent impedance and reactance from short-circuit test data?
 b. what specific single instrument provides useful information for calculating efficiency?
 c. is it necessary to calculate equivalent resistance? Why not? Explain. (Hint. See tabulations of copper loss in Exs. 13-15a and b.)
 d. is the equivalent copper loss referred to the high voltage side the same as the equivalent copper loss referred to the low voltage side? Explain.

13-24. With regard to the efficiency curve of a transformer shown in Fig. 13-12, explain
 a. why the efficiency is zero at zero load (open circuit)
 b. why the efficiency rises so rapidly

c. at what value of load the efficiency is a maximum (see Eqs. 13-34, 35)
d. why the efficiency is somewhat lower at heavy loads
e. why the efficiency is lower for lower PF loads
f. why a well-designed transformer is usually more efficient than the rotating machinery it serves.

13-25. With regard to all-day efficiency
a. what test data and information is necessary for its computation?
b. why is it computed on an energy rather than power basis?
c. why is it usually fairly high despite occasional low power factors and periods of relatively light use?

13-26. a. In addition to open- and short-circuit tests, why are phasing and polarity tests necessary before putting a transformer into service?
b. Define phasing.
c. How can a transformer have a polarity when it is used on ac? Explain.
d. What code letter determines high and low sides, respectively?
e. What is the significance of the number subscripts?
f. Is it possible to determine how coils are wound or the polarity from physical examination? Why not? Explain.

13-27. Given a transformer having several independent windings and a multi-tapped winding, explain
a. how each independent coil is identified
b. how the multi-tapped winding is identified
c. how the individual taps on the multi-tap winding are properly identified
d. how the polarity of each separate winding is determined.

13-28. For a transformer having 2 identical high voltage and 2 identical low voltage windings
a. how many possible combinations of voltage ratios may be obtained using all the windings, when connected as a conventional transformer?
b. repeat (a) above for an autotransformer, with all 4 coils series-connected.
c. repeat using high voltage and low voltage coils in parallel, but series-connected, as an autotransformer.

13-29. a. Is it possible to connect coils of unequal voltage rating in series? Explain.
b. Is it possible to connect coils of unequal voltage rating in parallel? Explain.
c. Are precautions required with regard to polarity in (a) above? Explain.
d. In connecting coils of equal voltage rating in parallel, what precautions are necessary in regard to instantaneous polarity? Explain.

13-30. a. Define an autotransformer.
b. Explain why autotransformers are more efficient than conventional transformers, having the same voltage ratios and primary, secondary load currents.
c. Can any transformer having two or more isolated windings be converted to an autotransformer? Explain how.

d. Why is the rating of an autotransformer higher than the rating of a conventional transformer of the same physical size and winding capacity?

13-31. a. Define power *transformed* from primary to secondary. Give equations.
b. Define power *transferred conductively* from primary to secondary. Give equations.
c. Which of the two powers is responsible for the increase in kVA of an autotransformer over a conventional isolation transformer?

13-32. For the same kVA rating, explain why an autotransformer
a. has less copper losses than a conventional transformer
b. has lower core loss
c. is smaller in size
d. has a higher efficiency.

13-33. In view of your answers to the previous question explain
a. why autotransformers are not used exclusively in comparison to conventional transformers
b. why autotransformers are limited to low voltage power distribution and usually restricted to use in reduced voltage motor starting.

13-34. a. What is the advantage, in terms of possible damage and continuity of service, of 3 single-phase transformers connected in a bank, over one polyphase transformer?
b. When connecting 3 transformers in either delta or wye, what precautions are required in regard to polarity, kVA rating, voltage and current ratings?

13-35. For a Y-Y transformer bank
a. what is the relation between the line and phase voltages, both primary and secondary sides?
b. what is the relation between the line and phase currents, both primary and secondary sides?
c. what is the relation between primary and secondary phase voltages and primary and secondary line voltages?
d. what is the phase shift from primary lines to secondary lines?

13-36. For a Y-Δ transformer bank, repeat all parts (a) through (d) of Quest. 13-35.

13-37. a. Is it possible to parallel a Y-Δ bank to a Δ-Y bank providing the line-to-line voltages are the same? Justify your answer by means of a phasor diagram.
b. For the 7 transformer banks listed in Sec. 13-14 (just prior to Ex. 13-23), list the parallel combinations which are *not* possible.
c. Explain why instantaneous polarity is so important in connecting transformers in delta. What is the effect of reversing one winding on the line voltages?
d. Repeat (c) for a Y-connection, illustrating result by phasor diagram.

13-38. a. Explain why the secondaries of mesh-or delta-connected transformers evidence a small voltage recorded on a voltmeter prior to closing the delta.

b. Explain the source of this voltage.
c. How is the voltage made to disappear, if the delta is closed?
d. On the basis of your answer to (c), does a Δ-Δ, Y-Δ or Δ-Y system evidence harmonic voltages? Explain fully.
e. On the basis of your answer to (d), explain why neutrals are absolutely necessary for Y-Y systems.

13-39.
a. Give 3 reasons for a neutral connection whenever a Y-bank is present.
b. Is it possible to provide a neutral in a Δ-Δ system?
c. What is the danger of center-tapping more than *one* secondary Δ winding and grounding it?
d. Why is the primary of a delta bank never grounded to a 3ϕ-4-wire source?

13-40. It is customary to use Y-Δ transformations to supply industrial and residential areas from high voltage transmission lines. Explain
a. advantages of using Y-Δ banks over Y-Y or Δ-Δ
b. how single-phase loads are accommodated
c. how a neutral is provided for 3-wire single-phase service
d. what effect an unbalance of load currents produces on the voltages of the delta secondary when the single-phase loading is higher than the 3-phase load supplied by the secondary transformer bank.

13-41. Assuming the load is constant, what effect is produced in a Δ-Δ bank when one transformer is removed and V-V operation is provided. Give
a. the power supplied by each transformer
b. the increase in load on each transformer
c. the total kVA rating in V-V as a fraction of the Δ-Δ total
d. the derating factor of each transformer, compared to its single phase rating.

13-42. Assuming 2 transformers are supplying rated capacity to a load in V-V. If a third transformer is added to provide a Δ-Δ bank, give
a. the total increased capacity over original rated capacity
b. cost of the third transformer
c. increase in rating of each transformer over original derating (part d, Ques. 41 above).

13-43.
a. Using Table 13-1, list all the possible transformer bank combinations both in Column A and Column B, beyond the 7 combinations originally begun in Sec. 13-14, using only two banks in parallel.
b. Why is it impossible to parallel a bank in Column B with that in Column A of Table 13-1?

13-44. Using transformers only, is it possible to transform
a. a single-phase system into a polyphase system? Explain.
b. a polyphase into a single-phase system? Explain.
c. a 2ϕ system into a 6ϕ system? Explain.
d. What is the advantage of using transformers only, for such transformations? Explain.

13-45.
a. How does a 2ϕ system differ from a 3ϕ system?
b. What is the relation between line and phase voltages in a 2ϕ system?

c. What is meant by a quarter-phase 5-wire system?
d. How many phase and line voltages are available in a system such as (c)?
e. Draw a Scott transformation showing 3ϕ to 2ϕ-5 wire transformation in which the phase to phase voltage ratios are 120 V/120 V.
f. What are the line voltages in (e) above, both primary and secondary?

13-46. a. What is meant by a true 6ϕ system?
b. Which transformations produce a true 6ϕ system?
c. What is one advantage of using higher order polyphase systems in rectification?
d. Name 5 different types of 6ϕ connections using 3ϕ to 6ϕ transformations.
e. Which transformation will supply a 6ϕ load using conventional single winding transformers having no center taps.

13-47. a. Why is the star connection preferred for 3ϕ to 6ϕ rectification compared to the mesh or double-delta?
b. List 4 advantages for polyphase rectification over single-phase.
c. To obtain the greatest ratio of dc output voltage to ac phase voltage, how many phases should be used, theoretically? How many phases produce this value practically within 1 percent?
d. How is the amplitude of the major harmonic affected by the number of phases rectified? How many phases reduce this amplitude to less than 2%?
e. How many phases appear to produce the maximum utilization of rectified power to voltampere rating of the transformer secondary?
f. What compromises should be made in determining optimum number of phases for polyphase rectification?

13-48. a. What two methods are employed to adjust the output dc voltage of polyphase rectifiers?
b. What is the advantage of using SCRs in place of diodes?
c. Why are interphase transformers used with SCRs?
d. What is the advantage of double-Y transformation shown in Fig. 13-45 over conventional 6ϕ half-wave transformation in terms of utilization factor?

PROBLEMS

13-1. The primary of a transformer, closely coupled, has an inductance of 20 H, a coefficient of coupling of 0.98 and mutual inductance of 9.8 H. Calculate the inductance of the secondary winding.

13-2. A commercial 400 Hz, 220 V/20 V transformer has 50 turns on its low voltage side. Calculate:
a. number of turns on its high side
b. ratio of transformation, α, when used as a step-down transformer
c. repeat (b) when used as a step-up transformer

d. volts/turn ratio of high side
e. volts/turn ratio of low side.

13-3. The high voltage side of a transformer has 750 turns and the low voltage side 50 turns. When the high side is connected to a rated voltage of 120 V, 60 Hz and rated load of 40 A is connected to the low side, calculate
a. transformation ratio, α
b. secondary voltage, assuming no internal transformer impedance voltage drops
c. resistance of the load
d. volts/turn ratio of secondary and primary, respectively
e. voltampere rating of the transformer.

13-4. A commercial 220 V/30 V, 3 kVA, 60 Hz transformer has a ratio of 3 V/turn. Calculate.
a. turns of high voltage side
b. turns of low voltage side
c. transformation ratio if used as a step-down transformer
d. transformation ratio if used as a step-up transformer
e. rated high-side current
f. rated low-side current.

13-5. A 10 Ω load draws 20 A from the high voltage side of a transformer whose $\alpha = 1/8$. Assuming no internal transformer voltage drops, calculate
a. secondary voltage
b. primary voltage
c. primary current
d. voltamperes transferred from primary to secondary
e. transformation ratio when used as a step-down transformer.

13-6. Using the volts/turn ratio, calculate the peak value of the mutual flux, ϕ_{pm}, for the transformers of
a. Prob. 13-2
b. Prob. 13-3
c. Prob. 13-4.

13-7. A 600 V/20 V, 1 kVA, 400 Hz, 3000 turn/100 turn transformer is to be used from a 60-Hz supply. Maintaining the same permissible flux density, calculate
a. maximum voltage which may be applied to the high side at 60 Hz
b. maximum voltage which may be applied to the low side at 60 Hz
c. original volts per turn ratio at 400 Hz
d. volts per turn ratio at 60 Hz
e. kVA rating of transformer at 60 Hz.

13-8. A 110 V/6 V, 60 Hz, 20 VA filament transformer is tested to withstand rms voltages as high as 1000 V on both its primary and secondary windings. If used at 400 Hz, and maintaining the same maximum permissible flux density, calculate
a. rating of high voltage side
b. rating of low voltage side
c. rating of the transformer, in VA.

13-9. Having solved the above problems, John Smith, a student, originates a brilliant scheme. Transformers are generally priced in proportion to kVA rating. The development of solid-state devices has rendered most vacuum tube filament transformers obsolete. At the same time, the introduction of 400-Hz solid-state circuitry has created a need for power-supply transformers for high voltage dc power supplies. Why not buy 60-Hz filament transformers, change the nameplate ratings and sell them at a higher kVA at a profit? He discovers he can purchase at discount quantity lots, a number of 60 Hz, 200 VA, 2000 V breakdown voltage tested transformers having a 120 V high-side and two low-side windings, 6 V and 12 V, respectively, at a cost of $5, each. John connects the low sides in series to use the full rating of the transformer secondaries. Calculate
a. the low-side voltage rating at 400 Hz
b. the high-side voltage rating at 400 Hz
c. the kVA rating at 400 Hz
d. selling price per transformer at a cost of $50/kVA
e. profit per transformer, if the new nameplates cost John $1 each.
f. Is John Smith misrepresenting his product in any way? Explain in terms of transformer efficiency.

13-10. The primary of a transformer, consisting of two 120 V windings in parallel, serves a fixed load and draws 6 A from a 120 V, 60 Hz supply. Calculate the current drawn from the supply when
a. only one coil is connected across the line
b. both 120 V windings are series-connected to a 240 V, 60 Hz supply
c. using a 120 V, 50 Hz supply
d. using a 120 V, 25 Hz supply.

13-11. If the permissible maximum flux density of a 220 V, 60 Hz transformer should not exceed 60 kilolines per square inch, how many turns should be used for the 220 V side? The cross-sectional area of the transformers is 22.5 in².

13-12. There are 1000 turns on the high voltage side of a 10 kVA, 10/1 transformer.
a. When 1000 V at 60 Hz is connected across the high voltage side, the maximum flux density is 5000 gauss (maxwells/square centimeter). What is the area of the core in cm² and in²?
b. If the applied voltage is raised to 1500 V, find the maximum flux density.
c. Repeat (b) at a frequency of 50 Hz at the same voltage (1500 V).

13-13. Assume that eddy current loss is a function of $(f \times B_m)^2$ but hysteresis loss is a function of $f^1 \times B_m^{1.75}$ for tests made on two transformers having the same weight and quality of iron. Calculate:
a. the ratio of their eddy current losses when operating at the same frequency, if their flux densities are 6000 and 4000 gauss, respectively
b. the ratio of their hysteresis losses at the flux densities given in (a).

13-14. A given 60 Hz transformer has a hysteresis loss of 200 W and an eddy current loss of 100 W at a maximum value of flux density of 200 webers per square meter, when rated voltage of 120 V is applied across its primary. Calculate:
  a. hysteresis and eddy current losses when the voltage is decreased to 110 V at the same frequency. Use assumptions of Prob. 13-13.
  b. maximum flux density hysteresis and eddy current losses if rated voltage is applied at a frequency of 50 Hz
  c. maximum flux density, hysteresis and eddy current loss when 60 V is applied at 30 Hz.

13-15. A 20 kVA, 660 V/120 V transformer has a no-load loss of 250 W and a high voltage side resistance of 0.2 Ω. Assuming that the load losses of the windings are equal, calculate
  a. resistance of the low voltage side
  b. full load equivalent copper loss
  c. transformer efficiencies at load values of 25, 50, 75, 100 and 125 per cent, unity PF, assuming that the regulation of the transformer is zero per cent.

13-16. The efficiency of a 20 kVA, 1200 V/120 V transformer is a maximum of 98.0% at 50% of rated load. Calculate:
  a. core loss
  b. efficiency at rated load
  c. efficiency at loads of 75% and 125%.

13-17. A 20 kVA, 1200/120 V transformer which is continuously excited is loaded at unity PF over a 24 hour period as follows: 5 hr at full load, 5 hr at half load, 5 hr at quarter load. Maximum efficiency occurs at full load and is 97%. Calculate the all-day transformer efficiency.

13-18. A 10 kVA, 60 Hz, 4800/240 V transformer is tested by the open-circuit and short-circuit tests, respectively. The test data is

	voltage	current	power	side used
Open-circuit test	240 V	1.5 A	60 W	low voltage
Short-circuit test	180 V	rated	180 W	high voltage

Using the above data, calculate
  a. equivalent resistance and reactance referred to the high voltage side
  b. equivalent resistance and reactance referred to the low voltage side
  c. voltage regulation of the step-down transformer at unity PF, full load
  d. repeat (c) at a PF of 0.8 lagging, full load.

13-19. From the data of Prob. 13-18, calculate
  a. core loss of the transformer
  b. full load copper loss of the transformer
  c. full load efficiency at a PF of 0.9 lagging
  d. all-day efficiency when the transformer is loaded as follows:
    6 hrs at full load, unity PF; 4 hrs at half load, 0.8 PF lagging; 6 hrs at one-quarter load, 0.6 lagging PF and 8 hrs at no load.

13-20. A 100 kVA, 60 Hz, 12000/240 V transformer is tested by the open- and short-circuit tests, respectively, and the test data is as follows:

test	voltage(V)	current(A)	power(W)	side used
Open-circuit	240	8.75	480	low voltage
Short-circuit	600	rated	1200	high voltage

Calculate from the above data, as a step-down transformer
a. regulation at 0.8 PF lagging
b. efficiencies at 0.8 PF lagging for $\frac{1}{8}, \frac{1}{4}, \frac{1}{2}, \frac{3}{4}, \frac{4}{4}$ and $\frac{5}{4}$ rated load
c. fraction of rated load at which maximum efficiency occurs
d. maximum efficiency at 0.8 PF lagging load.

13-21. Repeat Prob. 13-20b, for efficiencies at a PF of 0.6 lagging.

13-22. Repeat Prob. 13-20a, for regulation at a PF of 0.7 leading.

13-23. A 50 kVA, 600/240 V, 25 Hz transformer has a core loss of 200 W (of which 30 % is eddy-current loss) and a full-load copper loss of 650 W. If the transformer is operated at 600 V, 60 Hz, what should be the new rating of the transformer if the total losses must be kept the same?

13-24. A step-up autotransformer is used to supply 3 kV from a 2.4 kV supply line. If the secondary load is 50 A, calculate (neglecting losses and magnetizing current)
a. current in each part of the transformer
b. current drawn from the 2.4 kV supply line
c. kVA rating of the transformer
d. kVA rating of a comparable conventional two winding transformer necessary to accomplish the same transformation.

13-25. A step-down autotransformer is used to supply 100 A at 2 kV from a a 2.4 kV supply line. Calculate as in parts (a) through (d) of Prob. 13-24.

13-26. For the transformer in Prob. 13-24 calculate
a. power transformed from primary to secondary, at rated load, unity PF
b. power transferred conductively from primary to secondary, at rated load, unity PF.

13-27. For the transformer in Prob. 13-25 repeat parts (a) and (b) of Prob. 13-26.

13-28. Given 3 identical transformers with 7500 turns each on the high voltage sides, calculate the number of turns to be used on the low voltage sides, when the primaries are delta-connected to a 26,400 V, 3ϕ supply and the secondaries are connected
a. in Y to produce 4160 V between lines
b. in Δ to produce 4160 V between lines.

13-29. A Y-Δ transformation is used to convert 60 Hz, 13,200 V, 3ϕ, to 208 V, 3ϕ, using 3 identical single-phase transformers. If the maximum permissible flux density is 40,000 lines/in², and the transformer cross-sectional area is 40 in², calculate
a. number of turns in the high voltage sides
b. number of turns in the low voltage sides.

13-30. A 50 kVA, 220 V, three phase load must be supplied from a 13,200 V 3ϕ supply. Specify the voltage, current and kVA ratings of the single-phase transformers required for the following connections:
  a. Y-Y
  b. Y-Δ
  c. Δ-Y
  d. Δ-Δ
  e. V-V

13-31. A balanced 3ϕ load of 1.5 MVA is supplied from two identical transformers connected in V-V. For a load at unity PF, calculate
  a. the minimum capacity of each transformer in kVA
  b. the minimum capacity of each transformer if a Δ-Δ connection is used.

13-32. Repeat the calculations of Prob. 13-31 for the same MVA load at 0.8 PF lagging.

13-33. A transformer vault supplies a factory drawing 693.0 kW at unity PF through 2 V-V transformers. The incoming primary line voltage is 26,600 V and the secondary line voltage is 2300 V. Calculate, neglecting losses
  a. minimum kVA rating of each transformer
  b. voltage and current rating of each winding
  c. PF at which each winding operates, neglecting excitation current
  d. increased rating of the vault when a third identical transformer is added, in kW at unity PF
  e. per cent increase in kW and in transformer investment, under conditions of (d), above.

13-34. A transformer substation serves 2-phase, 5-wire power of 12,000 kW at unity PF at 13.8 kV from high voltage 3-phase 132 kV transmission lines. It is desired to use 2 Scott-connected transformers to accomplish the transformation. Specify for the main and teaser transformers, primary voltages and currents, secondary voltages currents, and MVA rating of each transformer.

13-35. An industrial plant consumes a total of 300 kW at 0.8 PF lagging from a 2400 V, 3ϕ, 3 wire supply which is the secondary of a Scott transformer whose primary is a balanced 240 V, quarter-phase, 5 wire system. For the two transformers, below, calculate
  a. main transformer primary and secondary currents
  b. teaser transformer primary and secondary currents.

13-36. An industrial plant is fed from a 230 V, 2ϕ 3-wire supply. It is desired to use a synchronous motor drawing 50 kW at unity PF rated at 230 V, 3-phase, 3 wire. (The sychronous motor is necessary for PF correction but a 2ϕ synchronous motor in this capacity is not available.) Draw a diagram showing the Scott transformation, 2ϕ to 3ϕ showing all primary and secondary currents in the main and teaser transformers, respectively.

ANSWERS

13-1 5 H   13-2(a) 550 t (b) 11 (c) 1/11 (d) 0.4 V/t (e) 0.4 V/t   13-3 (a) 15 (b) 8 V (c) 0.2 Ω (d) 0.16 V/t (e) 320 VA   13-4 (a) 73 t (b) 10 t (c) 7.33 (d) 0.136 (e) 13.62 A (f) 100 A   13-5 (a) 200 V (b) 25 V (c) 160 A (d) 4 kVA (e) 8   13-6 (a) 2.25×10^4 Mx (b) 6.0×10^4 Mx (c) 1.125×10^6 Mx   13-7 (a) 90 V (b) 3 V (c) 0.2 V/t (d) 0.03 V/t (e) 150 VA   13-8 (a) 733 V (b) 40 V (c) 133 VA   13-9 (a) 120 V (b) 800 V (c) 1.333 kVA (d) $66.67 (e) $60.67 (f) Yes   13-10 (a) 6 A (b) 3 A (c) 7.2 A (d) 14.4 A   13-11 61 t.   13-12 (a) 75 cm², 11.62 in² (b) 7.5 kG (c) 9 kG   13-13 (a) 2.25/1 (b) 2.06/1   13-14 (a) 172 W, 84.15 W (b) 240 Wb/m², 229.5 W, 100 W (c) 200 Wb/m², 100 W, 25 W   13-15 (a) 0.0066 Ω (b) 367 W (c) 94.75, 96.7, 97.1, 97.0 and 96.8%   13-16 (a) 100 W (b) 97.5% (c) 97.8%, 97.25%   13-17. 95.1%   13-18 (a) 41.6 Ω, 76 Ω; (b) 0.104 Ω, 0.19 Ω (c) 1.875% (d) 3.33%   13-19 (a) 60 W (b) 180 W (c) 97.4% (d) 98.3%   13-20 (a) 4.17% (b) 95.25; 97.3; 97.9; 98.1; 97.7; 97.6% (c) 0.633 (d) 98.2%   13-21 93.8; 96.5; 97.5; 97.5; 97.3; 97.0%   13-22 -4.167% 13-23 25.5 kVA   13-24 (a) 50 A, 12.5 A (b) 62.5 A (c) 60 kVA (d) 150 kVA   13-25 (a) 83.3 A, 16.7 A (b) 83.3 A (c) 66.7 kVA (d) 200 kVA   13-26 (a) 30 kW (b) 120 kW   13-27 (a) 33.3 kW (b) 166.7 kW   13-28 (a) 682 t (b) 1182 t   13-29 (a) 1790 t (b) 49 t   13-30 (a) Y-Y: 16.67 kVA; 7625/127.2 V; 2.19/131 A (b) Y-Δ: 16.67 kVA; 7625/220 V; 2.19/75.8 A (c) Δ-Y: 16.67 kVA; 13,200/127.2 V; 1.26/131 A (d) Δ-Δ: 16.67 kVA; 13,200/220 V; 1.26/75.8 A (e) V-V: 28.85 kVA; 13,200/220 V; 2.19/131 A.   13-31 (a) 866.7 kVA (b) 500 kVA.   13-32 (a) 866.7 kVA (b) 500 kVA   13-33 (a) 400 kVA (b) 26.6 kV at 15.05 A, 2.3 kV at 174 A (c) 0.866 (d) 1.2 MW (e) 73.2%, 50%.   13-34. Main transformer: $V_1 = 132$ kV, $I_1 = 52.5$ A, $V_2 = 13.8$ kV, $I_2 = 435$ A, rating $= 6.93$ MVA Teaser transformer: $V_1 = 114.2$ kV, $I_1 = 52.5$ A, $V_2 = 13.8$ kV, $I_2 = 435$ A, rating $= 6.0$ MVA   13-35 both (a) and (b), $I_2 = 90.2$ A and $I_1 = 781$ A   13-36 Main transformer: $V_2 = 230$ V, $I_2 = 125.5$ A, $V_1 = 230$ V, $I_1 = 108.7$ A Teaser transformer: $V_2 = 199.2$ V, $I_2 = 125.5$ A, $V_1 = 230$ V, $I_1 = 108.7$ A

appendix

TABLE A-1. NATURAL TRIGONOMETRIC FUNCTIONS

Angle, °	sin	tan	cot	cos	Angle, °	Angle, °	sin	tan	cot	cos	Angle, °
0.0	.00000	.00000	∞	1.00000	90.0	6.0	.10453	.10510	9.5144	.99452	84.0
.1	.00175	.00175	572.96	1.00000	.9	.1	.10626	.10687	9.3572	.99434	.9
.2	.00349	.00349	286.48	0.99999	.8	.2	.10800	.10863	9.2052	.99415	.8
.3	.00524	.00524	190.98	.99999	.7	.3	.10973	.11040	9.0579	.99396	.7
.4	.00698	.00698	143.24	.99993	.6	.4	.11147	.11217	8.9152	.99377	.6
.5	.00873	.00873	114.59	.99996	.5	.5	.11320	.11394	8.7769	.99357	.5
.6	.01047	.01047	95.489	.99995	.4	.6	.11494	.11570	8.6427	.99337	.4
.7	.01222	.01222	81.847	.99993	.3	.7	.11667	.11747	8.5126	.99317	.3
.8	.01396	.01396	71.615	.99990	.2	.8	.11840	.11924	8.3863	.99297	.2
.9	.01571	.01571	63.657	.99988	.1	.9	.12014	.12101	8.2636	.99276	.1
1.0	.01745	.01746	57.290	.99985	89.0	7.0	.12187	.12278	8.1443	.99255	83.0
.1	.01920	.01920	52.081	.99982	.9	.1	.12360	.12456	8.0285	.99233	.9
.2	.02094	.02095	47.740	.99978	.8	.2	.12533	.12633	7.9158	.99211	.8
.3	.02269	.02269	44.066	.99974	.7	.3	.12706	.12810	7.8062	.99189	.7
.4	.02443	.02444	40.917	.99970	.6	.4	.12880	.12988	7.6996	.99167	.6
.5	.02618	.02619	38.188	.99966	.5	.5	.13053	.13165	7.5958	.99144	.5
.6	.02792	.02793	35.801	.99961	.4	.6	.13226	.13343	7.4947	.99122	.4
.7	.02967	.02968	33.694	.99956	.3	.7	.13399	.13521	7.3962	.99098	.3
.8	.03141	.03143	31.821	.99951	.2	.8	.13572	.13698	7.3002	.99075	.2
.9	.03316	.03317	30.145	.99945	.1	.9	.13744	.13876	7.2066	.99051	.1
2.0	.03490	.03492	28.636	.99939	88.0	8.0	.13917	.14054	7.1154	.99027	82.0
.1	.03664	.03667	27.271	.99933	.9	.1	.14090	.14232	7.0264	.99002	.9
.2	.03839	.03842	26.031	.99926	.8	.2	.14263	.14410	6.9395	.98978	.8
.3	.04013	.04016	24.898	.99919	.7	.3	.14436	.14588	6.8548	.98953	.7
.4	.04188	.04191	23.859	.99912	.6	.4	.14608	.14767	6.7720	.98927	.6
.5	.04362	.04366	22.904	.99905	.5	.5	.14781	.14945	6.6912	.98902	.5
.6	.04536	.04541	22.022	.99897	.4	.6	.14954	.15124	6.6122	.98876	.4
.7	.04711	.04716	21.205	.99889	.3	.7	.15126	.15302	6.5350	.98849	.3
.8	.04885	.04891	20.446	.99881	.2	.8	.15299	.15481	6.4596	.98823	.2
.9	.05059	.05066	19.740	.99872	.1	.9	.15471	.15660	6.3859	.98796	.1
3.0	.05234	.05241	19.081	.99863	87.0	9.0	.15643	.15838	6.3138	.98769	81.0
.1	.05408	.05416	18.464	.99854	.9	.1	.15816	.16017	6.2432	.98741	.9
.2	.05582	.05591	17.886	.99844	.8	.2	.15988	.16196	6.1742	.98714	.8
.3	.05756	.05766	17.343	.99834	.7	.3	.16160	.16376	6.1066	.98686	.7
.4	.05931	.05941	16.832	.99824	.6	.4	.16333	.16555	6.0405	.98657	.6
.5	.06105	.06116	16.350	.99813	.5	.5	.16505	.16734	5.9758	.98629	.5
.6	.06279	.06291	15.895	.99803	.4	.6	.16677	.16914	5.9124	.98600	.4
.7	.06453	.06467	15.464	.99792	.3	.7	.16849	.17093	5.8502	.98570	.3
.8	.06627	.06642	15.056	.99780	.2	.8	.17021	.17273	5.7894	.98541	.2
.9	.06802	.06817	14.669	.99768	.1	.9	.17193	.17453	5.7297	.98511	.1
4.0	.06976	.06993	14.301	.99756	86.0	10.0	.17365	.17633	5.6713	.98481	80.0
.1	.07150	.07168	13.951	.99744	.9	.1	.17537	.17813	5.6140	.98450	.9
.2	.07324	.07344	13.617	.99731	.8	.2	.17708	.17993	5.5578	.98420	.8
.3	.07498	.07519	13.300	.99719	.7	.3	.17880	.18173	5.5026	.98389	.7
.4	.07672	.07695	12.996	.99705	.6	.4	.18052	.18353	5.4486	.98357	.6
.5	.07846	.07870	12.706	.99692	.5	.5	.18224	.18534	5.3955	.98325	.5
.6	.08020	.08046	12.429	.99678	.4	.6	.18395	.18714	5.3435	.98294	.4
.7	.08194	.08221	12.163	.99664	.3	.7	.18567	.18895	5.2924	.98261	.3
.8	.08368	.08397	11.909	.99649	.2	.8	.18738	.19076	5.2422	.98229	.2
.9	.08542	.08573	11.664	.99635	.1	.9	.18910	.19257	5.1929	.98196	.1
5.0	.08716	.08749	11.430	.99619	85.0	11.0	.19081	.19438	5.1446	.98163	79.0
.1	.08889	.08925	11.205	.99604	.9	.1	.19252	.19619	5.0970	.98129	.9
.2	.09063	.09101	10.988	.99588	.8	.2	.19423	.19801	5.0504	.98096	.8
.3	.09237	.09277	10.780	.99572	.7	.3	.19595	.19982	5.0045	.98061	.7
.4	.09411	.09453	10.579	.99556	.6	.4	.19766	.20164	4.9594	.98027	.6
.5	.09585	.09629	10.385	.99540	.5	.5	.19937	.20345	4.9152	.97992	.5
.6	.09758	.09805	10.199	.99523	.4	.6	.20108	.20527	4.8716	.97958	.4
.7	.09932	.09981	10.019	.99506	.3	.7	.20279	.20709	4.8288	.97922	.3
.8	.10106	.10158	9.8448	.99488	.2	.8	.20450	.20891	4.7867	.97887	.2
.9	.10279	.10334	9.6768	.99470	.1	.9	.20620	.21073	4.7453	.97851	.1
6.0	.10453	.10510	9.5144	.99452	84.0	12.0	.20791	.21256	4.7046	.97815	78.0
Angle, °	cos	cot	tan	sin	Angle, °	Angle, °	cos	cot	tan	sin	Angle, °

Appendix

TABLE A-1 (CONT.)

Angle, °	sin	tan	cot	cos	Angle, °	Angle, °	sin	tan	cot	cos	Angle, °
12.0	.20791	.21256	4.7046	.97815	78.0	18.0	.30902	.32492	3.0777	.95106	72.0
.1	.20962	.21438	4.6646	.97778	.9	.1	.31068	.32685	3.0595	.95052	.9
.2	.21132	.21621	4.6252	.97742	.8	.2	.31233	.32878	3.0415	.94997	.8
.3	.21303	.21804	4.5864	.97705	.7	.3	.31399	.33072	3.0237	.94943	.7
.4	.21474	.21986	4.5483	.97667	.6	.4	.31565	.33266	3.0061	.94888	.6
.5	.21644	.22169	4.5107	.97630	.5	.5	.31730	.33460	2.9887	.94832	.5
.6	.21814	.22353	4.4737	.97592	.4	.6	.31896	.33654	2.9714	.94777	.4
.7	.21985	.22536	4.4373	.97553	.3	.7	.32061	.33848	2.9544	.94721	.3
.8	.22155	.22719	4.4015	.97515	.2	.8	.32227	.34043	2.9375	.94665	.2
.9	.22325	.22903	4.3662	.97476	.1	.9	.32392	.34238	2.9208	.94609	.1
13.0	.22495	.23087	4.3315	.97437	77.0	19.0	.32557	.34433	2.9042	.94552	71.0
.1	.22665	.23271	4.2972	.97398	.9	.1	.32722	.34628	2.8878	.94495	.9
.2	.22835	.23455	4.2635	.97358	.8	.2	.32887	.34824	2.8716	.94438	.8
.3	.23005	.23639	4.2303	.97318	.7	.3	.33051	.35020	2.8556	.94380	.7
.4	.23175	.23823	4.1976	.97278	.6	.4	.33216	.35216	2.8397	.94322	.6
.5	.23345	.24008	4.1653	.97237	.5	.5	.33381	.35412	2.8239	.94264	.5
.6	.23514	.24193	4.1335	.97196	.4	.6	.33545	.35608	2.8083	.94206	.4
.7	.23684	.24377	4.1022	.97155	.3	.7	.33710	.35805	2.7929	.94147	.3
.8	.23853	.24562	4.0713	.97113	.2	.8	.33874	.36002	2.7776	.94088	.2
.9	.24023	.24747	4.0408	.97072	.1	.9	.34038	.36199	2.7625	.94029	.1
14.0	.24192	.24933	4.0108	.97030	76.0	20.0	.34202	.36397	2.7475	.93969	70.0
.1	.24362	.25118	3.9812	.96987	.9	.1	.34366	.36595	2.7326	.93909	.9
.2	.24531	.25304	3.9520	.96945	.8	.2	.34530	.36793	2.7179	.93849	.8
.3	.24700	.25490	3.9232	.96902	.7	.3	.34694	.36991	2.7034	.93789	.7
.4	.24869	.25676	3.8947	.96858	.6	.4	.34857	.37190	2.6889	.93728	.6
.5	.25038	.25862	3.8667	.96815	.5	.5	.35021	.37388	2.6746	.93667	.5
.6	.25207	.26048	3.8391	.96771	.4	.6	.35184	.37588	2.6605	.93606	.4
.7	.25376	.26235	3.8118	.96727	.3	.7	.35347	.37787	2.6464	.93544	.3
.8	.25545	.26421	3.7848	.96682	.2	.8	.35511	.37986	2.6325	.93483	.2
.9	.25713	.26608	3.7583	.96638	.1	.9	.35674	.38186	2.6187	.93420	.1
15.0	.25882	.26795	3.7321	.96593	75.0	21.0	.35837	.38386	2.6051	.93358	69.0
.1	.26050	.26982	3.7062	.96547	.9	.1	.36000	.38587	2.5916	.93295	.9
.2	.26219	.27169	3.6806	.96502	.8	.2	.36162	.38787	2.5782	.93232	.8
.3	.26387	.27357	3.6554	.96456	.7	.3	.36325	.38988	2.5649	.93169	.7
.4	.26556	.27545	3.6305	.96410	.6	.4	.36488	.39190	2.5517	.93106	.6
.5	.26724	.27732	3.6059	.96363	.5	.5	.36650	.39391	2.5386	.93042	.5
.6	.26892	.27921	3.5816	.96316	.4	.6	.36812	.39593	2.5257	.92978	.4
.7	.27060	.28109	3.5576	.96269	.3	.7	.36975	.39795	2.5129	.92913	.3
.8	.27228	.28297	3.5339	.96222	.2	.8	.37137	.39997	2.5002	.92849	.2
.9	.27396	.28486	3.5105	.96174	.1	.9	.37299	.40200	2.4876	.92784	.1
16.0	.27564	.28675	3.4874	.96126	74.0	22.0	.37461	.40403	2.4751	.92718	68.0
.1	.27731	.28864	3.4646	.96078	.9	.1	.37622	.40606	2.4627	.92653	.9
.2	.27899	.29053	3.4420	.96029	.8	.2	.37784	.40809	2.4504	.92587	.8
.3	.28067	.29242	3.4197	.95981	.7	.3	.37946	.41013	2.4383	.92521	.7
.4	.28234	.29432	3.3977	.95931	.6	.4	.38107	.41217	2.4262	.92455	.6
.5	.28402	.29621	3.3759	.95882	.5	.5	.38268	.41421	2.4142	.92388	.5
.6	.28569	.29811	3.3544	.95832	.4	.6	.38430	.41626	2.4023	.92321	.4
.7	.28736	.30001	3.3332	.95782	.3	.7	.38591	.41831	2.3906	.92254	.3
.8	.28903	.30192	3.3122	.95732	.2	.8	.38752	.42036	2.3789	.92186	.2
.9	.29070	.30382	3.2914	.95681	.1	.9	.38912	.42242	2.3673	.92119	.1
17.0	.29237	.30573	3.2709	.95630	73.0	23.0	.39073	.42447	2.3559	.92050	67.0
.1	.29404	.30764	3.2506	.95579	.9	.1	.39234	.42654	2.3445	.91982	.9
.2	.29571	.30955	3.2305	.95528	.8	.2	.39394	.42860	2.3332	.91914	.8
.3	.29737	.31147	3.2106	.95476	.7	.3	.39555	.43067	2.3220	.91845	.7
.4	.29904	.31338	3.1910	.95424	.6	.4	.39715	.43274	2.3109	.91775	.6
.5	.30071	.31530	3.1716	.95372	.5	.5	.39875	.43481	2.2998	.91706	.5
.6	.30237	.31722	3.1524	.95319	.4	.6	.40035	.43689	2.2889	.91636	.4
.7	.30403	.31914	3.1334	.95266	.3	.7	.40195	.43897	2.2781	.91566	.3
.8	.30570	.32106	3.1146	.95213	.2	.8	.40355	.44105	2.2673	.91496	.2
.9	.30736	.32299	3.0961	.95159	.1	.9	.40514	.44314	2.2566	.91425	.1
18.0	.30902	.32492	3.0777	.95106	72.0	24.0	.40674	.44523	2.2460	.91355	66.0
Angle, °	cos	cot	tan	sin	Angle, °	Angle, °	cos	cot	tan	sin	Angle, °

613

Appendix

TABLE A-1 (CONT.)

Angle, °	sin	tan	cot	cos	Angle, °	Angle, °	sin	tan	cot	cos	Angle, °
24.0	.40674	.44523	2.2460	.91355	66.0	30.0	.50000	.57735	1.7321	.86603	60.0
.1	.40833	.44732	2.2355	.91283	.9	.1	.50151	.57968	1.7251	.86515	.9
.2	.40992	.44942	2.2251	.91212	.8	.2	.50302	.58201	1.7182	.86427	.8
.3	.41151	.45152	2.2148	.91140	.7	.3	.50453	.58435	1.7113	.86340	.7
.4	.41310	.45362	2.2045	.91068	.6	.4	.50603	.58670	1.7045	.86251	.6
.5	.41469	.45573	2.1943	.90996	.5	.5	.50754	.58905	1.6977	.86163	.5
.6	.41628	.45784	2.1842	.90924	.4	.6	.50904	.59140	1.6909	.86074	.4
.7	.41787	.45995	2.1742	.90851	.3	.7	.51054	.59376	1.6842	.85985	.3
.8	.41945	.46206	2.1642	.90778	.2	.8	.51204	.59612	1.6775	.85896	.2
.9	.42104	.46418	2.1543	.90704	.1	.9	.51354	.59849	1.6709	.85806	.1
25.0	.42262	.46631	2.1445	.90631	65.0	31.0	.51504	.60086	1.6643	.85717	59.0
.1	.42420	.46843	2.1348	.90557	.9	.1	.51653	.60324	1.6577	.85627	.9
.2	.42578	.47056	2.1251	.90483	.8	.2	.51803	.60562	1.6512	.85536	.8
.3	.42736	.47270	2.1155	.90408	.7	.3	.51952	.60801	1.6447	.85446	.7
.4	.42894	.47483	2.1060	.90334	.6	.4	.52101	.61040	1.6383	.85355	.6
.5	.43051	.47698	2.0965	.90259	.5	.5	.52250	.61280	1.6319	.85264	.5
.6	.43209	.47912	2.0872	.90183	.4	.6	.52399	.61520	1.6255	.85173	.4
.7	.43366	.48127	2.0778	.90108	.3	.7	.52547	.61761	1.6191	.85081	.3
.8	.43523	.48342	2.0686	.90032	.2	.8	.52696	.62003	1.6128	.84989	.2
.9	.43680	.48557	2.0594	.89956	.1	.9	.52844	.62245	1.6066	.84897	.1
26.0	.43837	.48773	2.0503	.89879	64.0	32.0	.52992	.62487	1.6003	.84805	58.0
.1	.43994	.48989	2.0413	.89803	.9	.1	.53140	.62730	1.5941	.84712	.9
.2	.44151	.49206	2.0323	.89726	.8	.2	.53288	.62973	1.5880	.84619	.8
.3	.44307	.49423	2.0233	.89649	.7	.3	.53435	.63217	1.5818	.84526	.7
.4	.44464	.49640	2.0145	.89571	.6	.4	.53583	.63462	1.5757	.84433	.6
.5	.44620	.49858	2.0057	.89493	.5	.5	.53730	.63707	1.5697	.84339	.5
.6	.44776	.50076	1.9970	.89415	.4	.6	.53877	.63953	1.5637	.84245	.4
.7	.44932	.50295	1.9883	.89337	.3	.7	.54024	.64199	1.5577	.84151	.3
.8	.45088	.50514	1.9797	.89259	.2	.8	.54171	.64446	1.5517	.84057	.2
.9	.45243	.50733	1.9711	.89180	.1	.9	.54317	.64693	1.5458	.83962	.1
27.0	.45399	.50953	1.9626	.89101	63.0	33.0	.54464	.64941	1.5399	.83867	57.0
.1	.45554	.51173	1.9542	.89021	.9	.1	.54610	.65189	1.5340	.83772	.9
.2	.45710	.51393	1.9458	.88942	.8	.2	.54756	.65438	1.5282	.83676	.8
.3	.45865	.51614	1.9375	.88862	.7	.3	.54902	.65688	1.5224	.83581	.7
.4	.46020	.51835	1.9292	.88782	.6	.4	.55048	.65938	1.5166	.83485	.6
.5	.46175	.52057	1.9210	.88701	.5	.5	.55194	.66189	1.5108	.83389	.5
.6	.46330	.52279	1.9128	.88620	.4	.6	.55339	.66440	1.5051	.83292	.4
.7	.46484	.52501	1.9047	.88539	.3	.7	.55484	.66692	1.4994	.83195	.3
.8	.46639	.52724	1.8967	.88458	.2	.8	.55630	.66944	1.4938	.83098	.2
.9	.46793	.52947	1.8887	.88377	.1	.9	.55775	.67197	1.4882	.83001	.1
28.0	.46947	.53171	1.8807	.88295	62.0	34.0	.55919	.67451	1.4826	.82904	56.0
.1	.47101	.53395	1.8728	.88213	.9	.1	.56064	.67705	1.4770	.82806	.9
.2	.47255	.53620	1.8650	.88130	.8	.2	.56208	.67960	1.4715	.82708	.8
.3	.47409	.53844	1.8572	.88048	.7	.3	.56353	.68215	1.4659	.82610	.7
.4	.47562	.54070	1.8495	.87965	.6	.4	.56497	.68471	1.4605	.82511	.6
.5	.47716	.54296	1.8418	.87882	.5	.5	.56641	.68728	1.4550	.82413	.5
.6	.47869	.54522	1.8341	.87798	.4	.6	.56784	.68985	1.4496	.82314	.4
.7	.48022	.54748	1.8265	.87715	.3	.7	.56928	.69243	1.4442	.82214	.3
.8	.48175	.54975	1.8190	.87631	.2	.8	.57071	.69502	1.4388	.82115	.2
.9	.48328	.55203	1.8115	.87546	.1	.9	.57215	.69761	1.4335	.82015	.1
29.0	.48481	.55431	1.8040	.87462	61.0	35.0	.57358	.70021	1.4281	.81915	55.0
.1	.48634	.55659	1.7966	.87377	.9	.1	.57501	.70281	1.4229	.81815	.9
.2	.48786	.55888	1.7893	.87292	.8	.2	.57643	.70542	1.4176	.81714	.8
.3	.48938	.56117	1.7820	.87207	.7	.3	.57786	.70804	1.4124	.81614	.7
.4	.49090	.56347	1.7747	.87121	.6	.4	.57928	.71066	1.4071	.81513	.6
.5	.49242	.56577	1.7675	.87036	.5	.5	.58070	.71329	1.4019	.81412	.5
.6	.49394	.56808	1.7603	.86949	.4	.6	.58212	.71593	1.3968	.81310	.4
.7	.49546	.57039	1.7532	.86863	.3	.7	.58354	.71857	1.3916	.81208	.3
.8	.49697	.57271	1.7461	.86777	.2	.8	.58496	.72122	1.3865	.81106	.2
.9	.49849	.57503	1.7391	.86690	.1	.9	.58637	.72388	1.3814	.81004	.1
30.0	.50000	.57735	1.7321	.86603	60.0	36.0	.58779	.72654	1.3764	.80902	54.0
Angle, °	cos	cot	tan	sin	Angle, °	Angle, °	cos	cot	tan	sin	Angle, °

614

Appendix

TABLE A-1 (CONT.)

Angle, °	sin	tan	cot	cos	Angle, °	Angle, °	sin	tan	cot	cos	Angle, °
36.0	.58779	.72654	1.3764	.80902	54.0	40.5	.64945	.85408	1.1708	.76041	**49.5**
.1	.58920	.72921	1.3713	.80799	.9	.6	.65077	.85710	1.1667	.75927	.4
.2	.59061	.73189	1.3663	.80696	.8	.7	.65210	.86014	1.1626	.75813	.3
.3	.59201	.73457	1.3613	.80593	.7	.8	.65342	.86318	1.1585	.75700	.2
.4	.59342	.73726	1.3564	.80489	.6	.9	.65474	.86623	1.1544	.75585	.1
.5	.59482	.73996	1.3514	.80386	.5	41.0	.65606	.86929	1.1504	.75471	**49.0**
.6	.59622	.74267	1.3465	.80232	.4	.1	.65738	.87236	1.1463	.75356	.9
.7	.59763	.74538	1.3416	.80178	.3	.2	.65869	.87543	1.1423	.75241	.8
.8	.59902	.74810	1.3367	.80073	.2	.3	.66000	.87852	1.1383	.75126	.7
.9	.60042	.75082	1.3319	.79968	.1	.4	.66131	.88162	1.1343	.75011	.6
37.0	.60182	.75355	1.3270	.79864	53.0	.5	.66262	.88473	1.1303	.74896	.5
.1	.60321	.75629	1.3222	.79758	.9	.6	.66393	.88784	1.1263	.74780	.4
.2	.60460	.75904	1.3175	.79653	.8	.7	.66523	.89097	1.1224	.74664	.3
.3	.60599	.76180	1.3127	.79547	.7	.8	.66653	.89410	1.1184	.74548	.2
.4	.60738	.76456	1.3079	.79441	.6	.9	.66783	.89725	1.1145	.74431	.1
.5	.60876	.76733	1.3032	.79335	.5	42.0	.66913	.90040	1.1106	.74314	**48.0**
.6	.61015	.77010	1.2985	.79229	.4	.1	.67043	.90357	1.1067	.74198	.9
.7	.61153	.77289	1.2938	.79122	.3	.2	.67172	.90674	1.1028	.74080	.8
.8	.61291	.77568	1.2892	.79016	.2	.3	.67301	.90993	1.0990	.73963	.7
.9	.61429	.77848	1.2846	.78908	.1	.4	.67430	.91313	1.0951	.73846	.6
38.0	.61566	.78129	1.2799	.78801	52.0	.5	.67559	.91633	1.0913	.73728	.5
.1	.61704	.78410	1.2753	.78694	.9	.6	.67688	.91955	1.0875	.73610	.4
.2	.61841	.78692	1.2708	.78586	.8	.7	.67816	.92277	1.0837	.73491	.3
.3	.61978	.78975	1.2662	.78478	.7	.8	.67944	.92601	1.0799	.73373	.2
.4	.62115	.79259	1.2617	.78369	.6	.9	.68072	.92926	1.0761	.73254	.1
.5	.62251	.79544	1.2572	.78261	.5	43.0	.68200	.93252	1.0724	.73135	**47.0**
.6	.62388	.79829	1.2527	.78152	.4	.1	.68327	.93578	1.0686	.73016	.9
.7	.62524	.80115	1.2482	.78043	.3	.2	.68455	.93906	1.0649	.72897	.8
.8	.62660	.80402	1.2437	.77934	.2	.3	.68582	.94235	1.0612	.72777	.7
.9	.62796	.80690	1.2393	.77824	.1	.4	.68709	.94565	1.0575	.72657	.6
39.0	.62932	.80978	1.2349	.77715	51.0	.5	.68835	.94896	1.0538	.72537	.5
.1	.63068	.81268	1.2305	.77605	.9	.6	.68962	.95229	1.0501	.72417	.4
.2	.63203	.81558	1.2261	.77494	.8	.7	.69088	.95562	1.0464	.72297	.3
.3	.63338	.81849	1.2218	.77384	.7	.8	.69214	.95897	1.0428	.72176	.2
.4	.63473	.82141	1.2174	.77273	.6	.9	.69340	.96232	1.0392	.72055	.1
.5	.63608	.82434	1.2131	.77162	.5	44.0	.69466	.96569	1.0355	.71934	**46.0**
.6	.63742	.82727	1.2088	.77051	.4	.1	.69591	.96907	1.0319	.71813	.9
.7	.63877	.83022	1.2045	.76940	.3	.2	.69717	.97246	1.0283	.71691	.8
.8	.64011	.83317	1.2002	.76828	.2	.3	.69842	.97586	1.0247	.71569	.7
.9	.64145	.83613	1.1960	.76717	.1	.4	.69966	.97927	1.0212	.71447	.6
40.0	.64279	.83910	1.1918	.76604	50.0	.5	.70091	.98270	1.0176	.71325	.5
.1	.64412	.84208	1.1875	.76492	.9	.6	.70215	.98613	1.0141	.71203	.4
.2	.64546	.84507	1.1833	.76380	.8	.7	.70339	.98958	1.0105	.71080	.3
.3	.64679	.84806	1.1792	.76267	.7	.8	.70463	.99304	1.0070	.70957	.2
.4	.64812	.85107	1.1750	.76154	.6	.9	.70587	.99652	1.0035	.70834	.1
40.5	.64945	.85408	1.1708	.76041	49.5	45.0	.70711	1.00000	1.0000	.70711	45.0
Angle, °	cos	cot	tan	sin	Angle, °	Angle, °	cos	cot	tan	sin	Angle, °

615

Appendix

APPENDIX A-2. GRAPHICAL SYMBOLS

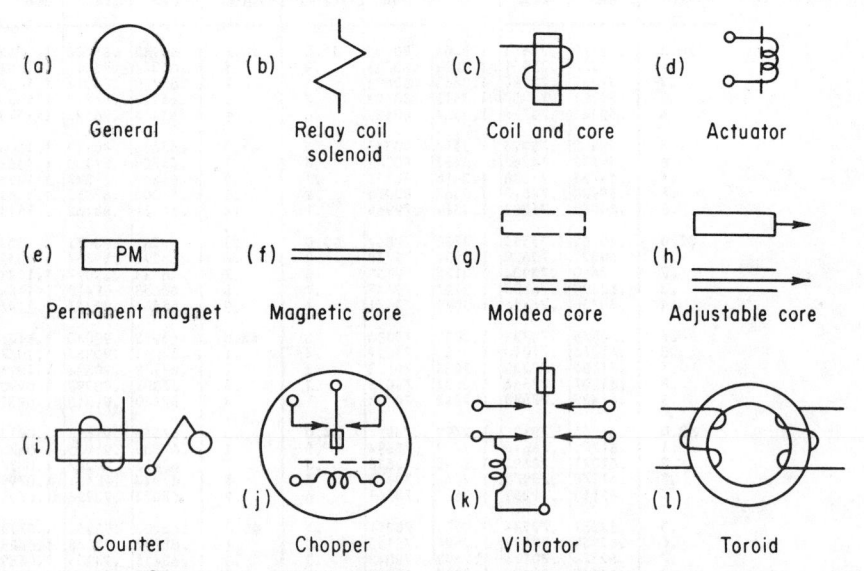

(a) General
(b) Relay coil solenoid
(c) Coil and core
(d) Actuator
(e) PM — Permanent magnet
(f) Magnetic core
(g) Molded core
(h) Adjustable core
(i) Counter
(j) Chopper
(k) Vibrator
(l) Toroid

Figure A-1. Symbols for windings, cores, and magnetic devices.

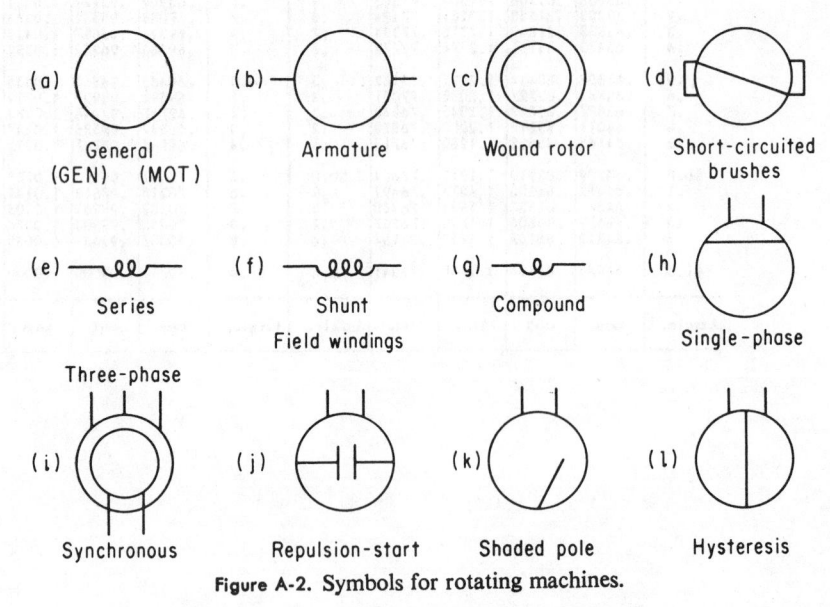

(a) General (GEN) (MOT)
(b) Armature
(c) Wound rotor
(d) Short-circuited brushes
(e) Series
(f) Shunt
(g) Compound
Field windings
(h) Single-phase
(i) Three-phase Synchronous
(j) Repulsion-start
(k) Shaded pole
(l) Hysteresis

Figure A-2. Symbols for rotating machines.

* Excerpted from G. Shiers, *Electronic Drafting,* Prentice-Hall, Inc., 1962.

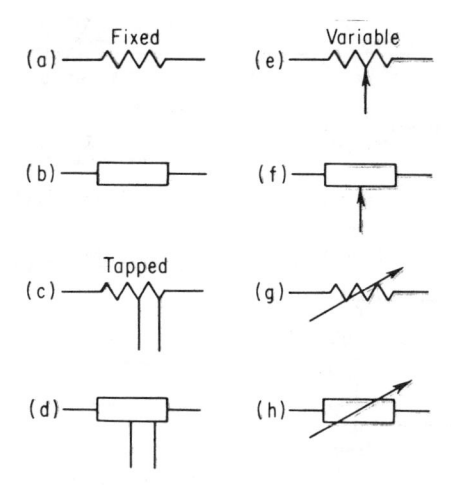

Figure A-3. Resistor symbols.

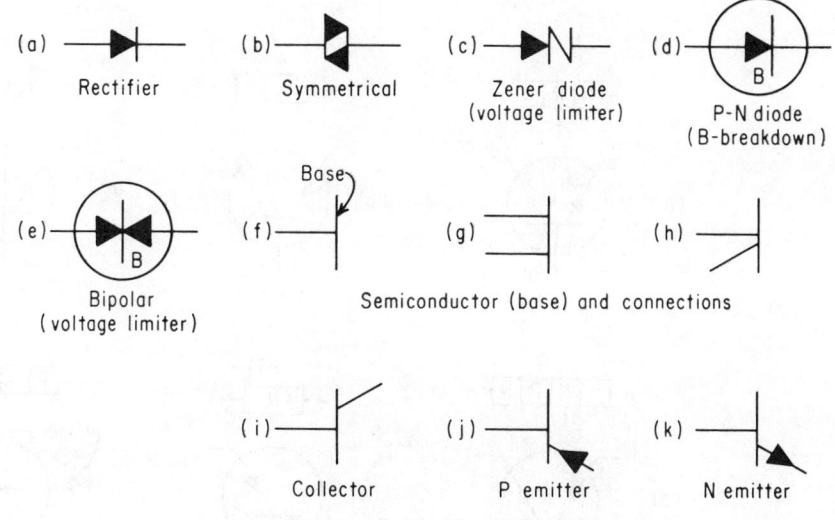

Figure A-4. Semiconductor symbols.

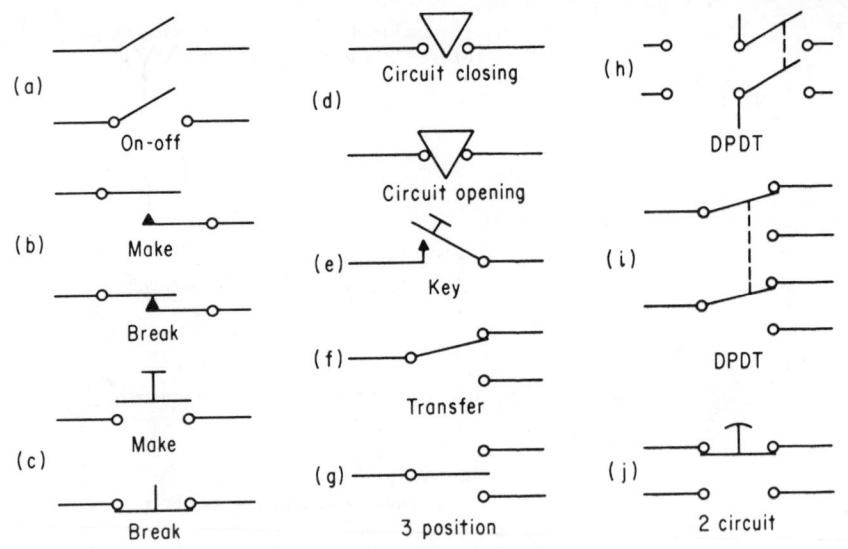

Figure A-5. Basic switch symbols.

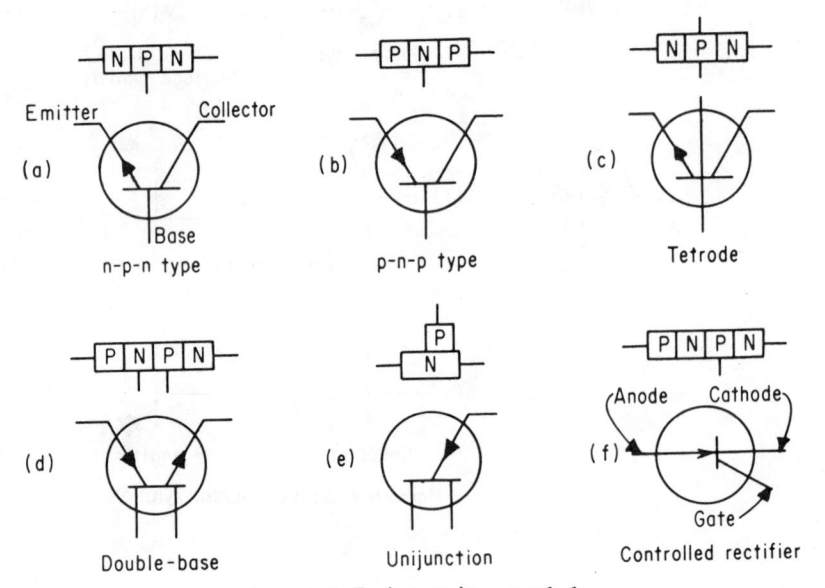

Figure A-6. Basic transistor symbols.

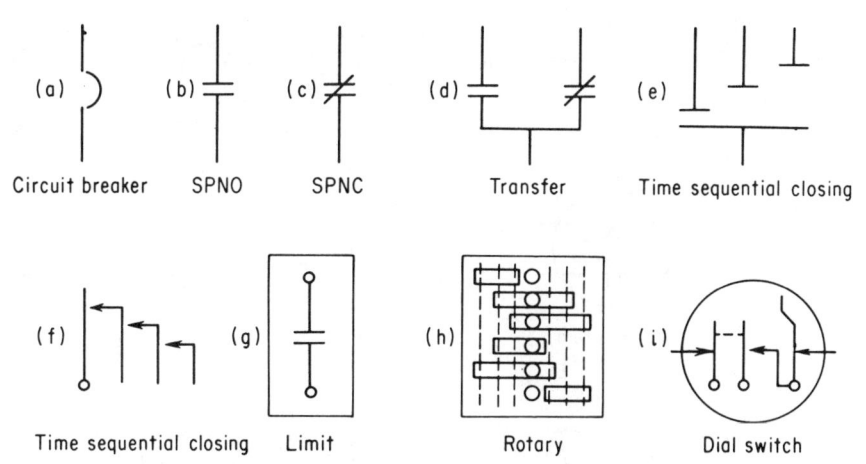

Figure A-7. Symbols for special-purpose switches and contactors.

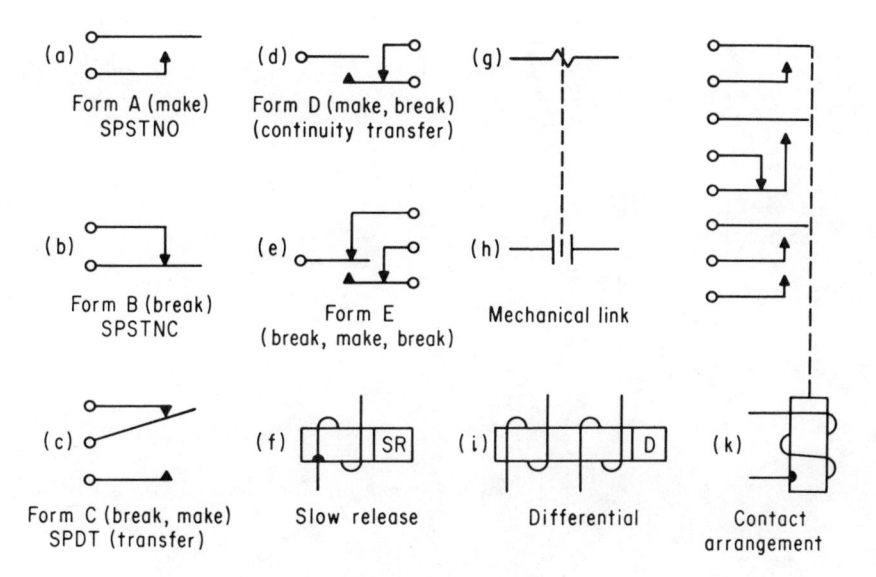

(a) Circuit breaker (b) SPNO (c) SPNC (d) Transfer (e) Time sequential closing (f) Time sequential closing (g) Limit (h) Rotary (i) Dial switch

Figure A-8. Relay symbols.

(a) Form A (make) SPSTNO
(b) Form B (break) SPSTNC
(c) Form C (break, make) SPDT (transfer)
(d) Form D (make, break) (continuity transfer)
(e) Form E (break, make, break)
(f) Slow release
(g)
(h) Mechanical link
(i) Differential
(k) Contact arrangement

619

Appendix

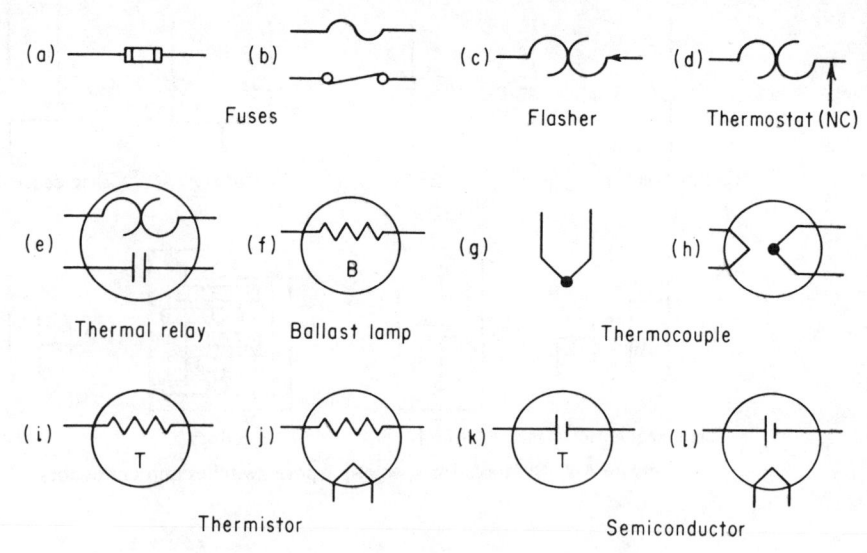

Figure A-9. Symbols used for thermally-operated devices.

620

Appendix

**TABLE A-3. FULL-LOAD CURRENTS IN AMPERES
DIRECT-CURRENT MOTORS (NEC 430-147)**

The following values of full-load currents are for motors running at base speed.

HP	120 V	240 V
1/4	2.9	1.5
1/3	3.6	1.8
1/2	5.2	2.6
3/4	7.4	3.7
1	9.4	4.7
1½	13.2	6.6
2	17	8.5
3	25	12.2
5	40	20
7½	58	29
10	76	38
15		55
20		72
25		89
30		106
40		140
50		173
60		206
75		255
100		341
125		425
150		506
200		675

TABLE A-4. FULL-LOAD CURRENTS IN AMPERES
SINGLE-PHASE ALTERNATING-CURRENT MOTORS (NEC 430-148)

The following values of full-load currents are for motors running at usual speeds and motors with normal torque characteristics. Motors built for especially low speeds or high torques may have higher full-load currents, in which case the nameplate current ratings should be used.

To obtain full-load currents of 208 and 200 V motors, increase corresponding 230 V motor full-load currents by 10 and 15 per cent, respectively.

The voltages listed are rated motor voltages. Corresponding nominal system voltages are 110 to 120, 220 to 240, 440 to 480.

HP	115 V	230 V	440 V
$\frac{1}{6}$	4.4	2.2	
$\frac{1}{4}$	5.8	2.9	
$\frac{1}{3}$	7.2	3.6	
$\frac{1}{2}$	9.8	4.9	
$\frac{3}{4}$	13.8	6.9	
1	16	8	
$1\frac{1}{2}$	20	10	
2	24	12	
3	34	17	
5	56	28	
$7\frac{1}{2}$	80	40	21
10	100	50	26

TABLE A-5. FULL-LOAD CURRENT*
THREE-PHASE A-C MOTORS (NEC 430-150)

	Induction Type Squirrel-Cage and Wound Rotor Amperes					Synchronous Type Unity Power Factor Amperes†			
HP	110 V	220 V	440 V	550 V	2300 V	220 V	440 V	550 V	2300 V
½	4	2	1	.8					
¾	5.6	2.8	1.4	1.1					
1	7	3.5	1.8	1.4					
1½	10	5	2.5	2.0					
2	13	6.5	3.3	2.6					
3		9	4.5	4					
5		15	7.5	6					
7½		22	11	9					
10		27	14	11					
15		40	20	16					
20		52	26	21					
25		64	32	26	7	54	27	22	5.4
30		78	39	31	8.5	65	33	26	6.5
40		104	52	41	10.5	86	43	35	8
50		125	63	50	13	108	54	44	10
60		150	75	60	16	128	64	51	12
75		185	93	74	19	161	81	65	15
100		246	123	98	25	211	106	85	20
125		310	155	124	31	264	132	106	25
150		360	180	144	37		158	127	30
200		480	240	192	48		210	168	40

For full-load currents of 208 and 200 volt motors, increase the corresponding 220 volt motor full-load current by 6 and 10 per cent, respectively.

* These values of full-load current are for motors running at speeds usual for belted motors and motors with normal torque characteristics. Motors built for especially low speeds or high torques may require more running current, in which case the nameplate current rating should be used.

† For 90 and 80 per cent P. F., the above figures should be multiplied by 1.1 and 1.25 respectively.

The voltages listed are rated motor voltages. Corresponding nominal system voltages are 110 to 120, 220 to 240, 440 to 480 and 550 to 600 volts.

TABLE A-6. LOCKED-ROTOR INDICATING CODE LETTERS (NEC 430-76)

Code letter	Kilovolt-amperes per horsepower with locked rotor
A	0 — 3.14
B	3.15 — 3.54
C	3.55 — 3.99
D	4.0 — 4.49
E	4.5 — 4.99
F	5.0 — 5.59
G	5.6 — 6.29
H	6.3 — 7.09
J	7.1 — 7.99
K	8.0 — 8.99
L	9.0 — 9.99
M	10.0 — 11.19
N	11.2 — 12.49
P	12.5 — 13.99
R	14.0 — 15.99
S	16.0 — 17.99
T	18.0 — 19.99
U	20.0 — 22.39
V	22.4 — and up

The above table is an adopted standard of the National Electrical Manufacturers Association.

The code letter indicating motor input with locked rotor must be in an individual block on the nameplate, properly designated. This code letter is to be used for determining branch-circuit overcurrent protection by reference to Table A-7.

TABLE A-7. MAXIMUM RATING OR SETTING OF MOTOR-BRANCH-CIRCUIT PROTECTIVE DEVICES FOR MOTORS MARKED WITH A CODE LETTER INDICATING LOCKED-ROTOR KVA (NEC 430-152)

Type of motor	Per cent of full-load current		
	Fuse rating	Circuit-breaker Setting Instantaneous type	Time limit type
All AC single-phase and polyphase squirrel cage and synchronous motors with full-voltage, resistor, or reactor starting:			
Code Letter A	150		150
Code Letters B to E	250		200
Code Letters F to V	300		250
All AC squirrel-cage and synchronous motors with auto-transformer starting:			
Code Letter A	150		150
Code Letters B to E	200		200
Code Letters F to V	250		200

For certain exceptions to the values specified see Sections 430-52 and 430-54 (NEC). The values given in the last column also cover the ratings of nonadjustable, time-limit types of circuit-breakers which may also be modified as in Section 430-52 (NEC).

Synchronous motors of the low-torque, low-speed type (usually 450 rpm or lower), such as are used to drive reciprocating compressors, pumps, etc., which start up unloaded, do not require a fuse rating or circuit-breaker setting in excess of 200 per cent of full-load current.

For motors not marked with a Code Letter, see Table A-8.

TABLE A-8. MAXIMUM RATING OR SETTING OF MOTOR-BRANCH-CIRCUIT PROTECTIVE DEVICES FOR MOTORS NOT MARKED WITH A CODE LETTER INDICATING LOCKED-ROTOR KVA (NEC 430-153)

	Per cent of full-load current		
		Circuit-breaker Setting	
Type of motor	Fuse rating	Instantaneous type	Time limit type
Single-phase, all types	300		250
Squirrel-cage and synchronous (full-voltage, resistor and reactor starting)	300		250
Squirrel-cage and synchronous (auto-transformer starting)			
Not more than 30 amperes	250		200
More than 30 amperes	200		200
High-reactance squirrel-cage			
Not more than 30 amperes	250		250
More than 30 amperes	200		200
Wound-rotor	150		150
Direct-current			
Not more than 50 H.P.	150	250	150
More than 50 H.P.	150	175	150
Sealed (Hermetic Type) Refrigeration Compressor*			
400 KVA lacked-rotor or less	175†		†175

For certain exceptions to the values specified see Sections 430-52, and 430-59 (NEC). The values given in the last column also cover the ratings of non-adjustable, time-limit types of circuit-breakers which may also be modified as in Section 430-52 (NEC).

Synchronous motors of the low-torque low-speed type (usually 450 rpm or lower) such as are used to drive reciprocating compressors, pumps, etc., which start up unloaded, do not require a fuse rating or circuit-breaker setting in excess of 200 per cent of full-load current.

For motors marked with a Code Letter, see Table A-7.

* The locked-rotor KVA is the product of the motor voltage and the motor locked-rotor current (LRA), given on the motor nameplate, divided by 1000 for single-phase motors, or divided by 580 for 3-phase motors.

† This value may be increased to 225 per cent if necessary to permit starting.

TABLE A-9. RATED, STARTING CURRENTS AND TORQUES OF VARIOUS CLASSES OF 220V* INDUCTION MOTORS†

Motor hp	Rated current amps	Starting current, amps Classes B, C, D	Starting current, amps Class F	Classes A and B 4 pole	Classes A and B 6 pole	Classes A and B 8 pole	Class C 4 pole	Class C 6 pole	Class C 8 pole
0.5	2.0	12				150			
1.0	3.5	24		275	175	150			
1.5	5.0	35		265	175	150			
2	6.5	45		250	175	150			
3	9.0	60		250	175	150		250	225
5	15	90		185	160	130	250	250	225
7.5	22	120		175	150	125	250	225	200
10	27	150		175	150	125	250	225	200
15	40	220		165	140	125	225	200	200
20	52	290		150	135	125	200	200	200
25	64	365		150	135	125	200	200	200
30	78	435	270	150	135	125	200	200	200
40	104	580	360	150	135	125	200	200	200
50	125	725	450	150	135	125	200	200	200
60	150	870	540	150	135	125	200	200	200
75	185	1085	675	150	135	125	200	200	200
100	246	1450	900	125	125	125	200	200	200
125	310	1815	1125	125	125	125	200	200	200
150	360	2170	1350	125	125	125	200	200	200
200	480	2900	1800	125	125	125	200	200	200

* For changes in voltage, use the following equations:

a. Starting current $= \dfrac{V_1}{220} \times I_s$,

where $V_1 =$ new voltage applied to stator and $I_s =$ starting current in above table.

b. Starting torque $= \left(\dfrac{V_1}{220}\right)^2 \times T_s$,

where $T_s =$ starting torque in above table.

† Starting currents of class A motors are usually *higher* than corresponding class B, C, D motors.

Starting torques of class D motors are usually *higher* than corresponding class A, B, C motors.

Starting torques of class F motors are usually *lower* than corresponding class A, B, C motors.

index

AC generators (*see* Alternators)
AC motors (*see* Motors, ac)
Alternator:
   armature reactance, 47, 171, 176
   armature reaction, 48, 158, 161, 251 ff
   armature windings, 51 ff, 58 ff, 170
   characteristics, 176
   construction, 43, 170, 173
   distribution factor, 62 ff
   efficiency, 476 ff, 479 ff
   elementary, 169
   equivalent circuit, 174, 182
   field construction, 170
   frequency, 68, 170
   generated voltage, 66, 176
   hunting, 220
   hydrogen cooling, 173, 478
   losses in, 476, 465
   open-circuit test, 184
   parallel operation, 206
   pitch factor, 60
   power factor, 176, 180
   prime movers, 173
   rating, 489 ff
   regulation, 176, 180, 183
   resistance test, 183
   short-circuit currents, 189
   short-circuit test, 184
   speed-frequency relations, 173
   synchronization of, 222
   synchronous impedance, 176, 182, 188
   voltage drops in, 176
   voltage regulators, 182
   waveform, 64
Amortisseur winding, 222
Amplidyne, 441
Armature:
   construction, 41
   core, 41
   flux, 145
   reactance, 47, 157, 171, 176
   resistance, 55, 77

Armature (*cont.*):
   reversing, 135
   shunting, 77
   voltage control 127 ff
   windings, 51 ff
      belt factor, 62
      chorded, 49, 60
      coil span, 51
      distribution factor, 62
      duplex, 54
      fractional pitch, 52, 64
      gramme ring, 18
      lap, 52, 57
      multiplex, 55
      pitch factor, 51, 60 ff
      simplex lap, 54
      simplex wave, 55
      types of, 51, 59
      wave, 52, 55, 57
Armature reaction:
   alternator, 48, 158, 161, 251 ff
   compensating for, 150
   cross-magnetizing, 161
   dc generator, 90
   dc motor, 136, 149
   demagnetizing, 161
   shift of neutral caused by, 149
   summary, 161
   synchronous motor, 158 ff
ASA method, 183
Asynchronous induction dynamo, 45, 300 ff
Autotransformer:
   connections, 554
   efficiency, 562 ff
   hazards in, 564
   polyphase motor starting, 336
   rating, 559 ff
   single phase speed control, 372
   starting, dual-capacitor motor, 374
   variable, 555

BALANCE coil, 419

Balancer set, 417
Bearings, 43, 498 ff
Belt factor, 62
Booster, 97, 280 ff
Brush
 drop (voltage) 77 ff, 89, 91, 117 ff, 123
 rigging, 43
 shifting, 149
 shifting repulstion motor, 380 ff
Brushless dc motor, 447 ff
Buildup, conditions for, 87 ff

CAPACITOR motor, 371 ff
Capacitor-start motor, 369
Capacitor, synchronous, 274 ff
Centrifugal switch, 129, 366 ff, 499
Changer, frequency, 281
Circuital law of magnetic circuit, 48
Coil:
 balance (*see* Balance coil)
 reactance, 47, 171, 176
 span, 51
Commutating pole, 41, 42, 47, 157
Commutation process, 154, 157 ff
Commutator:
 segment, 42, 57
 sparking of, 57, 154, 157 ff
Compensating winding, 152
Compound motor (*see* Motor, compound)
Compounding, degree of, 98 ff
Compounding, machine, 80 ff, 97 ff, 100 ff
Concatenation, 302
Conductor:
emf, induced in, 4 ff
force on, 26 ff
per turn, 4 ff, 24 ff, 51
Consequent pole, 345 ff
Control:
 amplidyne, 441 ff
 direction of rotation, 135
 field, 127 ff
 speed, 118, 345, 446
 voltage, 446 ff
Converters:
 armature reaction in, 407
 conversion ratios, 412
 currents in, 411 ff
 frequency, 348
 phase, 423 ff
 polyphase 411 ff
 rotary, 408 ff
 single phase, 408 ff
Cooling:
 duty cycle, 478, 493
 hydrogen, 173, 478 ff
Copper losses:
 alternator, 464 ff, 476 ff
 dynamo, 464 ff
 induction motor, 480 ff
 rotor, 481 ff
 stator, 481 ff

Core losses:
 alternator, 464 ff, 476 ff
 dynamo, 465, 469 ff
 induction motor, 486 ff
Counter emf, 26 ff, 29, 31, 117 ff
Critical field resistance, 88

DAMPER windings, 222
DC generator (*see* Generator, dc)
DC motor (*see* Motor, dc)
Differential compound:
 generator, 97, 98, 101 ff
 motor, 126, 129 ff, 133
Distortion flux (see Flux)
Distribution factor, 62
Diverter-pole generator, 402
Diverter (series field), 99
Dobrowolsky generator, 43, 419
Double excitation, 45, 146, 235, 301
Double squirrel-cage motor, 324
Duty cycle, 493
Dynamo:
 acyclic, 15, 405 ff
 basic principles, 29 ff
 comparison of motor and generator, 29 ff
 current ratings, 23
 construction, 40 ff
 energy conversion, 2 ff, 467 ff
 energy relations, 467 ff
 field structure, 41
 homopolar, 15, 405 ff
 induction, 45
 possibilities, 40
 power conversion, 467 ff
 universal, 144, 301
 voltage ratings, 23, 492
Dynamotor, 406

EDDY current losses, 465 ff, 539 ff
Effective resistance:
 alternator armature, 173 ff, 183 ff
 induction motor rotor, 327 ff
 induction motor stator, 481 ff
Efficiency:
 ac synchronous dynamo, 476, 479 ff
 alternator, 477, 479 ff
 asynchronous induction, 480
 dynamo, 464 ff, 469 ff
 generator, 463 ff
 induction motor, 483 ff
 maximum, 471 ff
 motor, 463 ff
 single phase motor, 489 ff
 transformer, 540 ff
Electromagnetic:
 energy conversion, 2
 force, 3, 27
 induction, 3, 4, 10
Emf:
 average, 24, 66
 counter, 26, 29, 31, 117 ff

630

INDEX

Emf (*cont.*):
  direction of induced, 9
  induced, 4, 5, 66
  rotor, 311 ff
  self-induced, 47, 50
  sinusoidal, 14
  square-shaped, 17
Enclosures, 493 ff
Equalizer, 204
Equivalent circuit:
  dc generator, 77 ff
  induction motor, 328
Excitation:
  armature and field, 47 ff, 145 ff
  double, 45, 146, 235, 301
  separate, 82 ff
Exciters, multified, 444 ff

Factor:
  distribution, 62
  pitch, 60
Faraday's law, 4
Feedback:
  loop, 443 ff
  negative, 443 ff
Field:
  coils, 49
  control, 443 ff
  discharge, 50
  flashing, 88
  reference, 441
  resultant, 147
  reversing, 88 ff, 135
  revolving, 237 ff
  rheostat, 49
  series, 42 ff
  shunt, 42 ff
Flashover, commutator, 57
Flat-compounding (*see* Compounding)
Fleming's rule:
  generator, 9
  left-hand (motor), 28 ff
  motor, 28 ff
  right-hand (generator), 9
Flux:
  armature, 145
  cross-magnetizing, 148
  demagnetizing, 147
  distortion, 147, 149
  leakage, 46 ff, 515
  main field, 146
  mutual, 46, 147, 515, 520 ff
Force:
  on a conductor, 26 ff
  electromotive, 3 ff
  reluctance, 3
Fractional pitch (*see* Pitch)
Frequency:
  changer, 281
  converter, 348

Frequency (*cont.*):
  rotor, 309
Friction, losses, 465 ff

GENERATOR
  ac (see Alternator)
  action, 29 ff
  dc:
    buildup, 87
    characteristics, 82 ff, 90 ff
    compound, 80, 97
    critical field resistance, 88
    degree of compounding, 98 ff
    differential, 97, 98, 101 ff
    diverter pole, 402
    effect of speed on, 93, 103
    efficiency, 464, 469 ff
    equalizer connections, 204
    equivalent circuit, 77 ff
    excitation, 82
    failure to buildup, 87 ff
    field resistance, 86
    flat-compound, 98
    flux distribution, 146
    homopolar, 405
    interpoles, 157 ff
    magnetization curve, 82 ff
    overcompound, 98
    parallel operation, 201 ff, 203 ff
    polarity, 13
    principles, 76
    rating of, 489 ff
    regulation of, 100
    Regulex, 444 ff
    Rosenberg, 439
    Rototrol, 444 ff
    self-excited, 88
    separately-excited, 82
    series, 96
    shunt, 86 ff
    special types, 401 ff
    third-brush, 403
    three-wire, 417 ff
    unbuilding, 89, 93
    undercompound, 98 ff
    voltage regulation, 94
  double current, 416
  homopolar, 405
  induction, 347
  systems, 3-wire, 417 ff
Gramme-ring armature, 18

HARMONICS, 64, 148
Homopolar dynamo, 405 ff
Horsepower:
  calculation 498
  motor, 134 ff
  ratings, 489 ff
Hunting:
  alternator, 220, 409
  synchronous converter, 409, 416

Hunting (*cont.*):
  synchronous motor, 235, 236
Hydrogen cooling (*see* Cooling)
Hysteresis:
  effect, 84, 93, 465 ff, 529 ff
  loss, 465 ff
  motor, 286

IMPEDANCE:
  equivalent, 174 ff
  synchronous, 176 ff, 182 ff
Induced emf, 4, 5, 66
Inductance, 47
Induction:
  electromagnetic, 3, 4, 10
  frequency converter, 348
  generator, 347
  phase converter, 423 ff
Induction motor (*see* Motor, induction)
Induction regulator, 483
Instability, 220, 235, 416
Insulating materials, 491
Interpoles, 41, 42, 47, 157
Inverted converter, 408

LEAKAGE coefficient, 49
Leakage flux, 46 ff, 515, 520
Leakage reactance, 47
Lenz's law, 10
Losses:
  alternator, 476 ff
  dynamo, 464 ff
  eddy current, 465 ff, 539 ff
  generator, dc, 464 ff, 469 ff
  hysteresis, 465 ff, 529 ff
  motor, dc, 473 ff
  motor, induction, 481 ff
  motor, single phase, 489

MAGNETIC:
  circuit, 47 ff
  circuital law, 48
  field, 47 ff
  leakage, 49
  neutral, 149
Magnetism, residual, 84 ff, 88
Magnetization curve:
  alternator, 184
  dc generator, 82 ff
Maintenance, 498 ff
Motor:
  action, 29
  adjustable-speed, 495
  adjustable varying-speed, 495
  compound, 126
  constant speed, 495
  counter emf, 26, 29, 117 ff
  dc:
    armature current in, 123
    armature resistance control, 122 ff
    braking, 135, 495

Motor (*cont.*):
  brush shifting, 149
  brushless, 447 ff
  characteristics, 110 ff
  commutation, 57, 154
  compensating windings, 152
  compound, 106
  constant speed, 127 ff
  differential compound, 126, 129
  dynamic braking, 135, 495
  effect of armature reaction, 136
  efficiency, 463, 469
  enclosures, 493 ff
  horsepower, 134, 498
  maintenance, 498 ff
  mechanical power developed, 119 ff
  plugging, 135, 495
  principles, 110
  reversing, 135
  rated output, 134, 498
  series, 125
  servo, 121, 434 ff
  shunt, 125
  speed characteristics, 117 ff, 127 ff
  speed regulation, 133
  starters, 122 ff
  starting, 122 ff
  torque, 111, 115, 121, 125, 134
  torque characteristics, 125 ff, 134
  types, 125 ff
definitions (speed) 495 ff
enclosures, 493 ff
multispeed, 495
non-reversible, 495
polyphase, induction:
  blocked rotor test, 328, 423
  classes of, 342
  construction, 493 ff
  double-squirrel-cage, 341 ff
  dynamic braking, 347, 495
  efficiency, 329 ff, 480, 483 ff
  enclosures, 493 ff
  frequency, 309 ff
  maintenance, 498 ff
  maximum torque, 314 ff
  no-load test, 483
  operating characteristics, 315 ff
  plugging, 495
  pole-changing, 345
  power factor, 320 ff
  principle, 305
  rotating field, 238, 302
  rotor current, 312 ff
  rotor emf, 311
  rotor frequency, 309
  rotor torque, 311 ff
  rotor voltage, 311
  running characteristics, 315 ff
  secondary resistance control, 319 ff
  slip, 307 ff
  starting, 320, 335

Motor (*cont.*):
  starting current, 312
  starting torque, 320
  stator resistance test, 481 ff
  synchronous-induction, 284
  two-speed, 495
  winding, 301
  wound rotor, 319
  wound rotor speed control, 326
 rating, 359, 498
 reversible, 495
 reversing, 495
 single phase:
  capacitor, 371 ff
  capacitor-start, 369
  classification, 391
  commutator types, 380
  construction, 361
  cross-field theory, 365
  double revolving field theory, 363
  dual value capacitor, 373
  enclosures, 493 ff
  hysteresis, 286
  induction, 391 ff
  maintenance, 498 ff
  reluctance, 285
  reluctance-start, 379
  repulsion, 380 ff, 384
  repulsion-induction, 386
  repulsion-start induction, 385
  series, 389
  shaded pole, 375
  speed control, 371 ff
  split-phase, 366
  subsynchronous, 287
  summary of types, 391 ff
  synchronous, 283 ff
  two-value capacitor, 373
  universal, 388
 supersynchronous, 282
 synchro (*see* Synchros)
 synchronous-induction, 284
 synchronous, polyphase:
  advantages, 235 ff
  amortisseur windings, 235 ff, 239
  applications, 235, 268, 271
  armature reaction, 251
  brushless, 289
  construction, 236
  damper windings, 235 ff, 239
  enclosures, 493 ff
  excitation, 249 ff
  exciters, 236, 239
  field current variation 253 ff, 256 ff
  frequency changer, 281
  high starting torque, 239 ff
  hunting, 235
  loading, 247 ff
  maintenance, 498 ff
  nonexcited, 283 ff
  normal excitation, 247

Motor (*cont.*):
  operation, 242
  pole slipping, 240
  power factor adjustment, 253 ff
  power factor correction, 268 ff, 276 ff
  principle, 236
  rating, 274
  reluctance torque, 273
  simplex rotor, 241
  speed of, 235, 239
  starting, 239 ff, 241 ff
  supersynchronous, 282
  synchronous-induction, 284
  torque, 244 ff, 270
  torque angle, 260 ff, 263 ff
  torque maximum, 272
  V-curves, 256 ff
 universal:
  applications, 388
  characteristics, 388
  speed, 388
 varying-speed, 495
 wound rotor induction (*see* Motor, polyphase, induction, wound rotor)
 multispeed, 495

NEUMANN's law, 5
No-load test (*see* Open circuit or Running-light test)
Nonexcited synchronous motor, 283 ff

OPEN circuit test:
 alternator, 184
 polyphase induction motor, 483
 transformer, 540 ff
Overcompound generator (*see* Generator, dc, overcompound)

PARALLEL operation:
 advantages, 197
 alternator, 207
 alternator load division, 218
 dc generator:
  compound, 203 ff
  shunt, 201 ff
 procedure, 204 ff
 requirements:
  for alternators, 207
  for dc generators, 202, 204
 synchronization, 207 ff
 synchronizing lamps, 209, 222
 synchronizing power, 213, 219
 synchroscope, 224
Permanent split-capacitor-motor (*see* Motor, single phase)
Phase converters, 423
Phase sequence indicator, 226
Pitch, coil 60
Pitch factor, 61
Plugging, 135, 495
Polarity, 13

Poles:
  alternator field, 170, 173
  commutating field, 41, 47, 157, ff
  dc field, 42
  nonsalient and salient, 173
  series field, 42, 79
  shunt field, 42, 77
Polyphase induction motor (*see* Motor, polyphase, induction)
Polyphase synchronous motor (*see* Motor, polyphase, synchronous)
Power:
  conversion, 462
  developed, 119
  flow diagrams, 467
  internal, 120
  losses, 465
  relations, 462
  rotor:
    asynchronous 328
    synchronous, 272
Power factor:
  correction, 268, 276
  synchronous capacitor, 274
  synchronous motor, 253, 268
  effect of, on:
    alternator regulation, 180 ff
    synchronous motor, 256 ff
  polyphase induction motor, 317 ff
  single phase motor, 367 ff
Pullout torque, 259, 297 ff

RATING:
  continuous duty, 493
  duty cycle, 493
  dynamo, 23, 489
  factors affecting, 489
  generator, 492, 497
  horsepower, 489, 493, 498
  intermittent duty, 493
  motor, 492, 497
  speed, 489, 494
  temperature, 490
  voltage, 492
Reactance, armature, 47, 157, 171, 176
  equivalent, 174
  leakage, 47
  synchronous, 186
Reactance coil, 419
Reaction, armature (*see* Armature reaction)
Reference field:
  input, 444
  voltage, 446
Regenerative braking, 135, 495
Regulation:
  alternator, 176, 180, 183, 443
  generator, 100
  speed, 133
  voltage, 95
Regulex, 444

Reluctance motor, 285
Reluctance-start motor, 379
Repulsion-induction motor, 386
Repulsion motor, 380 ff, 384
Repulsion-start induction motor, 385
Residual magnetism, 84 ff, 88
Resistance:
  armature circuit, 55, 77
  equivalent, 481
  field, 86
  losses, 465
Reversing, armature rotation:
  dc motor, 135
Revolving field, 238, 302
Rheostat, field, 49
Right-hand rule, 9
Rosenberg generator, 439
Rotary amplifier, 441
Rotary converter, 408
Rotating field:
  induction motor, 302 ff
  single-phase motor, 361 ff
  synchronous motor, 238 ff
Rotor:
  current, 312 ff
  double-cage, 341 ff
  emf, 311
  frequency, 309
  squirrel-cage, 301
  wound, 319
Rototrol, 444
Running light test, 469

SALIENT pole, 173
Saturation curve, 82 ff
Segment, commutator, 42, 57
Selection, motor, 497
Series:
  generator, 96
  motor:
    ac, 389
    dc, 125
    universal, 388
Servomotor:
  ac, 437 ff
  dc, 434 ff
Shunt generator (*see* Generator)
Shunt motor (*see* Motor)
Single-phase motor (*see* Motor)
Slip:
  definition, 307
  measurement, 332 ff
Slot lock, 436
Speed regulation, 133
Split-phase motor (*see* Motor, single-phase)
Squirrel-cage motor (*see* Motor, polyphase induction)
Starters:
  across-the-line, 335
  wound rotor, 320

Starting methods:
  autotransformer, 336
  full voltage, 335
  induction motor methods, 341
  reactance, 338
  reduced voltage, 338
  resistance, 338
  synchronous motor wye-delta, 338
Starting torque (*see* Torque)
Stator, 481
Stray load loss, 465 ff
Stray power loss, 465 ff
Strobe light, 334
Stroboscopic effect, 334
Subsynchronous motor, 287
Supersynchronous motor, 282
Switch:
  centrifugal, 129, 366 ff, 499
Symbols, Appendix A-2
Synchronizing devices (*see* Synchros)
Synchronizing procedure, 222
Synchronous:
  capacitor, 274 ff
  dynamo, 43 ff
  impedance method, 176 ff, 182 ff
  motor (*see* Motor, synchronous)
  reactance, 186
  reactor, 280
  speed:
    alternator, 173
    induction motor, 306
    synchronous motor, 235, 238
Synchronous-induction motor, 284
Synchros:
  construction, 425
  control transformer, 429
  differential receiver, 428
  differential transmitter, 427
  power, 432
  receiver, 426
  synchro-tie system, 407
  transmitter, 426
  zeroing, 431
Synchro-tie system, 433
Synchroscope, 224

TEMPERATURE:
  ambient, 490, 493
  rating, 490, 493
  rise, 490, 493
Testing:
  load, 487, 489
  locked-rotor, 392, 483
  no load, 468, 483
  open circuit, 184, 540
  resistance, 183
  short-circuit, 184
  stator resistance, 183
  synchronous impedance, 178, 182, 186
Third-brush generator, 403 ff
Three-wire generator, 417 ff

Torque:
  angle, 242 ff, 247 ff, 260 ff
  breakdown, 270, 314
  dc motor, 111 ff, 121 ff, 134
  defined, 111 ff
  developed, 116
  induction motor, 311 ff
  maximum, 270, 314
  reluctance, 3, 271, 285
  starting, 311, 313, 320, 324, 342 ff
  synchronous motor, 270 ff
Transformer:
  autotransformer (*see* Autotransformer)
  all-day efficiency, 546
  connections, 551 ff
  definitions, 514
  efficiency, 540 ff
  equivalent circuit, 529
  frequency, 522
  ideal, 517
  neutral, 574
  permissible flux, 522
  phasing, 547 ff
  polarity, 547 ff
  polyphase, 565 ff, 582 ff
  power conversion, 592
  practical, 526
  reflected impedance, 524 ff
  six phase, 585
  starting, 336
  tests, 536 ff, 540 ff
  transformation ratio, 519 ff
  T-T, 578
  V-V, 576
  Voltage regulation, 533 ff
Troubleshooting, motor, 498 ff
Tuning resistor, 444
Two revolving field theory, 362 ff
Two value capacitor motor (*see* Motor, single-phase)

UNDERCOMPOUND generator (*see* Generator, dc)
Unidirectional current, 16
Universal motor, 388 ff

V-CURVES, synchronous motor, 256
Voltage:
  alternator, 66, 176
  dc generator, 82 ff
  generated, 29, 66
  induced, 4
  ratings for generators, 492, 497
  ratings for motors, 492, 497
  reactance, 47 ff
  transmission line, 280

WINDAGE loss, 465, 478
Windings, armature (*see* Armature windings)
Wound rotor induction motor, 320
Wye-connected armature, 183
Wye-delta starting, 338